CONVERSION FACTORS FROM BG TO SI UNITS (Continued)

	To convert from	To	Multiply by
Pressure	lbf/ft^2	Pa	4.7880 E + 1
	lbf/in^2	Pa	6.8948 E + 3
	atm	Pa	1.0133 E + 5
	mm Hg	Pa	1.3332 E + 2
Specific weight	lbf/ft^3	N/m^3	1.5709 E + 2
Specific heat	$ft^2/(s^2 \cdot {}^\circ R)$	$m^2/(s^2 \cdot K)$	1.6723 E − 1
Temperature	°F	°C	$t_C = \frac{5}{9}(t_F - 32^\circ)$
	°R	K	0.5556
Velocity	ft/s	m/s	0.3048
	mi/h	m/s	4.4704 E − 1
	knot	m/s	5.1444 E − 1
Viscosity	$lbf \cdot s/ft^2$	$N \cdot s/m^2$	4.7880 E + 1
	$g/(cm \cdot s)$	$N \cdot s/m^2$	0.1
Volume	ft^3	m^3	2.8317 E − 2
	L	m^3	0.001
	gal (U.S.)	m^3	3.7854 E − 3
	fluid ounce (U.S.)	m^3	2.9574 E − 5
Volume flow	ft^3/s	m^3/s	2.8317 E − 2
	gal/min	m^3/s	6.3090 E − 5

Fluid
Mechanics

Fluid Mechanics

Second Edition

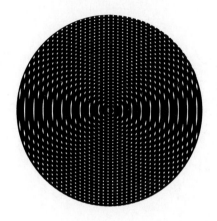

Frank M. White

Professor of Mechanical Engineering
University of Rhode Island

McGraw-Hill Publishing Company

New York   St. Louis   San Francisco   Auckland   Bogotá
Caracas   Hamburg   Lisbon   London   Madrid   Mexico   Milan
Montreal   New Delhi   Oklahoma City   Paris   San Juan
São Paulo   Singapore   Sydney   Tokyo   Toronto

Fluid Mechanics

Copyright © 1986, 1979 by McGraw-Hill, Inc. All rights reserved.
Printed in the United States of America. Except as permitted under the
United States Copyright Act of 1976, no part of this publication may be
reproduced or distributed in any form or by any means, or stored in a data
base or retrieval system, without the prior written permission of the
publisher.

7 8 9 10 11 12 DOC/DOC 9 9 8 7 6 5 4 3 2 1 0

ISBN 0-07-069673-X

This book was set in Times Roman.
The editors were Anne Murphy and David A. Damstra;
the designer was Elliot Epstein;
the production supervisor was Marietta Breitwieser.
New drawings were done by Fine Line Illustrations, Inc.
R. R. Donnelley & Sons Company was printer and binder.

Library of Congress Cataloging in Publication Data

White, Frank M.
    Fluid mechanics.

    Includes bibliographies and index.
    1. Fluid mechanics.   I. Title.
TA357.W48   1986   620.1'06   85-14939
ISBN 0-07-069673-X

To Jeanne

Contents

* These sections may be omitted without loss of continuity.

Preface

The second edition of this textbook sees additions and deletions but no philosophical change. The basic outline of eleven chapters and five appendixes remains the same. The triad of differential, integral, and experimental approaches is retained. There are now more problem exercises and fully worked examples. The informal, student-oriented style is retained. A number of new photographs and figures have been added.

A distinct improvement is the new breakdown of problem exercises by numbered topic, so that assignments can fit the class material more easily. In the first edition there were 1089 problem exercises. Of these, 410 have been deleted, and 490 entirely new problems have been added, for a total of 1169. Many of these new problems are more challenging and analytical in nature. I am indebted to many friends for suggesting new problems.

Although there are some revisions in every chapter, the most significant changes are in Chapters 2, 5, 8, 10, and 11. In Chapter 2 the two sections on pressure distribution in irrotational and arbitrary viscous flows have been removed. They were well meaning but premature. The chapter is now confined to hydrostatic and rigid-body conditions.

In Chapter 5 the power-product method has been eliminated in favor of exclusive use of the pi theorem. I like the power-product method, but it was confusing to present two competing methods in a first course.

In Chapter 8 a new section has been added on numerical finite-difference analysis of potential flow. There is an extensive numerical example showing how computer flow problems are set up and used along with other techniques such as boundary-layer theory. Although many modern potential-flow computer codes use the distributed-source or panel techniques, I feel that the finite-difference method is easier to understand in a first course.

In Chapter 10 the Moody chart has been deemphasized in favor of the Manning roughness correlation. Again it is felt that methods should not compete in a first course, and without a doubt the Manning formulation is the most popular method in open-channel flows. There is also some added material on backwater computations behind a weir or obstruction in a channel, a type of computation that occurs frequently in planning for hydropower projects.

Finally, some new material has been added in Chap. 11 on axial-flow pumps and on wind turbines. This broadens the coverage, but it is still difficult to give a full survey in one chapter of such a large subject as turbomachinery.

At the request of numerous reviewers and readers, the appendix material has been extended. The figures and tables in Appendix A now emphasize metric units. The compressible-flow tables in Appendix B now have very small Mach number increments of 0.02. Two new charts have been added for the property changes across an oblique-shock wave. Some new films have been added to Appendix C, and angular velocity components have been added to Appendix E.

McGraw-Hill now offers IBM-compatible software in fluid mechanics as a supplement to this textbook. The programs were designed by Prof. Daniel Olfe of the University of California at San Diego.

So many people have now helped me that it is impossible to list them all. I am especially indebted to the following reviewers who read these revised chapters and offered many helpful suggestions: Edward C. Chiang, Michigan Technological University; Rod W. Douglass, University of Nebraska, Lincoln; Charles L. Merkle, Pennsylvania State University; Ted Okiishi, Iowa State University; Philip J. Pritchard, Manhattan College; and James K. Strozier, United States Military Academy. Finally, I continue to enjoy the support of my family in these writing efforts.

Frank M. White

Fluid Mechanics

Introduction

1.1 PRELIMINARY REMARKS

Fluid mechanics is the study of fluids in motion or at rest and the subsequent effects of the fluid on the boundaries, which may be either solid surfaces or other fluids. The essence of the subject of fluid flow is that of a judicious compromise between theory and experiment. Since fluid flow is a branch of mechanics, it satisfies a set of well-documented basic conservation laws and thus a great deal of theoretical treatment is available. The theory is often frustrating, however, because it applies mainly to certain idealized situations which may be invalid in practical problems. The two chief obstacles to a workable theory are geometry and viscosity. The general theory of fluid motion (Chap. 4) is too difficult to enable the user to attack arbitrary geometric configurations, so that most textbooks concentrate on flat plates, circular pipes and other easy geometries. It is possible to apply numerical techniques to arbitrary geometries, and specialized textbooks are now appearing which explain these digital-computer approximations [1, 19].[1] This book will present many theoretical results, keeping their limitations in mind.

The second obstacle to theory is the action of viscosity, which can be neglected only in certain idealized flows (Chap. 8). First, viscosity increases the difficulty of the basic equations, although the boundary-layer approximation found by Ludwig Prandtl in 1904 (Chap. 7) has greatly simplified viscous-flow analyses. Second, viscosity has a destabilizing effect on all fluids, giving rise, at frustratingly small velocities, to a disorderly, random phenomenon called *turbulence*. The theory of turbulent flow is crude and heavily backed up by experiment (Chap. 6), yet it can be quite serviceable as an engineering estimate. Textbooks are beginning to present digital-computer techniques for turbulent-flow analysis [2], but they are based strictly upon empirical assumptions regarding the time mean of the turbulent stress field.

[1] Numbered references appear at the end of each chapter.

Thus there is theory available for fluid-flow problems, but in all cases it should be backed up by experiment. Often the experimental data provide the main source of information about specific flows, such as the drag and lift of immersed bodies (Chap. 7). Fortunately, fluid mechanics is a highly visual subject, with good instrumentation [3, 4], and the use of dimensional analysis and modeling concepts (Chap. 5) is widespread. Thus experimentation provides a natural and easy complement to the theory. Appendix C lists a variety of interesting films which have been prepared to visualize fluid-flow phenomena. You should keep in mind that theory and experiment should go hand in hand in all studies of fluid mechanics.

1.2 THE CONCEPT OF A FLUID

From the point of view of fluid mechanics, all matter consists of only two states, fluid and solid. The difference between the two is perfectly obvious to the layman, and it is an interesting exercise to ask a lay person to put this difference into words. The technical distinction lies with the reaction of the two to an applied shear or tangential stress. A solid can resist a shear stress by a static deformation; a fluid cannot. Any shear stress applied to a fluid, no matter how small, will result in motion of that fluid. The fluid moves and deforms continuously as long as the shear stress is applied. As a corollary, we can say that a fluid at rest must be in a state of zero shear stress, a state often called the hydrostatic stress condition in structural analysis. In this condition, Mohr's circle for stress reduces to a point, and there is no shear stress on any plane cut through the element under stress.

Given the definition of a fluid above, every layman also knows that there are two classes of fluids, *liquids* and *gases*. Again the distinction is a technical one concerning the effect of cohesive forces. A liquid, being composed of relatively close-packed molecules with strong cohesive forces, tends to retain its volume and will form a free surface in a gravitational field if unconfined from above. Free-surface flows are dominated by gravitational effects and are studied in Chaps. 5 and 10. Since gas molecules are widely spaced with negligible cohesive forces, a gas is free to expand until it encounters confining walls. A gas has no definite volume and left to itself without confinement forms an atmosphere which is essentially hydrostatic. The hydrostatic behavior of liquids and gases is taken up in Chap. 2. Gases cannot form a free surface, and thus gas flows are rarely concerned with gravitational effects other than buoyancy.

Figure 1.1 illustrates a solid block resting on a rigid plane and stressed by its own weight. The solid sags into a static deflection, shown as a highly exaggerated dashed line, resisting shear without flow. A free-body diagram of element A on the side of the block shows that there is shear in the block along a plane cut at an angle θ through A. Since the block sides are unsupported, element A has zero stress on the left and right sides and compression stress $\sigma = -p$ on the top and bottom. Mohr's circle does not reduce to a point, and there is nonzero shear stress in the block.

By contrast, the liquid and gas at rest in Fig. 1.1 require the supporting walls in order to eliminate shear stress. The walls exert a compression stress $-p$ and reduce Mohr's circle to a point with zero shear everywhere, i.e., the hydrostatic condition.

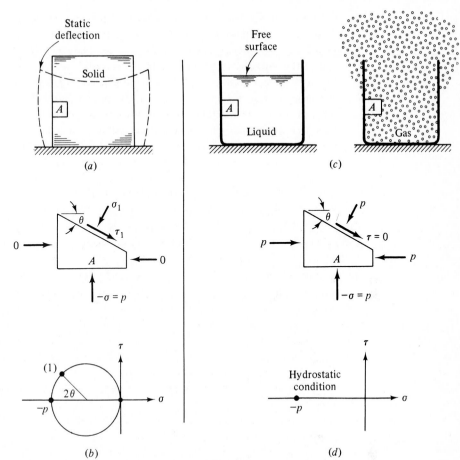

Fig. 1.1 A solid at rest can resist shear. (*a*) Static deflection of the solid; (*b*) equilibrium and Mohr's circle for solid element *A*. A fluid cannot resist shear. (*c*) Containing walls are needed; (*d*) equilibrium and Mohr's circle for fluid element *A*.

The liquid retains its volume and forms a free surface in the container. If the walls are removed, shear develops in the liquid and a big splash results. If the container is tilted, shear again develops, waves form, and the free surface seeks a horizontal configuration, pouring out over the lip if necessary. Meanwhile, the gas is unrestrained and expands out of the container, filling all available space. Element *A* in the gas is also hydrostatic and exerts a compression stress $-p$ on the walls.

In the above discussion clear decisions could be made about solids, liquids, and gases. Most engineering fluid-mechanics problems deal with these clear cases, i.e., the common liquids, such as water, oil, mercury, gasoline, and alcohol, and the common gases, such as air, helium, hydrogen, and steam, in their common temperature and pressure ranges. There are many borderline cases, however, of which you should be aware. Some apparently "solid" substances such as asphalt and lead resist shear stress for short periods but actually deform slowly and exhibit definite fluid behavior over long periods of time. Other substances, notably colloid and slurry mixtures, resist small shear stresses but "yield" at large stress and begin to flow like fluids. Specialized textbooks are devoted to this study of more general deformation and flow, a field called *rheology* [5]. Also, liquids and gases can coexist in two-phase mixtures, such as steam-water mixtures or water with entrapped air bubbles. Specialized textbooks present the analysis of such *two-phase flows* [6]. Finally, there are situations where the distinction between a liquid and a gas blurs. This is the case at temperatures and pressures above the so-called *critical point* of a substance, where only a single phase exists, primarily resembling a gas. As pressure increases far above the critical point, the gaslike substance becomes so dense that there is some resemblance to a liquid and the usual thermodynamic approximations like the perfect-gas law become inaccurate. The critical temperature and pressure of water are $T_c = 647$ K and $p_c = 219$ atm,[1] so that typical problems involving water and steam are below the critical point. Air, being a mixture of gases, has no distinct critical point, but its principal component, nitrogen, has $T_c = 126$ K and $p_c = 34$ atm. Thus typical problems involving air are in the range of high temperature and low pressure where air is distinctly and definitely a gas. This text will be concerned solely with clearly identifiable liquids and gases, and the borderline cases discussed above will be beyond our scope.

1.3 THE FLUID AS A CONTINUUM

We have already used technical terms such as fluid pressure and density without a rigorous discussion of their definition. As far as we know, fluids are aggregations of molecules, widely spaced for a gas, closely spaced for a liquid. The distance between molecules is very large compared with the molecular diameter. The molecules are not fixed in a lattice but move about freely relative to each other. Thus fluid density, or mass per unit volume, has no precise meaning because the number of molecules occupying a given volume continually changes. This effect becomes unimportant if the unit volume is large compared with, say, the cube of the molecular spacing, when the number of molecules within the volume will remain nearly constant in spite of the enormous interchange of particles across the boundaries. If, however, the chosen unit volume is too large, there could be a noticeable variation in the bulk aggregation of the particles. This situation is illustrated in Fig. 1.2, where the "density" as calculated from molecular mass δm within a given volume $\delta \mathcal{V}$ is plotted versus the size of the unit volume. There is a limiting volume $\delta \mathcal{V}^*$ below which molecular variations may be important and above which aggregate variations may be important. The density ρ of a fluid is best defined as

$$\rho = \lim_{\delta \mathcal{V} \to \delta \mathcal{V}^*} \frac{\delta m}{\delta \mathcal{V}} \tag{1.1}$$

The limiting volume $\delta \mathcal{V}^*$ is about 10^{-9} mm³ for all liquids and for gases at atmospheric pressure. For example, 10^{-9} mm³ of air at standard conditions contains approximately 3×10^7 molecules, which is sufficient to define a nearly constant density according to Eq. (1.1). Most engineering problems are concerned with physical dimensions much larger than this limiting volume, so that density is essentially a point function and fluid properties can be thought of as varying continually in space, as sketched in Fig. 1.2a. Such a fluid is called a *continuum*, which simply

---

[1] One atmosphere equals 2116 lbf/ft² = 101,300 Pa.

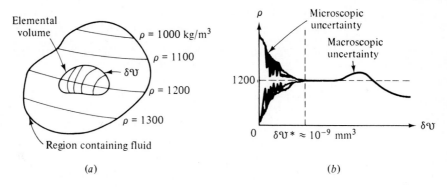

Fig. 1.2 The limit definition of continuum fluid density: (*a*) an elemental volume in a fluid region of variable continuum density; (*b*) calculated density versus size of the elemental volume.

means that its variation in properties is so smooth that the differential calculus can be used to analyze the substance. We shall assume that continuum calculus is valid for all the analyses in this book. Again there are borderline cases for gases at such low pressures that molecular spacing and mean free path[1] are comparable to, or larger than, the physical size of the system. This requires that the continuum approximation be dropped in favor of a molecular theory of rarefied-gas flow [7]. In principle, all fluid-mechanics problems can be attacked from the molecular viewpoint, but no such attempt will be made here. Note that the use of continuum calculus does not preclude the possibility of discontinuous jumps in fluid properties across a free surface or fluid interface or across a shock wave in a compressible fluid (Chap. 9). Our calculus in Chap. 4 must be flexible enough to handle discontinuous boundary conditions.

1.4 DIMENSIONS AND UNITS

A *dimension* is the measure by which a physical variable is expressed quantitatively. A *unit* is a particular way of attaching a number to the quantitative dimension. Thus length is a dimension associated with such variables as distance, displacement, width, deflection, and height, while centimeters and inches are both numerical units for expressing length. Dimension is a powerful concept about which a splendid tool called *dimensional analysis* has been developed (Chap. 5), while units are the nitty-gritty, the number which the customer wants as the final answer.

Systems of units have always varied widely from country to country, even after international agreements have been reached. Engineers need numbers and therefore need unit systems, and the numbers must be accurate because the safety of the public is at stake. You cannot design and build a piping system whose diameter is D and whose length is L. American engineers have persisted too long in clinging to British systems of units. There is too much margin for error in most British systems, and many an engineering student has flunked a test because of a missing or improper conversion factor of 12 or 144 or 32.2 or 60 or 1.8. Practicing engineers can make the same errors. The writer is aware from personal experience of a serious preliminary error in the design of an aircraft due to a missing factor of 32.2 to convert pounds of mass into slugs.

---

[1] The mean distance traveled by molecules between collisions.

Table 1.1

PRIMARY DIMENSIONS IN THE SI AND BG SYSTEMS

Primary dimension	SI unit	BG unit	Conversion factor
Mass $\{M\}$	Kilogram (kg)	Slug	1 slug = 14.5939 kg
Length $\{L\}$	Meter (m)	Foot (ft)	1 ft = 0.3048 m
Time $\{T\}$	Second (s)	Second (s)	1 s = 1 s
Temperature $\{\Theta\}$	Kelvin (K)	Rankine (°R)	1 K = 1.8°R

In 1872 an international meeting in France proposed a treaty called the Metric Convention, which was signed in 1875 by 17 countries including the United States. It was an improvement over British systems because its use of base 10 is the foundation of our number system, learned from childhood by all. Problems still remained because even the metric countries differed in their use of kiloponds instead of dynes or newtons, kilograms instead of grams, or calories instead of joules. To standardize the metric system, a General Conference of Weights and Measures attended in 1960 by 40 countries proposed the *International System of Units* (SI). We are now undergoing a painful period of transition to the SI system, an adjustment which may take the remainder of this century to complete. The professional societies have led the way. Since July 1, 1974 SI units have been required by all papers published by the American Society of Mechanical Engineers, which prepared a useful booklet explaining the SI system [8]. The present text will use SI units together with British gravitational (BG) units.

In fluid mechanics there are only four *primary dimensions* from which all other dimensions can be derived. They are mass, length, time, and temperature.[1] These dimensions and their units in both systems are given in Table 1.1. Note that the kelvin unit uses no degree symbol. The braces around a symbol like $\{M\}$ mean "the dimension of" mass. All other variables in fluid mechanics can be expressed in terms of $\{M\}$, $\{L\}$, $\{T\}$, and $\{\Theta\}$. For example, acceleration has the dimensions $\{LT^{-2}\}$. The most crucial of these secondary dimensions is that of force, which is directly related to mass, length, and time by Newton's second law

$$\mathbf{F} = m\mathbf{a} \tag{1.2}$$

From this we see that, dimensionally, $\{F\} = \{MLT^{-2}\}$. A constant of proportionality is avoided by defining the force unit exactly in terms of the primary units. Thus we define the newton and the pound of force

$$1 \text{ newton of force} = 1 \text{ N} \equiv 1 \text{ (kg} \cdot \text{m)/s}^2$$
$$1 \text{ pound of force} = 1 \text{ lbf} \equiv 1 \text{ (slug} \cdot \text{ft)/s}^2 = 4.4482 \text{ N} \tag{1.3}$$

In this book the abbreviation lbf is used for pound force and lb for pound mass. If instead one adopts other force units such as the dyne or the poundal or kilopond or

---

[1] If electromagnetic effects are important, a fifth primary dimension must be included, electric current $\{I\}$, the SI unit of which is the ampere (A).

Table 1.2

SECONDARY DIMENSIONS IN FLUID MECHANICS

Secondary dimension	SI unit	BG unit	Conversion factor
Area $\{L^2\}$	m²	ft²	1 m² = 10.764 ft²
Volume $\{L^3\}$	m³	ft³	1 m³ = 35.315 ft³
Velocity $\{LT^{-1}\}$	m/s	ft/s	1 ft/s = 0.3048 m/s
Acceleration $\{LT^{-2}\}$	m/s²	ft/s²	1 ft/s² = 0.3048 m/s²
Pressure or stress $\{ML^{-1}T^{-2}\}$	Pa = N/m²	lbf/ft²	1 lbf/ft² = 47.88 Pa
Angular velocity $\{T^{-1}\}$	s⁻¹	s⁻¹	1 s⁻¹ = 1 s⁻¹
Energy, heat, work $\{ML^2T^{-2}\}$	J = N · m	ft · lbf	1 ft · lbf = 1.3558 J
Power $\{ML^2T^{-3}\}$	W = J/s	(ft · lbf)/s	1(ft · lbf)/s = 1.3558 W
Density $\{ML^{-3}\}$	kg/m³	slugs/ft³	1 slug/ft³ = 515.4 kg/m³
Viscosity $\{ML^{-1}T^{-1}\}$	kg/(m · s)	slugs/(ft · s)	1 slug/(ft · s) = 47.88 kg/(m · s)
Specific heat $\{L^2T^{-2}\Theta^{-1}\}$	m²/(s² · K)	ft²/(s² · °R)	1 m²/(s² · K) = 5.980 ft²/(s² · °R)

adopts other mass units such as the gram or pound of mass, a constant of proportionality called g_c must be included in Eq. (1.2). We shall not use g_c in this book since it is not necessary in the SI and BG systems.

A list of some important secondary variables in fluid mechanics, with dimensions derived as combinations of the four primary dimensions, is given in Table 1.2. A more complete list of conversion factors is given in Appendix D.

EXAMPLE 1.1 A body weighs 1000 lbf when exposed to a standard earth gravity $g = 32.174$ ft/s². (*a*) What is its mass in kilograms? (*b*) What will the weight of this body be in newtons if it is exposed to the moon's standard acceleration $g_{\text{moon}} = 1.62$ m/s²? (*c*) How fast will the body accelerate if a net force of 400 lbf is applied to it on the moon or on the earth?

Solution

Part (a) Equation (1.2) holds with F = weight and $a = g_{\text{earth}}$

$$F = W = mg = 1000 \text{ lbf} = (m \text{ slugs})(32.174 \text{ ft/s}^2)$$

or

$$m = \frac{1000}{32.174} = (31.08 \text{ slugs})(14.5939 \text{ kg/slug}) = 453.6 \text{ kg} \qquad Ans. (a)$$

The change from 31.08 slugs to 453.6 kg illustrates the proper use of the conversion factor 14.5939 kg/slug.

Part (b) The mass of the body remains 453.6 kg regardless of its location. Equation (1.2) applies with a new value of a and hence a new force

$$F = W_{\text{moon}} = mg_{\text{moon}} = (453.6 \text{ kg})(1.62 \text{ m/s}^2) = 734.8 \text{ N} \qquad Ans. (b)$$

Part (c) This problem does not involve weight or gravity or position and is simply a direct application of Newton's law with an unbalanced force

$$F = 400 \text{ lbf} = ma = (31.08 \text{ slugs})(a \text{ ft/s}^2)$$

or $\qquad\qquad a = \dfrac{400}{31.08} = 12.43 \text{ ft/s}^2 = 3.79 \text{ m/s}^2$ *Ans. (c)*

This acceleration would be the same on the moon or earth or anywhere.

Many data in the literature are reported in inconvenient or arcane units suitable only to some industry or specialty or country. The engineer should convert these data into the SI or BG system before using them. This requires the systematic application of conversion factors, as in the following example.

EXAMPLE 1.2 An early viscosity unit in the cgs system is the poise, or g/(cm·s), named after J. L. M. Poiseuille, a French physician who performed pioneering experiments in 1840 on water flow in pipes. The viscosity of water (fresh or salt) at 293.16 K = 20°C is approximately $\mu = 0.01$ poise. Express this value in (a) SI and (b) BG units.

Solution

Part (a) $\qquad\qquad \mu = [0.01 \text{ g/(cm} \cdot \text{s)}] \dfrac{1 \text{ kg}}{1000 \text{ g}} (100 \text{ cm/m}) = 0.001 \text{ kg/(m} \cdot \text{s)}$ *Ans. (a)*

Part (b) $\qquad\qquad \mu = [0.001 \text{ kg/(m} \cdot \text{s)}] \dfrac{1 \text{ slug}}{14.59 \text{ kg}} (0.3048 \text{ m/ft})$

$$= 2.09 \times 10^{-5} \text{ slug/(ft} \cdot \text{s)}$$ *Ans. (b)*

Note: Result (b) could have been found directly from (a) by dividing (a) by the viscosity conversion factor 47.88 listed in Table 1.2.

We repeat our advice: faced with data in unusual units, convert them immediately into either SI or BG units because (1) it is more professional and (2) theoretical equations in fluid mechanics are *dimensionally consistent* and require no further conversion factors when these two fundamental unit systems are used, as the following example shows.

EXAMPLE 1.3 A useful theoretical equation for computing the relation between pressure, velocity, and altitude in a steady flow of a nearly inviscid, nearly incompressible fluid with negligible heat transfer and shaft work[1] is the Bernoulli relation, named after Daniel Bernoulli, who published a hydrodynamics textbook in 1738:

$$p_0 = p + \tfrac{1}{2}\rho V^2 + \rho g Z \tag{1.4}$$

where p_0 = stagnation pressure
$\qquad p$ = pressure in moving fluid
$\qquad V$ = velocity
$\qquad \rho$ = density
$\qquad Z$ = altitude
$\qquad g$ = gravitational acceleration

[1] That's an awful lot of assumptions, which need further study in Chap. 3.

(*a*) Show that Eq. (1.4) satisfies the principle of dimensional homogeneity, which states that all additive terms in a physical equation must have the same dimensions. (*b*) Show that consistent units result without additional conversion factors in SI units. (*c*) Repeat (*b*) for BG units.

Solution

Part (a)   We can express Eq. (1.4) dimensionally, using braces by entering the dimensions of each term from Table 1.2:

$$\{ML^{-1}T^{-2}\} = \{ML^{-1}T^{-2}\} + \{ML^{-3}\}\{L^2T^{-2}\} + \{ML^{-3}\}\{LT^{-2}\}\{L\}$$
$$= \{ML^{-1}T^{-2}\} \qquad \text{for all terms} \qquad\qquad Ans. \ (a)$$

Part (b)   Enter the SI units for each quantity from Table 1.2:

$$\{N/m^2\} = \{N/m^2\} + \{kg/m^3\}\{m^2/s^2\} + \{kg/m^3\}\{m/s^2\}\{m\}$$
$$= \{N/m^2\} + \{kg/(m \cdot s^2)\}$$

The right-hand side looks bad until we remember from Eq. (1.3) that $1 \ kg = 1 \ (N \cdot s^2)/m$. Substituting, we obtain

$$\{kg/(m \cdot s^2)\} = \frac{\{(N \cdot s^2)/m\}}{\{m \cdot s^2\}} = \{N/m^2\} \qquad\qquad Ans. \ (b)$$

Thus all terms in Bernoulli's equation will have units of pascals, or newtons per square meter, when SI units are used. No conversion factors are needed, which is true of all theoretical equations in fluid mechanics.

Part (c)   Introducing BG units for each term, we have

$$\{lbf/ft^2\} = \{lbf/ft^2\} + \{slugs/ft^3\}\{ft^2/s^2\} + \{slugs/ft^3\}\{ft/s^2\}\{ft\}$$
$$= \{lbf/ft^2\} + \{slugs/(ft \cdot s^2)\}$$

But, from Eq. (1.3), $1 \ slug = 1 \ lbf \cdot s^2/ft$, so that

$$\{slugs/(ft \cdot s^2)\} = \frac{\{(lbf \cdot s^2)/ft\}}{\{ft \cdot s^2\}} = \{lbf/ft^2\} \qquad\qquad Ans. \ (c)$$

All terms have the unit pounds force per square foot. No conversion factors are needed in the BG system either.

There is still a tendency in English-speaking countries to use pounds force per square inch as a pressure unit because the numbers are more manageable. For example, standard atmospheric pressure is $14.7 \ lbf/in^2 = 2116 \ lbf/ft^2 = 101{,}300 \ Pa$. The pascal is a small unit because the newton is less than $\frac{1}{4}$ lbf and a square meter is a very large area. It is felt nevertheless that the pascal will gradually gain universal acceptance; e.g., repair manuals for American automobiles now specify pressure measurements in pascals.

A final warning regarding dimensions and units: empirical formulas in engineering are sometimes dimensionally inconsistent, particularly those proposed by early workers. Consider this example.

EXAMPLE 1.4   In 1890 Robert Manning, an Irish engineer, proposed the following empirical formula for the average velocity V in uniform flow due to gravity down an open channel (BG units)

$$V = \frac{1.49}{n} R^{2/3} S^{1/2} \tag{1.5}$$

where R = hydraulic radius of the channel (Chaps. 6 and 10)
  S = channel slope (tangent of angle bottom makes with horizontal)
  n = Manning's roughness factor (Chap. 10)

n is a constant for a given surface condition for the walls and bottom of the channel. (*a*) Is Manning's formula dimensionally consistent? (*b*) Equation (1.5) is commonly taken to be valid in BG units with n taken as dimensionless. Rewrite it in SI units.

Solution
Part (a)   Introduce dimensions for each term. The slope S, being a tangent or ratio, is dimensionless, denoted by {unity} or {1}. Equation (1.5) in dimensional form is

$$\left\{\frac{L}{T}\right\} = \left\{\frac{1.49}{n}\right\}\{L^{2/3}\}\{1\}$$

This formula cannot be consistent unless $\{1.49/n\} = \{L^{1/3}/T\}$. If n is dimensionless (and it is never listed with units in textbooks), then the numerical value 1.49 must have units. This can be tragic to an engineer working in a different unit system unless the discrepancy is properly documented. In fact, Manning's formula, though popular, is inconsistent both dimensionally and physically and does not properly account for channel-roughness effects except in a narrow range of parameters, for water only.

Part (b)   From part (*a*), the number 1.49 must have dimensions $\{L^{1/3}/T\}$ and thus in BG units equals 1.49 ft$^{1/3}$/s. By using the SI conversion factor for length we have

$$(1.49 \text{ ft}^{1/3}/\text{s})(0.3048 \text{ m/ft})^{1/3} = 1.00 \text{ m}^{1/3}/\text{s}$$

Therefore Manning's formula in SI units becomes

$$V = \frac{1.0}{n} R^{2/3} S^{1/2} \qquad\qquad \textit{Ans. (b)}\quad (1.6)$$

with R in meters and V in meters per second. Actually, we misled you: this is the way Manning, a metric user, first proposed the formula. It was later converted into BG units. Such dimensionally inconsistent formulas are dangerous and should either be reanalyzed or treated as having very limited application.

1.5   PROPERTIES OF THE VELOCITY FIELD

In a given flow situation, the determination, by experiment or theory, of the properties of the fluid as a function of position and time is considered to be the *solution* to the problem. In almost all cases, the emphasis is on the space-time distribution of the fluid properties. One rarely keeps track of the actual fate of the specific fluid particles.[1] This treatment of properties as continuum-field functions distinguishes fluid mechanics from solid mechanics, where we are more likely to be interested in the trajectories of individual particles or systems.

[1] One example where fluid-particle paths are important is in water-quality analysis of the fate of contaminant discharges.

Eulerian and Lagrangian Descriptions

There are two different points of view in analyzing problems in mechanics. The first view, appropriate to fluid mechanics, is concerned with the field of flow and is called the *eulerian* method of description. In the eulerian method we compute the pressure field $p(x, y, z, t)$ of the flow pattern, not the pressure changes $p(t)$ which a particle experiences as it moves through the field.

The second method, which follows an individual particle moving through the flow, is called the *lagrangian* description. The lagrangian approach, which is more appropriate to solid mechanics, will not be treated in this book. However, certain numerical analyses of sharply bounded fluid flows, such as the motion of isolated fluid droplets, are very conveniently computed in lagrangian coordinates [1].

Fluid-dynamic measurements are also suited to the eulerian system. For example, when a pressure probe is introduced into a laboratory flow, it is fixed at a specific position (x, y, z). Its output thus contributes to the description of the eulerian pressure field $p(x, y, z, t)$. To simulate a lagrangian measurement, the probe would have to move downstream at the fluid particle speeds; this is sometimes done in oceanographic measurements, where flowmeters drift along with the prevailing currents.

The two different descriptions can be contrasted in the analysis of traffic flow along a freeway. A certain length of freeway may be selected for study and called the field of flow. Obviously, as time passes, various cars will enter and leave the field, and the identity of the specific cars within the field will constantly be changing. The traffic engineer ignores specific cars and concentrates on their average velocity as a function of time and position within the field, plus the flow rate or number of cars per hour passing a given section of the freeway. This engineer is using an eulerian description of the traffic flow. Other investigators, such as the police or social scientists, may be interested in the path or speed or destination of specific cars in the field. By following a specific car as a function of time they are using a lagrangian description of the flow.

The Velocity Field

Foremost among the properties of a flow is the velocity field $\mathbf{V}(x, y, z, t)$. In fact, determining the velocity is often tantamount to solving a flow problem, since other properties follow directly from the velocity field. Chapter 2 is devoted to the calculation of the pressure field once the velocity field is known. Books on heat transfer [e.g. 9] are essentially devoted to finding the temperature field from known velocity fields.

In general, velocity is a vector function of position and time and thus has three components u, v, and w, each a scalar field in itself

$$\mathbf{V}(x, y, z, t) = \mathbf{i}u(x, y, z, t) + \mathbf{j}v(x, y, z, t) + \mathbf{k}w(x, y, z, t) \tag{1.7}$$

The use of u, v, and w instead of the more logical component notation V_x, V_y, and V_z is the result of an almost unbreakable custom in fluid mechanics.

Several other quantities, called *kinematic properties*, can be derived by mathematically manipulating the velocity-field function. Examples are the displacement vector, the acceleration vector, the local angular-velocity vector, and the volume flux through a surface. The most important of these is the acceleration vector.

The Acceleration of a Particle

The acceleration **a** is fundamental to fluid mechanics since it occurs in Newton's law of dynamics. We require the total time derivative of **V**, which depends upon four independent variables x, y, z, and t. Therefore the acceleration vector will contain four separate derivative terms

$$\mathbf{a} = \frac{d\mathbf{V}}{dt} = \frac{\partial \mathbf{V}}{\partial t} + \frac{\partial \mathbf{V}}{\partial x}\frac{dx}{dt} + \frac{\partial \mathbf{V}}{\partial y}\frac{dy}{dt} + \frac{\partial \mathbf{V}}{\partial z}\frac{dz}{dt} \tag{1.8}$$

But the infinitesimal changes in position of a particle must be directly related by definition to the local velocity components

$$dx = u\,dt \qquad dy = v\,dt \qquad dz = w\,dt \tag{1.9}$$

Substituting for dx, dy, and dz in Eq. (1.8), we obtain the desired expression for the acceleration of a particle

$$\frac{d\mathbf{V}}{dt} = \frac{\partial \mathbf{V}}{\partial t} + \left(u\frac{\partial \mathbf{V}}{\partial x} + v\frac{\partial \mathbf{V}}{\partial y} + w\frac{\partial \mathbf{V}}{\partial z} \right) \tag{1.10}$$

The first term on the right-hand side is called the *local acceleration*, which vanishes if the flow is steady, i.e., independent of time. The last three terms in the parentheses are called the *convective acceleration*, which arises when the particle moves through regions of varying velocity, e.g., a nozzle or diffuser.

As a memory device, the convective-acceleration terms can be written as a dot product involving the vector **V** and the *gradient operator* ∇

$$\nabla = \mathbf{i}\frac{\partial}{\partial x} + \mathbf{j}\frac{\partial}{\partial y} + \mathbf{k}\frac{\partial}{\partial z} \tag{1.11}$$

The short notation for convective acceleration is thus

$$u\frac{\partial \mathbf{V}}{\partial x} + v\frac{\partial \mathbf{V}}{\partial y} + w\frac{\partial \mathbf{V}}{\partial z} \equiv (\mathbf{V}\cdot\nabla)\mathbf{V} \tag{1.12}$$

In this way the total acceleration [Eq. (1.10)] of a particle in the eulerian system can be written in the short form

$$\mathbf{a} = \frac{d\mathbf{V}}{dt} = \frac{\partial \mathbf{V}}{\partial t} + (\mathbf{V}\cdot\nabla)\mathbf{V} \tag{1.13}$$

Local   Convective

Only an unsteady flow can induce the local-acceleration term. But a steady flow, where **V** does not vary with time, will cause a convective acceleration which may be quite large, as in Example 1.6. Because the convective term $(\mathbf{V}\cdot\nabla)\mathbf{V}$ contains products of variables, it is a nonlinear term and causes some mathematical difficulties with exact differential analyses of flow problems (Chap. 4). On the other hand, it properly models fluid convective effects, and so we must keep it and be glad we got the term right.

Note that Eq. (1.13) contains a general expression for the eulerian time-derivative operator following a particle

$$\frac{d}{dt} = \frac{\partial}{\partial t} + (\mathbf{V} \cdot \nabla) \tag{1.14}$$

This operator can be applied to any fluid property, scalar or vector, e.g., the pressure

$$\frac{dp}{dt} = \frac{\partial p}{\partial t} + (\mathbf{V} \cdot \nabla)p = \frac{\partial p}{\partial t} + u\frac{\partial p}{\partial x} + v\frac{\partial p}{\partial y} + w\frac{\partial p}{\partial z} \tag{1.15}$$

We emphasize that this total time derivative follows a particle of fixed identity, making it convenient for expressing laws of particle mechanics in the eulerian fluid-field description. The operator d/dt is sometimes given a special name such as *substantial derivative* or *material derivative* and often assigned a special symbol such as D/Dt as a further reminder that it contains four terms and follows a fixed particle.

EXAMPLE 1.5   Given the eulerian velocity-vector field

$$\mathbf{V} = 3t\mathbf{i} + xz\mathbf{j} + ty^2\mathbf{k}$$

find the acceleration of a particle.

Solution   First note the specific given components

$$u = 3t \qquad v = xz \qquad w = ty^2$$

Then evaluate the vector derivatives required for Eq. (1.10)

$$\frac{\partial \mathbf{V}}{\partial t} = \mathbf{i}\frac{\partial u}{\partial t} + \mathbf{j}\frac{\partial v}{\partial t} + \mathbf{k}\frac{\partial w}{\partial t} = 3\mathbf{i} + y^2\mathbf{k}$$

$$\frac{\partial \mathbf{V}}{\partial x} = z\mathbf{j} \qquad \frac{\partial \mathbf{V}}{\partial y} = 2ty\mathbf{k} \qquad \frac{\partial \mathbf{V}}{\partial z} = x\mathbf{j}$$

This could have been worse: there are only five terms in all, whereas there could have been as many as twelve. Substitute directly into Eq. (1.10)

$$\frac{d\mathbf{V}}{dt} = (3\mathbf{i} + y^2\mathbf{k}) + (3t)(z\mathbf{j}) + (xz)(2ty\mathbf{k}) + (ty^2)(x\mathbf{j})$$

Collect terms for the final result

$$\frac{d\mathbf{V}}{dt} = 3\mathbf{i} + (3tz + txy^2)\mathbf{j} + (2xyzt + y^2)\mathbf{k} \qquad\qquad Ans.$$

Assuming that **V** is valid everywhere as given, this acceleration applies to all positions and times within the flow field.

Written out in full, the acceleration in Eq. (1.10) contains twelve separate terms, four associated with the time derivative of each scalar velocity component

$$\frac{d\mathbf{V}}{dt} = \mathbf{i}\,\frac{du}{dt} + \mathbf{j}\,\frac{dv}{dt} + \mathbf{k}\,\frac{dw}{dt} \tag{1.16}$$

where

$$\frac{du}{dt} = \frac{\partial u}{\partial t} + u\,\frac{\partial u}{\partial x} + v\,\frac{\partial u}{\partial y} + w\,\frac{\partial u}{\partial z}$$

$$\frac{dv}{dt} = \frac{\partial v}{\partial t} + u\,\frac{\partial v}{\partial x} + v\,\frac{\partial v}{\partial y} + w\,\frac{\partial v}{\partial z}$$

$$\frac{dw}{dt} = \frac{\partial w}{\partial t} + u\,\frac{\partial w}{\partial x} + v\,\frac{\partial w}{\partial y} + w\,\frac{\partial w}{\partial z}$$

The nine convective terms are all nonlinear. This is what we mean by stating that acceleration can cause mathematical difficulty in a flow analysis; however, many problems, e.g., duct flows, are concerned with only one or two of these terms.

Even in steady flow the convective acceleration can be very large if spatial changes in velocity are large. The following example will illustrate.

EXAMPLE 1.6   Flow through a converging nozzle can be approximated by a one-dimensional velocity distribution $u = u(x)$. For the nozzle shown, assume that the velocity varies linearly from $u = V_0$ at the entrance to $u = 3V_0$ at the exit

$$u(x) = V_0\!\left(1 + \frac{2x}{L}\right) \qquad \frac{\partial u}{\partial x} = \frac{2V_0}{L}$$

(a) Compute the acceleration du/dt as a general function of x; (b) evaluate du/dt at the entrance and exit if $V_0 = 10$ ft/s and $L = 1$ ft.

Solution
Part (a)   From Eq. (1.16) with $v = w = 0$ and u independent of time we have an acceleration

$$\frac{du}{dt} = u\,\frac{\partial u}{\partial x} = \left[V_0\!\left(1 + \frac{2x}{L}\right)\right]\frac{2V_0}{L} = \frac{2V_0^2}{L}\left(1 + \frac{2x}{L}\right) \qquad Ans.\ (a)$$

The fluid acceleration is definitely not zero even though this is a steady flow.

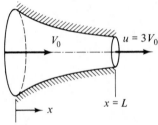

Fig. E1.6 $x = 0$

Part (b)   Introducing the numerical values $V_0 = 10$ ft/s and $L = 1$ ft, we have, at the entrance, $x = 0$,

$$\left(\frac{du}{dt}\right)_{x=0} = \frac{2V_0^2}{L} = \frac{2(10 \text{ ft/s})^2}{1 \text{ ft}} = 200 \text{ ft/s}^2 \qquad \textit{Ans. (b)}$$

This is more than 6 times the acceleration of gravity. At the exit, $x = L$, we have

$$\left(\frac{du}{dt}\right)_{x=L} = \frac{2V_0^2}{L}(1 + 2) = \frac{6V_0^2}{L} = 3\left(\frac{du}{dt}\right)_{x=0} = 600 \text{ ft/s}^2 \qquad \textit{Ans. (b)}$$

The exit acceleration is 18.6 times that of gravity. Thus very large dynamic forces can be involved in what are otherwise rather innocent-appearing fluid motions.

Volume and Mass Rate of Flow

Probably the second most important kinematic property is the volume rate of flow Q passing through an (imaginary) surface in the flow field.

Suppose that the surface S in Fig. 1.3a is a sort of magic wire mesh through which the flow passes without resistance. How much volume of fluid passes through S in unit time? If, as usual, $\mathbf{V}$ varies with position, we must integrate over the elemental surface area dA in Fig. 1.3a. Also, $\mathbf{V}$ may pass through dA at some angle θ off the normal. Let $\mathbf{n}$ be defined as the unit vector normal to dA. Then the volume swept out of dA in time dt is the volume of the slanted parallelepiped in Fig. 1.3b

$$d\mathcal{V} = V \, dt \, dA \cos \theta = (\mathbf{V} \cdot \mathbf{n}) \, dA \, dt \qquad (1.17)$$

The integral of $d\mathcal{V}/dt$ is the total volume rate of flow Q through the surface S

$$Q = \int_S (\mathbf{V} \cdot \mathbf{n}) \, dA = \int_S V_n \, dA \qquad (1.18)$$

We could replace $\mathbf{V} \cdot \mathbf{n}$ by its equivalent, V_n, the component of $\mathbf{V}$ normal to dA, but the use of the dot product allows Q to have a sign to distinguish between inflow and outflow. By convention throughout this book we consider $\mathbf{n}$ to be the *outward*

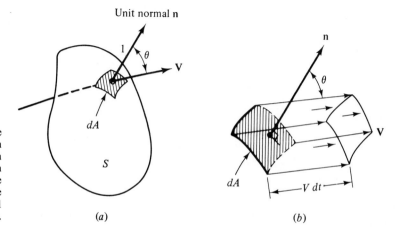

Fig. 1.3   Volume rate of flow through an arbitrary surface: (*a*) an elemental area dA on the surface; (*b*) the incremental volume swept through dA equals $V \, dt \, dA \cos \theta$.

Unit normal **n**

(*a*)

(*b*)

normal unit vector. Therefore $\mathbf{V} \cdot \mathbf{n}$ denotes outflow if it is positive and inflow if negative. This will be an extremely useful housekeeping device when computing volume and mass flow in the basic control-volume relations of Chap. 3.

Volume flow can be multiplied by density to obtain the mass flow $\dot{m}$. If density varies over the surface, it must be part of the surface integral

$$\dot{m} = \int_S \rho (\mathbf{V} \cdot \mathbf{n}) \, dA = \int_S \rho V_n \, dA \tag{1.19}$$

If density is constant, it comes out of the integral and a direct proportionality results:

Constant density: $\qquad\qquad\qquad \dot{m} = \rho Q \tag{1.20}$

The volume flow is often used, particularly in duct flow, to define an average velocity passing through the surface

$$V_{\text{av}} = \frac{Q}{A} = \frac{\int_S V_n \, dA}{\int_S dA} \tag{1.21}$$

This provides a useful reference velocity which can be used to scale the results of the analysis to other flows.

EXAMPLE 1.7   Using the velocity vector of Example 1.5, evaluate the volume flow and the average velocity through the square surface whose vertices are at $(0, 1, 0)$, $(0, 1, 2)$, $(2, 1, 2)$, and $(2, 1, 0)$.

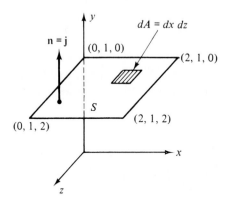

Fig. E1.7

Solution   The surface S is shown in the sketch and is such that $\mathbf{n} = \mathbf{j}$ and $dA = dx\, dz$ everywhere. The velocity field is $\mathbf{V} = 3t\mathbf{i} + xz\mathbf{j} + ty^2\mathbf{k}$. The normal component to S is $\mathbf{V} \cdot \mathbf{n} = \mathbf{V} \cdot \mathbf{j} = v$, the y component, which equals xz. The limits on the integral for Q are 0 to 2 for both dx and dz. The volume flow is thus

$$Q = \int_S V_n \, dA = \int_0^2 \int xz \, dx \, dz = 4.0 \text{ units} \qquad\qquad Ans.$$

The area of the surface is $2(2) = 4$ units. Then the average velocity from Eq. (1.21) is

$$V_{av} = \frac{Q}{A} = \frac{4.0}{4.0} = 1.0 \text{ unit} \qquad\qquad Ans.$$

The problem is too artificial to assign units to these results. The next example is more practical.

EXAMPLE 1.8   At low velocities, the flow through a long circular tube has a paraboloid velocity distribution

$$u = u_{max}\left(1 - \frac{r^2}{R^2}\right)$$

where R is the tube radius and u_{max} is the maximum velocity, which occurs at the tube centerline. (*a*) Find a general expression for volume flow and average velocity through the tube; (*b*) compute the volume flow if $R = 3$ cm and $u_{max} = 8$ m/s; and (*c*) compute the mass flow if $\rho = 1000$ kg/m³.

Solution

Part (a)   The area S is the cross section of the tube, and $\mathbf{n} = \mathbf{i}$. The normal component $\mathbf{V} \cdot \mathbf{n} = \mathbf{V} \cdot \mathbf{i} = u$. Since u varies only with r, the element dA can be taken to be the annular strip $dA = 2\pi r\, dr$. The volume flow becomes

$$Q = \int_S u\, dA = \int_0^R u_{max}\left(1 - \frac{r^2}{R^2}\right)2\pi r\, dr$$

Carrying out the integration over r, we obtain

$$Q = \tfrac{1}{2}u_{max}\pi R^2 \qquad\qquad Ans.\ (a)$$

The average velocity is

$$u_{av} = \frac{Q}{A} = \frac{\tfrac{1}{2}u_{max}\pi R^2}{\pi R^2} = \tfrac{1}{2}u_{max} \qquad\qquad Ans.\ (a)$$

The average velocity is half the maximum, which is an accepted result for low-speed, or *laminar*, flow through a long tube. The theory of tube flow is given in Sec. 6.4.

Part (b)   For the given numerical values

$$Q = \tfrac{1}{2}(8 \text{ m/s})\pi(0.03 \text{ m})^2 = 0.0113 \text{ m}^3/\text{s} \qquad\qquad Ans.\ (b)$$

Part (c)   For the given density, assumed constant,

$$\dot{m} = \rho Q = (1000 \text{ kg/m}^3)(0.0113 \text{ m}^3/\text{s}) = 11.3 \text{ kg/s} \qquad\qquad Ans.\ (c)$$

Other Kinematic Properties

Several other kinematic properties are derived from the velocity field by spatial differentiation. They will be treated in detail in Chap. 4. The vector operators of divergence and curl both have important physical meaning when applied to the velocity vector. We discuss these two operations here to illustrate the scope of the velocity-field properties.

The divergence of velocity equals the rate of volume expansion of a fluid element per unit initial volume

$$\frac{1}{\mathcal{V}}\frac{d\mathcal{V}}{dt} = \text{div } \mathbf{V} = \nabla \cdot \mathbf{V} \tag{1.22}$$

In like manner, we shall show in Chap. 4 that the curl of the velocity equals twice the local angular velocity $\boldsymbol{\omega}$ of a fluid element

$$\boldsymbol{\omega} = \tfrac{1}{2}\text{curl }\mathbf{V} = \tfrac{1}{2}\nabla \times \mathbf{V} \tag{1.23}$$

Neither of these two properties is of any great direct interest in flow analysis, but it happens that for many important flows both or either is negligibly small. If the volume expansion is negligible, the flow is termed *incompressible*

$$\nabla \cdot \mathbf{V} = \frac{\partial u}{\partial x} + \frac{\partial v}{\partial y} + \frac{\partial w}{\partial z} = 0 \tag{1.24}$$

In like manner, a flow with negligible local angular velocity is called *irrotational*

$$\nabla \times \mathbf{V} = \begin{vmatrix} \mathbf{i} & \mathbf{j} & \mathbf{k} \\ \dfrac{\partial}{\partial x} & \dfrac{\partial}{\partial y} & \dfrac{\partial}{\partial z} \\ u & v & w \end{vmatrix} = 0 \tag{1.25}$$

The theory of incompressible and/or irrotational flows is considerably simplified from the general case. If a flow is both incompressible and irrotational, we can treat it with the technique of linear potential analysis borrowed from electromagnetic theory. Such idealized problems, called *potential flows*, are discussed in Chap. 8.

EXAMPLE 1.9   Determine whether the velocity field of Example 1.5 is incompressible, irrotational, both, or neither.

Solution   The given velocity field is

$$\mathbf{V} = 3t\mathbf{i} + xz\mathbf{j} + ty^2\mathbf{k}$$

The divergence of this, from Eq. (1.24), is

$$\nabla \cdot \mathbf{V} = \frac{\partial}{\partial x}(3t) + \frac{\partial}{\partial y}(xz) + \frac{\partial}{\partial z}(ty^2) \equiv 0$$

Therefore this velocity field is incompressible. To be truthful, though, it is not a very realistic flow field. The curl of the velocity field is, from Eq. (1.25),

$$\nabla \times \mathbf{V} = \begin{vmatrix} \mathbf{i} & \mathbf{j} & \mathbf{k} \\ \dfrac{\partial}{\partial x} & \dfrac{\partial}{\partial y} & \dfrac{\partial}{\partial z} \\ 3t & xz & ty^2 \end{vmatrix} = (2ty - x)\mathbf{i} + z\mathbf{k}$$

This is not zero; hence the flow field is rotational, not irrotational. Potential theory cannot be used to analyze this flow.

1.6 THERMODYNAMIC PROPERTIES OF A FLUID

While the velocity field **V** is the most important fluid property, it interacts closely with the thermodynamic properties of the fluid. We have already introduced into the discussion the three most common such properties

1. Pressure p
2. Density ρ
3. Temperature T

These three are constant companions of the velocity vector in flow analyses. Four other thermodynamic properties become important when work, heat, and energy balances are treated (Chaps. 3 and 4):

4. Internal energy e
5. Enthalpy $h = \hat{u} + p/\rho$
6. Entropy s
7. Specific heats c_p and c_v

In addition, friction and heat-conduction effects are governed by the two so-called *transport properties*:

8. Coefficient of viscosity μ
9. Thermal conductivity k

All nine of these quantities are true thermodynamic properties which are determined by the thermodynamic condition or *state* of the fluid. For example, for a single-phase substance such as water or oxygen, two basic properties such as pressure and temperature are sufficient to fix the value of all the others:

$$\rho = \rho(p, T) \qquad h = h(p, T) \qquad \mu = \mu(p, T) \tag{1.26}$$

and so on for every quantity in the list. Note that the specific volume, so important in thermodynamic analyses, is omitted here in favor of its inverse, the density ρ.

Recall that thermodynamic properties describe the state of a *system*, i.e., a collection of matter of fixed identity which interacts with its surroundings. In most cases here the system will be a small fluid element, and all properties will be assumed to be continuum properties of the flow field: $\rho = \rho(x, y, z, t)$, etc.

Recall also that thermodynamics is normally concerned with *static* systems, whereas fluids are usually in variable motion with constantly changing properties. Do the static properties retain their meaning in a fluid flow which is technically not in equilibrium? The answer is yes, from a statistical argument. In gases at normal pressure (and even more so for liquids) an enormous number of molecular collisions occurs over a very short distance of the order of 1 μm, so that a fluid subjected to sudden changes rapidly adjusts itself toward equilibrium. We therefore assume that all the thermodynamic properties listed above exist as point functions in a flowing fluid and follow all the laws and state relations of ordinary equilibrium thermodynamics. There are, of course, important nonequilibrium effects such as chemical and nuclear reactions in flowing fluids which are not treated in this text.

Potential and Kinetic Energy

In thermostatics the only energy in a substance is that stored in a system by molecular activity and molecular bonding forces. This is commonly denoted as *internal energy* $\hat{u}$. A commonly accepted adjustment to this static situation for fluid flow is to add two more energy terms which arise from newtonian mechanics, the potential energy and kinetic energy.

The potential energy equals the work required to move the system of mass m from the origin to a position vector $\mathbf{r} = \mathbf{i}x + \mathbf{j}y + \mathbf{k}z$ against a gravity field $\mathbf{g}$. Its value is $-m\mathbf{g} \cdot \mathbf{r}$, or $-\mathbf{g} \cdot \mathbf{r}$ per unit mass. The kinetic energy equals the work required to change the speed of the mass from zero to velocity V. Its value is $\frac{1}{2}mV^2$ or $\frac{1}{2}V^2$ per unit mass. Then by common convention the internal energy e per unit mass in fluid mechanics is the sum of three terms

$$e = \hat{u} + \tfrac{1}{2}V^2 + (-\mathbf{g} \cdot \mathbf{r}) \tag{1.27}$$

Also, throughout this book we shall define z as upward, so that $\mathbf{g} = -g\mathbf{k}$ and $\mathbf{g} \cdot \mathbf{r} = -gz$. Then Eq. (1.27) becomes

$$e = \hat{u} + \tfrac{1}{2}V^2 + gz \tag{1.28}$$

The molecular internal energy $\hat{u}$ is a function of T and p for the single-phase pure substance, whereas the potential and kinetic energy are kinematic properties.

State Relations for Gases

Thermodynamic properties are found both theoretically and experimentally to be related to each other by state relations which differ for each substance. As mentioned, we shall confine ourselves here to single-phase pure substances, e.g., water in its liquid phase. The second most common fluid, air, is a mixture of gases, but since the mixture ratios remain nearly constant between 160 and 2200 K, in this temperature range air can be considered to be a pure substance.

All gases at high temperatures and low pressures (relative to their critical point) are in good agreement with the *perfect-gas law*

$$p = \rho R T \qquad R = c_p - c_v = \text{gas constant} \tag{1.29}$$

Since Eq. (1.29) is dimensionally consistent, R has the same dimensions as specific heat, $\{L^2 T^{-2} \Theta^{-1}\}$, or velocity squared per temperature unit (kelvin or degree Rankine). Each gas has its own constant R equal to a universal constant Λ divided by the molecular weight

$$R_{\text{gas}} = \frac{\Lambda}{M_{\text{gas}}} \tag{1.30}$$

where $\Lambda = 49{,}700 \text{ ft}^2/(\text{s}^2 \cdot {}^\circ\text{R}) = 8310 \text{ m}^2/(\text{s}^2 \cdot \text{K})$. Most applications in this book are for air, with $M = 28.97$

$$R_{\text{air}} = 1717 \text{ ft}^2/(\text{s}^2 \cdot {}^\circ\text{R}) = 287 \text{ m}^2/(\text{s}^2 \cdot \text{K}) \tag{1.31}$$

Standard atmospheric pressure is 2116 lbf/ft^2, and standard temperature is 60°F = 520°R. Thus standard air density is

$$\rho_{air} = \frac{2116}{(1717)(520)} = 0.00237 \text{ slug/ft}^3 = 1.22 \text{ kg/m}^3 \tag{1.32}$$

This is a nominal value suitable for problems.

One proves in thermodynamics that Eq. (1.29) requires that the internal molecular energy $\hat{u}$ of a perfect gas vary only with temperature: $\hat{u} = \hat{u}(T)$. Therefore the specific heat c_v also varies only with temperature

$$c_v = \left(\frac{\partial \hat{u}}{\partial T}\right)_\rho = \frac{d\hat{u}}{dT} = c_v(T)$$

or

$$d\hat{u} = c_v(T)\, dT \tag{1.33}$$

In like manner h and c_p of a perfect gas also vary only with temperature

$$h = \hat{u} + \frac{p}{\rho} = \hat{u} + RT = h(T)$$

$$c_p = \left(\frac{\partial h}{\partial T}\right)_p = \frac{dh}{dT} = c_p(T) \tag{1.34}$$

$$dh = c_p(T)\, dT$$

The ratio of specific heats of a perfect gas is an important dimensionless parameter in compressible-flow analysis (Chap. 9)

$$\gamma = \frac{c_p}{c_v} = \gamma(T) \geq 1 \tag{1.35}$$

As a first approximation in airflow analysis we commonly take c_p, c_v, and γ to be constant

$$\gamma_{air} \approx 1.4$$

$$c_v = \frac{R}{\gamma - 1} \approx 4293 \text{ ft}^2/(\text{s}^2 \cdot °\text{R}) = 718 \text{ m}^2/(\text{s}^2 \cdot \text{K})$$

$$c_p = \frac{\gamma R}{\gamma - 1} \approx 6010 \text{ ft}^2/(\text{s}^2 \cdot °\text{R}) = 1005 \text{ m}^2/(\text{s}^2 \cdot \text{K})$$

Actually, for all gases, c_p and c_v increase gradually with temperature, and γ decreases gradually. Experimental values of the specific-heat ratio for eight common gases are shown in Fig. 1.4.

Many flow problems involve steam. Typical steam operating conditions are relatively close to the critical point, so that the perfect-gas approximation is inaccurate. The properties of steam are therefore available in tabular form [12], but the error of using the perfect-gas law is not usually great, as the following example shows.

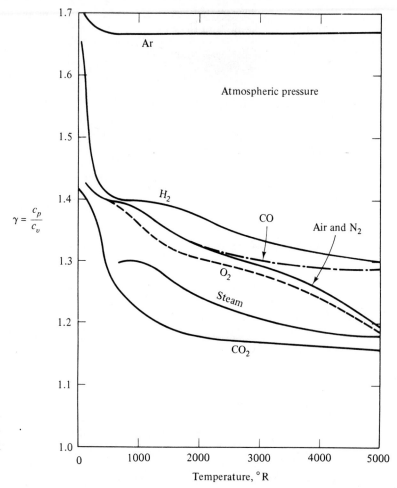

$$\gamma = \frac{c_p}{c_v}$$

Ar

Atmospheric pressure

H_2

CO

Air and N_2

O_2

Steam

CO_2

Fig. 1.4 Specific-heat ratio of eight common gases as a function of temperature. (*Data from Ref. 11.*)

Temperature, °R

EXAMPLE 1.10   Estimate ρ and c_p of steam at $100\ \mathrm{lbf/in^2}$ amd $400°F$ (*a*) by a perfect-gas approximation and (*b*) from the ASME Steam Tables [12].

Solution

Part (*a*)   First convert to BG units: $p = 100\ \mathrm{lbf/in^2} = 14{,}400\ \mathrm{lb/ft^2}$, $T = 400°F = 860°R$. From Appendix Table A.4 the molecular weight of H_2O is $2M_H + M_O = 2(1.008) + 16.0 = 18.016$. Then from Eq. (1.30) the gas constant of steam is approximately

$$R = \frac{49{,}700}{18.016} = 2759\ \mathrm{ft^2/(s^2 \cdot °R)}$$

whence, from the perfect-gas law,

$$\rho \approx \frac{p}{RT} = \frac{14{,}400}{2759(860)} = 0.00607\ \mathrm{slug/ft^3} \qquad\qquad Ans.\ (a)$$

From Fig. 1.4, γ of steam at 860°R is approximately 1.30. Then, from Eq. (1.36)

$$c_p \approx \frac{\gamma R}{\gamma - 1} = \frac{1.30(2759)}{1.30 - 1} = 12{,}000 \text{ ft}^2/(\text{s}^2 \cdot {}^\circ\text{R}) \qquad \textit{Ans. (a)}$$

Part (b) From Ref. 12 (table 3, p. 153) the specific volume of steam at 100 lbf/in² and 400°F is $(4.935 \text{ ft}^3/\text{lb})(32.174 \text{ lb/slug}) = 159 \text{ ft}^3/\text{slug}$. Then the inverse is

$$\rho = \frac{1}{159} = 0.00630 \text{ slug/ft}^3 \qquad \textit{Ans. (b)}$$

The perfect-gas estimate above is about 3.6 percent low.

Reference 12 (table 9, p. 278) lists the value of c_p at 100 lbf/in² and 400°F as 0.536 Btu/(lb·°R). Convert this into BG units

$$c_p = [0.536 \text{ Btu/(lb} \cdot {}^\circ\text{R)}][778.2(\text{ft} \cdot \text{lbf})/\text{Btu}](32.174 \text{ lb/slug})$$
$$= 13{,}400(\text{ft} \cdot \text{lbf})/(\text{slug} \cdot {}^\circ\text{R}) = 13{,}400 \text{ ft}^2/(\text{s}^2 \cdot {}^\circ\text{R}) \qquad \textit{Ans. (b)}$$

The perfect-gas estimate above is about 10.5 percent low. The chief reason for the discrepancy is because 100 lbf/in² and 400°F are very close to the critical point and saturation line of steam. At higher temperatures and lower pressures, say 800°F and 50 lbf/in², the perfect-gas law gives ρ and c_p with an accuracy within 1 percent.

Note that the use of the steam tables in fluid-mechanics applications gives more accurate state relations but requires a continual use of conversion factors between the pound mass–Btu type of thermodynamic unit and the BG system.

Specific Weight and Specific Gravity

The weight per unit volume of a fluid is called its *specific weight* and equals ρg, the product of its density and the acceleration of gravity. For example, in standard earth gravity, where $g = 32.174 \text{ ft/s}^2 = 9.807 \text{ m/s}^2$, the specific weights of air and water at 20°C and 1 atm are

$$\rho g_{\text{air}} = (1.204 \text{ kg/m}^3)(9.807 \text{ m/s}^2) = 11.8 \text{ N/m}^3 = 0.0752 \text{ lbf/ft}^3$$
$$\rho g_{\text{water}} = (998 \text{ kg/m}^3)(9.807 \text{ m/s}^2) = 9790 \text{ N/m}^3 = 62.4 \text{ lbf/ft}^3$$

The specific weight is very useful in the hydrostatic-pressure problems of Chap. 2. In the engineering literature specific weight is often given the special symbol $\gamma = \rho g$, but here we reserve γ to mean the specific-heat ratio from Eq. (1.35).

Another useful concept is the *specific gravity* (SG), which is the ratio of density to the standard density of some reference fluid at 20°C and 1 atm. It is customary to relate gases to standard air and liquids to water

$$\text{SG}_{\text{gas}} = \frac{\rho_{\text{gas}}}{\rho_{\text{air}}} = \frac{\rho_{\text{gas}}}{1.204 \text{ kg/m}^3}$$
$$\text{SG}_{\text{liquid}} = \frac{\rho_{\text{liquid}}}{\rho_{\text{water}}} = \frac{\rho_{\text{liquid}}}{998 \text{ kg/m}^3} \qquad (1.36)$$

For example, the specific gravity of mercury, $\rho = 13,580\,kg/m^3$, is $13,580/998 = 13.6$. The specific gravity of helium at $100°C$ and 1 atm is approximately $0.131/1.204 = 0.11$. Engineers find these dimensionless ratios easier to remember than the actual density of a variety of fluids.

State Relations for Liquids

The writer knows of no "perfect-liquid law" comparable to that for gases. Liquids are nearly incompressible and have a single reasonably constant specific heat. Thus an idealized state relation for a liquid is

$$p \approx \text{const} \qquad c_p \approx c_v \approx \text{const} \qquad dh \approx c_p\, dT \qquad (1.37)$$

Most of the flow problems in this book can be attacked with these simple assumptions. Water is normally taken to have a density of 1.94 slugs/ft^3 and a specific heat $c_p = 25,200\,ft^2/(s^2 \cdot °R)$. The Steam Tables may be used if more accuracy is required.

The density of a liquid usually decreases slightly with temperature and increases moderately with pressure. If we neglect the temperature effect, an empirical pressure-density relation for a liquid is

$$\frac{p}{p_a} \approx (B + 1)\left(\frac{\rho}{\rho_a}\right)^n - B \qquad (1.38)$$

where B and n are dimensionless parameters which vary slightly with temperature and p_a and ρ_a are standard atmospheric values. Water can be fit approximately to the values $B \approx 3000$ and $n \approx 7$.

Seawater is a variable mixture of water and salt and thus requires three thermodynamic properties to define its state. These are normally taken as pressure, temperature, and the *salinity* $\hat{S}$, defined as the weight of the dissolved salt divided by the weight of the mixture. The average salinity of seawater is 0.035, usually written as 35 parts per thousand or 35 ‰. The average density of seawater is 2.00 slugs/ft^3. Strictly speaking, seawater has three specific heats, all approximately equal to the value for pure water of $25,200\,ft^2/(s^2 \cdot °R) = 4210\,m^2/(s^2 \cdot K)$.

EXAMPLE 1.11

The pressure at the deepest part of the ocean is approximately 1100 atm. Estimate the density of seawater at this pressure.

Solution

Equation (1.38) holds for either water or seawater. The ratio p/p_a is given as 1100

$$1100 \approx (3001)\left(\frac{\rho}{\rho_a}\right)^7 - 3000$$

or

$$\frac{\rho}{\rho_a} = \left(\frac{4100}{3001}\right)^{1/7} = 1.046$$

Assuming an average surface seawater density $\rho_a = 2.00$ slugs/ft^3, we compute

$$\rho \approx 1.046(2.00) = 2.09 \text{ slugs/ft}^3 \qquad \qquad Ans.$$

Even at these immense pressures, the density increase is less than 5 percent, which justifies the treatment of a liquid flow as essentially incompressible.

1.7 VISCOSITY AND OTHER SECONDARY PROPERTIES

The quantities such as pressure, temperature, and density discussed in the previous section are *primary* thermodynamic variables characteristic of any system. There are also certain secondary variables which characterize specific fluid-mechanical behavior. The most important of these is viscosity, which relates the local stresses in a moving fluid to the strain rate of the fluid element.

Viscosity

When a fluid is sheared, it begins to move at a strain rate inversely proportional to a property called its *coefficient of viscosity* μ. Consider a fluid element sheared in one plane by a single shear stress τ, as in Fig. 1.5a. The shear strain angle $\delta\theta$ will continuously grow with time as long as the stress τ is maintained, the upper surface moving at speed δu larger than the lower. Such common fluids as water, oil, and air show a linear relation between applied shear and resulting strain rate

$$\tau \propto \frac{\delta\theta}{\delta t} \tag{1.39}$$

From the geometry of Fig. 1.5a we see that

$$\tan \delta\theta = \frac{\delta u\, \delta t}{\delta y} \tag{1.40}$$

In the limit of infinitesimal changes, this becomes a relation between shear strain rate and velocity gradient

$$\frac{d\theta}{dt} = \frac{du}{dy} \tag{1.41}$$

From Eq. (1.39), then, the applied shear is also proportional to the velocity gradient for the common linear fluids. The constant of proportionality is the viscosity coefficient μ

$$\tau = \mu \frac{d\theta}{dt} = \mu \frac{du}{dy} \tag{1.42}$$

Fig. 1.5 Shear stress causes continuum shear deformation in a fluid: (*a*) a fluid element straining at a rate $\delta\theta/\delta t$; (*b*) newtonian shear distribution in a shear layer near a wall.

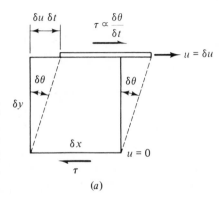

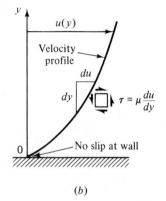

(*a*)   (*b*)

Table 1.3

**VISCOSITY AND KINEMATIC VISCOSITY OF EIGHT FLUIDS
AT 1 ATM AND 20°C**

Fluid	μ, kg/(m·s)†	Ratio $\mu/\mu(H_2)$	ρ, kg/m³	ν, m²/s†	Ratio $\nu/\nu(Hg)$
Hydrogen	8.8 E−6	1.0	0.084	1.05 E−4	920
Air	1.8 E−5	2.1	1.20	1.51 E−5	130
Gasoline	2.9 E−4	33	680	4.22 E−7	3.7
Water	1.0 E−3	114	998	1.01 E−6	8.7
Ethyl alcohol	1.2 E−3	135	789	1.52 E−6	13
Mercury	1.5 E−3	170	13,580	1.16 E−7	1.0
SAE 30 oil	0.29	33,000	891	3.25 E−4	2,850
Glycerin	1.5	170,000	1,264	1.18 E−3	10,300

† 1 kg/(m·s) = 0.0209 slug/(ft·s); 1 m²/s = 10.76 ft²/s.

Equation (1.42) is dimensionally consistent; therefore μ has dimensions of stress-time: $\{FT/L^2\}$ or $\{M/LT\}$. The BG unit is slugs per foot-second and the SI unit is kilograms per meter-second. The linear fluids which follow Eq. (1.42) are called *newtonian fluids*, after Sir Isaac Newton, who first postulated this resistance law in 1687.

We do not really care about the strain angle $\theta(t)$ in fluid mechanics, concentrating instead on the velocity distribution $u(y)$, as in Fig. 1.5b. We shall use Eq. (1.42) in Chap. 4 to derive a differential equation for finding the velocity distribution $u(y)$—and, more generally, $\mathbf{V}(x, y, z, t)$—in a viscous fluid. Figure 1.5b illustrates a shear layer, or *boundary layer*, near a solid wall. The shear stress is proportional to the slope of the velocity profile and is greatest at the wall. Further, at the wall, the velocity u is zero relative to the wall: this is called the *no-slip condition* and is characteristic of all viscous-fluid flows.

The viscosity of newtonian fluids is a true thermodynamic property and varies with temperature and pressure. At a given state (p, T) there is a vast range of values among the common fluids. Table 1.3 lists the viscosity of eight fluids at standard pressure and temperature. There is a variation of six orders of magnitude from hydrogen up to glycerin. Thus there will be wide differences between fluids subjected to the same applied stresses.

As we shall see in Chaps. 5 and 6, the primary parameter correlating the viscous behavior of all newtonian fluids is the dimensionless Reynolds number

$$\text{Re} = \frac{\rho VL}{\mu} = \frac{VL}{\nu} \qquad (1.43)$$

where V and L represent the characteristic velocity and length scales of the flow. Since ρ and μ occur as a ratio in this parameter, this ratio is significant and is called the kinematic viscosity

$$\nu = \frac{\mu}{\rho}$$

The mass units cancel, and v has units of square meters per second or square feet per second; hence the term kinematic viscosity.

Table 1.3 also lists values of v for the same eight fluids. The pecking order changes considerably, and mercury, the heaviest, has the smallest viscosity relative to its own weight. All gases have high v relative to thin liquids such as gasoline, water, and alcohol. Oil and glycerin still have the highest v, but the ratio is smaller. For a given value of V and L in a flow, these fluids exhibit a spread of four orders of magnitude in Reynolds number.

Flow between Plates

A classic problem is the flow induced between a fixed lower plate and an upper plate moving steadily at velocity $\mathbf{V}$, as shown in Fig. 1.6. The clearance between plates is h, and the fluid is newtonian and does not slip at either plate. If the plates are large, this steady shearing motion will set up a velocity distribution $u(y)$, as shown, with $v = w = 0$. Substituting this distribution into Eq. (1.16), we find that the fluid acceleration is zero everywhere.

With zero acceleration and assuming no pressure variation in the flow direction, you should show that a force balance on a small fluid element leads to the result that the shear stress is constant throughout the fluid. Then Eq. (1.42) becomes

$$\frac{du}{dy} = \frac{\tau}{\mu} = \text{const}$$

which we can integrate to obtain

$$u = a + by$$

The velocity distribution is linear, as shown in Fig. 1.6, and the constants a and b can be evaluated from the no-slip condition at the upper and lower walls

$$u = \begin{cases} 0 = a + b(0) & \text{at } y = 0 \\ V = a + b(h) & \text{at } y = h \end{cases}$$

Hence $a = 0$ and $b = V/h$. Then the velocity profile between the plates is given by

$$u = V \frac{y}{h} \tag{1.44}$$

as indicated in Fig. 1.6. Turbulent flow (Chap. 6) does not have this shape.

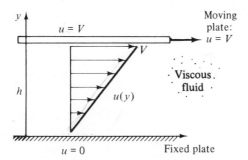

Fig. 1.6 Viscous flow induced by relative motion between two parallel plates.

Although viscosity has a profound effect on fluid motion, the actual viscous stresses are quite small in magnitude even for oils, as shown in the following example.

EXAMPLE 1.12 Suppose that the fluid being sheared in Fig. 1.6 is SAE 30 oil at 20°C. Compute the shear stress in the oil if $V = 3$ m/s and $h = 2$ cm.

Solution The shear stress is found from Eq. (1.42) by differentiating Eq. (1.44)

$$\tau = \mu \frac{du}{dy} = \frac{\mu V}{h} \tag{1}$$

From Table 1.3 for SAE 30 oil, $\mu = 0.29$ kg/(m · s). Then, for the given values of V and h, Eq. (1) predicts

$$\tau = \frac{[0.29 \text{ kg/(m · s)}](3 \text{ m/s})}{0.02 \text{ m}} = 43 \text{ kg/(m · s}^2)$$

$$= 43 \text{ N/m}^2 = 43 \text{ Pa} \qquad\qquad Ans.$$

Although oil is very viscous, this is a modest shear stress, about 2400 times less than atmospheric pressure. Viscous stresses in gases and thin liquids are even smaller.

Variation of Viscosity with Temperature

Temperature has a strong effect and pressure a moderate effect on viscosity. The viscosity of gases and most liquids increases slowly with pressure. Water is anomalous in showing a very slight decrease below 30°C. Since the change in viscosity is only a few percent up to 100 atm, we shall neglect pressure effects in this book.

Gas viscosity increases with temperature. Two common approximations are the power law and the Sutherland law:

$$\frac{\mu}{\mu_0} \approx \begin{cases} \left(\dfrac{T}{T_0}\right)^n & \text{power law} \\[2mm] \dfrac{(T/T_0)^{3/2}(T_0 + S)}{T + S} & \text{Sutherland law} \end{cases} \tag{1.45}$$

where μ_0 is a known viscosity at a known absolute temperature T_0 (usually 273 K). The constants n and S are fit to the data, and both formulas are adequate over a wide range of temperatures. For air, $n \approx 0.7$ and $S \approx 110$ K $= 199°$R. Other values are given in Ref. 5 of Chap. 4.

Liquid viscosity decreases with temperature and is roughly exponential, $\mu \approx ae^{-bT}$; but a better fit is the empirical result that $\ln \mu$ is quadratic in $1/T$, where T is absolute temperature

$$\ln \frac{\mu}{\mu_0} \approx a + b\left(\frac{T_0}{T}\right) + c\left(\frac{T_0}{T}\right)^2 \tag{1.46}$$

For water, with $T_0 = 273.16$ K, $\mu_0 = 0.001792$ kg/(m · s), suggested values are $a = -1.94$, $b = -4.80$, and $c = 6.74$, with accuracy about ± 1 percent. The viscosity of water is tabulated in Appendix Table A.1.

Thermal Conductivity

Just as viscosity relates applied stress to resulting strain rate, there is a property called *thermal conductivity k* which relates the vector rate of heat flow per unit area $\mathbf{q}$ to the vector gradient of temperature ∇T. This proportionality, observed experimentally for fluids and solids, is known as *Fourier's law of heat conduction*

$$\mathbf{q} = -k\nabla T \tag{1.47a}$$

which can also be written as three scalar equations

$$q_x = -k\frac{\partial T}{\partial x} \qquad q_y = -k\frac{\partial T}{\partial y} \qquad q_z = -k\frac{\partial T}{\partial z} \tag{1.47b}$$

The minus sign satisfies the convention that heat flux is positive in the direction of decreasing temperature. Fourier's law is dimensionally consistent, and k has SI units of joules per second-meter-kelvin. Thermal conductivity k is a thermodynamic property and varies with temperature and pressure in much the same way as viscosity. The ratio k/k_0 can be correlated with T/T_0 in the same manner as Eqs. (1.45) and (1.46) for gases and liquids, respectively.

Further data on viscosity and thermal-conductivity variations can be found in Ref. 10.

Nonnewtonian Fluids

Fluids which do not follow the linear law of Eq. (1.42) are called nonnewtonian and are treated in books on rheology [5]. Figure 1.7a compares four examples with a newtonian fluid. A *dilatant*, or shear-thickening, fluid increases resistance with increasing applied stress. Alternately, a *pseudoplastic*, or shear-thinning, fluid

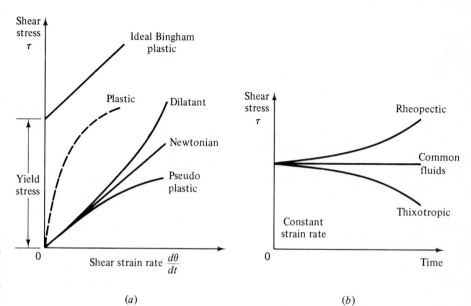

Fig. 1.7 Rheological behavior of various viscous materials: (*a*) stress versus strain rate; (*b*) effect of time on applied stress.

decreases resistance with increasing stress. If the thinning effect is very strong, as with the dashed-line curve, the fluid is termed *plastic*. The limiting case of a plastic substance is one which requires a finite yield stress before it begins to flow. The linear-flow *Bingham-plastic* idealization is shown, but the flow behavior after yield may also be nonlinear. An example of a yielding fluid is toothpaste, which will not flow out of the tube until a finite stress is applied by squeezing.

A further complication of nonnewtonian behavior is the transient effect shown in Fig. 1.6*b*. Some fluids require a gradually increasing shear stress to maintain a constant strain rate and are called *rheopectic*. The opposite case of a fluid which thins out with time and requires decreasing stress is termed *thixotropic*. We shall neglect nonnewtonian effects in this book; see Ref. 5 for further study.

Surface Tension

A liquid, being unable to expand freely, will form an *interface* with a second liquid or gas. The physical chemistry of such interfacial surfaces is quite complex, and whole textbooks are devoted to this specialty [13]. Molecules deep within the liquid repel each other because of their close packing. Molecules at the surface are less dense and attract each other. Since half of their neighbors are missing, the mechanical effect is that the surface is in tension. We can account adequately for surface effects in fluid mechanics with the concept of surface tension.

If a cut of length dL is made in an interfacial surface, equal and opposite forces of magnitude $\Upsilon\, dL$ are exposed normal to the cut and parallel to the surface, where Υ is called the *coefficient of surface tension*. The dimensions of Υ are $\{F/L\}$, with SI units of newtons per meter and BG units of pounds force per foot. An alternate concept is to open up the cut to an area dA; this requires work to be done of amount $\Upsilon\, dA$. Thus the coefficient Υ can also be regarded as the surface energy per unit area of the interface, in newton-meters per square meter or foot-pounds force per square foot.

The two most common interfaces are water-air and mercury-air. For a clean surface at $20°C = 68°F$, the measured surface tension is

$$\Upsilon = \begin{cases} 0.0050 \text{ lbf/ft} = 0.073 \text{ N/m} & \text{air-water} \\ 0.033 \text{ lbf/ft} = 0.48 \text{ N/m} & \text{air-mercury} \end{cases} \qquad (1.48)$$

These are design values and can change considerably if the surface contains contaminants like detergents or slicks. Generally Υ decreases with liquid temperature and is zero at the critical point.

If the interface is curved, a mechanical balance shows that there is a pressure difference across the interface, the pressure being higher on the concave side, as illustrated in Fig. 1.8. In Fig. 1.8*a*, the pressure increase in the interior of a liquid cylinder is balanced by two surface-tension forces

$$2RL\,\Delta p = 2\Upsilon L$$

or

$$\Delta p = \frac{\Upsilon}{R} \qquad (1.49)$$

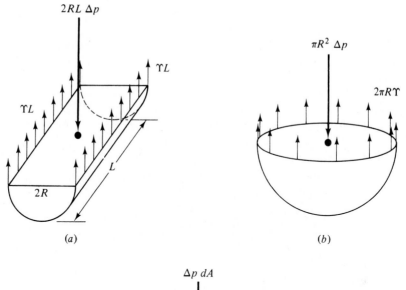

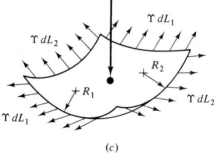

Fig. 1.8 Pressure change across a curved interface due to surface tension: (*a*) interior of a liquid cylinder; (*b*) interior of a spherical droplet; (*c*) general curved interface.

We are not considering the weight of the liquid in this calculation. In Fig. 1.8*b*, the pressure increase in the interior of a spherical droplet balances a ring of surface-tension force

$$\pi R^2 \, \Delta p = 2\pi R \Upsilon$$

or

$$\Delta p = \frac{2\Upsilon}{R} \tag{1.50}$$

We can use this result to predict the pressure increase inside a soap bubble, which has *two* interfaces with air, an inner and outer surface of nearly the same radius R

$$\Delta p_{\text{bubble}} \approx 2 \, \Delta p_{\text{droplet}} = \frac{4\Upsilon}{R} \tag{1.51}$$

Figure 1.8*c* shows the general case of an arbitrarily curved interface whose principal radii of curvature are R_1 and R_2. A force balance normal to the surface will show that the pressure increase on the concave side is

$$\Delta p = \Upsilon(R_1^{-1} + R_2^{-1}) \tag{1.52}$$

Gas

Liquid

Fig. 1.9 Contact-angle effects at liquid-gas-solid interface. If $\theta < 90°$, the liquid "wets" the solid; if $\theta > 90°$, the liquid is nonwetting.

Gas

Liquid

Nonwetting

θ θ

Solid

Equations (1.49) to (1.51) can all be derived from this general relation; e.g., in Eq. (1.49) $R_1 = R$ and $R_2 = \infty$.

A second important surface effect is the *contact angle* θ which appears when a liquid interface intersects with a solid surface, as in Fig. 1.9. The force balance would then involve both Υ and θ. If the contact angle is less than 90°, the liquid is said to *wet* the solid; if $\theta > 90°$, the liquid is termed *nonwetting*. For example, water wets soap but does not wet wax. Water is extremely wetting to a clean glass surface, with $\theta \approx 0°$. Like Υ, the contact angle θ is sensitive to the actual physicochemical conditions of the solid-liquid interface. For a clean mercury-air-glass interface, $\theta = 130°$.

Example 1.13 illustrates how surface tension causes a fluid interface to rise or fall in a capillary tube.

EXAMPLE 1.13   Derive an expression for the change in height h in a circular tube of a liquid with surface tension Υ and contact angle θ, as in Fig. E1.13.

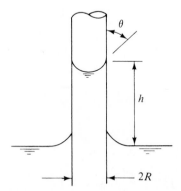

θ

h

Fig. E1.13

$2R$

Solution   The vertical component of the ring surface-tension force at the interface in the tube must balance the weight of the column of fluid of height h

$$2\pi R \Upsilon \cos \theta = \rho g \pi R^2 h$$

Solving for h, we have the desired result

$$h = \frac{2\Upsilon \cos \theta}{\rho g R} \qquad\qquad Ans.$$

Thus the capillary height decreases inversely with tube radius R and is positive if $\theta < 90°$ (wetting liquid) and negative (capillary depression) if $\theta > 90°$.

Suppose that $R = 1$ mm. Then the capillary rise for a water-air-glass interface, $\theta \approx 0°$, $\Upsilon = 0.073$ N/m, and $\rho = 1000$ kg/m^3 is

$$h = \frac{2(0.073 \text{ N/m})(\cos 0°)}{(1000 \text{ kg/m}^3)(9.81 \text{ m/s}^2)(0.001 \text{ m})} = 0.015 \text{ (N} \cdot \text{s}^2)/\text{kg} = 0.015 \text{ m} = 1.5 \text{ cm}$$

For a mercury-air-glass interface, with $\theta = 130°$, $\Upsilon = 0.48$ N/m, and $\rho = 13,600$ kg/m^3. the capillary rise is

$$h = \frac{2(0.48)(\cos 130°)}{13,600(9.81)(0.001)} = -0.46 \text{ cm}$$

When a small-diameter tube is used to make pressure measurements (Chap. 2), these capillary effects must be corrected for.

Vapor Pressure

Vapor pressure is the pressure at which a liquid boils and is in equilibrium with its own vapor. For example, the vapor pressure of water at 68°F is 49 lbf/ft^2, while that of mercury is only 0.0035 lbf/ft^2. If the liquid pressure is greater than the vapor pressure, the only exchange between liquid and vapor is evaporation at the interface. If, however, the liquid pressure falls below the vapor pressure, vapor bubbles begin to appear in the liquid. If water is heated to 212°F, its vapor pressure rises to 2116 lbf/ft^2 and thus water at normal atmospheric pressure will boil. When the liquid pressure is dropped below the vapor pressure due to a flow phenomenon, we call the process *cavitation*. As we shall see in Chap. 2, if water is accelerated from rest to about 50 ft/s, its pressure drops by about 15 lbf/in^2, or 1 atm. This can cause cavitation. Figure 1.10 shows cavitation occurring in the low-pressure region associated with the tip vortices shed from a marine propeller.

The dimensionless parameter describing flow-induced boiling is the *cavitation number*

$$\text{Ca} = \frac{p_a - p_v}{\frac{1}{2}\rho V^2} \tag{1.53}$$

where p_a = ambient pressure
p_v = vapor pressure
V = characteristic flow velocity

Depending upon the geometry, a given flow has a critical value of Ca below which the flow will begin to cavitate. Values of surface tension and vapor pressure of water are given in Table 1.4.

No-Slip and No-Temperature-Jump Conditions

When a fluid flow is bounded by a solid surface, molecular interactions cause the fluid in contact with the surface to seek momentum and energy equilibrium with that surface. All liquids essentially are in equilibrium with the surface they contact.

Fig. 1.10 Cavitation may occur in a high-velocity liquid flow. In the photograph bubbles form a helical pattern in the tip vortices shed from a marine propeller. (*Courtesy of the Garfield Thomas Water Tunnel, Pennsylvania State University.*)

Table 1.4

SURFACE TENSION† AND VAPOR PRESSURE OF PURE WATER

T, °C	Υ, N/m	p_v, kPa
0	0.0756	0.611
10	0.0742	1.227
20	0.0728	2.337
30	0.0712	4.242
40	0.0696	7.375
50	0.0679	12.34
60	0.0662	19.92
70	0.0644	31.16
80	0.0626	47.35
90	0.0608	70.11
100	0.0589	101.33

† In contact with air.

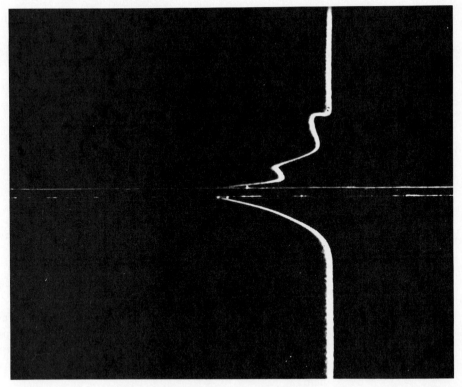

Fig. 1.11 The no-slip condition in water flow past a thin fixed plate. The upper flow is turbulent; the lower flow is laminar. The velocity profile is made visible by a line of hydrogen bubbles discharged from the wire across the flow. [*From* Illustrated Experiments in Fluid Mechanics (*The NCFMF Book of Film Notes*), National Committee for Fluid Mechanics Films, Education Development Center, Inc., copyright 1972.]

All gases are, too, except under the most rarefied conditions [7]. Excluding rarefied gases, then, all fluids at a point of contact with a solid take on the velocity and temperature of that surface

$$\mathbf{V}_{\text{fluid}} \equiv \mathbf{V}_{\text{wall}} \qquad T_{\text{fluid}} \equiv T_{\text{wall}} \qquad (1.54)$$

These are called the *no-slip* and *no-temperature-jump conditions*, respectively. They serve as *boundary conditions* for analysis of fluid flow past a solid surface (Chap. 6). Figure 1.11 illustrates the no-slip condition for water flow past the top and bottom surfaces of a fixed thin plate. The flow past the upper surface is disorderly, or turbulent, while the lower surface flow is smooth, or laminar.[1] In both cases there is clearly no slip at the wall, where the water takes on the zero velocity of the fixed plate. The velocity profile is made visible by the discharge of a line of hydrogen bubbles from the wire shown stretched across the flow.

To decrease the mathematical difficulty, the no-slip condition is partially relaxed in the analysis of inviscid flow (Chap. 8). The flow is allowed to "slip" past the surface but not to permeate through the surface

$$V_{\text{normal}}(\text{fluid}) \equiv V_{\text{normal}}(\text{solid}) \qquad (1.55)$$

while the tangential velocity V_t is allowed to be independent of the wall. The analysis is much simpler, but the flow patterns are highly idealized.

---

[1] Laminar and turbulent flows are studied in Chaps. 6 and 7.

1.8 BASIC FLOW-ANALYSIS TECHNIQUES

There are three basic ways to attack a fluid-flow problem. They are equally important for a student learning the subject, and this book tries to give adequate coverage to each method:

1. Control volume, or *integral* analysis (Chap. 3)
2. Infinitesimal system, or *differential* analysis (Chap. 4)
3. Experimental study, or *dimensional* analysis (Chap. 5)

In all cases, the flow must satisfy the three basic conservation laws of mechanics[1] plus a thermodynamic state relation and associated boundary conditions:

1. Conservation of mass (continuity)
2. Conservation of (linear) momentum (Newton's second law)
3. Conservation of energy (first law of thermodynamics)
4. A state relation like $\rho = \rho(p, T)$
5. Appropriate boundary conditions at solid surfaces, interfaces, inlets, and exits

In integral and differential analyses, these five relations are modeled mathematically and solved by computational methods. In an experimental study, the fluid itself performs this task without using any mathematics. In other words, these laws are believed to be fundamental to physics, and no fluid flow is known to violate them.

A control volume is a finite region, chosen carefully by the analyst, with open boundaries which mass, momentum, and energy are allowed to cross. The analyst makes a budget, or balance, between the incoming and outgoing fluid and the resultant changes within the control volume. The result is a powerful tool but a crude one. Details of the flow are normally washed out or ignored in control-volume analyses. Nevertheless, the control-volume technique of Chap. 3 never fails to yield useful and quantitative information to the engineering analyst.

When the conservation laws are written for an infinitesimal system of fluid in motion, they become the basic differential equations of fluid flow. To apply them to a specific problem, one must integrate these equations mathematically subject to the boundary conditions of the particular problem. Exact analytic solutions are often possible only for very simple geometries and boundary conditions (Chap. 4). Otherwise, one attempts numerical integration on a digital computer, i.e., a summing procedure for finite-sized systems which one hopes will approximate the exact integral calculus [1]. Even computer analysis often fails to provide an accurate simulation, because of either inadequate storage or inability to model the finely detailed flow structure characteristic of irregular geometries or turbulent-flow patterns. Thus differential analysis sometimes promises more than it delivers, although we can successfully study a number of classic and useful solutions.

---

[1] In fluids which are variable mixtures of components, such as seawater, a fourth basic law is required, *conservation of species*. For an example of salt-conservation analysis, see Ref. 15, p. 397.

A properly planned experiment is very often the best way to study a practical engineering flow problem. Guidelines for planning flow experiments are given in Chap. 5. For example, no theory presently available, whether differential or integral, calculus or computer, is able to make an accurate computation of the aerodynamic drag and side force of an automobile moving down a highway with crosswinds. One must solve the problem by experiment. The experiment may be *full scale*: one can test a real automobile on a real highway in real crosswinds. For that matter, there are wind tunnels in existence large enough to hold a full-scale car without significant blockage effects. Normally, however, in the design stage, one tests a small model automobile in a small wind tunnel. Without proper interpretation, the model results may be poor and mislead the designer (Chap. 5). For example, the model may lack important details such as surface finish or underbody protuberances. The "wind" produced by the tunnel propellers may lack the turbulent gustiness of real winds. It is the job of the fluid-flow analyst, using such techniques as dimensional analysis, to plan an experiment which gives an accurate estimate of full-scale or *prototype* results expected in the final product.

It is possible to classify flows, but there is no general agreement on how to do it. Most classifications deal with the assumptions made in the proposed flow analysis. They come in pairs, and we normally assume that a given flow is either

$$\text{Steady} \quad \text{or} \quad \text{unsteady} \qquad (1.56a)$$

$$\text{Inviscid} \quad \text{or} \quad \text{viscous} \qquad (1.56b)$$

$$\text{Incompressible} \quad \text{or} \quad \text{compressible} \qquad (1.56c)$$

$$\text{Gas} \quad \text{or} \quad \text{liquid} \qquad (1.56d)$$

As Fig. 1.12 indicates, we choose one assumption from each pair. We may have a steady viscous compressible gas flow, or an unsteady inviscid ($\mu = 0$) incompressible liquid flow. Although there is no such thing as a truly inviscid fluid, the assumption $\mu = 0$ gives adequate results in many analyses (Chap. 8). Often the assumptions overlap: a flow may be viscous in the boundary layer near a solid surface (Fig. 1.11) and effectively inviscid away from the surface. The viscous part of the flow may be laminar or transitional or turbulent or combine patches of all three types of viscous flow. A flow may involve both a gas and a liquid and the free surface, or interface, between them (Chap. 10). A flow may be compressible in one region and have nearly constant density in another. Nevertheless, Eq. (1.56) and Fig. 1.12 give the basic binary assumptions of flow analysis, and Chaps. 6 to 10 try to separate them and isolate the basic effect of each assumption.

Fig. 1.12 Ready for a flow analysis? Then choose one assumption out of each box.

Steady / Unsteady — Inviscid / Viscous — Incompressible / Compressible — Gas / Liquid

1.9 FLOW PATTERNS: STREAMLINES, STREAKLINES, AND PATHLINES

Fluid mechanics is a highly visual subject. The patterns of flow can be visualized in a dozen different ways, and you can view these sketches or photographs and learn a great deal qualitatively and often quantitatively about the flow. Appendix C lists a number of fine films which have been made to illustrate various flow phenomena.

Four basic types of line patterns are used to visualize flows:

1. A *streamline* is a line everywhere tangent to the velocity vector at a given instant.
2. A *pathline* is the actual path traversed by a given fluid particle.
3. A *streakline* is the locus of particles which have earlier passed through a prescribed point.
4. A *timeline* is a set of fluid particles that form a line at a given instant.

The streamline is convenient to calculate mathematically, while the other three are easier to generate experimentally. Note that a streamline and a timeline are instantaneous lines, while the pathline and streakline are generated by the passage of time. The velocity profile shown in Fig. 1.11 is really a timeline generated earlier by a single discharge of bubbles from the wire. A pathline can be found by a time exposure of a single marked particle moving through the flow. Streamlines are difficult to generate experimentally in unsteady flow unless one marks a great many particles and notes their direction of motion during a very short time interval [14, p. 35]. In steady flow the situation simplifies greatly:

Streamlines, pathlines, and streaklines are identical in steady flow.

In fluid mechanics the most common mathematical result for visualization purposes is the streamline pattern. Figure 1.13a shows a typical set of streamlines, and Fig. 1.13b shows a closed pattern called a *streamtube*. By definition the fluid within a streamtube is confined there because it cannot cross the streamlines; thus the streamtube walls need not be solid but may be fluid surfaces.

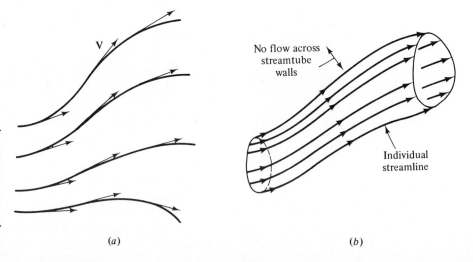

Fig. 1.13 The most common method of flow-pattern presentation: (*a*) streamlines are everywhere tangent to the local velocity vector; (*b*) a streamtube is formed by a closed collection of streamlines.

V

No flow across streamtube walls

Individual streamline

(*a*)

(*b*)

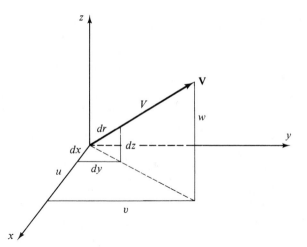

Fig. 1.14 Geometric relations for defining a streamline.

Streamlines can be calculated from the velocity field by the geometric relationships in Fig. 1.14. Since every vector arc length $d\mathbf{r}$ along a streamline must be tangent to $\mathbf{V}$, their respective components must be in exact proportion:

Streamline:
$$\frac{dx}{u} = \frac{dy}{v} = \frac{dz}{w} = \frac{dr}{V} \qquad (1.57)$$

If the components u, v, and w are known functions of position and time, Eq. (1.57) can be integrated to find the streamline passing through (x_0, y_0, z_0, t_0). The integration may be rather laborious. Another idea is to introduce a parameter ds equal to the ratios in Eq. (1.57). Thus

$$\frac{dx}{ds} = u \qquad \frac{dy}{ds} = v \qquad \frac{dz}{ds} = w \qquad (1.58)$$

Integrate Eq. (1.58) with respect to s, holding time constant, with the initial condition (x_0, y_0, z_0, t_0) at $s = 0$. Then eliminate s to obtain the desired function $f(x, y, z, t)$ representing the streamline.

The pathline is defined by integration of the relation between velocity and displacement in Eq. (1.9)

$$\frac{dx}{dt} = u(x, y, z, t) \qquad \frac{dy}{dt} = v(x, y, z, t) \qquad \frac{dz}{dt} = w(x, y, z, t) \qquad (1.59)$$

Integrate with respect to time using the condition (x_0, y_0, z_0, t_0). Then eliminate time to give the pathline function $f(x, y, z, t)$. Again the integration may be laborious.

Finally, to compute the streakline, take the integrated result of Eq. (1.59) and retain time as a parameter. Find the integration constants which cause the pathlines to pass through (x_0, y_0, z_0) for a sequence of times $\xi < t$. Then eliminate ξ from the result to obtain the streaklines.

EXAMPLE 1.14   Given the velocity distribution
$$u = Kx \qquad v = -Ky \qquad w = 0 \qquad (1)$$
where K is constant, compute and plot the streamlines of the flow, including directions, and give some possible interpretations of the pattern.

Solution   Since time does not appear explicitly in Eq. (1), the motion is steady, so that streamlines, pathlines, and streaklines will coincide. Since $w = 0$ everywhere, the motion is two-dimensional, in the xy plane. The reader can check from Eqs. (1.24) and (1.25) that the motion is incompressible and irrotational. The streamlines can be computed by substituting the expressions for u and v into Eq. (1.57)

$$\frac{dx}{Kx} = -\frac{dy}{Ky}$$

or

$$\int \frac{dx}{x} = -\int \frac{dy}{y}$$

Integrating, we obtain $\ln x = -\ln y + \ln C$, or

$$xy = C \qquad\qquad\qquad\qquad Ans. \quad (2)$$

This is the general expression for the streamlines, which are hyperbolas. The complete pattern is plotted in Fig. E1.14 by assigning various values to the constant C. The arrowheads can be determined only by returning to Eqs. (1) to ascertain the velocity component directions, assuming K is positive. For example, in the upper right quadrant ($x > 0$, $y > 0$), u is positive and v is negative; hence the flow moves down and to the right, establishing the arrowheads as shown.

Note that the streamline pattern is entirely independent of the constant K. It could represent the impingement of two opposing streams, or the upper half could simulate the flow of a single downward stream against a flat wall. Taken in isolation, the upper right quadrant is similar to the flow in a 90° corner. This is definitely a realistic flow pattern and is discussed again in Chap. 8 (Figs. 8.18*b* and 8.19*b*).

Finally note the peculiarity that the two streamlines ($C = 0$) have opposite directions and intersect each other. This is possible only at a point where $u = v = w = 0$, which occurs at the origin in this case. Such a point of zero velocity is called a *stagnation point*.

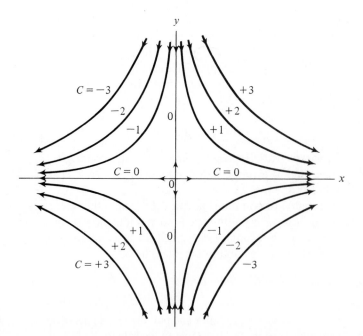

Fig. E1.14 Streamlines for the velocity distribution given by Eq. (1), for $K > 0$.

A streakline can be produced experimentally by the continuous release of marked particles (dye, smoke, or bubbles) from a given point. Figure 1.15 shows two examples. The flow in Fig. 1.15b is unsteady and periodic due to the flapping of the plate against the oncoming stream. We see that the dash-dot streakline does not coincide with either the streamline or pathline passing through the same release point. This is characteristic of unsteady flow, but in Fig. 1.15a the smoke filaments form streaklines which are identical to the streamlines and pathlines. We noted earlier that this coincidence of lines is always true of steady flow: since the velocity never changes magnitude or direction at any point, every particle which comes along repeats the behavior of its earlier neighbors.

Methods of experimental flow visualization are:

1. Dye, smoke, or bubble discharges
2. Surface powder or flakes on liquid flows
3. Floating or neutral-density particles
4. Optical techniques which detect density changes in gas flows: shadowgraph, schlieren, and interferometer
5. Tufts of yarn attached to boundary surfaces
6. Evaporative coatings on boundary surfaces
7. Luminescent fluids or additives

These methods are illustrated in film 3 in Appendix C. The mathematical implications of flow-pattern analysis are discussed in detail in Ref. 16.

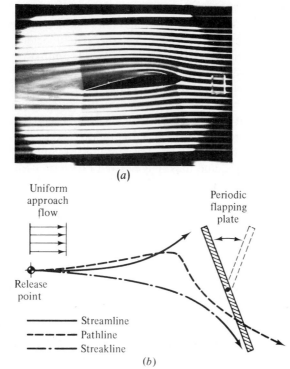

Fig. 1.15 Experimental visualization of steady and unsteady flow: (a) steady flow past an airfoil visualized by smoke filaments (*C. A. A. SCIENTIFIC—Prime Movers Laboratory Systems*); (b) unsteady flow past an oscillating plate with a point bubble release (from an experiment in Ref. 14).

1.10 HISTORY AND SCOPE OF FLUID MECHANICS

Like most scientific disciplines, fluid mechanics has a history of erratically occurring early achievements, then an intermediate era of steady fundamental discoveries in the eighteenth and nineteenth centuries, leading to the twentieth-century era of "modern practice," as we self-centeredly term our limited but up-to-date knowledge. Ancient civilizations had enough knowledge to solve certain flow problems. Sailing ships with oars and irrigation systems were both known in prehistoric times. The Greeks produced quantitative information. Archimedes and Hero of Alexandria both postulated the parallelogram law for addition of vectors in the third century B.C. Archimedes (285–212 B.C.) formulated the laws of buoyancy and applied them to floating and submerged bodies, actually deriving a form of the differential calculus as part of the analysis. The Romans built extensive aqueduct systems in the fourth century B.C. but left no records showing any quantitative knowledge of design principles.

From the birth of Christ to the Renaissance there was a steady improvement in the design of such flow systems as ships and canals and water conduits but no recorded evidence of fundamental improvements in flow analysis. Then Leonardo da Vinci (1452–1519) derived the equation of conservation of mass in one-dimensional steady flow. Leonardo was an excellent experimentalist, and his notes contain accurate descriptions of waves, jets, hydraulic jumps, eddy formation, and both low-drag (streamlined) and high-drag (parachute) designs. A Frenchman, Edme Mariotte (1620–1684), built the first wind tunnel and tested models in it.

Problems involving the momentum of fluids could finally be analyzed after Isaac Newton (1642–1727) postulated his laws of motion and the law of viscosity of the linear fluids now called newtonian. The theory first yielded to the assumption of a "perfect" or frictionless fluid, and eighteenth-century mathematicians (Daniel Bernoulli, Leonhard Euler, Jean d'Alembert, Joseph-Louis Lagrange, and Pierre-Simon Laplace) produced many beautiful solutions of frictionless-flow problems. Euler developed both the differential equations of motion and their integrated form, now called the Bernoulli equation. D'Alembert used them to show his famous paradox: that a body immersed in a frictionless fluid has zero drag. These beautiful results amounted to overkill, since perfect-fluid assumptions have very limited application in practice, and most engineering flows are dominated by the effects of viscosity. Engineers began to reject what they regarded as a totally unrealistic theory and developed the science of *hydraulics*, relying almost entirely on experiment. Such experimentalists as Chézy, Pitot, Borda, Weber, Francis, Hagen, Poiseuille, Darcy, Manning, Bazin, and Weisbach produced data on a variety of flows such as open channels, ship resistance, pipe flows, waves, and turbines. All too often the data were used in raw form without regard to the fundamental physics of flow.

At the end of the nineteenth century unification between experimental *hydraulics* and theoretical *hydrodynamics* finally began. William Froude (1810–1879) and his son Robert (1846–1924) developed laws of model testing, Lord Rayleigh (1842–1919) proposed the technique of dimensional analysis, and Osborne Reynolds (1842–1912) published the classic pipe experiment in 1883 which showed the importance of the dimensionless Reynolds number named after him. Meanwhile, viscous-flow theory was available but unexploited, since Navier (1785–1836) and Stokes (1819–1903) had successfully added newtonian viscous terms to the equations of motion. The resulting Navier-Stokes equations were too difficult to analyze for arbitrary flows. Then, in 1904, a German engineer, Ludwig Prandtl (1875–1953), published perhaps the most important paper ever written on fluid mechanics. Prandtl pointed out that fluid flows with small viscosity, e.g., waterflows and airflows, can be divided into a thin viscous layer, or *boundary layer*, near solid surfaces and interfaces, patched onto a nearly inviscid outer layer, where the Euler and Bernoulli equations apply. Boundary-layer theory has proved to be the single most important tool in modern flow analysis. The twentieth-century foundations for the present state-of-the-art in fluid mechanics were laid in a series of broad-based experiments and theories by Prandtl and this two chief friendly competitors, Theodore von Kármán (1881–1963) and Sir Geoffrey I. Taylor (1886–1975). Many of the results sketched here from a historical point of view will of course be discussed in this textbook. More historical details can be found in Ref. 17.

Since the earth is 75 percent covered with water and 100 percent covered with air, the scope of fluid mechanics is vast and touches nearly every human endeavor. The sciences of meteorology, physical oceanography, and hydrology are concerned with naturally occurring fluid flows, as are medical studies of breathing and blood circulation. All transportation problems involve fluid motion, with well-developed specialties in aerodynamics of aircraft and rockets and in naval hydrodynamics of ships and submarines. Almost all of our electric energy is developed either from water flow or from steam flow through turbine generators. All combustion problems involve fluid motion, as do the more classic problems of irrigation, flood control, water supply, sewage disposal, projectile motion, and oil and gas pipelines. The aim of this book is to present enough fundamental concepts and practical applications in fluid mechanics to prepare you to move smoothly into any of these specialized fields of the science of flow—and then be prepared to move out again as new technologies develop.

REFERENCES

1. P. J. Roache, "Computational Fluid Dynamics," Hermosa, Albuquerque, N.M., 1972.
2. T. Cebeci and A. M. O. Smith, "Analysis of Turbulent Boundary Layers," Academic, New York, 1974.
3. V. A. Sandborn, "Class Notes for Experimental Methods in Fluid Mechanics," Colorado State University, Civil Engineering Department, Fort Collins, Colo., 1972.
4. P. Bradshaw, "Experimental Fluid Mechanics," Macmillan, New York, 1964.
5. M. Reiner, "Deformation, Strain and Flow: An Elementary Introduction to Rheology," 3d ed., Lewis, London, 1969.
6. G. B. Wallis, "One-Dimensional Two-Phase Flow," McGraw-Hill, New York, 1969.
7. G. N. Patterson, "Introduction to the Kinetic Theory of Gas Flows," University of Toronto Press, Toronto, 1971.
8. "ASME Orientation and Guide for Use of SI Units," Guide No. SI-1, 7th ed., American Society of Mechanical Engineers, 1976.

9. J. P. Holman, "Heat Transfer," 5th ed., McGraw-Hill, 1981.

10. R. C. Reid, J. M. Pravsnitz, and T. K. Sherwood, "The Properties of Gases and Liquids," 3d ed., McGraw-Hill, New York, 1977.

11. J. Hilsenrath et al., Tables of Thermodynamic and Transport Properties, *U.S. Nat. Bur. Stand. Circ. 564*, 1955; reprinted Pergamon, New York, 1960.

12. "ASME Steam Tables," American Society of Mechanical Engineers, New York, 1967.

13. A. W. Adamson, "Physical Chemistry of Surfaces," Interscience, New York, 1960.

14. National Committee for Fluid Mechanics Films, "Illustrated Experiments in Fluid Mechanics," M.I.T. Press, Cambridge, Mass., 1972.

15. G. Neumann and W. J. Pierson Jr., "Principles of Physical Oceanography," Prentice-Hall, Englewood Cliffs, N.J., 1966.

16. I. G. Currie, "Fundamental Mechanics of Fluids," McGraw-Hill, New York, 1974.

17. H. Rouse and S. Ince, "History of Hydraulics," Iowa Institute of Hydraulic Research, State University of Iowa, Ames, 1957.

18. W. Merzkirch, "Flow Visualization," Academic, New York, 1974.

19 C. Y. Chow, "An Introduction to Computational Fluid Mechanics," Wiley, New York, 1979.

RECOMMENDED FILMS

Films numbered 3, 4, 6, 13, 16, and 17 in Appendix C.

Problems

PROBLEM DISTRIBUTION

Section	Topic	Problems
1.1, 2, 3	Fluid-continuum concepts	1.1–1.8
1.4	Dimensions and units	1.9–1.20
1.5	The velocity field	1.21–1.39
1.6	Thermodynamic properties	1.40–1.58
1.7	Viscosity: no-slip condition	1.59–1.75
1.7	Surface tension, cavitation	1.76–1.92
1.8	Flow-analysis classification	1.93
1.9	Streamlines and pathlines	1.94–1.98
1.10	History of fluid mechanics	1.99–1.100

1.1 If Avogadro's number is 6.023×10^{23} molecules per mole, how many molecules are there in 1 cm³ of (*a*) air, (*b*) water, (*c*) hydrogen, and (*d*) helium at 1 atm pressure and 20°C?

1.2 Estimate the number of molecules in the entire atmosphere of the earth.

1.3 A gas begins to deviate from the continuum concept when it contains less than about 10^{12} molecules per cubic millimeter. At 20°C what absolute pressure in pascals does this limit for air correspond to?

1.4 If a transparent closed container is filled with a clear fluid with no visible meniscus or free surface, how can one tell without opening the container, whether the fluid is liquid or gas?

1.5 A cylindrical beaker of diameter 3 in and height 5 in weighs 35 oz when filled with liquid and 9 oz when empty. What is the density of the liquid in SI and BG units?

1.6 How high will the free surface be if 1 ft³ of water is poured into a container that is a right circular cone 18 in high with base diameter of 20 in? How much additional water is required to fill the container?

1.7 If the conical container in Prob. 1.6 can be filled with 27 kg of a certain oil, what is the density of the oil in SI and BG units?

1.8 Which of the following laws of physics would be appropriate for analysis of fluid motion: (*a*) Boyle's law, (*b*) Charles' law, (*c*) Newton's second law, (*d*) Ohm's law, (*e*) first law of thermodynamics, (*f*) Hooke's law, (*g*) second law of thermodynamics, (*h*) perfect-gas law, (*i*) Gibbs-Dalton law?

1.9 If Eq. (1.2) is written with a conversion constant g_c, that is, $F = ma/g_c$, what is the numerical value of g_c (*a*) when F is in pounds force, m in pounds, and a in feet per second squared; (b) when F is in dynes, m in grams, and a in centimeters per second squared; (c) when F is in kilograms (force), m in kilograms (mass), and a in meters per second squared?

1.10 In the $\{MLT\Theta\}$ system, what is the dimensional representation of (a) moment of a force, (b) modulus of elasticity, (c) kinematic viscosity, (d) mass rate of flow, (e) angular acceleration, (f) rate of heat flow, (g) entropy?

1.11 A mass of 3 kg is placed on a planet with acceleration of gravity of 5.0 m/s². (a) What is its mass in kilograms? (b) What is its weight in newtons? (c) What force is required to accelerate the mass at 2.0 m/s²?

1.12 SAE 70 lubricating oil at 100°F has a viscosity of 0.0088 slug/(ft · s). What is its viscosity in kilograms per meter-second? If the oil weighs 55 lbf/ft³, what is its kinematic viscosity, $v = \mu/\rho$, in square meters per second?

1.13 For low-speed (laminar) flow through a circular pipe, as shown in Fig. Pl.13, the velocity distribution takes the form

$$u = \frac{B}{\mu}(r_0^2 - r^2)$$

where μ is the fluid viscosity. What are the units of the constant B?

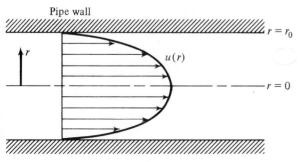

Pipe wall

$r = r_0$

$u(r)$

$r = 0$

Fig. P1.13

1.14 The *mean free path* l of a gas is defined as the mean distance traveled by molecules between collisions. According to kinetic theory [7], the mean free path of an ideal gas is given by

$$l = 1.26\frac{\mu}{\rho}(RT)^{-1/2}$$

where R is the gas constant and T the absolute temperature. Is the constant 1.26 dimensionless?

1.15 The Stokes–Oseen formula [16] for the drag force F on a sphere of diameter D in a fluid stream of low velocity V is

$$F = 3\pi\mu DV + \frac{9\pi}{16}\rho V^2 D^2$$

Is this formula dimensionally consistent?

1.16 The speed of propagation C of waves traveling at the interface between two fluids is given by

$$C = \left(\frac{\pi\Upsilon}{\rho_a\lambda}\right)^{1/2}$$

where λ is the wavelength and ρ_a the average density of the two fluids. If the formula is dimensionally consistent, what are the units of Υ? What might it represent?

1.17 The specific heat c_p of air at room conditions is approximately 0.24 Btu/(lb · °R), which is a useful unit in thermodynamic calculations. In fluid mechanics c_p is more conveniently expressed as a velocity squared per temperature unit, as in Table 1.2. What is the specific heat of air (a) in ft²/(s² · °R) and (b) in m²/(s² · K)?

1.18 We have seen that algebraic physical equations such as that of Bernoulli, Eq. (1.4), are dimensionally consistent. Are differential equations in physics also consistent? For example, determine whether the heat-conduction equation

$$\rho c_p \frac{\partial T}{\partial t} = k\frac{\partial^2 T}{\partial x^2}$$

is dimensionally consistent. Can you draw a general conclusion from this result?

1.19 The horsepower (hp) is a unit of power equal to 550 (ft · lbf)/s. How many kilowatthours of energy are expended by a 25-hp motor operating for 12 h?

1.20 A popular formula in the hydraulics literature is the Hazen-Williams formula for volume flow rate Q in a pipe of diameter D and pressure gradient dp/dx

$$Q = 61.9D^{2.63}\left(\frac{dp}{dx}\right)^{0.54}$$

What are the dimensions of the constant 61.9? Can this formula be used with confidence for a variety of liquids and gases?

1.21 A velocity field is given by $u = 3y^2$, $v = 2x$, $w = 0$, in arbitrary units. Is this flow steady or unsteady? Is it two- or three-dimensional? At $(x, y, z) = (2, 1, 0)$ compute (a) the velocity, (b) the local acceleration, and (c) the convective acceleration.

1.22   For the velocity field of Prob. 1.21, at $(2, 1, 0)$, compute (a) the acceleration component parallel to the velocity vector and (b) the component normal to the velocity vector. The answer in part (b) is not zero. What does it represent?

1.23   An idealized velocity field is given by the formula

$$\mathbf{V} = 3tx\mathbf{i} - t^2y\mathbf{j} + 2xz\mathbf{k}$$

Is this flow steady or unsteady? Is it two- or three-dimensional? At the point $(x, y, z) = (1, -1, 0)$ compute (a) the total acceleration vector and (b) the unit vector normal to the acceleration.

1.24   For the pipe-flow velocity field of Prob. 1.13 determine (a) the maximum velocity in terms of B, μ, and r_0 and (b), using Eq. (1.19), the mass flow rate in terms of B, μ, and r_0.

1.25   For steady flow through a conical nozzle the axial velocity is approximately

$$u = U_0\left(1 - \frac{x}{L}\right)^{-2}$$

where U_0 is the entrance velocity and L is the distance to the apparent vertex of the cone. Compute (a) a general expression for the axial acceleration du/dt and (b) the acceleration in g's at the entrance and at $x = 1$ m if $U_0 = 5$ m/s and $L = 2$ m.

1.26   A two-dimensional velocity field is given by

$$\mathbf{V} = (x^2 - y^2 + x)\mathbf{i} - (2xy + y)\mathbf{j}$$

in arbitrary units. At $x = 2$ and $y = 1$ compute (a) the accelerations a_x and a_y, (b) the velocity component in the direction $\theta = 30°$, and (c) the directions of maximum acceleration and maximum velocity.

1.27   The two-dimensional pressure field, in arbitrary units, $p = 4x^3 - 2y^2$ is associated with the velocity field in Prob. 1.26. Compute the rate of change dp/dt at $x = 2$ and $y = 1$.

1.28   The velocity field in the neighborhood of a stagnation point is given by $u = U_0x/L$, $v = -U_0y/L$, $w = 0$. (a) Show that the acceleration vector is purely radial. (b) If $L = 2$ ft, what is the magnitude of U_0 if the total acceleration at $(x, y) = (L, L)$ is 30 ft/s^2?

1.29   At higher speeds the flow in the pipe of Fig. P11.3 becomes turbulent and has a velocity distribution approximated by

$$u \approx u_{max}\left(1 - \frac{r}{r_0}\right)^n$$

where $n \approx 0.14$. Sketch this profile and compare with Prob. 1.13. Using Eq. (1.21), compute the volume flow and average velocity in the pipe.

1.30   Suppose that a particle moves around the circular path $x^2 + y^2 = 4$ m^2 at a uniform tangential velocity of 3 m/s. Express this motion in terms of u and v components. Compute the tangential and radial acceleration at the point $(x, y) = (2, 0)$.

1.31   The velocity field in a diffuser is $u = U_0 e^{-2x/L}$, and the density field is $\rho = \rho_0 e^{-x/L}$. Find the rate of change of density at $x = L$.

1.32   A velocity field in arbitrary units is given by

$$\mathbf{V} = 2x^2\mathbf{i} - xy\mathbf{j} - 3xz\mathbf{k}$$

Find the volume flow Q passing through the unit square bounded by $(x, y, z) = (1, 0, 0)$, $(1, 1, 0)$, $(1, 1, 1)$, and $(1, 0, 1)$.

1.33   For the velocity field of Prob. 1.32 find the volume flow passing through the triangle whose vertices are at $(x, y, z) = (1, 0, 0)$, $(0, 1, 0)$, and $(0, 0, 1)$.

1.34   The velocity field of Prob. 1.21 is associated with the temperature field $T = 4xy^2$, in arbitrary units. Compute the rate of change dT/dt at the point $(x, y) = (2, 1)$.

1.35   If the fluid in Prob. 1.13 is water at 20°C and 1 atm, what is the centerline velocity U_0 if the tube radius is 1 cm and the mass flow through the tube is 1.2 kg/s?

1.36   If the volume flow Q across the upper surface in Fig. P1.36 equals the difference between inlet and outlet flows past the plate, what is this flow Q in terms of the inlet velocity U_0 and the height δ of the fluid region shown in the figure?

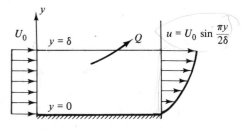

Plate (width b into paper)

Fig. P1.36

1.37 The velocity profile in water flow down a spillway is given approximately by

$$u = U_0 \left(\frac{y}{h}\right)^{1/7}$$

where $y = 0$ denotes the bottom and the depth is h. If $U_0 = 1.5$ m/s, $h = 2$ m, and the width is 20 m, how many hours will it take 10^6 m^3 of water to pass this section of the spillway?

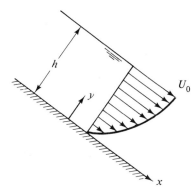

Fig. P1.37

1.38 True or false? (*a*) In flow through a tube the mass flow and volume flow occur everywhere in the same ratio. (*b*) In steady flow there can be no fluid acceleration. (*c*) In tube flow the average velocity is always less than or equal to the maximum velocity. (*d*) In circular vortex motion there is no fluid acceleration. (*e*) In unsteady flow there is always a fluid acceleration.

1.39 A nearly frictionless liquid flows from the bottom of a large tank through a 1-cm-diameter hole at a rate of 800 cm^3/s. If the fluid flows radially toward the hole with the same volume flow across every section, compute the convective acceleration at points 20 and 40 cm from the hole.

1.40 Using the perfect-gas law, estimate the specific gravity of (*a*) CO_2 and (*b*) H_2 at 100°C and 150 kPa.

1.41 The heaviest gas known to the author is uranium hexafluoride, $^{238}UF_6$. Estimate its specific gravity at 20°C and 1 atm.

1.42 A tank contains 45,000 cm^3 of helium gas at 200 kPa and 60°C. Estimate the total weight of gas (*a*) on earth and (*b*) on the moon.

1.43 How much heat transfer in joules is required to expand the helium in Prob. 1.42 at constant pressure to a new volume of 65,000 cm^3?

1.44 Some values of the density of mercury versus pressure at 20°C are as follows:

p, atm	1	1000	2000
ρ, kg/m^3	13,545	13,599	13,652

Fit these data to the empirical approximation, Eq. (1.38), to find best values of B and n for mercury.

1.45 As will be shown in Sec. 9.2, the speed of sound of a liquid equals the square root of its pressure-density gradient, $dp/d\rho$. From the data in Prob. 1.44 estimate the sound speed, in meters per second, of mercury at 1 atm and compare with Table 9.1.

1.46 Air at 1 atm and 60°F has an internal energy of approximately $\hat{u} = 3.12 \times 10^6$ (ft · lb)/slug. If the air is moving at $V = 400$ ft/s and has zero potential energy, what is its total energy in foot-pounds per slug?

1.47 If nitrogen has a molecular weight of 28, what is its density in kilograms per cubic meter according to the perfect-gas law when $p = 200,000$ Pa and $T = 50$°C?

1.48 Using the perfect-gas approximation and Table A.4, estimate the total heat required to raise 2 kg of NO gas from 60 to 200°C at a constant pressure of 200 kPa.

1.49 In Table A.4, most common gases (air, nitrogen, oxygen, hydrogen) have a specific-heat ratio of about 1.40. Why do argon and helium have such high values? Why does NH_4 have a low value?

1.50 If 3 kg of NH_4 is confined in a rigid tank at 180°C and 140 kPa, estimate how much heat must be removed to lower the gas pressure to 90 kPa.

1.51 If a gas occupies 1 m^3 at 1 atm pressure, what pressure is required to reduce the volume of the gas by 1 percent under isothermal conditions if the fluid is (*a*) air, (*b*) argon, (*c*) hydrogen?

1.52 If water occupies 1 m^3 at 1 atm pressure, what pressure is required to reduce its volume by 1 percent according to Eq. (1.38)?

1.53 Using the perfect-gas law, estimate the density of steam in slugs per cubic foot at 1800°R and 4 atm and compare with the value listed in the ASME Steam Tables.

1.54 The *isentropic bulk modulus* of a fluid is defined as the change in pressure per fractional change in density

$$B = \rho \left(\frac{\partial p}{\partial \rho}\right)_s$$

and has dimensions of pressure or stress. Estimate the bulk modulus of (a) a perfect gas and (b) water at standard conditions as a multiple of atmospheric pressure.

1.55 Using the curve fit to the data for mercury in Prob. 1.44, estimate its bulk modulus in megapascals. How does one compute the speed of sound from this value?

1.56 A gas has a molecular weight of 56 and a specific heat $c_v = 3000$ (ft·lb)/(slug·°R). What is its specific-heat ratio?

1.57 Compute the enthalpy change, assuming a perfect gas, of CO_2 in joules per kilogram for a temperature rise from 0 to 100°C.

1.58 Estimate how much the density of water is reduced from its value at 1 atm if the pressure is reduced to zero.

1.59 In Fig. 1.6, if the fluid is glycerin at 20°C and the width between plates is 6 mm, what shear stress is required to move the upper plate at 2.5 m/s? What is the Reynolds number if L is taken to be the distance between plates?

1.60 Carbon tetrachloride at 20°C has a viscosity of 0.00097 kg/(m·s). What shear stress in pascals is required to deform this fluid at a strain rate of 5000 s^{-1}?

1.61 SAE 10 oil at 20°C is sheared between two parallel plates 0.01 in apart with the lower plate fixed and the upper plate moving at 15 ft/s. Compute the shear stress in the oil.

1.62 Knowing μ of air at 20°C from Table 1.3, estimate its viscosity at 1500°F from (a) the power law and (b) the Sutherland law, and compare with the accepted value of 9.1×10^{-7} slug/(ft·s).

1.63 Knowing μ of water at 20°C from Table 1.3, estimate its viscosity at 90°C from Eq. (1.46) and compare with the accepted value of 0.000317 kg/(m·s).

1.64 Some measured values of the viscosity of ethyl alcohol at 1 atm are as follows:

T, °C	μ, kg/(m·s)
−40	0.00481
0	0.00177
40	0.000834
80	0.000430

Fit these values to Eq. (1.46) and assess the accuracy of the formula. Take $T_0 = 273.16$ K and compare also with Table A.3.

1.65 An 8-kg flat block of metal slides down a 20° inclined plane while lubricated by a 2-mm-thick film of SAE 30 oil at 20°C. The contact area is 0.2 m². What is the terminal velocity of the block?

1.66 A shaft 8.00 cm in diameter is being pushed through a bearing sleeve 8.02 cm in diameter and 30 cm long. The clearance, assumed uniform, is filled with oil with $v = 0.005$ m²/s and SG = 0.9. If the shaft moves axially at 0.5 m/s, estimate the resistance force exerted by the oil on the shaft.

1.67 The shaft in Prob. 1.66 is now fixed axially and rotated inside the sleeve at 1800 r/min. Estimate (a) the resisting torque exerted by the oil in newton-meters and (b) the power in kilowatts required to rotate the shaft.

1.68 A steel (7850 kg/m³) shaft 3 cm in diameter and 40 cm long falls of its own weight inside a vertical open tube 3.02 cm in diameter. The clearance, assumed uniform, is a film of glycerin at 20°C. How fast will the cylinder fall at terminal conditions?

1.69 In Prob. 1.13 if $u_{max} = 7$ m/s, the fluid is glycerin at 20°C, and the pipe diameter 3 cm, what is the shear stress in pascals (a) at the centerline and (b) at the wall?

1.70 Air at 20°C forms a boundary layer near a solid wall of sine-wave-shaped velocity profile. The boundary-layer thickness is 6 mm, and the peak velocity is 10 m/s. Compute the shear stress in the boundary layer in pascals at y equal to (a) 0, (b) 3 mm, and (c) 6 mm.

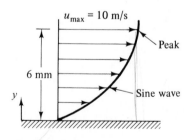

Fig. P1.70

1.71 A solid cone of angle 2θ, base radius r_0, and density ρ_c is rotating with initial angular velocity ω_0 inside a conical seat. The clearance h is filled with oil of viscosity μ. Neglecting air drag, derive an expression for the time required to reduce the cone's angular velocity from ω_0 to $0.1\omega_0$.

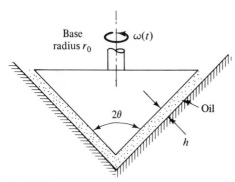

Fig. P1.71

1.72 A disk of radius R rotates at angular velocity Ω inside an oil bath of viscosity μ, as shown. Assuming a linear velocity profile and neglecting shear on the outer disk edges, derive an expression for the viscous torque on the disk.

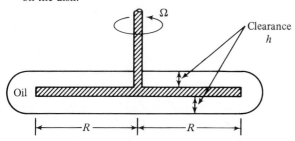

Fig. P1.72

1.73 What is the Reynolds number Re of the boundary-layer flow in Prob. 1.70, based on maximum velocity, boundary-layer thickness, and the properties of air?

1.74 A dimensionless fluid property called the Prandtl number is important in heat-transfer analysis and is formed from the viscosity, the thermal conductivity, and the specific heat. What form does this dimensionless group have: (a) $\mu/c_p k$; (b) $\mu k/c_p$; (c) $\mu c_p/k$; (d) $\mu c_p/k^2$; (e) none of these?

1.75 Repeat Example 1.12 for a fluid heated unevenly to cause the viscosity to vary linearly from $\frac{1}{2}\mu$ at the lower surface to μ at the upper surface. What is the ratio of the shear stress in this case to the stress in Example 1.12?

1.76 Assuming that a soda-water bubble is equivalent to an air-water interface with $\Upsilon = 0.005$ lbf/ft, what is the pressure difference in pounds per square foot between the inside and outside of a bubble whose diameter is 0.004 in?

1.77 Derive Eq. (1.52) by making a force balance of the fluid interface sketched in Fig. 1.7c.

1.78 If the interface in a surface-tension problem has the two-dimensional form $z = \eta(x, t)$, show that Eq. (1.52) takes the form

$$\Delta p = \frac{\Upsilon(\partial^2\eta/\partial x^2)}{[1 + (\partial\eta/\partial x)^2]^{3/2}}$$

Under what conditions can this expression be linearized by setting the denominator equal to unity?

1.79 Make an analysis of the shape of the water-air interface near a plane wall, assuming that the slope is small, $R^{-1} \approx d^2\eta/dx^2$, and that the pressure difference across the interface is balanced by the specific weight and the interface height, $\Delta p \approx \rho g \eta$. The boundary conditions are a wetting contact angle θ at $x = 0$ and a horizontal surface $\eta = 0$ as $x \to \infty$. What is the maximum height h at the wall?

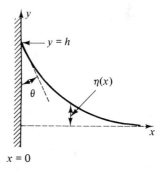

Fig. P1.79

1.80 Repeat Prob. 1.79 using the exact curvature expression $R^{-1} = \eta''/(1 + \eta'^2)^{3/2}$. Find the value of h.

1.81 At low speeds the jet issuing from a faucet approximates a liquid cylinder with an air-water interface. What is the pressure difference between inside and outside of the jet when the diameter is 2 mm and the temperature is 10°C?

1.82 At 60°C the surface tension of mercury and water is 0.47 and 0.0662 N/m, respectively. What capillary-height changes will occur in these two fluids when they are in contact with air in a glass tube of diameter 0.5 mm?

1.83 At 30°C what diameter glass tube is necessary to keep the capillary-height change of water less than 1 mm?

1.84 A 1-in diameter soap bubble has an internal pressure 0.004 lbf/in² greater than that of the outside atmosphere. Compute the surface tension of the soap-air interface in pounds force per foot.

1.85 Derive an expression for the capillary-height change h for a fluid of surface tension Υ and contact angle θ between two vertical parallel plates a distance W apart. What will h be for water if $W = 1$ mm?

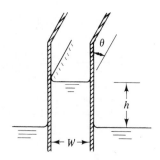

Fig. P1.85

1.86 Return to Prob. 1.16, which asked only for dimensional reasoning, and discuss whether Υ is indeed the coefficient of surface tension. If so, is the formula physically realistic; i.e., should wave speed increase with surface tension?

1.87 Neglecting the weight of the wire, what force is required to lift a thin wire ring 4 cm in diameter from a water surface at 20°C? Is this a good way to measure surface tension? Should the wire be of any particular material?

1.88 A spherical soap bubble of diameter d_1 coalesces with another bubble of diameter d_2 to form a single bubble d_3 containing the same amount of air. Assuming an isothermal process, derive an analytic expression for finding d_3 as a function of d_1, d_2, ambient pressure p_a, and interfacial tension Υ.

1.89 The glass tube in Fig. P1.89 is used to measure the pressure p_1 in the water tank. The tube diameter is

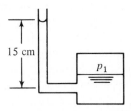

Fig. P1.89

1 mm, and the water is at 30°C. After correcting for surface tension, what is the true water height in the tube? What percent error is made if no correction is computed?

1.90 An atomizer forms water droplets with a diameter of 50 μm, or 5×10^{-5} m. What excess pressure exists in the interior of these droplets for water at 30°C?

1.91 In pump flow the lowest pressure usually occurs at the pump inlet. For a pump handling water at 30°C, what inlet pressure in pounds force per square inch absolute is liable to cause cavitation?

1.92 Early mountaineers used boiling water to estimate their height. If they reach the top and find that water boils at 80°C, approximately how high is the mountain?

1.93 Examine the photographs of the flows in Figures 1.11, 5.2, and 9.28 and classify them with the boxes in Fig. 1.12.

1.94 A velocity field is given by $u = V \cos\theta$, $v = V \sin\theta$, and $w = 0$, where V and θ are constants. Find an expression for the streamlines of this flow.

1.95 A two-dimensional steady velocity field is given by $u = x^2 - y^2$, $v = -2xy$. Derive the streamline pattern and sketch a few streamlines in the upper half plane. *Hint*: The differential equation is exact.

1.96 Problems 1.94 and 1.95 illustrate the determination of the streamline pattern $f(x, y) = $ const from the known velocity components $u(x, y)$ and $v(x, y)$. Is the inverse of this problem possible; i.e., can one compute u and v from a given streamline pattern $f(x, y) = $ const?

1.97 A two-dimensional unsteady velocity field is given by $u = x(1 + 2t)$, $v = y$. Find the equation of the time-varying streamlines which all pass through the point (x_0, y_0) at some time t. Sketch some.

1.98 Repeat Prob. 1.97 to find the equation of the pathline which passes through (x_0, y_0) at time $t = 0$. Sketch it.

1.99 Do some outside reading on the British physicist Sir George Stokes and cite at least three things in fluid mechanics which are named after him.

1.100 List some of the important contributions to twentieth-century fluid mechanics made by Ludwig Prandtl, Theodore von Kármán, and Sir Geoffrey I. Taylor.

Pressure Distribution in a Fluid

2.1 PRESSURE AND PRESSURE GRADIENT

In Fig. 1.1 we saw that a fluid at rest cannot support shear stress and thus Mohr's circle reduces to a point. In other words, the normal stress on any plane through a fluid element at rest is equal to a unique value called the *fluid pressure p*, taken positive for compression by common convention. This is such an important concept that we shall review it with another approach.

Figure 2.1 shows a small wedge of fluid at rest of size Δx by Δz by Δs and depth b into the paper. There is no shear by definition, but we postulate that the pressures p_x, p_z, and p_n may be different on each face. The weight of the element may also be important. Summation of forces must equal zero (no acceleration) in both the x and z directions

$$\sum F_x = 0 = p_x b\,\Delta z - p_n b\,\Delta s \sin\theta$$
$$\sum F_z = 0 = p_z b\,\Delta x - p_n b\,\Delta s \cos\theta - \tfrac{1}{2}\rho g b\,\Delta x\,\Delta z \tag{2.1}$$

but the geometry of the wedge is such that

$$\Delta s \sin\theta = \Delta z \qquad \Delta s \cos\theta = \Delta x \tag{2.2}$$

Substitution into Eq. (2.1) and rearrangement gives

$$p_x = p_n \qquad p_z = p_n + \tfrac{1}{2}\rho g\,\Delta z \tag{2.3}$$

These relations illustrate two important principles of the hydrostatic, or shear-free, condition: (1) there is no pressure change in the horizontal direction, and (2) there is a vertical change in pressure proportional to density, gravity, and the depth change. We shall exploit these results to the fullest, starting in Sec. 2.3.

In the limit as the fluid wedge shrinks to a "point," $\Delta z \to 0$ and Eqs. (2.3) become

$$p_x = p_z = p_n = p \tag{2.4}$$

51

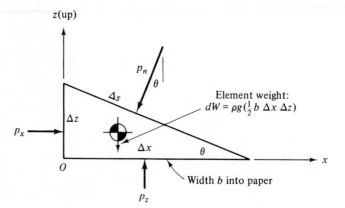

Fig. 2.1 Equilibrium of a small wedge of fluid at rest.

Since θ is arbitrary, we conclude that the pressure p at a point in a static fluid is independent of orientation.

What about the pressure at a point in a moving fluid? As discussed in Sec. 1.7, if there are strain rates in a moving fluid, there will be viscous stresses, both shear and normal in general (Sec. 4.3). In that case (Chap. 4) the pressure is defined as the average of the three normal stresses σ_{ii} on the element

$$p = -\tfrac{1}{3}(\sigma_{xx} + \sigma_{yy} + \sigma_{zz}) \tag{2.5}$$

The negative sign occurs because a compression stress is considered to be negative whereas p is positive. Equation (2.5) is subtle and rarely needed since the great majority of viscous flows have negligible viscous normal stresses (Chap. 4).

Pressure Force on a Fluid Element

Pressure (or any other stress, for that matter) causes no net force on a fluid element unless it varies *spatially*.[1] To see this, consider the pressure acting on the two x faces in Fig. 2.2. Let the pressure vary arbitrarily

$$p = p(x, y, z, t) \tag{2.6}$$

The net force in the x direction on the element in Fig. 2.2 is given by

$$dF_x = p\, dy\, dz - \left(p + \frac{\partial p}{\partial x}\, dx\right) dy\, dz = -\frac{\partial p}{\partial x}\, dx\, dy\, dz \tag{2.7}$$

In like manner the net force dF_y involves $-\partial p/\partial y$ and the net force dF_z concerns $-\partial p/\partial z$. The total net-force vector on the element due to pressure is

$$d\mathbf{F}_{\text{press}} = \left(-\mathbf{i}\frac{\partial p}{\partial x} - \mathbf{j}\frac{\partial p}{\partial y} - \mathbf{k}\frac{\partial p}{\partial z}\right) dx\, dy\, dz \tag{2.8}$$

We recognize the term in parentheses as the negative vector gradient of p. Denoting $\mathbf{f}$ as the net force per unit element volume, we rewrite Eq. (2.8) as

$$\mathbf{f}_{\text{press}} = -\nabla p \tag{2.9}$$

[1] An interesting application for a large element is in Fig. 3.7.

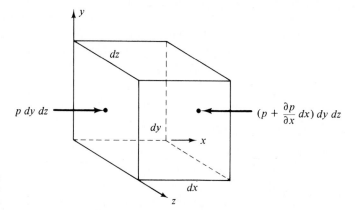

Fig. 2.2 Net x force on an element due to pressure variation.

Thus it is not the pressure but the pressure *gradient* causing a net force which must be balanced by gravity or acceleration or some other effect in the fluid.

2.2 EQUILIBRIUM OF A FLUID ELEMENT

The pressure gradient is a *surface* force which acts on the sides of the element. There may also be a *body* force, due to electromagnetic or gravitational potentials, acting on the entire mass of the element. Here we consider only the gravity force, or weight of the element

$$d\mathbf{F}_{\text{grav}} = \rho\mathbf{g}\, dx\, dy\, dz$$

or

$$\mathbf{f}_{\text{grav}} = \rho\mathbf{g}$$

(2.10)

In general there may also be a surface force due to the gradient, if any, of the viscous stresses. For completeness, we write this term here without derivation and consider it more thoroughly in Chap. 4. For an incompressible fluid with constant viscosity, the net viscous force is

$$\mathbf{f}_{\text{VS}} = \mu\left(\frac{\partial^2\mathbf{V}}{\partial x^2} + \frac{\partial^2\mathbf{V}}{\partial y^2} + \frac{\partial^2\mathbf{V}}{\partial z^2}\right) = \mu\,\nabla^2\mathbf{V}$$

(2.11)

where VS stands for viscous stresses and μ is the coefficient of viscosity from Chap. 1. Note that the term $\mathbf{g}$ in Eq. (2.10) denotes the acceleration of gravity, a vector acting toward the center of the earth. On earth the average magnitude of $\mathbf{g}$ is $32.174\ \text{ft/s}^2 = 9.807\ \text{m/s}^2$.

The total vector resultant of these three forces, pressure, gravity, and viscous stress, must either keep the element in equilibrium or cause it to move with acceleration $\mathbf{a}$. From Newton's law, Eq. (1.2), we have

$$\rho\mathbf{a} = \sum\mathbf{f} = \mathbf{f}_{\text{press}} + \mathbf{f}_{\text{grav}} + \mathbf{f}_{\text{visc}} = -\nabla p + \rho\mathbf{g} + \mu\,\nabla^2\mathbf{V}$$

(2.12)

This is one form of the differential momentum equation for a fluid element and will be studied further in Chap. 4. Vector addition is implied by Eq. (2.12): the acceleration reflects the local balance of forces and is not necessarily parallel to the local-velocity vector, which reflects the direction of motion at that instant.

The present chapter is concerned with cases where the velocity and acceleration are known, leaving one to solve for the pressure variation in the fluid. Later chapters will take up the more general problem where pressure, velocity, and acceleration are all unknown. Rewrite Eq. (2.12) as

$$\nabla p = \rho(\mathbf{g} - \mathbf{a}) + \mu \nabla^2 \mathbf{V} = \mathbf{B}(x, y, z, t) \tag{2.13}$$

where $\mathbf{B}$ is a short notation for the vector sum on the right-hand side. If $\mathbf{V}$ and $\mathbf{a} = d\mathbf{V}/dt$ are known functions of space and time and the density and viscosity are known, we can solve Eq. (2.13) for $p(x, y, z, t)$ by direct integration. By components, Eq. (2.13) is equivalent to three simultaneous first-order differential equations

$$\frac{\partial p}{\partial x} = B_x(x, y, z, t) \qquad \frac{\partial p}{\partial y} = B_y(x, y, z, t) \qquad \frac{\partial p}{\partial z} = B_z(x, y, z, t) \tag{2.14}$$

Since the right-hand sides are known functions, they can be integrated systematically to obtain the distribution $p(x, y, z, t)$ except for an unknown function of time, which remains because we have no relation for $\partial p/\partial t$. This extra function is found from a condition of known time variation $p_0(t)$ at some point (x_0, y_0, z_0). If the flow is steady (independent of time), the unknown function is a constant and is found from knowledge of a single known pressure p_0 at a point (x_0, y_0, z_0). If this sounds complicated, it is not: we shall illustrate with many examples. Finding the pressure distribution from a known velocity distribution is one of the easiest problems in fluid mechanics, which is why we put it in Chap. 2.

Examining Eq. (2.13), we can single out at least four special cases:

1. Flow at rest or at constant velocity: the acceleration and viscous terms vanish identically, and p depends only upon gravity and density. This is the *hydrostatic* condition. See Sec. 2.3.
2. Rigid-body translation and rotation: the viscous term vanishes identically, and p depends only upon the term $\rho(\mathbf{g} - \mathbf{a})$. See Sec. 2.9.
3. Irrotational motion ($\nabla \times \mathbf{V} \equiv 0$): the viscous term vanishes identically, and an exact integral called *Bernoulli's equation* can be found for the pressure distribution. See Sec. 4.9.
4. Arbitrary viscous motion: nothing helpful happens, no general rules apply, but still the integration is quite straightforward. See Sec. 6.4.

Let us consider cases 1 and 2 here.

Absolute, Gage, and Vacuum Pressure

Before embarking on examples, we should note that engineers are likely to specify a pressure measurement in two ways, either *absolute* pressure with respect to a zero pressure reference, or else with respect to local atmospheric pressure. In the latter case the measured pressure may be higher or lower than the local atmosphere. If higher, the difference is called *gage* pressure; if lower, the difference is called *vacuum*

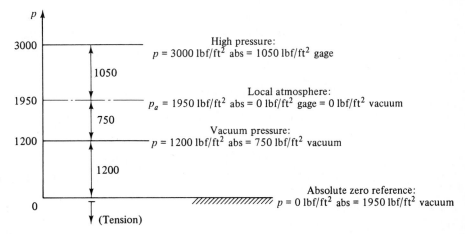

Fig. 2.3 Illustration of absolute, gage, and vacuum pressure readings.

pressure. This state of affairs is illustrated in Fig. 2.3. The local atmospheric pressure is, say, 1950 lbf/ft², which might reflect a storm condition in a sea-level city or normal conditions in a city at an altitude of 2300 ft. We can write this as $p_a =$ 1950 lbf/ft² absolute or 0 lbf/ft² gage or 0 lbf/ft² vacuum. Suppose a laboratory pressure gage reads an absolute pressure of 3000 lbf/ft². This is higher than local atmospheric pressure, and the value may be reported as $3000 - 1950 = 1050$ lbf/ft² gage pressure. (It may then be *your* problem later to determine what p_a was on that day.) Suppose another gage reads 1200 lbf/ft² absolute. Locally this would be a vacuum pressure and might be reported as $1950 - 1200 = 750$ lbf/ft² vacuum. (Again you might later need to know p_a to carry on your analysis.) Gage pressure is the opposite of vacuum pressure: 750 lbf/ft² vacuum $= -750$ lbf/ft² gage, but negative values are not normally reported. From time to time we shall specify gage or vacuum pressure in problems to keep you on your toes.

2.3 HYDROSTATIC PRESSURE DISTRIBUTIONS

If the fluid is at rest or at constant velocity, $\mathbf{a} = 0$ and $\nabla^2 \mathbf{V} = 0$. Equation (2.13) for the pressure distribution reduces to

$$\nabla p = \rho \mathbf{g} \qquad (2.15)$$

This is a *hydrostatic* distribution and is correct for all fluids at rest, regardless of their viscosity, because the viscous term vanishes identically.

Recall from vector analysis that the vector ∇p expresses the magnitude and direction of the maximum spatial rate of increase of the scalar property p. As a result, ∇p is perpendicular everywhere to surfaces of constant p. Thus Eq. (2.15) states that a fluid in hydrostatic equilibrium will align its constant-pressure surfaces everywhere normal to the local-gravity vector. The maximum pressure increase will be in the direction of gravity, i.e., "down." If the fluid is a liquid, its free surface, being at atmospheric pressure, will be normal to local gravity, or "horizontal." You probably knew all this before, but Eq. (2.15) is the proof of it.

In our customary coordinate system z is "up." Thus the local-gravity vector for small-scale problems is

$$\mathbf{g} = -g\mathbf{k} \tag{2.16}$$

where g is the magnitude of local gravity, for example, 9.807 m/s². For these coordinates Eq. (2.15) has the components

$$\frac{\partial p}{\partial x} = 0 \qquad \frac{\partial p}{\partial y} = 0 \qquad \frac{\partial p}{\partial z} = -\rho g \tag{2.17}$$

the first two of which tell us that p is independent of x and y. Hence $\partial p/\partial z$ can be replaced by the total derivative dp/dz, and the hydrostatic condition reduces to

$$\frac{dp}{dz} = -\rho g$$

or

$$p_2 - p_1 = -\int_1^2 \rho g \, dz \tag{2.18}$$

Equation (2.18) is the solution to the hydrostatic problem. The integration requires an assumption about the density and gravity distribution. Gases and liquids are usually treated differently.

We state the following conclusions about a hydrostatic condition:

Pressure in a continuously distributed uniform static fluid varies only with vertical distance and is independent of the shape of the container. The pressure is the same at all points on a given horizontal plane in the fluid. The pressure increases with depth in the fluid.

An illustration of this is shown in Fig. 2.4. The free surface of the container is atmospheric and forms a horizontal plane. Points a, b, c, and d are at equal depth in a horizontal plane and are interconnected by the same fluid, water; therefore all points have the same pressure. The same is true of points A, B, and C on the bottom, which all have the same higher pressure than at a, b, c, and d. However, point D, although at the same depth as A, B, and C, has a different pressure because it lies beneath a different fluid, mercury.

Fig. 2.4 Hydrostatic-pressure distribution. Points a, b, c, and d are at equal depths in water and therefore have identical pressures. Points A, B, and C are also at equal depths in water and have identical pressures higher than a, b, c, and d. Point D has a different pressure from A, B, and C because it is not connected to them by a water path.

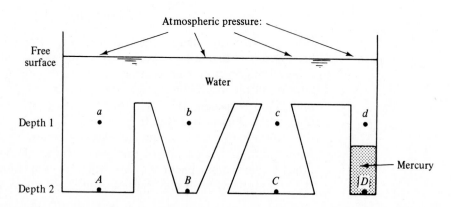

Effect of Variable Gravity

For a spherical planet of uniform density, the acceleration of gravity varies inversely as the square of the radius from its center

$$g = g_0 \left(\frac{r_0}{r}\right)^2 \tag{2.19}$$

where r_0 is the planet radius and g_0 is the surface value of g. For earth, $r_0 \approx 3960$ statute miles ≈ 6400 km. In typical engineering problems the deviation from r_0 extends from the deepest ocean, about 11 km, to the atmospheric height of supersonic transport operation, about 20 km. This gives a maximum variation in g of $(6400/6420)^2$ or 0.6 percent. We therefore neglect the variation of g in most problems.

Hydrostatic Pressure in Liquids

Liquids are so nearly incompressible that we can neglect their density variation in hydrostatics. In Example 1.11 we saw that water density increases only 4.6 percent at the deepest part of the ocean. Its effect on hydrostatics would be about half of this, or 2.3 percent. Thus we assume constant density in liquid hydrostatic calculations, for which Eq. (2.18) integrates to

Liquids:
$$p_2 - p_1 = -\rho g(z_2 - z_1) \tag{2.20}$$

or
$$z_1 - z_2 = \frac{p_2}{\rho g} - \frac{p_1}{\rho g}$$

We use the first form in most problems. The quantity ρg is called the *specific weight* of the fluid, with dimensions of weight per unit volume; some values are tabulated in Table 2.1. The quantity $p/\rho g$ is a length called the *pressure head* of the fluid.

For lakes and oceans, the coordinate system is usually chosen as in Fig. 2.5, with $z = 0$ at the free surface, where p equals the surface atmospheric pressure p_a. When

Table 2.1

SPECIFIC WEIGHT OF SOME COMMON FLUIDS

Fluid	Specific weight ρg at $68°F = 20°C$	
	lbf/ft^3	N/m^3
Air (at 1 atm)	0.0752	11.8
Ethyl alcohol	49.2	7,733
SAE 30 oil	57.3	8,996
Water	62.4	9,790
Seawater	64.0	10,050
Glycerin	78.7	12,360
Carbon tetrachloride	99.1	15,570
Mercury	846	133,100

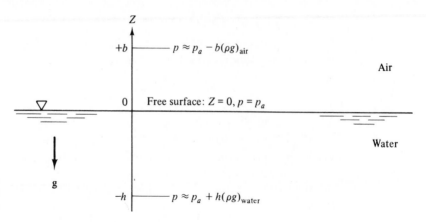

Fig. 2.5 Hydrostatic-pressure distribution in oceans and atmospheres.

we introduce the reference value $(p_1, z_1) = (p_a, 0)$, Eq. (2.20) becomes, for p at any (negative) depth z,

Lakes and oceans: $$p = p_a - \rho g z \qquad (2.21)$$

where ρg is the average specific weight of the lake or ocean. As we shall see, Eq. (2.21) holds in the atmosphere also with an accuracy of 2 percent for heights z up to 1000 ft.

EXAMPLE 2.1 Newfound Lake, a freshwater lake near Bristol, New Hampshire, has a maximum depth of 60 m, and mean atmospheric pressure is 91 kPa. Estimate the absolute pressure at this maximum depth in kilopascals.

Solution From Table 2.1, take $\rho g \approx 9790$ N/m³. With $p_a = 91$ kPa and $z = -60$ m, Eq. (2.21) predicts that the pressure at this depth will be

$$p = 91 \text{ kN/m}^2 - (9790 \text{ N/m}^3)(-60 \text{ m}) \frac{1 \text{ kN}}{1000 \text{ N}}$$

$$= 91 \text{ kPa} + 587 \text{ kN/m}^2 = 678 \text{ kPa} \qquad \qquad Ans.$$

By omitting p_a we could state the result as $p = 587$ kPa (gage).

The Mercury Barometer

The simplest practical application of the hydrostatic formula (2.20) is the barometer (Fig. 2.6), which measures atmospheric pressure. A tube is filled with mercury and inverted while submerged in a reservoir. This causes a near vacuum in the closed upper end because mercury has an extremely small vapor pressure at room temperatures (0.16 Pa at 20°C). Since atmospheric pressure forces a mercury column to rise a distance h into the tube, the upper mercury surface is at zero pressure.

From Fig. 2.6, Eq. (2.20) applies with $p_1 = 0$ at $z_1 = h$ and $p_2 = p_a$ at $z_2 = 0$:

$$p_a - 0 = -\rho_M g(0 - h)$$

or $$h = \frac{p_a}{\rho_M g} \qquad (2.22)$$

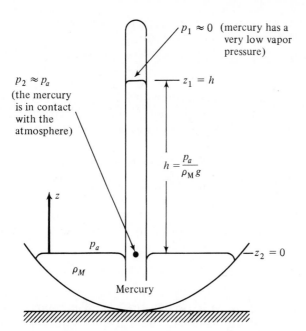

$p_1 \approx 0$ (mercury has a very low vapor pressure)

$z_1 = h$

$p_2 \approx p_a$ (the mercury is in contact with the atmosphere)

$h = \dfrac{p_a}{\rho_M g}$

z

p_a

ρ_M

$z_2 = 0$

Mercury

Fig. 2.6 The mercury barometer: the height of the mercury column is a measure of atmospheric pressure.

At sea-level standard, with $p_a = 101{,}350$ Pa and $\rho_M g = 133{,}100$ N/m^2 from Table 2.1, the barometric height is $h = 101{,}350/133{,}100 = 0.761$ m or 761 mm. In the United States the weather service reports this as an atmospheric "pressure" of 29.96 in Hg (inches of mercury). Mercury is used because it is the heaviest common liquid. A water barometer would be 34 ft high.

Hydrostatic Pressure in Gases

Gases are compressible, with density nearly proportional to pressure. Thus density must be considered as a variable in Eq. (2.18) if the integration carries over large pressure changes. It is sufficiently accurate to introduce the perfect-gas law $p = \rho R T$ in Eq. (2.18)

$$\frac{dp}{dz} = -\rho g = -\frac{p}{RT} g$$

Separate the variables and integrate between points 1 and 2:

$$\int_1^2 \frac{dp}{p} = \ln \frac{p_2}{p_1} = -\frac{g}{R} \int_1^2 \frac{dz}{T} \tag{2.23}$$

The integral over z requires an assumption about the temperature variation $T(z)$. One common approximation is the *isothermal atmosphere*, where $T = T_0$:

$$p_2 = p_1 \exp\left[-\frac{g(z_2 - z_1)}{RT_0} \right] \tag{2.24}$$

The quantity in brackets is dimensionless. (Think that over; it must be dimensionless, right?) Equation (2.24) is a fair approximation for the earth, but actually the earth's mean atmospheric temperature drops off nearly linearly with z up to an altitude of about 36,000 ft (11,000 m):

$$T \approx T_0 - Bz \qquad (2.25)$$

where T_0 is sea-level temperature (absolute) and B is the *lapse rate*, both of which vary somewhat from day to day. By international agreement [1] the following standard values are assumed to apply from 0 to 36,000 ft:

$$T_0 = 518.69°R = 288.16 \text{ K} = 15°C$$

$$B = 0.003566 \text{ °R/ft} = 0.00650 \text{ K/m} \qquad (2.26)$$

This lower portion of the atmosphere is called the *troposphere*. Introducing Eq. (2.25) into (2.23) and integrating, we obtain the more accurate relation

$$p = p_a \left(1 - \frac{Bz}{T_0}\right)^{g/RB} \qquad \text{where } \frac{g}{RB} = 5.26 \text{ (air)} \qquad (2.27)$$

in the troposphere, with $z = 0$ at sea level. The exponent g/RB is dimensionless (again it must be) and has the standard value of 5.26 for air, with $R = 1717 \text{ ft}^2/(\text{s}^2 \cdot °R)$.

The U.S. Standard Atmosphere [1] is sketched in Fig. 2.7. The pressure is seen to be nearly zero at $z = 30$ km. For tabulated properties see Appendix Table A.6.

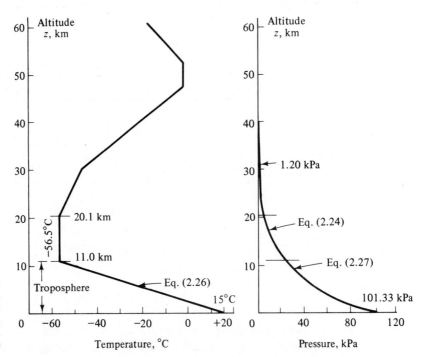

Fig. 2.7 Temperature and pressure distribution in the U.S. Standard Atmosphere. (*From Ref. 1.*)

EXAMPLE 2.2 If standard atmospheric pressure is 101,300 Pa, compute the standard pressure at an altitude of 3000 m (*a*) by the exact formula and compare with (*b*) an isothermal assumption and (*c*) a constant-density formula.

Solution

Part (a) From the exact relation (2.27)

$$p = p_a \left[1 - \frac{(0.00650 \text{ K/m})(3000 \text{ m})}{(288.16 \text{ K})} \right]^{5.26} = p_a(0.9323)^{5.26}$$

$$= 101,300(0.6917) = 70,070 \text{ Pa} = 70.07 \text{ kPa} \qquad\qquad Ans. (a)$$

Part (b) For the isothermal assumption, Eq. (2.24) applies with $p_1 = p_a$ and $z_1 = 0$:

$$p \approx p_a \exp \left\{ -\frac{(9.807 \text{ m/s}^2)(3000 \text{ m})}{[287 \text{ m}^2/(\text{s}^2 \cdot \text{K})](288.16 \text{ K})} \right\}$$

$$= p_a \exp(-0.3557) = 101,300(0.7006) = 70.98 \text{ kPa} \qquad\qquad Ans. (b)$$

This is 1.3 percent higher than the exact result (*a*).

Part (c) The simple linear approximation (2.21) is surprisingly accurate even over the distance of 3000 m. Taking $\rho g \approx 11.8 \text{ N/m}^3$ from Table 2.1, we get

$$p \approx p_a - \rho g z = 101,300 - (11.8 \text{ N/m}^3)(3000 \text{ m})$$

$$= 101,300 - 35,400 = 65,900 \text{ Pa} = 65.90 \text{ kPa} \qquad\qquad Ans. (c)$$

This is only 6.0 percent low compared with the exact result (*a*).

The error involved in using the linear approximation (2.21) can be evaluated by expanding the exact formula (2.27) into a series

$$\left(1 - \frac{Bz}{T_0} \right)^n = 1 - n\frac{Bz}{T_0} + \frac{n(n-1)}{2!}\left(\frac{Bz}{T_0} \right)^2 - \cdots \qquad\qquad (2.28)$$

where $n = g/RB$. Introducing these first three terms of the series into Eq. (2.27) and rearranging, we obtain

$$p = p_a - \rho_a g z \left(1 - \frac{n-1}{2}\frac{Bz}{T_0} + \cdots \right) \qquad\qquad (2.29)$$

Thus the error in using the linear formula (2.21) is small if the second term in parentheses in (2.29) is small compared with unity. This is true if

$$z \ll \frac{2T_0}{(n-1)B} = 20,800 \text{ m} \qquad\qquad (2.30)$$

We thus expect errors of less than 5 percent if z or δz is less than 1000 m. This explains why the error in part (*c*) of Example 2.2 was rather small.

2.4 APPLICATION TO MANOMETRY

From the hydrostatic formula (2.20), a change in elevation $z_2 - z_1$ of a liquid is equivalent to a change in pressure $(p_2 - p_1)/\rho g$. Thus a static column of one or more liquids or gases can be used to measure pressure differences between two

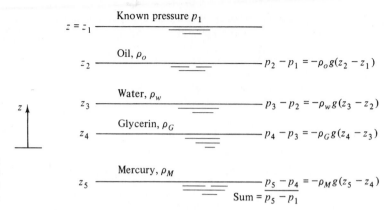

Known pressure p_1

$z = z_1$

Oil, ρ_o

z_2 $p_2 - p_1 = -\rho_o g(z_2 - z_1)$

Water, ρ_w

z_3 $p_3 - p_2 = -\rho_w g(z_3 - z_2)$

Glycerin, ρ_G

z_4 $p_4 - p_3 = -\rho_G g(z_4 - z_3)$

Mercury, ρ_M

z_5 $p_5 - p_4 = -\rho_M g(z_5 - z_4)$

Sum $= p_5 - p_1$

Fig. 2.8 Evaluating pressure changes through a column of multiple fluids.

points. Such a device is called a *manometer*. If multiple fluids are used, we must change the density in the formula as we move from one fluid to another. Figure 2.8 illustrates the use of the formula with a column of multiple fluids. The pressure change through each fluid is calculated separately. If we wish to know the total change $p_5 - p_1$, we add the successive changes $p_2 - p_1$, $p_3 - p_2$, $p_4 - p_3$, and $p_5 - p_4$. The intermediate values of p cancel, and we have, for the example of Fig. 2.8,

$$
\begin{aligned}
p_5 - p_1 = &-\rho_0 g(z_2 - z_1) - \rho_w g(z_3 - z_2) - \rho_G g(z_4 - z_3) \\
&- \rho_M g(z_5 - z_4)
\end{aligned}
\tag{2.31}
$$

No additional simplification is possible on the right-hand side because of the different densities. Notice that we have placed the fluids in order from the lightest on top to the heaviest at bottom. This is the only stable configuration. If we attempt to layer them in any other manner, the fluids will overturn and seek the stable arrangement.

Figure 2.9 shows a simple open manometer for measuring p_A in a closed chamber relative to atmospheric pressure p_a, in other words, measuring the gage pressure. The chamber fluid ρ_1 is combined with a second fluid ρ_2, perhaps for two reasons: (1) to protect the environment from a corrosive chamber fluid or (2) a heavier fluid ρ_2 will keep z_2 small and the open tube can be shorter. The basic

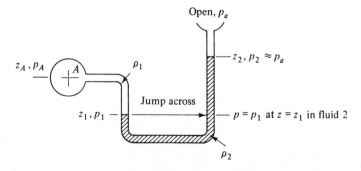

Open, p_a

z_A, p_A A ρ_1 $z_2, p_2 \approx p_a$

Jump across

z_1, p_1 $p = p_1$ at $z = z_1$ in fluid 2

Fig. 2.9 Simple open manometer for measuring p_A relative to atmospheric pressure.

ρ_2

hydrostatic relation (2.20) is applied in two steps, first from z_A to z_1 and then from z_1 to z_2:

$$p_A - p_1 = -\rho_1 g(z_A - z_1) \qquad p_1 - p_2 = -\rho_2 g(z_1 - z_2) \qquad (2.32)$$

The small amount of air between z_2 and the open end is neglected and we assume that $p_2 \approx p_a$. When we add the two relations above, p_1 cancels and we obtain the desired result

$$p_A = p_a - \rho_1 g(z_A - z_1) - \rho_2 g(z_1 - z_2) \qquad (2.33)$$

Notice that the pressure at level z_1 is equal to p_1 on both sides of the U-tube in Fig. 2.9. This is because a continuous length of the same fluid connects the two equal levels. The hydrostatic relation (2.20) requires this as a form of Pascal's law:

Any two points at the same elevation in a continuous mass of the same static fluid will be at the same pressure.

This rule can be used to facilitate calculations in multiple-U-tube problems. We can jump across the U-tube if we stay within the same fluid.

EXAMPLE 2.3   (a) Find the pressure p_A in Fig. 2.9 if $p_a = 2116$ lbf/ft^2, $z_A = 7$ in, $z_1 = 4$ in, $z_2 = 13$ in, fluid 1 is water, and fluid 2 is mercury. (b) What would the height z_2 be for the same pressure p_A if the mercury were replaced by glycerin?

Solution

Part (a)   The pressure relation has already been derived in Eq. (2.33). From Table 2.1, $\rho_1 g = 62.4$ and $\rho_2 g = 846$ lbf/ft^3. We need z_A, z_1, and z_2 in feet to produce p in pounds force per square foot. Substitute into (2.33)

$$p_A = 2116 - 62.4(\tfrac{7}{12} - \tfrac{4}{12}) - 846(\tfrac{4}{12} - \tfrac{13}{12})$$

$$= 2116 - 16 + 635 = 2735 \text{ lbf/ft}^2 \qquad \textit{Ans. (a)}$$

Part (b)   The specific weight of glycerin is 78.7 lbf/ft^3. The balance in Eq. (2.33) would now be

$$2735 = 2116 - 16 - 78.7(\tfrac{4}{12} - z_2)$$

or

$$z_2 = \frac{635}{78.7} + \frac{4}{12} = 8.4 \text{ ft} = 101 \text{ in} \qquad \textit{Ans. (b)}$$

This would require a much taller U-tube and is not a good idea in the long run because glycerin tends to mix with water over a period of time.

Instead of attempting to remember manometer formulations, it is simpler to remember Eq. (2.20)[1] and derive the pressure-change relation for each new situation. Figure 2.10 illustrates a multiple-fluid manometer problem for finding the

---

[1] The author prefers as a memory device to keep the minus sign with ρg so that the subscripts on p and z will be the same on both sides.

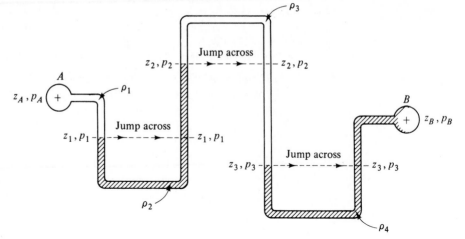

Fig. 2.10 A
complicated multiple-
fluid manometer to
relate p_A to p_B. This
system is not especially
practical but makes a
good homework or
examination problem.

difference in pressure between the two chambers A and B. Repeated application of Eq. (2.20) results in

$$p_A - p_1 = -\rho_1 g(z_A - z_1) \qquad p_1 - p_2 = -\rho_2 g(z_1 - z_2)$$
$$p_2 - p_3 = -\rho_3 g(z_2 - z_3) \qquad p_3 - p_B = -\rho_4 g(z_3 - z_B) \qquad (2.34)$$

Summing these four gives the desired pressure difference $p_A - p_B$.

EXAMPLE 2.4   Find the pressure difference $p_A - p_B$ in Fig. 2.10 if $z_A = 1.6$ m, $z_1 = 0.7$ m, $z_2 = 2.1$ m, $z_3 = 0.9$ m, $z_B = 1.8$ m, fluids 1 and 3 are water, and fluids 2 and 4 are mercury.

Solution   The specific weights of water and mercury are 9790 and 133,100 N/m³, respectively. The pressure relations have already been derived in Eq. (2.34). Summing them and substituting numerical values, we have

$$p_A - p_B = -9790(1.6 - 0.7) - 133,100(0.7 - 2.1) - 9790(2.1 - 0.9)$$
$$- 133,100(0.9 - 1.8) = -8811 + 186,340 - 11,748 + 119,790$$
$$= 285,571 \text{ Pa} = 286 \text{ kPa} \qquad \qquad \textit{Ans.}$$

The intermediate six-figure result of 285,571 Pa is utterly fatuous since the measurements cannot be made that accurately.

In making these manometer calculations we have neglected the capillary-height changes due to surface tension, which were discussed in Example 1.13. These effects cancel out if there is a fluid interface, or *meniscus*, on both sides of the U-tube, as in Fig. 2.9. Otherwise, as in the right-hand U-tube of Fig. 2.10, a capillary correction can be made or the effect can be made negligible by using large-bore tubes.

2.5 HYDROSTATIC FORCES ON PLANE SURFACES

A common problem in the design of structures which interact with fluids is the computation of the hydrostatic force on a plane surface. If we neglect density changes in the fluid, Eq. (2.20) applies and the pressure on any submerged surface varies linearly with depth. For a plane surface, the linear stress distribution is exactly analogous to combined bending and compression of a beam in strength-of-materials theory. The hydrostatic problem thus reduces to simple formulas involving the centroid and moments of inertia of the plate cross-sectional area.

Figure 2.11 shows a plane panel of arbitrary shape completely submerged in a liquid. The panel plane makes an arbitrary angle θ with the horizontal free surface, so that the depth varies over the panel surface. If h is the depth to any element area dA of the plate, from Eq. (2.20) the pressure there is $p = p_a + \rho g h$.

To derive formulas involving the plate shape, establish an xy coordinate system in the plane of the plate with the origin at its centroid, plus a dummy coordinate ξ down from the surface in the plane of the plate. Then the total hydrostatic force on one side of the plate is given by

$$F = \int p \, dA = \int (p_a + \rho g h) \, dA = p_a A + \rho g \int h \, dA \qquad (2.35)$$

The remaining integral is evaluated by noticing from Fig. 2.11 that $h = \xi \sin \theta$ and, by definition, the centroidal slant distance from the surface to the plate is

$$\xi_{CG} = \frac{1}{A} \int \xi \, dA \qquad (2.36)$$

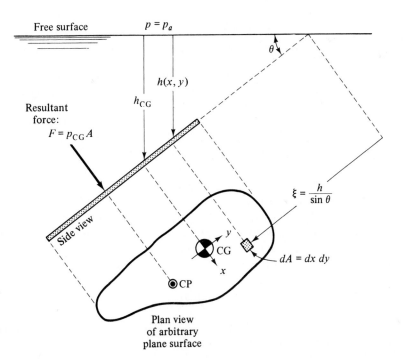

Fig. 2.11 Hydrostatic force and center of pressure on an arbitrary plane surface of area A inclined at an angle θ below the free surface.

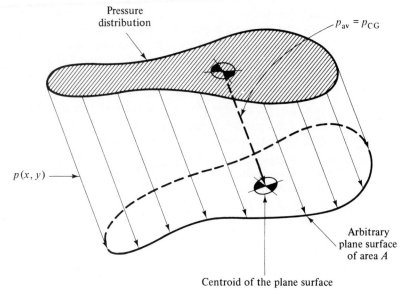

Pressure distribution

$p_{av} = p_{CG}$

$p(x, y)$

Arbitrary plane surface of area A

Centroid of the plane surface

Fig. 2.12 The hydrostatic-pressure force on a plane surface is equal, regardless of its shape, to the resultant of the three-dimensional linear pressure distribution on that surface, $F = p_{CG}A$.

Therefore, since θ is constant along the plate, Eq. (2.35) becomes

$$F = p_a A + \rho g \sin \theta \int \xi \, dA = p_a A + \rho g \sin \theta \, \xi_{CG} A \qquad (2.37)$$

Finally, unravel this by noticing that $\xi_{CG} \sin \theta = h_{CG}$, the depth straight down from the surface to the plate centroid. Thus

$$F = p_a A + \rho g h_{CG} A = (p_a + \rho g h_{CG})A = p_{CG} A \qquad (2.38)$$

The force on one side of any plane submerged surface in a uniform fluid equals the pressure at the plate centroid times the plate area, independent of the shape of the plate or the angle θ at which it is slanted.

Equation (2.38) can be visualized physically in Fig. 2.12 as the resultant of a linear stress distribution over the plate area. This simulates combined compression and bending of a beam of the same cross section. It follows that the "bending" portion of the stress causes no force if its "neutral axis" passes through the plate centroid of area. Thus the remaining "compression" part must equal the centroid stress times the plate area. This is the result of Eq. (2.38).

However, to balance the bending-moment portion of the stress, the resultant force F does not act through the centroid but below it toward the high-pressure side. Its line of action passes through the *center of pressure* CP of the plate, as sketched in Fig. 2.11. To find the coordinates (x_{CP}, y_{CP}), we sum moments of the elemental force $p \, dA$ about the centroid and equate to the moment of the resultant F. To compute y_{CP}, we equate

$$F y_{CP} = \int y p \, dA = \int y(p_a + \rho g \xi \sin \theta) \, dA = \rho g \sin \theta \int y \xi \, dA \qquad (2.39)$$

The term $\int p_a y \, dA$ vanishes by definition of centroidal axes. Introducing $\xi = \xi_{CG} - y$, we obtain

$$F y_{CP} = \rho g \sin \theta \left(\xi_{CG} \int y \, dA - \int y^2 \, dA \right) = -\rho g \sin \theta \, I_{xx} \qquad (2.40)$$

where again $\int y \, dA = 0$ and I_{xx} is the area moment of inertia of the plate area about its centroidal x axis, computed in the plane of the plate. Substituting for F gives the result

$$y_{CP} = -\rho g \sin \theta \frac{I_{xx}}{p_{CG} A} \qquad (2.41)$$

The negative sign in Eq. (2.41) shows that y_{CP} is below the centroid at a deeper level and, unlike F, depends upon the angle θ. If we move the plate deeper, y_{CP} approaches the centroid because every term in Eq. (2.41) remains constant except p_{CG}, which increases.

The determination of x_{CP} is exactly similar:

$$F x_{CP} = \int x p \, dA = \int x [p_a + \rho g (\xi_{CG} - y) \sin \theta] \, dA$$

$$= -\rho g \sin \theta \int x y \, dA = -\rho g \sin \theta \, I_{xy} \qquad (2.42)$$

where I_{xy} is the product of inertia of the plate, again computed in the plane of the plate. Substituting for F gives

$$x_{CP} = -\rho g \sin \theta \frac{I_{xy}}{p_{CG} A} \qquad (2.43)$$

For positive I_{xy}, x_{CP} is negative because the dominant pressure force acts in the third, or lower left, quadrant of the panel. If $I_{xy} = 0$, usually implying symmetry, $x_{CP} = 0$ and the center of pressure lies directly below the centroid on the y axis.

Gage-Pressure Formulas

In most cases the ambient pressure p_a is neglected because it acts on both sides of the plate; e.g., the other side of the plate is inside a ship or on the dry side of a gate or dam. In this case $p_{CG} = \rho g h_{CG}$, and the center of pressure becomes independent of specific weight

$$F = \rho g h_{CG} A \qquad y_{CP} = -\frac{I_{xx} \sin \theta}{h_{CG} A} \qquad x_{CP} = -\frac{I_{xy} \sin \theta}{h_{CG} A} \qquad (2.44)$$

Figure 2.13 gives the area and moments of inertia of several common cross sections for use with these formulas.

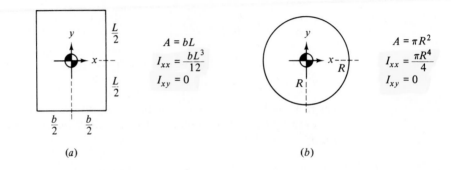

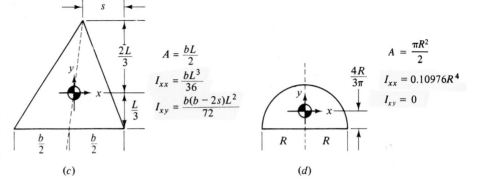

Fig. 2.13 Centroidal moments of inertia for various cross sections: (*a*) rectangle, (*b*) circle, (*c*) triangle, and (*d*) semicircle.

EXAMPLE 2.5 The gate in Fig. E2.5*a* is 5 ft wide, is hinged at point *B*, and rests against a smooth wall at point *A*. Compute (*a*) the force on the gate due to seawater pressure; (*b*) the horizontal force *P* exerted by the wall at point *A*; and (*c*) the reactions at the hinge *B*.

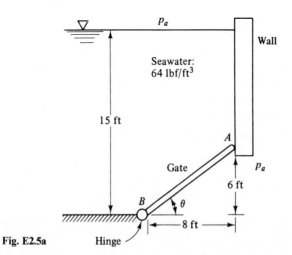

Fig. E2.5a

Solution

Part (a) By geometry the gate is 10 ft long from A to B, and its centroid is halfway between, or at elevation 3 ft above point B. The depth h_{CG} is thus $15 - 3 = 12$ ft. The gate area is $5(10) = 50$ ft^2. Neglect p_a as acting on both sides of the gate. From Eq. (2.38) the hydrostatic force on the gate is

$$F = p_{CG}A = \rho g h_{CG} A = (64 \text{ lbf/ft}^3)(12 \text{ ft})(50 \text{ ft}^2) = 38{,}400 \text{ lbf} \qquad \textit{Ans. (a)}$$

Part (b) First we must find the center of pressure of F. A free-body diagram of the gate is shown in Fig. E2.5b. The gate is a rectangle, and hence

$$I_{xy} = 0 \quad \text{and} \quad I_{xx} = \frac{bL^3}{12} = \frac{(5 \text{ ft})(10 \text{ ft})^3}{12} = 417 \text{ ft}^4$$

The distance l from the CG to the CP is given by Eq. (2.44) since p_a is neglected

$$l = -y_{CP} = +\frac{I_{xx} \sin \theta}{h_{CG} A} = \frac{(417 \text{ ft}^4)(\frac{6}{10})}{(12 \text{ ft})(50 \text{ ft}^2)} = 0.417 \text{ ft}$$

The distance from point B to force F is thus $10 - l - 5 = 4.583$ ft. Summing moments counterclockwise about B gives

$$PL \sin \theta - F(5 - l) = P(6 \text{ ft}) - (38{,}400 \text{ lbf})(4.583 \text{ ft}) = 0$$

or $\qquad\qquad\qquad\qquad\qquad\qquad P = 29{,}300 \text{ lbf} \qquad\qquad\qquad\qquad\qquad \textit{Ans. (b)}$

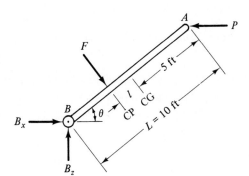

Fig. E2..5b

Part (c) With F and P known, the reactions B_x and B_z are found by summing forces on the gate

$$\sum F_x = 0 = B_x + F \sin \theta - P = B_x + 38{,}400(0.6) - 29{,}300$$

or $\qquad\qquad\qquad\qquad\qquad\qquad B_x = 6300 \text{ lbf}$

$$\sum F_z = 0 = B_z - F \cos \theta = B_z - 38{,}400(0.8)$$

or $\qquad\qquad\qquad\qquad\qquad\qquad B_z = 30{,}700 \text{ lbf} \qquad\qquad\qquad\qquad\qquad \textit{Ans. (c)}$

This example should have reviewed your knowledge of statics.

EXAMPLE 2.6 A tank of oil has a right triangular panel near the bottom, as in Fig. E2.6. Omitting p_a, find the (a) hydrostatic force and (b) CP on the panel.

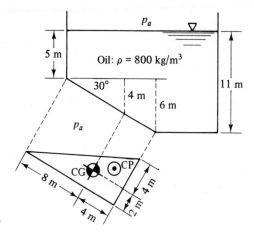

Fig. E2.6

Solution

Part (a) The triangle has properties given in Fig. 2.13c. The centroid is one-third up (4 m) and one-third over (2 m) from the lower left corner, as shown. The area is

$$\tfrac{1}{2}(6 \text{ m})(12 \text{ m}) = 36 \text{ m}^2$$

The moments of inertia are

$$I_{xx} = \frac{bL^3}{36} = \frac{(6 \text{ m})(12 \text{ m})^3}{36} = 288 \text{ m}^4$$

and

$$I_{xy} = \frac{b(b - 2s)L^2}{72} = \frac{(6 \text{ m})[6 \text{ m} - 2(6 \text{ m})](12 \text{ m})^2}{72} = -72 \text{ m}^4$$

The depth to the centroid is $h_{CG} = 5 + 4 = 9$ m; thus the hydrostatic force from Eq. (2.44) is

$$F = \rho g h_{CG} A = (800 \text{ kg/m}^3)(9.807 \text{ m/s}^2)(9 \text{ m})(36 \text{ m}^2)$$
$$= 2.54 \times 10^6 \text{ (kg} \cdot \text{m)/s}^2 = 2.54 \times 10^6 \text{ N} = 2.54 \text{ MN} \qquad \textit{Ans. (a)}$$

Part (b) The CP position is given by Eqs. (2.44):

$$y_{CP} = -\frac{I_{xx} \sin \theta}{h_{CG} A} = -\frac{(288 \text{ m}^4)(\sin 30°)}{(9 \text{ m})(36 \text{ m}^2)} = -0.444 \text{ m}$$

$$x_{CP} = -\frac{I_{xy} \sin \theta}{h_{CG} A} = -\frac{(-72 \text{ m}^4)(\sin 30°)}{(9 \text{ m})(36 \text{ m}^2)} = +0.111 \text{ m} \qquad \textit{Ans. (b)}$$

The resultant force $F = 2.54$ MN acts through this point, which is down and to the right of the centroid.

2.6 HYDROSTATIC FORCES ON CURVED SURFACES

The resultant pressure force on a curved surface is most easily computed by separating into horizontal and vertical components. Consider the arbitrary curved surface sketched in Fig.. 2.14a. The incremental pressure forces, being normal to the local area element, vary in direction along the surface and thus cannot be added numerically. We could sum the separate three components of these elemental

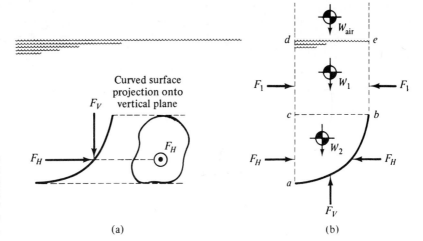

Fig. 2.14 Computation of hydrostatic force on a curved surface: (*a*) submerged curved surface; (*b*) free-body diagram of fluid above the curved surface.

(a)

(b)

pressure forces, but it turns out that we need not perform a laborious three-way integration.

Figure 2.14*b* shows a free-body diagram of the column of fluid contained in the vertical projection above the curved surface. The desired forces F_H and F_V are exerted by the surface on the fluid column. Other forces are shown due to fluid weight and horizontal pressure on the vertical sides of this column. The column of fluid must be in static equilibrium. On the upper part of the column, *bcde*, the horizontal components F_1 exactly balance and are not relevant to the discussion. On the lower, irregular portion of fluid *abc* adjoining the surface, summation of horizontal forces shows that the desired force F_H due to the curved surface is exactly equal to the force F_H on the vertical left side of the fluid column. This left-side force can be computed by the plane-surface formula, Eq. (2.38), based on a vertical projection of the area of the curved surface. This is a general rule and simplifies the analysis:

> The horizontal component of force on a curved surface equals the force on the plane area formed by the projection of the curved surface onto a vertical plane normal to the component.

If there are two horizontal components, both can be computed by this scheme.

Summation of vertical forces on the fluid free body then shows that

$$F_V = W_1 + W_2 + W_{\text{air}} \tag{2.45}$$

We can state this in words as our second general rule:

> The vertical component of pressure force on a curved surface equals in magnitude and direction the weight of the entire column of fluid, both liquid and atmosphere, above the curved surface.

Thus the calculation of F_V involves little more than finding centers of mass of a column of fluid—perhaps a little integration if the lower portion *abc* has a particularly vexing shape.

EXAMPLE 2.7   A dam has a parabolic shape $z/z_0 = (x/x_0)^2$ as shown in Fig. E2.7a, with $x_0 = 10$ ft and $z_0 = 24$ ft. The fluid is water, $\rho g = 62.4$ lbf/ft³, and atmospheric pressure may be omitted. Compute the forces F_H and F_V on the dam and the position CP where they act. The width of the dam is 50 ft.

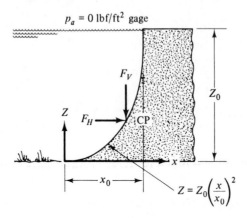

$p_a = 0$ lbf/ft² gage

$Z = Z_0 \left(\dfrac{x}{x_0}\right)^2$

Fig. E2.7a

Solution   The vertical projection of this curved surface is a rectangle 24 ft high and 50 ft wide, with its centroid halfway down, or $h_{CG} = 12$ ft. The force F_H is thus

$$F_H = \rho g h_{CG} A_{proj} = (62.4 \text{ lbf/ft}^3)(12 \text{ ft})(24 \text{ ft})(50 \text{ ft})$$
$$= 899,000 \text{ lbf} = 899 \times 10^3 \text{ lbf} \qquad\qquad\qquad Ans.$$

The line of action of F_H is below the centroid by an amount

$$y_{CP} = -\frac{I_{xx} \sin\theta}{h_{CG} A_{proj}} = -\frac{\frac{1}{12}(50 \text{ ft})(24 \text{ ft})^3(\sin 90°)}{(12 \text{ ft})(24 \text{ ft})(50 \text{ ft})} = -4 \text{ ft}$$

Thus F_H is $12 + 4 = 16$ ft, or two-thirds, down from the free surface or 8 ft from the bottom, as might have been evident by inspection of the triangular pressure distribution.

The vertical component F_V equals the weight of the parabolic portion of fluid above the curved surface. The geometric properties of a parabola are shown in Fig. E2.7b. The weight of this amount of water is

$$F_V = \rho g(\tfrac{2}{3}x_0 z_0 b) = (62.4 \text{ lbf/ft}^3)(\tfrac{2}{3})(10 \text{ ft})(24 \text{ ft})(50 \text{ ft})$$
$$= 499,000 \text{ lbf} = 499 \times 10^3 \text{ lbf} \qquad\qquad\qquad Ans.$$

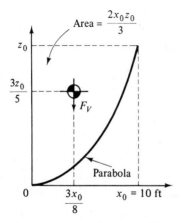

Area $= \dfrac{2x_0 z_0}{3}$

z_0

$\dfrac{3z_0}{5}$

F_V

Parabola

$\dfrac{3x_0}{8}$

$x_0 = 10$ ft

Fig. E2.7b

This acts downward on the surface at a distance $3x_0/8 = 3.75$ ft over from the origin of coordinates. Note that the vertical distance $3z_0/5$ in Fig. E2.7b is irrelevant.

The total resultant force acting on the dam is

$$F = (F_H^2 + F_V^2)^{1/2} = [(499)^2 + (899)^2]^{1/2} = 1028 \times 10^3 \text{ lbf}$$

As seen in Fig. E2.7c, this force acts down and to the right at an angle of $29° = \tan^{-1}\frac{499}{899}$. The force F passes through the point $(x, z) = (3.75 \text{ ft}, 8 \text{ ft})$. If we move down along the $29°$ line until we strike the dam, we find an equivalent center of pressure on the dam at

$$x_{CP} = 5.43 \text{ ft} \qquad z_{CP} = 7.07 \text{ ft} \qquad\qquad Ans.$$

This definition of CP is rather artificial, but this is an unavoidable complication of dealing with a curved surface.

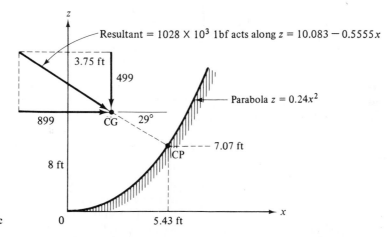

Fig. E2.7c

2.7 HYDROSTATIC FORCES IN LAYERED FLUIDS

The formulas for plane and curved surfaces in Secs. 2.5 and 2.6 are valid only for a fluid of uniform density. If the fluid is layered with different densities, as in Fig. 2.15, a single formula cannot resolve the problem because the slope of the linear pressure distribution changes between layers. However, the formulas apply separately to each layer, and thus the appropriate remedy is to compute and sum the separate layer forces and moments.

Consider the slanted plane surface immersed in a two-layer fluid in Fig. 2.15. The slope of the pressure distribution becomes steeper as we move down into the denser second layer. The total force on the plate does *not* equal the pressure at the centroid times the plate area, but the plate portion in each layer does satisfy the formula, so that we can sum forces to find the total

$$F = \sum F_i = \sum p_{CG_i} A_i \qquad\qquad (2.46)$$

Similarly, the centroid of the plate portion in each layer can be used to locate the center of pressure on that portion

$$y_{CP_i} = -\frac{\rho_i g \sin \theta_i I_{xx_i}}{p_{CG_i} A_i} \qquad x_{CP_i} = -\frac{\rho_i g \sin \theta_i I_{xy_i}}{p_{CG_i} A_i} \qquad (2.47)$$

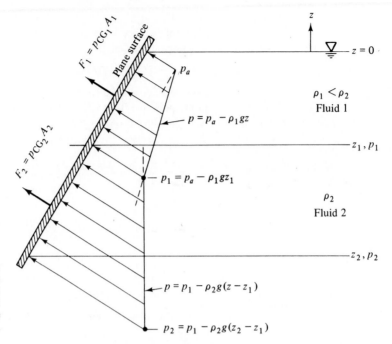

Fig. 2.15 Hydrostatic forces on a surface immersed in a layered fluid must be summed together in separate pieces.

These formulas locate the center of pressure of that particular F_i with respect to the centroid of that particular portion of plate in the layer, not with respect to the centroid of the entire plate. The center of pressure of the total force $F = \sum F_i$ can then be found by summing moments about some convenient point such as the surface. The following example will illustrate.

EXAMPLE 2.8   A tank 20 ft deep and 7 ft wide is layered with 8 ft of oil, 6 ft of water, and 4 ft of mercury. Compute (*a*) the total hydrostatic force and (*b*) the resultant center of pressure of the fluid on the right-hand side of the tank.

Solution

Part (a)   Divide the end panel into three parts as sketched in Fig. E2.8 and find the hydrostatic pressure at the centroid of each part, using the relation (2.38) in steps as in Fig. E2.8:

$$p_{CG_1} = (55.0 \text{ lbf/ft}^3)(4 \text{ ft}) = 220 \text{ lbf/ft}^2$$

$$p_{CG_2} = (55.0)(8) + 62.4(3) = 627 \text{ lbf/ft}^2$$

$$p_{CG_3} = (55.0)(8) + 62.4(6) + 846(2) = 2506 \text{ lbf/ft}^2$$

These pressures are then multiplied by the respective panel areas to find the force on each portion:

$$F_1 = p_{CG_1} A_1 = (220 \text{ lbf/ft}^2)(8 \text{ ft})(7 \text{ ft}) = \quad 12,300 \text{ lbf}$$

$$F_2 = p_{CG_2} A_2 = 627(6)(7) = \quad 26,300 \text{ lbf}$$

$$F_3 = p_{CG_3} A_3 = 2506(4)(7) = \quad 70,200 \text{ lbf}$$

$$F = \sum F_i = \overline{108,800 \text{ lbf}} \qquad Ans. (a)$$

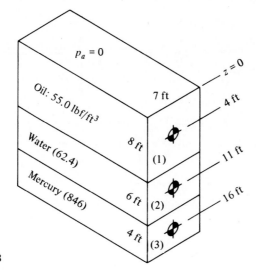

Fig. E2.8

Part (b)   Equations (2.47) can be used to locate the CP of each force F_i, noting that $\theta = 90°$ and $\sin\theta = 1$ for all parts. The moments of inertia are $I_{xx_1} = (7\text{ ft})(8\text{ ft})^3/12 = 298.7\text{ ft}^4$, $I_{xx_2} = 7(6)^3/12 = 126.0\text{ ft}^4$, and $I_{xx_3} = 7(4)^3/12 = 37.3\text{ ft}^4$. The centers of pressure are thus at

$$y_{CP_1} = -\frac{\rho_1 g I_{xx_1}}{F_1} = -\frac{(55.0\text{ lbf/ft}^3)(298.7\text{ ft}^4)}{12,300\text{ lbf}} = -1.33\text{ ft}$$

$$y_{CP_2} = -\frac{62.4(126.0)}{26,300} = -0.30\text{ ft} \qquad y_{CP_3} = -\frac{846(37.3)}{70,200} = -0.45\text{ ft}$$

This locates $z_{CP_1} = -4 - 1.33 = -5.33\text{ ft}$, $z_{CP_2} = -11 - 0.30 = -11.30\text{ ft}$, and $z_{CP_3} = -16 - 0.45 = -16.45\text{ ft}$. Summing moments about the surface then gives

$$\sum F_i z_{CP_i} = F z_{CP}$$

or $\qquad 12,300(-5.33) + 26,300(-11.30) + 70,200(-16.45) = 108,800 z_{CP}$

or $$z_{CP} = -\frac{1,518,000}{108,800} = -13.95\text{ ft} \qquad\qquad Ans. (b)$$

The center of pressure of the total resultant force on the right side of the tank lies 13.95 ft below the surface.

2.8   BUOYANCY AND STABILITY

The same principles used to compute hydrostatic forces on surfaces can be applied to the net pressure force on a completely submerged or floating body. The results are the two laws of buoyancy discovered by Archimedes in the third century B.C.:

1. A body immersed in a fluid experiences a vertical buoyant force equal to the weight of the fluid it displaces.

2. A floating body displaces its own weight in the fluid in which it floats.

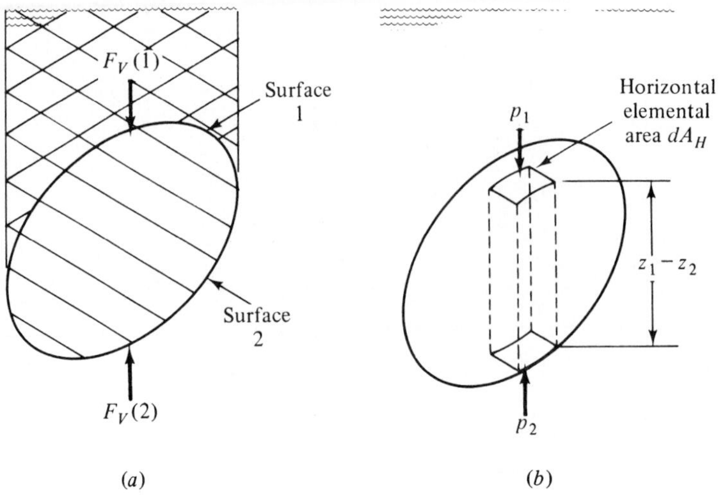

Fig. 2.16 Two different approaches to the buoyant force on an arbitrary immersed body: (*a*) forces on upper and lower curved surfaces; (*b*) summation of elemental vertical-pressure forces.

These two laws are easily derived by referring to Fig. 2.16. In Fig. 2.16*a*, the body lies between an upper curved surface 1 and a lower curved surface 2. From Eq. (2.45) for vertical force, the body experiences a net upward force

$$F_B = F_V(2) - F_V(1)$$

$$= \text{(fluid weight above 2)} - \text{(fluid weight above 1)}$$

$$= \text{weight of fluid equivalent to body volume} \tag{2.48}$$

Alternatively, from Fig. 2.16*b*, we can sum the vertical forces on elemental vertical slices through the immersed body

$$F_B = \int_{\text{body}} (p_2 - p_1)\, dA_H = -\rho g \int (z_2 - z_1)\, dA_H = (\rho g)(\text{body volume}) \tag{2.49}$$

These are identical results and equivalent to law 1 above.

Equation (2.49) assumes that the fluid has uniform specific weight. The line of action of the buoyant force passes through the center of volume of the displaced body; i.e., its center of mass is computed as if it had uniform density. This point through which F_B acts is called the *center of buoyancy*, commonly labeled B or CB on a drawing. Of course the point B may or may not correspond to the actual center of mass of the body's own material, which may have variable density.

Equation (2.49) can be generalized to a layered fluid (LF) by summing the weights of each layer of density ρ_i displaced by the immersed body

$$(F_B)_{\text{LF}} = \sum \rho_i g (\text{displaced volume})_i \tag{2.50}$$

Each displaced layer would have its own center of volume, and one would have to sum moments of the incremental buoyant forces in order to find the center of buoyancy of the immersed body.

Since liquids are relatively heavy, we are conscious of their buoyant forces, but gases also exert buoyancy on any body immersed in them. For example, human beings have an average specific weight of about 60 lbf/ft^3. We may record the

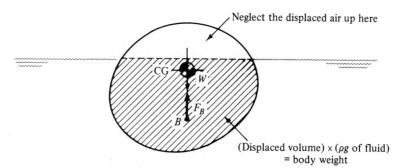

Neglect the displaced air up here

(Displaced volume) × (ρg of fluid)
= body weight

Fig. 2.17 Static equilibrium of a floating body.

weight of a person as 180 lbf and thus estimate his total volume as 3.0 ft³. However, in so doing we are neglecting the buoyant force of the air surrounding the person. At standard conditions, the specific weight of air is 0.0763 lbf/ft³; hence the buoyant force is approximately 0.23 lbf. If measured in vacuo, the person would weigh about 0.23 lbf more. For balloons and blimps the buoyant force of air instead of being negligible is the controlling factor in the design. Also, many flow phenomena, e.g., natural convection of heat and vertical mixing in the ocean, are strongly dependent upon seemingly small buoyant forces.

Floating bodies are a special case; only a portion of the body is submerged, the remainder poking up out of the free surface. This is illustrated in Fig. 2.17, where the shaded portion is the displaced volume. Equation (2.49) is modified to apply to this smaller volume

$$F_B = (\rho g)(\text{displaced volume}) = \text{floating-body weight} \qquad (2.51)$$

Not only does the buoyant force equal the body weight but they are *collinear* since there can be no net moments for static equilibrium. Equation (2.51) is the mathematical equivalent of Archimedes' law 2 previously stated.

EXAMPLE 2.9   A block of concrete weighs 100 lbf in air and "weighs" only 60 lbf when immersed in fresh water (62.4 lbf/ft³). What is the average specific weight of the block?

Solution   A free-body diagram of the submerged block shows a balance between the apparent weight, the buoyant force, and the actual weight

$$\sum F_z = 0 = 60 + F_B - 100$$

or
$$F_B = 40 \text{ lbf} = (62.4)(\text{block volume in ft}^3)$$

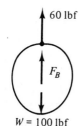

60 lbf

F_B

Fig. E2.9 $W = 100$ lbf

Solving gives the volume of the block as $40/62.4 = 0.641$ ft^3. Therefore the specific weight of the block is

$$(\rho g)_{\text{block}} = \frac{100 \text{ lbf}}{0.641 \text{ ft}^3} = 156 \text{ lbf/ft}^3 \qquad Ans.$$

Occasionally, a body will have exactly the right weight and volume for its ratio to equal the specific weight of the fluid. If so, the body will be *neutrally buoyant* and will remain at rest at any point where it is immersed in the fluid. Small neutrally buoyant particles are sometimes used in flow visualization, and a neutrally buoyant body called a *Swallow float* [2] is used to track oceanographic currents. A submarine can achieve positive, neutral, or negative buoyancy by pumping water in or out of its ballast tanks.

Stability

A floating body as in Fig. 2.17 may not approve of the position in which it is floating. If so, it will overturn at the first opportunity and is said to be statically *unstable*, like a pencil balanced upon its point. The least disturbance will cause it to seek another equilibrium position which is stable. Engineers must design to avoid floating instability. The only way to tell for sure whether a floating position is stable is to "disturb" the body a slight amount mathematically and see whether it develops a restoring moment which will return it to its original position. If so, it is stable; if not, unstable. Such calculations for arbitrary floating bodies have been honed to a fine art by naval architects [3], but we can at least outline the basic principle of the static-stability calculation. Figure 2.18 illustrates the computation for the usual case of a symmetric floating body. The steps are as follows:

1. The basic floating position is calculated from Eq. (2.51). The body's center of mass G and center of buoyancy B are computed.

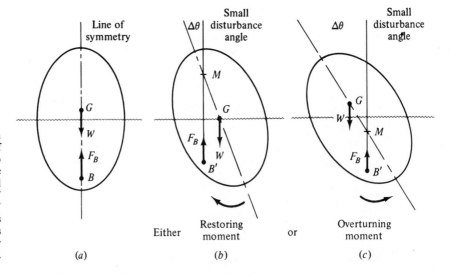

Fig. 2.18 Calculation of the metacenter M of a floating body to determine its static stability: (*a*) initial floating position; (*b*) B' moves far out (point M above G denotes stability); (*c*) B' moves slightly (point M below G denotes instability).

2. The body is tilted a small angle $\Delta\theta$ and a new waterline established for the body to float at this angle. The new position B' of the center of buoyancy is calculated. A vertical line drawn upward from B' intersects the line of symmetry at a point M, called the *metacenter*, which is independent of $\Delta\theta$ for small angles.

3. If point M is above G, that is, if the *metacentric height* $\overline{MG}$ is positive, a restoring moment is present and the original position is stable. If M is below G (negative $\overline{MG}$), the body is unstable and will overturn if disturbed. Stability increases with increasing $\overline{MG}$.

Thus the metacentric height is a property of the cross section for the given weight, and its value gives an indication of the stability of the body. For a body of varying cross section and draft, such as a ship, the computation of the metacenter can be very involved. We shall use a two-dimensional, or uniform, cross section as an example.

EXAMPLE 2.10   A barge has a uniform rectangular cross section of width $2L$ and vertical draft of height H. Determine (*a*) the metacentric height for a small tilt angle and (*b*) the range of ratio L/H for which the barge is statically stable. Assume that the center of mass G of the barge is exactly at the waterline, as shown in Fig. E2.10*a*.

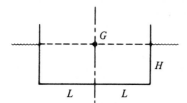

Fig. E2.10a

Solution   The barge in its tilted position $\Delta\theta$ is shown in Fig. E2.10*b*. Because of symmetry the tilted body floats, again with G at the waterline, and triangles aeG and fbG are identical in size. The new center of buoyancy B' can be found by summing moments of the submerged areas about, say, the point G and setting this equal to the total submerged area ($2LH$) times the

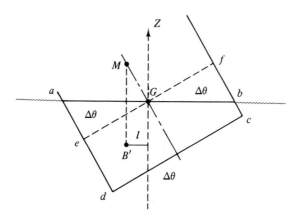

Fig. E2.10b

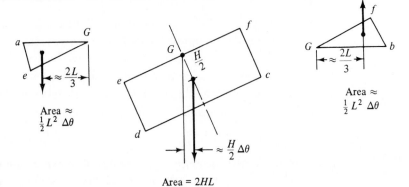

Fig. E2.10c

Area = $2HL$

distance l. The submerged area $aGbcde$ is the sum of rectangle $efdc$ plus triangle aeG minus triangle fbG, and the centroids of each of these pieces can easily be found for small angles of tilt (Fig. E2.10c). Summing moments of area counterclockwise about G, we have

$$2(\tfrac{1}{2}L^2 \, \Delta\theta)\frac{2L}{3} - 2HL\left(\frac{H}{2}\Delta\theta\right) \equiv 2HLl$$

We can solve this for the distance $l = (L^2/3H - H/2)\,\Delta\theta$. From the tilted position showing M above, the distance $\overline{MG} \approx l/\Delta\theta$, for small angles. Thus we have the approximate result for metacentric height, independent of $\Delta\theta$,

$$\overline{MG} = \frac{L^2}{3H} - \frac{H}{2} \qquad\qquad Ans.\,(a)$$

This can be positive or stable only if $L^2 > 3H^2/2$ or

$$L > 1.225H \qquad\qquad Ans.\,(b)$$

The wider the barge relative to its draft, the more stable it is.

2.9 PRESSURE DISTRIBUTION IN RIGID-BODY MOTION

In rigid-body motion, all particles are in combined translation and rotation and there is no relative motion between particles. With no relative motion, there are no strains or strain rates, so that the viscous term $\mu \, \nabla^2 \mathbf{V}$ in Eq. (2.13) vanishes, leaving a balance between pressure, gravity, and particle acceleration

$$\nabla p = \rho(\mathbf{g} - \mathbf{a}) \qquad\qquad (2.52)$$

The pressure gradient acts in the direction $\mathbf{g} - \mathbf{a}$, and lines of constant pressure (including the free surface, if any) are perpendicular to this direction. The general case of combined translation and rotation of a rigid body is shown in Fig. 2.19. If the center of rotation is at point O and the translational velocity is V_0 at this point, the velocity of an arbitrary point P on the body is given by[1]

$$\mathbf{V} = \mathbf{V}_0 + \mathbf{\Omega} \times \mathbf{r}_0 \qquad\qquad (2.53)$$

[1] For a more detailed derivation of rigid-body motion, see Ref. 4, sec. 2.7.

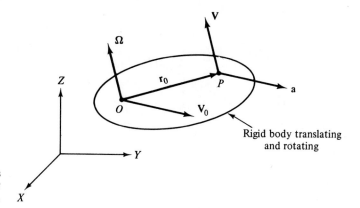

Fig. 2.19 Kinematics of rigid-body velocity and acceleration.

where $\boldsymbol{\Omega}$ is the angular-velocity vector and $\mathbf{r}_0$ is the position of point P. Differentiating, we obtain the most general form of the acceleration of a rigid body

$$\mathbf{a} = \frac{d\mathbf{V}}{dt} + \boldsymbol{\Omega} \times (\boldsymbol{\Omega} \times \mathbf{r}_0) + \frac{d\boldsymbol{\Omega}}{dt} \times \mathbf{r}_0 \tag{2.54}$$

Looking at the right-hand side, we see that the first term is the translational acceleration, the second term is the *centripetal acceleration*, whose direction is from point P perpendicular toward the axis of rotation, and the third term is the linear acceleration due to changes in the angular velocity. It is rare for all three of these terms to apply to any one fluid flow. In fact, fluids can rarely move in rigid-body motion unless restrained by confining walls for a long time. For example, suppose a tank of water is in a car which starts a constant acceleration. The water in the tank would begin to slosh about, and that sloshing would damp out very slowly until finally the particles of water would be in approximately rigid-body acceleration. This would take so long that the car would have reached hypersonic speeds. Nevertheless, we can at least discuss the pressure distribution in a tank of rigidly accelerating water. The following is an example where the water in the tank will reach uniform acceleration rapidly.

EXAMPLE 2.11 A tank of water 1 m deep is in free fall under gravity with negligible drag. Compute the pressure at the bottom of the tank if $p_a = 101$ kPa.

Solution Being unsupported in this condition, the water particles tend to fall downward as a rigid hunk of fluid. In free fall with no drag, the downward acceleration is $\mathbf{a} = \mathbf{g}$. Thus Eq. (2.52) for this situation gives $\nabla p = \rho(\mathbf{g} - \mathbf{g}) = 0$. The pressure in the water is thus *constant everywhere* and equal to the atmospheric pressure 101 kPa. In other words, the walls are doing no service in sustaining the pressure distribution which would normally exist.

Uniform Linear Acceleration

In this general case of uniform rigid-body acceleration, Eq. (2.52) applies, **a** having the same magnitude and direction for all particles. With reference to Fig. 2.20, the parallelogram sum of **g** and $-\mathbf{a}$ gives the direction of the pressure gradient or

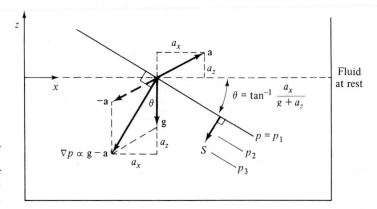

Fig. 2.20 Tilting of constant-pressure surfaces in a tank of liquid in rigid-body acceleration.

greatest rate of increase of p. The surfaces of constant pressure must be perpendicular to this and are thus tilted at an angle θ such that

$$\theta = \tan^{-1} \frac{a_x}{g + a_z} \qquad (2.55)$$

One of these tilted lines is the free surface, which is found by the requirement that the fluid retain its volume unless it spills out. The rate of increase of pressure in the direction $\mathbf{g} - \mathbf{a}$ is greater than in ordinary hydrostatics and is given by

$$\frac{dp}{ds} = \rho G \qquad \text{where } G = [a_x^2 + (g + a_z)^2]^{1/2} \qquad (2.56)$$

These results are independent of the size or shape of the container as long as the fluid is continuously connected throughout the container.

EXAMPLE 2.12   A drag racer rests her coffee mug on a horizontal tray while she accelerates at $7\,\text{m/s}^2$. The mug is 10 cm deep and 6 cm in diameter and contains coffee 7 cm deep at rest. (*a*) Assuming rigid-body acceleration of the coffee, determine whether it will spill out of the mug. (*b*) Calculate the gage pressure in the corner at point A if the density of coffee is $1010\,\text{kg/m}^3$.

Solution
Part (a)   The free surface tilts at the angle θ given by Eq. (2.55) regardless of the shape of the mug. With $a_z = 0$ and standard gravity

$$\theta = \tan^{-1} \frac{a_x}{g} = \tan^{-1} \frac{7.0}{9.81} = 35.5°$$

If the mug is symmetric about its central axis, the volume of coffee is conserved if the tilted surface intersects the original rest surface exactly at the centerline, as shown in Fig. E2.12. Thus the deflection at the left side of the mug is

$$z = (3\,\text{cm})(\tan \theta) = 2.14\,\text{cm} \qquad \textit{Ans. (a)}$$

This is less than the 3-cm clearance available, so the coffee will not spill unless it was sloshed during the start-up of acceleration.

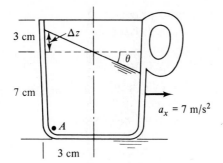

Fig. E2.12

Part (b)   When at rest, the gage pressure at point A is given by Eq. (2.20)

$$p_A = \rho g(z_{\text{surf}} - z_A) = (1010 \text{ kg/m}^3)(9.81 \text{ m/s}^2)(0.07 \text{ m}) = 694 \text{ N/m}^2 = 694 \text{ Pa}$$

During acceleration, Eq. (2.56) applies, with $G = [(7.0)^2 + (9.81)^2]^{1/2} = 12.05 \text{ m/s}^2$. The distance Δs down normal from the tilted surface to point A is

$$\Delta s = (7.0 + 2.14) \cos \theta = 7.44 \text{ cm}$$

Thus the pressure at point A becomes

$$p_A = \rho G \, \Delta s = 1010(12.05)(0.0744) = 906 \text{ Pa} \qquad\qquad Ans. (b)$$

which is an increase of 31 percent over the pressure when at rest.

Rigid-Body Rotation

As a second special case, consider rotation of the fluid about the z axis without any translation, as sketched in Fig. 2.21. We assume that the container has been rotating long enough at constant Ω for the fluid to have attained rigid-body rotation. The fluid acceleration will then be the centripetal term in Eq. (2.54). In the coordinates of Fig. 2.21 the angular-velocity and position vectors are given by

$$\boldsymbol{\Omega} = \mathbf{k}\Omega \qquad \mathbf{r}_0 = \mathbf{i}_r r \qquad\qquad (2.57)$$

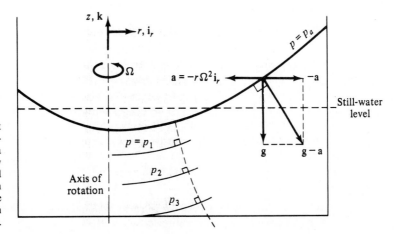

Fig. 2.21 Development of paraboloid constant-pressure surfaces in a fluid in fluid-body rotation. The dashed line along the direction of maximum pressure increase is an exponential curve.

Then the acceleration is given by

$$\mathbf{\Omega} \times (\mathbf{\Omega} \times \mathbf{r}_0) = -r\Omega^2 \mathbf{i}_r \tag{2.58}$$

as marked in the figure, and Eq. (2.52) for the force balance becomes

$$\nabla p = \mathbf{i}_r \frac{\partial p}{\partial r} + \mathbf{k} \frac{\partial p}{\partial z} = \rho(\mathbf{g} - \mathbf{a}) = \rho(r\Omega^2 \mathbf{i}_r - g\mathbf{k}) \tag{2.59}$$

Equating like components, we find the pressure field by solving two first-order partial differential equations

$$\frac{\partial p}{\partial r} = \rho r \Omega^2 \qquad \frac{\partial p}{\partial z} = -\rho g \tag{2.60}$$

This is our first specific example of the generalized three-dimensional problem described by Eqs. (2.14) for more than one independent variable. The right-hand sides of (2.60) are known functions of r and z. One can proceed as follows: integrate the first equation "partially," i.e., holding z constant, with respect to r. The result is

$$p = \tfrac{1}{2}\rho r^2 \Omega^2 + f(z) \tag{2.61}$$

where the "constant" of integration is actually a function $f(z)$.[1] Now differentiate this with respect to z and compare with the second relation of (2.60):

$$\frac{\partial p}{\partial z} = 0 + f'(z) = -\rho g$$

or

$$f(z) = -\rho g z + C \tag{2.62a}$$

where C is a constant. Thus Eq. (2.61) now becomes

$$p = \text{const} - \rho g z + \tfrac{1}{2}\rho r^2 \Omega^2 \tag{2.62b}$$

This is the pressure distribution in the fluid. The value of C is found by specifying the pressure at one point. If $p = p_0$ at $(r, z) = (0, 0)$, then $C = p_0$. The final desired distribution is

$$p = p_0 - \rho g z + \tfrac{1}{2}\rho r^2 \Omega^2 \tag{2.63}$$

The pressure is linear in z and parabolic in r. If we wish to plot a constant-pressure surface, say, $p = p_1$, Eq. (2.63) becomes

$$z = \frac{p_0 - p_1}{\rho g} + \frac{r^2 \Omega^2}{2g} = a + br^2 \tag{2.64}$$

Thus the surfaces are paraboloids of revolution, concave upward, with their minimum point on the axis of rotation. Some examples are sketched in Fig. 2.21.

As in the previous example of linear acceleration, the position of the free surface is found by conserving the volume of fluid. For a noncircular container with the axis of rotation off center, as in Fig. 2.21, a lot of laborious mensuration is required, and a single problem will take you all weekend. However, the calculation is easy for

---

[1] This is because $f(z)$ vanishes when differentiated with respect to r. If you don't see this, you should review your calculus.

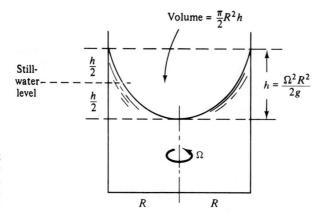

Volume = $\frac{\pi}{2}R^2 h$

Still-water level

$\frac{h}{2}$

$\frac{h}{2}$

$h = \dfrac{\Omega^2 R^2}{2g}$

Ω

R R

Fig. 2.22 Determining the free-surface position for rotation of a cylinder of fluid about its central axis.

a cylinder rotating about its central axis, as in Fig. 2.22. Since the volume of a paraboloid is one-half the base area times its height, the still-water level is exactly halfway between the high and low points of the free surface. The center of the fluid drops an amount $h/2 = \Omega^2 R^2/4g$, and the edges rise an equal amount.

EXAMPLE 2.13   The coffee cup in Example 2.12 is removed from the drag racer, placed on a turntable, and rotated about its central axis until a rigid-body mode occurs. Find (a) the angular velocity which will cause the coffee to just reach the lip of the cup and (b) the gage pressure at point A for this condition.

Solution

Part (a)   The cup contains 7 cm of coffee. The remaining distance of 3 cm up to the lip must equal the distance $h/2$ in Fig. 2.22. Thus

$$\frac{h}{2} = 0.03 \text{ m} = \frac{\Omega^2 R^2}{4g} = \frac{\Omega^2(0.03 \text{ m})^2}{4(9.81 \text{ m/s}^2)}$$

Solving, we obtain

$$\Omega^2 = 1308 \quad \text{or} \quad \Omega = 36.2 \text{ rad/s} = 345 \text{ r/min} \qquad \textit{Ans. (a)}$$

Part (b)   To compute the pressure, it is convenient to put the origin of coordinates r and z at the bottom of the free-surface depression, as shown in Fig. E2.13. The gage pressure here is

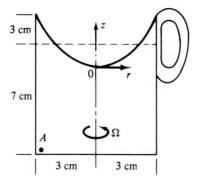

3 cm

z

0 r

7 cm

A Ω

Fig. E2.13   3 cm   3 cm

$p_0 = 0$, and point A is at $(r, z) = (3 \text{ cm}, -4 \text{ cm})$. Equation (2.63) can then be evaluated

$$p_A = 0 - (1010 \text{ kg/m}^3)(9.81 \text{ m/s}^2)(-0.04 \text{ m})$$
$$+ \tfrac{1}{2}(1010 \text{ kg/m}^3)(0.03 \text{ m})^2(1308 \text{ rad}^2/\text{s}^2)$$
$$= 396 \text{ N/m}^2 + 594 \text{ N/m}^2 = 990 \text{ Pa} \qquad\qquad Ans. (b)$$

This is about 43 percent greater than the still-water pressure $p_A = 694$ Pa.

Here, as in the linear-acceleration case, it should be emphasized that the paraboloid pressure distribution (2.63) sets up in *any* fluid under rigid-body rotation, regardless of the shape or size of the container. The container may even be closed and filled with fluid. It is only necessary that the fluid be continuously interconnected throughout the container. The following example will illustrate a peculiar case in which one can visualize an imaginary free surface extending outside the walls of the container.

EXAMPLE 2.14   A U-tube with a radius of 10 in and containing mercury to a height of 30 in is rotated about its center at 180 r/min until a rigid-body mode is achieved. The diameter of the tubing is negligible. Atmospheric pressure is 2116 lbf/ft². Find the pressure at point A in the rotating condition.

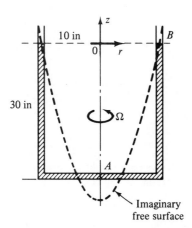

Fig. E2.14

Solution   Convert the angular velocity into radians per second

$$\Omega = (180 \text{ r/min})\frac{2\pi}{60} = 18.85 \text{ rad/s}$$

From Table 2.1 we find for mercury that $\rho g = 846 \text{ lbf/ft}^3$ and hence $\rho = 846/32.2 = 26.3 \text{ slugs/ft}^3$. At this high rotation rate the free surface will slant upward at a fierce angle [about 84°; check this from Eq. (2.64)], but the tubing is so thin that the free surface will remain at approximately the same 30-in height, point B. Placing our origin of coordinates at this height, we can calculate the constant C in Eq. (2.62b) from the condition $p_B = 2116 \, lbf/ft^2$ at $(r, z) = (10 \, in, 0)$:

$$p_B = 2116 \, lbf/ft^2 = C - 0 + \tfrac{1}{2}(26.3 \, slugs/ft^3)(\tfrac{10}{12} \, ft)^2 (18.85 \, rad/s)^2$$

or
$$C = 2116 - 3245 = -1129 \, lbf/ft^2$$

We then obtain p_A by evaluating Eq. (2.63) at $(r, z) = (0, -30 \, in)$

$$p_A = -1129 - (846 \, lbf/ft^3)(-\tfrac{30}{12} \, ft) = -1129 + 2115 = 986 \, lbf/ft^2 \qquad Ans.$$

This is less than atmospheric pressure, and we can see why if we follow the free-surface paraboloid down from point B along the dashed line in the figure. It will cross the horizontal portion of the U-tube (where p will be atmospheric) and fall *below* point A. From Fig. 2.22 the actual drop from point B will be

$$h = \frac{\Omega^2 R^2}{2g} = \frac{(18.85)^2 (\tfrac{10}{12})^2}{2(32.2)} = 3.83 \, ft = \text{`}46 \, in$$

Thus p_A is about 16 inHg below atmospheric pressure, or about $\tfrac{16}{12}(846) = 1128 \, lbf/ft^2$ below $p_a = 2116 \, lbf/ft^2$, which checks with the answer above. When the tube is at rest,

$$p_A = 2116 - 846(-\tfrac{30}{12}) = 4231 \, lbf/ft^2$$

Hence rotation has reduced the pressure at point A by 77 percent. Further rotation can reduce p_A to near-zero pressure, and cavitation can occur.

An interesting by-product of this analysis for rigid-body rotation is that the lines everywhere parallel to the pressure gradient form a family of curved surfaces, as sketched in Fig. 2.21. They are everywhere orthogonal to the constant-pressure surfaces, and hence their slope is the negative inverse of the slope computed from Eq. (2.64):

$$\frac{dz}{dr}\bigg|_{GL} = -\frac{1}{(dz/dr)_{p=const}} = -\frac{1}{r\Omega^2/g}$$

where GL stands for gradient line

or
$$\frac{dz}{dr} = -\frac{g}{r\Omega^2} \tag{2.65}$$

Separating the variables and integrating, we find the equation of the pressure-gradient surfaces

$$r = C_1 \exp\left(-\frac{\Omega^2 z}{g}\right) \tag{2.66}$$

Notice that this result and Eq. (2.64) are independent of the density of the fluid. In the absence of friction and Coriolis effects, Eq. (2.66) defines the lines along which the apparent net gravitational field would act on a particle. Depending upon its density, a small particle or bubble would tend to rise or fall in the fluid along these exponential lines, as demonstrated experimentally in Ref. 5. Also, buoyant

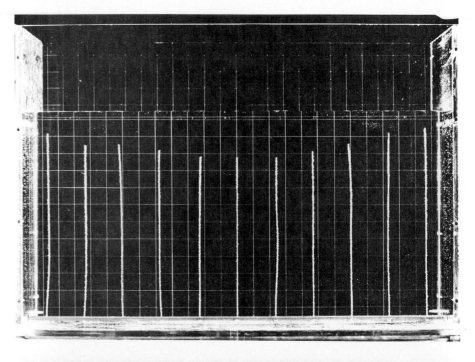

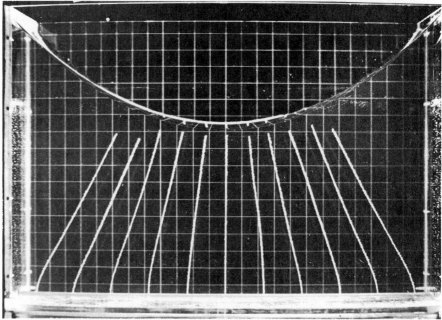

Fig. 2.23 Experimental demonstration with buoyant streamers of the fluid force field in rigid-body rotation: (*top*) fluid at rest (streamers hang vertically upward); (*bottom*) rigid-body rotation (streamers are aligned with the direction of maximum pressure gradient). (*From Ref. 5, courtesy of R. Ian Fletcher.*)

streamers would align themselves with these exponential lines, thus avoiding any stress other than pure tension. Figure 2.23 shows the configuration of such streamers before and during rotation.

2.10 PRESSURE MEASUREMENT

There are many devices available for measuring pressure, both in a static and in a moving stream. All take advantage of the fact that pressure applied to a finite area of material causes a force and stress and displacement in the material. These mechanical effects can then be quantified in various ways:

1. A force balance
2. The height of a liquid column (manometer)
3. Direct displacement measurement
4. Indirect (electrical) measurement of displacement

Force balances are commonly used to *calibrate* pressure instruments rather than as devices for routine measurement. The manometer, analyzed in Sec. 2.4, is a simple and inexpensive device with no moving parts except the liquid column itself. Manometers can be constructed with extreme accuracy by using fluids of small density differences, tilted liquid columns, and a micrometer-mounted pointer or eyepiece to locate the meniscus accurately. Examples are given in Fig. 2.24.

When measuring pressure in a moving fluid (the so-called *static pressure*), care must be taken not to disturb the flow. The best way to do this is to take the measurement through a *static hole* in the wall of the flow, as illustrated for the two instruments in Fig. 2.24. The hole should be normal to the wall, and burrs should be avoided. If the hole is small enough (typically 1 mm diameter), there will be no flow into the measuring tube once the pressure has adjusted to a steady value. Thus the flow is almost undisturbed. An oscillating flow pressure, however, can cause a large error due to possible dynamic response of the tubing. Other devices of smaller

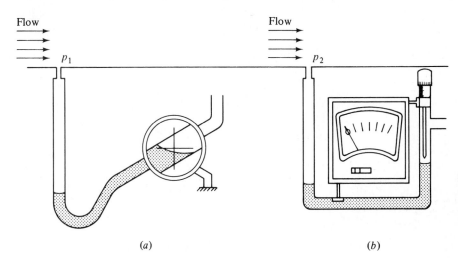

Fig. 2.24 Two types of accurate manometers for precise measurements: (*a*) tilted tube with eyepiece; (*b*) micrometer pointer with ammeter detector.

(*a*)   (*b*)

dimensions are used for dynamic-pressure measurements (see Fig. 2.26d). Note that the manometers in Fig. 2.24 are arranged to measure the absolute pressures p_1 and p_2. If the pressure difference $p_1 - p_2$ is desired, a significant error is incurred by subtracting two independent measurements, and it would be far better to connect both ends of one instrument to the two static holes p_1 and p_2 so that one manometer reads the difference directly.

The third type of pressure gage is a direct displacement measurement. An inexpensive and reliable device is the *bourdon tube*, sketched in Fig. 2.25. A curved tube with a flattened cross section will deflect outward when pressurized internally. The deflection can be measured by a linkage attached to a calibrated dial-gage pointer. Extreme accuracy can be obtained through proper design, and commercial bourdon gages are available with an accuracy of ± 0.1 percent of full scale.

The fourth type of pressure gage measures the displacement of the gage sensing element by electrical means. The common techniques are (1) resistive, (2) capacitive, (3) inductive, (4) reluctive, (5) piezoelectric, and (6) eddy currents. Some examples are sketched in Fig. 2.26.

The reluctive device in Fig. 2.26a is called a *linear variable differential transformer* (LVDT). Note that it uses a bourdon tube for convenience as a source of mechanical displacement from the applied pressure. The strain gages in Fig. 2.26b are attached to a membrane which deflects under pressure, while the potentiometer gage (Fig. 2.26c) has a small bellows which deflects the pickup on the variable resistor. The piezoelectric transducer (Fig. 2.26d) uses the discovery of Pierre and Jacques Curie in 1880 that certain quartz crystals generate an electric charge when stressed. The piezoelectric gage has no internal fluid volume; i.e., the diaphragm is installed flush with the wall and reacts almost instantaneously to a change in local pressure. Thus the dynamic response is excellent, and piezoelectric sensors are in wide use for rapid transient-pressure measurements such as blast waves or turbulent-pressure fluctuations.

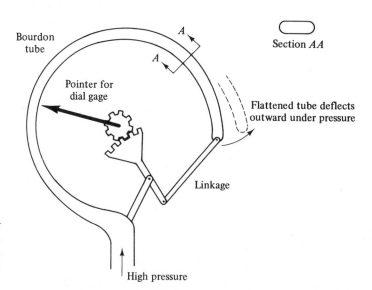

Fig. 2.25 Schematic of a bourdon-tube device for mechanical measurement of high pressures.

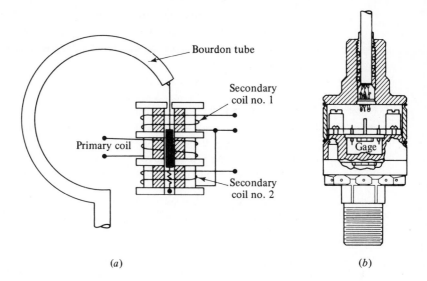

(a)   (b)

Fig. 2.26 Four types of pressure transducers with electric output: (a) differential transformer (core motion causes a change in transformer reluctance); (b) strain-gage transducer (membrane deflection causes gage strain); (c) potentiometer transducer (capsule deflection causes variable-resistance change); (d) piezoelectric transducer (diaphragm deflection causes induced charge in crystals).

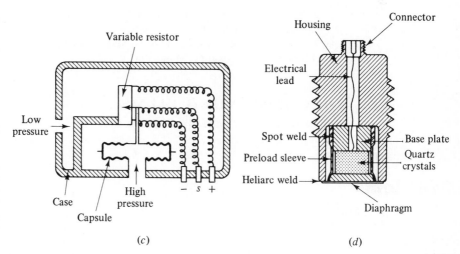

(c)   (d)

All the pressure devices with electric sensors have the great advantage that their output can be recorded on a paper chart or magnetic tape or other such permanent storage. This advantage serves to offset the relatively high cost ($100 to $2000) of electromechanical transducers.

Further discussion of pressure-measuring devices can be found in Refs. 7 to 9.

SUMMARY

This chapter has been devoted entirely to the computation of pressure distributions and the resulting forces and moments in a static fluid or a fluid with a known velocity field. All hydrostatic (Secs. 2.3 to 2.8) and rigid-body (Sec. 2.9) problems are solved in this manner and are classic cases which every student should understand. In many irrotational and almost all viscous flows, both pressure and velocity are unknowns and are solved together as a system of equations in the chapters which follow.

REFERENCES

1. "U.S. Standard Atmosphere, 1962," U.S. Government Printing Office, Washington, 1962.
2. G. Neumann and W. J. Pierson, Jr., "Principles of Physical Oceanography," Prentice-Hall, Englewood Cliffs, N.J., 1966.
3. T. C. Gillmer, "Modern Ship Design," chap. 5, United States Naval Institute, Annapolis, Md., 1970.
4. D. T. Greenwood, "Principles of Dynamics," Prentice-Hall, Englewood Cliffs, N.J., 1965.
5. R. I. Fletcher, The Apparent Field of Gravity in a Rotating Fluid System, *Am. J. Phys.* vol. 40, pp. 959–965, July 1972.
6. National Committee for Fluid Mechanics Films, "Illustrated Experiments in Fluid Mechanics," M.I.T. Press, Cambridge, Mass., 1972.
7. J. P. Holman, "Experimental Methods for Engineers," 4th ed., McGraw-Hill, New York, 1983.
8. R. P. Benedict, "Fundamentals of Temperature, Pressure, and Flow Measurement," 3d ed., Wiley, New York, 1984.
9. T. G. Beckwith, N. L. Buck, and R. D. Marangoni, "Mechanical Measurements," 3d ed., Addison-Wesley, Reading, Mass., 1982.

RECOMMENDED FILMS

Films numbered 5, 17, and 24 in Appendix C.

Problems

PROBLEM DISTRIBUTION

Section	Topic	Problems
2.1	Pressure and pressure gradient	2.1–2.5
2.2	Static equilibrium, gage pressure	2.6–2.8
2.3	Hydrostatic pressure; barometers	2.9–2.24
2.3	The atmosphere	2.25–2.33
2.4	Manometers	2.34–2.47
2.5	Force on plane surfaces	2.48–2.81
2.6	Force on curved surfaces	2.82–2.101
2.7	Forces in layered fluids	2.102–2.103
2.8	Buoyancy, Archimedes' principle	2.104–2.126
2.8	Stability of floating bodies	2.127–2.136
2.9	Uniform acceleration	2.137–2.151
2.9	Rigid-body rotation	2.152–2.160
2.10	Pressure measurements	None

2.1 Figure 2.1 and Eqs. (2.3) and (2.4) show that the pressure is the same in all directions at a point in a fluid at rest. How does this analysis differ from the discussion in Sec. 1.2 and Fig. 1.1c and d, which showed that Mohr's circle shrinks to a point in a hydrostatic fluid?

2.2 For small changes Δz the hydrostatic relation Eq. (2.20) becomes $\Delta p = \rho g \, \Delta z$, yet in Eq. (2.3) we would appear to get the relation $\Delta p = \frac{1}{2}\rho g \, \Delta z$. Explain the factor of $\frac{1}{2}$ in this second relation.

2.3 For the two-dimensional stress field shown in Fig. P2.3 it is found that

$$\sigma_{xx} = 2000 \text{ lbf/ft}^2 \qquad \sigma_{yy} = 1800 \text{ lbf/ft}^2$$

$$\sigma_{xy} = \sigma_{yx} = 100 \text{ lbf/ft}^2$$

Find the shear and normal stress in pounds per square foot on the plane AA cutting through at a $30°$ angle as shown.

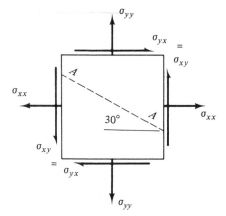

Fig. P2.3

2.4 For the stress field of Fig. P2.3 we are given

$$\sigma_{xx} = 2000 \text{ lbf/ft}^2 \qquad \sigma_{yy} = 1800 \text{ lbf/ft}^2$$

$$\sigma_n(AA) = 2100 \text{ lbf/ft}^2$$

Compute the shear stress on plane AA and the shear stress σ_{xy}.

2.5 Equation (2.13) is a vector equation which has three components. Write out its y component in full.

2.6 Atlanta, Georgia has an average altitude of 1100 ft. On a day when atmospheric pressure equals the U.S. standard value, gage A in a laboratory experiment reads 99 kPa and gage B reads 97 kPa. Express these readings in gage pressure or vacuum pressure as the case may be.

2.7 Any pressure reading can be expressed as a head or length $h = p/\rho g$. What is standard sea-level pressure expressed in (a) meters of water, (b) feet of water, (c) inches of mercury, and (d) millimeters of mercury?

2.8 On a day when the barometer reads 29.5 inHg a pressure gage reads 1.3 lbf/in² vacuum. Express the absolute pressure in the gage in (a) inches of mercury and (b) pascals.

2.9 Express the absolute pressure at point C in Fig. 2.4 in pascals if it lies 85 cm below the surface and the barometer reads 29.7 inHg.

2.10 Express a pressure of 2 kPa in terms of a column height of 20°C fluid for (a) benzene, (b) methanol, and (c) seawater.

2.11 What depth change in feet in (a) water, (b) seawater, and (c) mercury will cause a pressure increase of 1 atm?

2.12 The deepest known point in the ocean is 11,034 m in the Mariana Trench in the Pacific. Assuming seawater to have a constant specific weight of 10,050 N/m³, what is the pressure at this point in atmospheres?

2.13 Repeat Prob. 2.12 by integrating the more exact nonlinear pressure-density relation for water, Eq. (1.38), from the surface to the sea bottom. What is the percent change in the computed bottom pressure?

2.14 The closed tank in Fig. P2.14 is at 20°C. If the pressure at point A is 90 kPa absolute, what is the absolute pressure at point B in kilopascals? What percent error do you make by neglecting the specific weight of the air?

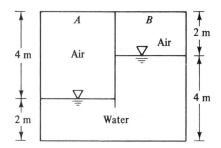

Fig. P2.14

2.15 The answer to Prob. 1.44 is that mercury best fits $B = 41,350$ and $n = 6$. Consider this a reward if you read ahead this far. Now use this formula to determine the pressure in the Mariana Trench (Prob. 2.12) if the ocean were filled with mercury instead of seawater.

2.16 The air-oil-water system in Fig. P2.16 is at 20°C. Knowing that gage A reads 15 lbf/in² absolute and gage B reads 1.25 lbf/in² less than gage C, compute (a) the specific weight of the oil in pounds force per cubic foot and (b) the actual reading of gage C in pounds force per square inch absolute.

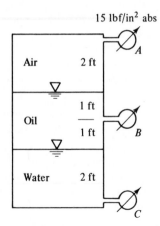

Fig. P2.16

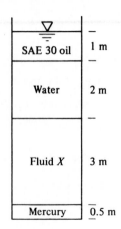

Fig. P2.18

2.17 The system in Fig. P2.17 is at 20°C. If the pressure at point A is 2000 lbf/ft^2, determine the pressures at points B, C, and D in pounds force per square foot.

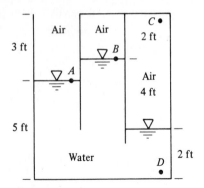

Fig. P2.17

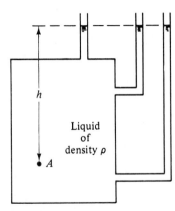

Fig. P2.19

2.18 The system in Fig. P2.18 is at 20°C. If atmospheric pressure is 101.33 kPa and the pressure at the bottom of the tank is 237 kPa, what is the specific gravity of fluid X?

2.19 When open tubes called *piezometers* are connected to a tank of liquid under pressure, the liquid rises to the *piezometer head*, or *pressure head*, of the liquid. If p_A is the pressure at point A, show that all three piezometers rise to the same level $h = p_A/\rho g$.

2.20 The hydraulic jack in Fig. P2.20 is filled with oil at 56 lbf/ft^3. Neglecting the weight of the two pistons, what force F on the handle is required to support the 2000-lbf weight for this design?

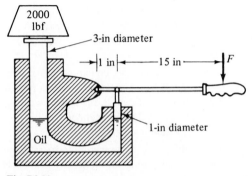

Fig. P2.20

2.21 The Red Sea, being in a region of high evaporation, is very salty. The average pressure at a depth of 20 m in the Red Sea is 202 kPa above atmospheric pressure. What is the density of Red Sea water in kilograms per cubic meter? What is its percentage salt content by weight?

2.22 The fuel gage for a gasoline tank in a car reads proportional to the bottom gage pressure as in Fig. P2.22. If the tank is 30 cm deep and accidentally contains 2 cm of water plus gasoline, how many centimeters of air remain at the top when the gage erroneously reads "full"?

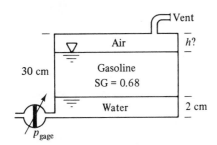

Fig. P2.22

2.23 At 20°C gage A reads 300 kPa absolute. What is the height h of the water in centimeters? What should gage B read in kilopascals absolute?

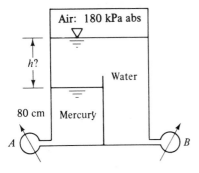

Fig. P2.23

2.24 The U-tube in Fig. P2.24 has a 1-cm ID and contains mercury as shown. If 10 cm³ of water is poured into the right-hand leg, what will the free-surface height in each leg be after the sloshing has died down?

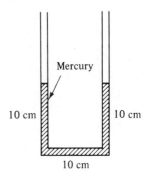

Fig. P2.24

2.25 If air were incompressible at its nominal specific weight of 11.8 N/m³, how high in meters would the atmosphere have to be to cause sea-level pressure to be its nominal value of 2116 lbf/ft²?

2.26 The atmosphere of Venus is 90 percent CO_2, but assume it to be 100 percent. Its surface pressure is about 2 MPa, and its average surface temperature is 190°C. For an isothermal atmosphere, estimate the pressure at an altitude of 20 km. Take $g = 28.5$ ft/s².

2.27 From the standard atmosphere in Fig. 2.7, estimate the mass of the entire atmosphere in kilograms, assuming the radius of the earth to be 6400 km.

2.28 Investigate the effect of a doubling of the lapse rate on atmospheric pressure. If T_0 is 20°C and $p_a = $ 99 kPa, compute the pressure at 3500 m for a lapse rate of (*a*) 0.0005 K/m and (*b*) 0.001 K/m. What do you conclude?

2.29 Sea-level temperature and pressure and the atmospheric lapse rate vary from day to day. At a certain time sea-level temperature is 50°F and sea-level pressure is 29 inHg. An airplane overhead registers a temperature of 20°F and a pressure of 11 lbf/in² absolute. Estimate the altitude of the plane in feet.

2.30 What percentage error do you make in Prob. 2.29 if you simply assume a standard lapse rate of 0.003566 °R/ft?

2.31 From Fig. 2.7 standard pressure at $z = 30$ km is 1.20 kPa. From 30 to 40 km the temperature rises approximately linearly from −45 to −20°C. Estimate the pressure at $z = 40$ km in pascals.

2.32 On a cold day at the North Pole sea-level temperature is −50°C and remains approximately isothermal up to 3000 m. Estimate the atmospheric pressure there at $z = 3000$ m if sea-level pressure is 100 kPa.

2.33 On a summer day in Narragansett, Rhode Island, sea-level temperature is 80°F, sea-level pressure is 29.5 inHg, and the lapse rate is 0.0036 °R/ft. Estimate the specific weight of the local atmosphere in pounds force per cubic foot at an altitude of 5000 ft.

2.34 In Fig. P2.34 fluid 2 is carbon tetrachloride and fluid 1 is glycerin. If $p_a = 101$ kPa, determine the absolute pressure at point A.

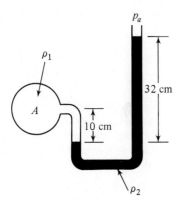

Fig. P2.34

2.35 In Fig. P2.35 all fluids are at 20°C. Determine the pressure difference between points A and B.

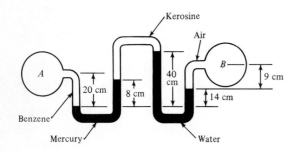

Fig. P2.35

2.36 In Fig. P2.36 all fluids are at 20°C. If $p_B - p_A = 99$ kPa, what must the height H be in centimeters?

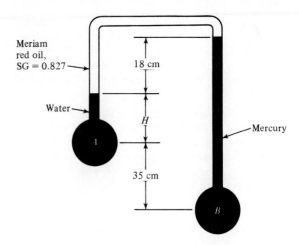

Fig. P2.36

2.37 For Fig. 2.9 if fluid 1 is water and fluid 2 is mercury, and $z_A = 0$, $z_1 = -10$ cm, what is the level z_2 for which $p_A = p_{atm}$?

2.38 For Fig. 2.9 fluid 1 is air and fluid 2 is mercury. The right leg is fitted with a scale graduated to read p_A in kilopascals gage. What should the difference between scale graduations in millimeters be?

2.39 The inclined manometer in Fig. P2.39 contains Meriam red manometer oil, SG = 0.827. Assume that the reservoir is very large. If the inclined arm is fitted with graduations 1 in apart, what should the angle θ be if each graduation corresponds to 1 lbf/ft² gage pressure for p_A?

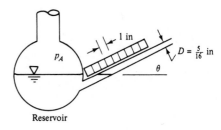

Fig. P2.39

2.40 Repeat Prob. 2.39 for a finite reservoir approximating a half-full hemisphere with a diameter of 6 in.

2.41 The system in Fig. P2.41 is at 20°C. Compute the pressure at point A in pounds force per square foot absolute.

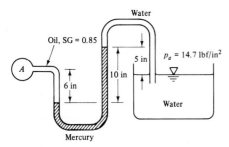

Fig. P2.41

2.42 Very small pressure differences $p_A - p_B$ can be measured accurately by the two-fluid differential manometer in Fig. P2.42. Density ρ_2 is only slightly larger than the upper fluid ρ_1. Derive an expression for the proportionality between h and $p_A - p_B$ if the reservoirs are very large.

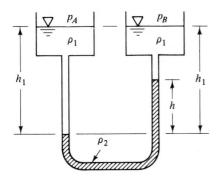

Fig. P2.42

2.43 To illustrate the effectiveness of the system in Fig. P2.42, consider that a height difference of 1 cm in a fluid with SG = 0.82 corresponds to a pressure difference of 80.3 Pa. In Fig. P2.42, what difference $p_A - p_B$ corresponds to $h = 1$ cm when $SG_1 = 0.81$ and $SG_2 = 0.83$? What is the increase in accuracy?

2.44 Water flows downward in a pipe at 45°, as shown in Fig. P2.44. The pressure drop $p_1 - p_2$ is partly due to gravity and partly due to friction. The mercury manometer reads a 6-in height difference. What is the total pressure drop $p_1 - p_2$ in pounds force per square inch? What is the pressure drop due to friction only between

1 and 2 in pounds force per square inch? Does the manometer reading correspond only to friction drop? Why?

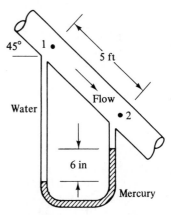

Fig. P2.44

2.45 Determine the gage pressure at point A in pascals. Is it higher or lower than atmospheric?

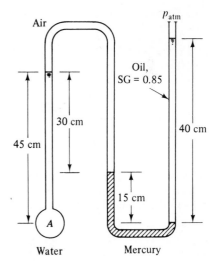

Fig. P2.45

2.46 In Fig. P2.46, what will the level h of the oil in the right-hand tube be in centimeters? Both tubes are open to the atmosphere.

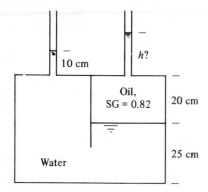

Fig. P2.46

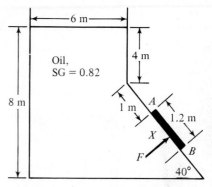

Fig. P2.51

2.47 The cylindrical tank in Fig. P2.47 is being filled with water at 20°C by a pump which produces an absolute pressure of 150 kPa at the inlet. At the instant shown the air pressure is 100 kPa and $H = 25$ cm. If the air compresses isothermally, what will H be when the flow ceases because the tank becomes hydrostatic?

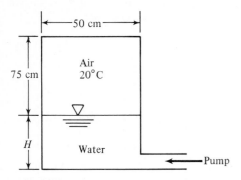

Fig. P2.47

2.48 The tank in Prob. 2.46 is 30 cm wide into the paper. Compute the force in newtons on the left-hand side panel of the tank and its center-of-pressure position.

2.49 Repeat Prob. 2.48 but instead compute the force and center of pressure on the right-hand side panel.

2.50 Repeat Example 2.5 but instead let the hinge be at point A and let point B rest against a smooth bottom.

2.51 Gate AB in Fig. P2.51 is 1.2 m long and 0.8 m into the paper. Neglecting atmospheric pressure, compute the force F on the gate and its center-of-pressure position X.

2.52 A vertical submerged gate 8 ft wide and 10 ft high is hinged at the top and held closed by water whose level is 5 ft above the gate top. What horizontal force applied at the bottom of the gate is required to open it?

2.53 A vat filled with oil (SG = 0.82) is 7 m long and 3 m deep and has a trapezoidal cross section 4 m wide at the bottom and 6 m wide at the top. Compute (*a*) the weight of oil in the vat, (*b*) the force on the vat bottom, and (*c*) the force on the trapezoidal end panel.

2.54 The gate AB in Fig. P2.54 is 6 ft wide into the paper, hinged at point A, and restrained by a stop at point B. Compute the force on the stop and the reactions at A if the water depth h is 10 ft.

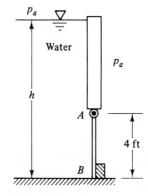

Fig. P2.54

2.55 In Fig. P2.54 the stop B will break if the force on it equals 10,000 lbf. For what water depth h is this condition reached?

2.56 In Fig. P2.54 the hinge A will break if its horizontal reaction equals 9000 lbf. For what water depth h is this condition reached?

2.57 Neglecting p_a, find the hydrostatic force on the triangular gate ABC in Fig. P2.57 and the position of its center of pressure.

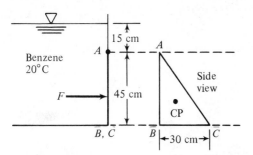

Fig. P2.57

2.58 In Fig. P2.58 the cover gate AB closes a circular opening 80 cm in diameter. The gate is held closed by a 200-kg mass as shown. Assume standard gravity at 20°C. At what water level h will the gate be dislodged? Neglect the weight of the gate.

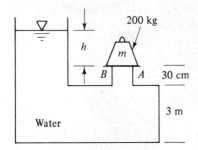

Fig. P2.58

2.59 When freshly poured between wooden forms, concrete approximates a fluid with SG = 2.40. Figure P2.59 shows an 8- by 10-ft by 6-in slab poured between wooden forms which are connected by four corner bolts A, B, C, D as shown. Neglecting end effects, compute the forces in all four bolts.

2.60 Find the net hydrostatic force per unit width on the rectangular gate AB in Fig. P2.60 and its line of action.

2.61 A cylindrical wooden-stave barrel is 4 ft in diameter and 6 ft high. It is held together by steel hoops at the top and bottom, each with a cross section of 0.35 in². If the barrel is filled with apple juice (SG = 1.02), compute the tension stress in each hoop in pounds force per square inch absolute.

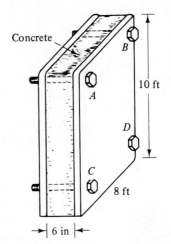

Fig. P2.59

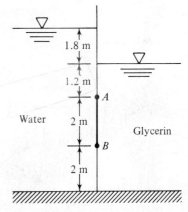

Fig. P2.60

2.62 The gate AB in Fig. P2.62 is 15 ft long and 10 ft wide into the paper and hinged at B with a stop at A. Neglecting the weight of the gate, compute the water level h for which the gate will start to fall.

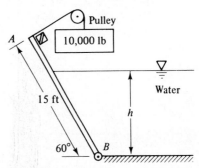

Fig. P2.62

2.63 Repeat Prob. 2.62 by including the weight of the gate, which is 1-in-thick steel, SG = 7.85.

2.64 The turbine inlet duct coming from a large water dam is 3 m in diameter. It is closed by a vertical circular gate, whose center point is 50 m below the dam water level. Compute the force on the gate and its center of pressure.

2.65 Gate AB in Fig. P2.65 is semicircular, hinged at B, and held by a horizontal force P at A. What force P is required for equilibrium?

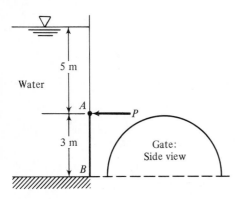

Fig. P2.65

2.66 Dam ABC in Fig. P2.66 is 40 m wide into the paper and made of concrete weighing 22 kN/m³. Find the hydrostatic force on surface AB and its moment about point C. Could this force tip the dam over?

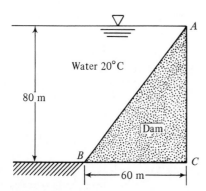

Fig. P2.66

2.67 Isosceles triangle gate AB in Fig. P2.67 is hinged at A and weightless. What horizontal force P is required at point B for equilibrium?

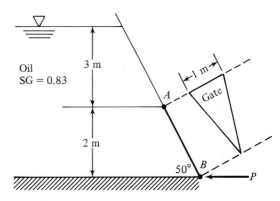

Fig. P2.67

2.68 The tank in Fig. P2.68 is 50 cm wide into the paper. Compute the hydrostatic force on (a) the lower panel BC and (b) the upper panel AD. Neglect atmospheric pressure.

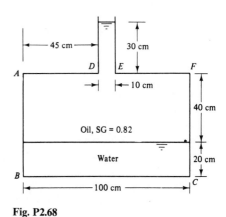

Fig. P2.68

2.69 The water tank in Fig. P2.69 is pressurized, as shown by the mercury-manometer reading. Determine the hydrostatic force per unit depth on gate AB.

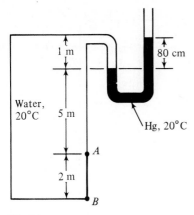

Fig. P2.69

ocean level h will the gate first open? Neglect the gate weight.

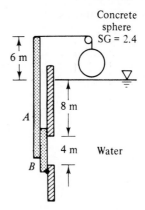

Fig. P2.71

2.70 Calculate the force and center of pressure on one side of the vertical triangular panel *ABC* in Fig. P2.70. Neglect p_{atm}.

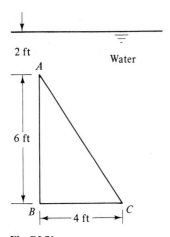

Fig. P2.70

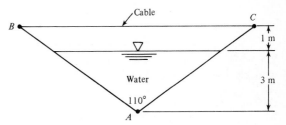

Fig. P2.72

2.71 In Fig. P2.71 gate *AB* is 3 m wide into the paper and is connected by a rod and pulley to a concrete sphere (SG = 2.40). What diameter of the sphere is just sufficient to keep the gate closed?

2.72 The V-shaped container in Fig. P2.72 is hinged at *A* and held together by cable *BC* at the top. If cable spacing is 1 m into the paper, what is the cable tension?

2.73 Gate *AB* is 5 ft wide into the paper and opens to let fresh water out when the ocean tide is dropping. The hinge at *A* is 2 ft above the freshwater level. At what

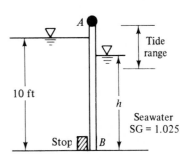

Fig. P2.73

2.74 In Prob. 2.73 investigate analytically whether the required height h is independent of the gate width b into the paper.

2.75 Compute the force on one side of the parabolic panel *ABC* in Fig. P2.75, neglecting atmospheric pressure. Compute the vertical distance down to the center of pressure knowing that $I_{xx} = 2bh^3/7$ for a parabola about a horizontal axis through *A*.

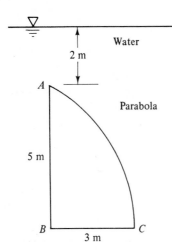

Fig. P2.75

2.76 The huge conical beaker in Fig. P2.76 is filled with kerosine (SG = 0.62). An evacuated hemisphere *ABC* of diameter 2 m rests at the bottom as shown. Compute the total force on the circular base *AB* of the hemisphere. Don't ask what this problem really means.

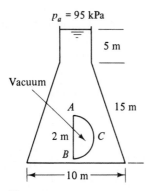

Fig. P2.76

2.77 The circular gate *ABC* in Fig. P2.77 has a 1-m radius and is hinged at *B*. Compute the force *P* just sufficient to keep the gate from opening when $h = 10$ m. Neglect atmospheric pressure.

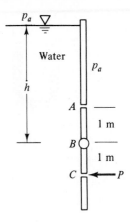

Fig. P2.77

2.78 Repeat Prob. 2.77 to derive an analytic expression for *P* as a function of *h*. Is there anything unusual about your solution?

2.79 Gate *ABC* in Fig. P2.79 is 1 m square and is hinged at *B*. It will open automatically when the water level *h* becomes high enough. Determine the lowest height for which the gate will open. Neglect atmospheric pressure.

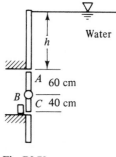

Fig. P2.79

2.80 Repeat Prob. 2.79 to determine *h* analytically for fluid of arbitrary density ρ. Is your solution unusual?

2.81 Gate *AB* is 7 ft into the paper and weighs 3000 lbf when submerged. It is hinged at *B* and rests against a smooth wall at *A*. Determine the water level *h* at the left which will just cause the gate to open.

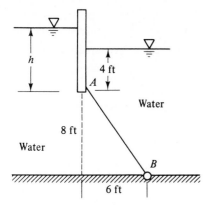

Fig. P2.81

2.82 The dam in Fig. P2.82 is a quarter-circle 50 m wide into the paper. Determine the horizontal and vertical components of the hydrostatic force against the dam and the point CP where the resultant strikes the dam.

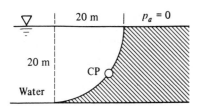

Fig. P2.82

2.83 Gate *AB* is a quarter circle 10 ft wide into the paper and hinged at *B*. Find the force *F* just sufficient to keep the gate from opening. Neglect the weight of the gate.

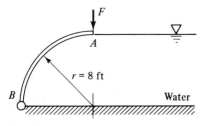

Fig. P2.83

2.84 Repeat Prob. 2.83 by including the 4000-lbf weight of the gate, which is a steel plate of uniform thickness.

2.85 Compute the horizontal and vertical components of the hydrostatic force on the quarter-circle panel at the bottom of the water tank in Fig. P2.85.

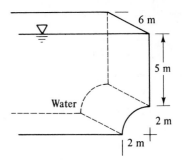

Fig. P2.85

2.86 Compute the horizontal and vertical components of the hydrostatic force on the hemispherical bulge at the bottom of the tank in Fig. P2.86.

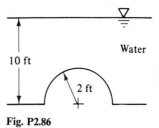

Fig. P2.86

2.87 The bottle of champagne (SG = 0.96) in Fig. P2.87 is under pressure as shown by the mercury-manometer reading. Compute the net force on the 2-in-radius hemispherical end cap at the bottom of the bottle.

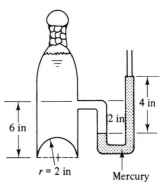

Fig. P2.87

2.88 The half-cylinder *ABC* in Fig. P2.88 is 10 ft wide into the paper. Calculate the net moment of the hydrostatic oil forces on the cylinder about point *C*.

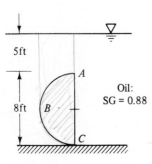

Fig. P2.88

2.89 Compute the hydrostatic force and its line of action on the semicylindrical bulge *ABC* in Fig. P2.89.

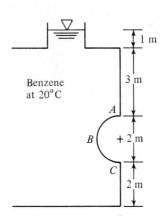

Fig. P2.89

2.90 A 1-ft-diameter hole in the bottom of the tank in Fig. P2.90 is closed by a conical 45° plug. Neglecting the weight of the plug, compute the force *F* required to keep the plug in the hole.

2.91 The hemispherical dome in Fig. P2.91 weighs 30 kN and is filled with water and attached to the floor by six equally spaced bolts. What is the force in each bolt required to hold the dome down?

2.92 A 4-m-diameter water tank consists of two half-cylinders, each weighing 4.5 kN/m, bolted together as shown in Fig. P2.92. If the support of the end caps is neglected, determine the force induced in each bolt.

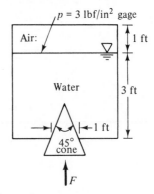

Fig. P2.90

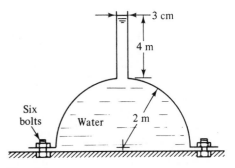

Fig. P2.91

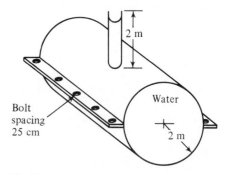

Fig. P2.92

2.93 The 2-ft-radius cylinder in Fig. P2.93 is 8 ft wide into the paper. Compute the horizontal and vertical components of the water against the cylinder.

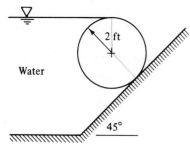

Fig. P2.93

2.94 The 4-ft-diameter log (SG = 0.80) in Fig. P2.94 is 10 ft long into the paper and dams water as shown. Compute the net vertical and horizontal reactions at point C.

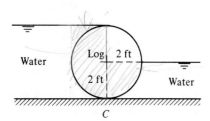

Fig. P2.94

2.95 The 2-m-diameter cylinder in Fig. P2.95 is 5 m long into the paper and rests in static equilibrium against the smooth wall at point B. Compute (a) the weight and (b) the specific gravity of the cylinder. Assume zero wall friction at point B.

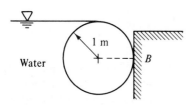

Fig. P2.95

2.96 The tank in Fig. P2.96 is 2 m wide into the paper. Neglecting atmospheric pressure, compute the hydrostatic (a) horizontal force, (b) vertical force, and (c) resultant force on quarter-circle panel BC.

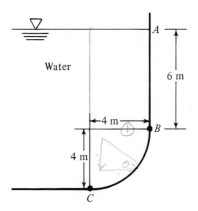

Fig. P2.96

2.97 Gate AB is a quarter-circle 8 ft wide, hinged at B and resting against a smooth wall at A. Compute the reaction forces at point A and at point B.

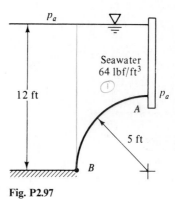

Fig. P2.97

2.98 Gate ABC in Fig. P2.98 is a quarter-circle 10 ft wide into the paper. Compute the horizontal and vertical hydrostatic forces on the gate and the line of action of the resultant force.

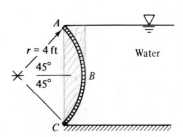

Fig. P2.98

2.99 A 2-ft-diameter sphere weighing 500 lbf closes a 1-ft-diameter hole in the bottom of the tank in Fig. P2.99. Compute the force F required to dislodge the sphere from the hole.

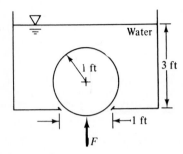

Fig. P2.99

2.100 Pressurized water fills the tank in Fig. P2.100. Compute the net hydrostatic force on the conical surface ABC.

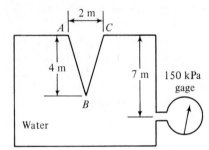

Fig. P2.100

2.101 Gate AB in Fig. P2.101 is 10 m wide into the paper, parabolic in shape, and hinged at B. Compute the force F required to hold the gate in equilibrium. Neglect atmospheric pressure.

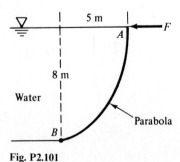

Fig. P2.101

2.102 A horizontal cylindrical tank 10 ft long and 6 ft in diameter is half-filled with water, and the remainder is filled with oil (SG = 0.85). Compute the force and center of pressure on the end circular panel of the tank.

2.103 The cylindrical tank in Fig. P2.89 has a hemispherical end cap ABC and contains oil and water as shown. Compute the resultant force and line of action of the fluids on the end cap ABC.

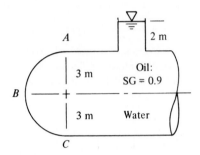

Fig. P2.103

2.104 The can in Fig. P2.104 floats in the position shown. What is its weight in newtons?

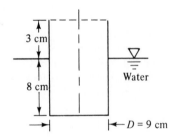

Fig. P2.104

2.105 It is said that Archimedes discovered the buoyancy laws when asked by King Hiero of Syracuse to determine whether or not his new crown was gold (SG = 19.3). Archimedes found the weight of the crown in air to be 13.0 N and its weight in water to be 11.8 N. Was it gold or not?

2.106 Repeat Prob. 2.105 by assuming that the crown is an alloy of gold (SG = 19.3) and silver (SG = 10.5). For the same measured weights, compute the percentage silver in the crown.

2.107 A sphere of buoyant solid molded foam is immersed in seawater (64 lbf/ft³) and moored at the bottom. The sphere radius is 14 in. The mooring line has a tension of 150 lbf. What is the specific weight of the sphere?

2.108 Repeat Prob. 2.62 assuming that the 10,000-lbf weight is concrete (SG = 2.40) and is hanging submerged in the water.

2.109 A *hydrometer* floats at a level which is a measure of the specific gravity of the liquid. The stem is of constant diameter D, and a weight in the bottom bulb stabilizes the body, as shown in Fig. P2.109. If the total hydrometer weight is 0.03 lbf and the stem diameter is 0.3 in, compute the height h where it will float when the liquid has SG = 1.3.

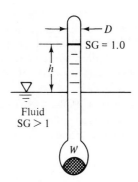

Fig. P2.109

2.110 A hydrometer has a weight of 0.16 N and a stem diameter of 1 cm. What is the distance between scale markings for SG = 1.0 and SG = 1.1? Between SG = 1.1 and SG = 1.2?

2.111 For the hydrometer of Fig. P2.109 derive an analytic formula for the float position h as a function of SG, W, D, and the specific weight $\rho_0 g$ of pure water. Are the scale markings linear or nonlinear as a function of SG?

2.112 A wooden pole (SG = 0.65), 9 cm by 9 cm by 5 m long, hangs vertically from a string in such a way that 3 m is submerged in water and 2 m is above the surface. What is the tension in the string?

2.113 A spar buoy is a buoyant rod weighted at the bottom so that it floats upright and can be used for measurements or markers. The spar in Fig. P2.113 is wood (SG = 0.6), 2 by 2 in by 12 ft and floats in sea-

water (SG = 1.025). How many pounds of steel (SG = 7.85) should be added to the bottom so that exactly $h = 2$ ft of the spar is exposed?

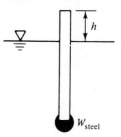

Fig. P2.113

2.114 A right circular cone is 8 cm in diameter and 15 cm high and weighs 1 N in air. How much force is required to push this cone vertex-downward into benzene so that its base is exactly at the surface? How much additional force will push the base 10 cm below the surface?

2.115 The 2- by 2-in by 12-ft spar buoy of Fig. P2.113 has 6 lb of steel weight attached and has gone aground on a rock 8 ft deep (Fig. P2.115). Compute the angle θ at which the buoy will lean, assuming that the rock exerts no moment on the buoy.

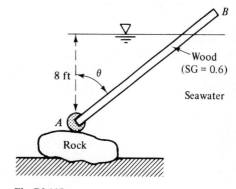

Fig. P2.115

2.116 The solid cube 12 cm on a side in Fig. P2.116 is balanced by a 2-kg mass on the beam scale when the cube is immersed in water (SG = 1.0). What is the specific weight of the cube material?

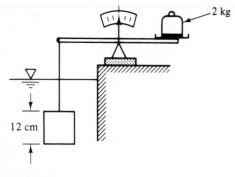

Fig. P2.116

2.117 The balloon in Fig. P2.117 is filled with helium (molecular weight = 4.00) and pressurized to 110 kPa. The atmosphere conditions are shown. Compute the tension in the mooring line.

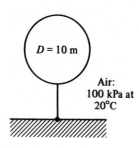

Fig. P2.117

2.118 A 1-ft-diameter hollow sphere is made of steel (SG = 7.85) with 0.16-in wall thickness. How high will this sphere float in water (SG = 1.0)? How much weight must be added inside to make the sphere neutrally buoyant?

2.119 When a 5-lb weight is placed on the end of a floating 4- by 4-in by 9-ft wooden beam, the beam tilts at $1.5°$ with its right upper corner at the surface, as in Fig. P2.119. What is the specific weight of the wood?

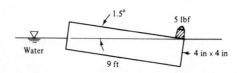

Fig. P2.119

2.120 A wooden beam (SG = 0.6) is 15 by 15 cm by 4 m and is hinged at A. At what angle θ will the beam float in the water?

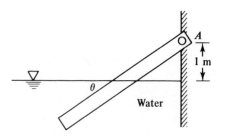

Fig. P2.120

2.121 A barge weighs 50 tons empty and is 20 ft wide, 50 ft long, and 8 ft high. What will be its *draft*, i.e., its depth below the waterline, when loaded with 130 tons of gravel and floating in seawater (SG = 1.025)? One ton = 2000 lbf.

2.122 A block of steel (SG = 7.85) will "float" at a mercury-water interface as in Fig. P2.122. What will be the ratio of the distances a and b for this condition?

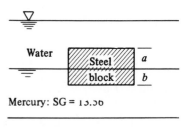

Mercury: SG = 13.56

Fig. P2.122

2.123 In an estuary where fresh water meets and mixes with seawater there often occurs a stratified salinity condition with fresh water on top and salt water on the bottom, as in Fig. P2.123. The interface is called a *halocline*. An idealization of this would be constant density on each side of the halocline as shown. A 35-cm-diameter sphere weighing 50 lbf would "float" near such a halocline. Compute the sphere position for the idealization in Fig. P2.123.

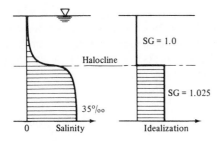

Fig. P2.123

2.124 A balloon weighing 3 lbf is 5 ft in diameter. It is filled with hydrogen (molecular weight 2.02) at 15 lbf/in² absolute and 60°F and released. At what altitude in the U.S. Standard Atmosphere will this balloon be neutrally buoyant?

2.125 A rectangular barge 20 ft wide by 50 ft long by 8 ft deep floats empty with a draft of 3 ft in a canal lock 30 ft wide by 60 ft long and water depth 6 ft when the empty barge is present. If 200,000 lbf of steel is loaded onto the barge, what are the new draft of the barge and the new water depth in the lock?

2.126 An open can 15 cm in diameter and 30 cm long (Fig. P2.126) is eased downward so that the air within is trapped and compressed by the water. The can weighs 4 N and is made of sheet steel (SG = 7.85). Assuming isothermal compression of the air in the can, compute the height of water intrusion h and the height L of the can above the water when the can is floating. Assume that the position is stable.

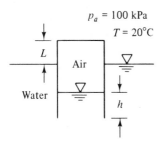

Fig. P2.126

2.127 A spar buoy 2 by 2 in by 10 ft is made of wood (40 lbf/ft³) and weighted with 4 lbf of steel at its bottom end. For this assumed vertical floating position compute the metacentric height $\overline{MG}$. Is the buoy stable?

2.128 A cube of side length L and specific gravity $S = 0.8$ floats in water (SG = 1.0) as shown in Fig. P2.128. Compute the metacentric height $\overline{MG}$. Is the cube stable?

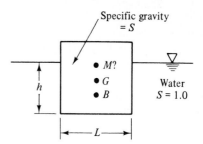

Fig. P2.128

2.129 Make a complete analytic study of Prob. 2.128 and determine the range of values of S between 0 and 1.0 for which the cube is stable.

2.130 Investigate the stability of the wooden pole in Prob. 2.112. Is it stable when hanging vertically?

2.131 A rectangular barge is 6 by 15 by 30 ft long and is piled so high with gravel that its center of gravity is 2 ft above the waterline, as in Fig. P2.131. Is the barge stable for this configuration?

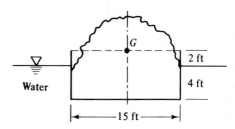

Fig. P2.131

2.132 A solid right circular cone has SG = 0.99 and floats vertically as in Fig. P2.132. Is this a stable position for the cone?

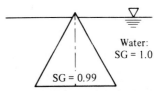

Fig. P2.132

2.133 Suppose that the barge in Fig. P2.131 has no gravel but instead is filled with oil (SG = 0.83) to a depth of 5 ft. Is the barge stable? What complicates the analysis?

2.134 When floating in water (SG = 1.0), an equilateral triangular body (SG = 0.9) might take one of the two positions shown in Fig. P2.134. Which is the more stable position? Assume large width into the paper.

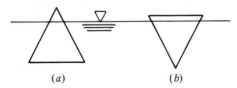

(a)   (b)

Fig. P2.134

2.135 A solid cylinder is 1 m in diameter and 5 m long, with SG = 0.5. When floating in water (SG = 1.0), what is its stable position? What is its metacentric height?

2.136 Investigate the stability of the floating can in Prob. 2.126. Make simplifying assumptions if necessary.

2.137 An open tank of water 3 m deep is being accelerated upward at $4 \, \text{m/s}^2$. Compute the gage pressure at the bottom of the tank in pascals.

2.138 The liquid fuel (SG = 0.72) in a rocket is 8 ft deep and vented to the atmosphere. The rocket is accelerating upward at 3 g's. Compute the pressure at the bottom of the fuel tank in pounds force per square inch gage.

2.139 The tank of water in Fig. P2.139 accelerates to the right with the fluid in rigid-body motion. Compute a_x in meters per second squared. Does the solution change if the fluid is mercury?

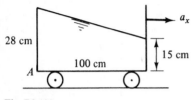

Fig. P2.139

2.140 Compute the gage pressure at point A in Fig. P2.139 if the fluid is mercury (SG = 13.56).

2.141 The same water tank from Prob. 2.139 is now moving in rigid-body motion up a 30° inclined plane with constant acceleration. Compute a in meters per second squared. Is the acceleration up or down? Compute gage p_A.

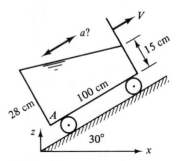

Fig. P2.141

2.142 The tank of water in Fig. P2.142 is 10 cm wide into the paper. If it is accelerated uniformly to the right at $5.0 \, \text{m/s}^2$, what will the water depth be on side AB? What will the pressure force be on panel AB?

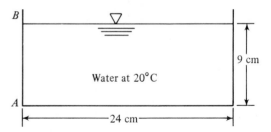

Fig. P2.142

2.143 The tank of water in Fig. P2.143 is full and open to the atmosphere at point A. For what acceleration a_x in feet per second squared will the pressure at point B be atmospheric?

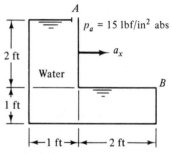

Fig. P2.143

2.144 For the water tank of Fig. P2.143, what acceleration a_x in feet per second squared will cause the pressure at point B to be zero absolute? Neglect cavitation.

2.145 A full can of water is 30 cm in diameter and 60 cm high. What gradually applied rigid-body acceleration will cause one-quarter of the water to spill out?

2.146 A bucket 1 ft deep contains 6 in of water and is being twirled in a vertical circle on a 3 ft rope at 1 r/s. Estimate the gage pressure at the bottom of the bucket when it is at (*a*) its highest and (*b*) its lowest position.

2.147 The tank of water in Fig. P2.147 accelerates uniformly by freely rolling down a 30° incline. If the wheels are frictionless, what is the angle θ? Can you explain this interesting result?

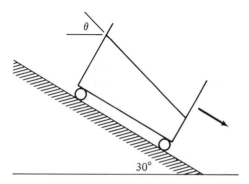

Fig. P2.147

2.148 A fish tank 12 in deep by 15 by 30 in is to be carried in a station wagon which may experience accelerations of $\frac{1}{2}g$. What is the maximum water depth which will avoid spilling in rigid-body motion? What is the best alignment of the tank with respect to car motion? What is a proper engineering safety factor on water depth to avoid sloshing spills?

2.149 The 6-ft-radius waterwheel in Fig. P2.149 is being used to lift water with its 1-ft-diameter half-

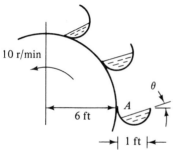

Fig. P2.149

cylinder blades. If the wheel rotates at 10 r/min and rigid-body motion is assumed, what is the water surface angle θ at position A?

2.150 A cheap accelerometer, probably worth the price, can be made from a U-tube as in Fig. P2.150. The water-level change h is a measure of a_x. If $L = 20$ cm and $D = 1$ cm, what will h be if $a_x = 5$ m/s²? Can the scale markings on the tube be linear multiples of a_x? What are the disadvantages of this design?

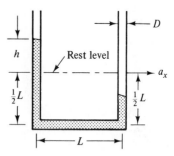

Fig. P2.150

2.151 The U-tube in Fig. P2.151 is open at A and closed at D. If accelerated to the right at uniform a_x, what acceleration will cause the pressure at point C to be atmospheric? The fluid is water (SG = 1.0).

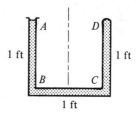

Fig. P2.151

2.152 A 6-in-diameter cylinder 1 ft high is full of water. If it is rotated about its central axis in rigid-body motion, compute the rate in revolutions per minute for which one-third of the water will spill out.

2.153 For the cylinder of Prob. 2.152 compute the rotation rate in revolutions per minute for which the bottom will be barely exposed.

2.154 If the U-tube in Fig. P2.150 is rotated about the right leg at 80 r/min, what will the level h in the left leg be if $L = 20$ cm and $D = 1$ cm?

2.155 For what uniform rotation rate in revolutions per minute about axis C will the U-tube in Fig. P2.155 take the configuration shown? The fluid is mercury at 20°C.

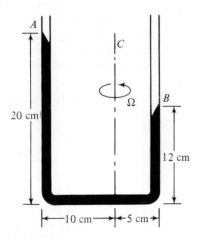

Fig. P2.155

2.156 Suppose that the U-tube of Fig. P2.151 is rotated about axis DC. If the fluid is water at 122°F and atmospheric pressure is 2116 lbf/ft² absolute at what rotation rate will the fluid within the tube begin to vaporize? At what point will this occur?

2.157 The 45° V-tube in Fig. P2.157 contains water and is open at A and closed at C. What uniform rotation rate in revolutions per minute about axis AB will cause the pressure to be equal at points B and C? For this condition, at what point in leg BC will the pressure be a minimum?

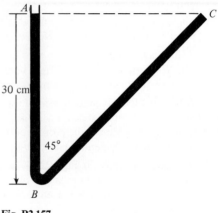

Fig. P2.157

2.158 A 9-in-high, 6-in-diameter can is filled with SAE 30 oil at 20°C and the lid is sealed at standard sea-level pressure. The can is then rotated about its central axis at 3000 r/min. Assuming rigid-body motion, what will the maximum oil pressure in the can be in pounds per square inch absolute?

2.159 For Prob. 2.158, compute the total hydrostatic force on the bottom of the can in pounds force.

2.160 A very deep 8-in-diameter can contains 6 in of water overlaid with 6 in of oil (SG = 0.86). If the can is rotated about its central axis at 300 r/min and rigid-body motion ensues, what will the shape of the air-oil and oil-water interfaces be? What will the maximum fluid pressure in the can be in pounds force per square inch gage?

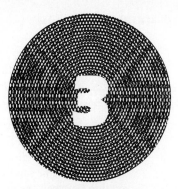

Integral Relations
for a
Control Volume

3.1 BASIC PHYSICAL LAWS OF FLUID MECHANICS

It is time now to really get serious about flow problems. The fluid-statics applications of Chap. 2 were more like fun than work, at least in the writer's opinion. Statics problems basically require only the density of the fluid and knowledge of the position of the free surface, but most flow problems require the analysis of an arbitrary state of variable fluid motion defined by the geometry, the boundary conditions, and the laws of mechanics. This chapter and the next two outline the three basic approaches to the analysis of arbitrary flow problems:

1. Control-volume, or large-scale, analysis (Chap. 3)
2. Differential, or small-scale, analysis (Chap. 4)
3. Experimental, or dimensional, analysis (Chap. 5)

The three approaches are roughly equal in importance, but control-volume analysis is more equal, being the single most valuable tool available to the engineer for flow analysis. It gives "engineering" answers, sometimes gross and crude but always useful. In principle, the differential approach of Chap. 4 can be used for any problem, but in practice the lack of mathematical tools and the inability of the digital computer to model small-scale processes make the differential approach rather limited. Similarly, although the dimensional analysis of Chap. 5 can be applied to any problem, the lack of time and money and generality often makes experimentation a limited approach. But a control-volume analysis takes about half an hour and gives useful results. Thus, in a trio of approaches, the control volume is king. Oddly enough, it is the newest of the three. Differential analysis began with Euler and Lagrange in the eighteenth century, and dimensional analysis was pioneered by Lord Rayleigh in the late nineteenth century, but the control volume, although proposed by Euler, was not developed on a rigorous basis as an analytical tool until the 1940s.

Systems versus Control Volumes

All the laws of mechanics are written for a *system*, which is defined as an arbitrary quantity of mass of fixed identity. Everything external to this system is denoted by the term *surroundings*, and the system is separated from its surroundings by its *boundaries*. The laws of mechanics then state what happens when there is an interaction between the system and its surroundings.

First, the system is a fixed quantity of mass, denoted by m. Thus the mass of the system is conserved and does not change.[1] This is a law of mechanics and has a very simple mathematical form called *conservation of mass*:

$$m_{\text{syst}} = \text{const}$$

or
$$\frac{dm}{dt} = 0 \tag{3.1}$$

This is so obvious in solid-mechanics problems that we often forget about it. In fluid mechanics, we must pay a lot of attention to mass conservation, and it takes a little analysis to make it hold.

Second, if the surroundings exert a net force $\mathbf{F}$ on the system, Newton's second law states that the mass will begin to accelerate[2]

$$\mathbf{F} = m\mathbf{a} = m\frac{d\mathbf{V}}{dt} = \frac{d}{dt}(m\mathbf{V}) \tag{3.2}$$

In Eq. (2.12) we saw this relation applied to a differential element of viscous incompressible fluid. In fluid mechanics Newton's law is called the conservation of linear momentum or, alternately, the momentum principle. Note that it is a vector law which implies the three scalar equations $F_x = ma_x$, $F_y = ma_y$, and $F_z = ma_z$.

Third, if the surroundings exert a net moment $\mathbf{M}$ about the center of mass of the system, there will be a rotation effect

$$\mathbf{M} = \frac{d\mathbf{H}}{dt} \tag{3.3}$$

where $\mathbf{H} = \sum (\mathbf{r} \times \mathbf{V})\, \delta m$ is the angular momentum of the system about its center of mass. Here we shall call Eq. (3.3) conservation of angular momentum or, alternately, the angular-momentum principle. Note that it is also a vector equation implying three scalar equations such as $M_x = dH_x/dt$.

For an arbitrary mass and arbitrary moment, $\mathbf{H}$ is quite complicated and contains nine terms (see, for example, Ref. 1, p. 285). In elementary dynamics we commonly treat only a rigid body rotating about a fixed x axis, for which Eq. (3.3) reduces to

$$M_x = I_x \frac{d}{dt}(\omega_x) \tag{3.4}$$

---

[1] We are neglecting nuclear reactions, where mass can be changed into energy.

[2] We are neglecting relativistic effects, where Newton's law must be modified.

where ω_x is the angular velocity of the body and I_x is its mass moment of inertia about the x axis. Unfortunately, fluid systems are not rigid and rarely reduce to such a simple relation, as we shall see in Sec. 3.6.

Fourth, if heat dQ is added to the system or work dW is done by the system, the system energy dE must change according to the energy relation, or first law of thermodynamics,

$$dQ - dW = dE$$

or $\qquad\qquad\qquad \dfrac{dQ}{dt} - \dfrac{dW}{dt} = \dfrac{dE}{dt} \qquad\qquad\qquad$ (3.5)

Like mass conservation, Eq. (3.1), this is a scalar relation having only a single component.

Finally, the second law of thermodynamics relates entropy change dS to heat added dQ and absolute temperature T

$$dS \geq \frac{dQ}{T} \qquad\qquad (3.6)$$

This is valid for a system and can be written in control-volume form, but there are almost no practical applications in fluid mechanics except to analyze flow-loss details (see Sec. 9.5).

All these laws involve thermodynamic properties, and thus we must supplement them with state relations $p = p(\rho, T)$ and $e = e(\rho, T)$ for the particular fluid being studied, as in Sec. 1.6.

The purpose of this chapter is to put our four basic laws into the control-volume form suitable for large-scale flows:

1. Conservation of mass (Sec. 3.3)
2. Conservation of linear momentum (Sec. 3.4)
3. Conservation of angular momentum (Sec. 3.6)
4. The energy equation (Sec. 3.7)

Wherever necessary to complete the analysis we shall also introduce a state relation such as the perfect-gas law.

Equations (3.1) to (3.6) apply to either fluid or solid systems. They are ideal for solid mechanics, where we follow the same system forever because it represents the product we are designing and building. For example, we follow a beam as it deflects under load. We follow a piston as it oscillates back and forth. We follow a rocket system all the way to Mars.

But fluid systems do not demand this concentrated attention. It is rare that we wish to follow the ultimate path of a specific particle of fluid. Instead it is likely that the fluid forms the environment whose effect on our product we wish to know. For the three examples cited above, we wish to know the wind loads on the beam, the fluid pressures on the piston, and the drag and lift loads on the rocket. This requires that the basic laws be rewritten to apply to a specific *region* in the neighborhood of our product. In other words, where the fluid particles in the wind go after they leave

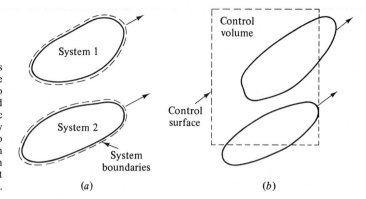

Fig. 3.1 System versus control-volume analysis: (*a*) two systems passing by and satisfying all the basic laws while on their way to infinity; (*b*) two systems passing through a control volume, which enables their local effect to be analyzed.

the beam is of little interest to a beam designer. The user's point of view underlies the need for the control-volume analysis of this chapter.

Although thermodynamics is not at all the main topic of this book, it would be a shame if the student did not review at least the first law and the state relations, as discussed for example in Refs. 6 and 7.

The control-volume approach is further illustrated in Fig. 3.1. If we are to analyze the systems in Fig. 3.1*a*, we must apply the basic laws (3.1) to (3.5) to each system and account for all interactions with the surroundings of each system. We must follow them forever or else to boundaries where known interactions occur. Since fluid systems constantly deform and interact with each other, this means in practice that the basic laws must be applied to differential fluid elements and patched together by analytic or numerical integration to form the resulting flow field. Thus system analysis in fluid mechanics is *differential analysis*, to be discussed in Chap. 4.

In contrast, the imaginary control volume in Fig. 3.1*b* happens to have the same two systems passing through it at the moment. They will soon be gone and other systems will come along, but no matter. The basic laws are rewritten to apply to the effect of the flow on this local region called a control volume. All we need to know is the flow field in this region; it is usually sufficient simply to make a reasonable assumption about the local field. Flow conditions away from this local region are then irrelevant. The technique for doing this is the subject of the present chapter.

3.2 THE REYNOLDS TRANSPORT THEOREM

In order to convert a system analysis into a control-volume analysis we must convert our mathematics to apply to a specific region rather than to individual masses. This conversion, called the *Reynolds transport theorem*, can be applied to all the basic laws. Examining the basic laws (3.1) to (3.3) and (3.5), we see that they are all concerned with the time derivative of fluid properties m, $\mathbf{V}$, $\mathbf{H}$, and E. Therefore what we need is to relate the time derivative of a system property to the rate of change of that property within a certain region.

The desired conversion formula differs slightly according as the control volume is fixed, moving, or deformable. Figure 3.2 illustrates these three cases. The fixed

Fig. 3.2 Fixed, moving, and deformable control volumes: (*a*) fixed control volume for nozzle-stress analysis; (*b*) control volume moving at ship speed for drag-force analysis; (*c*) control volume deforming within cylinder for transient pressure-variation analysis.

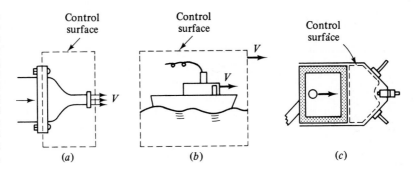

control volume in Fig. 3.2*a* encloses a stationary region of interest to a nozzle designer. The control surface is an abstract concept and does not hinder the flow in any way. It slices through the jet leaving the nozzle, circles around through the surrounding atmosphere, and slices through the flange bolts and the fluid within the nozzle. This particular control volume exposes the stresses in the flange bolts, which contribute to applied forces in the momentum analysis. In this sense the control volume resembles the *free-body* concept, which is applied to systems in solid-mechanics analyses.

Figure 3.2*b* illustrates a moving control volume. Here the ship is of interest, not the ocean, so that the control surface chases the ship at ship speed V. The control volume is of fixed volume, but the relative motion between water and ship must be considered. If V is constant, this relative motion is a steady-flow pattern, which simplifies the analysis.[1] If V is variable, the relative motion is unsteady, so that the computed results are time-variable and certain terms enter the momentum analysis to reflect the noninertial frame of reference.

Figure 3.2*c* shows a deforming control volume. Varying relative motion at the boundaries becomes a factor, and the rate of change of shape of the control volume enters the analysis. We shall begin by deriving the fixed-control-volume case and consider the other cases as advanced topics.

One-Dimensional Fixed Control Volume

As a simple first example, consider a duct or streamtube with a nearly one-dimensional flow $V = V(x)$, as shown in Fig. 3.3. The selected control volume is a portion of the duct which happens to be filled exactly by system 2 at a particular instant t. At time $t + dt$, system 2 has begun to move out and a sliver of system 1 has entered from the left. The shaded areas show an outflow sliver of volume $A_b V_b \, dt$ and an inflow volume $A_a V_a \, dt$.

Now let B be any property of the fluid (energy, momentum, etc.) and let $\beta = dB/dm$ be the *intensive* value or the amount of B per unit mass in any small portion of the fluid. The total amount of B in the control volume is thus

$$B_{CV} = \int_{CV} \beta \rho \, d\mathcal{V} \qquad \beta = \frac{dB}{dm} \qquad (3.7)$$

[1] A *wind tunnel* with a fixed model simulates a moving control volume with a moving body, whereas a *tow tank* simulates a fixed control volume with a moving body.

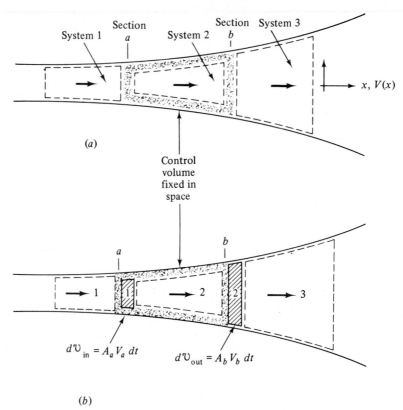

Section *a* — System 1, Section — System 2, *b* Section — System 3

x, V(x)

(a)

Control volume fixed in space

a

b

1 | 1 → 2 | 2 → 3

$d\mathcal{V}_{in} = A_a V_a \, dt$

$d\mathcal{V}_{out} = A_b V_b \, dt$

Fig. 3.3 Example of inflow and outflow as three systems pass through a control volume: (*a*) system 2 fills the control volume at time *t*; (*b*) at time *t + dt* system 2 begins to leave and system 1 enters.

(b)

where $\rho \, d\mathcal{V}$ is a differential mass of the fluid. We want to relate the rate of change of B_{CV} to the rate of change of the amount of B in system 2 which happens to coincide with the control volume at time t. The time derivative of B_{CV} is defined by the calculus limit

$$\frac{d}{dt}(B_{CV}) = \frac{1}{dt} B_{CV}(t + dt) - \frac{1}{dt} B_{CV}(t)$$

$$= \frac{1}{dt} [B_2(t + dt) - (\beta\rho \, d\mathcal{V})_{out} + (\beta\rho \, d\mathcal{V})_{in}] - \frac{1}{dt} [B_2(t)]$$

$$= \frac{1}{dt} [B_2(t + dt) - B_2(t)] - (\beta\rho A V)_{out} + (\beta\rho A V)_{in} \tag{3.8}$$

The first term on the right is the rate of change of B within system 2 at the instant it occupies the control volume. By rearranging Eq. (3.8) we have the desired conversion formula relating changes in any property B of a local system to one-dimensional computations concerning a fixed control volume which instantaneously encloses the system.

$$\frac{d}{dt}(B_{syst}) = \frac{d}{dt}\left(\int_{CV} \beta\rho \, d\mathcal{V}\right) + (\beta\rho A V)_{out} - (\beta\rho A V)_{in} \tag{3.9}$$

This is the one-dimensional Reynolds transport theorem for a fixed volume. The three terms on the right-hand side are, respectively,

1. The rate of change of B within the control volume
2. The flux of B passing out of the control surface
3. The flux of B passing into the control surface

If the flow pattern is steady, the first term vanishes. Equation (3.9) can readily be generalized to an arbitrary flow pattern, as follows.

Arbitrary Fixed Control Volume

Figure 3.4 shows a generalized fixed control volume with an arbitrary flow pattern passing through. The only additional complication is that there are variable slivers of inflow and outflow of fluid all about the control surface. In general, each differential area dA of surface will have a different velocity $\mathbf{V}$ making a different angle θ with the local normal to dA. Some elemental areas will have inflow volume $(VA\cos\theta)_{\text{in}}\,dt$, and others will have outflow volume $(VA\cos\theta)_{\text{out}}\,dt$, as seen in Fig. 3.4. Some surfaces might correspond to streamlines ($\theta = 90°$) or solid walls ($\mathbf{V} = 0$) with neither inflow nor outflow. Equation (3.9) generalizes to

$$\frac{d}{dt}(B_{\text{syst}}) = \frac{d}{dt}\left(\iiint_{\text{CV}} \beta\rho\,d\mathcal{V}\right) + \iint_{\text{CS}} \beta\rho V\cos\theta\,dA_{\text{out}} - \iint_{\text{CS}} \beta\rho V\cos\theta\,dA_{\text{in}} \quad (3.10)$$

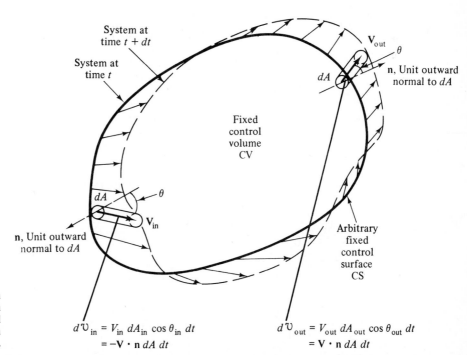

$$d\mathcal{V}_{\text{in}} = V_{\text{in}}\,dA_{\text{in}}\cos\theta_{\text{in}}\,dt$$
$$= -\mathbf{V}\cdot\mathbf{n}\,dA\,dt$$

$$d\mathcal{V}_{\text{out}} = V_{\text{out}}\,dA_{\text{out}}\cos\theta_{\text{out}}\,dt$$
$$= \mathbf{V}\cdot\mathbf{n}\,dA\,dt$$

This is the Reynolds transport theorem for an arbitrary fixed control volume. By letting the property B be mass, momentum, angular momentum, or energy we can rewrite all the basic laws in control-volume form. Note that all three of the control-volume integrals are concerned with the intensive property β. Since the control volume is fixed in space, the elemental volumes $d\mathcal{V}$ do not vary with time, so that the time derivative of the volume integral vanishes unless either β or ρ varies with time (unsteady flow).

Equation (3.10) expresses the basic formula that a system derivative equals the rate of change of B within the control volume plus the flux of B out of the control surface minus the flux of B into the control surface. The quantity B (or β) may be any vector or scalar property of the fluid. Two alternate forms are possible for the flux terms. First we may notice that $V \cos \theta$ is the component of V normal to the area element of the control surface. Thus we can write

$$\text{Flux terms} = \iint_{\text{CS}} \beta \rho V_n \, dA_{\text{out}} - \iint_{\text{CS}} \beta \rho V_n \, dA_{\text{in}} = \iint_{\text{CS}} \beta \, d\dot{m}_{\text{out}} - \iint_{\text{CS}} \beta \, d\dot{m}_{\text{in}} \quad (3.11a)$$

where $d\dot{m} = \rho V_n \, dA$ is the differential mass flux through the surface. Form (3.11a) helps visualize what is being calculated.

A second alternate form offers elegance and compactness as advantages. If $\mathbf{n}$ is defined as the *outward* normal unit vector everywhere on the control surface, then $\mathbf{V} \cdot \mathbf{n} = V_n$ for outflow and $-V_n$ for inflow. Therefore the flux terms can be represented by a single integral involving $\mathbf{V} \cdot \mathbf{n}$ which accounts for both positive outflow and negative inflow

$$\text{Flux terms} = \iint_{\text{CS}} \beta \rho (\mathbf{V} \cdot \mathbf{n}) \, dA \quad (3.11b)$$

The compact form of the Reynolds transport theorem is thus

$$\frac{d}{dt}(B_{\text{syst}}) = \frac{d}{dt}\left(\iiint_{\text{CV}} \beta \rho \, d\mathcal{V} \right) + \iint_{\text{CS}} \beta \rho (\mathbf{V} \cdot \mathbf{n}) \, dA \quad (3.12)$$

This is beautiful but only occasionally useful, when the coordinate system is ideally suited to the control volume selected. Otherwise the computations are easier when the flux of B out is added and the flux of B in is subtracted, according to (3.10) or (3.11a).

The time-derivative term can be written in the equivalent form

$$\frac{d}{dt}\left(\iiint_{\text{CV}} \beta \rho \, d\mathcal{V} \right) = \iiint_{\text{CV}} \frac{\partial}{\partial t}(\beta \rho) \, d\mathcal{V} \quad (3.13)$$

for the fixed control volume since the volume elements do not vary.

Control Volume Moving at Constant Velocity

If the control volume is moving uniformly at velocity V_s, as in Fig. 3.2b, an observer fixed to the control volume will see a relative velocity V_r of fluid crossing the control surface, defined by

$$V_r = V - V_s \tag{3.14}$$

where V is the fluid velocity relative to the same coordinate system in which the control volume motion V_s is observed. Note that Eq. (3.14) is a vector subtraction. The flux terms will be proportional to V_r, but the volume integral is unchanged because the control volume moves as a fixed shape without deforming. The Reynolds transport theorem for this case of a uniformly moving control volume is

$$\frac{d}{dt}(B_{\text{syst}}) = \frac{d}{dt}\left(\iiint_{\text{CV}} \beta\rho \, d\mathcal{V}\right) + \iint_{\text{CS}} \beta\rho(V_r \cdot n) \, dA \tag{3.15}$$

which reduces to Eq. (3.12) if $V_s \equiv 0$.

Control Volume of Constant Shape but Variable Velocity[1]

If the control volume moves with a velocity $V_s(t)$ which retains its shape, the volume elements do not change with time but the boundary relative velocity $V_r = V(r, t) - V_s(t)$ becomes a somewhat more complicated function. Equation (3.15) is unchanged in form, but the area integral may be more laborious to evaluate.

Arbitrarily Moving and Deformable Control Volume[2]

The most general situation is when the control volume is both moving and deforming arbitrarily, as illustrated in Fig. 3.5. The flux of volume across the control surface is again proportional to the relative normal velocity component $V_r \cdot n$, as in Eq. (3.15). However, since the control surface has a deformation, its velocity $V_s = V_s(r, t)$, so that the relative velocity $V_r = V(r, t) - V_s(r, t)$ is or can be a complicated function, even though the flux integral is the same as in Eq. (3.15). Meanwhile, the volume integral in Eq. (3.15) must allow the volume elements to distort with time. Thus the time derivative must be applied *after* integration. For the deforming control volume, then, the transport theorem takes the form

$$\frac{d}{dt}(B_{\text{syst}}) = \frac{d}{dt}\left(\iiint_{\text{CV}} \beta\rho \, d\mathcal{V}\right) + \iint_{\text{CS}} \beta\rho(V_r \cdot n) \, dA \tag{3.16}$$

This is the most general case, which we can compare with the equivalent form for a fixed control volume

$$\frac{d}{dt}(B_{\text{syst}}) = \iiint_{\text{CV}} \frac{\partial}{\partial t}(\beta\rho) \, d\mathcal{V} + \iint_{\text{CS}} \beta\rho(V \cdot n) \, dA \tag{3.17}$$

---

[1] This section may be omitted without loss of continuity.

[2] This section may be omitted without loss of continuity.

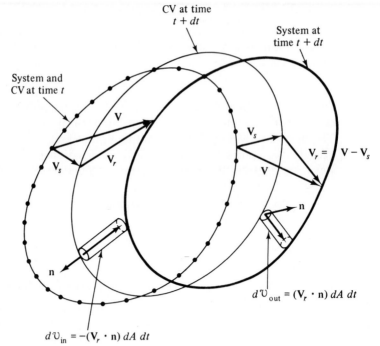

Fig. 3.5 Relative-velocity effects between a system and a control volume when both move and deform. The system boundaries move at velocity **V**, and the control surface moves at velocity $\mathbf{V}_s$.

The moving and deforming control volume, Eq. (3.16), contains only two complications: (1) the time derivative of the triple integral must be taken outside, and (2) the double integral involves the *relative* velocity $\mathbf{V}_r$ between the fluid system and the control surface. These differences and mathematical subtleties are best shown by examples.

One-Dimensional Flux-Term Approximations

In many applications, the flow crosses the boundaries of the control surface only at certain simplified inlets and exits which are approximately *one-dimensional*; i.e., the flow properties are nearly uniform over the cross section of the inlet or exit. Then the double-integral flux terms required in Eq. (3.16) reduce to a simple sum of positive (exit) and negative (inlet) product terms involving the flow properties at each cross section

$$\iint_{\text{CS}} \beta \rho (\mathbf{V}_r \cdot \mathbf{n}) \, dA = \sum (\beta_i \rho_i V_{ri} A_i)_{\text{out}} - \sum (\beta_i \rho_i V_{ri} A_i)_{\text{in}} \qquad (3.18)$$

An example of this situation is shown in Fig. 3.6. There are inlet flows at sections 1 and 4 and outflow at sections 2, 3, and 5. For this particular problem Eq. (3.18) would be

$$\iint_{\text{CS}} \beta \rho (\mathbf{V}_r \cdot \mathbf{n}) \, dA = \beta_2 \rho_2 V_{r2} A_2 + \beta_3 \rho_3 V_{r3} A_3$$

$$+ \beta_5 \rho_5 V_{r5} A_5 - \beta_1 \rho_1 V_{r1} A_1 - \beta_4 \rho_4 V_{r4} A_4 \qquad (3.19)$$

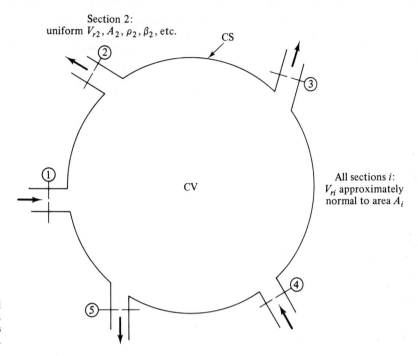

Section 2:
uniform $V_{r2}, A_2, \rho_2, \beta_2$, etc.

CS

CV

All sections i:
V_{ri} approximately
normal to area A_i

Fig. 3.6 A control
volume with simplified
one-dimensional inlets
and exits.

with no contribution from any other portion of the control surface because there is
no flow across the boundary.

EXAMPLE 3.1 A fixed control volume has three one-dimensional boundary sections, as shown in Fig. E3.1.
The flow within the control volume is steady. The flow properties at each section are tabu-

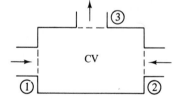

Fig. E3.1

lated below. Find the rate of change of energy of the system which occupies the control
volume at this instant.

Section	Type	ρ, kg/m^3	V, m/s	A, m^2	e, J/kg
1	Inlet	800	5.0	2.0	300
2	Inlet	800	8.0	3.0	100
3	Outlet	800	17.0	2.0	150

Solution   The property under study here is energy, and so $B = E$ and $\beta = dE/dm = e$, the energy per unit mass. Since the control volume is fixed, Eq. (3.17) applies

$$\left(\frac{dE}{dt}\right)_{\text{syst}} = \iiint_{\text{CV}} \frac{\partial}{\partial t}(e\rho)\, d\mathcal{V} + \iint_{\text{CS}} e\rho(\mathbf{V} \cdot \mathbf{n})\, dA$$

The flow within is steady, so that $\partial(e\rho)/\partial t \equiv 0$ and the volume integral vanishes. The area integral consists of two inlet and one outlet sections, as given in the table

$$\left(\frac{dE}{dt}\right)_{\text{syst}} = -e_1\rho_1 A_1 V_1 - e_2\rho_2 A_2 V_2 + e_3\rho_3 A_3 V_3$$

Introducing the numerical values from the table, we have

$$\left(\frac{dE}{dt}\right)_{\text{syst}} = -(300 \text{ J/kg})(800 \text{ kg/m}^3)(2 \text{ m}^2)(5 \text{ m/s})$$

$$- 100(800)(3)(8) + 150(800)(2)(17)$$

$$= (-2{,}400{,}000 - 1{,}920{,}000 + 4{,}080{,}000) \text{ J/s}$$

$$= -240{,}000 \text{ J/s} = -0.24 \text{ MJ/s} \qquad\qquad Ans.$$

Thus the system is losing energy at the rate of $0.24 \text{ MJ/s} = 0.24 \text{ MW}$. Since we have accounted for all fluid energy crossing the boundary, we conclude from the first law that there must be heat loss through the control surface or the system must be doing work on the environment through some device not shown. Notice that the use of SI units leads to a consistent result in joules per second without any conversion factors. We promised in Chap. 1 that this would be the case.

   Note: This problem involves energy, but suppose we check the balance of mass also. Then $B = $ mass, m, and $\beta = dm/dm = $ unity. Again the volume integral vanishes for steady flow, and Eq. (3.17) reduces to

$$\left(\frac{dm}{dt}\right)_{\text{syst}} = \iint_{\text{CS}} \rho(\mathbf{V} \cdot \mathbf{n})\, dA = -\rho_1 A_1 V_1 - \rho_2 A_2 V_2 + \rho_3 A_3 V_3$$

$$= -(800 \text{ kg/m}^3)(2 \text{ m}^2)(5 \text{ m/s}) - 800(3)(8) + 800(17)(2)$$

$$= (-8000 - 19{,}200 + 27{,}200) \text{ kg/s} = 0 \text{ kg/s}$$

Thus the system mass does not change, which correctly expresses the law of conservation of system mass, Eq. (3.1).

EXAMPLE 3.2   The balloon in Fig. E3.2 is being filled through section 1, where the area is A_1, velocity is V_1, and fluid density is ρ_1. The average density within the balloon is $\rho_b(t)$. Find an expression for the rate of change of system mass within the balloon at this instant.

Solution   It is convenient to define a deformable control surface just outside the balloon, expanding at the same rate $R(t)$. Equation (3.16) applies with $V_r = 0$ on the balloon surface and $V_r = V_1$ at the pipe entrance. For mass change, we take $B = m$ and $\beta = dm/dm = 1$. Equation (3.16) becomes

$$\left(\frac{dm}{dt}\right)_{\text{syst}} = \frac{d}{dt}\left(\iiint_{\text{CV}} \rho\, d\mathcal{V}\right) + \iint_{\text{CS}} \rho(\mathbf{V}_r \cdot \mathbf{n})\, dA$$

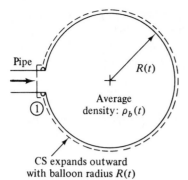

Fig. E3.2

Mass flux occurs only at the inlet, so that the double integral reduces to the single negative term $-\rho_1 A_1 V_1$. The fluid mass within the control volume is approximately the average density times the volume of a sphere. The equation thus becomes

$$\left(\frac{dm}{dt}\right)_{\text{syst}} = \frac{d}{dt}\left(\rho_b \tfrac{4}{3}\pi R^3\right) - \rho_1 A_1 V_1 \qquad\qquad Ans.$$

This is the desired result for system mass rate of change. Actually, by the conservation law (3.1), this change must be zero. Thus the balloon density and radius are related to the inlet mass flux by

$$\frac{d}{dt}\left(\rho_b R^3\right) = \frac{3}{4\pi}\rho_1 A_1 V_1$$

This is a first-order differential equation which could form part of an engineering analysis of balloon inflation. It cannot be solved without further use of mechanics and thermodynamics to relate the four unknowns ρ_b, ρ_1, V_1, and R. The pressure and temperature and the elastic properties of the balloon would also have to be brought into the analysis.

For advanced study, many more details of the analysis of deformable control volumes can be found in Hansen [4] and Potter and Foss [5].

3.3 CONSERVATION OF MASS

The Reynolds transport theorem, Eq. (3.16) or (3.17), establishes a relation between system rates of change and control-volume surface and volume integrals. But system derivatives are related to the basic laws of mechanics, Eqs. (3.1) to (3.5). Eliminating system derivatives between the two gives the control-volume, or *integral*, forms of the laws of mechanics of fluids. The dummy variable B becomes, respectively, mass, linear momentum, angular momentum, and energy.

For conservation of mass, as discussed in Examples 3.1 and 3.2, $B = m$ and $\beta = dm/dm = 1$. Equation (3.1) becomes

$$\left(\frac{dm}{dt}\right)_{\text{syst}} = 0 = \frac{d}{dt}\left(\iiint_{\text{CV}} \rho\, d\mathcal{V}\right) + \iint_{\text{CS}} \rho(\mathbf{V}_r \cdot \mathbf{n})\, dA \qquad (3.20)$$

This is the integral mass-conservation law for a deformable control volume. For a fixed control volume, we have

$$\iiint_{\text{CV}} \frac{\partial \rho}{\partial t} \, d\mathcal{V} + \iint_{\text{CS}} \rho(\mathbf{V} \cdot \mathbf{n}) \, dA = 0 \tag{3.21}$$

If the control volume has only a number of one-dimensional inlets and outlets, we can write

$$\iiint_{\text{CV}} \frac{\partial \rho}{\partial t} \, d\mathcal{V} + \sum_i (\rho_i A_i V_i)_{\text{out}} - \sum_i (\rho_i A_i V_i)_{\text{in}} = 0 \tag{3.22}$$

Other special cases occur. Suppose that the flow within the control volume is steady; then $\partial \rho / \partial t \equiv 0$, and Eq. (3.21) reduces to

$$\iint_{\text{CS}} \rho(\mathbf{V} \cdot \mathbf{n}) \, dA = 0 \tag{3.23}$$

This states that in steady flow the mass flows entering and leaving the control volume must balance exactly.[1] If, further, the inlets and outlets are one-dimensional, we have for steady flow

$$\sum_i (\rho_i A_i V_i)_{\text{in}} = \sum_i (\rho_i A_i V_i)_{\text{out}} \tag{3.24}$$

This simple approximation is widely used in engineering analyses. For example, referring back to Fig. 3.6, if the flow in that control volume is steady, the three outlet mass fluxes balance the two inlets

$$\text{Outflow} = \text{inflow}$$
$$\rho_2 A_2 V_2 + \rho_3 A_3 V_3 + \rho_5 A_5 V_5 = \rho_1 A_1 V_1 + \rho_4 A_4 V_4 \tag{3.25}$$

The quantity $\rho A V$ is called the *mass flow* $\dot{m}$ passing through the one-dimensional cross section and has consistent units of kilograms per second (or slugs per second) when using SI (or BG) units. Equation (3.25) can be rewritten in the short form

$$\dot{m}_2 + \dot{m}_3 + \dot{m}_5 = \dot{m}_1 + \dot{m}_4 \tag{3.26}$$

and, in general, the steady-flow–mass-conservation relation (3.23) can be written as

$$\sum_i (\dot{m}_i)_{\text{out}} = \sum_i (\dot{m}_i)_{\text{in}} \tag{3.27}$$

---

[1] Throughout this section we are neglecting *sources* or *sinks* of mass which might be embedded in the control volume. Equations (3.20) and (3.21) can readily be modified to add source and sink terms, but this is rarely necessary.

If the inlets and outlets are not one-dimensional, one has to compute $\dot{m}$ by integration over the section

$$\dot{m}_{cs} = \iint\limits_{cs} \rho(\mathbf{V} \cdot \mathbf{n})\, dA \tag{3.28}$$

where cs stands for cross section. An illustration of this is given in Example 3.4.

Incompressible Flow

Still further simplification is possible if the fluid is incompressible, which we may define as having density variations which are negligible in the mass-conservation requirement.[1] As we saw in Chap. 1, all liquids are nearly incompressible, and gas flows can *behave* as if they were incompressible, particularly if the gas velocity is less than about 30 percent of the speed of sound of the gas.

Again consider the fixed control volume. If the fluid is nearly incompressible, $\partial\rho/\partial t$ is negligible and the volume integral in Eq. (3.21) may be neglected, after which the density can be slipped outside the surface integral and divided out since it is nonzero. The result is a conservation law for incompressible flows, whether steady or unsteady

$$\iint\limits_{cs} (\mathbf{V} \cdot \mathbf{n})\, dA = 0 \tag{3.29}$$

If the inlets and outlets are one-dimensional, we have

$$\sum_i (V_i A_i)_{out} = \sum_i (V_i A_i)_{in} \tag{3.30}$$

or

$$\sum Q_{out} = \sum Q_{in}$$

where $Q_i = V_i A_i$ is called the *volume flow* passing through the given cross-section. Again, if consistent units are used, $Q = VA$ will have units of cubic meters per second (SI) or cubic feet per second (BG). If the cross section is not one-dimensional, we have to integrate

$$Q_{cs} = \iint\limits_{cs} (\mathbf{V} \cdot \mathbf{n})\, dA \tag{3.31}$$

Equation (3.31) allows us to define an *average velocity* V_{av} which when multiplied by the section area gives the correct volume flow

$$V_{av} = \frac{Q}{A} = \frac{1}{A} \iint (\mathbf{V} \cdot \mathbf{n})\, dA \tag{3.32}$$

---

[1] Be warned that there is subjectivity in specifying incompressibility. Oceanographers consider a 0.1 percent density variation very significant, while aerodynamicists often neglect density variations in highly compressible, even hypersonic, gas flows. Your task is to justify the incompressible approximation when you make it.

This could be called the *volume-average velocity*. If the density varies across the section, we can define an average density in the same manner

$$\rho_{av} = \frac{1}{A} \iint \rho \, dA \tag{3.33}$$

But the mass flow would contain the product of density and velocity, and the average product $(\rho V)_{av}$ would in general have a different value from the product of the averages

$$(\rho V)_{av} = \frac{1}{A} \iint \rho (\mathbf{V} \cdot \mathbf{n}) \, dA \approx \rho_{av} V_{av} \tag{3.34}$$

We shall illustrate this in Example 3.4. We can often neglect the difference or, if necessary, use a correction factor between mass average and volume average.

EXAMPLE 3.3   Write the conservation-of-mass relation for steady flow through a stream-tube (flow everywhere parallel to the walls) with a single one-dimensional inlet 1 and exit 2 (Fig. E3.3).

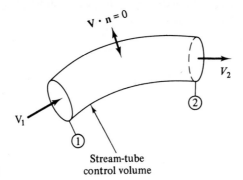

Fig. E3.3

Stream-tube
control volume

Solution   For steady flow Eq. (3.24) applies with the single inlet and exit

$$\dot{m} = \rho_1 A_1 V_1 = \rho_2 A_2 V_2 = \text{const}$$

Thus, in a streamtube in steady flow, the mass flow is constant across every section of the tube. If the density is constant, then

$$Q = A_1 V_1 = A_2 V_2 = \text{const} \quad \text{or} \quad V_2 = \frac{A_1}{A_2} V_1$$

The volume flow is constant in the tube in steady incompressible flow, and the velocity increases as the section area decreases. This relation was derived by Leonardo da Vinci in 1500.

EXAMPLE 3.4   For steady viscous flow through a circular tube (Fig. E3.4) the axial velocity profile is given approximately by

$$u = U_0 \left(1 - \frac{r}{R} \right)^m$$

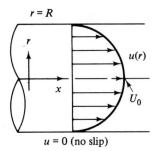

$r = R$

$u(r)$

x

U_0

Fig. E3.4 $u = 0$ (no slip)

so that u varies from zero at the wall ($r = R$), or no slip, and up to a maximum $u = U_0$ at the centerline $r = 0$. For highly viscous (laminar) flow, $m \approx \frac{1}{2}$, while for slightly viscous (turbulent) flow $m \approx \frac{1}{7}$. (a) Compute the average velocity if the density is constant. (b) Suppose the density also varies, according to the formula

$$\rho = \tfrac{1}{2}\rho_0 \left[1 + \left(1 - \frac{r}{R} \right)^n \right]$$

i.e., it increases from $0.5\rho_0$ at the walls to ρ_0 at the centerline. This simulates high-speed gas flow through a tube with hot walls. Compute the average density and the average product of density and velocity, and compare.

Solution

Part (a)   The average velocity is defined by Eq. (3.32). Here $\mathbf{V} = \mathbf{i}u$ and $\mathbf{n} = \mathbf{i}$, and thus $\mathbf{V} \cdot \mathbf{n} = u$. Since the flow is symmetric, the differential area can be taken as a circular strip $dA = 2\pi r \, dr$. Equation (3.32) becomes

$$V_{\mathrm{av}} = \frac{1}{A} \iint u \, dA = \frac{1}{\pi R^2} \int_0^R U_0 \left(1 - \frac{r}{R} \right)^m 2\pi r \, dr$$

or $$V_{\mathrm{av}} = U_0 \frac{2}{(1+m)(2+m)} \qquad \qquad Ans. (a)$$

For the laminar-flow approximation, $m \approx \frac{1}{2}$, $V_{\mathrm{av}} \approx 0.53 U_0$. (The exact laminar theory in Chap. 6 gives $V_{\mathrm{av}} = 0.50 U_0$.) For turbulent flow, $m \approx \frac{1}{7}$, $V_{\mathrm{av}} \approx 0.82 U_0$. (There is no exact turbulent theory, and so we accept this approximation.) The turbulent velocity profile is more uniform across the section, and thus the average velocity is only slightly less than maximum.

Part (b)   The average density is given by Eq. (3.33)

$$\rho_{\mathrm{av}} = \frac{1}{\pi R^2} \int_0^R \tfrac{1}{2}\rho_0 \left[1 + \left(1 - \frac{r}{R} \right)^n \right] 2\pi r \, dr = \rho_0 \left[\frac{1}{2} + \frac{1}{(1+n)(2+n)} \right] \qquad Ans. (b)$$

For the laminar approximation, $n \approx \frac{1}{2}$, this gives $\rho_{\mathrm{av}} = 0.77\rho_0$. The turbulent value $n \approx \frac{1}{7}$ gives $\rho_{\mathrm{av}} = 0.91\rho_0$, which is again closer to the maximum value. (There is no exact theory for either laminar or turbulent density distribution in a tube.) The average product of velocity and density is obtained from Eq. (3.34)

$$(\rho V)_{\mathrm{av}} = \frac{1}{\pi R^2} \int_0^R \tfrac{1}{2}\rho_0 \left[1 + \left(1 - \frac{r}{R} \right)^n \right] U_0 \left(1 - \frac{r}{R} \right)^m 2\pi r \, dr$$

$$= \rho_0 U_0 \left[\frac{2 + 2m + n}{(1+m)(1+m+n)} - \frac{4 + 2m + n}{(2+m)(2+m+n)} \right] \qquad Ans. (b)$$

This does *not* equal the product of ρ_{av} and V_{av}. To illustrate, try our two approximations:

$$(\rho V)_{av} = \begin{cases} 0.433\rho_0 U_0 = 1.06\rho_{av} V_{av} & \text{laminar, } m \approx n \approx \frac{1}{2} \\ 0.749\rho_0 U_0 = 1.01\rho_{av} V_{av} & \text{turbulent, } m \approx n \approx \frac{1}{7} \end{cases}$$

Thus, if we use average density and average velocity to compute the mass flux in this example, we make a 6 percent error in laminar flow and a 1 percent error in turbulent flow. Close but no cigar: although engineering analyses of variable-density flow have an uncertainty of about 10 percent, we sometimes account for this correction.

EXAMPLE 3.5 Consider the constant-density velocity field

$$u = \frac{V_0 x}{L} \qquad v = 0 \qquad w = -\frac{V_0 z}{L}$$

similar to Example 1.14. Use the triangular control volume in Fig. E3.5, bounded by $(0, 0)$, (L, L), and $(0, L)$, with depth b into paper. Compute the volume flow through sections 1, 2, and 3 and compare to see whether mass is conserved.

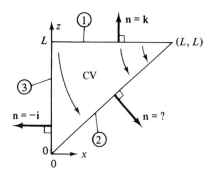

Fig. E3.5 Depth b into paper

Solution The velocity field everywhere has the form $\mathbf{V} = \mathbf{i}u + \mathbf{k}w$. This must be evaluated along each section. We shall save section 2 until last because it looks tricky. Section 1 is the plane $z = L$ with depth b. The unit outward normal is $\mathbf{n} = \mathbf{k}$, as shown. The differential area is a strip of depth b varying with x: $dA = b\,dx$. The normal velocity is

$$(\mathbf{V} \cdot \mathbf{n})_1 = (\mathbf{i}u + \mathbf{k}w) \cdot \mathbf{k} = w|_1 = -\frac{V_0 z}{L}\bigg|_{z=L} = -V_0$$

The volume flow through section 1 is thus, from Eq. (3.31),

$$Q_1 = \int_1 (\mathbf{V} \cdot \mathbf{n})\,dA = \int_0^L (-V_0)b\,dx = -V_0 bL \qquad \text{Ans. 1}$$

Since this is negative, section 1 is a net inflow. Check the units: $V_0 bL$ is a velocity times an area; OK.

Section 3 is the plane $x = 0$ with depth b. The unit normal is $\mathbf{n} = -\mathbf{i}$, as shown, and $dA = b\,dz$. The normal velocity is

$$(\mathbf{V} \cdot \mathbf{n})_3 = (i u + k w) \cdot (-\mathbf{i}) = -u|_3 = -\left. \frac{V_0 x}{L} \right|_{x=0} = 0 \qquad\qquad \textit{Ans. 3}$$

Thus $V_n \equiv 0$ all along section 3; hence $Q_3 = 0$.

Finally, section 2 is the plane $x = z$ with depth b. The normal direction is to the right $\mathbf{i}$ and down $-\mathbf{k}$ but must have *unit* value; thus $\mathbf{n} = (1/\sqrt{2})(\mathbf{i} - \mathbf{k})$. The differential area is either $dA = \sqrt{2}\,b\,dx$ or $dA = \sqrt{2}\,b\,dz$. The normal velocity is

$$(\mathbf{V} \cdot \mathbf{n})_2 = (i u + k w) \cdot \frac{1}{\sqrt{2}}(\mathbf{i} - \mathbf{k}) = \frac{1}{\sqrt{2}}(u - w)_2$$

$$= \frac{1}{\sqrt{2}}\left[V_0 \frac{x}{L} - \left(-V_0 \frac{z}{L} \right) \right]_{x=z} = \frac{\sqrt{2}\,V_0 x}{L} \ \text{or} \ \frac{\sqrt{2}\,V_0 z}{L}$$

Then the volume flow through section 2 is

$$Q_2 = \int_2 (\mathbf{V} \cdot \mathbf{n})\,dA = \int_0^L \frac{\sqrt{2}\,V_0 x}{L}(\sqrt{2}\,b\,dx) = V_0 b L \qquad\qquad \textit{Ans. 2}$$

This answer is positive, indicating an outflow. These are the desired results. We should note that the volume flow is zero through the front and back triangular faces of the prismatic control volume because $V_n = v = 0$ on those faces.

The sum of the three volume flows is

$$Q_1 + Q_2 + Q_3 = -V_0 b L + V_0 b L + 0 = 0$$

Mass is conserved in this constant-density flow, and there are no net sources or sinks within the control volume. This is a very realistic flow, as described in Example 1.14.

EXAMPLE 3.6   The tank in Fig. E3.6 is being filled with water by two one-dimensional inlets. Air is trapped at the top of the tank. The water height is h. (*a*) Find an expression for the change in water height dh/dt. (*b*) Compute dh/dt if $D_1 = 1$ in, $D_2 = 3$ in, $V_1 = 3$ ft/s, $V_2 = 2$ ft/s, and $A_t = 2$ ft^2, assuming water at 20°C.

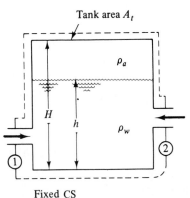

Fig. E3.6   Fixed CS

Solution

Part (a)   A suggested control volume encircles the tank and cuts through the two inlets. The flow within is unsteady, and Eq. (3.22) applies with no outlets and two inlets

$$\frac{\partial}{\partial t}\left(\iiint_{CV} \rho \, d\mathcal{V}\right) - \rho_1 A_1 V_1 - \rho_2 A_2 V_2 = 0 \tag{1}$$

Now, if A_t is the tank cross-sectional area, the unsteady term can be evaluated as follows:

$$\frac{\partial}{\partial t}\left(\iiint_{CV} \rho \, d\mathcal{V}\right) = \frac{\partial}{\partial t}(\rho_w A_t h) + \frac{\partial}{\partial t}[\rho_a A_t(H-h)] = \rho_w A_t \frac{dh}{dt} \tag{2}$$

The ρ_a term vanishes because it is the rate of change of air mass and is zero because the air is trapped at the top. Substituting (2) into (1), we find the change of water height

$$\frac{dh}{dt} = \frac{\rho_1 A_1 V_1 + \rho_2 A_2 V_2}{\rho_w A_t} \qquad\qquad\qquad Ans.\ (a)$$

For water, $\rho_1 = \rho_2 = \rho_w$, and this result reduces to

$$\frac{dh}{dt} = \frac{A_1 V_1 + A_2 V_2}{A_t} = \frac{Q_1 + Q_2}{A_t} \tag{3}$$

Part (b)   The two inlet volume flows are

$$Q_1 = A_1 V_1 = \tfrac{1}{4}\pi(\tfrac{1}{12}\ \text{ft})^2(3\ \text{ft/s}) = 0.016\ \text{ft}^3/\text{s}$$

$$Q_2 = A_2 V_2 = \tfrac{1}{4}\pi(\tfrac{3}{12}\ \text{ft})^2(2\ \text{ft/s}) = 0.098\ \text{ft}^3/\text{s}$$

Then, from Eq. (3),

$$\frac{dh}{dt} = \frac{0.016 + 0.098\ \text{ft}^3/\text{s}}{2\ \text{ft}^2} = 0.057\ \text{ft/s} \qquad\qquad Ans.\ (b)$$

Suggestion: Repeat this problem with the top of the tank open.

An illustration of a mass balance with a deforming control volume has already been given in Example 3.2.

The control-volume mass relations, Eq. (3.20) or (3.21), are fundamental to all fluid-flow analyses. They involve only velocity and density. Vector directions are of no consequence except to determine the normal velocity at the surface and hence whether the flow is *in* or *out*. Although your specific analysis may concern forces or moments or energy, you must always make sure that mass is balanced as part of the analysis; otherwise the results will be unrealistic and probably rotten. We shall see in the examples which follow how mass conservation is constantly checked while performing an analysis of other fluid properties.

3.4   CONSERVATION OF LINEAR MOMENTUM

In Newton's law, Eq. (3.2), the property being differentiated is the linear momentum $m\mathbf{V}$. Therefore our dummy variable is $\mathbf{B} = m\mathbf{V}$ and $\beta = d\mathbf{B}/dm = \mathbf{V}$, and application of the Reynolds transport theorem gives the linear-momentum relation for a deformable control volume

$$\frac{d}{dt}(m\mathbf{V})_{\text{syst}} = \sum \mathbf{F} = \frac{d}{dt}\left(\iiint_{\text{CV}} \mathbf{V}\rho \, d\mathcal{V}\right) + \iint_{\text{CS}} \mathbf{V}\rho(\mathbf{V}_r \cdot \mathbf{n}) \, dA \qquad (3.35)$$

The following points concerning this relation should be strongly emphasized:

1. The term $\mathbf{V}$ is the fluid velocity relative to an *inertial* (nonaccelerating) coordinate system; otherwise Newton's law must be modified to include noninertial relative-acceleration terms (see the end of this section).
2. The term $\sum \mathbf{F}$ is the *vector* sum of all forces acting on the control-volume material considered as a free body; i.e., it includes surface forces on all fluids and solids cut by the control surface plus all body forces (gravity and electromagnetic) acting on the masses within the control volume.
3. The entire equation is a vector relation; both the integrals are vectors due to the term $\mathbf{V}$ in the integrands. The equation thus has three components. If we want only, say, the x component, the equation reduces to

$$\sum F_x = \frac{d}{dt}\left(\iiint_{\text{CV}} u\rho \, d\mathcal{V}\right) + \iint_{\text{CS}} u\rho(\mathbf{V}_r \cdot \mathbf{n}) \, dA \qquad (3.36)$$

and similarly, $\sum F_y$ and $\sum F_z$ would involve v and w, respectively. Failure to account for the vector nature of the linear-momentum relation (3.35) is probably the greatest source of student error in control-volume analyses.

For a fixed control volume, the relative velocity $\mathbf{V}_r \equiv \mathbf{V}$, and we can use the partial derivative

$$\sum \mathbf{F} = \frac{\partial}{\partial t}\left(\iiint_{\text{CV}} \mathbf{V}\rho \, d\mathcal{V}\right) + \iint_{\text{CS}} \mathbf{V}\rho(\mathbf{V} \cdot \mathbf{n}) \, dA \qquad (3.37)$$

Again we stress that this is a vector relation and that $\mathbf{V}$ must be an inertial-frame velocity. Most of the momentum analyses in this text are concerned with Eq. (3.37).

One-Dimensional Momentum Flux

By analogy with the term mass flow used in Eq. (3.28), the surface integral in Eq. (3.37) is called the *momentum-flux term*. If we denote momentum by $\mathbf{M}$, then

$$\dot{\mathbf{M}}_{\text{cs}} = \iint_{\text{sec}} \mathbf{V}\rho(\mathbf{V} \cdot \mathbf{n}) \, dA \qquad (3.38)$$

Because of the dot product, the result will be negative for inlet momentum flux and positive for outlet flux. If the cross section is one-dimensional, $\mathbf{V}$ and ρ are uniform over the area and the integrated result is

$$\dot{\mathbf{M}}_{\text{sec } i} = \mathbf{V}_i(\rho_i V_{ni} A_i) = \dot{m}_i \mathbf{V}_i \tag{3.39}$$

for outlet flux and $-\dot{m}_i \mathbf{V}_i$ for inlet flux. Thus if the control volume has only one-dimensional inlets and outlets, Eq. (3.37) reduces to

$$\sum \mathbf{F} = \frac{\partial}{\partial t} \left(\iiint_{\text{CV}} \mathbf{V}\rho \, d\mathcal{V} \right) + \sum (\dot{m}_i \mathbf{V}_i)_{\text{out}} - \sum (\dot{m}_i \mathbf{V}_i)_{\text{in}} \tag{3.40}$$

This is a commonly used approximation in engineering analyses. It is crucial to realize that we are dealing with vector sums. Equation (3.40) states that the net vector force on a fixed control volume equals the rate of change of vector momentum within the control volume plus the vector sum of outlet momentum fluxes minus the vector sum of inlet fluxes.

Net Pressure Force on a Closed Control Surface

Generally speaking, the surface forces on a control volume are due to (1) forces exposed by cutting through solid bodies which protrude through the surface and (2) forces due to pressure and viscous stresses of the surrounding fluid. The computation of pressure force is relatively simple, as shown in Fig. 3.7. Recall from Chap. 2 that the external pressure force on a surface is normal to the surface and *inward*. Since the unit vector $\mathbf{n}$ is defined as *outward*, one way to write the pressure force is

$$\mathbf{F}_{\text{press}} = \iint_{\text{CS}} p(-\mathbf{n}) \, dA \tag{3.41}$$

Fig. 3.7 Pressure-force computation by subtracting out a uniform distribution: (*a*) uniform pressure, $\mathbf{F} = -p_a \iint \mathbf{n} \, dA \equiv 0$; (*b*) nonuniform pressure, $\mathbf{F} = -\iint (p - p_a)\mathbf{n} \, dA$.

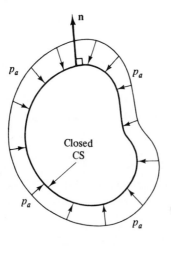

(*a*)

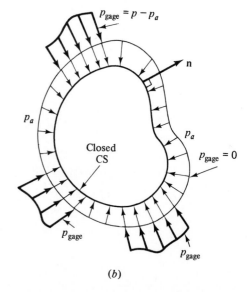

(*b*)

Now if the pressure has a uniform value p_a all around the surface, as in Fig. 3.7a, the net pressure force is zero[1]

$$\mathbf{F}_{\text{UP}} = \iint p_a(-\mathbf{n})\, dA = -p_a \iint \mathbf{n}\, dA \equiv 0 \qquad (3.42)$$

where the subscript UP stands for uniform pressure. This result is *independent of the shape of the surface*[1] as long as the surface is closed, and all our control volumes are closed. Thus a seemingly complicated pressure-force problem can be simplified by subtracting out any convenient uniform pressure p_a and working only with the pieces of gage pressure which remain, as illustrated in Fig. 3.7b. Thus Eq. (3.41) is entirely equivalent to

$$\mathbf{F}_{\text{press}} = \iint_{\text{CS}} (p - p_a)(-\mathbf{n})\, dA = \iint_{\text{CS}} p_{\text{gage}}(-\mathbf{n})\, dA \qquad (3.43)$$

This trick can mean quite a saving in computation.

EXAMPLE 3.7   A control volume of a nozzle section has surface pressures of 40 lbf/in² absolute at section 1 and atmospheric pressure of 15 lbf/in² absolute at section 2 and on the external rounded part of the nozzle, as in Fig. E3.7a. Compute the net pressure force if $D_1 = 3$ in and $D_2 = 1$ in.

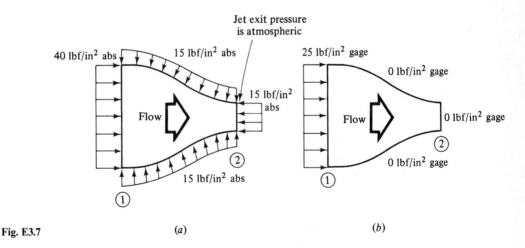

Fig. E3.7   (a)   (b)

Solution   We do not have to bother with the outer surface if we subtract 15 lbf/in² from all surfaces. This leaves 25 lbf/in² gage at section 1 and 0 lbf/in² gage everywhere else, as in Fig. E3.7b. Then the net pressure force is computed from section 1 only

$$\mathbf{F} = p_{g1}(-\mathbf{n})_1 A_1 = (25 \text{ lbf/in}^2)\frac{\pi}{4}(3 \text{ in})^2\mathbf{i} = 177\mathbf{i} \text{ lbf} \qquad Ans.$$

[1] Can you prove this? It is a consequence of Gauss' theorem from vector analysis.

Notice that we did not change inches to feet in this case because with pressure in pounds force per square inch and area in square inches the product gives force directly in pounds. More often, though, the change back to standard units is necessary and desirable. *Note*: This problem computes pressure force only. There are probably other forces involved in Fig. E3.7, e.g., nozzle-wall stresses in the cuts through sections 1 and 2 and the weight of the fluid within the control volume.

Pressure Condition at a Jet Exit

Figure E3.7 illustrates a pressure boundary condition commonly used for jet exit-flow problems. When a fluid flow leaves a confined internal duct and exits into an ambient "atmosphere," its free surface is exposed to that atmosphere. Therefore the jet itself will essentially be at atmospheric pressure also. This condition was used at section 2 in Fig. E3.7.

Only two effects could maintain a pressure difference between the atmosphere and a free exit jet. The first is surface tension, Eq. (1.49), which is usually negligible. The second effect is a *supersonic* jet, which can separate itself from an atmosphere with expansion or compression waves (Chap. 9). For the majority of applications, therefore, we shall set the pressure in an exit jet as atmospheric.

EXAMPLE 3.8   A fixed control volume of a streamtube in steady flow has a uniform inlet flow (ρ_1, A_1, V_1) and a uniform exit flow (ρ_2, A_2, V_2), as shown in Fig. 3.8. Find an expression for the net force on the control volume.

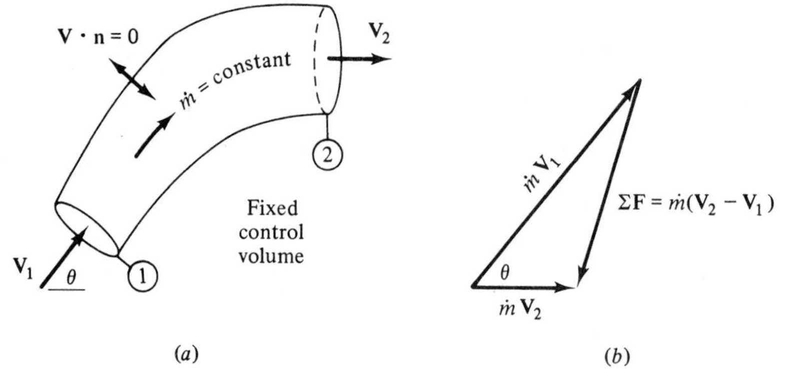

Fig. 3.8   Net force on a one-dimensional streamtube in steady flow: (a) streamtube in steady flow; (b) vector diagram for computing net force.

Solution   Equation (3.40) applies with one inlet and exit

$$\sum \mathbf{F} = \dot{m}_2 \mathbf{V}_2 - \dot{m}_1 \mathbf{V}_1 = (\rho_2 A_2 V_2)\mathbf{V}_2 - (\rho_1 A_1 V_1)\mathbf{V}_1$$

The volume-integral term vanishes for steady flow, but from conservation of mass in Example 3.3 we saw that

$$\dot{m}_1 = \dot{m}_2 = \dot{m} = \text{const}$$

Therefore a simple form for the desired result is

$$\sum \mathbf{F} = \dot{m}(\mathbf{V}_2 - \mathbf{V}_1) \qquad\qquad Ans.$$

This is a *vector* relation and is sketched in Fig. 3.8*b*. The term $\sum \mathbf{F}$ represents the net force acting on the control volume due to all causes; it is needed to balance the change in momentum of the fluid as it turns and decelerates while passing through the control volume.

EXAMPLE 3.9   As shown in Fig. 3.9*a*, a fixed vane turns a water jet of area A through an angle θ without changing its velocity magnitude. The flow is steady, pressure is p_a everywhere, and friction on the vane is negligible. (*a*) Find the components F_x and F_y of the applied vane force. (*b*) Find expressions for the force magnitude F and the angle ϕ between F and the horizontal. Plot versus θ.

Solution

Part (a)   The control volume selected in Fig. 3.9*a* cuts through the inlet and exit of the jet and through the vane support, exposing the vane force **F**. Since there is no cut along the vane-jet interface,

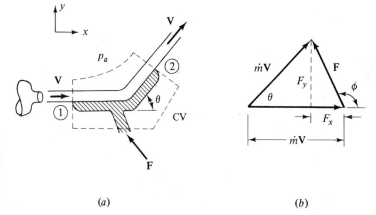

Fig. 3.9   Net applied force on a fixed jet-turning vane: (*a*) geometry of the vane turning the water jet; (*b*) vector diagram for the net force.

(*a*)   (*b*)

vane friction is internally self-canceling. The pressure force is zero in the uniform atmosphere. We neglect the weight of fluid and the vane weight within the control volume. Then Eq. (3.40) reduces to

$$\mathbf{F}_{\text{vane}} = \dot{m}_2 \mathbf{V}_2 - \dot{m}_1 \mathbf{V}_1$$

But the magnitude $V_1 = V_2 = V$ as given, and conservation of mass for the streamtube requires $\dot{m}_1 = \dot{m}_2 = \dot{m} = \rho A V$. The vector diagram for force and momentum change becomes an isosceles triangle with legs $\dot{m}V$ and base **F**, as in Fig. 3.9*b*. We can readily find the force components from this diagram

$$F_x = \dot{m}V(\cos \theta - 1) \qquad F_y = \dot{m}V \sin \theta \qquad \textit{Ans. (a)}$$

where $\dot{m}V = \rho A V^2$ for this case. This is the desired result.

Part (b) ·   The force magnitude is obtained from part (*a*)

$$F = (F_x^2 + F_y^2)^{1/2} = \dot{m}V[\sin^2 \theta + (\cos \theta - 1)^2]^{1/2} = 2\dot{m}V \sin \frac{\theta}{2} \qquad \textit{Ans. (b)}$$

From the geometry of Fig. 3.9*b* we obtain

$$\phi = 180° - \tan^{-1} \frac{F_y}{F_x} = 90° + \frac{\theta}{2} \qquad \textit{Ans. (b)}$$

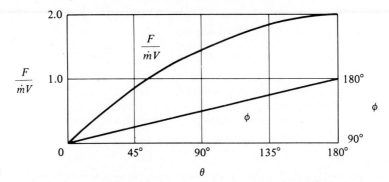

Fig. E3.9

These can be plotted versus θ as shown in Fig. E3.9. Two special cases are of interest. First, the maximum force occurs at $\theta = 180°$, that is, when the jet is turned around and thrown back in the opposite direction with its momentum completely reversed. This force is $2\dot{m}V$ and acts to the *left*; that is, $\phi = 180°$. Second, at very small turning angles ($\theta < 10°$) we obtain approximately

$$F \approx \dot{m}V\theta \qquad \phi \approx 90°$$

The force is linearly proportional to the turning angle and acts nearly normal to the jet. This is the principle of a lifting vane, or airfoil, which causes a slight change in the oncoming flow direction and thereby creates a lift force normal to the basic flow.

EXAMPLE 3.10 A water jet of velocity V_j impinges normal to a flat plate which moves to the right at velocity V_c, as shown in Fig. 3.10a. Find the force required to keep the plate moving at constant velocity if the jet density is 1000 kg/m^3, the jet area is 3 cm^2, and V_j and V_c are 20 and 15 m/s, respectively. Neglect the weight of the jet and plate and assume steady flow with respect to the moving plate with the jet splitting into an equal upward and downward half-jet.

Solution The suggested control volume in Fig. 3.10a cuts through the plate support to expose the desired forces R_x and R_y. This control volume moves at speed V_c and thus is fixed relative to the plate, as in Fig. 3.10b. We must satisfy both mass and momentum conservation for the

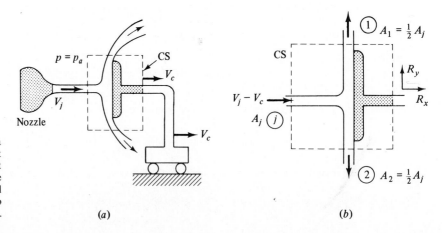

Fig. 3.10 Force on a plate moving at constant velocity: (a) jet striking a moving plate normally; (b) control volume fixed relative to the plate.

(a)

(b)

assumed steady-flow pattern in Fig. 3.10b. There are two outlets and one inlet, and Eq. (3.30) applies for mass conservation

$$\dot{m}_{out} = \dot{m}_{in}$$

or

$$\rho_1 A_1 V_1 + \rho_2 A_2 V_2 = \rho_j A_j (V_j - V_c) \tag{1}$$

We assume that the water is incompressible, $\rho_1 = \rho_2 = \rho_j$, and we are given that $A_1 = A_2 = \frac{1}{2} A_j$. Therefore Eq. (1) reduces to

$$V_1 + V_2 = 2(V_j - V_c) \tag{2}$$

Strictly speaking, this is all that mass conservation tells us. However, from the symmetry of the jet deflection and the neglect of fluid weight, we conclude that the two velocities V_1 and V_2 must be equal, and hence (2) becomes

$$V_1 = V_2 = V_j - V_c \tag{3}$$

For the given numerical values, we have

$$V_1 = V_2 = 20 - 15 = 5 \text{ m/s}$$

Now we can compute R_x and R_y from the two components of momentum conservation. Equation (3.40) applies with the unsteady term zero

$$\sum F_x = R_x = \dot{m}_1 u_1 + \dot{m}_2 u_2 - \dot{m}_j u_j \tag{4}$$

where from the mass analysis, $\dot{m}_1 = \dot{m}_2 = \frac{1}{2}\dot{m}_j = \frac{1}{2}\rho_j A_j(V_j - V_c)$. Now check the flow directions at each section: $u_1 = u_2 = 0$, and $u_j = V_j - V_c = 5$ m/s. Thus Eq. (4) becomes

$$R_x = -\dot{m}_j u_j = -[\rho_j A_j(V_j - V_c)](V_j - V_c) \tag{5}$$

For the given numerical values we have

$$R_x = -(1000 \text{ kg/m}^3)(0.0003 \text{ m}^2)(5 \text{ m/s})^2 = -7.5 \text{ (kg·m)/s}^2 = -7.5 \text{ N} \quad Ans. (a)$$

This acts to the *left*; i.e., it requires a restraining force to keep the plate from accelerating to the right due to the continuous impact of the jet. The vertical force is

$$F_y = R_y = \dot{m}_1 v_1 + \dot{m}_2 v_2 - \dot{m}_j v_j$$

Check directions again: $v_1 = V_1, v_2 = -V_2, v_j = 0$. Thus

$$R_y = \dot{m}_1(V_1) + \dot{m}_2(-V_2) = \frac{1}{2}\dot{m}_j(V_1 - V_2) \tag{6}$$

But since we found earlier that $V_1 = V_2$, this means that $R_y = 0$, as we could expect from the symmetry of the jet deflection.[1] Two other results are of interest. First, the relative velocity at section 1 was found to be 5 m/s up, from Eq. (3). If we convert this to absolute motion by adding on the control-volume speed $V_c = 15$ m/s to the right, we find that the absolute velocity $\mathbf{V}_1 = 15\mathbf{i} + 5\mathbf{j}$ m/s, or 15.8 m/s at an angle of 18.4° upward, as indicated in Fig. 3.10a. Thus the absolute jet speed changes after hitting the plate. Second, the computed force R_x does not change if we assume the jet deflects in all radial directions along the plate surface rather than just up and down. Since the plate is normal to the x axis, there would still be zero outlet x-momentum flux when Eq. (4) was rewritten for a radial-deflection condition.

---

[1] Symmetry can be a powerful tool if used properly. Try to learn more about the uses and misuses of symmetry conditions. Here we doggedly computed the results without invoking symmetry.

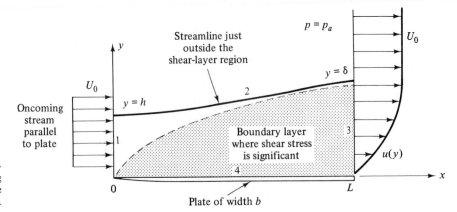

Fig. 3.11 Control-
volume analysis of drag
force on a flat plate due
to boundary shear.

EXAMPLE 3.11   The previous example treated a plate at normal incidence to an oncoming flow. In Fig. 3.11
the plate is parallel to the flow. The stream is not a jet but a broad river, or *free stream*, of
uniform velocity $\mathbf{V} = U_0\mathbf{i}$. The pressure is assumed uniform, and so it has no net force on the
plate. The plate does not block the flow as in Fig. 3.10, so that the only effect is due to
boundary shear, which was neglected in the previous example. The no-slip condition at the
wall brings the fluid there to a halt, and these slowly moving particles retard their neighbors
above, so that at the end of the plate there is a significant retarded shear layer, or *boundary
layer*, of thickness $y = \delta$. The viscous stresses along the wall can sum to a finite drag force on
the plate. These effects are illustrated in Fig. 3.11. The problem is to make an integral analysis and find the drag force D in terms of the flow properties ρ, U_0, and δ and the plate
dimensions L and b.[1]

Solution   Like most practical cases, this problem requires a combined mass and momentum balance.
A proper selection of control volume is essential, and we select the four-sided region from O
to h to δ to L and back to the origin O, as shown in Fig. 3.11. Had we chosen to cut across
horizontally from left to right along the height $y = h$, we would have cut through the shear
layer and exposed unknown shear stresses. Instead we follow the streamline passing
through $(x, y) = (0, h)$, which is outside the shear layer and also has no mass flow across it.
The four control-volume sides are thus

1. From $(0, 0)$ to $(0, h)$: a one-dimensional inlet, $\mathbf{V} \cdot \mathbf{n} = -U_0$
2. From $(0, h)$ to (L, δ): a streamline, no shear, $\mathbf{V} \cdot \mathbf{n} \equiv 0$
3. From (L, δ) to $(L, 0)$: a two-dimensional outlet, $\mathbf{V} \cdot \mathbf{n} = +u(y)$
4. From $(L, 0)$ to $(0, 0)$: a streamline just above the plate surface, $\mathbf{V} \cdot \mathbf{n} = 0$, shear forces
   summing to the drag force $-D\mathbf{i}$ acting from the plate onto the retarded fluid

The pressure is uniform, and so there is no net pressure force. Since the flow is assumed
incompressible and steady, Eq. (3.37) applies with no unsteady term and fluxes only across
sections 1 and 3:

$$\sum F_x = -D = \rho \iint_1 u(\mathbf{V} \cdot \mathbf{n}) \, dA + \rho \iint_3 u(\mathbf{V} \cdot \mathbf{n}) \, dA$$

$$= \rho \int_0^h U_0(-U_0)b \, dy + \rho \int_0^\delta u(+u)b \, dy$$

[1] The general analysis of such wall-shear problems, called *boundary-layer theory*, is treated in Sec. 7.3.

Evaluating the first integral and rearranging gives

$$D = \rho U_0^2 bh - \rho b \int_0^\delta u^2 \, dy \qquad (1)$$

This could be considered the answer to the problem, but it is not useful because the height h is not known with respect to the shear-layer thickness δ. This is found by applying mass conservation, since the control volume forms a streamtube

$$\rho \iint_{CS} (\mathbf{V} \cdot \mathbf{n}) \, dA = 0 = \rho \int_0^h (-U_0) b \, dy + \rho \int_0^\delta ub \, dy$$

or

$$U_0 h = \int_0^\delta u \, dy \qquad (2)$$

after canceling b and ρ and evaluating the first integral. Introduce this value of h into Eq. (1) for a much cleaner result

$$D = \rho b \int_0^\delta u(U_0 - u) \, dy \Big|_{x=L} \qquad (3) \quad Ans.$$

This result was first derived by Theodore von Kármán in 1921.[1] It relates the friction drag on one side of a flat plate to the integral of the *momentum defect* $u(U_0 - u)$ across the trailing cross section of the flow past the plate. Since $U_0 - u$ vanishes as y increases, the integral has a finite value. Equation (3) is an example of *momentum-integral theory* for boundary layers, which is treated in Chap. 7. To illustrate the magnitude of this drag force, we can use a simple approximation for the outlet-velocity profile $u(y)$ which simulates low-speed, or *laminar*, shear flow

$$u \approx U_0 \left(\frac{2y}{\delta} - \frac{y^2}{\delta^2} \right) \qquad \text{for } 0 \le y \le \delta \qquad (4)$$

Substituting into Eq. (3) and letting $\eta = y/\delta$ for convenience, we obtain

$$D = \rho b U_0^2 \delta \int_0^1 (2\eta - \eta^2)(1 - 2\eta + \eta^2) \, d\eta = \tfrac{2}{15} \rho U_0^2 b \delta \qquad (5)$$

This is within 1 percent of the accepted result from laminar boundary-layer theory (Chap. 7) in spite of the crudeness of the Eq. (4) approximation. This is a happy situation and has led to the wide use of Kármán's integral theory in the analysis of viscous flows. Note that D increases with the shear-layer thickness δ, which itself increases with plate length and the viscosity of the fluid (see Sec. 7.4).

Momentum-Flux Correction Factor

For flow in a duct, the axial velocity is usually nonuniform, as in Example 3.4. For this case the simple momentum-flux calculation $\int u\rho(\mathbf{V} \cdot \mathbf{n}) \, dA = \dot{m}V = \rho AV^2$ is somewhat in error and should be corrected to $\beta\rho AV^2$, where β is the dimensionless momentum-flux correction factor, $\beta \ge 1$.

---

[1] The autobiography of this great twentieth-century engineer and teacher [2] is recommended for its historical and scientific insight.

The factor β accounts for the variation of u^2 across the duct section. That is, we compute the exact flux and set it equal to a flux based on average velocity in the duct

$$\rho \int u^2 \, dA = \beta \dot{m} V_{av} = \beta \rho A V_{av}^2$$

or

$$\beta = \frac{1}{A} \int \left(\frac{u}{V_{av}} \right)^2 dA \qquad (3.43a)$$

Values of β can be computed based on typical duct velocity profiles similar to those in Example 3.4. The results are:

Laminar flow:

$$u = U_0 \left(1 - \frac{r^2}{R^2} \right) \qquad \beta = \tfrac{4}{3} \qquad (3.43b)$$

Turbulent flow:

$$u \approx U_0 \left(1 - \frac{r}{R} \right)^m \qquad \tfrac{1}{9} \leq m \leq \tfrac{1}{5}$$

$$\beta = \frac{(1 + m)^2 (2 + m)^2}{2(1 + 2m)(2 + 2m)} \qquad (3.43c)$$

The turbulent correction factors have the following range of values:

m	$\tfrac{1}{5}$	$\tfrac{1}{6}$	$\tfrac{1}{7}$	$\tfrac{1}{8}$	$\tfrac{1}{9}$
β	1.037	1.027	1.020	1.016	1.013

Turbulent flow:

These are so close to unity that they are normally neglected. The laminar correction may sometimes be important.

To illustrate a typical use of these correction factors, the solution to Example 3.8 for nonuniform velocities at sections 1 and 2 would be given as

$$\sum \mathbf{F} = \dot{m}(\beta_2 \mathbf{V}_2 - \beta_1 \mathbf{V}_1) \qquad (3.43d)$$

Note that the basic parameters and vector character of the result are not changed at all by this correction.

Noninertial Reference Frame[1]

All previous derivations and examples in this section have assumed that the coordinate system is inertial, i.e., at rest or moving at constant velocity. In this case the rate of change of velocity equals the absolute acceleration of the system, and Newton's law applies directly in the form of Eqs. (3.2) and (3.35).

---

[1] This section may be omitted without loss of continuity.

In many cases it is convenient to use a *noninertial*, or accelerating, coordinate system. An example would be coordinates fixed to a rocket during takeoff. A second example is any flow on the earth's surface, which is accelerating relative to the fixed stars because of the rotation of the earth. Atmospheric and oceanographic flows experience the so-called *Coriolis acceleration*, outlined below. It is typically less than $10^{-5}g$, where g is the acceleration of gravity, but its accumulated effect over distances of many kilometers can be dominant in geophysical flows. By contrast, the Coriolis acceleration is negligible in small-scale problems like pipe or airfoil flows.

Suppose that the fluid flow has velocity $\mathbf{V}$ relative to a noninertial xyz coordinate system, as shown in Fig. 3.12. Then $d\mathbf{V}/dt$ will represent a noninertial acceleration which must be added vectorially to a relative acceleration $\mathbf{a}_{rel}$ to give the absolute acceleration $\mathbf{a}_i$ relative to some inertial coordinate system XYZ, as in Fig. 3.12. Thus

$$\mathbf{a}_i = \frac{d\mathbf{V}}{dt} + \mathbf{a}_{rel} \tag{3.44}$$

Since Newton's law applies to the absolute acceleration,

$$\sum \mathbf{F} = m\mathbf{a}_i = m\left(\frac{d\mathbf{V}}{dt} + \mathbf{a}_{rel}\right)$$

or

$$\sum \mathbf{F} - m\mathbf{a}_{rel} = m\frac{d\mathbf{V}}{dt} \tag{3.45}$$

Thus Newton's law in noninertial coordinates xyz is equivalent to adding additional "force" terms $-m\mathbf{a}_{rel}$ to account for noninertial effects. In the most general case, sketched in Fig. 3.12, the term $\mathbf{a}_{rel}$ contains four parts, three of which account

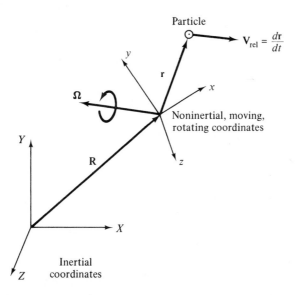

Fig. 3.12 Geometry of fixed versus accelerating coordinates.

for the angular velocity $\mathbf{\Omega}(t)$ of the noninertial coordinates. By inspection of Fig. 3.12, the absolute displacement of a particle is

$$\mathbf{S}_i = \mathbf{r} + \mathbf{R} \tag{3.46}$$

Differentiation gives the absolute velocity

$$\mathbf{V}_i = \mathbf{V} + \frac{d\mathbf{R}}{dt} + \mathbf{\Omega} \times \mathbf{r} \tag{3.47}$$

A second differentiation gives the absolute acceleration:

$$\mathbf{a}_i = \frac{d\mathbf{V}}{dt} + \frac{d^2\mathbf{R}}{dt^2} + \frac{d\mathbf{\Omega}}{dt} \times \mathbf{r} + 2\mathbf{\Omega} \times \mathbf{V} + \mathbf{\Omega} \times (\mathbf{\Omega} \times \mathbf{r}) \tag{3.48}$$

By comparison with Eq. (3.44), we see that the last four terms on the right represent the additional relative acceleration:

1. $d^2\mathbf{R}/dt^2$ is the acceleration of the noninertial origin of the coordinates xyz.
2. $(d\mathbf{\Omega}/dt) \times \mathbf{r}$ is the angular-acceleration effect.
3. $2\mathbf{\Omega} \times \mathbf{V}$ is the Coriolis acceleration.
4. $\mathbf{\Omega} \times (\mathbf{\Omega} \times \mathbf{r})$ is the centripetal acceleration, directed from the particle normal to the axis of rotation with magnitude $\Omega^2 L$, where L is the normal distance to the axis.[1]

Equation (3.45) differs from Eq. (3.2) only in the added inertia forces on the left-hand side. Thus the control-volume formulation of linear momentum in noninertial coordinates merely adds inertia terms by integrating the added relative acceleration over each differential mass in the control volume

$$\sum \mathbf{F} - \iiint_{\mathrm{CV}} \mathbf{a}_{\mathrm{rel}}\, dm = \frac{d}{dt}\left(\iiint_{\mathrm{CV}} \mathbf{V}\rho\, d\mathcal{V} \right) + \iint_{\mathrm{CS}} \mathbf{V}\rho(\mathbf{V}_r \cdot \mathbf{n})\, dA \tag{3.49}$$

where $$\mathbf{a}_{\mathrm{rel}} = \frac{d^2\mathbf{R}}{dt^2} + \frac{d\mathbf{\Omega}}{dt} \times \mathbf{r} + 2\mathbf{\Omega} \times \mathbf{V} + \mathbf{\Omega} \times (\mathbf{\Omega} \times \mathbf{r})$$

This is the noninertial equivalent to the inertial form given in Eq. (3.35). In order to analyze such problems one must have knowledge of the displacement $\mathbf{R}$ and angular velocity $\mathbf{\Omega}$ of the noninertial coordinates.

If the control volume is nondeformable, Eq. (3.49) reduces to

$$\sum \mathbf{F} - \iiint_{\mathrm{CV}} \mathbf{a}_{\mathrm{rel}}\, dm = \frac{\partial}{\partial t}\left(\iiint_{\mathrm{CV}} \mathbf{V}\rho\, d\mathcal{V} \right) + \iint_{\mathrm{CS}} \mathbf{V}\rho(\mathbf{V} \cdot \mathbf{n})\, dA \tag{3.50}$$

In other words, the right-hand side reduces to that of Eq. (3.37).

---

[1] A complete discussion of these noninertial coordinate terms is given, for example, in Ref. 4, pp. 49–51.

EXAMPLE 3.12   Repeat Example 3.10 with the plate and its cart unrestrained horizontally and thus allowed to accelerate to the right. Derive (*a*) the equation of motion for cart velocity $V_c(t)$ and (*b*) the time required for the cart to accelerate from rest to 95 percent of the jet velocity and (*c*) compute actual numerical values from (*b*) for the conditions of Example 3.10 and a cart mass of 3 kg. Neglect cartwheel friction.

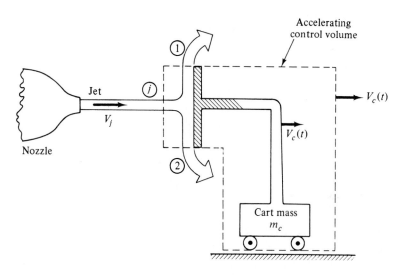

Fig. E3.12

Solution

Part (a)   The control volume is shown in Fig. E3.12 to include the cart and its wheel. The control volume moves at speed $V_c(t)$. The flow within the control volume is unsteady because the inlet velocity $V_j - V_c$ varies with time. The mass-conservation relation is Eq. (3.22)

$$\frac{\partial}{\partial t}\left(\iiint_{\text{CV}} \rho \, d\mathcal{V}\right) + \rho_1 A_1 V_1 + \rho_2 A_2 V_2 - \rho_j A_j (V_j - V_c) = 0 \qquad (1)$$

Again assume incompressible flow so that the density cancels out. The triple integral is the mass within the control volume, which remains constant if the jet cross section is constant. Therefore we conclude that the time derivative of the triple integral is zero, and the previous result holds

$$V_1 = V_2 = V_j - V_c \qquad (2)$$

These speeds V_1 and V_2 are relative to the moving cart. The momentum relation (3.50) is now applied to the *x* direction with the coordinates *xyz* fixed to the accelerating cart

$$\sum F_x - \iiint_{\text{CV}} a_{x,\,\text{rel}} \, dm = \frac{\partial}{\partial t}\left(\iiint_{\text{CV}} u\rho \, d\mathcal{V}\right) + \iint_{\text{CS}} u\rho(\mathbf{V} \cdot \mathbf{n}) \, dA$$

Now the cart is unrestrained, and the pressure is constant; hence $\sum F_x = 0$. The relative acceleration of the control volume is none other than the rate of change of cart speed since the cart is not rotating relative to inertial coordinates

$$a_{x,\,\text{rel}} = \frac{d}{dt}(V_c) \qquad (3)$$

The area integrals on the right-hand side are the same as found in Example 3.10

$$\iint_{CS} u\rho(\mathbf{V} \cdot \mathbf{n}) \, dA = -\dot{m}_j u_j = -\rho_j A_j (V_j - V_c)^2 \tag{4}$$

with only the jet inlet flux having nonzero u velocity. Finally, evaluation of the two remaining (triple) integrals requires an assumption about the mass and x momentum of the fluid within the control volume. Since these are probably small with respect to the *cart* mass and momentum, we apply a little art: we neglect the contribution of fluid mass to the volume integrals,[1] so that the integrals are, approximately,

$$\iiint_{CV} a_{x,\,\text{rel}} \, dm \approx m_c \frac{dV_c}{dt} \qquad \frac{\partial}{\partial t}\left(\iiint_{CV} u\rho \, d\mathcal{V}\right) \approx 0$$

where m_c is the mass of the cart and plate. Substituting back gives

$$0 - m_c \frac{dV_c}{dt} = 0 - \rho A_j (V_j - V_c)^2$$

or

$$\frac{dV_c}{dt} = K(V_j - V_c)^2 \qquad K = \frac{\rho A_j}{m_c} \qquad \textit{Ans. (a)} \quad (5)$$

This is the desired differential equation for the cart speed.

Part (b) We can find the time to reach 95 percent of the jet speed by integrating Eq. (5), assuming that K and V_j are constant. Separate the variables and use a lower limit starting from rest, that is, $V_c = 0$ at $t = 0$

$$\int_0^{V_c} (V_j - V_c)^{-2} \, dV_c = \int_0^t K \, dt$$

Evaluate the integrals to get

$$\frac{1}{V_j - V_c} - \frac{1}{V_j} = Kt$$

or

$$\frac{V_c}{V_j} = \frac{V_j Kt}{1 + V_j Kt} \tag{6}$$

This is the general solution for the cart speed $V_c(t)$ when starting from rest. The time t^* when $V_c/V_j = 0.95$ is thus given by

$$0.95 = \frac{V_j Kt^*}{1 + V_j Kt^*}$$

or

$$t^* = \frac{19}{KV_j} = \frac{19 m_c}{\rho A_j V_j} \qquad \textit{Ans. (b)}$$

Part (c) The constant $1/KV_j$ is an indication of the response time of the cart when struck by the jet. For the given conditions,

$$t^* = \frac{19(3 \text{ kg})}{(1000 \text{ kg/m}^3)(0.0003 \text{ m}^2)(20 \text{ m/s})} = 9.5 \text{ s} \qquad \textit{Ans. (c)}$$

[1] Have your instructor assign you to analyze the conditions under which these assumptions are valid.

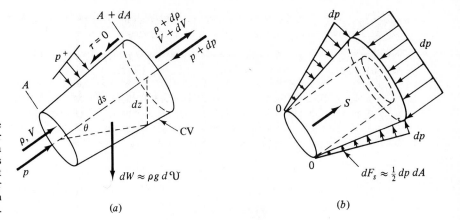

Fig. 3.13 The Bernoulli equation for frictionless flow along a streamline: (*a*) forces and fluxes; (*b*) net pressure force after uniform subtraction of *p*.

3.5 THE BERNOULLI EQUATION

A widely used relation between pressure, velocity, and elevation, dating back to Daniel Bernoulli and Leonhard Euler in the eighteenth century, can be derived by considering frictionless flow through an infinitesimal streamtube, as shown in Fig. 3.13*a*. The area A is small enough for the properties ρ, V, and p to be considered uniform over the section. All properties ρ, V, p, and A gradually change in the streamwise direction s. The streamtube is slanted at some arbitrary angle θ, so that the elevation change between sections is $dz = ds \sin \theta$. The figure shows the inevitable retarding friction on the walls of the streamtube, which we are neglecting.[1]

The conservation-of-mass requirement for this elemental control volume is

$$\frac{\partial}{\partial t} \left(\iiint_{\text{CV}} \rho \, d\mathcal{V} \right) + \dot{m}_{\text{out}} - \dot{m}_{\text{in}} = 0$$

or

$$\frac{\partial \rho}{\partial t} \, d\mathcal{V} + d\dot{m} = 0 \tag{3.51}$$

The differential volume of the element is

$$d\mathcal{V} \approx (A + \tfrac{1}{2} dA) \, ds \approx A \, ds \tag{3.52}$$

Equation (3.51) thus relates mass flux to density change

$$d\dot{m} = d(\rho A V) = -\frac{\partial \rho}{\partial t} A \, ds \tag{3.53}$$

Now write conservation of momentum in the streamwise direction

$$\sum F_s = \frac{\partial}{\partial t} \left(\iiint_{\text{CV}} V_s \, dm \right) + \iint_{\text{CS}} V_s \, d\dot{m} \tag{3.54}$$

[1] More subtly, the derivation also neglects heat and shaft work addition to the fluid.

The force terms are due to pressure and gravity. The gravity term is the streamwise weight component, which is negative

$$dF_{s,\,grav} = -dW \sin\theta \approx -\rho gA \, ds \sin\theta \approx -\rho gA \, dz \qquad (3.55)$$

The pressure force is more easily visualized by first subtracting a uniform value p from all surfaces, remembering from Fig. 3.7 that this causes no change in the net force. The result is shown in Fig. 3.13b. The pressure along the slanted tube walls has a streamwise component which acts not on A itself but on the outer ring of area increase, dA. The net force is, approximately,

$$dF_{s,\,press} \approx \tfrac{1}{2}dp \, dA - dp(A + dA) \approx -A \, dp \qquad (3.56)$$

to first order. The integral terms on the right-hand side of Eq. (3.54) are

$$\frac{\partial}{\partial t}\left(\iiint_{CV} V_s \, dm \right) \approx \frac{\partial}{\partial t}(\rho V_s A \, ds) \approx \frac{\partial}{\partial t}(\rho V)A \, ds$$

$$\iint_{CS} V_s \, d\dot{m} \approx (V + dV)(\dot{m} + d\dot{m}) - V\dot{m} \approx \dot{m} \, dV + V \, d\dot{m} \qquad (3.57)$$

again neglecting second-order differentials. All terms in Eq. (3.54) can now be evaluated

$$-A \, dp - \rho gA \, dz = \frac{\partial}{\partial t}(\rho V)A \, ds + \dot{m} \, dV + V \, d\dot{m} \qquad (3.58)$$

A nice cancellation occurs if we split up the derivative term

$$\frac{\partial}{\partial t}(\rho V) = \rho \frac{\partial V}{\partial t} + V \frac{\partial \rho}{\partial t} = \rho \frac{\partial V}{\partial t} + V \frac{-d\dot{m}}{A \, ds} \qquad (3.59)$$

where the last substitution is from continuity, Eq. (3.53). Now combine Eqs. (3.58) and (3.59)

$$-A \, dp - \rho gA \, dz = \frac{\partial V}{\partial t}\rho A \, ds + \rho AV \, dV \qquad (3.60)$$

divide by ρA, and rearrange

$$\frac{\partial V}{\partial t} ds + \frac{dp}{\rho} + V \, dV + g \, dz = 0 \qquad (3.61)$$

This is Bernoulli's equation for *unsteady frictionless flow along a streamline*. It is in differential form and can be integrated between any two points 1 and 2 on the streamline

$$\int_1^2 \frac{\partial V}{\partial t} ds + \int_1^2 \frac{dp}{\rho} + \tfrac{1}{2}(V_2^2 - V_1^2) + g(z_2 - z_1) = 0 \qquad (3.62)$$

To evaluate the two remaining integrals one must estimate the unsteady effect $\partial V/\partial t$ and the variation of density with pressure. At this time we consider only steady ($\partial V/\partial t = 0$) incompressible (constant-density) flow, for which Eq. (3.62) becomes

$$\frac{p_2 - p_1}{\rho} + \tfrac{1}{2}(V_2^2 - V_1^2) + g(z_2 - z_1) = 0$$

or

$$\frac{p_1}{\rho} + \tfrac{1}{2}V_1^2 + gz_1 = \frac{p_2}{\rho} + \tfrac{1}{2}V_2^2 + gz_2 = \text{const} \qquad (3.63)$$

This is the Bernoulli equation for steady frictionless incompressible flow along a streamline.

There are two additional rather subtle restrictions on Eq. (3.63):

> Equation (3.63) is not valid in a region where heat or work transfer occurs in a fluid, and usually the Bernoulli constant *changes* downstream of such a region.

The basic reason for this restriction is that heat and work are married to frictional effects which invalidate our assumption of frictionless flow. This becomes clearer in Sec. 3.7, when we write the first law (energy equation) in control-volume form and compare it with Bernoulli's equation.

As we shall see in Sec. 3.7, when the energy equation is written for the special case of steady flow through a streamtube with shaft work and friction, it takes the form

$$\frac{p_1}{\rho} + \tfrac{1}{2}V_1^2 + gz_1 = \left(\frac{p_2}{\rho} + \tfrac{1}{2}V_2^2 + gz_2\right) + w_s + w_f \qquad (3.64a)$$

or

$$\frac{p_1}{\rho g} + \frac{V_1^2}{2g} + z_1 = \left(\frac{p_2}{\rho g} + \frac{V_2^2}{2g} + z_2\right) + h_s + h_f \qquad (3.64b)$$

In Eq. (3.64a) all terms are energies per unit mass: w_s is the shaft work per unit mass done by the fluid (positive for a turbine, negative for a pump), and w_f is the loss of energy per unit mass due to friction between sections 1 and 2.

In the alternate form (3.46b) all terms are heads, or lengths, actually energies per unit weight = (ft · lbf)/lbf = ft. (In the SI system joules per newton equal meters.) The term $h_s = w_s/g$ is the work head done by the fluid, and $h_f = w_f/g$ is the friction head loss between 1 and 2. We see that the Bernoulli equation (3.63) is much more restrictive than the energy equation (3.64) because it contains only the "mechanical" terms of pressure work, kinetic energy, and potential energy which arise from the frictionless momentum-integral relation, Eq. (3.58).

Figure 3.14 illustrates some practical limitations on the use of Bernoulli's equation in the form (3.63). For the wind-tunnel model test of Fig. 3.14a, Eq. (3.63) is valid in the core flow of the tunnel but not in the tunnel-wall boundary layers, the model-surface boundary layers, or the wake of the model, all of which are frictional regions.

In Fig. 3.14b, Bernoulli's equation is valid upstream and downstream of the propeller but with a different constant $h_0 = z + p/\rho g + V^2/2g$ caused by the work addition to the fluid. The Bernoulli equation is invalid near the propeller blades

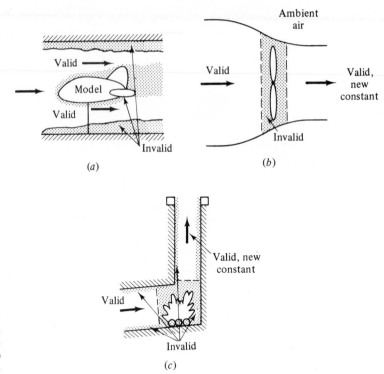

Fig. 3.14 Illustration of regions of validity and invalidity of the Bernoulli equation: (*a*) tunnel model, (*b*) propeller, (*c*) chimney.

and in the helical vortices (not shown) shed downstream of the blade edges. Also, the Bernoulli constants are higher in the flowing slipstream than in the ambient atmosphere because of the slipstream kinetic energy.

In Fig. 3.14*c*, Eq. (3.63) is valid before and after the fire in a chimney but with a change in Bernoulli constant caused by heat addition. The Bernoulli equation is not valid in the fire itself or in the chimney-wall boundary layers.

The moral is to apply Eq. (3.63) only for the very restrictive conditions for which it applies: steady incompressible flow along a streamline with no friction losses, no heat transfer, and no shaft work between sections 1 and 2.

Hydraulic and Energy Grade Lines

A useful visual interpretation of Bernoulli's equation is to sketch two grade lines of a flow. The energy grade line (EGL) shows the height of the total Bernoulli constant $h_0 = z + p/\rho g + V^2/2g$. In frictionless flow with no work or heat transfer, Eq. (3.63), the EGL has constant height. The hydraulic grade line (HGL) shows the height corresponding to elevation and pressure head $z + p/\rho g$, that is, the EGL minus the velocity head $V^2/2g$. The HGL is the height to which liquid would rise in a piezometer tube (see Prob. 2.19) attached to the flow. In an open-channel flow the HGL is identical to the free surface of the water.

Figure 3.15 illustrates the EGL and HGL for frictionless flow at sections 1 and 2 of a duct. The piezometer tubes measure the static-pressure head $z + p/\rho g$ and thus

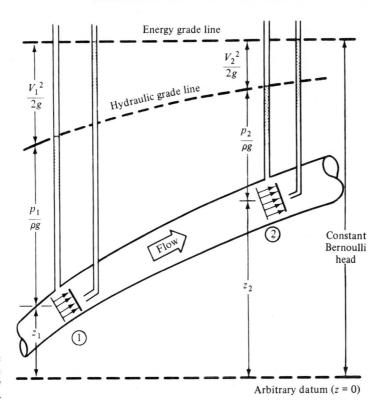

Fig. 3.15 Hydraulic and energy grade lines for frictionless flow in a duct.

outline the HGL. The pitot stagnation-velocity tubes measure the total head $z + p/\rho g + V^2/2g$, which corresponds to the EGL. In this particular case the EGL is constant, and the HGL rises due to a drop in velocity.

In more general flow conditions the EGL will drop slowly due to friction losses and will drop sharply due to a major loss (a valve or obstruction) or due to work extraction (to a turbine). The EGL can rise only if there is work addition (as from a pump or propeller). The HGL generally follows the behavior of the EGL with respect to losses or work transfer and it also rises and/or falls if the velocity decreases and/or increases. These details will be treated in Sec. 3.7.

As mentioned before, no conversion factors are needed in computations with the Bernoulli equation if consistent SI or BG units are used, as the following examples will show.

In all Bernoulli-type problems in this text we shall consistently take point 1 upstream and point 2 downstream.

EXAMPLE 3.13 Find a relation between nozzle discharge velocity V_2 and tank free-surface height h as in Fig. E3.13. Assume steady frictionless flow.

Solution As mentioned, we always choose point 1 upstream and point 2 downstream. Try to choose 1 and 2 where maximum information is known or desired. Here we select 1 as the tank free surface, where elevation and pressure are known, and point 2 as the nozzle exit, where again pressure and elevation are known. The two unknowns are V_1 and V_2.

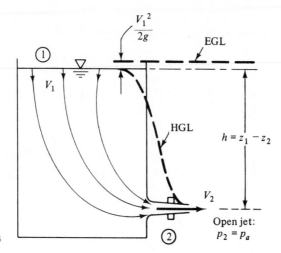

$\dfrac{V_1^2}{2g}$

EGL

V_1

HGL

$h = z_1 - z_2$

V_2

Open jet:
$p_2 = p_a$

Fig. E3.13

Mass conservation is usually a vital part of Bernoulli analyses. If A_1 is the tank cross section and A_2 the nozzle area, this is approximately a one-dimensional flow with constant density, Eq. (3.30),

$$A_1 V_1 = A_2 V_2 \qquad (1)$$

Bernoulli's equation (3.63) gives

$$\frac{p_1}{\rho} + \tfrac{1}{2}V_1^2 + g z_1 = \frac{p_2}{\rho} + \tfrac{1}{2}V_2^2 + g z_2$$

But since sections 1 and 2 are both exposed to atmospheric pressure, $p_1 = p_2 = p_a$, the pressure terms cancel, leaving

$$V_2^2 - V_1^2 = 2g(z_1 - z_2) = 2gh \qquad (2)$$

Eliminating V_1 between Eqs. (1) and (2), we obtain the desired result:

$$V_2^2 = \frac{2gh}{1 - A_2^2/A_1^2} \qquad \text{Ans.} \quad (3)$$

Generally the nozzle area A_2 is very much smaller than the tank area A_1, so that the ratio A_2^2/A_1^2 is doubly negligible, and an accurate approximation for the outlet velocity is

$$V_2 \approx (2gh)^{1/2} \qquad \text{Ans.} \quad (4)$$

This formula, discovered by Evangelista Torricelli in 1644, states that the discharge velocity equals the speed which a frictionless particle would attain if it fell freely from point 1 to 2. In other words, the potential energy of the surface fluid is entirely converted into kinetic energy of efflux, which is consistent with the neglect of friction and the fact that no net pressure work is done. Note that Eq. (4) is independent of the fluid density, a characteristic of gravity-driven flows.

Except for the wall boundary layers, the streamlines from 1 to 2 all behave the same way, and we can assume that the Bernoulli constant h_0 is the same for all the core flow, i.e., that it is irrotational. However, the outlet flow is likely to be nonuniform, not one-dimensional, so that the average velocity is only approximately equal to Torricelli's result. The engineer will then adjust the formula to include a dimensionless *discharge coefficient* c_d

$$(V_2)_{av} = \frac{Q}{A_2} = c_d(2gh)^{1/2} \tag{5}$$

As discussed in Sec. 6.10, the discharge coefficient of a nozzle varies from about 0.6 to 1.0 as a function of (dimensionless) flow conditions and nozzle shape.

Before proceeding with more examples, we should note carefully that a solution by Bernoulli's equation (3.63) does *not* require a control-volume analysis, only a selection of two points 1 and 2 along a given streamline. The control volume was used to derive the differential relation (3.61), but the integrated form (3.63) is valid all along the streamline for frictionless flow with no heat transfer or shaft work and a control volume is not necessary.

EXAMPLE 3.14   Rework Example 3.13 to account, at least approximately, for the unsteady-flow condition caused by the draining of the tank.

Solution   Essentially we are asked to include the unsteady integral term involving $\partial V/\partial t$ from Eq. (3.62). This will result in a new term added to Eq. (2) from Example 3.13:

$$2 \int_1^2 \frac{\partial V}{\partial t}\, ds + V_2^2 - V_1^2 = 2gh \tag{1}$$

Since the flow is incompressible, the continuity equation still retains the simple form $A_1 V_1 = A_2 V_2$ from Example 3.13. To integrate the unsteady term we must estimate the acceleration all along the streamline. Most of the streamline is in the tank region where $\partial V/\partial t \approx dV_1/dt$. The length of the average streamline is slightly longer than the nozzle depth h. A crude estimate for the integral is thus

$$\int_1^2 \frac{\partial V}{\partial t}\, ds \approx \int_1^2 \frac{dV_1}{dt}\, ds \approx \frac{dV_1}{dt} h \tag{2}$$

But since A_1 and A_2 are constant, $dV_1/dt \approx (A_2/A_1)(dV_2/dt)$. Substitution into Eq. (1) gives

$$2h \frac{A_2}{A_1} \frac{dV_2}{dt} + V_2^2 \left(1 - \frac{A_2^2}{A_1^2}\right) \approx 2gh \tag{3}$$

This is a first-order differential equation for $V_2(t)$. It is complicated by the fact that the depth h is variable, $h = h(t)$, as determined by the variation in $V_1(t)$

$$h(t) = h_0 - \int_0^t V_1\, dt \tag{4}$$

Equations (3) and (4) must be solved simultaneously, but the problem is well posed and can be handled analytically or numerically. We can also estimate the size of the first term in Eq. (3) by using the approximation $V_2 \approx (2gh)^{1/2}$ from the previous example. After differentiation, we obtain

$$2h \frac{A_2}{A_1} \frac{dV_2}{dt} \approx -\left(\frac{A_2}{A_1}\right)^2 V_2^2 \tag{5}$$

which is negligible if $A_2 \ll A_1$, as originally postulated.

EXAMPLE 3.15 A constriction in a pipe will cause the velocity to rise and the pressure to fall at section 2 in the throat. The pressure difference is a measure of the flow rate through the pipe. The smoothly necked-down system shown in Fig. E3.15 is called a *venturi tube*. Find an expression for the mass flux in the tube as a function of the pressure change.

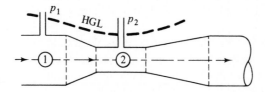

Fig. E3.15

Solution   Bernoulli's equation is assumed to hold along the center streamline

$$\frac{p_1}{\rho} + \tfrac{1}{2}V_1^2 + gz_1 = \frac{p_2}{\rho} + \tfrac{1}{2}V_2^2 + gz_2$$

If the tube is horizontal, $z_1 = z_2$ and we can solve for V_2

$$V_2^2 - V_1^2 = \frac{2\,\Delta p}{\rho} \qquad \Delta p = p_1 - p_2 \tag{1}$$

We relate the velocities from the incompressible continuity relation

$$A_1 V_1 = A_2 V_2$$

or
$$V_1 = \beta^2 V_2 \qquad \beta = \frac{D_2}{D_1} \tag{2}$$

Combining (1) and (2), we obtain a formula for the velocity in the throat

$$V_2 = \left[\frac{2\,\Delta p}{\rho(1 - \beta^4)} \right]^{1/2} \tag{3}$$

The mass flux is given by

$$\dot{m} = \rho A_2 V_2 = A_2 \left(\frac{2\rho\,\Delta p}{1 - \beta^4} \right)^{1/2} \tag{4}$$

This is the ideal frictionless mass flux. In practice, we measure $\dot{m}_{\text{actual}} = c_d \dot{m}_{\text{ideal}}$ and correlate the discharge coefficient c_d.

EXAMPLE 3.16 A 10-cm fire hose with a 3-cm nozzle discharges $1.5\ \text{m}^3/\text{min}$ to the atmosphere. Assuming frictionless flow, find the force F_B exerted by the flange bolts to hold the nozzle on the hose.

Solution   We use Bernoulli's equation and continuity to find the pressure p_1 upstream of the nozzle and then use a control-volume momentum analysis to compute the bolt force as in Fig. E3.16.

The flow from 1 to 2 is a constriction exactly similar in effect to the venturi in Example 3.15, for which Eq. (1) gave

$$p_1 = p_2 + \tfrac{1}{2}\rho(V_2^2 - V_1^2) \tag{1}$$

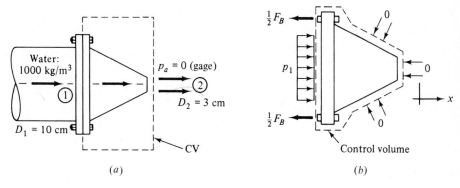

Fig. E3.16

(a)   (b)

The velocities are found from the known flow rate $Q = 1.5 \text{ m}^3/\text{min}$ or $0.025 \text{ m}^3/\text{s}$:

$$V_2 = \frac{Q}{A_2} = \frac{0.025 \text{ m}^3/\text{s}}{(\pi/4)(0.03 \text{ m})^2} = 35.4 \text{ m/s}$$

$$V_1 = \frac{Q}{A_1} = \frac{0.025 \text{ m}^3/\text{s}}{(\pi/4)(0.1 \text{ m})^2} = 3.2 \text{ m/s}$$

We are given $p_2 = p_a = 0$ gage pressure. Then Eq. (1) becomes

$$p_1 = \tfrac{1}{2}(1000 \text{ kg/m}^3)(35.4^2 - 3.2^2) \text{ m}^2/\text{s}^2$$
$$= 620{,}000 \text{ kg/(m} \cdot \text{s}^2) = 620{,}000 \text{ Pa gage}$$

The control-volume force balance is shown in Fig. E3.16b:

$$\sum F_x = -F_B + p_1 A_1$$

and the zero gage pressure on all other surfaces contributes no force. The x-momentum flux is $+\dot{m}V_2$ at the outlet and $-\dot{m}V_1$ at the inlet. The steady-flow momentum relation (3.40) thus gives

$$-F_B + p_1 A_1 = \dot{m}(V_2 - V_1)$$

or

$$F_B = p_1 A_1 - \dot{m}(V_2 - V_1) \qquad (2)$$

Substituting in the given numerical values, we find

$$\dot{m} = \rho Q = (1000 \text{ kg/m}^3)(0.025 \text{ m}^3/\text{s}) = 25 \text{ kg/s}$$

$$A_1 = \frac{\pi}{4} D_1^2 = \frac{\pi}{4}(0.1 \text{ m})^2 = 0.00785 \text{ m}^2$$

$$F_B = (620{,}000 \text{ N/m}^2)(0.00785 \text{ m}^2) - (25 \text{ kg/s})(35.4 - 3.2 \text{ m/s})$$
$$= 4872 \text{ N} - 805 \text{ (kg} \cdot \text{m)/s}^2 = 4067 \text{ N (915 lbf)} \qquad \qquad Ans.$$

This gives an idea of why it takes more than one firefighter to hold a fire hose at full discharge.

Notice from these examples that the solution of a typical problem involving Bernoulli's equation almost always leads to a consideration of the continuity equation as an equal partner in the analysis. The only exception is when the complete velocity distribution is already known from a previous or given analysis, but that means that the continuity relation has already been used to obtain the given information. The point is that the continuity relation is always an important element in a flow analysis.

3.6 THE ANGULAR-MOMENTUM THEOREM[1]

A control-volume analysis can be applied to the angular-momentum relation, Eq. (3.3), by letting our dummy variable **B** be the angular-momentum vector **H**. However, since the system considered here is typically a group of nonrigid fluid particles of variable velocity, the concept of mass moment of inertia is of no help and we have to calculate the instantaneous angular momentum by integration over the elemental masses dm. If O is the point about which moments are desired, the angular momentum about O is given by

$$\mathbf{H}_O = \int_{syst} (\mathbf{r} \times \mathbf{V}) \, dm \tag{3.65}$$

where **r** is the position vector from O to the elemental mass dm and **V** is the velocity of that element. The amount of angular momentum per unit mass is thus seen to be

$$\beta = \frac{d\mathbf{H}_O}{dm} = \mathbf{r} \times \mathbf{V} \tag{3.66}$$

The Reynolds transport theorem (3.16) then tells us that

$$\left. \frac{d\mathbf{H}_O}{dt} \right|_{syst} = \frac{d}{dt} \left[\iiint_{CV} (\mathbf{r} \times \mathbf{V}) \rho \, d\mathcal{V} \right] + \iint_{CS} (\mathbf{r} \times \mathbf{V}) \rho (\mathbf{V}_r \cdot \mathbf{n}) \, dA \tag{3.67}$$

for the most general case of a deformable control volume. But from conservation of angular momentum (3.3) this must equal the sum of all the moments about point O applied to the control volume

$$\frac{d\mathbf{H}_O}{dt} = \sum \mathbf{M}_O = \sum (\mathbf{r} \times \mathbf{F})_O \tag{3.68}$$

Note that the total moment equals the summation of moments of all applied forces about point O. Recall, however, that this law, like Newton's law (3.2), assumes that the particle velocity **V** is relative to an *inertial* coordinate system. If not, the moments about point O of the relative acceleration terms $\mathbf{a}_{rel}$ in Eq. (3.49) must also be included

$$\sum \mathbf{M}_O = \sum (\mathbf{r} \times \mathbf{F})_O - \iiint_{CV} (\mathbf{r} \times \mathbf{a}_{rel}) \, dm \tag{3.69}$$

---

[1] This section may be omitted without loss of continuity.

where the four terms constituting $\mathbf{a}_{\text{rel}}$ are given in Eq. (3.49). Thus the most general case of the angular-momentum theorem is for a deformable control volume associated with a noninertial coordinate system. We combine Eqs. (3.67) and (3.69) to obtain

$$\sum (\mathbf{r} \times \mathbf{F})_0 - \iiint\limits_{CV} (\mathbf{r} \times \mathbf{a}_{\text{rel}})\, dm = \frac{d}{dt}\left[\iiint\limits_{CV} (\mathbf{r} \times \mathbf{V})\rho\, d\mathcal{V}\right] + \iint\limits_{CS} (\mathbf{r} \times \mathbf{V})\rho(\mathbf{V}_r \cdot \mathbf{n})\, dA$$

(3.70)

For a nondeformable inertial control volume, this reduces to

$$\sum \mathbf{M}_O = \frac{\partial}{\partial t}\left[\iiint\limits_{CV} (\mathbf{r} \times \mathbf{V})\rho\, d\mathcal{V}\right] + \iint\limits_{CS} (\mathbf{r} \times \mathbf{V})\rho(\mathbf{V} \cdot \mathbf{n})\, dA \qquad (3.71)$$

Further, if there are only one-dimensional inlets and exits, the angular-momentum flux terms evaluated on the control surface become

$$\iint\limits_{CS} (\mathbf{r} \times \mathbf{V})\rho(\mathbf{V} \cdot \mathbf{n})\, dA = \sum (\mathbf{r} \times \mathbf{V})_{\text{out}}\dot{m}_{\text{out}} - \sum (\mathbf{r} \times \mathbf{V})_{\text{in}}\dot{m}_{\text{in}} \qquad (3.72)$$

Although at this stage the angular-momentum theorem can be considered to be a supplementary topic, it has direct application to many important fluid-flow problems involving torques or moments. A particularly important case is the analysis of rotating fluid-flow devices, usually called *turbomachines* (Chap. 11).

EXAMPLE 3.17   As shown in Fig. E3.17a a pipe bend is supported at point A and connected to a flow system by flexible couplings at sections 1 and 2. The fluid is incompressible, and ambient pressure p_a is zero. (a) Find an expression for the torque T which must be resisted by the support at A, in terms of the flow properties at sections 1 and 2 and the distances h_1 and h_2. (b) Compute this torque if $D_1 = D_2 = 3$ in, $p_1 = 100\ \text{lbf/in}^2$ gage, $p_2 = 80\ \text{lbf/in}^2$ gage, $V_1 = 40\ \text{ft/s}$, $h_1 = 2$ in, $h_2 = 10$ in, and $\rho = 1.94\ \text{slugs/ft}^3$.

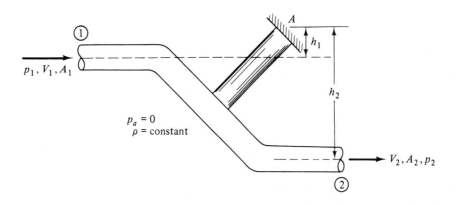

Fig. E3.17a

Solution

Part (a) The control volume chosen in Fig. E3.17b cuts through sections 1 and 2 and through the support at A, where the torque T_A is desired. The flexible-couplings description specifies that there is no torque at either section 1 or 2, and so the cuts there expose no moments. For the

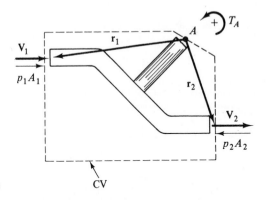

Fig. E3.17b

CV

angular-momentum terms $\mathbf{r} \times \mathbf{V}$, $\mathbf{r}$ should be taken from point A to sections 1 and 2. Note that the gage pressure forces $p_1 A_1$ and $p_2 A_2$ both have moments about A. Equation (3.71) with one-dimensional flux terms becomes

$$\sum \mathbf{M}_A = \mathbf{T}_A + \mathbf{r}_1 \times (-p_1 A_1 \mathbf{n}_1) + \mathbf{r}_2 \times (-p_2 A_2 \mathbf{n}_2)$$
$$= (\mathbf{r}_2 \times \mathbf{V}_2)(+\dot{m}_{\text{out}}) + (\mathbf{r}_1 \times \mathbf{V}_1)(-\dot{m}_{\text{in}}) \tag{1}$$

Figure E3.17c shows that all the cross products are associated either with $r_1 \sin \theta_1 = h_1$ or $r_2 \sin \theta_2 = h_2$, the perpendicular distances from point A to the pipe axes at 1 and 2. Remem-

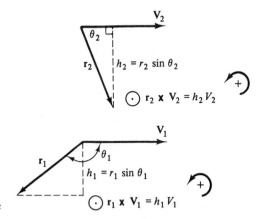

Fig. E3.17c

ber that $\dot{m}_{\text{in}} = \dot{m}_{\text{out}}$ from the steady-flow continuity relation. In terms of counterclockwise moments, Eq. (1) then becomes

$$T_A + p_1 A_1 h_1 - p_2 A_2 h_2 = \dot{m}(h_2 V_2 - h_1 V_1) \tag{2}$$

Rewriting this, we find the desired torque to be

$$T_A = h_2(p_2 A_2 + \dot{m} V_2) - h_1(p_1 A_1 + \dot{m} V_1) \qquad \text{Ans. (a)} \quad (3)$$

counterclockwise. The quantities p_1 and p_2 are gage pressures. Note that this result is independent of the shape of the pipe bend and varies only with the properties at sections 1 and 2 and the distances h_1 and h_2.[1]

Part (b) The inlet and exit areas are the same:

$$A_1 = A_2 = \frac{\pi}{4}(3)^2 = 7.07 \text{ in}^2 = 0.0491 \text{ ft}^2$$

Since the density is constant, we conclude from continuity that $V_2 = V_1 = 40$ ft/s. The mass flow is

$$\dot{m} = \rho A_1 V_1 = 1.94(0.0491)(40) = 3.81 \text{ slugs/s}$$

Equation (3) can be evaluated as

$$T_A = (\tfrac{10}{12} \text{ ft})[80(7.07) \text{ lbf} + 3.81(40) \text{ lbf}]$$
$$- (\tfrac{2}{12} \text{ ft})[100(7.07) \text{ lbf} + 3.81(40) \text{ lbf}]$$
$$= 598 - 143 = 455 \text{ ft} \cdot \text{lbf counterclockwise} \qquad \textit{Ans. (b)}$$

We got a little daring there and multiplied p in pounds force per square inch gage times A in square inches to get pounds without changing units to pounds force per square foot and square feet.

EXAMPLE 3.18 Figure 3.16 shows a schematic of a centrifugal pump. The fluid enters axially and passes through the pump blades, which rotate at angular velocity ω; the velocity of the fluid is changed from V_1 to V_2 and its pressure from p_1 to p_2. (*a*) Find an expression for the torque T_O which must be applied to these blades in order to maintain this flow. (*b*) The power supplied to the pump would be $P = \omega T_O$. To illustrate numerically, suppose $r_1 = 0.2$ m, $r_2 = 0.5$ m, and $b = 0.15$ m. Let the pump rotate at 600 r/min and deliver water at 2.5 m³/s with a density of 1000 kg/m³. Compute the idealized torque and power supplied.

Solution

Part (a) The control volume is chosen to be the annular region between sections 1 and 2 where the flow passes through the pump blades (see Fig. 3.16). The flow is steady and assumed incompressible. The contribution of pressure to the torque about axis O is zero since the pressure forces at 1 and 2 act radially through O. Equation (3.71) becomes

$$\sum \mathbf{M}_O = \mathbf{T}_O = (\mathbf{r}_2 \times \mathbf{V}_2)\dot{m}_{\text{out}} - (\mathbf{r}_1 \times \mathbf{V}_1)\dot{m}_{\text{in}} \qquad (1)$$

where steady-flow continuity tells us that

$$\dot{m}_{\text{in}} = \rho V_{n1} 2\pi r_1 b = \dot{m}_{\text{out}} = \rho V_{n2} 2\pi r_2 b = \rho Q$$

The cross product $\mathbf{r} \times \mathbf{V}$ is found to be clockwise about O at both sections:

$$\mathbf{r}_2 \times \mathbf{V}_2 = r_2 V_{t2} \sin 90° \, \mathbf{k} = r_2 V_{t2} \mathbf{k} \qquad \text{clockwise}$$
$$\mathbf{r}_1 \times \mathbf{V}_1 = r_1 V_{t1} \mathbf{k} \qquad \text{clockwise}$$

Equation (1) thus becomes the desired formula for torque

$$\mathbf{T}_O = \rho Q(r_2 V_{t2} - r_1 V_{t1})\mathbf{k} \qquad \text{clockwise} \qquad \textit{Ans.} \quad (2a)$$

---

[1] Indirectly, the pipe-bend shape probably affects the pressure change from p_1 to p_2.

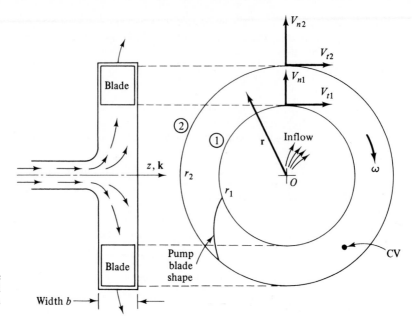

Fig. 3.16 Schematic of a simplified centrifugal pump.

This relation is called *Euler's turbine formula*. In an idealized pump, the inlet and outlet tangential velocities would match the blade rotational speeds, $V_{t1} = \omega r_1$, $V_{t2} = \omega r_2$. Then the formula for torque supplied becomes

$$T_O = \rho Q \omega(r_2^2 - r_1^2) \qquad \text{clockwise} \tag{2b}$$

Part (b)   Convert ω to $600(2\pi/60) = 62.8$ rad/s. The normal velocities are not needed here but follow from the flow rate

$$V_{n1} = \frac{Q}{2\pi r_1 b} = \frac{2.5 \text{ m}^3/\text{s}}{2\pi(0.2 \text{ m})(0.15 \text{ m})} = 13.3 \text{ m/s}$$

$$V_{n2} = \frac{Q}{2\pi r_2 b} = \frac{2.5}{2\pi(0.5)(0.15)} = 5.3 \text{ m/s}$$

For the idealized inlet and outlet tangential velocity equals tip speed

$$V_{t1} = \omega r_1 = (62.8 \text{ rad/s})(0.2 \text{ m}) = 12.6 \text{ m/s}$$

$$V_{t2} = \omega r_2 = 62.8(0.5) = 31.4 \text{ m/s}$$

Equation (2a) predicts the required torque to be

$$T_O = (1000 \text{ kg/m}^3)(2.5 \text{ m}^3/\text{s})[(0.5 \text{ m})(31.4 \text{ m/s}) - (0.2 \text{ m})(12.6 \text{ m/s})]$$
$$= 33,000 \text{ (kg} \cdot \text{m}^2)/\text{s}^2 = 33,000 \text{ N} \cdot \text{m} \qquad \qquad Ans.$$

The power required is

$$P = \omega T_O = (62.8 \text{ rad/s})(33,000 \text{ N} \cdot \text{m}) = 2,070,000 \text{ (N} \cdot \text{m})/\text{s}$$
$$= 2.07 \text{ MW} \quad (2780 \text{ hp}) \qquad \qquad Ans.$$

In actual practice the tangential velocities are considerably less than the impeller-tip speeds, and the design power requirements for this pump may be only 1 MW or less.

EXAMPLE 3.19

Figure 3.17 shows a lawn-sprinkler arm viewed from above. The arm rotates about O at constant angular velocity ω. The volume flux entering the arm at O is Q, and the fluid is incompressible. There is a retarding torque at O, due to bearing friction, of amount $-T_O\mathbf{k}$. Find an expression for the rotation rate ω in terms of the arm and flow properties.

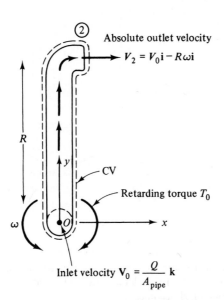

Absolute outlet velocity
$V_2 = V_0\mathbf{i} - R\omega\mathbf{i}$

Fig. 3.17 View from above of a single arm of a rotating lawn sprinkler.

Inlet velocity $V_0 = \dfrac{Q}{A_{\text{pipe}}}\mathbf{k}$

Solution

The entering velocity is $V_0\mathbf{k}$, where $V_0 = Q/A_{\text{pipe}}$. Equation (3.71) applies to the control volume sketched in Fig. 3.17 only if $\mathbf{V}$ is the absolute velocity relative to an inertial frame. Thus the exit velocity at section 2 is

$$\mathbf{V}_2 = V_0\mathbf{i} - R\omega\mathbf{i}$$

Equation (3.71) then predicts that, for steady flow,

$$\sum \mathbf{M}_O = -T_O\mathbf{k} = (\mathbf{r}_2 \times \mathbf{V}_2)\dot{m}_{\text{out}} - (\mathbf{r}_1 \times \mathbf{V}_1)\dot{m}_{\text{in}} \qquad (1)$$

where, from continuity, $\dot{m}_{\text{out}} = \dot{m}_{\text{in}} = \rho Q$. The cross products with reference to point O are

$$\mathbf{r}_2 \times \mathbf{V}_2 = R\mathbf{j} \times (V_0 - R\omega)\mathbf{i} = (R^2\omega - RV_0)\mathbf{k}$$
$$\mathbf{r}_1 \times \mathbf{V}_1 = 0\mathbf{j} \times V_0\mathbf{k} = 0$$

Equation (1) thus becomes

$$-T_O\mathbf{k} = \rho Q(R^2\omega - RV_0)\mathbf{k}$$

or
$$\omega = \frac{V_0}{R} - \frac{T_O}{\rho Q R^2} \qquad\qquad Ans.$$

The result may surprise you: even if the retarding torque T_O is negligible, the arm rotational speed is limited to the value V_0/R imposed by the inlet speed and the arm length.

3.7 THE ENERGY EQUATION[1]

As our fourth and final basic law we apply the Reynolds transport theorem (3.12) to the first law of thermodynamics, Eq. (3.5). The dummy variable B becomes energy E, and the energy per unit mass is $\beta = dE/dm = e$. Equation (3.5) can then be written for a fixed control volume as follows:[2]

$$\frac{dQ}{dt} - \frac{dW}{dt} = \frac{dE}{dt} = \frac{\partial}{\partial t}\left(\iiint_{CV} e\rho \, d\mathcal{V}\right) + \iint_{CS} e\rho(\mathbf{V} \cdot \mathbf{n}) \, dA \qquad (3.73)$$

Recall that positive Q denotes heat added to the system and positive W denotes work done by the system.

The system energy per unit mass e may be of several types:

$$e = e_{\text{internal}} + e_{\text{kinetic}} + e_{\text{potential}} + e_{\text{other}} \qquad (3.74)$$

where e_{other} could encompass chemical reactions, nuclear reactions, and electrostatic or magnetic field effects. We neglect e_{other} here and consider only the first three terms as discussed in Eq. (1.28), with z defined as "up":

$$e = \hat{u} + \tfrac{1}{2}V^2 + gz \qquad (3.75)$$

The heat and work terms could be examined in detail. If this were a heat-transfer book, dQ/dT would be broken down into conduction, convection, and radiation effects and whole chapters written on each (see, for example, Ref. 3). Here we leave the term untouched and consider it only occasionally.

Using for convenience the overdot to denote the time derivative, we shall divide the work term into three parts

$$\dot{W} = \dot{W}_{\text{shaft}} + \dot{W}_{\text{press}} + \dot{W}_{\text{viscous stresses}} = \dot{W}_s + \dot{W}_p + \dot{W}_v \qquad (3.76)$$

The work of gravitational forces has already been included as potential energy in Eq. (3.75). Other types of work, e.g., those due to electromagnetic forces, are excluded here.

The shaft work isolates that portion of the work which is deliberately done by a machine (pump impeller, fan blade, piston, etc.) protruding through the control surface into the control volume. No further specification other than $\dot{W}_s$ is desired at this point, but calculations of the work done by turbomachines will be performed in Chap. 11.

The rate of work $\dot{W}_p$ done on pressure forces occurs at the surface only; all work on internal portions of the material in the control volume is by equal and opposite forces and is self-canceling. The pressure work equals the pressure force on a small surface element dA times the normal velocity component into the control volume

$$d\dot{W}_p = -(p \, dA)V_{n,\text{in}} = -p(-\mathbf{V} \cdot \mathbf{n}) \, dA \qquad (3.77)$$

---

[1] This section should be read for information and enrichment even if you lack formal background in thermodynamics.

[2] The energy equation for a deformable control volume is rather complicated and will not be discussed here. See Refs. 4 and 5 for further details.

The total pressure work is the integral over the control surface

$$\dot{W}_p = \iint\limits_{\text{CS}} p(\mathbf{V} \cdot \mathbf{n}) \, dA \tag{3.78}$$

A cautionary remark: if part of the control surface is the surface of a machine part, we prefer to delegate that portion of the pressure to the *shaft-work* term $\dot{W}_s$, not to $\dot{W}_p$, which is primarily meant to isolate the fluid-flow pressure-work terms.

Finally, the shear work due to viscous stresses occurs at the control surface, the internal work terms again being self-canceling, and consists of the product of each viscous stress (one normal and two tangential) and the respective velocity component

$$d\dot{W}_v = -\boldsymbol{\tau} \cdot \mathbf{V} \, dA$$

or

$$\dot{W}_v = - \iint\limits_{\text{CS}} \boldsymbol{\tau} \cdot \mathbf{V} \, dA \tag{3.79}$$

where $\boldsymbol{\tau}$ is the stress vector on the elemental surface dA. This term may vanish or be negligible according to the particular type of surface at that part of the control volume:

Solid surface. For all parts of the control surface which are solid confining walls, $\mathbf{V} = 0$ from the viscous no-slip condition; hence $\dot{W}_v =$ zero identically.

Surface of a machine. Here the viscous work is contributed by the machine, and so we absorb this work in the term $\dot{W}_s$.

An inlet or outlet. At an inlet or outlet, the flow is approximately normal to the element dA; hence the only viscous-work term comes from the normal stress $\tau_{nn} V_n \, dA$. Since viscous normal stresses are extremely small in all but rare cases, e.g., the interior of a shock wave, it is customary to neglect viscous work at inlets and outlets of the control volume.

Streamline surface. If the control surface is a streamline such as the upper curve in the boundary-layer analysis of Fig. 3.11, the viscous-work term must be evaluated and retained if shear stresses are significant along this line. In the particular case of Fig. 3.11, the streamline is outside the boundary layer, and viscous work is negligible.

The net result of the above discussion is that the rate-of-work term in Eq. (3.73) consists essentially of

$$\dot{W} = \dot{W}_s + \iint\limits_{\text{CS}} p(\mathbf{V} \cdot \mathbf{n}) \, dA - \iint\limits_{\text{CS}} (\boldsymbol{\tau} \cdot \mathbf{V})_{\text{SS}} \, dA \tag{3.80}$$

where the subscript SS stands for stream surface. When we introduce (3.80) and (3.75) into (3.73), we find that the pressure-work term can be combined with the energy-flux term since both involve surface integrals of $\mathbf{V} \cdot \mathbf{n}$. The control-volume energy equation thus becomes

$$\dot{Q} - \dot{W}_s - (\dot{W}_v)_{SS} = \frac{\partial}{\partial t} \left(\iiint_{CV} e\rho \, d\mho \right) + \iint_{CS} \left(e + \frac{p}{\rho} \right) \rho (\mathbf{V} \cdot \mathbf{n}) \, dA \quad (3.81)$$

Using e from (3.75), we see that the enthalpy $\hat{h} = \hat{u} + p/\rho$ occurs in the control-surface integral. The final general form for the energy equation for a fixed control volume becomes

$$\dot{Q} - \dot{W}_s - \dot{W}_v = \frac{\partial}{\partial t} \left[\iiint_{CV} (\hat{u} + \tfrac{1}{2}V^2 + gz)\rho \, d\mho \right] + \iint_{CS} (\hat{h} + \tfrac{1}{2}V^2 + gz)\rho (\mathbf{V} \cdot \mathbf{n}) \, dA$$

$$(3.82)$$

As mentioned above, the shear-work term $\dot{W}_v$ is rarely important.

One-Dimensional Energy-Flux Terms

If the control volume has a series of one-dimensional inlets and outlets, as in Fig. 3.6, the surface integral in (3.82) reduces to a summation of outlet fluxes minus inlet fluxes

$$\iint_{CS} (\hat{h} + \tfrac{1}{2}V^2 + gz)\rho (\mathbf{V} \cdot \mathbf{n}) \, dA$$

$$= \sum (\hat{h} + \tfrac{1}{2}V^2 + gz)_{out} \dot{m}_{out} - \sum (\hat{h} + \tfrac{1}{2}V^2 + gz)_{in} \dot{m}_{in} \quad (3.83)$$

where the values of $\hat{h}$, $\tfrac{1}{2}V^2$, and gz are taken to be averages over each cross section.

EXAMPLE 3.20   A steady-flow machine (Fig. E3.20) takes in air at section 1 and discharges it at sections 2 and 3. The properties at each section are as follows:

Section	A, ft^2	Q, ft^3/s	T, °F	p, lbf/in^2 abs	z, ft
1	0.4	100	70	20	1.0
2	1.0	40	100	30	4.0
3	0.25	50	200	??	1.5

Work is provided to the machine at the rate of 150 hp. Find the pressure p_3 in pounds force per square inch absolute and the heat transfer $\dot{Q}$ in Btu per second. Assume that air is a perfect gas with $R = 1715$ and $c_p = 6003$ (ft·lbf)/(slug·°R).

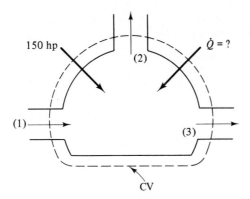

Fig. E3.20

CV

Solution   The control volume chosen cuts across the three desired sections and otherwise follows the solid walls of the machine. Therefore the shear-work term W_v is negligible. We have enough information to compute $V_i = Q_i/A_i$ immediately

$$V_1 = \frac{100}{0.4} = 250 \text{ ft/s} \qquad V_2 = \frac{40}{1.0} = 40 \text{ ft/s} \qquad V_3 = \frac{50}{0.25} = 200 \text{ ft/s}$$

and also the densities $\rho_i = p_i/RT_i$

$$\rho_1 = \frac{20(144)}{1715(70 + 460)} = 0.00317 \text{ slug/ft}^3$$

$$\rho_2 = \frac{30(144)}{1715(560)} = 0.00450 \text{ slug/ft}^3$$

but ρ_3 is determined from the steady-flow continuity relation:

$$\dot{m}_1 = \dot{m}_2 + \dot{m}_3$$

$$\rho_1 Q_1 = \rho_2 Q_2 + \rho_3 Q_3 \qquad \qquad (1)$$

$$0.00317(100) = 0.00450(40) + \rho_3(50)$$

or

$$50\rho_3 = 0.317 - 0.180 = 0.137 \text{ slug/s}$$

$$\rho_3 = \frac{0.137}{50} = 0.00274 \text{ slug/ft}^3 = \frac{144 p_3}{1715(660)}$$

$$p_3 = 21.5 \text{ lbf/in}^2 \text{ absolute} \qquad \qquad \textit{Ans.}$$

Note that the volume flux $Q_1 \neq Q_2 + Q_3$ because of the density changes.

For steady flow, the volume integral in (3.82) vanishes, and we have agreed to neglect viscous work. With one inlet and two outlets, we obtain

$$\dot{Q} - \dot{W}_s = -\dot{m}_1(\hat{h}_1 + \tfrac{1}{2}V_1^2 + gz_1) + \dot{m}_2(\hat{h}_2 + \tfrac{1}{2}V_2^2 + gz_2) + \dot{m}_3(\hat{h}_3 + \tfrac{1}{2}V_3^2 + gz_3) \quad (2)$$

where $\dot{W}_s$ is given in horsepower and can be quickly converted into consistent BG units:

$$\dot{W}_s = -150 \text{ hp } [550 \text{ (ft} \cdot \text{lbf)/(s} \cdot \text{hp)}]$$
$$= -82{,}500 \text{ (ft} \cdot \text{lbf)/s} \qquad \text{negative work on the system}$$

For a perfect gas with constant c_p, $\hat{h} = c_p T$ plus an arbitrary constant. It is instructive to separate the flux terms in Eq. (2) above to examine their magnitudes:

Enthalpy flux:

$$c_p(-\dot{m}_1 T_1 + \dot{m}_2 T_2 + \dot{m}_3 T_3) = [6003 \ (\text{ft} \cdot \text{lbf})/(\text{slug} \cdot {}^\circ\text{R})][(-0.317 \ \text{slug/s})(530{}^\circ\text{R})$$
$$+ \ 0.180(560) + 0.137(660)]$$
$$= \ -1{,}009{,}000 + 605{,}000 + 543{,}000$$
$$= \ +139{,}000 \ (\text{ft} \cdot \text{lbf})/\text{s}$$

Kinetic-energy flux:

$$-\dot{m}_1(\tfrac{1}{2}V_1^2) + \dot{m}_2(\tfrac{1}{2}V_2^2) + \dot{m}_3(\tfrac{1}{2}V_3^2) = \tfrac{1}{2}[-0.317(250)^2 + 0.180(40)^2 + 0.137(200)^2]$$
$$= -9900 + 150 + 2750 = -7000 \ (\text{ft} \cdot \text{lbf})/\text{s}$$

Potential-energy flux:

$$g(-\dot{m}_1 z_1 + \dot{m}_2 z_2 + \dot{m}_3 z_3) = 32.2[-0.317(1.0) + 0.180(4.0) + 0.137(1.5)]$$
$$= -10 + 23 + 7 = +20 \ (\text{ft} \cdot \text{lbf})/\text{s}$$

These are typical effects: the potential-energy flux is negligible in gas flows, the kinetic-energy flux is small in low-speed flows, and the enthalpy flux is dominant. It is only when we neglect heat-transfer effects that the kinetic and potential energy become important. Anyway, we can now solve for the heat flux

$$\dot{Q} = -82{,}500 + 139{,}000 - 7000 + 20 = 49{,}520 \ (\text{ft} \cdot \text{lbf})/\text{s} \qquad (3)$$

Converting, we get

$$\dot{Q} = \frac{49{,}520}{778.2 \ (\text{ft} \cdot \text{lbf})/\text{Btu}} = +63.6 \ \text{Btu/s} \qquad\qquad Ans.$$

The Steady-Flow Energy Equation

For steady flow with one inlet and one outlet, both assumed one-dimensional, Eq. (3.82) reduces to a celebrated relation used in many engineering analyses. Let section 1 be the inlet and section 2 be the outlet. Then

$$\dot{Q} - \dot{W}_s - \dot{W}_v = -\dot{m}_1(\hat{h}_1 + \tfrac{1}{2}V_1^2 + gz_1) + \dot{m}_2(\hat{h}_2 + \tfrac{1}{2}V_2^2 + gz_2) \qquad (3.84)$$

But, from continuity, $\dot{m}_1 = \dot{m}_2 = \dot{m}$, and we can rearrange (3.84) as follows:

$$\hat{h}_1 + \tfrac{1}{2}V_1^2 + gz_1 = (\hat{h}_2 + \tfrac{1}{2}V_2^2 + gz_2) - q + w_s + w_v \qquad (3.85)$$

where $q = \dot{Q}/\dot{m} = dQ/dm$, the heat transferred to the fluid per unit mass. Similarly, $w_s = \dot{W}_s/\dot{m} = dW_s/dm$ and $w_v = \dot{W}_v/\dot{m} = dW_v/dm$. Equation (3.85) is a general form of the *steady-flow energy equation*, which states that the upstream *stagnation enthalpy* $H_1 = (\hat{h} + \tfrac{1}{2}V^2 + gz)_1$ differs from the downstream value H_2 only if there is heat transfer, shaft work, or viscous work as the fluid passes between sections 1 and 2. Recall that q is positive if heat is added to the control volume and w_s and w_v are positive if work is done by the fluid on the surroundings.

Each term in Eq. (3.85) has the dimensions of energy per unit mass, or velocity squared, which is a form commonly used by mechanical engineers. If we divide through by g, each term becomes a length, or head, which is a form preferred by civil engineers. Since both enthalpy and head use the same symbol h, we break up enthalpy into $\hat{u} + p/\rho$ to avoid confusion. Equation (3.85) then becomes

$$\frac{p_1}{\rho g} + \frac{\hat{u}_1}{g} + \frac{V_1^2}{2g} + z_1 = \left(\frac{p_2}{\rho g} + \frac{\hat{u}_2}{g} + \frac{V_2^2}{2g} + z_2\right) - h_q + h_s + h_v \quad (3.86)$$

where $h_q = q/g$, $h_s = w_s/g$, and $h_v = w_v/g$ are the head changes due to heat transfer, shaft work, and viscous work.

Comparison with Bernoulli's Equation

When Bernoulli's equation (3.63) was derived in the previous section, it was pointed out that it neglected friction, heat transfer, and all work except that due to pressure forces. The steady-flow energy equation (3.86) also corresponds effectively to streamline flow (one inlet and one exit) and therefore allows a comparison. Equation (3.86) can be rearranged as follows:

$$\frac{p_1}{\rho} + \tfrac{1}{2}V_1^2 + gz_1 = \left(\frac{p_2}{\rho} + \tfrac{1}{2}V_2^2 + gz_2\right) + w_s + w_v + (\hat{u}_2 - \hat{u}_1 - q) \quad (3.87)$$

The equivalent head-length form would be

$$h_{0_1} = h_{0_2} + h_s + h_v + \frac{\hat{u}_2 - \hat{u}_1 - q}{g} \quad (3.88)$$

where $h_0 = z + p/\rho g + V^2/2g$ is the total Bernoulli head or height of the EGL. Thus the Bernoulli head can change between 1 and 2 due to shaft work, viscous work, heat transfer, and dissipation. Equation (3.88) is much less restrictive than the momentum-type Bernoulli equation (3.63).

The last term in (3.88) may contain both reversible and irreversible (dissipative) portions

$$\frac{\hat{u}_2 - \hat{u}_1 - q}{g} = \Delta h_{\text{rev}} + \Delta h_{\text{losses}} \geq 0 \quad (3.89)$$

The reversible head changes are due to gradually applied heat transfer or to interchange between mechanical and internal energy during compression or expansion of the fluid. This is a common occurrence in gas-dynamics problems and is treated in Chap. 9.

The irreversible loss changes occur in all real flows and are the result of viscous dissipation converting mechanical energy into nonrecoverable internal energy plus heat transfer. We shall denote Δh_{losses} by h_f, meaning the head loss due to friction. The second law of thermodynamics requires that h_f always be positive in a real fluid flow. Methods of correlating losses with flow conditions are discussed in Sec. 6.7.

Figure 3.18 illustrates some of the effects which can occur in a generalized flow system with heat, work, and friction effects. The Bernoulli constant h_0, or EGL, begins at the reservoir height and slowly drops through the first pipe due to friction loss h_f. The HGL drops at the same rate because the velocity in the pipe is constant from the continuity requirement.

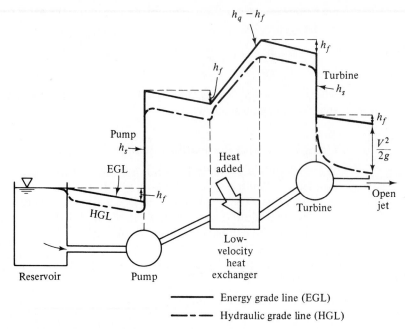

Fig. 3.18 Hypothetical example showing the effect of friction, work, and heat transfer on the Bernoulli constant or EGL and the pressure head (HGL).

The EGL then rises sharply through the pump by the amount h_s corresponding to the pump input. There is a similar rise in pressure reflected in the HGL. Both lines then drop slightly due to friction in the second pipe. The heat exchanger adds heat to the fluid and causes the EGL and HGL to rise more or less linearly through the exchanger. The rise is labeled $h_q - h_f$ to indicate that there are some losses in the exchanger. Then there is friction in the third pipe.

The EGL and HGL drop sharply through the turbine by an amount h_s corresponding to turbine output. The flow finally proceeds through a pipe to the open atmosphere. The EGL in this last pipe shows only the slight drop due to friction loss, but the HGL drops by a large amount, reflecting the high velocity and large pressure change in this last pipe. The HGL ends exactly at the open-jet elevation as the jet drops to atmospheric pressure and gains kinetic energy.

Before proceeding with examples, you should be cautioned that the steady-flow energy equation (3.87), although powerful and useful, is only a simplified special case of the general control-volume energy relation, Eq. (3.82).

EXAMPLE 3.21   Air [$R = 1715$, $c_p = 6003$ (ft · lbf)/(slug · °R)] flows steadily, as shown in Fig. E3.21, through a turbine which produces 700 hp. For the inlet and exit conditions shown, estimate (*a*) the exit velocity V_2 and (*b*) the heat transferred $\dot{Q}$ in Btu per hour.

Solution

Part (*a*)   The inlet and exit densities can be computed from the perfect-gas law

$$\rho_1 = \frac{p_1}{RT_1} = \frac{150(144)}{1715(460 + 300)} = 0.0166 \text{ slug/ft}^3$$

$$\rho_2 = \frac{p_2}{RT_2} = \frac{40(144)}{1715(460 + 35)} = 0.00679 \text{ slug/ft}^3$$

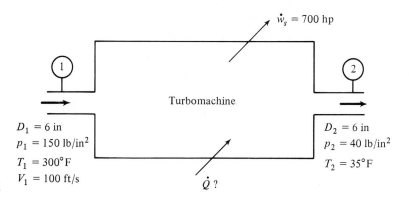

Fig. E3.21

The mass flow is determined by the inlet conditions

$$\dot{m} = \rho_1 A_1 V_1 = (0.0166)\frac{\pi}{4}\left(\frac{6}{12}\right)^2(100) = 0.325 \text{ slug/s}$$

Knowing mass flow, we compute the exit velocity

$$\dot{m} = 0.325 = \rho_2 A_2 V_2 = (0.00679)\frac{\pi}{4}\left(\frac{6}{12}\right)^2 V_2$$

or
$$V_2 = 244 \text{ ft/s} \qquad \qquad Ans. (a)$$

Part (b)   The steady-flow energy equation (3.84) applies with $\dot{W}_v = 0$, $z_1 = z_2$, and $\hat{h} = c_p T$

$$\dot{Q} - \dot{W}_s = \dot{m}(c_p T_2 + \tfrac{1}{2}V_2^2 - c_p T_1 - \tfrac{1}{2}V_1^2)$$

Convert the turbine work to foot-pounds force per second with the conversion factor 1 hp = 550 (ft · lbf)/s. The turbine work is positive

$$\dot{Q} - 700(550) = 0.325[6003(495) + \tfrac{1}{2}(244)^2 - 6003(760) - \tfrac{1}{2}(100)^2]$$
$$= -510,000 \text{ (ft · lbf)/s}$$

or
$$\dot{Q} = -125,000 \text{ (ft · lbf)/s}$$

Convert this to British thermal units as follows:

$$\dot{Q} = -125,000 \text{ (ft · lbf)/s} \frac{3600 \text{ s/h}}{778.2 \text{ (ft · lbf)/Btu}}$$
$$= -576,000 \text{ Btu/h} \qquad \qquad Ans. (b)$$

The negative sign indicates that this heat transfer is a *loss* from the control volume.

EXAMPLE 3.22   A hydroelectric power plant (Fig. E3.22) takes in 30 m³/s of water through its turbine and discharges it at $V_2 = 2$ m/s at atmospheric pressure. The head loss in the turbine and penstock system is $h_f = 20$ m. Estimate the power extracted by the turbine in megawatts.

Solution   We neglect viscous work and heat transfer and take section 1 at the reservoir surface, where $V_1 \approx 0$ and $p_1 = p_a$. Section 2 is at the turbine outlet. Equation (3.88) becomes

$$\frac{p_a}{\rho g} + \tfrac{1}{2}(0)^2 + 100 \text{ m} = \frac{p_a}{\rho g} + \frac{\tfrac{1}{2}(2 \text{ m/s})^2}{9.81 \text{ m/s}^2} + 0 \text{ m} + h_f + h_s \qquad (1)$$

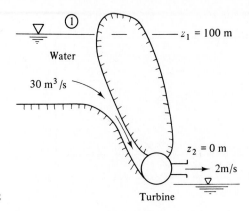

Fig. E3.22

With h_f given as 20 m, we solve for

$$h_s = 100 - 20 - 0.2 = 79.8 \text{ m}$$

or $\quad w_s = h_s g = (79.8 \text{ m})(9.81 \text{ m/s}^2) = 783 \text{ m}^2/\text{s}^2 = 783 \text{ (N} \cdot \text{m)/kg} = 783 \text{ J/kg}$ (2)

The result is positive, as expected from our sign convention that work done by the fluid on a turbine is positive. The power extracted is

$$P = \dot{m}w_s = \rho Q w_s = (1000 \text{ kg/m}^3)(30 \text{ m}^3/\text{s})(783 \text{ J/kg})$$
$$= 23.5 \times 10^6 \text{ J/s} = 23.5 \times 10^6 \text{ W} = 23.5 \text{ MW} \qquad \textit{Ans.}$$

The turbine drives an electric generator which itself has losses of about 15 percent and probably generates about 20 MW of power.

EXAMPLE 3.23 The pump in Fig. E3.23 delivers water at $3 \text{ ft}^3/\text{s}$ to a machine at section 2, which is 20 ft higher than the reservoir surface. The losses between 1 and 2 are given by $h_f = KV_2^2/2g$, where $K = 7.5$ is a dimensionless loss coefficient (see Sec. 6.7). Find the power required to drive the pump if it is 80 percent efficient.

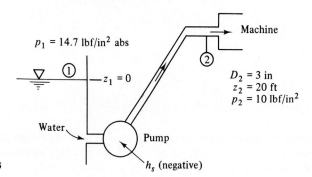

Fig. E3.23

Solution   The flow is steady except for the slow decrease in reservoir depth, which we neglect, taking $V_1 \approx 0$. We can compute V_2 from the given flow rate and diameter

$$V_2 = \frac{Q}{A_2} = \frac{3 \text{ ft}^3/\text{s}}{(\pi/4)(\tfrac{3}{12} \text{ ft})^2} = 61.1 \text{ ft/s}$$

Because of the solid walls and one-dimensional ports, the viscous work is zero. Equation (3.88) becomes

$$\frac{p_1}{\rho g} + \frac{V_1^2}{2g} + z_1 = \frac{p_2}{\rho g} + \frac{V_2^2}{2g} + z_2 + h_s + h_f$$

With $V_1 = z_1 = 0$ and $h_f = KV_2^2/2g$, we can solve for

$$h_s = \frac{p_1 - p_2}{\rho g} - z_2 - (1 + K)\frac{V_2^2}{2g} \tag{1}$$

The pressures must be in pounds force per square foot for consistent units, with $\rho g = 62.4 \, \text{lbf/ft}^3$ for water. Introducing numerical values, we have

$$h_s = \frac{(14.7 - 10)(144) \, \text{lbf/ft}^2}{62.4 \, \text{lbf/ft}^3} - 20 \, \text{ft} - \frac{(1 + 7.5)(61.1 \, \text{ft/s})^2}{2(32.2 \, \text{ft/s}^2)}$$

$$= 11 - 20 - 493 = -502 \, \text{ft} \tag{2}$$

The pump head is negative, indicating work done on the fluid. The power delivered is

$$\boxed{P = \dot{m}w_s} = \rho Q g h_s = (1.94 \, \text{slugs/ft}^3)(3 \, \text{ft}^3/\text{s})(32.2 \, \text{ft/s}^2)(-502 \, \text{ft})$$

$$= -94{,}100 \, (\text{ft} \cdot \text{lbf})/\text{s}$$

$$\text{hp} = \frac{94{,}100 \, (\text{ft} \cdot \text{lbf})/\text{s}}{550 \, (\text{ft} \cdot \text{lbf})/(\text{s} \cdot \text{hp})} = 171 \, \text{hp}$$

The input power thus required to drive the 80 percent efficient pump is

$$P_{\text{input}} = \frac{P}{\text{efficiency}} = \frac{171}{0.8} = 214 \, \text{hp} \qquad\qquad Ans.$$

Kinetic-Energy Correction Factor

Often the flow entering or leaving a port is not strictly one-dimensional. In particular, the velocity may vary over the cross section, as in Fig. E3.4. In this case the kinetic-energy term in Eq. (3.83) for a given port should be modified by a dimensionless correction factor α so that the integral can be proportional to the square of the average velocity through the port

$$\iint\limits_{\text{port}} (\tfrac{1}{2}V^2)\rho(\mathbf{V} \cdot \mathbf{n}) \, dA \equiv \alpha(\tfrac{1}{2}V_{\text{av}}^2)\dot{m} \tag{3.90}$$

where $\qquad\qquad V_{\text{av}} = \frac{1}{A}\iint u \, dA \qquad$ for incompressible flow

If the density is also variable, the integration is very cumbersome; we shall not treat this complication. Letting u be the velocity normal to the port, Eq. (3.90) becomes, for incompressible flow,

$$\tfrac{1}{2}\rho \iint u^3 \, dA = \tfrac{1}{2}\rho\alpha V_{\text{av}}^3 A$$

or $\qquad\qquad \alpha = \frac{1}{A}\iint \left(\frac{u}{V_{\text{av}}}\right)^3 dA \tag{3.91}$

This is the kinetic-energy correction factor, having a value of about 2.0 for fully developed laminar pipe flow and from 1.04 to 1.11 for turbulent pipe flow. The incompressible steady-flow energy equation (3.87) would become

$$\frac{p_1}{\rho} + \tfrac{1}{2}\alpha_1 V_1^2 + gz_1 = \left(\frac{p_2}{\rho} + \tfrac{1}{2}\alpha_2 V_2^2 + gz_2\right) + w_s + \Delta H_{\text{losses}} \qquad (3.92)$$

To compute numerical values, we can use these approximations:

Laminar flow:
$$u = U_0\left[1 - \left(\frac{r}{R}\right)^2\right]$$

from which
$$V_{\text{av}} = 0.5U_0$$

and
$$\alpha = 2.0 \qquad (3.93)$$

Turbulent flow:
$$u \approx U_0\left(1 - \frac{r}{R}\right)^m \qquad m \approx \tfrac{1}{7}$$

from which, in Example 3.4,

$$V_{\text{av}} = \frac{2U_0}{(1 + m)(2 + m)}$$

Substitution into Eq. (3.91) gives

$$\alpha = \frac{(1 + m)^3(2 + m)^3}{4(1 + 3m)(2 + 3m)} \qquad (3.94)$$

and numerical values are as follows:

Turbulent flow: m	$\frac{1}{5}$	$\frac{1}{6}$	$\frac{1}{7}$	$\frac{1}{8}$	$\frac{1}{9}$
α	1.106	1.077	1.058	1.046	1.037

These values are only slightly different from unity and are normally neglected in elementary turbulent-flow analyses.

SUMMARY

This chapter has analyzed the four basic equations of fluid mechanics: conservation of (1) mass, (2) linear momentum, (3) angular momentum, and (4) energy. The equations were attacked "in the large," i.e., applied to whole regions of a flow. As such, the typical analysis will involve an approximation of the flow field within the region, giving somewhat crude but always instructive quantitative results. However, the basic control-volume relations are rigorous and correct and will give exact results if applied to the exact flow field.

There are two main points to a control-volume analysis. The first is the selection of a proper, clever, workable control volume. There is no substitute for experience, but the following guidelines apply. The control volume should cut through the place where the information or solution is desired. It should cut through places where maximum information is already known. If the momentum equation is to be used, it should *not* cut through solid walls unless absolutely necessary, since this will expose possible unknown stresses and forces and moments which make the solution for the desired force difficult or impossible. Finally, every attempt should be made to place the control volume in a frame of reference where the flow is steady or quasi-steady, since the steady formulation is much simpler to evaluate.

The second main point to a control-volume analysis is that of reducing the analysis down to a case which applies to the problem at hand. The 23 examples in this chapter give only an introduction to the search for appropriate simplifying assumptions. You will need to solve 23 or 123 more examples in order to become truly experienced in simplifying the problem just enough and no more. In the meantime, it would be wise for the beginner to adopt a very general form of the control-volume conservation laws and then make a series of simplifications to achieve the final analysis. Starting with the general form, one can ask a series of questions:

1. Is the control volume nondeforming or nonaccelerating?
2. Is the flow field steady? Can we change to a steady-flow frame?
3. Can friction be neglected?
4. Is the fluid incompressible? If not, is the perfect-gas law applicable?
5. Are gravity or other body forces negligible?
6. Is there heat transfer, shaft work, or viscous work?
7. Are the inlet and outlet flows approximately one-dimensional?
8. Is atmospheric pressure important to the analysis? Is the pressure hydrostatic on any portions of the control surface?
9. Are there reservoir conditions which change so slowly that the velocity and time rates of change can be neglected?

In this way, by approving or rejecting a list of basic simplifications like those above, one can avoid pulling, say, Bernoulli's equation off the shelf when it does not apply.

REFERENCES

1. D. T. Greenwood, "Principles of Dynamics," Prentice-Hall, Englewood Cliffs, N.J., 1965.
2. T. von Kármán, "The Wind and Beyond," Little, Brown, Boston, 1967.
3. J. P. Holman, "Heat Transfer," 5th ed., McGraw-Hill, New York, 1981.
4. A. G. Hansen, "Fluid Mechanics," Wiley, New York, 1967.
5. M. C. Potter and J. F. Foss, "Fluid Mechanics," Ronald, New York, 1975.
6. G. J. Van Wylen and R. E. Sonntag, "Fundamentals of Classical Thermodynamics," 3d ed., Wiley, New York, 1985.
7. W. C. Reynolds and H. C. Perkins, "Engineering Thermodynamics," 2d ed., McGraw-Hill, New York, 1977.

Problems

PROBLEM DISTRIBUTION

3.1 In thermodynamics [6, 7] the basic equations are often analyzed for a "control mass" or an "open system." How do these two terms compare with the control volume used in this chapter?

3.2 For the ship steaming at constant speed in Fig. 3.2b, draw the air and water streamlines in the vicinity of the ship for (a) the control volume moving at ship speed and (b) a control volume fixed to the earth. Show directions.

3.3 Write down the intensive, or per-unit-mass, properties β for each of the following fluid properties B: (a) mass, (b) linear momentum, (c) angular momentum, (d) energy, (e) entropy, (f) enthalpy, and (g) salt dissolved in water.

3.4 Discuss whether the following flows are steady or unsteady: (a) flow near an airplane moving at 900 km/h, (b) flow in a pipe as the downstream valve is opened at a uniform rate, (c) flow of a river past a bridge pier, (d) water motion near a beach as a uniform train of waves passes by, (e) flow over the spillway of a dam. Elaborate if these questions seem ambiguous.

3.5 A marine ecosystem test tank contains water of salinity S and density ρ. Seawater enters the tank at conditions (S_i, ρ_i, A_i, V_i) and mixes instantly. Water leaves the tank at velocity V_0 and area A_0. If system salt mass is conserved, write an expression for the rate of change of salt mass M_{CV} within the control volume (tank).

3.6 Air at 30°C and 100 kPa flows through a 15- by 30-cm rectangular duct at 15 N/s. Compute (a) the volume flux in cubic meters per second and (b) the average velocity in meters per second.

3.7 Oil (SG = 0.86) flows through a 30-in-diameter pipeline at 8000 gal/min (1 U.S. gal = 231 in^3). Compute (a) the average velocity in feet per second, (b) the volume flux in cubic feet per second, and (c) the mass flux in slugs per second.

3.8 Water flows steadily through a box at three sections in Fig. P3.8. Section 1 has a diameter of 3 in, and the flow in is 1 ft^3/s. Section 2 has a diameter of 2 in, and the flow out is 30 ft/s average velocity. Compute the average velocity and volume flux at section 3 if $D_3 = 1$ in. Is the flow at 3 in or out?

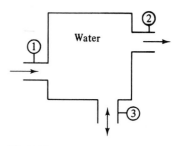

Fig. P3.8

3.9 The water tank in Fig. P3.9 is being filled through section 1 at $V_1 = 4$ m/s and through section 3 at $Q_3 = 0.01$ m^3/s. If the water level h is constant, determine the exit velocity V_2.

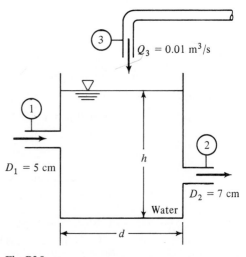

Fig. P3.9

3.10 If the water level varies in Prob. 3.9 and $V_2 = 6$ m/s, find the rate of change dh/dt if the tank diameter $d = 95$ cm.

3.11 For the general case of the flow depicted in Fig. 3.9 derive an expression for dh/dt in terms of tank size and the volume flows Q_1, Q_2, and Q_3 at the three ports.

3.12 Water flows steadily through the nozzle in Fig. P3.12 at 50 kg/s. The diameters are $D_1 = 20$ cm and $D_2 = 6$ cm. Compute the average velocities at sections 1 and 2 in meters per second.

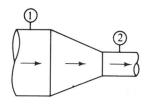

Fig. P3.12

3.13 The hypodermic needle in Fig. P3.13 contains some sort of serum (SG = 1.02). If the plunger is pushed in steadily at 0.8 in/s, what is the exit velocity V_2 in feet per second? Assume no leakage past the plunger.

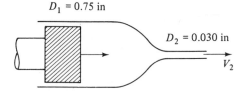

$D_1 = 0.75$ in

$D_2 = 0.030$ in

V_2

Fig. P3.13

3.14 Repeat Prob. P3.13 assuming that there is leakage back past the plunger equal to one-third of the volume flux out of the needle. Compute V_2 in feet per second and the average leakage velocity relative to the needle walls if the plunger diameter is 0.746 in.

3.15 A full gasoline tank has an 8-cm-diameter steel plunger at section 1, which is being pushed into the tank at 5 cm/s. Assuming the fluid is incompressible (SG = 0.68), how many pounds of gasoline per second are being forced out at section 2, $D_2 = 2$ cm?

3.16 A gasoline pump fills a 75-L tank in 1 min 10 s. If the pump exit diameter is 3 cm, what is the average pump-flow exit velocity in centimeters per second?

3.17 An incompressible fluid passes an impermeable plate with a uniform inlet flow and a parabolic exit flow, as shown in Fig. P3.17. The parabola has a maximum ($u = U_0$) at $y = \delta$. Compute the volume flow Q across the top surface of the control volume.

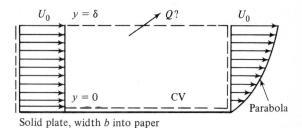

U_0 $y = \delta$ $Q?$ U_0

$y = 0$ CV

Parabola

Solid plate, width b into paper

Fig. P3.17

3.18 A tank filled with air at 20°C and 100 kPa is to be evacuated by a vacuum pump. The tank volume is 1 m^3, and the pump evacuates 80 L/min of air regardless of the pressure. Assuming a perfect gas and an isothermal process, compute the time in minutes to pump the tank down to 1 kPa pressure. *Hint*: This problem results in a first-order differential equation.

3.19 As shown in Fig. P3.19, an incompressible fluid is squeezed outward between two large circular disks by the uniform downward motion V_0 of the upper disk. Using the cylindrical control volume shown and assuming one-dimensional radial outflow, derive an expression for the outflow velocity $V(r)$.

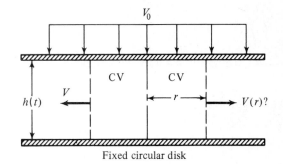

V_0

CV CV

$h(t)$ V r $V(r)?$

Fixed circular disk

Fig. P3.19

3.20 The tank in Fig. P3.20 is admitting water at 90 N/s weight flow and ejecting gasoline at 50 N/s. If all fluids are incompressible, how much air in newtons per hour is passing through the vent? In which direction?

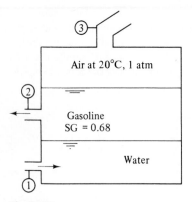

Fig. P3.20

3.21 Oil (SG = 0.86) flows in a pipe so that the axial velocity has the laminar-flow shape $u = U_0(1 - r^2/R^2)$. If the mass flux is 0.1 kg/s and $R = 1$ cm, what is the centerline velocity U_0 in meters per second?

3.22 Water flows in a wide smooth channel with a turbulent velocity profile $u \approx U_0(z/z_0)^{1/7}$, as shown in Fig. P3.22. If $U_0 = 2.8$ ft/s and $z_0 = 8$ ft, compute the volume flux and weight flux in the channel per unit width into the paper.

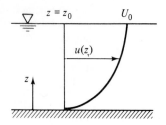

Fig. P3.22

3.23 The flow in the inlet between parallel plates in Fig. P3.23 is uniform, $u = U_0 = 4$ cm/s, while downstream the flow develops into the parabolic laminar profile $u = az(z_0 - z)$, where a is a constant. If $z_0 = 1$ cm

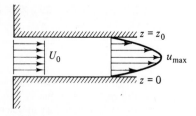

Fig. P3.23

and the fluid is glycerin at 20°C, for steady flow what is the value of u_{max} in centimeters per second?

3.24 Air at 70°F and 15 lbf/in² absolute enters a chamber at section 1 at a velocity of 200 ft/s. It leaves section 2 at 1200°F and 200 lbf/in² absolute. What is the exit velocity in feet per second if $D_1 = 6$ in and $D_2 = 2$ in? Assume the flow is steady.

3.25 Compressed air at 20°C is exhausting from a 1-cm-diameter hole in a 35-cm-diameter rigid spherical tank. Assume an isothermal process with exhaust-jet density equal to tank density. If the initial pressure is 250 kPa and the initial exhaust rate is 0.05 kg/s, estimate the tank pressure after 0.3 s.

3.26 The V-shaped tank in Fig. P3.26 has width b into the paper and is filled from the inlet pipe at volume flow rate Q. Derive expressions for (a) the rate of change dh/dt and (b) the time Δt for the surface to rise from h_1 to h_2.

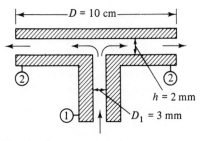

Fig. P3.26

3.27 Oil (SG = 0.86) enters at section 1 at a weight flux of 0.06 N/s to lubricate a thrust bearing. The 10-cm-diameter bearing plates are 2 mm apart. Assuming steady flow, compute (a) the inlet average velocity V_1, (b) the outlet average velocity V_2 assuming radial flow, and (c) the outlet volume flux in milliliters per second.

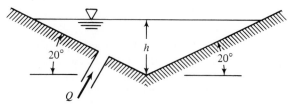

Fig. P3.27

3.28 When it's raining and you have on your new clothes and no umbrella, some people say it's better to run and some say you should walk to keep drier. Suppose that it is raining straight down at a volume flux of

10^{-5} m³/s per square meter of ground area. You have to go 100 m in this rain. Assume an average droplet size of 1 mm³. If a human adult approximates a box 2 m high by 1 m wide by 0.5 m deep, analyze this situation to see whether you should walk at 1 m/s or run at 4 m/s to stay drier. Are there any other physical processes or circumstances to be considered?

3.29 In Fig. P3.29 pipes 1 and 2 are of diameter 2 cm; $D_3 = 3$ cm. Alcohol (SG = 0.79) enters section 1 at 8 m/s while water enters section 2 at 12 m/s. Assuming ideal mixing of incompressible fluids, compute the exit velocity and density of the mixture at section 3.

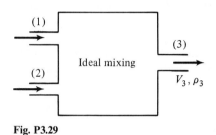

Fig. P3.29

3.30 Traffic engineers have long noted that on any crowded highway the "number flux" of cars passing any given section increases with car speed up to about 80 km/h and then decreases. The reason apparently is that drivers keep more separation distance between cars at high speeds. Analyze this problem to determine whether there is indeed a maximum car flux when separation distance is proportional to (a) car speed; (b) speed squared.

3.31 A bellows may be modeled as a deforming wedge-shaped volume as in Fig. P3.31. The check valve on the left (pleated) end is closed during the stroke. If b is the bellows width into the paper, derive an expression for outlet mass flow $\dot{m}_0$ as a function of stroke $\theta(t)$.

3.32 A 10-m³ tank is filled with fresh water. At time zero a salt solution containing 500 N of salt per cubic meter of solution begins entering at section 1 at 0.5 m³/s, while an equal volume flux 0.5 m³/s of tank mixture begins exiting at section 2. Assume that a mixer immediately disperses the salt solution in the tank. How many newtons of salt will be dissolved in the tank after 1 min? How long will it take the tank solution to have 200 N/m³ of salt?

3.33 In some wind tunnels the test-section wall is porous or perforated; fluid is sucked out to provide a thin viscous boundary layer. The wall in Fig. P3.33 is 4 m long and contains 800 holes of 6-mm diameter per square meter of area. The suction velocity out each hole is $V_s = 10$ m/s, and the test section entrance velocity is $V_1 = 45$ m/s. Assuming incompressible flow of air at 20°C and 1 atm, compute (a) V_0; (b) the total wall suction volume flow; (c) V_2; and (d) V_f.

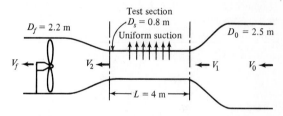

Fig. P3.33

3.34 A rocket motor is operating steadily, as shown in Fig. P3.34. The products of combustion flowing out the

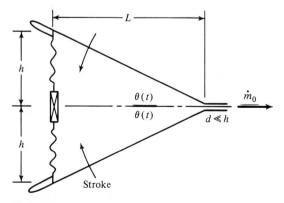

Fig. P3.31

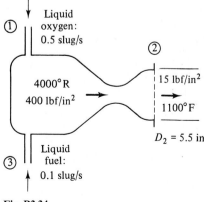

Fig. P3.34

exhaust nozzle approximate a perfect gas with a molecular weight of 26. For the given conditions calculate V_2 in feet per second.

3.35 In contrast to the liquid rocket in Fig. P3.34, the solid-propellant rocket in Fig. P3.35 is self-contained and has no entrance ducts. Using a control-volume analysis for the conditions shown in Fig. P3.35, compute the rate of mass loss of the propellant, assuming that the exit gas has a molecular weight of 31.

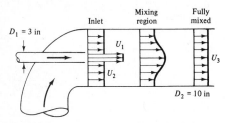

Fig. P3.35

3.36 The water-jet pump in Fig. P3.36 injects water at $U_1 = 100$ ft/s through a 3-in pipe and entrains a secondary flow of water $U_2 = 10$ ft/s in the annular region around the small pipe. The two flows become fully mixed downstream, where U_3 is approximately constant. If the flow is steady and incompressible, compute U_3 in feet per second.

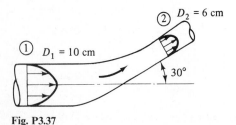

Fig. P3.36

3.37 Water enters section 1 in Fig. P3.37 at 200 N/s and exits at a 30° angle at section 2. Section 1 has the

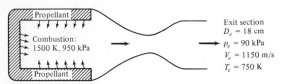

Fig. P3.37

laminar profile $u = u_{m1}(1 - r^2/R^2)$, while section 2 has changed to a turbulent profile

$$u = u_{m2}\left(1 - \frac{r}{R}\right)^{1/7}$$

For steady incompressible flow, what are the maximum velocities u_{m1} and u_{m2} in meters per second?

3.38 Compute the force and its direction of the water on the 30° duct elbow in Prob. 3.37 assuming (a) uniform velocities at both sections and (b) variable velocity profiles as given in Prob. 3.37. Assume steady flow with constant pressure and no friction.

3.39 The cart in Fig. P3.39 is supported by wheels and a linear spring as shown. The water jet is deflected 50° by the cart. Compute (a) the force on the wheels caused by the jet and (b) the spring deflection compared with its unstressed position.

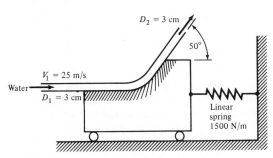

Fig. P3.39

3.40 The water jet in Fig. P3.40 strikes normal to a fixed plate. Neglect gravity and friction and compute the force F required to hold the plate fixed in newtons.

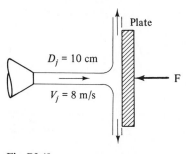

Fig. P3.40

3.41 In Fig. P3.41 the vane turns the water jet completely around. Find an expression for the maximum jet velocity V_0 if the maximum possible support force is F_0.

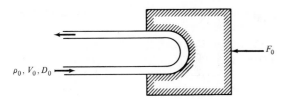

Fig. P3.41

3.42 Make an analysis of the case shown in Fig. P3.42, where the vane moves to the right at constant velocity V_c on a cart. The jet velocity and area are V_j and A_j. The pressure is uniform. Compute the forces F_x and F_y required to restrain the cart and keep it in this nonaccelerating condition. Neglect cart weight and friction.

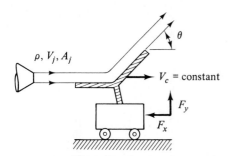

Fig. P3.42

3.43 For the moving cart and vane of Fig. P3.42, let the jet be water of 3 in diameter at 35 ft/s. If the vane angle is $\theta = 120°$ and the cart is moving at 15 ft/s, compute the required F_x.

3.44 For the moving-vane analysis of Prob. 3.42, find the cart velocity V_c as a fraction of jet velocity V_j for which (a) the force F_x is a maximum and (b) the power transmitted $P = V_c F_x$ is a maximum.

3.45 In Fig. P3.45 a perfectly balanced weight and platform are supported by a steady water jet. If the total

weight supported is 900 N, what is the proper jet velocity?

3.46 When a jet strikes an inclined fixed plate, as in Fig. P3.46, it breaks into two jets at 2 and 3 of equal velocity $V = V_{jet}$ but unequal fluxes αQ at 2 and $(1 - \alpha)Q$ at section 3, α being a fraction. The reason is that for frictionless flow the fluid can exert no tangential force F_t on the plate. The condition $F_t = 0$ enables us to solve for α. Perform this analysis and find α as a function of the plate angle θ. Why doesn't the answer depend upon the properties of the jet?

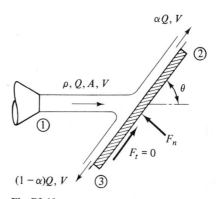

Fig. P3.46

3.47 For Prob. 3.46 evaluate the normal force F_n required to hold the plate fixed for $\theta = 60°$ and a water jet with $V = 40$ ft/s and $A = 0.1$ ft^2.

3.48 The boat in Fig. P3.48 is being driven at a steady speed V_0 by a jet of compressed air issuing from a 3-cm diameter hole at $V_e = 350$ m/s. Jet exit conditions are $p_e = 100$ kPa and $T_e = 20°$C. The hull drag force has the form kV_0^2, where $k = 35$ N·s^2/m^2. Air drag on the upper part of the boat is negligible. Estimate the steady boat speed V_0.

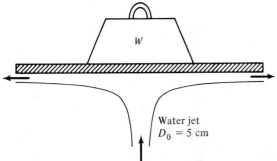

Fig. P3.45

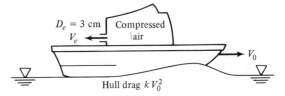

Fig. P3.48

3.49 The horizontal nozzle in Fig. P3.49 has $D_1 = 8$ in and $D_2 = 4$ in. The inlet pressure $p_1 = 50$ lbf/in^2 absolute, and the exit velocity $V_2 = 72$ ft/s. Compute the force provided by the flange bolts to hold the nozzle on. Assume incompressible steady flow.

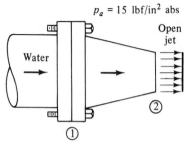

Fig. P3.49

3.50 The jet engine on a test stand in Fig. P3.50 takes in air at 20°C and 1 atm at section 1, where $V_1 = 200$ m/s and $A_1 = 0.3$ m². The fuel air ratio is 1:40. The air leaves at atmospheric pressure and a higher temperature at section 2, where $V_2 = 1000$ m/s and $A_2 = 0.25$ m². Compute the test-stand reaction R_x in newtons which balances the thrust of this engine in steady flow.

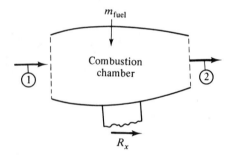

Fig. P3.50

3.51 In Prob. 3.50 what is the exit temperature T_2 in degrees Celsius? If you made this calculation in Prob. 3.50, you are ahead of the game. If the combustion effciency is reduced so that $T_2 = 600$°C with no other changes, what will be the thrust of the engine?

3.52 In the jet engine of Fig. P3.50 both the entrance and exit are open and at atmospheric pressure. So how is the thrust *applied* to the engine? This is, what physical effect causes the momentum change of the jet flow to be transmitted to the support structure of the engine?

3.53 The steady entrance flow in Fig. P3.53 develops from uniform flow U_0 at section 1 to the laminar paraboloid

$$u = u_{max}\left(1 - \frac{r^2}{R^2}\right)$$

at section 2. Find the wall drag F as a function of $p_1, p_2, \rho, U_0,$ and R.

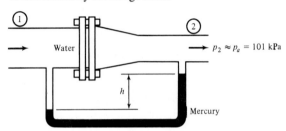

Friction drag on fluid

Fig. P3.53

3.54 For the pipe-flow reducing section of Fig. P3.54, $D_1 = 6$ cm, $D_2 = 4$ cm, and p_2 is approximately atmospheric. If the entrance velocity is $V_1 = 4$ m/s and the manometer reading is $h = 28$ cm, estimate the total force resisted by the flange bolts.

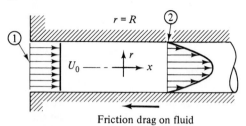

Fig. P3.54

3.55 For the shear flow past a flat plate as in Fig. 3.11 it is found for sea-level standard airflow at 10 m/s past a plate with $L = 1$ m and $b = 2$ m that the drag on one side of the plate is 0.2 N. If the flow is laminar, as in Eq. (4) of Example 3.11, estimate the boundary-layer thickness δ at the trailing edge in millimeters.

3.56 Repeat the analysis of Example 3.11 to find D as a function of $\rho, b, \delta,$ and U_0 for two alternate velocity-profile estimates: (*a*) a linear profile $u \approx U_0 y/\delta$ and (*b*) a sine-wave profile $u \approx U_0 \sin(\pi y/2\delta)$. Which of these should give a more accurate estimate?

3.57 A water jet drives a 180° impulse turbine at 40 ft/s to the right, as in Fig. P3.57. Compute (*a*) the

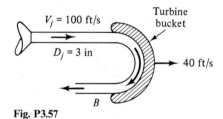

Fig. P3.57

force imparted to the bucket in pounds and (*b*) the horsepower imparted to the bucket.

3.58 The water tank in Fig. P3.58 stands on a frictionless cart and feeds a jet of diameter 4 cm and velocity 8 m/s, which is deflected 60° by a vane. Compute the tension in the supporting cable.

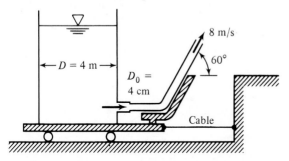

Fig. P3.58

3.59 When a pipe flow suddenly expands from A_1 to A_2, as in Fig. P3.59, low-speed, low-friction eddies appear in the corners and the flow gradually expands to A_2 downstream. Using the suggested control volume

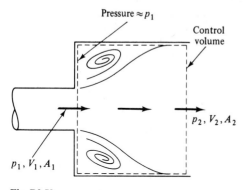

Fig. P3.59

for incompressible steady flow and assuming that $p \approx p_1$ on the corner annular ring as shown, show that the downstream pressure is given by

$$p_2 = p_1 + \rho V_1^2 \frac{A_1}{A_2}\left(1 - \frac{A_1}{A_2}\right)$$

Neglect wall friction.

3.60 Repeat Prob. 3.59 for the assumption that the average pressure on the corner annular ring is $\frac{1}{2}(p_1 + p_2)$, which is an extreme case corresponding to a

certain eddy pattern. What is the percentage difference in the change $p_2 - p_1$?

3.61 A vane of turning angle $180° - \theta$ is mounted on a water tank, as shown in Fig. P3.61. It is struck by a 2-in-diameter, 50 ft/s water jet, which turns and falls into the water without spilling. What is the force F required to hold the water tank stationary if $\theta = 90°$?

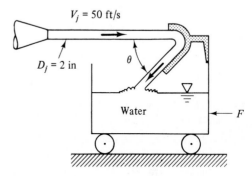

Fig. P3.61

3.62 Water exits to the atmosphere ($p_a = 101$ kPa) through a split nozzle as in Fig. P3.62. Duct areas are $A_1 = 0.01$ m^2 and $A_2 = A_3 = 0.005$ m^2. The flow rate is $Q_2 = Q_3 = 150$ m^3/h, and inlet pressure $p_1 = 140$ kPa absolute. Compute the force on the flange bolts at section 1.

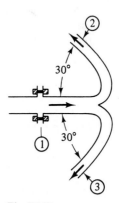

Fig. P3.62

3.63 Water flow in open channels can be controlled and measured with the sluice gate in Fig. P3.63. At a moderate distance upstream and downstream of the gate, sections 1 and 2, the flow is uniform and the pressure is hydrostatic. Derive an expression for the force F required to hold the gate as a function of ρ, V_1, g, h_1, and h_2. Why does the water level rise at point A on the gate? Neglect bottom friction in your analysis.

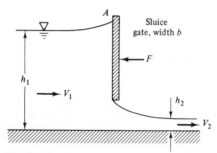

Fig. P3.63

3.64 Compute the force F to hold the gate in Fig. P3.63 if $h_1 = 3$ m, $h_2 = 0.5$ m, and $V_1 = 1.15$ m/s.

3.65 The box in Fig. P3.65 has three 1-in-diameter holes on the right side. The center hole admits water at 0.2 ft^3/s, and the upper and lower holes eject water at 0.1 ft^3/s each. The details of the box interior are not known. Compute the force, if any, which this water flow causes on the box.

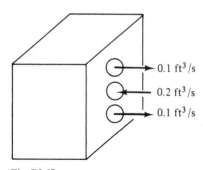

Fig. P3.65

3.66 The water tank in Fig. P3.66 weighs 1000 N empty and contains 1 m^3 of water at 20°C. The entrance and exit pipes are identical, $D_1 = D_2 = 5$ cm, and both

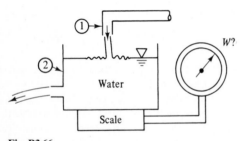

Fig. P3.66

flow at 0.06 m^3/s. What should the scale reading W be in newtons?

3.67 For the rocket engine of Prob. 3.34 compute the thrust in pounds force if atmospheric pressure is 15 lbf/in^2 absolute.

3.68 The rocket in Fig. P3.68 has a supersonic exhaust, and the exit pressure p_e is not necessarily equal to p_a. Show that the force F required to hold this rocket on the test stand is $F = \rho_e A_e V_e^2 + A_e(p_e - p_a)$. Is this force F what we term the thrust of the rocket?

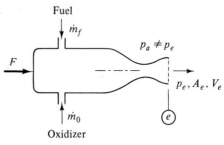

Fig. P3.68

3.69 A water jet 9 cm in diameter and $Q = 20{,}000$ cm^3/s flows over a fixed cone of half-angle θ, as in Fig. P3.69, with a base diameter of 40 cm. If the water leaves as a conical sheet with the same velocity as the jet, what is the force F in newtons required to hold the cone fixed? How does this force vary with the base diameter?

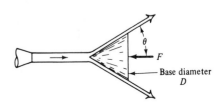

Fig. P3.69

3.70 A dredger is loading sand (SG = 2.65) onto a moored barge, as in Fig. P3.70. The sand leaves the

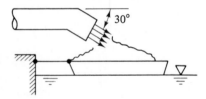

Fig. P3.70

dredger pipe at 5 ft/s with a weight flux of 800 lbf/s. What is the tension on the mooring line caused by this sand loading?

3.71 Suppose that a deflector is deployed at the exit of the jet engine of Prob. 3.50, as shown in Fig. P3.71. What will the reaction R_x on the test stand be now? How does this reaction relate to the braking ability of the deflector?

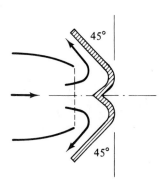

Fig. P3.71

3.72 When immersed in a uniform stream, a thick cylinder will create a broad, low-velocity wake indicative of a large *drag* force in the downstream direction. An idealization of the flow pattern is shown in Fig. P3.72. If the pressures are equal at the inlet and outlet

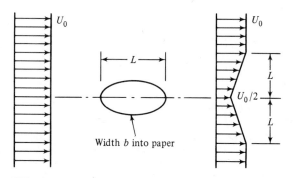

Fig. P3.72

sections, estimate the drag force F in terms of U_0, ρ, L, and width b of the cylinder. Compute the drag coefficient

$$C_D = \frac{F}{\frac{1}{2}\rho U_0^2 bL}$$

3.73 A pump in a water tank directs a jet at 30 ft/s and 0.4 ft³/s against a vane, as in Fig. P3.73. Compute the force F to hold the cart stationary if the jet follows the path A.

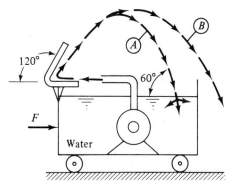

Fig. P3.73

3.74 Compute the force F in Prob. 3.73 if the fluid jet follows path B. The tank holds 35 ft³ of water at this instant.

3.75 Consider a steady adiabatic airflow in a 15-cm-diameter pipe from $p_1 = 350$ kPa and $T_1 = 20°$C to a downstream section, where $p_2 = 113$ kPa and $T_2 = 2.5°$C. If $V_1 = 69$ m/s and the pipe is 70 m long from 1 to 2, what is the average wall shear stress between 1 and 2?

3.76 For the inclined-plate flow geometry of Fig. P3.46, find the angle θ for which the flow rate Q_2 is exactly $2Q_3$.

3.77 Water flows steadily through a reducing pipe bend, as in Fig. P3.77. Known conditions are $p_1 = 300$ kPa, $D_1 = 30$ cm, $V_1 = 2$ m/s, $p_2 = 150$ kPa, and

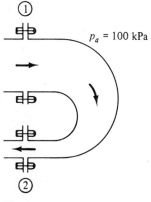

Fig. P3.77

$D_2 = 10$ cm. Compute the axial force which must be resisted by the flange bolts, neglecting the bend and water weight.

3.78 An air jet of diameter D_1 enters a series of moving blades at absolute velocity V_1 and angle β_1 and leaves at absolute velocity V_2 and angle β_2, as in Fig. P3.78. The blades move at speed u. If $V_1 = 60$ m/s, $V_2 = 40$ m/s. $\beta_1 = 30°$, and $\beta_2 = 60°$, find the velocity of the blades u assuming uniform incompressible flow. If $\rho = 1.2$ kg/m^3 and $D_1 = 5$ cm, find the power in watts applied to the blades.

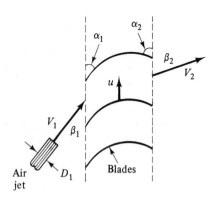

Fig. P3.78

3.79 Derive an analytic formula for the power P applied to the blades in Prob. 3.78 as a function of the jet mass flux $\dot{m}$, the velocities V_1 and V_2, and the angles β_1 and β_2.

3.80 A river of width b and depth h_1 passes over a submerged obstacle, or "drowned weir," in Fig. P3.80, emerging at a new flow condition (V_2, h_2). Neglect atmospheric pressure and assume that the water pressure is hydrostatic at both sections 1 and 2. Derive an expression for the force exerted by the river on the obstacle in terms of V_1, h_1, h_2, b, and ρg. Neglect water friction on the river bottom.

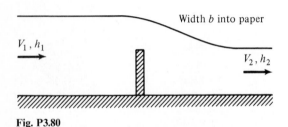

Fig. P3.80

3.81 According to Torricelli's theorem in Example 3.12, the ideal efflux velocity from a hole in a tank is $V = (2gh)^{1/2}$. For the circular tank of Fig. P3.81 the hole diameter is 10 cm and $h = 3$ m. The tank weighs 1000 N empty. What is the coefficient of static friction μ' for which the tank will just begin to move to the right?

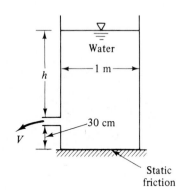

Fig. P3.81

3.82 The cart in Fig. P3.82 is struck by a water jet moving at 50 ft/s and 5 ft^3/s. The jet is turned around completely by the curved vane in the cart. What force F is required to hold the cart stationary? Neglect wheel friction.

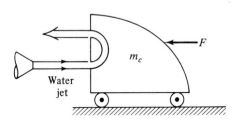

Fig. P3.82

3.83 The system of blades driven by a free jet in Fig. P3.78 is called an *impulse turbine* (Chap. 11). Suppose that the jet is water with $V_1 = 200$ ft/s and $\beta_1 = 40°$. If the blades move at $u = 80$ ft/s and develop a power of 180 hp [1 hp = 550 (ft · lbf)/s], compute the exit conditions V_2 and β_2 and the blade angles α_1 and α_2 for $\dot{m} = 10$ slugs/s.

3.84 Air (1.2 kg/m^3) flows in the 25-cm duct at 10 m/s. It is choked by a 90° cone in the exit, as in Fig. P3.84. Neglect wall friction and estimate the force of the airflow on the cone.

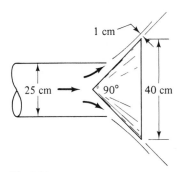

Fig. P3.84

3.85 The thin-plate orifice in Fig. P3.85 causes a large pressure drop. For water flow at $4\,\text{ft}^3/\text{s}$ with pipe $D = 8\,\text{in}$ and orifice $d = 5\,\text{in}$, $p_1 - p_2 = 13\,\text{lbf/in}^2$. If wall friction is negligible, estimate the force on the orifice plate.

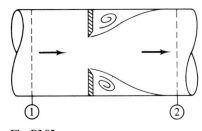

Fig. P3.85

3.86 For the water-jet pump of Prob. 3.36 suppose that the inlet pressure $p_1 = p_2 = 20\,\text{lbf/in}^2$ gage and the distance from inlet to section 3 is 90 in. If the average wall shear stress between 1 and 3 is $3\,\text{lbf/ft}^2$, what is the pressure p_3? Why is it higher than p_1?

3.87 Figure P3.87 simulates a *manifold* flow, where fluid is removed through a porous or perforated section of pipe. Assume that wall friction is negligible and the

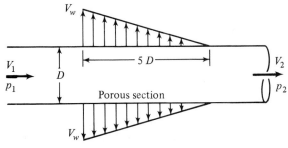

Fig. P3.87

suction velocity V_w is much smaller than V_1 or V_2. For incompressible flow, compute downstream pressure p_2 as a function of V_w and fluid density.

3.88 Water at 68°F flows at $0.1\,\text{ft}^3/\text{s}$ through a straight pipe of diameter 1 in and length 10 ft. The inlet pressure is $4500\,\text{lbf/ft}^2$, and the exit pressure is $3200\,\text{lbf/ft}^2$. What is the average wall shear stress in the pipe?

3.89 Consider a water-jet pump similar to Fig. P3.36 with $D_1 = 2\,\text{in}$ and $D_3 = D_2 = 8\,\text{in}$. If $V_1 = 100\,\text{ft/s}$ and the pressure rises by an amount $p_1 - p_3 = 5\,\text{lbf/in}^2$, what is the exit flow rate Q_3 in cubic feet per second? How much flow Q_2 is entrained by the jet? Neglect wall friction.

3.90 As shown in Fig. P3.90, a liquid column of height h is confined in a vertical tube of cross-sectional area A by a stopper. At $t = 0$ the stopper is suddenly removed, exposing the bottom of the liquid to atmospheric pressure. Using a control-volume analysis of mass and vertical momentum, derive the differential equation for the downward motion $V(t)$ of the liquid. Assume one-dimensional, incompressible, frictionless flow.

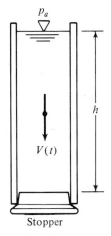

Fig. P3.90

3.91 For Fig. P3.40 compute the force F and power P required to push the plate steadily toward the jet at a speed of 5 m/s.

3.92 For the moving cart of Fig. P3.42 assume that wheel friction is zero and that there is no restraining force F_x. Derive the differential equation for the acceleration of the cart if m_c is the mass of the cart plus vane. Solve the equation with the initial condition $V_c = 0$ at

$t = 0$, assuming that the jet continues to touch the vane throughout the motion.

3.93 For Prob. 3.92 with $V_j = 35$ ft/s, $D_j = 3$ in, $\theta = 120°$, and $m_c = 3$ slugs with a water jet, how long will it take the cart to accelerate from rest to $V_c = 15$ ft/s?

3.94 Extend Prob. 3.90 to include a linear (laminar) wall shear-stress resistance of the form $\tau = kV$, where k is a constant. Solve for the resulting velocity $V(t)$.

3.95 For Prob. 3.61 suppose the cart is unrestrained and wheel friction is zero. If the initial mass of the cart plus the water it contains is M_0 at $t = 0$, derive and solve the differential equation of the cart motion $V_c(t)$ as it accelerates to the right.

3.96 Solve Prob. 3.95 for the numerical data of Prob. 3.61 and an initial mass $M_0 = 40$ slugs. How long does it take the cart to accelerate from rest to $V_c = 10$ ft/s?

3.97 Extend Prob. 3.90 to the case in Fig. P3.97, where the stoppered liquid has both a vertical and a horizontal leg, of equal cross section A. Neglect wall friction and derive the differential equation for liquid velocity. *Hint*: Combine two control volumes, one for each leg of the tube.

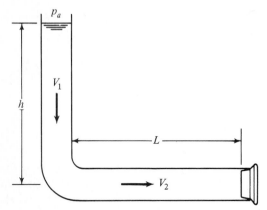

Fig. P3.97

3.98 Consider a rocket of initial mass M_0, steady exhaust mass flux $\dot{m}$, and exit velocity V_e relative to the rocket. The rocket moves straight up, as in Fig. P3.98. Set up the differential equation for rocket motion $V(t)$ and show that the solution is

$$V = -V_e \ln\left(1 - \frac{\dot{m}t}{M_0}\right) - gt$$

if air resistance is neglected and the flow pattern within the rocket is steady.

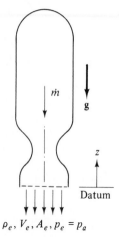

Fig. P3.98

3.99 Suppose that the rocket motor of Fig. P3.34 starts from rest and moves straight up, as in Fig. P3.98. If the initial weight is 2000 lbf, what will be the velocity of the rocket after 10 s? How high will it have risen?

3.100 Extend Prob. 3.90 to the case of the liquid motion in a frictionless U-tube whose liquid column is displaced a distance Z upward and then released, as in Fig. P3.100. Neglect the short horizontal leg and combine control volume analyses for the left and right legs to derive a single differential equation for $V(t)$ of the liquid column.

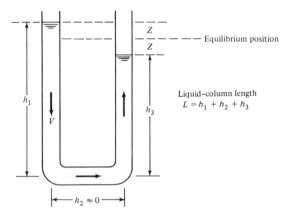

Fig. P3.100

3.101 A rocket ship is moving at 9000 m/s in outer space, where gravity is negligible. It is desired to slow the rocket down to 8900 m/s by firing a retrorocket forward. The retrorocket burns 7 kg/s of fuel and oxidizer at an exhaust velocity of 1500 m/s relative to the rocket. If the initial mass of the rocket ship is 1600 kg, how long should the retrorocket burn, and how much fuel will be burned? Assume that the exhaust pressure is nearly equal to ambient conditions.

3.102 Suppose that an airplane has two jet engines with specifications the same as in Prob. 3.50, each equipped with a deflector as in Fig. P3.71. The airplane lands at a speed of 50 m/s and has a mass of 90,000 kg. If the engines maintain thrust and the deflectors are deployed with no other brakes or air resistance, how much will the plane slow down in 20 s? How far down the landing strip will it have traveled?

3.103 Suppose that the cart in Prob. 3.82 is unrestrained and has zero wheel friction. If the cart velocity to the right is $V_c = 0$ at $t = 0$, how long will it take to accelerate to 10 ft/s? What will the cart speed be after 3 s? Let $m_c = 15$ slugs.

3.104 A rocket is attached to a rigid horizontal rod hinged at the origin as in Fig. P3.104. Its initial mass is M_0, and its exit properties are $\dot{m}$ and V_e relative to the rocket. Set up the differential equation for rocket motion and solve for the angular velocity $\omega(t)$ of the rod. Neglect gravity, air drag, and the rod mass.

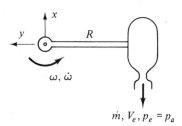

Fig. P3.104

3.105 For Fig. P3.104 suppose $R = 2$ m, $M_0 = 3$ kg, $\dot{m} = 0.08$ kg/s, and $V_e = 1600$ m/s. What will the angular velocity be after 10 s of burning? What will the tension in the rod be?

3.106 Extend Prob. 3.104 to the case where the rocket has a linear drag force $F = kV$, where k is constant. What will the *terminal* angular velocity be, i.e., the long-term motion when the angular acceleration is zero?

3.107 The cart in Fig. P3.107 moves at constant velocity $V_0 = 35$ m/s and takes on water with a scoop 1.5 m wide which dips $h = 3$ cm into a pond. How much water is added to the cart per second and what is the drag force on the scoop?

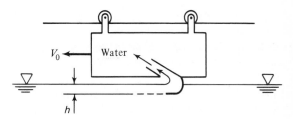

Fig. P3.107

3.108 Assume that a rocket sled is to be decelerated by a scoop, as in Fig. P3.108, which dips into water a distance h and deflects a water jet upward at 60°. Neglect air drag. If the sled and scoop weigh 2500 lbf and $h = 4$ in, what will the acceleration of the sled be when $V = 500$ ft/s? The sled has zero thrust. The scoop has a width $b = 1$ ft into the paper.

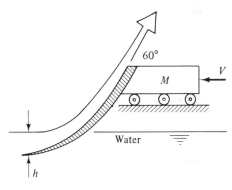

Fig. P3.108

3.109 In Fig. P3.108 derive an analytic expression for the motion $V(t)$ of the sled assuming $V = V_0$ at $t = 0$. How long will it take the sled to slow down to $0.1V_0$?

3.110 For Prob. 3.108 compute the time it takes the sled to decelerate from 500 to 50 ft/s.

3.111 In jet-momentum problems like Fig. P3.46 we have assumed that the various velocities $V_1 = V_2 = V_3 = V$. Can Bernoulli's equation (3.63) be used to establish this condition? What are the relevant assumptions?

3.112 Repeat Prob. 3.49 by assuming that p_1 is unknown and using Bernoulli's equation with no losses. Compute the new bolt force for this assumption. What is the head loss h_f between 1 and 2 for the given data of Prob. 3.49?

3.113 Reanalyze Prob. 3.54 as an application of Bernoulli's equation. For the given manometer reading of 28 cm, what is the head loss between sections 1 and 2? What would the manometer reading be if there were no head losses?

3.114 The nozzle in Fig. P3.114 creates a horizontal jet of alcohol (SG = 0.79) which strikes a vertical plate. If the jet causes a force of 500 N on the plate, what is (a) the mass flow rate of alcohol and (b) the pressure at section 1 if there are no losses in the nozzle?

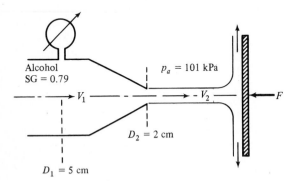

Fig. P3.114

3.115 Utilizing Prob. 3.59 as an application of Bernoulli's equation with losses, find an expression for the head loss h_f between sections 1 and 2 as a function of V_1 and A_1/A_2.

3.116 In Prob. 3.63 the upstream velocity V_1 was assumed known. However, if losses are neglected we can compute V_1 in terms of h_1 and h_2. Derive such an expression for V_1. Are all streamlines (top, bottom, middle) equally applicable? Compute V_1 in Prob. 3.64 and compare with the given value of 1.15 m/s. If $V_1 = 1.15$ is accurate, what is the head loss h_f in meters?

3.117 Utilizing Prob. 3.77 as an application of Bernoulli's equation with losses, compute the head loss h_f between 1 and 2 assuming that $z_1 - z_2 = 1$ m. Is this a realistic result?

3.118 For the data of Prob. 3.85 what is the head loss in meters between sections 1 and 2? What is the dimensionless head loss $h_f/(V_1^2/2g)$? Is this a large or small loss?

3.119 A free liquid jet, as in Fig. P3.119, has constant ambient pressure and small losses; hence from Bernoulli's equation $z + V^2/2g$ is constant along the jet. For the fire nozzle with $V_{\text{exit}} = 100$ ft/s in Fig. P3.119, what is (a) the minimum and (b) the maximum value of

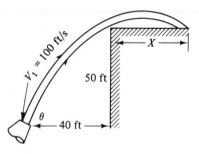

Fig. P3.119

θ for which the water jet will clear the corner of the building? For these two cases what is (c) the jet velocity as it nears the corner and (d) the distance X where it strikes the roof?

3.120 Assuming that the container in Fig. P3.120 is large and losses are negligible, derive an expression for the distance X where the free jet leaving horizontally will strike the floor, as a function of h and H. For what ratio h/H will X be maximum? Sketch the three trajectories for $h/H = 0.25, 0.5,$ and 0.75.

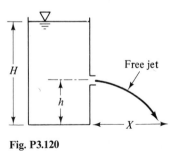

Fig. P3.120

3.121 In Fig. P3.121 what should the water level h be in centimeters for the free jet to just clear the wall?

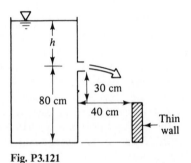

Fig. P3.121

3.122 Neglecting losses, compute the water level h in Fig. P3.122 for which the water will begin to form vapor cavities at the throat of the nozzle.

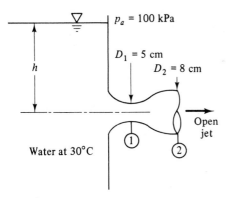

Fig. P3.122

3.123 For the water flow of Fig. P3.123 atmospheric pressure is 15 lbf/in^2 absolute, and the vapor pressure is 2 lbf/in^2 absolute. Neglecting losses, for what nozzle diameter D will cavitation just occur? To avoid cavitation should you increase or decrease D from this critical value?

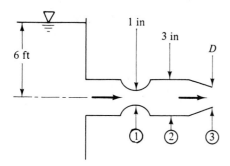

Fig. P3.123

3.124 A necked-down, or *venturi*, section of a pipe flow develops a low pressure which can be used to aspirate fluid upward from a reservoir, as shown in Fig. P3.124. Using Bernoulli's equation with no losses, derive an expression for the exit velocity V_2 which is just sufficient to cause the reservoir fluid to rise in the tube up to section 1.

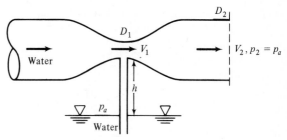

Fig. P3.124

3.125 The liquid in Fig. P3.125 is gasoline (SG = 0.85). The losses in the exit pipe equal $KV^2/2g$, where $K = 6.0$. The tank reservoir is large. Compute the flow rate in cubic feet per minute.

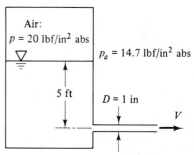

Fig. P3.125

3.126 The manometer fluid in Fig. P3.126 is mercury. Neglecting losses, calculate the flow rate in the tube in cubic feet per second if the flowing fluid is (*a*) water, 62.4 lbf/ft^3; (*b*) air, 0.076 lbf/ft^3.

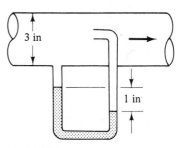

Fig. P3.126

3.127 In Fig. P3.127 the flowing fluid is air, 12 N/m^3, and the manometer fluid is Meriam red oil, SG = 0.827. Assuming no losses, compute the flow rate in cubic meters per second.

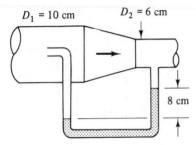

Fig. P3.127

3.128 In Fig. P3.128 the fluid is water, and the pressure gage reads $p_1 = 170\,\text{kPa}$ gage. If the mass flux is 15 kg/s, what is the head loss h_f between 1 and 2 in meters?

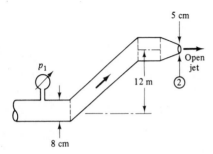

Fig. P3.128

3.129 Once it has been started by sufficient suction, the *siphon* in Fig. P3.129 will run continuously as long as reservoir fluid is available. Using Bernoulli's equation with no losses, show (*a*) that the exit velocity V_2 depends only upon gravity and the distance H and (*b*) that the lowest (vacuum) pressure occurs at point 3 and depends on the distance $L + H$.

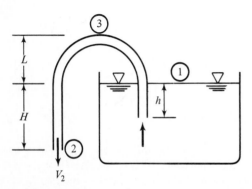

Fig. P3.129

3.130 If losses are neglected for the water-flow system in Fig. P3.130, what should the mercury manometer reading h be in feet? The left manometer leg is open to the atmosphere and $V_1 = 2\,\text{ft/s}$.

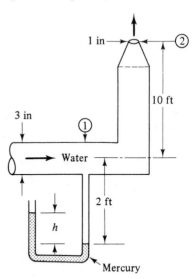

Fig. P3.130

3.131 In Fig. P3.131 the pipe exit losses are $KV^2/2g$, where V is the exit velocity and $K = 1.5$. What is the exit weight flux of water in pounds force per second? Be careful in applying Bernoulli's equation across two different fluids like gasoline and water.

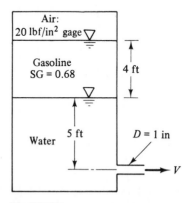

Fig. P3.131

3.132 For the water flow between two reservoirs in Fig. P3.132, the flow rate is found to be 0.016 m³/s. What is the head loss in the pipe in meters? If atmospheric pressure is 100 kPa and the vapor pressure is

8 kPa, for what constriction diameter D in centimeters will cavitation occur? Assume no additional losses due to changes in the constriction.

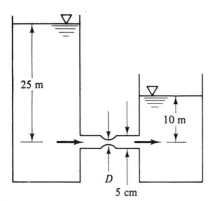

Fig. P3.132

3.133 The horizontal wye fitting in Fig. P3.133 splits Q_1 into two equal flow rates. At section 1 $Q_1 = 4 \text{ ft}^3/\text{s}$ and $p_1 = 20 \text{ lbf/in}^2$ gage. Neglecting losses, compute the pressures p_2 and p_3 and the force required (two components) to keep the wye in place.

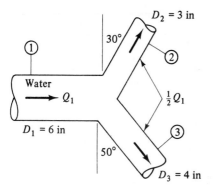

Fig. P3.133

3.134 A cylindrical tank of diameter D contains liquid to an initial height h_0. At time $t = 0$ a small stopper of diameter d is removed from the bottom. Using Bernoulli's equation with no losses, derive (a) a differential equation for the free-surface height $h(t)$ during draining and (b) an expression for the time t_0 to drain the entire tank.

3.135 In the water flow over the spillway in Fig. P3.135 the velocity is uniform at sections 1 and 2 and the pressure approximately hydrostatic. Neglecting

losses, compute V_1 and V_2 and the horizontal force exerted by the water on the spillway. Assume unit width.

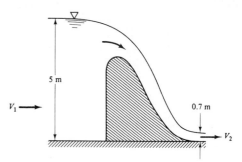

Fig. P3.135

3.136 For the water channel flow down the sloping ramp of Fig. P3.136, $h_1 = 1 \text{ m}$, $H = 3 \text{ m}$, and $V_1 = 4 \text{ m/s}$. The flow is uniform at 1 and 2. Neglecting losses, find the downstream depth h_2 and show that three solutions are possible, of which only two are realistic.

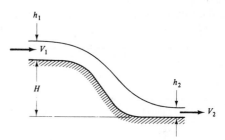

Fig. P3.136

3.137 For water flow up the sloping channel in Fig. P3.137, $h_1 = 0.5 \text{ ft}$, $V_1 = 15 \text{ ft/s}$, and $H = 2 \text{ ft}$. Neglect

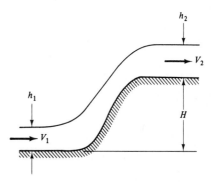

Fig. P3.137

losses and assume uniform flow at 1 and 2. Find the downstream depth h_2 and show that three solutions are possible, of which only two are realistic.

3.138   A large tank in Fig. P3.138 is connected to a pipe of length L closed by a valve. At $t = 0$ the valve is opened. Use the unsteady Bernoulli equation, assuming $h \approx$ constant and negligible velocities in the tank, derive a differential equation for $V(t)$ in the pipe and solve it analytically.

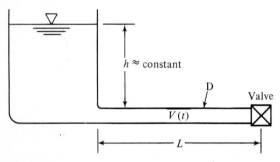

Fig. P3.138

3.139   If a 10-lb force is applied steadily to the piston in Fig. P3.139 and flow losses are negligible, what will the water-jet exit velocity V_2 be in feet per second?

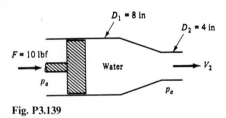

Fig. P3.139

3.140   A hump in the bottom of a water channel as in Fig. P3.140 is a measure of flow rate, sometimes called a

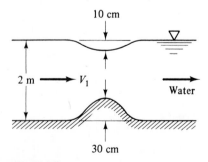

Fig. P3.140

venturi flume. If the water surface dips down 10 cm when the hump is 30 cm high, what is the volume flux Q_1 per unit width, assuming no losses? Is the surface dip proportional to Q_1?

3.141   For a single vane being driven by a jet as in Fig. P3.42, the power transmitted to the vane is

$$P = \rho A_j (V_j - V_c)^2 V_c (1 - \cos \theta)$$

whereas for a series of similar vanes being driven by the same jet the power is more, namely,

$$P = \rho A_j V_j (V_j - V_c) V_c (1 - \cos \theta)$$

Verify that this is true and explain the difference.

3.142   As can often be seen in a kitchen sink when the faucet is running, a high-speed channel flow (V_1, h_1) may "jump" to a low-speed, low-energy condition (V_2, h_2) as in Fig. P3.142. The pressure at sections 1 and 2 is approximately hydrostatic, and wall friction is negligible. Use the continuity and momentum relations to find h_2 and V_2 in terms of (h_1, V_1). The Bernoulli equation (3.63) is invalid because of losses in the jump.

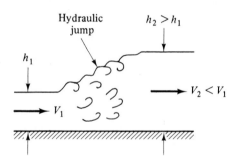

Fig. P3.142

3.143   For the hydraulic jump of Fig. P3.142 compute the head loss h_f in terms of the upstream and downstream depths h_1, h_2. Is the head loss always positive if $h_2 > h_1$?

3.144   A channel flow 2 ft deep moving at 15 ft/s suffers a hydraulic jump. Compute the velocity and depth downstream of the jump and the head loss in feet.

3.145   A *venturi meter* (Fig. P3.145) is a carefully designed constriction whose pressure difference is a measure of the flow rate in a pipe. Using Bernoulli's equation for steady incompressible flow with no losses, show that the flow rate Q is related to the manometer reading h by

$$Q = \frac{A_2}{\sqrt{1 - (D_2/D_1)^4}} \sqrt{\frac{2gh(\rho_M - \rho)}{\rho}}$$

where ρ_M is the density of the manometer fluid.

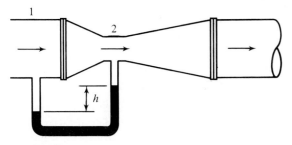

Fig. P3.145

3.146 The horizontal lawn sprinkler in Fig. P3.146 has a flow rate of $0.01 \text{ ft}^3/\text{s}$ introduced vertically through the center. Assuming no friction, what torque in foot-pounds force is required to keep the arms from rotating?

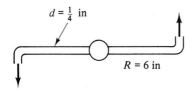

Fig. P3.146

3.147 If the sprinkler in Fig. P3.146 is released, what will its steady rotation rate be if its retarding friction torque is (a) zero; (b) $0.05 \text{ ft} \cdot \text{lbf}$?

3.148 The wye joint in Fig. P3.148 splits the pipe flow into equal amounts $Q/2$, which exit at a distance R_0

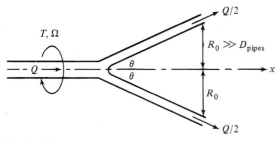

Fig. P3.148

from the x axis. The system rotates about the x axis at a rate Ω. Neglecting gravity and friction, what torque T is needed to maintain the rotation?

3.149 The three-arm lawn sprinkler of Fig. P3.149 receives water through the center at 1200 mL/s. If collar friction is negligible, what is the steady rotation rate in revolutions per minute for (a) $\theta = 0°$; (b) $\theta = 30°$?

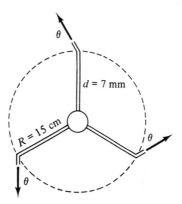

Fig. P3.149

3.150 Repeat Example 3.19 for the case where the arm starts from rest and spins up to its final rotation speed. The mass moment of inertia of the empty arm about O is I_O. Derive the differential equation for angular acceleration $d\omega/dt$ of the arm and solve it for initial condition $\omega = 0$ at $t = 0$.

3.151 Water flows at $0.15 \text{ ft}^3/\text{s}$ in a double pipe bend in Fig. P3.151. The pipe diameter is 1 in throughout. The pressure $p_1 = 25 \text{ lbf/in}^2$ gage, and the head loss between 1 and 2 is $h_f = 1.2V_1^2/2g$. Compute the torque T at point B necessary to keep the pipe from rotating.

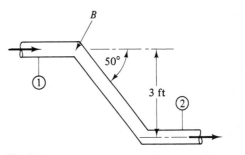

Fig. P3.151

3.152 Water flows at $0.3 \text{ m}^3/\text{s}$ through the pipe bend and nozzle in Fig. P3.152. Atmospheric pressure is

100 kPa, and $D_1 = 30$ cm, $D_2 = 15$ cm. The head loss between 1 and 2 is $h_f = 1.9 V_1^2/2g$. Compute the torque T at point B required to hold the bend stationary.

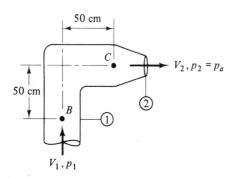

Fig. P3.152

3.153 A liquid flows through a 90° bend as shown in Fig. P3.153 and issues at velocity V_w from a series of small holes in the top of the pipe near a closed valve. Assuming that pipe size is much smaller than the lengths R and L, derive an expression for the torque at point O required to hold the system stationary.

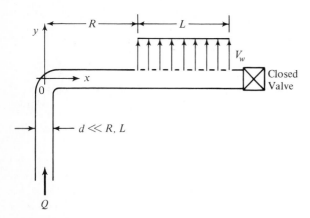

Fig. P3.153

3.154 Extend Prob. 3.46 to the problem of computing the center of pressure of the normal force F_n. Assume that the jets 1, 2, and 3 are two-dimensional sheets of liquid of thicknesses h_1, h_2, and h_3. Find the distance L in terms of (h_1, h_2, h_3) for which no moments are needed to hold the plate at rest.

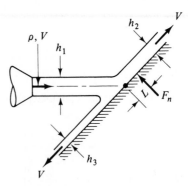

Fig. P3.154

3.155 A centrifugal pump impeller has $r_1 = 3$ in and $r_2 = 12$ in with $b = 2$ in. When running at 1800 r/min it pumps 8 ft³/s water, with $V_{t1} \approx 0$ and $V_{t2} = 100$ ft/s. Losses are negligible. Compute (a) absolute velocities V_1 and V_2, (b) the horsepower required, (c) the total head change H across the impeller, and (d) the pressure change $p_2 - p_1$ in pounds force per square inch. How does the power compare with the idealized flow, where V_{t1} and V_{t2} equal the impeller-tip speeds?

3.156 A radial-flow turbine has $r_1 = 15$ cm, $r_2 = 40$ cm, $b_1 = 5$ cm, $b_2 = 3$ cm. It admits water at $V_{r2} = 12$ m/s inward and develops tangential velocities $V_{t1} = 5$ m/s and $V_{t2} = 30$ m/s at 1200 r/min. Neglecting losses, compute (a) the power developed in kilowatts, (b) the total head change H, and (c) $p_2 - p_1$.

3.157 The waterwheel in Fig. P3.157 is being driven at 150 r/min by a 150 ft/s water jet 3 in in diameter. Assuming no losses, what is the power developed by the wheel? For what speed Ω in revolutions per minute would the horsepower developed be a maximum?

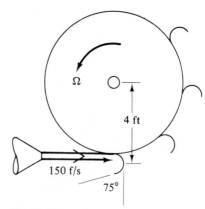

Fig. P3.157

3.158 For steady flow of water through the device in Fig. P3.158, we are given inlet conditions $D_1 = 10$ cm, $Q_1 = 200$ m^3/h, and $P_1 = 170$ kPa. Outlet conditions are $D_2 = 8$ cm, $Q_2 = 100$ m^3/h, $p_2 = 205$ kPa, $D_3 = 5$ cm, and $p_3 = 240$ kPa. Heat transfer, temperature, and gravity effects are negligible. Compute the rate of shaft work done in this device. Is it done on or by the fluid?

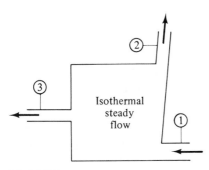

Fig. P3.158

3.159 The pump in Fig. P3.159 draws water from a reservoir through a 12-cm-diameter pipe and discharges it at high velocity through a 5-cm-diameter nozzle. Total friction head loss is 7 m. If the pump delivers 35 kW of power to the water, what are (a) the exit velocity V_e and (b) the flow rate Q?

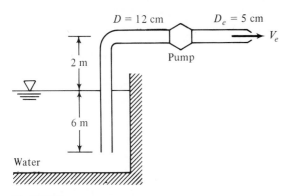

Fig. P3.159

3.160 A power plant on a river, as in Fig. P3.160, must eliminate 40 MW of waste heat to the river. The river inlet flow is $Q_i = 2$ m^3/s, and $T_i = 20°$C. For steady flow with negligible heat losses to atmosphere and ground, what will the downstream river temperature T_0 be? Take $c_p = 4200$ J/(kg · K) for water.

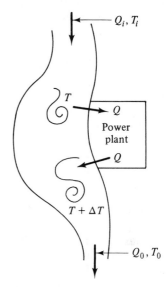

Fig. P3.160

3.161 For the conditions of Prob. 3.161, if the power plant is to heat the nearby river water by $\Delta T = 10°$C, what should the flow rate Q in cubic meters per second be through the plant heat exchanger? How will the value of Q affect the temperature T_0?

3.162 Air flows steadily through a pipe 1500 m long and 20 cm in diameter. At the inlet $p_1 = 120$ kPa, $T_1 = 20°$C, and $V_1 = 17$ m/s, while at the outlet $V_2 = 41$ m/s. Estimate (a) $\dot{m}$, (b) p_2, and (c) T_2. Is there any appropriate way to estimate the head loss h_f? Neglect heat transfer and shaft work.

3.163 Water flows at 10 ft/s through a pipe 1000 ft long with diameter 1 in. The inlet pressure $p_1 = 200$ lbf/in^2 gage, and the exit section is 100 ft higher than the inlet. What is the exit pressure p_2 if the friction head loss is $h_f = 350$ ft?

3.164 A 30-in-diameter pipeline carries oil (SG = 0.86) at 500,000 barrels per day (1 barrel = 42 US gal, 1 US gal = 231 in^3). The friction head loss is 8.4 ft per 1000 ft of pipe. It is planned to place pumping stations every 12 statute miles along the pipe. Compute (a) the pressure drop in pounds per square inch between pumping stations and (b) the horsepower which must be delivered to the oil by each pump.

3.165 It is proposed to build a dam in a river where the flow rate is $10 \text{ m}^3/\text{s}$ and a 30-m drop in elevation can be achieved for flow through a turbine. If the turbine is 80 percent efficient, what is the maximum power in kilowatts which can be achieved?

3.166 A pump delivers $6 \text{ m}^3/\text{h}$ from a pond to an elevated storage tank. The pump is 2 m above the pond, and the discharge is 15 m above the pump. The friction head loss is 2.5 m. If the pump is 65 percent efficient, what power is needed to drive it?

3.167 The long pipe in Fig. P3.167 is filled with water. When valve A is closed, $p_2 - p_1 = 10 \text{ lbf/in}^2$. When the valve is open and water flows at $10 \text{ ft}^3/\text{s}$, $p_1 - p_2 = 34 \text{ lbf/in}^2$. What is the friction head loss between 1 and 2 in ft for the flowing condition?

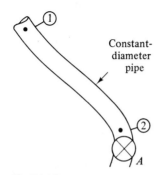

Fig. P3.167

3.168 Air enters a steady-flow compressor at 100 kPa, 20°C, and 10 m/s and is discharged at 350 kPa, 100°C, and 80 m/s. The weight flux is 5.5 N/s, and heat transfer is negligible. Compute the rate of shaft work in kilowatts done on the fluid.

3.169 A fireboat pump draws seawater (SG = 1.025) from a 6-in submerged pipe and discharges it at 120 ft/s through a 2-in nozzle, as in Fig. P3.169. Total head loss is 8 ft. If the pump is 70 percent efficient, how much horsepower is required to drive it?

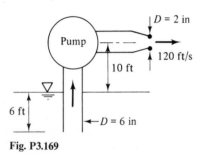

Fig. P3.169

3.170 The pump in Fig. P3.170 discharges water at $0.02 \text{ m}^3/\text{s}$. Neglecting losses and elevation changes, what power in kilowatts is delivered to the water by the pump?

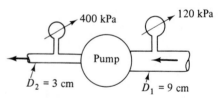

Fig. P3.170

3.171 The pump in Fig. P3.171 draws water from a reservoir and delivers it through a 5-cm-diameter nozzle to a maximum height of 25 m above the nozzle. Pipe-friction head losses are 3.5 m. What power must be delivered to the water by the pump?

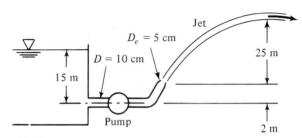

Fig. P3.171

3.172 Steam enters a turbine at 400 lbf/in² absolute, 600°F, and 10 ft/s and is discharged at 100 ft/s and 20 lbf/in² absolute saturated conditions. The mass flux is 2.0 lb/s, and the heat loss from the turbine is 6 Btu per pound of steam. Head losses and elevation changes are negligible. How much horsepower does the turbine develop?

3.173 The pump in Fig. P3.173 delivers water at 4 ft³/s from the lower reservoir to the upper. The pipe friction

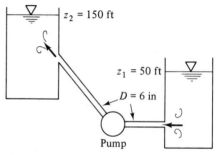

Fig. P3.173

loss equals $35V^2/2g$. If the pump is 70 percent efficient, what horsepower is needed to drive it?

3.174 The air-cushion vehicle in Fig. P3.174 brings in air through a fan and discharges it at high velocity through an annular skirt with small ground clearance h. If the vehicle weighs 50 kN and $p_a = 101$ kPa, $T_a = 20°C$, estimate the airflow rate and power delivered by the fan. *Hint*: The air in region 1 is nearly stagnant and supports the weight of the vehicle.

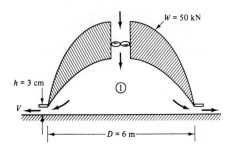

Fig. P3.174

3.175 Gasoline (SG = 0.68) flows through the pump in Fig. P3.175 at 3 ft³/s. Head losses between 1 and 2 are 10 ft, and the pump delivers 30 hp to the flow. What should the mercury manometer reading h be in feet?

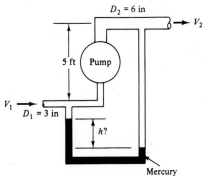

Fig. P3.175

3.176 The large turbine in Fig. P3.176 takes in 100 m³/s of water and discharges it 10 m below the lower surface. Pipe losses are $2.5V^2/2g$. What is the power in kilowatts developed by the turbine?

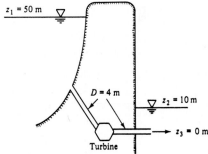

Fig. P3.176

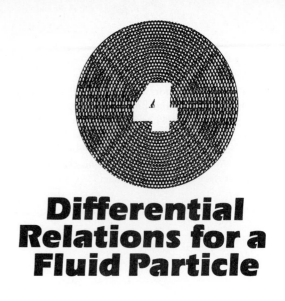

Differential Relations for a Fluid Particle

4.1 DIFFERENTIAL SYSTEMS VERSUS CONTROL VOLUMES

This chapter treats the second in our trio of techniques for analyzing fluid motion, small-scale, or *differential*, analysis. That is, we apply our four basic conservation laws to an infinitesimally small control volume or, alternately, to an infinitesimal fluid system. In either case the results yield the basic *differential equations* of fluid motion. Appropriate *boundary conditions* are also developed.

In their most basic form, these differential equations of motion are quite difficult to solve, and very little is known about their general mathematical properties. However, certain things can be done which have great educational value. For example, as shown in Chap. 5, the equations (even if unsolved) reveal the basic dimensionless parameters which govern fluid motion. Second, as shown in Chap. 6, a great number of useful solutions can be found if one makes two simplifying assumptions: (1) steady flow and (2) incompressible flow. A third and rather drastic simplification, frictionless flow, makes our old friend the Bernoulli equation valid and yields a wide variety of idealized, or *perfect-fluid*, solutions possible. These idealized flows are treated in Chap. 8, and we must be careful to ascertain whether such solutions are in fact realistic when compared with actual fluid motion. Finally, even the difficult general differential equations are now yielding to the approximating technique known as *numerical analysis*, whereby the derivatives are simulated by algebraic relations between a finite number of grid points in the flow field which are then solved on a digital computer. Reference 1 is an example of a textbook devoted entirely to numerical analysis of fluid motion.

The present chapter will derive the basic differential equations of motion and attempt through discussion and examples to illustrate the meaning and use of these equations. Subsequent chapters will then be concerned with specific applications and particular solutions of the equations.

4.2 THE DIFFERENTIAL EQUATION OF MASS CONSERVATION

All the basic differential equations can be derived by considering either an elemental control volume or an elemental system. Here we choose an infinitesimal fixed control volume (dx, dy, dz), as in Fig. 4.1, and use our basic control-volume relations from Chap. 3. The flow through each side of the element is approximately one-dimensional, and so the appropriate mass-conservation relation to use here is

$$\iiint_{CV} \frac{\partial \rho}{\partial t} \, d\mathcal{V} + \sum_i (\rho_i A_i V_i)_{\text{out}} - \sum_i (\rho_i A_i V_i)_{\text{in}} = 0 \qquad (3.22)$$

The element is so small that the volume integral simply reduces to a differential term

$$\iiint_{CV} \frac{\partial \rho}{\partial t} \, d\mathcal{V} \approx \frac{\partial \rho}{\partial t} \, dx \, dy \, dz \qquad (4.1)$$

The mass-flow terms occur on all six faces, three inlets and three outlets. We make use of the field or continuum concept from Chap. 1, where all fluid properties are considered to be uniformly varying functions of time and position, such as $\rho = \rho(x, y, z, t)$. Thus, if T is the temperature on the left face of the element in Fig. 4.1, the right face will have a slightly different temperature $T + (\partial T/\partial x) \, dx$. For mass conservation, if ρu is known on the left face, the value of this product on the right face is $\rho u + (\partial \rho u/\partial x) \, dx$.

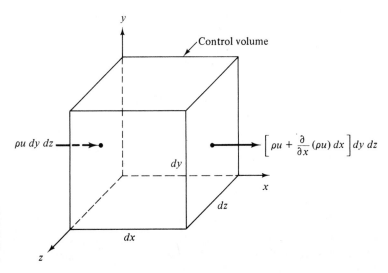

Fig. 4.1 Elemental cartesian fixed control volume showing the inlet and outlet mass flows on the x faces.

Figure 4.1 shows only the mass flows on the x or left and right faces. The flows on the y (bottom and top) and the z (back and front) faces have been omitted to avoid cluttering up the drawing. We can list all of these six flows as follows:

Face	Inlet mass flow	Outlet mass flow
x	$\rho u \, dy \, dz$	$\left[\rho u + \dfrac{\partial}{\partial x}(\rho u)\, dx \right] dy \, dz$
y	$\rho v \, dx \, dz$	$\left[\rho v + \dfrac{\partial}{\partial y}(\rho v)\, dy \right] dx \, dz$
z	$\rho w \, dx \, dy$	$\left[\rho w + \dfrac{\partial}{\partial z}(\rho w)\, dz \right] dx \, dy$

Introduce these terms and Eq. (4.1) into Eq. (3.22) above and we have

$$\frac{\partial \rho}{\partial t} dx \, dy \, dz + \frac{\partial}{\partial x}(\rho u)\, dx \, dy \, dz + \frac{\partial}{\partial y}(\rho v)\, dx \, dy \, dz + \frac{\partial}{\partial z}(\rho w)\, dx \, dy \, dz = 0$$

(4.2)

The element volume cancels out of all terms, leaving a partial differential equation involving the derivatives of density and velocity

$$\frac{\partial \rho}{\partial t} + \frac{\partial}{\partial x}(\rho u) + \frac{\partial}{\partial y}(\rho v) + \frac{\partial}{\partial z}(\rho w) = 0 \qquad (4.3)$$

This is the desired result: conservation of mass for an infinitesimal control volume. It is often called the *equation of continuity* because it requires no assumptions except that the density and velocity are continuum functions. That is, the flow may be either steady or unsteady, viscous or frictionless, compressible or incompressible.[1] However, the equation does not allow for any source or sink singularities within the element.

The vector-gradient operator

$$\mathbf{V} = \mathbf{i}\frac{\partial}{\partial x} + \mathbf{j}\frac{\partial}{\partial y} + \mathbf{k}\frac{\partial}{\partial z} \qquad (4.4)$$

enables us to rewrite the equation of continuity in a compact form, not that it helps much in finding a solution. The last three terms of Eq. (4.3) are equivalent to the divergence of the vector $\rho\mathbf{V}$

$$\frac{\partial}{\partial x}(\rho u) + \frac{\partial}{\partial y}(\rho v) + \frac{\partial}{\partial z}(\rho w) \equiv \mathbf{V} \cdot (\rho\mathbf{V}) \qquad (4.5)$$

[1] One case where Eq. (4.3) might need special care is *two-phase flow*, where the density is discontinuous between the phases. For further details on this case, see, for example, Ref. 2.

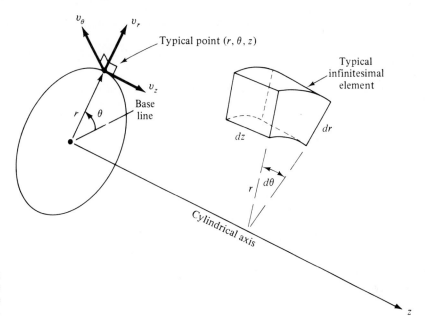

Fig. 4.2 Definition sketch for the cylindrical coordinate system.

so that the compact form of the continuity relation is

$$\frac{\partial \rho}{\partial t} + \nabla \cdot (\rho \mathbf{V}) = 0 \tag{4.6}$$

In this vector form the equation is still quite general and can readily be converted into other than cartesian coordinate systems.

Cylindrical Polar Coordinates

The most common alternative to the cartesian system is the *cylindrical polar* coordinate system, sketched in Fig. 4.2. An arbitrary point P is defined by a distance z along the axis, a radial distance r from the axis, and a rotation angle θ about the axis. The three independent velocity components are an axial velocity v_z, a radial velocity v_r, and a circumferential velocity v_θ, which is positive counterclockwise, i.e., in the direction of increasing θ. In general, all components, as well as pressure and density and other fluid properties, are continuous functions of r, θ, z, and t.

The divergence of any vector function $\mathbf{A}(r, \theta, z, t)$ is found by making the transformation of coordinates,

$$r = (x^2 + y^2)^{1/2} \qquad \theta = \tan^{-1}\frac{y}{x} \qquad z = z \tag{4.7}$$

and the result will be given here without proof[1]

$$\nabla \cdot \mathbf{A} = \frac{1}{r}\frac{\partial}{\partial r}(rA_r) + \frac{1}{r}\frac{\partial}{\partial \theta}(A_\theta) + \frac{\partial}{\partial z}(A_z) \tag{4.8}$$

[1] See, for example, Ref. 3, p. 783.

The general continuity equation (4.6) in cylindrical polar coordinates is thus

$$\frac{\partial \rho}{\partial t} + \frac{1}{r}\frac{\partial}{\partial r}(r\rho v_r) + \frac{1}{r}\frac{\partial}{\partial \theta}(\rho v_\theta) + \frac{\partial}{\partial z}(\rho v_z) = 0 \qquad (4.9)$$

There are other orthogonal curvilinear coordinate systems, notably *spherical polar* coordinates, which occasionally merit use in a fluid-mechanics problem. We shall not treat these systems here except in Prob. 4.12.

There are also other ways to derive the basic continuity equation (4.6) which are interesting and instructive. Ask your instructor about these alternate approaches.

Steady Compressible Flow

If the flow is steady, $\partial/\partial t \equiv 0$ and all properties are functions of position only. Equation (4.6) reduces to

Cartesian:
$$\frac{\partial}{\partial x}(\rho u) + \frac{\partial}{\partial y}(\rho v) + \frac{\partial}{\partial z}(\rho w) = 0$$

Cylindrical:
$$\frac{1}{r}\frac{\partial}{\partial r}(r\rho v_r) + \frac{1}{r}\frac{\partial}{\partial \theta}(\rho v_\theta) + \frac{\partial}{\partial z}(\rho v_z) = 0 \qquad (4.10)$$

Since density and velocity are both variables, these are still nonlinear and rather formidable, but a number of special-case solutions have been found.

Incompressible Flow

A special case which affords great simplification is incompressible flow, where the density changes are negligible. Then $\partial \rho/\partial t \approx 0$ regardless of whether the flow is steady or unsteady, and the density can be slipped out of the divergence in Eq. (4.6) and divided out. The result is

$$\nabla \cdot \mathbf{V} = 0 \qquad (4.11)$$

valid for steady or unsteady incompressible flow. The two coordinate forms are

Cartesian:
$$\frac{\partial u}{\partial x} + \frac{\partial v}{\partial y} + \frac{\partial w}{\partial z} = 0 \qquad (4.12a)$$

Cylindrical:
$$\frac{1}{r}\frac{\partial}{\partial r}(rv_r) + \frac{1}{r}\frac{\partial}{\partial \theta}(v_\theta) + \frac{\partial}{\partial z}(v_z) = 0 \qquad (4.12b)$$

These are *linear* differential equations, and a wide variety of solutions are known, as discussed in Chaps. 6 to 8. Since no author or instructor can resist a wide variety of solutions, it follows that a great deal of time is spent studying incompressible flows. Fortunately, this is exactly what should be done, because most practical engineering flows are approximately incompressible, the chief exception being the high-speed gas flows treated in Chap. 9.

When is a given flow approximately incompressible? We can derive a nice criterion by playing a little fast and loose with density approximations. In essence, we wish to slip the density out of the divergence in Eq. (4.6) and approximate a typical term as, for example,

$$\frac{\partial}{\partial x}(\rho u) \approx \rho \frac{\partial u}{\partial x} \tag{4.13}$$

This is equivalent to the strong inequality

$$\left| u \frac{\partial \rho}{\partial x} \right| \ll \left| \rho \frac{\partial u}{\partial x} \right|$$

or

$$\left| \frac{\delta \rho}{\rho} \right| \ll \left| \frac{\delta V}{V} \right| \tag{4.14}$$

As we shall see in Chap. 9, the pressure change is approximately proportional to the density change and the square of the speed of sound a of the fluid

$$\delta p \approx a^2 \, \delta \rho \tag{4.15}$$

Meanwhile, if elevation changes are negligible, the pressure is related to the velocity change by Bernoulli's equation (3.63)

$$\delta p \approx -\rho V \, \delta V \tag{4.16}$$

Combining Eqs. (4.14) to (4.16), we obtain an explicit criterion for incompressible flow:

$$\frac{V^2}{a^2} = \text{Ma}^2 \ll 1 \tag{4.17}$$

where $\text{Ma} = V/a$ is the dimensionless Mach number of the flow. How small is small? The commonly accepted limit is

$$\text{Ma} \leq 0.3 \tag{4.18}$$

For air at standard conditions, a flow can thus be considered incompressible if the velocity is less than about 100 m/s (330 ft/s). This encompasses a wide variety of airflows: automobile and train motions, light aircraft, landing and takeoff of high-speed aircraft, most pipe flows, and turbomachinery at moderate rotational speeds. Further, it is clear that almost all liquid flows are incompressible, since flow velocities are small and the speed of sound very large.[1]

Before attempting to analyze the continuity equation, we shall proceed with the derivation of the momentum and energy equations, so that we can analyze them as a group. A very clever device called the *stream function* can often make short work of the continuity equation, but we shall save it until Sec. 4.7.

One further remark is appropriate: the continuity equation is always important and must always be satisfied for a rational analysis of a flow pattern. Any newly

---

[1] An exception occurs in geophysical flows, where a density change is imposed thermally or mechanically rather than by the flow conditions themselves. An example is fresh water layered upon salt water or warm air layered upon cold air in the atmosphere. We say that the fluid is *stratified*, and we must account for vertical density changes in Eq. (4.6) even if the velocities are small.

discovered momentum or energy "solution" will ultimately crash in flames when subjected to critical analysis if it does not also satisfy the continuity equation.

EXAMPLE 4.1   Under what conditions does the velocity field

$$\mathbf{V} = (a_1 x + b_1 y + c_1 z)\mathbf{i} + (a_2 x + b_2 y + c_2 z)\mathbf{j} + (a_3 x + b_3 y + c_3 z)\mathbf{k}$$

where a_1, b_1, etc. = const, represent an incompressible flow which conserves mass?

Solution   Recalling that $\mathbf{V} = u\mathbf{i} + v\mathbf{j} + w\mathbf{k}$, we see that $u = (a_1 x + b_1 y + c_1 z)$, etc. Substituting into Eq. (4.12a) for incompressible continuity, we obtain

$$\frac{\partial}{\partial x}(a_1 x + b_1 y + c_1 z) + \frac{\partial}{\partial y}(a_2 x + b_2 y + c_2 z) + \frac{\partial}{\partial z}(a_3 x + b_3 y + c_3 z) = 0$$

or $a_1 + b_2 + c_3 = 0$ *Ans.*

At least two of the constants a_1, b_2, and c_3 must have opposite signs. Continuity imposes no restrictions whatever on the constants, b_1, c_1, a_2, c_2, a_3, and b_3, which do not contribute to a mass increase or decrease of a differential element.

EXAMPLE 4.2   An incompressible velocity field is given by

$$u = a(x^2 - y^2) \qquad v \text{ unknown} \qquad w = b$$

where a and b are constants. What must the form of the velocity component v be?

Solution   Again Eq. (4.12a) applies

$$\frac{\partial}{\partial x}(ax^2 - ay^2) + \frac{\partial v}{\partial y} + \frac{\partial b}{\partial z} = 0$$

or $\dfrac{\partial v}{\partial y} = -2ax$ (1)

This is easily integrated partially with respect to y

$$v(x, y, z, t) = -2axy + f(x, z, t) \qquad\qquad Ans.$$

This is the only possible form for v which satisfies the incompressible continuity equation. The function of integration f is entirely arbitrary since it vanishes when v is differentiated with respect to y.[1]

EXAMPLE 4.3   A centrifugal impeller of 40 cm diameter is used to pump hydrogen at 15°C and 1 atm pressure. What is the maximum allowable impeller rotational speed to avoid compressibility effects at the blade tips?

Solution   The speed of sound of hydrogen for these conditions is $a = 1300$ m/s. Assume that the gas velocity leaving the impeller is approximately equal to the impeller-tip speed

$$V = \Omega r = \tfrac{1}{2}\Omega D$$

---

[1] This is a very realistic flow which simulates the turning of an inviscid fluid through a 60° angle; see Examples 4.6 and 4.8.

Our rule of thumb, Eq. (4.18), neglects compressibility if

$$V = \tfrac{1}{2}\Omega D \le 0.3a = 390 \text{ m/s}$$

or $\qquad\qquad \tfrac{1}{2}\Omega(0.4 \text{ m}) \le 390 \text{ m/s} \qquad \Omega \le 1950 \text{ rad/s}$

Thus we estimate the allowable speed to be quite large

$$\Omega \le 310 \text{ r/s (18,600 r/min)} \qquad\qquad Ans.$$

An impeller moving at this speed in air would create shock waves at the tips but not in a light gas like hydrogen.

4.3 THE DIFFERENTIAL EQUATION OF LINEAR MOMENTUM

Having done it once in Sec. 4.2 for mass conservation, we can move along a little faster this time. We use the same elemental control volume as in Fig. 4.1, for which the appropriate form of the linear-momentum relation is

$$\sum \mathbf{F} = \frac{\partial}{\partial t}\left(\iiint\limits_{\text{CV}} \mathbf{V}\rho \, d\mathcal{V} \right) + \sum (\dot{m}_i \mathbf{V}_i)_{\text{out}} - \sum (\dot{m}_i \mathbf{V}_i)_{\text{in}} \qquad (3.40)$$

Again the element is so small that the volume integral simply reduces to a derivative term

$$\frac{\partial}{\partial t}\left(\iiint\limits_{\text{CV}} \mathbf{V}\rho \, d\mathcal{V} \right) \approx \frac{\partial}{\partial t}(\rho\mathbf{V}) \, dx \, dy \, dz \qquad (4.19)$$

The momentum fluxes occur on all six faces, three inlets and three outlets. Referring again to Fig. 4.1, we can form a table of momentum fluxes by exact analogy with the discussion which led up to Eq. (4.2) for net mass flux:

Faces	Inlet momentum flux	Outlet momentum flux
x	$\rho u\mathbf{V} \, dy \, dz$	$\left[\rho u\mathbf{V} + \dfrac{\partial}{\partial x}(\rho u\mathbf{V}) \, dx \right] dy \, dz$
y	$\rho v\mathbf{V} \, dx \, dz$	$\left[\rho v\mathbf{V} + \dfrac{\partial}{\partial y}(\rho v\mathbf{V}) \, dy \right] dx \, dz$
z	$\rho w\mathbf{V} \, dx \, dy$	$\left[\rho w\mathbf{V} + \dfrac{\partial}{\partial z}(\rho w\mathbf{V}) \, dz \right] dx \, dy$

Introduce these terms and Eq. (4.19) into Eq. (3.40) and get the intermediate result

$$\sum \mathbf{F} = dx \, dy \, dz\left[\frac{\partial}{\partial t}(\rho\mathbf{V}) + \frac{\partial}{\partial x}(\rho u\mathbf{V}) + \frac{\partial}{\partial y}(\rho v\mathbf{V}) + \frac{\partial}{\partial z}(\rho w\mathbf{V}) \right] \qquad (4.20)$$

Note that this is a vector relation. A simplification occurs if we split up the term in brackets as follows:

$$\frac{\partial}{\partial t}(\rho\mathbf{V}) + \frac{\partial}{\partial x}(\rho u\mathbf{V}) + \frac{\partial}{\partial y}(\rho v\mathbf{V}) + \frac{\partial}{\partial z}(\rho w\mathbf{V})$$

$$= \mathbf{V}\left[\frac{\partial\rho}{\partial t} + \nabla\cdot(\rho\mathbf{V})\right] + \rho\left(\frac{\partial\mathbf{V}}{\partial t} + u\frac{\partial\mathbf{V}}{\partial x} + v\frac{\partial\mathbf{V}}{\partial y} + w\frac{\partial\mathbf{V}}{\partial z}\right) \quad (4.21)$$

The term in brackets on the right-hand side is seen to be the equation of continuity, Eq. (4.6), which vanishes identically. The long term in parentheses on the right-hand side is seen from Eq. (1.10) to be the total acceleration of a particle which instantaneously occupies the control volume

$$\frac{\partial\mathbf{V}}{\partial t} + u\frac{\partial\mathbf{V}}{\partial x} + v\frac{\partial\mathbf{V}}{\partial y} + w\frac{\partial\mathbf{V}}{\partial z} = \frac{d\mathbf{V}}{dt} \quad (1.10)$$

Thus we have now reduced Eq. (4.20) to

$$\sum\mathbf{F} = \rho\frac{d\mathbf{V}}{dt}\,dx\,dy\,dz \quad (4.22)$$

It might be well for you to stop and rest now and think about what we have just done. What is the relation between Eq. (4.22) and (3.40) for an infinitesimal control volume? Could we have *begun* the analysis at Eq. (4.22)?

Equation (4.22) points out that the net force on the control volume must be of differential size and proportional to the element volume. These forces are of two types, *body* forces and *surface* forces. Body forces are due to external fields (gravity, magnetism, electric potential) which act upon the entire mass within the element. The only body force we shall consider in this book is gravity. The gravity force on the differential mass $\rho\,dx\,dy\,dz$ within the control volume is

$$d\mathbf{F}_{\text{grav}} = \rho\mathbf{g}\,dx\,dy\,dz \quad (4.23)$$

where $\mathbf{g}$ may in general have an arbitrary orientation with respect to the coordinate system. In many applications, such as Bernoulli's equation, we take z "up," and $\mathbf{g} = -g\mathbf{k}$.

The surface forces are due to the stresses on the sides of the control surface. These stresses, as discussed in Chap. 2, are the sum of hydrostatic pressure plus viscous stresses τ_{ij} which arise from motion with velocity gradients

$$\sigma_{ij} = \begin{vmatrix} -p+\tau_{xx} & \tau_{yx} & \tau_{zx} \\ \tau_{xy} & -p+\tau_{yy} & \tau_{zy} \\ \tau_{xz} & \tau_{yz} & -p+\tau_{zz} \end{vmatrix} \quad (4.24)$$

The subscript notation for stresses is given in Fig. 4.3.

It is not these stresses but their *gradients*, or differences, which cause a net force on the differential control surface. This is seen by referring to Fig. 4.4, which shows only the x-directed stresses to avoid cluttering up the drawing. For example, the leftward force $\sigma_{xx}\,dy\,dz$ on the left face is balanced by the rightward force $\sigma_{xx}\,dy\,dz$

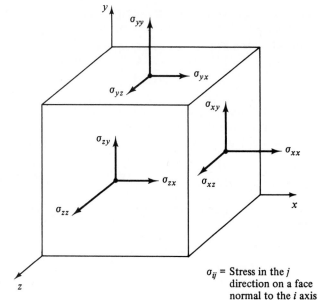

Fig. 4.3 Notation for stresses.

σ_{ij} = Stress in the j direction on a face normal to the i axis

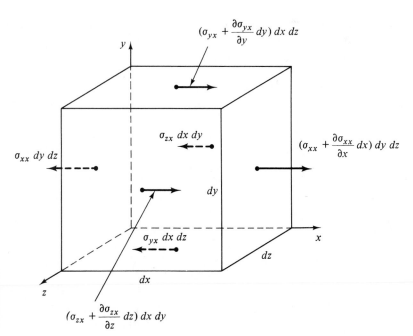

Fig. 4.4 Elemental cartesian fixed control volume showing the surface forces in the x direction only.

on the right face, leaving only the net rightward force $\partial\sigma_{xx}/\partial x\ dx\ dy\ dz$ on the right face. The same thing happens on the other four faces, so that the net surface force in the x direction is given by

$$dF_{x,\,\text{surf}} = \left[\frac{\partial}{\partial x}(\sigma_{xx}) + \frac{\partial}{\partial y}(\sigma_{yx}) + \frac{\partial}{\partial z}(\sigma_{zx})\right] dx\ dy\ dz \qquad (4.25)$$

We see that this force is proportional to the element volume. Notice that the stress terms are taken from the *top row* of the array in Eq. (4.24). Splitting this row into pressure plus viscous stresses, we can rewrite Eq. (4.25) as

$$\frac{dF_x}{d\mathcal{V}} = -\frac{\partial p}{\partial x} + \frac{\partial}{\partial x}(\tau_{xx}) + \frac{\partial}{\partial y}(\tau_{yx}) + \frac{\partial}{\partial z}(\tau_{zx}) \qquad (4.26)$$

In exactly similar manner, we can derive the y and z forces per unit volume on the control surface

$$\frac{dF_y}{d\mathcal{V}} = -\frac{\partial p}{\partial y} + \frac{\partial}{\partial x}(\tau_{xy}) + \frac{\partial}{\partial y}(\tau_{yy}) + \frac{\partial}{\partial z}(\tau_{zy})$$

$$\frac{dF_z}{d\mathcal{V}} = -\frac{\partial p}{\partial z} + \frac{\partial}{\partial x}(\tau_{xz}) + \frac{\partial}{\partial y}(\tau_{yz}) + \frac{\partial}{\partial z}(\tau_{zz}) \qquad (4.27)$$

Now we multiply Eqs. (4.26) and (4.27) by $\mathbf{i}$, $\mathbf{j}$, and $\mathbf{k}$, respectively, and add to obtain an expression for the net vector surface force

$$\left(\frac{d\mathbf{F}}{d\mathcal{V}}\right)_{\text{surf}} = -\nabla p + \left(\frac{d\mathbf{F}}{d\mathcal{V}}\right)_{\text{viscous}} \qquad (4.28)$$

where the viscous force has a total of nine terms:

$$\left(\frac{d\mathbf{F}}{d\mathcal{V}}\right)_{\text{viscous}} = \mathbf{i}\left(\frac{\partial\tau_{xx}}{\partial x} + \frac{\partial\tau_{yx}}{\partial y} + \frac{\partial\tau_{zx}}{\partial y}\right)$$

$$+ \mathbf{j}\left(\frac{\partial\tau_{xy}}{\partial x} + \frac{\partial\tau_{yy}}{\partial y} + \frac{\partial\tau_{zy}}{\partial z}\right)$$

$$+ \mathbf{k}\left(\frac{\partial\tau_{xz}}{\partial x} + \frac{\partial\tau_{yz}}{\partial y} + \frac{\partial\tau_{zz}}{\partial z}\right) \qquad (4.29)$$

Since each term in parentheses in (4.29) represents the divergence of a stress-component vector acting on the x, y, and z faces respectively, Eq. (4.29) is sometimes expressed in divergence form

$$\left(\frac{d\mathbf{F}}{d\mathcal{V}}\right)_{\text{viscous}} = \nabla \cdot \tau_{ij} \qquad (4.30)$$

where

$$\tau_{ij} = \begin{bmatrix} \tau_{xx} & \tau_{yx} & \tau_{zx} \\ \tau_{xy} & \tau_{yy} & \tau_{zy} \\ \tau_{xz} & \tau_{yz} & \tau_{zz} \end{bmatrix} \qquad (4.31)$$

is the viscous-stress tensor acting on the element. The surface force is thus the sum of the *pressure-gradient* vector and the divergence of the viscous-stress tensor. Substituting into Eq. (4.22) and utilizing Eq. (4.23), we have the basic differential momentum equation for an infinitesimal element

$$\rho\mathbf{g} - \nabla p + \nabla \cdot \tau_{ij} = \rho \frac{d\mathbf{V}}{dt} \tag{4.32}$$

where

$$\frac{d\mathbf{V}}{dt} = \frac{\partial \mathbf{V}}{\partial t} + u\frac{\partial \mathbf{V}}{\partial x} + v\frac{\partial \mathbf{V}}{\partial y} + w\frac{\partial \mathbf{V}}{\partial z} \tag{4.33}$$

We can also express Eq. (4.32) in words:

Gravity force per unit volume + pressure force per unit volume
+ viscous force per unit volume = density × acceleration (4.34)

Equation (4.32) is so brief and compact that its inherent complexity is almost invisible. It is a *vector* equation, each of whose component equations contains nine terms. Let us therefore write out the component equations in full to illustrate the mathematical difficulties inherent in the momentum equation

$$\rho g_x - \frac{\partial p}{\partial x} + \frac{\partial \tau_{xx}}{\partial x} + \frac{\partial \tau_{yx}}{\partial y} + \frac{\partial \tau_{zx}}{\partial z} = \rho\left(\frac{\partial u}{\partial t} + u\frac{\partial u}{\partial x} + v\frac{\partial u}{\partial y} + w\frac{\partial u}{\partial z}\right)$$

$$\rho g_y - \frac{\partial p}{\partial y} + \frac{\partial \tau_{xy}}{\partial x} + \frac{\partial \tau_{yy}}{\partial y} + \frac{\partial \tau_{zy}}{\partial z} = \rho\left(\frac{\partial v}{\partial t} + u\frac{\partial v}{\partial x} + v\frac{\partial v}{\partial y} + w\frac{\partial v}{\partial z}\right) \tag{4.35}$$

$$\rho g_z - \frac{\partial p}{\partial z} + \frac{\partial \tau_{xz}}{\partial x} + \frac{\partial \tau_{yz}}{\partial y} + \frac{\partial \tau_{zz}}{\partial z} = \rho\left(\frac{\partial w}{\partial t} + u\frac{\partial w}{\partial x} + v\frac{\partial w}{\partial y} + w\frac{\partial w}{\partial z}\right)$$

This is the differential momentum equation in its full glory, and it is valid for any fluid in any general motion, particular fluids being characterized by particular viscous-stress terms. Note that the last three terms on the right-hand side of each component equation in (4.35) are nonlinear, which complicates the general mathematical analysis.

Inviscid Flow: Euler's Equation

Equation (4.35) is not ready to use until we write the viscous stresses in terms of velocity components. The simplest assumption is frictionless flow, $\tau_{ij} = 0$, for which Eq. (4.35) reduces to

$$\rho\mathbf{g} - \nabla p = \rho\frac{d\mathbf{V}}{dt} \tag{4.36}$$

This is Euler's equation for inviscid flow. We show in Sec. 4.9 that Euler's equation can be integrated along a streamline to yield the frictionless Bernoulli equation, (3.61) or (3.63). The complete analysis of inviscid flow fields, using continuity and the Bernoulli relation, is given in Chap. 8.

Newtonian Fluid: Navier-Stokes Equations

For a newtonian fluid, as discussed in Sec. 1.7, the viscous stresses are proportional to the element strain rates and the coefficient of viscosity. For incompressible flow, the generalization of Eq. (1.42) to three-dimensional viscous flow is[1]

$$\tau_{xx} = 2\mu \frac{\partial u}{\partial x} \qquad \tau_{yy} = 2\mu \frac{\partial v}{\partial y} \qquad \tau_{zz} = 2\mu \frac{\partial w}{\partial z}$$

$$\tau_{xy} = \tau_{yx} = \mu\left(\frac{\partial u}{\partial y} + \frac{\partial v}{\partial x}\right) \qquad \tau_{xz} = \tau_{zx} = \mu\left(\frac{\partial w}{\partial x} + \frac{\partial u}{\partial z}\right) \qquad (4.37)$$

$$\tau_{yz} = \tau_{zy} = \mu\left(\frac{\partial v}{\partial z} + \frac{\partial w}{\partial y}\right)$$

where μ is the viscosity coefficient. Substitution into Eq. (4.35) gives the differential momentum equation for a newtonian fluid with constant density and viscosity

$$\rho g_x - \frac{\partial p}{\partial x} + \mu\left(\frac{\partial^2 u}{\partial x^2} + \frac{\partial^2 u}{\partial y^2} + \frac{\partial^2 u}{\partial z^2}\right) = \rho \frac{du}{dt}$$

$$\rho g_y - \frac{\partial p}{\partial y} + \mu\left(\frac{\partial^2 v}{\partial x^2} + \frac{\partial^2 v}{\partial y^2} + \frac{\partial^2 v}{\partial z^2}\right) = \rho \frac{dv}{dt} \qquad (4.38)$$

$$\rho g_z - \frac{\partial p}{\partial z} + \mu\left(\frac{\partial^2 w}{\partial x^2} + \frac{\partial^2 w}{\partial y^2} + \frac{\partial^2 w}{\partial z^2}\right) = \rho \frac{dw}{dt}$$

These are the Navier-Stokes equations, named after C. L. M. H. Navier (1785–1836) and Sir George G. Stokes (1819–1903), who are credited with the derivation. They are second-order nonlinear partial differential equations and quite formidable, but surprisingly many solutions have been found to a variety of interesting viscous-flow problems, some of which are discussed in Chap. 6 (see also Refs. 4 and 5). For compressible flow, see Eq. (2.29) of Ref. 5.

Equation (4.38) has four unknowns, p, u, v, and w. It should be combined with the incompressible continuity relation (4.12) to form four equations in these four unknowns. We shall discuss this again in Sec. 4.6, which presents the appropriate boundary conditions for these equations.

EXAMPLE 4.4   Take the velocity field of Example 4.2, with $b = 0$ for algebraic convenience

$$u = a(x^2 - y^2) \qquad v = -2axy \qquad w = 0$$

and determine under what conditions it is a solution to the Navier-Stokes momentum equation (4.38). Assuming that these conditions are met, determine the resulting pressure distribution when z is "up" ($g_x = 0, g_y = 0, g_z = -g$).

---

[1] When compressibility is significant, additional small terms arise containing the element volume expansion rate and a *second* coefficient of viscosity; see Refs. 4 and 5 for details.

Solution Make a direct substitution of u, v, w into Eq. (4.38)

$$\rho(0) - \frac{\partial p}{\partial x} + \mu(2a - 2a) = 2a^2\rho(x^3 + xy^2) \tag{1}$$

$$\rho(0) - \frac{\partial p}{\partial y} + \mu(0) = 2a^2\rho(x^2y + y^3) \tag{2}$$

$$\rho(-g) - \frac{\partial p}{\partial z} + \mu(0) = 0 \tag{3}$$

The viscous terms vanish identically (although μ is *not* zero). Equation (3) can be integrated partially to obtain

$$p = -\rho g z + f_1(x, y) \tag{4}$$

i.e., the pressure is hydrostatic in the z direction, which follows anyway from the fact that the flow is two-dimensional ($w = 0$). Now the question is: Do Eqs. (1) and (2) show that the given velocity field *is* a solution? One way to find out is to form the mixed derivative $\partial^2 p/(\partial x \, \partial y)$ from (1) and (2) separately and then compare them.
Differentiate Eq. (1) with respect to y

$$\frac{\partial^2 p}{\partial x \, \partial y} = -4a^2\rho xy \tag{5}$$

Now differentiate Eq. (2) with respect to x

$$\frac{\partial^2 p}{\partial x \, \partial y} = \frac{\partial}{\partial x}[2a^2\rho(x^2y + y^3)] = -4a^2\rho xy \tag{6}$$

Since these are identical, the given velocity field is an *exact* solution to the Navier-Stokes equation. *Ans.*
To find the pressure distribution, substitute Eq. (4) into Eqs. (1) and (2), which will enable us to find $f_1(x, y)$

$$\frac{\partial f_1}{\partial x} = -2a^2\rho(x^3 + xy^2) \tag{7}$$

$$\frac{\partial f_1}{\partial y} = -2a^2\rho(x^2y + y^3) \tag{8}$$

Integrate Eq. (7) partially with respect to x

$$f_1 = -\tfrac{1}{2}a^2\rho(x^4 + 2x^2y^2) + f_2(y) \tag{9}$$

Differentiate this with respect to y and compare with Eq. (8)

$$\frac{\partial f_1}{\partial y} = -2a^2\rho x^2y + f'_2(y) \tag{10}$$

Comparing (8) and (10), we see they are equivalent if

$$f'_2(y) = -2a^2\rho y^3$$

or $$f_2(y) = -\tfrac{1}{2}a^2\rho y^4 + C \tag{11}$$

where C is a constant. Combine Eqs. (4), (9), and (11) to give the complete expression for pressure distribution

$$p(x, y, z) = -\rho g z - \tfrac{1}{2} a^2 \rho (x^4 + y^4 + 2x^2 y^2) + C \qquad \text{Ans.} \quad (12)$$

This is the desired solution. Do you recognize it? Not unless you go back to the beginning and square the velocity components:

$$u^2 + v^2 + w^2 = V^2 = a^2 (x^4 + y^4 + 2x^2 y^2) \qquad (13)$$

Comparing with Eq. (12), we can rewrite the pressure distribution as

$$p + \tfrac{1}{2} \rho V^2 + \rho g z = C \qquad (14)$$

This is Bernoulli's equation (3.63). That is no accident, because the velocity distribution given in this problem is one of a family of flows which are solutions to the Navier-Stokes equation and which satisfy Bernoulli's incompressible equation everywhere in the flow field. They are called *irrotational flows*, for which curl $\mathbf{V} = \nabla \times \mathbf{V} \equiv 0$. This subject will be discussed again in Sec. 4.9.

4.4 THE DIFFERENTIAL EQUATION OF ANGULAR MOMENTUM

Having now been through the same approach for both mass and linear momentum, we can go rapidly through a derivation of the differential angular-momentum relation. The appropriate form of the integral angular-momentum equation for a fixed control volume is

$$\sum \mathbf{M}_O = \frac{\partial}{\partial t} \left[\iiint_{\text{CV}} (\mathbf{r} \times \mathbf{V}) \rho \, d\mathcal{V} \right] + \iint_{\text{CS}} (\mathbf{r} \times \mathbf{V}) \rho (\mathbf{V} \cdot \mathbf{n}) \, dA \qquad (3.71)$$

We shall confine ourselves to an axis O which is parallel to the z axis and passes through the centroid of the elemental control volume. This is shown in Fig. 4.5. Let θ be the angle of rotation about O of the fluid within the control volume. The only

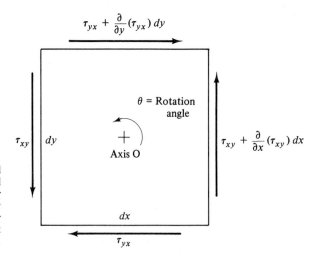

Fig. 4.5 Elemental cartesian fixed control volume showing shear stresses which may cause a net angular acceleration about axis O.

stresses which have moments about O are the shear stresses τ_{xy} and τ_{yx}. We can evaluate the moments about O and the angular-momentum terms about O. A lot of algebra is involved, and we give here only the result

$$\left[\tau_{xy} - \tau_{yx} + \frac{1}{2}\frac{\partial}{\partial x}(\tau_{xy})\,dx - \frac{1}{2}\frac{\partial}{\partial y}(\tau_{yx})\,dy\right]dx\,dy\,dz$$

$$= \tfrac{1}{12}\rho(dx\,dy\,dz)(dx^2 + dy^2)\frac{d^2\theta}{dt^2} \quad (4.39)$$

Assuming that the angular acceleration $d^2\theta/dt^2$ is not infinite, we can neglect all higher-order differential terms, which leaves a finite and interesting result

$$\tau_{xy} \approx \tau_{yx} \quad (4.40)$$

Had we summed moments about axes parallel to y or x, we would have obtained exactly analogous results

$$\tau_{xz} \approx \tau_{zx} \qquad \tau_{yz} \approx \tau_{zy} \quad (4.41)$$

There is *no* differential angular-momentum equation. Application of the integral theorem to a differential element gives the result, well known to students of stress analysis, that the shear stresses are symmetric: $\tau_{ij} = \tau_{ji}$. This is the only result of this section.[1] There is no differential equation to remember, which leaves room in your brain for the next topic, the differential energy equation.

4.5 THE DIFFERENTIAL EQUATION OF ENERGY [2]

We are now so used to this type of derivation that we can race through the energy equation at a bewildering pace. The appropriate integral relation for the fixed control volume of Fig. 4.1 is

$$\dot{Q} - \dot{W}_s - \dot{W}_v = \frac{\partial}{\partial t}\left(\iiint_{CV} e\rho\,d\mathcal{V}\right) + \iint_{CS}\left(e + \frac{p}{\rho}\right)\rho(\mathbf{V}\cdot\mathbf{n})\,dA \quad (3.81)$$

where $\dot{W}_s = 0$ because there can be no infinitesimal shaft protruding into the control volume. By analogy with Eq. (4.20), the right-hand side becomes, for this tiny element,

$$\dot{Q} - \dot{W}_v = \left[\frac{\partial}{\partial t}(\rho e) + \frac{\partial}{\partial x}(\rho u \xi) + \frac{\partial}{\partial y}(\rho v \xi) + \frac{\partial}{\partial z}(\rho w \xi)\right]dx\,dy\,dz \quad (4.42)$$

where $\xi = e + \dfrac{p}{\rho}$

---

[1] We are neglecting the possibility of a finite *couple* being applied to the element by some powerful external force field. See, for example, Ref. 6, p. 44.

[2] This section may be omitted without loss of continuity.

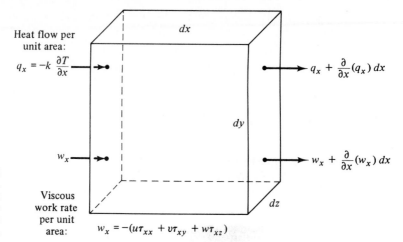

Heat flow per unit area:

$q_x = -k \dfrac{\partial T}{\partial x}$

$q_x + \dfrac{\partial}{\partial x}(q_x)\,dx$

$w_x + \dfrac{\partial}{\partial x}(w_x)\,dx$

Viscous work rate per unit area:

$w_x = -(u\tau_{xx} + v\tau_{xy} + w\tau_{xz})$

Fig. 4.6 Elemental cartesian control volume showing heat flow and viscous-work-rate terms in the x direction.

When we use the continuity equation by analogy with Eq. (4.21), this becomes

$$\dot{Q} - \dot{W}_v = \left(\rho\,\frac{de}{dt} + \mathbf{V}\cdot\nabla p\right) dx\,dy\,dz \tag{4.43}$$

To evaluate $\dot{Q}$, we neglect radiation and consider only heat conduction through the sides of the element. The heat flow by conduction follows Fourier's law from Chap. 1

$$\mathbf{q} = -k\,\nabla T \tag{1.47}$$

where k is the coefficient of thermal conductivity of the fluid. Figure 4.6 shows the heat flow passing through the x faces, the y and z heat flows being omitted for clarity. We can list these six heat-flux terms:

Faces	Inlet heat flux	Outlet heat flux
x	$q_x\,dy\,dz$	$\left[q_x + \dfrac{\partial}{\partial x}(q_x)\,dx\right]dy\,dz$
y	$q_y\,dx\,dz$	$\left[q_y + \dfrac{\partial}{\partial y}(q_y)\,dy\right]dx\,dz$
z	$q_z\,dx\,dy$	$\left[q_z + \dfrac{\partial}{\partial z}(q_z)\,dz\right]dx\,dy$

By adding the inlet terms and subtracting the outlet terms we obtain the net heat added to the element

$$\dot{Q} = -\left[\frac{\partial}{\partial x}(q_x) + \frac{\partial}{\partial y}(q_y) + \frac{\partial}{\partial z}(q_z)\right] dx\,dy\,dz = -\nabla\cdot\mathbf{q}\,dx\,dy\,dz \tag{4.44}$$

As expected, the heat flux is proportional to the element volume. Introducing Fourier's law from Eq. (1.47), we have

$$\dot{Q} = \nabla \cdot (k \, \nabla T) \, dx \, dy \, dz \tag{4.45}$$

The rate of work done by viscous stresses equals the product of the stress component times its corresponding velocity component times the area of the element face. Figure 4.6 shows that the work rate on the left x face is

$$\dot{W}_{v,\text{LF}} = w_x \, dy \, dz \qquad \text{where } w_x = -(u\tau_{xx} + v\tau_{xy} + w\tau_{xz}) \tag{4.46}$$

(where the subscript LF stands for left face) and a slightly different work on the right face due to the gradient in w_x. These work fluxes could be tabulated in exactly the same manner as the heat fluxes in the previous table, with w_x replacing q_x, etc. After subtracting outlet from inlet terms, the net viscous-work rate becomes

$$\dot{W}_v = -\left[\frac{\partial}{\partial x}(u\tau_{xx} + v\tau_{xy} + w\tau_{xz}) + \frac{\partial}{\partial y}(u\tau_{yx} + v\tau_{yy} + w\tau_{yz})\right.$$

$$\left. + \frac{\partial}{\partial z}(u\tau_{zx} + v\tau_{zy} + w\tau_{zz})\right] dx \, dy \, dz$$

$$= -\nabla \cdot (\mathbf{V} \cdot \tau_{ij}) \, dx \, dy \, dz \tag{4.47}$$

We now substitute Eqs. (4.45) and (4.47) into Eq. (4.43) to obtain one form of the differential energy equation

$$\rho \frac{de}{dt} + \mathbf{V} \cdot \nabla p = \nabla \cdot (k \, \nabla T) + \nabla \cdot (\mathbf{V} \cdot \tau_{ij}) \qquad \text{where } e = \hat{u} + \tfrac{1}{2}V^2 + gz \tag{4.48}$$

A more useful form is obtained if we split up the viscous-work term

$$\nabla \cdot (\mathbf{V} \cdot \tau_{ij}) \equiv \mathbf{V} \cdot (\nabla \cdot \tau_{ij}) + \Phi \tag{4.49}$$

where Φ is short for the *viscous-dissipation function*.[1] For a newtonian incompressible viscous fluid, this function has the form

$$\Phi = \mu\left[2\left(\frac{\partial u}{\partial x}\right)^2 + 2\left(\frac{\partial v}{\partial y}\right)^2 + 2\left(\frac{\partial w}{\partial z}\right)^2 + \left(\frac{\partial v}{\partial x} + \frac{\partial u}{\partial y}\right)^2\right.$$

$$\left. + \left(\frac{\partial w}{\partial y} + \frac{\partial v}{\partial z}\right)^2 + \left(\frac{\partial u}{\partial z} + \frac{\partial w}{\partial x}\right)^2\right] \tag{4.50}$$

Since all terms are quadratic, viscous dissipation is always positive, so that a viscous flow always tends to lose its available energy due to dissipation, in accordance with the second law of thermodynamics.

Now substitute Eq. (4.49) into Eq. (4.48), using the linear-momentum equation (4.33) to eliminate $\nabla \cdot \tau_{ij}$. This will cause the kinetic and potential energy to cancel out, leaving a more customary form of the general differential energy equation

$$\rho \frac{d\hat{u}}{dt} + p(\nabla \cdot \mathbf{V}) = \nabla \cdot (k \, \nabla T) + \Phi \tag{4.51}$$

[1] For further details, see, for example, Ref. 5, p. 75.

This equation is valid for a newtonian fluid under very general conditions of unsteady, compressible, viscous, heat-conducting flow, except that it neglects radiation heat transfer and internal *sources* of heat that might occur during a chemical or nuclear reaction.

Equation (4.51) is too difficult to analyze except on a digital computer [1]. It is customary to make the following approximations:

$$d\hat{u} \approx c_v \, dT \qquad c_v, \mu, k, \rho \approx \text{const} \tag{4.52}$$

Equation (4.51) then takes the simpler form

$$\rho c_v \frac{dT}{dt} = k \, \nabla^2 T + \Phi \tag{4.53}$$

which involves temperature T as the sole primary variable plus velocity as a secondary variable through the total time-derivative operator

$$\frac{dT}{dt} = \frac{\partial T}{\partial t} + u \frac{\partial T}{\partial x} + v \frac{\partial T}{\partial y} + w \frac{\partial T}{\partial z} \tag{4.54}$$

A great many interesting solutions to Eq. (4.53) are known for various flow conditions, and extended treatments are given in advanced books on viscous flow [4, 5] and books on heat transfer [7, 8].

One well-known special case of Eq. (4.53) occurs when the fluid is at rest or has negligible velocity, where the dissipation Φ and convective terms become negligible

$$\rho c_v \frac{\partial T}{\partial t} = k \, \nabla^2 T \tag{4.55}$$

This is called the *heat-conduction equation* in applied mathematics and is valid for solids and fluids at rest. The solution to Eq. (4.55) for various conditions is a large part of courses and books on heat transfer.

This completes the derivation of the basic differential equations of fluid motion.

4.6 BOUNDARY CONDITIONS FOR THE BASIC EQUATIONS

There are three basic differential equations of fluid motion, just derived. Let us summarize them here:

Continuity:
$$\frac{\partial \rho}{\partial t} + \nabla \cdot (\rho \mathbf{V}) = 0 \tag{4.56}$$

Momentum:
$$\rho \frac{d\mathbf{V}}{dt} = \rho \mathbf{g} - \nabla p + \nabla \cdot \tau_{ij} \tag{4.57}$$

Energy:
$$\rho \frac{d\hat{u}}{dt} + p(\nabla \cdot \mathbf{V}) = \nabla \cdot (k \, \nabla T) + \Phi \tag{4.58}$$

where Φ is given by Eq. (4.50). In general, the density is variable, so that these three equations contain five unknowns, ρ, V, p, $\hat{u}$, and T. Therefore we need two additional relations to complete the system of equations. These are provided by data or algebraic expressions for the state relations of the thermodynamic properties

$$\rho = \rho(p, T) \qquad \hat{u} = \hat{u}(p, T) \tag{4.59}$$

For example, for a perfect gas with constant specific heats we complete the system with

$$\rho = \frac{p}{RT} \qquad \hat{u} = \int c_v \, dT \approx c_v T + \text{const} \tag{4.60}$$

It is shown in advanced books [4, 5] that this system of equations, (4.56) to (4.59), is well posed and can be solved analytically or numerically, subject to the proper boundary conditions.

What are the proper boundary conditions? First, if the flow is unsteady, there must be an *initial condition* or initial spatial distribution known for each variable:

At $t = 0$: $\rho, V, p, \hat{u}, T = \text{known } f(x, y, z) \tag{4.61}$

Thereafter, for all times t to be analyzed, we must know something about the variables at each *boundary* enclosing the flow.

Figure 4.7 illustrates the three most common types of boundaries encountered in fluid-flow analysis: (1) a solid wall, (2) an inlet or outlet, and (3) a liquid-gas interface.

First, for a solid, impermeable wall, there is no slip and no temperature jump in a viscous heat-conducting fluid

$$\mathbf{V}_{\text{fluid}} = \mathbf{V}_{\text{wall}} \qquad T_{\text{fluid}} = T_{\text{wall}} \qquad \text{solid wall} \tag{4.62}$$

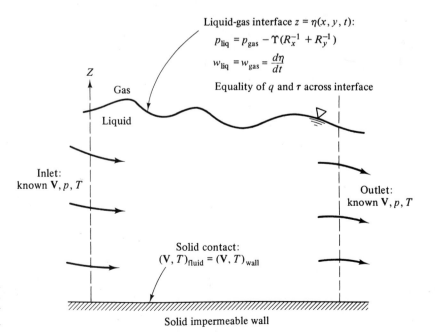

Liquid-gas interface $z = \eta(x, y, t)$:

$$p_{\text{liq}} = p_{\text{gas}} - \Upsilon(R_x^{-1} + R_y^{-1})$$

$$w_{\text{liq}} = w_{\text{gas}} = \frac{d\eta}{dt}$$

Equality of q and τ across interface

Gas

Liquid

Inlet:
known $\mathbf{V}, p, T$

Outlet:
known $\mathbf{V}, p, T$

Solid contact:
$(\mathbf{V}, T)_{\text{fluid}} = (\mathbf{V}, T)_{\text{wall}}$

Fig. 4.7 Typical boundary conditions in a viscous heat-conducting fluid-flow analysis.

Solid impermeable wall

The only exception to Eq. (4.62) occurs in an extremely rarefied gas flow, where slippage can be present [5].

Second, at any inlet or outlet section of the flow, the complete distribution of velocity, pressure, and temperature must be known for all times:

$$\text{Inlet or outlet:} \qquad \text{Known } \mathbf{V}, p, T \qquad (4.63)$$

These inlet and outlet sections can be and often are at $\pm\infty$, simulating a body immersed in an infinite expanse of fluid.

Finally, the most complex conditions occur at a liquid-gas interface, or free surface, as sketched in Fig. 4.7. Let us denote the interface by

$$\text{Interface:} \qquad z = \eta(x, y, t) \qquad (4.64)$$

Then there must be equality of vertical velocity across the interface, so that no holes appear between liquid and gas:

$$w_{\text{liq}} = w_{\text{gas}} = \frac{d\eta}{dt} = \frac{\partial\eta}{\partial t} + u\frac{\partial\eta}{\partial x} + v\frac{\partial\eta}{\partial y} \qquad (4.65)$$

This is called the *kinematic boundary condition*.

There must be mechanical equilibrium across the interface. The viscous-shear stresses must balance

$$(\tau_{zy})_{\text{liq}} = (\tau_{zy})_{\text{gas}} \qquad (\tau_{zx})_{\text{liq}} = (\tau_{zx})_{\text{gas}} \qquad (4.66)$$

Neglecting the viscous normal stresses, the pressures must balance at the interface except for surface-tension effects

$$p_{\text{liq}} = p_{\text{gas}} - \Upsilon(R_x^{-1} + R_y^{-1}) \qquad (4.67)$$

which is equivalent to Eq. (1.52). The radii of curvature can be written in terms of the free-surface position η

$$R_x^{-1} + R_y^{-1} = \frac{\partial}{\partial x}\left[\frac{\partial\eta/\partial x}{\sqrt{1 + (\partial\eta/\partial x)^2 + (\partial\eta/\partial y)^2}}\right]$$
$$+ \frac{\partial}{\partial y}\left[\frac{\partial\eta/\partial y}{\sqrt{1 + (\partial\eta/\partial x)^2 + (\partial\eta/\partial y)^2}}\right] \qquad (4.68)$$

Good luck if you have to use Eq. (4.68). You will probably want to set the denominators equal to unity for simplicity.

Finally, the heat transfer must be the same on both sides of the interface, since no heat can be stored in the infinitesimally thin interface

$$(q_z)_{\text{liq}} = (q_z)_{\text{gas}} \qquad (4.69)$$

Neglecting radiation, this is equivalent to

$$\left(k\frac{\partial T}{\partial z}\right)_{\text{liq}} = \left(k\frac{\partial T}{\partial z}\right)_{\text{gas}} \qquad (4.70)$$

This is as much detail as we wish to give at this level of exposition. Further and even more complicated details on fluid-flow boundary conditions are given in Refs. 5 and 9.

Simplified Free-Surface Conditions

In the introductory analyses given in this book, such as open-channel flows in Chap. 10, we shall back away from the exact conditions (4.65) to (4.69) and assume that the upper fluid is an "atmosphere" which merely exerts pressure upon the lower fluid, with shear and heat conduction negligible. We also neglect nonlinear terms involving the slopes of the free surface. We then have a much simpler and linear set of conditions at the surface

$$p_{\text{liq}} \approx p_{\text{gas}} - \Upsilon\left(\frac{\partial^2 \eta}{\partial x^2} + \frac{\partial^2 \eta}{\partial y^2}\right) \qquad w_{\text{liq}} \approx \frac{\partial \eta}{\partial t}$$

$$\left(\frac{\partial V}{\partial z}\right)_{\text{liq}} \approx 0 \qquad \left(\frac{\partial T}{\partial z}\right)_{\text{liq}} \approx 0 \qquad (4.71)$$

In many cases, such as open-channel flow, we can also neglect surface tension, so that

$$p_{\text{liq}} \approx p_{\text{atm}} \qquad (4.72)$$

These are the types of approximations which will be used in Chap. 10. The nondimensional forms of these conditions will also be useful in Chap. 5.

Incompressible Flow with Constant Properties

Flow with constant ρ, μ, and k is a basic simplification which will be used for example throughout Chap. 6. The basic equations of motion (4.56) to (4.58) reduce to:

Continuity:
$$\mathbf{V} \cdot \mathbf{V} = 0 \qquad (4.73)$$

Momentum:
$$\rho \frac{d\mathbf{V}}{dt} = \rho \mathbf{g} - \nabla p + \mu \nabla^2 \mathbf{V} \qquad (4.74)$$

Energy:
$$\rho c_v \frac{dT}{dt} = k \nabla^2 T + \Phi \qquad (4.75)$$

Since ρ is constant, there are only three unknowns, p, $\mathbf{V}$, and T. The system is closed.[1] Not only that, the system splits apart: continuity and momentum are independent of T. Thus we can solve Eqs. (4.73) and (4.74) entirely separately for the pressure and velocity, using such boundary conditions as:

Solid surface:
$$\mathbf{V} = \mathbf{V}_{\text{wall}} \qquad (4.76)$$

Inlet or outlet:
$$\text{Known } \mathbf{V}, p \qquad (4.77)$$

Free surface:
$$p \approx p_a \qquad w \approx \frac{\partial \eta}{\partial t} \qquad (4.78)$$

---

[1] For this system, what are the thermodynamic equivalents to Eq. (4.59)?

Later, entirely at our leisure,[1] we can solve for the temperature distribution from Eq. (4.75), which depends upon velocity $\mathbf{V}$ through the dissipation Φ and the total time-derivative operator d/dt.

Inviscid-Flow Approximations

Chapter 8 assumes inviscid flow throughout, for which the viscosity $\mu = 0$. The momentum equation (4.74) reduces to

$$\rho \frac{d\mathbf{V}}{dt} = \rho \mathbf{g} - \nabla p \tag{4.79}$$

This is *Euler's equation*; it can be integrated along a streamline to obtain Bernoulli's equation (see Sec. 4.8). By neglecting viscosity we have lost the second-order derivative of $\mathbf{V}$ in Eq. (4.74); therefore we must relax one boundary condition on velocity. The only mathematically sensible condition to drop is the no-slip condition at the wall. We let the flow slip parallel to the wall but do not allow it to flow into the wall. The proper inviscid condition is that the normal velocities must match at any solid surface:

Inviscid flow: $$(V_n)_{\text{fluid}} = (V_n)_{\text{wall}} \tag{4.80}$$

In most cases the wall is fixed; therefore the proper inviscid-flow condition is

$$V_n = 0 \tag{4.81}$$

There is *no* condition whatever on the tangential velocity component at the wall in inviscid flow. The tangential velocity will be part of the solution, and the correct value will appear after the analysis is completed (see Chap. 8).

EXAMPLE 4.5   For steady incompressible laminar flow through a long tube, the velocity distribution is given by

$$v_z = U\left(1 - \frac{r^2}{R^2}\right) \qquad v_r = v_\theta = 0$$

where U is the maximum, or centerline, velocity and R is the tube radius. If the wall temperature is constant at T_w and the temperature $T = T(r)$ only, find $T(r)$ for this flow.

Solution   With $T = T(r)$, Eq. (4.75) reduces for steady flow to

$$\rho c_v v_r \frac{dT}{dr} = \frac{k}{r} \frac{d}{dr}\left(r \frac{dT}{dr}\right) + \mu\left(\frac{dv_z}{dr}\right)^2 \tag{1}$$

But since $v_r = 0$ for this flow, the convective term on the left vanishes. Introduce v_z into Eq. (1) to obtain

$$\frac{k}{r} \frac{d}{dr}\left(r \frac{dT}{dr}\right) = -\mu\left(\frac{dv_z}{dr}\right)^2 = -\frac{4U^2 \mu r^2}{R^4} \tag{2}$$

---

[1] Since temperature is entirely *uncoupled* by this assumption, we may never get around to solving for it here and may ask you to wait until a course on heat transfer.

Multiply through by r/k and integrate once:

$$r\frac{dT}{dr} = -\frac{\mu U^2 r^4}{kR^4} + C_1 \tag{3}$$

Divide through by r and integrate once again:

$$T = -\frac{\mu U^2 r^4}{4kR^4} + C_1 \ln r + C_2 \tag{4}$$

Now we are in position to apply our boundary conditions to evaluate C_1 and C_2.

First, since the logarithm of zero is $-\infty$, the temperature at $r = 0$ will be infinite unless

$$C_1 = 0 \tag{5}$$

Thus we eliminate the possibility of a logarithmic singularity. The same thing will happen if we apply the *symmetry* condition $dT/dr = 0$ at $r = 0$ to Eq. (3). The constant C_2 is then found by the wall-temperature condition at $r = R$

$$T = T_w = -\frac{\mu U^2}{4k} + C_2$$

or

$$C_2 = T_w + \frac{\mu U^2}{4k} \tag{6}$$

The correct solution is thus

$$T(r) = T_w + \frac{\mu U^2}{4k}\left(1 - \frac{r^4}{R^4}\right) \qquad\qquad Ans. \tag{7}$$

which is a fourth-order parabolic distribution with a maximum value $T_0 = T_w + \mu U^2/4k$ at the centerline.

4.7 THE STREAM FUNCTION

We have seen in Sec. 4.6 that even if the temperature is uncoupled from our system of equations of motion, we must solve the continuity and momentum equations simultaneously for pressure and velocity. The *stream function* ψ is a clever device which allows us to wipe out the continuity equation and solve the momentum equation directly for the single variable ψ.

The stream-function idea works only if the continuity equation (4.56) can be reduced to *two* terms. In general, we have *four* terms:

Cartesian:
$$\frac{\partial \rho}{\partial t} + \frac{\partial}{\partial x}(\rho u) + \frac{\partial}{\partial y}(\rho v) + \frac{\partial}{\partial z}(\rho w) = 0$$

Cylindrical:
$$\frac{\partial \rho}{\partial t} + \frac{1}{r}\frac{\partial}{\partial r}(r\rho v_r) + \frac{1}{r}\frac{\partial}{\partial \theta}(\rho v_\theta) + \frac{\partial}{\partial z}(\rho v_z) = 0 \tag{4.82}$$

First, let us eliminate unsteady flow, which is a peculiar and unrealistic application of the stream-function idea. Reduce either of Eqs. (4.82) to any *two* terms. The most common application is incompressible flow in the xy plane

$$\frac{\partial u}{\partial x} + \frac{\partial v}{\partial y} = 0 \tag{4.83}$$

This equation is satisfied *identically* if a function $\psi(x, y)$ is defined such that Eq. (4.83) becomes

$$\frac{\partial}{\partial x}\left(\frac{\partial \psi}{\partial y}\right) + \frac{\partial}{\partial y}\left(-\frac{\partial \psi}{\partial x}\right) \equiv 0 \qquad (4.84)$$

Comparison of (4.83) and (4.84) shows that this new function ψ must be defined such that

$$u = \frac{\partial \psi}{\partial y} \qquad v = -\frac{\partial \psi}{\partial x} \qquad (4.85)$$

or

$$\mathbf{V} = \mathbf{i}\frac{\partial \psi}{\partial y} - \mathbf{j}\frac{\partial \psi}{\partial x}$$

Is this legitimate? Yes, it is just a mathematical trick of replacing two variables (u and v) by a single higher-order function ψ. The vorticity, or curl $\mathbf{V}$, is an interesting function

$$\text{curl } \mathbf{V} = 2\mathbf{k}\omega_z = -\mathbf{k}\,\nabla^2\psi \qquad \text{where } \nabla^2\psi = \frac{\partial^2\psi}{\partial x^2} + \frac{\partial^2\psi}{\partial y^2} \qquad (4.86)$$

Thus, if we take the curl of the momentum equation (4.74) and utilize Eq. (4.86), we obtain a single equation for ψ

$$\frac{\partial \psi}{\partial y}\frac{\partial}{\partial x}(\nabla^2\psi) - \frac{\partial \psi}{\partial x}\frac{\partial}{\partial y}(\nabla^2\psi) = v\,\nabla^2(\nabla^2\psi) \qquad (4.87)$$

where $v = \mu/\rho$ is the kinematic viscosity. This is partly a victory and partly a defeat: Eq. (4.87) is scalar and has only one variable, ψ, but it now contains *fourth*-order derivatives and probably will require computer analysis. There will be four boundary conditions required on ψ. For example, for the flow of a uniform stream in the x direction past a solid body, the four conditions would be

At infinity: $\qquad\qquad \dfrac{\partial \psi}{\partial y} = U_\infty \qquad \dfrac{\partial \psi}{\partial x} = 0$

$$\qquad\qquad\qquad\qquad\qquad\qquad\qquad\qquad\qquad\qquad\qquad (4.88)$$

At the body: $\qquad\qquad \dfrac{\partial \psi}{\partial y} = \dfrac{\partial \psi}{\partial x} = 0$

Many examples of numerical solution of Eqs. (4.87) and (4.88) are given in Ref. 1.

One important application is inviscid *irrotational* flow in the xy plane, where $\omega_z \equiv 0$. Equations (4.86) and (4.87) reduce to

$$\nabla^2\psi = \frac{\partial^2\psi}{\partial x^2} + \frac{\partial^2\psi}{\partial y^2} = 0 \qquad (4.89)$$

This is the second-order *Laplace equation* (Chap. 8), for which many solutions and analytical techniques are known. Also, boundary conditions like Eq. (4.88) reduce to

At infinity: $\qquad\qquad\qquad \psi = U_\infty y + \text{const}$

$$\qquad\qquad\qquad\qquad\qquad\qquad\qquad\qquad\qquad\qquad\qquad (4.90)$$

At the body: $\qquad\qquad\qquad \psi = \text{const}$

It is well within our capability to find some useful solutions to Eqs. (4.89) and (4.90), which we shall do in Chap. 8.

Geometric Interpretation of ψ

The fancy mathematics above would serve by itself to make the stream function immortal and always useful to engineers. Even better, though, ψ has a beautiful geometric interpretation: lines of constant ψ are *streamlines* of the flow. This can be shown as follows. From Eq. (1.57) the definition of a streamline in two-dimensional flow is

$$\frac{dx}{u} = \frac{dy}{v}$$

or $\qquad\qquad\qquad u\, dy - v\, dx = 0 \qquad \text{streamline} \qquad\qquad (4.91)$

Introducing the stream function from Eq. (4.85), we have

$$\frac{\partial \psi}{\partial x}\, dx + \frac{\partial \psi}{\partial y}\, dy = 0 = d\psi \qquad\qquad (4.92)$$

Thus the change in ψ is zero along a streamline, or

$$\psi = \text{const along a streamline} \qquad\qquad (4.93)$$

Having found a given solution $\psi(x, y)$, we can plot lines of constant ψ to give the streamlines of the flow.

There is also a physical interpretation which relates ψ to volume flow. From Fig. 4.8, we can compute the volume flow dQ through an element ds of control surface of unit depth

$$dQ = \mathbf{V} \cdot \mathbf{n}\, dA = \left(\mathbf{i}\frac{\partial \psi}{\partial y} - \mathbf{j}\frac{\partial \psi}{\partial x} \right) \cdot \left(\mathbf{i}\frac{dy}{ds} - \mathbf{j}\frac{dx}{ds} \right) ds(1)$$

$$= \frac{\partial \psi}{\partial x}\, dx + \frac{\partial \psi}{\partial y}\, dy = d\psi \qquad\qquad (4.94)$$

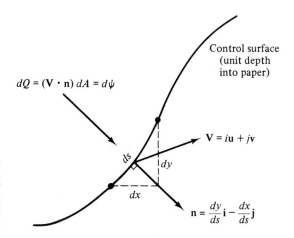

$dQ = (\mathbf{V} \cdot \mathbf{n})\, dA = d\psi$

Control surface (unit depth into paper)

$\mathbf{V} = i u + j v$

ds

dy

dx

$\mathbf{n} = \dfrac{dy}{ds}\mathbf{i} - \dfrac{dx}{ds}\mathbf{j}$

Fig. 4.8 Geometric interpretation of stream function: volume flow through a differential portion of a control surface.

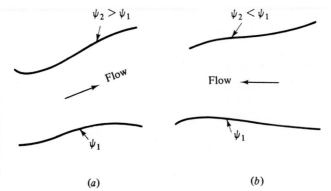

Fig. 4.9 Sign convention for flow in terms of change in stream function: (a) flow to the right if ψ_U is greater; (b) flow to the left if ψ_L is greater.

Thus the change in ψ across the element is numerically equal to the volume flow through the element. The volume flow between any two points in the flow field is equal to the change in stream function between those points:

$$Q_{1\to2} = \int_1^2 (\mathbf{V} \cdot \mathbf{n})\, dA = \int_1^2 d\psi = \psi_2 - \psi_1 \tag{4.95}$$

Further, the direction of the flow can be ascertained by noting whether ψ increases or decreases. As sketched in Fig. 4.9, the flow is to the right if ψ_U is greater than ψ_L, where the subscripts stand for upper and lower, as before; otherwise the flow is to the left.

Both the stream function and the velocity potential were invented by the French mathematician Joseph Louis Lagrange and published in his treatise on fluid mechanics in 1781.

EXAMPLE **4.6**   If a stream function exists for the velocity field of Example 4.4

$$u = a(x^2 - y^2) \qquad v = -2axy \qquad w = 0$$

find it, plot it, and interpret it.

Solution   Since this flow field was shown expressly in Example 4.2 to satisfy the equation of continuity, we are pretty sure that a stream function does exist. We can check again to see if

$$\frac{\partial u}{\partial x} + \frac{\partial v}{\partial y} = 0$$

Substitute: $2ax + (-2ax) = 0$ checks

Therefore we are certain that a stream function exists. To find ψ we simply set

$$u = \frac{\partial \psi}{\partial y} = ax^2 - ay^2 \tag{1}$$

$$v = -\frac{\partial \psi}{\partial x} = -2axy \tag{2}$$

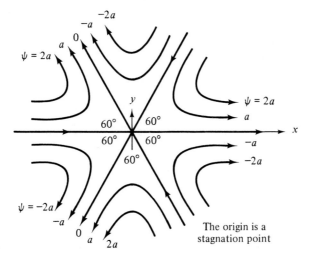

Fig. E4.6a

and work from either one toward the other. Integrate (1) partially

$$\psi = ax^2y - \frac{ay^3}{3} + f(x) \tag{3}$$

Differentiate (3) with respect to x and compare with (2)

$$\frac{\partial \psi}{\partial x} = 2axy + f'(x) = 2axy \tag{4}$$

Therefore $f'(x) = 0$, or $f = $ constant. The complete stream function is thus found

$$\psi = a\left(x^2y - \frac{y^3}{3}\right) + C \qquad\qquad Ans. \tag{5}$$

To plot this, set $C = 0$ for convenience and plot the function

$$3x^2y - y^3 = \frac{3\psi}{a} \tag{6}$$

for constant values of ψ. The result is shown in Fig. E4.6a to be six 60° wedges of circulating motion, each with identical flow patterns except for the arrows. Once the streamlines are labeled, the flow directions follow from the sign convention of Fig. 4.9. How can the flow be interpreted? Since there is slip along all streamlines, no streamline can truly represent a solid surface in a viscous flow. However, the flow could represent the impingement of three incoming streams at 60, 180, and 300°. This would be a rather unrealistic yet exact solution to the Navier-Stokes equation, as we showed in Example 4.4.

By allowing the flow to slip as a frictionless approximation, we could let any given streamline be a body shape. Some examples are shown in Fig. E4.6b.

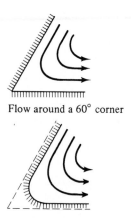

Flow around a 60° corner

Flow around a
rounded 60° corner

Fig. E4.6b

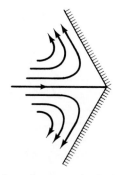

Incoming stream impinging
against a 120° corner

A stream function also exists in a variety of other physical situations where only two coordinates are needed to define the flow. Three examples are illustrated here.

Steady Plane Compressible Flow

Suppose now that the density is variable but that $w = 0$, so that the flow is in the xy plane. Then the equation of continuity becomes

$$\frac{\partial}{\partial x}(\rho u) + \frac{\partial}{\partial y}(\rho v) = 0 \tag{4.96}$$

We see that this is in exactly the same form as Eq. (4.84). Therefore a compressible-flow stream function can be defined such that

$$\rho u = \frac{\partial \psi}{\partial y} \qquad \rho v = -\frac{\partial \psi}{\partial x} \tag{4.97}$$

Again lines of constant ψ are streamlines of the flow, but the change in ψ is now equal to the *mass* flow, not the volume flow

$$d\dot{m} = \rho(\mathbf{V} \cdot \mathbf{n}) \, dA = d\psi$$

or

$$\dot{m}_{1 \to 2} = \int_1^2 \rho(\mathbf{V} \cdot \mathbf{n}) \, dA = \psi_2 - \psi_1 \tag{4.98}$$

The sign convention on flow direction is the same as in Fig. 4.9. This particular stream function combines density with velocity and must be substituted into not only momentum but also the energy and state relations (4.58) and (4.59) with pressure and temperature as companion variables. Thus the compressible stream function is not a great victory, and further assumptions must be made to effect an analytical solution to a typical problem (see, for example, Ref. 5, chap. 7).

Incompressible Plane Flow in Polar Coordinates

Suppose that the important coordinates are r and θ, with $v_z = 0$, and that the density is constant. Then Eq. (4.82b) reduces to

$$\frac{1}{r}\frac{\partial}{\partial r}(rv_r) + \frac{1}{r}\frac{\partial}{\partial \theta}(v_\theta) = 0 \tag{4.99}$$

After multiplying through by r, this is the same as the analogous form of Eq. (4.84)

$$\frac{\partial}{\partial r}\left(\frac{\partial \psi}{\partial \theta}\right) + \frac{\partial}{\partial \theta}\left(-\frac{\partial \psi}{\partial r}\right) = 0 \tag{4.100}$$

By comparison of (4.99) and (4.100) we deduce the form of the incompressible polar-coordinate stream function

$$v_r = \frac{1}{r}\frac{\partial \psi}{\partial \theta} \qquad v_\theta = -\frac{\partial \psi}{\partial r} \tag{4.101}$$

Once again lines of constant ψ are streamlines, and the change in ψ is the *volume flow*, $Q_{1\to2} = \psi_2 - \psi_1$. The sign convention is the same as in Fig. 4.9. This type of stream function is very useful in analyzing flows with cylinders, vortices, sources, and sinks (Chap. 8).

Incompressible Axisymmetric Flow

As a final example, suppose that the flow is three-dimensional (v_r, v_z) but with no circumferential variations, $v_\theta = \partial/\partial\theta = 0$ (see Fig. 4.2 for definition of coordinates). Such a flow is termed *axisymmetric*, and the flow pattern is the same when viewed on any meridional plane through the axis of revolution z. For incompressible flow, Eq. (4.82) becomes

$$\frac{1}{r}\frac{\partial}{\partial r}(rv_r) + \frac{\partial}{\partial z}(v_z) = 0 \tag{4.102}$$

This doesn't seem to work: can't we get rid of the one r outside? But when we realize that r and z are independent coordinates, Eq. (4.102) can be rewritten as

$$\frac{\partial}{\partial r}(rv_r) + \frac{\partial}{\partial z}(rv_z) = 0 \tag{4.103}$$

By analogy with Eq. (4.84), this has the form

$$\frac{\partial}{\partial r}\left(-\frac{\partial \psi}{\partial z}\right) + \frac{\partial}{\partial z}\left(\frac{\partial \psi}{\partial r}\right) = 0 \tag{4.104}$$

By comparing (4.103) and (4.104), we deduce the form of an incompressible axisymmetric stream function $\psi(r, z)$

$$v_r = -\frac{1}{r}\frac{\partial \psi}{\partial z} \qquad v_z = \frac{1}{r}\frac{\partial \psi}{\partial r} \tag{4.105}$$

Here again lines of constant ψ are streamlines, but there is a factor (2π) in the volume flow: $Q_{1 \to 2} = 2\pi(\psi_2 - \psi_1)$. The sign convention on flow is the same as in Fig. 4.9.

EXAMPLE 4.7 Investigate the stream function in polar coordinates

$$\psi = U \sin \theta \left(r - \frac{R^2}{r} \right) \tag{1}$$

where U and R are constants, a velocity and a length, respectively. Plot the streamlines. What does the flow represent? Is it a realistic solution to the basic equations?

Solution The streamlines are lines of constant ψ, which has units of square meters per second. Note that ψ/UR is dimensionless. Rewrite Eq. (1) in dimensionless form

$$\frac{\psi}{UR} = \sin \theta \left(\eta - \frac{1}{\eta} \right) \qquad \eta = \frac{r}{R} \tag{2}$$

Of particular interest is the special line $\psi = 0$. From Eq. (1) or (2) this occurs when (a) $\theta = 0°$ or $180°$ and (b) $r = R$. Case (a) is the x axis and case (b) is a circle of radius R, both of which are plotted in Fig. E4.7.

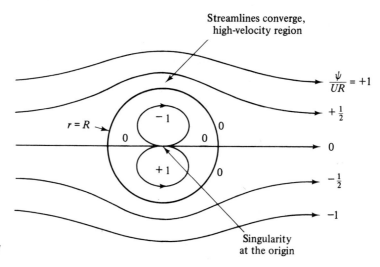

Fig. E4.7

For any other nonzero value of ψ it is easiest to pick a value of r and solve for θ:

$$\sin \theta = \frac{\psi/UR}{r/R - R/r} \tag{3}$$

In general, there will be two solutions for θ because of the symmetry about the y axis. For example take $\psi/UR = +1.0$:

Guess r/R	3.0	2.5	2.0	1.8	1.7	1.618
Compute θ	22°	28°	42°	54°	64°	90°
	158°	152°	138°	156°	116°	

This line is plotted in Fig. E4.7 and passes over the circle $r = R$. You have to watch it, though, because there is a second curve for $\psi/UR = +1.0$ for small $r < R$ below the x axis:

Guess r/R	0.618	0.6	0.5	0.4	0.3	0.2	0.1
Compute θ	$-90°$	$-70°$	$-42°$	$-28°$	$-19°$	$-12°$	$-6°$
		$-110°$	$-138°$	$-152°$	$-161°$	$-168°$	$-174°$

This second curve plots as a closed curve inside the circle $r = R$. There is a singularity of infinite velocity and indeterminate flow direction at the origin. Figure E4.7 shows the full pattern.

The given stream function, Eq. (1), is an exact and classic solution to the momentum equation (4.38) for frictionless flow. Outside the circle $r = R$ it represents two-dimensional inviscid flow of a uniform stream past a circular cylinder (Sec. 8.3). Inside the circle it represents a rather unrealistic trapped circulating motion of what is called a *line doublet*.

4.8 VORTICITY AND IRROTATIONALITY

The assumption of zero fluid angular velocity, or irrotationality, is a very useful simplification. Here we show that angular velocity is associated with the curl of the local-velocity vector.

The differential relations for deformation of a fluid element can be derived by examining Fig. 4.10. Two fluid lines AB and BC, initially perpendicular at time t,

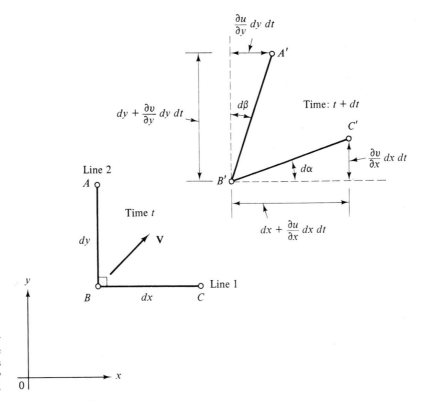

Fig. 4.10 Angular velocity and strain rate of two fluid lines deforming in the xy plane.

move and deform so that at $t + dt$ they have slightly different lengths $A'B'$ and $B'C'$ and are slightly off the perpendicular by angles $d\alpha$ and $d\beta$. Such deformation occurs kinematically because A, B, and C have slightly different velocities when the velocity field V has spatial gradients. All these differential changes in the motion of A, B, and C are noted in Fig. 4.10.

We define the angular velocity ω_z about the z axis as the average rate of counterclockwise turning of the two lines

$$\omega_z = \frac{1}{2}\left(\frac{d\alpha}{dt} - \frac{d\beta}{dt}\right) \tag{4.106}$$

But from Fig. 4.10 $d\alpha$ and $d\beta$ are each directly related to velocity derivatives in the limit of small dt

$$d\alpha = \lim_{dt \to 0}\left(\tan^{-1}\frac{\partial v/\partial x \, dx \, dt}{dx + \partial u/\partial x \, dx \, dt}\right) = \frac{\partial v}{\partial x} \, dt$$

$$d\beta = \lim_{dt \to 0}\left(\tan^{-1}\frac{\partial u/\partial y \, dy \, dt}{dy + \partial v/\partial y \, dy \, dt}\right) = \frac{\partial u}{\partial y} \, dt \tag{4.107}$$

Combining Eqs. (4.106) and (4.107) gives the desired result:

$$\omega_z = \frac{1}{2}\left(\frac{\partial v}{\partial x} - \frac{\partial u}{\partial y}\right) \tag{4.108}$$

In exactly similar manner we determine the other two rates

$$\omega_x = \frac{1}{2}\left(\frac{\partial w}{\partial y} - \frac{\partial v}{\partial z}\right) \qquad \omega_y = \frac{1}{2}\left(\frac{\partial u}{\partial z} - \frac{\partial w}{\partial x}\right) \tag{4.109}$$

The vector $\boldsymbol{\omega} = \mathbf{i}\omega_x + \mathbf{j}\omega_y + \mathbf{k}\omega_z$ is thus one-half the curl of the velocity vector

$$\boldsymbol{\omega} = \tfrac{1}{2}\operatorname{curl}\mathbf{V} = \frac{1}{2}\begin{vmatrix} \mathbf{i} & \mathbf{j} & \mathbf{k} \\ \dfrac{\partial}{\partial x} & \dfrac{\partial}{\partial y} & \dfrac{\partial}{\partial z} \\ u & v & w \end{vmatrix} \tag{4.110}$$

Since the factor of $\tfrac{1}{2}$ is annoying, many workers prefer to use a vector twice as large, called the *vorticity*:

$$\boldsymbol{\zeta} = 2\boldsymbol{\omega} = \operatorname{curl}\mathbf{V} \tag{4.111}$$

Many flows have negligible or zero vorticity and are called *irrotational*

$$\operatorname{curl}\mathbf{V} \equiv 0 \tag{4.112}$$

The next section expands on this idea. Such flows can be incompressible or compressible, steady or unsteady.

We may also note that Fig. 4.10 demonstrates the shear-strain rate of the element, which is defined as the rate of closure of the initially perpendicular lines

$$\dot{\epsilon}_{xy} = \frac{d\alpha}{dt} + \frac{d\beta}{dt} = \frac{\partial v}{\partial x} + \frac{\partial u}{\partial y} \tag{4.113}$$

When multiplied by viscosity μ, this equals the shear stress τ_{xy} in a newtonian fluid, as discussed earlier in Eqs. (4.37). Appendix E lists strain-rate and vorticity components in cylindrical coordinates.

4.9 FRICTIONLESS IRROTATIONAL FLOWS

When a flow is both frictionless and irrotational, pleasant things happen. First, the momentum equation (4.38) reduces to Euler's equation

$$\rho \frac{d\mathbf{V}}{dt} = \rho \mathbf{g} - \nabla p \tag{4.114}$$

Second, there is a great simplification in the acceleration term. Recall from Sec. 1.5 that acceleration has two terms

$$\frac{d\mathbf{V}}{dt} = \frac{\partial \mathbf{V}}{\partial t} + (\mathbf{V} \cdot \nabla)\mathbf{V} \tag{1.13}$$

A beautiful vector identity exists for the second term [11]:

$$(\mathbf{V} \cdot \nabla)\mathbf{V} \equiv \nabla(\tfrac{1}{2}V^2) + \boldsymbol{\zeta} \times \mathbf{V} \tag{4.115}$$

where $\boldsymbol{\zeta} = $ curl $\mathbf{V}$ from Eq. (4.111) is the fluid vorticity.

Now combine (4.114) and (4.115), divide by ρ, and rearrange on the left-hand side. Dot the entire equation into an arbitrary vector displacement $d\mathbf{r}$

$$\left[\frac{\partial \mathbf{V}}{\partial t} + \nabla(\tfrac{1}{2}V^2) + \boldsymbol{\zeta} \times \mathbf{V} + \frac{1}{\rho}\nabla p - \mathbf{g} \right] \cdot (d\mathbf{r}) = 0 \tag{4.116}$$

Nothing works right unless we can get rid of the third term. We want

$$(\boldsymbol{\zeta} \times \mathbf{V}) \cdot (d\mathbf{r}) \equiv 0 \tag{4.117}$$

This will be true under various conditions:

1. $\mathbf{V}$ is zero; trivial, no flow (hydrostatics).
2. $\boldsymbol{\zeta}$ is zero; irrotational flow.
3. $d\mathbf{r}$ is perpendicular to $\boldsymbol{\zeta} \times \mathbf{V}$; this is called *Beltrami flow* and is rather specialized and rare.
4. $d\mathbf{r}$ is parallel to $\mathbf{V}$; we integrate *along a streamline* (see Sec. 3.5).

Condition 4 is the common assumption. If we integrate along a streamline in frictionless compressible flow and take, for convenience, $\mathbf{g} = -g\mathbf{k}$, Eq. (4.116) reduces to

$$\frac{\partial \mathbf{V}}{\partial \mathbf{t}} \cdot d\mathbf{r} + d(\tfrac{1}{2}V^2) + \frac{dp}{\rho} + g\,dz = 0 \tag{4.118}$$

Except for the first term, these are exact differentials. Integrate between any two points 1 and 2 along the streamline:

$$\int_1^2 \frac{\partial V}{\partial t} \, ds + \int_1^2 \frac{dp}{\rho} + \tfrac{1}{2}(V_2^2 - V_1^2) + g(z_2 - z_1) = 0 \qquad (4.119)$$

where ds is the arc length along the streamline. Equation (4.119) is Bernoulli's equation for frictionless unsteady flow along a streamline and is identical to Eq. (3.62). For incompressible steady flow, it reduces to

$$\frac{p}{\rho} + \tfrac{1}{2}V^2 + gz = \text{const along streamline} \qquad (4.120)$$

The constant may vary from streamline to streamline unless the flow is also irrotational (assumption 2). For irrotational flow, $\zeta = 0$, the offending term Eq. (4.117) vanishes regardless of the direction of $d\mathbf{r}$, and Eq. (4.120) then holds all over the flow field with the same constant.

Velocity Potential

Irrotationality gives rise to a scalar function ϕ similar and complementary to the stream function ψ. From a theorem in vector analysis [11] a vector with zero curl must be the gradient of a scalar function

$$\text{If} \quad \nabla \times \mathbf{V} \equiv 0 \quad \text{then} \quad \mathbf{V} = \nabla \phi \qquad (4.121)$$

where $\phi = \phi(x, y, z, t)$ is called the *velocity potential function*. Knowledge of ϕ thus immediately gives the velocity components

$$u = \frac{\partial \phi}{\partial x} \qquad v = \frac{\partial \phi}{\partial y} \qquad w = \frac{\partial \phi}{\partial z} \qquad (4.122)$$

Lines of constant ϕ are called the *potential lines* of the flow.

Note that ϕ, unlike the stream function, is fully three-dimensional and not limited to two coordinates. It reduces a velocity problem with three unknowns u, v, and w to a single unknown potential ϕ; many examples will be given in Chap. 8. The velocity potential also simplifies the unsteady Bernoulli equation (4.118) because if ϕ exists, we obtain

$$\frac{\partial \mathbf{V}}{\partial t} \cdot d\mathbf{r} = \frac{\partial}{\partial t}(\nabla \phi) \cdot d\mathbf{r} = d\!\left(\frac{\partial \phi}{\partial t}\right) \qquad (4.123)$$

Equation (4.118) then becomes a relation between ϕ and p

$$\frac{\partial \phi}{\partial t} + \int \frac{dp}{\rho} + \tfrac{1}{2}|\nabla \phi|^2 + gz = \text{const} \qquad (4.124)$$

This is the unsteady irrotational Bernoulli equation. It is very important in the analysis of accelerating flow fields (see, for example, Refs. 10 and 15), but the only application in this text will be in Sec. 9.3 for steady flow.

Orthogonality of Streamlines and Potential Lines

If a flow is both irrotational and described by only two coordinates, ψ and ϕ both exist and the streamlines and potential lines are everywhere mutually perpendicular except at a stagnation point. For example, for incompressible flow in the xy plane, we would have

$$u = \frac{\partial \psi}{\partial y} = \frac{\partial \phi}{\partial x} \tag{4.125}$$

$$v = -\frac{\partial \psi}{\partial x} = \frac{\partial \phi}{\partial y} \tag{4.126}$$

Can you tell by inspection that these relations imply not only orthogonality but also that ϕ and ψ satisfy Laplace's equation?[1] A line of constant ϕ would be such that the change in ϕ is zero

$$d\phi = \frac{\partial \phi}{\partial x}\, dx + \frac{\partial \phi}{\partial y}\, dy = 0 = u\, dx + v\, dy \tag{4.127}$$

Solving, we have

$$\left(\frac{dy}{dx}\right)_{\phi = \text{const}} = -\frac{u}{v} = -\frac{1}{(dy/dx)_{\psi = \text{const}}} \tag{4.128}$$

Equation (4.128) is the mathematical condition that lines of constant ϕ and ψ be mutually orthogonal. It may not be true at a stagnation point, where both u and v are zero, so that their ratio in Eq. (4.128) is indeterminate.

Generation of Rotationality

This is the second time we have discussed Bernoulli's equation under different circumstances (the first was in Sec. 3.5). Such reinforcement is useful, since this is probably the most widely used equation in fluid mechanics. It requires frictionless flow with no shaft work or heat transfer between sections 1 and 2. The flow may or may not be irrotational, the latter being an easier condition, allowing a universal Bernoulli constant.

The only remaining question is: *When* is a flow irrotational? In other words, when does a flow have negligible angular velocity? The exact analysis of fluid rotationality under arbitrary conditions is a topic for advanced study, e.g., Ref. 10, sec. 8.5; Ref. 9, sec. 5.2; and Ref. 5, sec. 2.10. We shall simply state those results here without proof.

A fluid flow which is initially irrotational may become rotational if

1. There are significant viscous forces induced by jets, wakes, or solid boundaries. In this case Bernoulli's equation will not be valid in such viscous regions.

---

[1] Equations (4.125) and (4.126) are called the *Cauchy-Riemann equations* and are studied in complex-variable theory.

2. There are entropy gradients caused by curved shock waves (see Fig. 4.11*b*).

3. There are density gradients caused by *stratification* (uneven heating) rather than by pressure gradients.

4. There are significant *noninertial* effects such as the earth's rotation (the Coriolis acceleration).

In cases 2 to 4, Bernoulli's equation still holds along a streamline if friction is negligible. We shall not study cases 3 and 4 in this book. Case 2 will be treated briefly in Chap. 9 on gas dynamics. Primarily we are concerned with case 1, where rotation is induced by viscous stresses. This occurs near solid surfaces, where the no-slip condition creates a boundary layer through which the stream velocity drops to zero, and in jets and wakes, where streams of different velocities meet in a region of high shear.

Internal flows, such as pipes and ducts, are mostly viscous, and the wall layers grow to meet in the core of the duct. Bernoulli's equation does not hold in such flows unless modified for viscous losses.

External flows, such as a body immersed in a stream, are partly viscous and partly inviscid, the two regions being patched together at the edge of the shear layer or boundary layer. Two examples are shown in Fig. 4.11. Figure 4.11*a* shows a low-speed subsonic flow past a body. The approach stream is irrotational, i.e., the curl of a constant is zero, but viscous stresses create a rotational shear layer beside and downstream of the body. Generally speaking (see Chap. 6), the shear layer is laminar, or smooth, near the front of the body and turbulent, or disorderly, toward the rear. A separated, or deadwater, region usually occurs near the trailing edge, followed by an unsteady turbulent wake extending far downstream. Some sort of laminar or turbulent viscous theory must be applied to these viscous regions; they are then patched onto the outer flow, which is frictionless and irrotational. If the stream Mach number is less than about 0.3, we can combine Eq. (4.122) with the incompressible continuity equation (4.73)

$$\mathbf{\nabla} \cdot \mathbf{V} = \mathbf{\nabla} \cdot (\mathbf{\nabla}\phi) = 0$$

or
$$\nabla^2\phi = 0 = \frac{\partial^2\phi}{\partial x^2} + \frac{\partial^2\phi}{\partial y^2} + \frac{\partial^2\phi}{\partial z^2} \tag{4.129}$$

This is Laplace's equation in three dimensions, there being no restraint on the number of coordinates in potential flow. A great deal of Chap. 8 will be concerned with solving Eq. (4.129) for practical engineering problems; it holds in the entire region of Fig. 4.11*a* outside the shear layer.

Figure 4.11*b* shows a supersonic flow past a body. A curved shock wave generally forms in front, and the flow downstream is *rotational* due to entropy gradients (case 2). We can use Euler's equation (4.114) in this frictionless region but not potential theory. The shear layers have the same general character as in Fig. 4.11*a* except that the separation zone is slight or often absent and the wake is usually thinner. Theory of separated flow is presently qualitative, but we can make quantitative estimates of laminar and turbulent boundary layers and wakes.

Viscous regions where Bernoulli's equation fails:

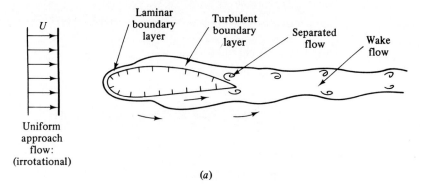

(a)

Fig. 4.11 Typical flow patterns illustrating viscous regions patched onto nearly frictionless regions: (a) low subsonic flow past a body ($U \ll a$); frictionless, irrotational potential flow outside the boundary layer (Bernoulli and Laplace equations valid); (b) supersonic flow past a body ($U > a$); frictionless, rotational flow outside the boundary layer (Bernoulli equation valid, potential flow invalid).

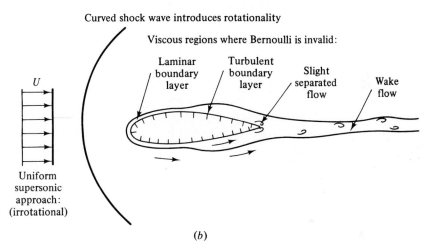

(b)

EXAMPLE 4.8 If a velocity potential exists for the velocity field of Example 4.4

$$u = a(x^2 - y^2) \qquad v = -2axy \qquad w = 0$$

find it, plot it, and compare with Example 4.6.

Solution Since $w = 0$, the curl of **V** has only one (z) component, and we must show that it is zero

$$(\mathbf{\nabla} \times \mathbf{V})_z = 2\omega_z = \frac{\partial v}{\partial x} - \frac{\partial u}{\partial y} = \frac{\partial}{\partial x}(-2axy) - \frac{\partial}{\partial y}(ax^2 - ay^2)$$

$$= -2ay + 2ay = 0 \qquad \text{checks}$$

The flow is indeed irrotational. A potential exists. *Ans.*

To find $\phi(x, y)$, set

$$u = \frac{\partial \phi}{\partial x} = ax^2 - ay^2 \tag{1}$$

$$v = \frac{\partial \phi}{\partial y} = -2axy \tag{2}$$

Integrate (1)

$$\phi = \frac{ax^3}{3} - axy^2 + f(y) \tag{3}$$

Differentiate (3) and compare with (2)

$$\frac{\partial \phi}{\partial y} = -2axy + f'(y) = -2axy \tag{4}$$

Therefore $f' = 0$, *or* $f = $ constant. The velocity potential is

$$\phi = \frac{ax^3}{3} - axy^2 + C \qquad\qquad Ans.$$

Letting $C = 0$, we can plot the ϕ lines in the same fashion as in Example 4.6. The result is shown in Fig. E4.8 (no arrows on ϕ). For this particular problem, the ϕ lines form the same

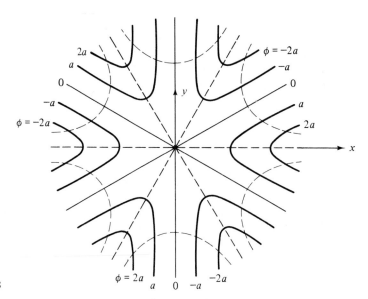

Fig. E4.8

pattern as the ψ lines of Example 4.6 (which are shown here as dashed lines) but are displaced 30°. The ϕ and ψ lines are everywhere perpendicular except at the origin, a stagnation point, where they are 30° apart. We expected trouble at the stagnation point, and there is no general rule for determining the behavior of the lines at that point.

SUMMARY

This chapter supplements Chap. 3 by using an infinitesimal control volume to derive the three basic partial differential equations of fluid motion: (1) the continuity equation (4.56), (2) the momentum equation (4.57), and (3) the energy equation (4.58). These equations, together with thermodynamic state relations for the fluid and appropriate boundary conditions, can in principle be solved for the complete flow field in a given engineering-fluid-mechanics problem.

In practice, most of our subsequent chapters will attack problems with constant density and viscosity, for which the energy equation (temperature) can be uncoupled and laid away on the shelf for later study. In such problems we consider only pressure and velocity variations. This will be the primary emphasis in dimensional analysis (Chap. 5), viscous flow (Chap. 6), inviscid flow (Chap. 8), open-channel flow (Chap. 10), and turbomachinery (Chap. 11). Only Chap. 9 on compressible flow will consider density and temperature variations and then mostly by neglecting friction.

This chapter has presented considerable *discussion* of the meaning of the basic equations, but there is more to be said. Whole books are written on the basic equations [11], and Ref. 12 contains 360 solved problems which relate fluid mechanics to the whole of continuum mechanics.

REFERENCES

1. P. J. Roache, "Computational Fluid Dynamics," Hermosa, Albuquerque, N.M., 1972.
2. G. B. Wallis, "One-dimensional Two-Phase Flow," McGraw-Hill, New York, 1969.
3. S. M. Selby, "CRC Handbook of Tables for Mathematics," 4th ed., CRC Press Inc., Cleveland, 1976.
4. H. Schlichting, "Boundary Layer Theory," 7th ed., McGraw-Hill, New York, 1979.
5. F. M. White, "Viscous Fluid Flow," McGraw-Hill, New York, 1974.
6. D. Frederick and T. S. Chang, "Continuum Mechanics," Allyn & Bacon, Boston, 1965.
7. J. P. Holman, "Heat Transfer," 5th ed., McGraw-Hill, New York, 1981.
8. W. M. Kays and M. Crawford, "Convective Heat and Mass Transfer," 2d ed., McGraw-Hill, New York, 1980.
9. G. K. Batchelor, "An Introduction to Fluid Dynamics," Cambridge University Press, Cambridge, 1967.
10. L. Prandtl and O. G. Tietjens, "Fundamentals of Hydro- and Aeromechanics," Dover, New York, 1957.
11. R. Aris, "Vectors, Tensors, and the Basic Equations of Fluid Mechanics," Prentice-Hall, Englewood Cliffs, N.J., 1962.
12. G. E. Mase, "Continuum Mechanics," Schaum's Outline Series, McGraw-Hill, New York, 1970.
13. W. F. Hughes and E. W. Gaylord, "Basic Equations of Engineering Science," Schaum's Outline Series, McGraw-Hill, New York, 1964.
14. G. Astarita and G. Marrucci, "Principles of Non-Newtonian Fluid Mechanics," McGraw-Hill, New York, 1974.
15. H. Lamb, "Hydrodynamics," 6th ed., Dover, New York, 1945.
16. A. Szeri, "Tribology: Friction, Lubrication, and Wear," McGraw-Hill, New York, 1980.

RECOMMENDED FILMS

Films numbered 2 to 5 and 16 in Appendix C.

Problems

PROBLEM DISTRIBUTION

4.1 Show that the general continuity equation (4.3) can be written in the equivalent form $d\rho/dt + \rho(\nabla \cdot \mathbf{V}) = 0$.

4.2 Write the special cases of the equation of continuity for (a) steady compressible flow in the yz plane, (b) unsteady incompressible flow in the xz plane, (c) unsteady compressible flow in the y direction only, (d) steady compressible flow in plane polar coordinates.

4.3 Derive Eq. (4.12b) for cylindrical coordinates by considering the flux of an incompressible fluid in and out of the elemental control volume in Fig. 4.2.

4.4 Derive Eq. (4.9) by considering the flux of an unsteady compressible fluid through the elemental control volume of Fig. 4.2.

4.5 An incompressible flow field is given by $u = Kxz^2$ and $w = Cy$, where K and C are constants. What does continuity tell us about the form of the velocity component v?

4.6 The linear velocity profile in Fig. 1.6 was obtained by a momentum argument regarding constant shear stress in the fluid. Suppose we know only that $u = u(y)$, $v = v(x)$, and $w = 0$ between the two plates. What does continuity alone tell us about the possible forms of u and v? How do the boundary conditions help resolve these forms?

4.7 An incompressible flow field has the cylindrical components $v_\theta = Cr$, $v_z = K(R^2 - r^2)$, $v_r = 0$, where C and K are constants and $r \le R$, $z \le L$. Does this flow satisfy continuity? What might it represent physically?

4.8 A two-dimensional incompressible velocity field has $u = K(1 - e^{-ay})$, for $x \le L$ and $0 \le y \le \infty$. What is the most general form of $v(x, y)$ for which continuity is satisfied and $v = v_0$ at $y = 0$? What are the proper dimensions for the constants K and a?

4.9 An incompressible flow in polar coordinates is given by

$$v_r = K \cos \theta \left(1 - \frac{b}{r^2} \right)$$

$$v_\theta = -K \sin \theta \left(1 + \frac{b}{r^2} \right)$$

Does this field satisfy continuity? For consistency, what should the dimensions of the constants K and b be? Sketch the surface where $v_r = 0$ and interpret.

4.10 Which of the following velocity fields satisfies conservation of mass for incompressible plane flow?
(a) $u = x, v = y$
(b) $u = y, v = x$
(c) $u = 2x, v = -2y$
(d) $u = 3xt, v = -3yt$
(e) $u = xy + y^2t, v = xy + x^2t$
(f) $u = 3x^2y^2, v = -2xy^3$
Ignore the vulgar dimensional inconsistencies.

4.11 If the radial velocity for incompressible flow is given by $v_r = (b \cos \theta)/r^2$, $b = $ constant, what is the most general form of $v_\theta(r, \theta)$ which satisfies continuity?

4.12 Spherical polar coordinates (r, θ, ϕ) are defined in Fig. P4.12. The cartesian transformations are

$$x = r \sin \theta \cos \phi$$

$$y = r \sin \theta \sin \phi$$

$$z = r \cos \theta$$

Show that the cartesian incompressible continuity relation (4.12a) can be transformed into the spherical polar form

$$\frac{1}{r^2} \frac{\partial}{\partial r} (r^2 v_r) + \frac{1}{r \sin \theta} \frac{\partial}{\partial \theta} (v_\theta \sin \theta) + \frac{1}{r \sin \theta} \frac{\partial}{\partial \phi} (v_\phi) = 0$$

What is the most general form of v_r when the flow is purely radial; that is, v_θ and v_ϕ are zero?

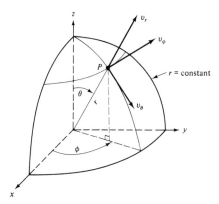

Fig. P4.12

4.13 A two-dimensional velocity field is given by

$$u = -\frac{Ky}{x^2 + y^2} \qquad v = \frac{Kx}{x^2 + y^2}$$

where K is constant. Does this field satisfy incompressible continuity? Transform these velocities into polar components v_r and v_θ. What might the flow represent?

4.14 For incompressible polar coordinate flow, what is the most general form of a purely circulatory motion, $v_\theta = v_\theta(r, \theta, t)$ and $v_r = 0$, which satisfies continuity?

4.15 What is the most general form of a purely radial polar coordinate incompressible flow pattern, $v_r = v_r(r, \theta, t)$ and $v_\theta = 0$, which satisfies continuity?

4.16 An incompressible steady flow pattern is given by $u = x^3 + 2z^2$ and $w = y^3 - 2yz$. What is the most general form of the third component, $v(x, y, z)$, which satisfies continuity?

4.17 A certain two-dimensional shear flow near a wall, as in Fig. P4.17, has the velocity component

$$u = U\left(\frac{2y}{ax} - \frac{y^2}{a^2 x^2}\right)$$

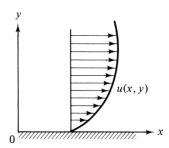

Fig. P4.17

where a is constant. Derive from continuity the velocity component $v(x, y)$ assuming that $v = 0$ at the wall, $y = 0$.

4.18 Prob. 4.17 became artificial and somewhat misleading as soon as we picked an algebraic expression for $u(x, y)$. Yet we can get many insights from continuity without substituting any numbers at all. Consider the flat-plate boundary-layer flow in Fig. P4.18 (recall Example 3.11). From the no-slip condition $v = 0$ all along the wall $y = 0$, and $u = U = $ constant outside the layer. If the layer thickness δ increases with x as shown, prove with incompressible two-dimensional continuity that (*a*) the component $v(x, y)$ is everywhere positive within the layer; (*b*) v increases parabolically with y very near the wall; and (*c*) v reaches a positive maximum at $y = \delta$.

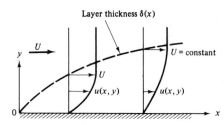

Fig. P4.18

4.19 From Prob. 3.53 the axial velocity field for fully developed laminar flow in a pipe is

$$v_z = u_{max}\left(1 - \frac{r^2}{R^2}\right)$$

and there is no swirl, $v_\theta = 0$. Determine the radial velocity field $v_r(r, z)$ from the incompressible relation if u_{max} is constant and $v_r = 0$ at $r = R$.

4.20 For steady compressible flow of gas through a nozzle, the velocity distribution is found to be approximately one-dimensional, $u = U_0(1 + x/L)$, $v = w = 0$, where U_0 and L are constants. Using continuity, find an expression for the density distribution $\rho(x)$, if $\rho = \rho_0$ at $x = 0$. At what position x has the density dropped 10 percent below ρ_0?

4.21 Air flows under steady, approximately one-dimensional conditions through the conical nozzle in Fig. P4.21. If the speed of sound is approximately 340 m/s, what is the minimum nozzle-diameter ratio D_e/D_0 for which we can safely neglect compressibility effects if $V_0 = $ (*a*) 10 m/s; (*b*) 30 m/s?

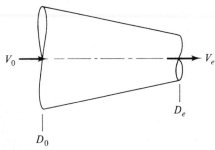

Fig. P4.21

4.22 Consider the plane polar coordinate flow field $v_r = Q/r$, $v_\theta = K/r$, where Q and K are constants. Does this flow satisfy continuity for an incompressible fluid? Is it irrotational? Sketch a few streamlines and give a physical interpretation of the flow.

4.23 Curvilinear, or streamline, coordinates are defined in Fig. P4.23, where n is normal to the streamline in the plane of the radius of curvature R. Show that Euler's frictionless momentum equation (4.36) in streamline coordinates becomes

$$\frac{\partial V}{\partial t} + V\frac{\partial V}{\partial s} = -\frac{1}{\rho}\frac{\partial p}{\partial s} + g_s \qquad (1)$$

$$-V\frac{\partial \theta}{\partial t} - \frac{V^2}{R} = -\frac{1}{\rho}\frac{\partial p}{\partial n} + g_n \qquad (2)$$

Further show that the integral of Eq. (1) with respect to s is none other than our old friend Bernoulli's equation (3.62).

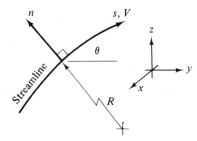

Fig. P4.23

4.24 A frictionless, incompressible steady flow field is given by

$$\mathbf{V} = 2xy\mathbf{i} - y^2\mathbf{j}$$

in arbitrary units. Let the density be $\rho_0 = $ constant and neglect gravity. Find an expression for the pressure gradient in the x direction and evaluate this gradient at $(1, 2, 0)$.

4.25 If z is "up," what are the conditions on the constants a and b for which the velocity field $u = ay$, $v = bx$, $w = 0$ is an exact solution to the continuity and Navier-Stokes equations for incompressible flow?

4.26 Consider the incompressible flow field $u = Ax + By$, $v = Cx + Dy$, $w = 0$. For what conditions on the constants (A, B, C, D) is this flow an exact solution to the continuity and Navier-Stokes equations with negligible gravity?

4.27 Show that the two-dimensional flow field of Example 1.14 is an exact solution to the incompressible Navier-Stokes equations (4.38). Neglecting gravity, compute the pressure field $p(x,y)$ and relate it to the absolute velocity $V^2 = u^2 + v^2$. Interpret the result.

4.28 Show that the Couette flow between parallel plates in Fig. 1.6, $u = Vy/h$, $v = w = 0$, is an exact solution to the incompressible Navier-Stokes equations (4.38). Find an expression for the only nonzero shear stress in the fluid.

4.29 The answer to Prob. 4.14 is $v_\theta = f(r)$ only. Do not reveal this to your friends if they are still working on Prob. 4.14. Show that this flow field is an exact solution to the Navier-Stokes equations (4.38) for only two special cases of the function $f(r)$. Neglect gravity. Interpret these two cases physically.

4.30 From Prob. 4.15 the purely radial polar coordinate flow which satisfies continuity is $v_r = f(\theta)/r$, where f is an arbitrary function. Determine what particular forms of $f(\theta)$ satisfy the full Navier-Stokes equations in polar coordinate form from Appendix E, Eqs. (E.5) and (E.6).

4.31 The fully developed laminar-pipe-flow solution of Prob. 3.53, $v_z = u_{max}(1 - r^2/R^2)$, $v_\theta = 0$, $v_r = 0$, is an exact solution to the cylindrical Navier-Stokes equations (Appendix E). Neglecting gravity, compute the pressure distribution in the pipe $p(r, z)$ and the shear-stress distribution $\tau(r, z)$, using R, u_{max}, and μ as parameters. Why does the maximum shear occur at the wall? Why does the density not appear as a parameter?

4.32 For fully developed laminar flow between parallel plates, as in Prob. 3.23, the velocity profile is $u = 4u_{max}z(z_0 - z)/z_0^2$, $v = 0$, $w = 0$, an exact solution to the Navier-Stokes equations. Neglecting gravity, compute the pressure distribution $p(x, z)$ and the shear-stress distribution $\tau(x, z)$, using z_0, u_{max}, and μ as parameters. Why does the maximum shear occur at the walls? Why does density not appear as a parameter?

4.33 An incompressible flow field is given by

$$\mathbf{V} = x^2\mathbf{i} - z^2\mathbf{j} - 2xz\mathbf{k}$$

with V in meters per second and (x, y, z) in meters. If the fluid viscosity is $0.05\ \text{kg/(m·s)}$, evaluate the entire viscous stress tensor, Eq. (4.31), at the point $(x, y\ z) = (1, 2, 3)$.

4.34 From the Navier-Stokes equations for incompressible flow in polar coordinates (Appendix E for cylindrical coordinates), find the most general case of purely circulating motion $v_\theta(r)$, $v_r = v_z = 0$, for flow with no slip between two fixed concentric cylinders, as in Fig. P4.34.

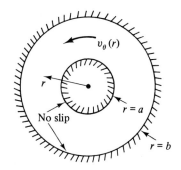

Fig. P4.34

4.35 A constant-thickness film of viscous liquid flows in laminar motion down a plate inclined at angle θ, as in Fig. P4.35. The velocity profile is

$$u = Cy(2h - y) \qquad v = w = 0$$

Find the constant C in terms of the specific weight and viscosity and the angle θ. Find the volume flux Q per unit width in terms of these parameters.

4.36 Show that the incompressible flow field $v_r = 0$, $v_\theta = K/r$, $v_z = 0$, where K is a constant, is an exact solution to the Navier-Stokes equations in cylindrical coordinates (see Appendix E). Neglecting gravity, compute the pressure distribution $p(r, \theta, z)$ in terms of density and v_θ^2 and interpret the result.

4.37 According to the theory in Chap. 8, near the front of a rounded two-dimensional body, as in Fig. P4.37, the velocity approaching the stagnation point is given by $u = U_0(1 - a^2/x^2)$, where a is the nose radius, as shown, and U_0 is the velocity far upstream. Compute the value and position of (a) the maximum deceleration of the fluid and (b) the maximum viscous normal stress in the fluid.

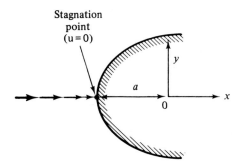

Fig. P4.37

4.38 For Prob. 4.37 assume that the fluid is water at $20°C$ and $U_0 = 8\ \text{m/s}$, $a = 7\ \text{cm}$. Compute, along the line approaching the stagnation point, (a) the maximum deceleration of the fluid in meters per second squared and (b) the maximum viscous normal stress in pascals.

4.39 A flat plate of essentially infinite width and breadth oscillates sinusoidally in its own plane beneath a viscous fluid, as in Fig. P4.39. The fluid is at rest far

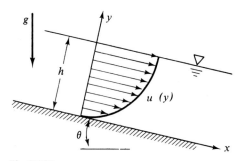

Fig. P4.35

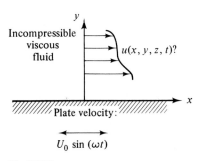

Fig. P4.39

above the plate. Making as many simplifying assumptions as you can, set up the governing differential equation and boundary conditions for finding the velocity field u in the fluid. Do not solve (if you *can* solve it immediately, you might be able to get exempted from the balance of this course with credit).

4.40   A solid circular cylinder of radius R rotates at angular velocity Ω in a viscous incompressible fluid which is at rest far from the cylinder, as in Fig. P4.40. Make simplifying assumptions and derive the governing differential equation and boundary conditions for the velocity field v_θ in the fluid. Do not solve unless obsessed with this problem.

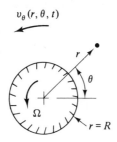

Fig. P4.40

4.41   Manipulate the general differential energy equation (4.51) to show that the rate of change of entropy s in a newtonian fluid with constant k is given by

$$\rho T \frac{ds}{dt} = k \nabla^2 T + \Phi$$

Hint: Recall from thermodynamics that $T\,ds = d\hat{h} - dp/\rho$.

4.42   Problems involving viscous dissipation of energy are dependent on viscosity μ, thermal conductivity k, stream velocity U_0, and stream temperature T_0. Group these parameters into the dimensionless *Brinkman number*, which is proportional to μ.

4.43   As mentioned in Prob. 4.32, the velocity profile for laminar flow between two plates, as in Fig. P4.43, is

$$u = \frac{4u_{\max}\,y(h - y)}{h^2} \qquad v = w = 0$$

If the wall temperature is T_w at both walls, use the incompressible-flow energy equation (4.75) to solve for the temperature distribution $T(y)$ between the walls for steady flow.

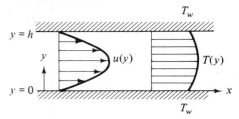

Fig. P4.43

4.44   Repeat the analysis of Prob. 4.43 if temperature of the lower wall is T_0 and of the upper wall is T_1, both constant.

4.45   In duct-flow problems with heat transfer one often defines an average fluid temperature. Consider the duct flow of Fig. P4.43 of width b into the paper. Using a control-volume integral analysis with constant density and specific heat, derive an expression for the temperature arising if the entire duct flow poured into a bucket and was stirred uniformly. Assume arbitrary $u(y)$ and $T(y)$. This average is called the *cup-mixing temperature* of the flow.

4.46   For the fluid between two concentric cylinders in Fig. P4.34 let the outer cylinder be fixed and the inner cylinder rotate counterclockwise at angular velocity Ω. What are the proper boundary conditions? Integrate to find the velocity distribution $v_\theta(r)$ if $v_r = v_z = 0$.

4.47   Consider a viscous film of liquid draining uniformly down the side of a vertical rod of radius a, as in Fig. P4.47. At some distance down the rod the film will approach a terminal or *fully developed* draining

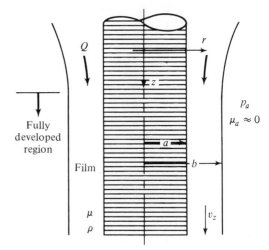

Fig. P4.47

flow of constant outer radius b, with $v_z = v_z(r)$, $v_\theta = v_r = 0$. Assume that the atmosphere offers no shear resistance to the film motion. Derive a differential equation for v_z, state the proper boundary conditions, and solve for the film velocity distribution. How does the film radius b relate to the total film volume flow rate Q?

4.48 Assume that there is no motion in the viscous fluid between cylinders in Fig. P4.34 and that the fluid has thermal conductivity k. If the inner and outer cylinder surface temperatures are T_a and T_b, respectively, derive a differential equation for $T(r)$ between cylinders, state the boundary conditions, and solve for the temperature.

4.49 The flow pattern in bearing lubrication can be illustrated by Fig. P4.49, where a viscous oil (ρ, μ) is forced into the gap $h(x)$ between a fixed slipper block and a wall moving at velocity U. If the gap is thin, $h \ll L$, it can be shown that the pressure and velocity distributions are of the form $p = p(x)$, $u = u(y)$, $v = w = 0$. Neglecting gravity, reduce the Navier-Stokes equations (4.38) to a single differential equation for $u(y)$. What are the proper boundary conditions? Integrate and show that

$$u = \frac{1}{2\mu} \frac{dp}{dx}(y^2 - yh) + U\left(1 - \frac{y}{h}\right)$$

where $h = h(x)$ may be an arbitrary slowly varying gap width. (For further information on lubrication theory, see Ref. 16.)

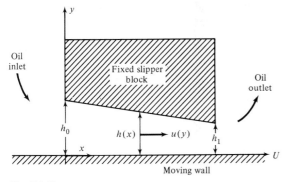

Fig. P4.49

4.50 The viscous oil in Fig. P4.50 is set into steady motion by a concentric inner cylinder moving axially at velocity U inside a fixed outer cylinder. Assuming constant pressure and density and a purely axial fluid motion, solve Eqs. (4.38) for the fluid velocity distribution $v_z(r)$. What are the proper boundary conditions?

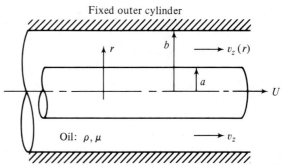

Fig. P4.50

4.51 Determine the incompressible two-dimensional stream function $\psi(x, y)$ which represents the flow field given in Example 1.14.

4.52 Investigate the stream function $\psi = K(x^2 - y^2)$, $K = $ constant. Plot the streamlines in the full xy plane, find any stagnation points, and interpret what the flow could represent.

4.53 Investigate the polar coordinate stream function $\psi = Kr^{1/2} \sin \frac{1}{2}\theta$, $K = $ constant. Plot the streamlines in the full xy plane, find any stagnation points, and interpret.

4.54 Investigate the polar coordinate stream function $\psi = Kr^{2/3} \sin(2\theta/3)$, $K = $ constant. Plot the streamlines in all except the bottom right quadrant and interpret.

4.55 In spherical polar coordinates, as in Fig. P4.12, the flow is called *axisymmetric* if $v_\phi \equiv 0$ and $\partial/\partial\phi \equiv 0$, so that $v_r = v_r(r, \theta)$ and $v_\theta = v(r, \theta)$. Show that a stream function $\psi(r, \theta)$ exists for this case and is given by

$$v_r = \frac{1}{r^2 \sin \theta} \frac{\partial \psi}{\partial \theta} \qquad v_\theta = -\frac{1}{r \sin \theta} \frac{\partial \psi}{\partial r}$$

This is called the *Stokes stream function* [5, p. 204].

4.56 Investigate the velocity potential $\phi = Kxy$, $K = $ constant. Sketch the potential lines in the full xy plane, find any stagnation points, and sketch in by eye the orthogonal streamlines. What could the flow represent?

4.57 Determine the incompressible two-dimensional velocity potential $\phi(x, y)$ which represents the flow field given in Example 1.14. Sketch a few potential and stream lines.

4.58 Show that the incompressible velocity potential in plane polar coordinates $\phi(r, \theta)$ is such that

$$v_r = \frac{\partial \phi}{\partial r} \qquad v_\theta = \frac{1}{r} \frac{\partial \phi}{\partial \theta}$$

Further show that the angular velocity about the z axis in such a flow would be given by

$$2\omega_z = \frac{1}{r}\frac{\partial}{\partial r}(rv_\theta) - \frac{1}{r}\frac{\partial}{\partial \theta}(v_r)$$

Finally show that ϕ as defined above satisfies Laplace's equation in polar coordinates for incompressible flow.

4.59 Investigate the polar coordinate velocity potential $\phi = Kr^{1/2}\cos\frac{1}{2}\theta$, $K =$ constant. Plot the potential lines in the full xy plane, sketch in by eye the orthogonal streamlines, and interpret.

4.60 Show that the linear Couette flow between plates in Fig. 1.6 has a stream function but no velocity potential. Why is this so?

4.61 Find the two-dimensional velocity potential $\phi(r, \theta)$ for the polar coordinate flow pattern of Prob. 4.22.

4.62 Show that the velocity potential $\phi(r, z)$ in axisymmetric cylindrical coordinates (see Fig. 4.2) is defined such that

$$v_r = \frac{\partial \phi}{\partial r} \qquad v_z = \frac{\partial \phi}{\partial z}$$

Further show that for incompressible flow this potential satisfies Laplace's equation in (r, z) coordinates.

4.63 A two-dimensional incompressible flow is defined by

$$u = -\frac{Ky}{x^2 + y^2} \qquad v = \frac{Kx}{x^2 + y^2}$$

where $K =$ constant. Is this flow irrotational? If so, find its velocity potential, sketch a few potential lines, and interpret the flow pattern.

4.64 Show that the parabolic laminar-duct-flow pattern of Prob. 4.43 and Fig. P4.43 has a stream function but no velocity potential. Explain why this is so.

4.65 Show that the fully developed laminar pipe flow of Prob. 3.53,

$$v_z = u_{max}\left(1 - \frac{r^2}{R^2}\right)$$

$v_r = 0$, has a stream function $\psi(r, z)$ but no velocity potential. Why is this so?

4.66 Find the polar coordinate velocity potential corresponding to the stream functions (a) $\psi = K\theta$ and (b) $\psi = K\ln r$, where $K =$ constant.

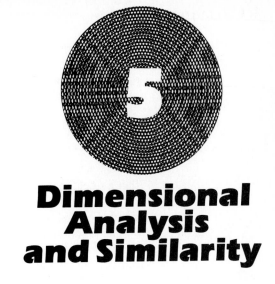

Dimensional Analysis and Similarity

5.1 INTRODUCTION

This chapter treats the third and final method in our trio of techniques for studying fluid flows, dimensional analysis. The emphasis here is on the use of dimensional analysis to plan experiments and present data compactly, but many workers also use it in theoretical studies.

Basically, dimensional analysis is a method for reducing the number and complexity of experimental variables which affect a given physical phenomenon, using a sort of compacting technique. If a phenomenon depends upon n dimensional variables, dimensional analysis will reduce the problem to only k *dimensionless* variables, where the reduction $n - k = 1, 2, 3,$ or 4, depending upon the problem complexity. Generally $n - k$ equals the number of different dimensions (sometimes called basic or primary or fundamental dimensions) which govern the problem. In fluid mechanics, the four basic dimensions are usually taken to be mass M, length L, time T, and temperature Θ, or an $MLT\Theta$ system for short. Sometimes one uses an $FLT\Theta$ system, with force F replacing mass.

Although its purpose is to reduce variables and group them in dimensionless form, dimensional analysis has several side benefits. The first is an enormous saving in time and money. Suppose one knew that the force F on a particular body immersed in a stream of fluid depended only on the body length L, the stream velocity V, the fluid density ρ, and the fluid viscosity μ; that is,

$$F = f(L, V, \rho, \mu) \tag{5.1}$$

Suppose further that the geometry and flow conditions are so complicated that our integral theories (Chap. 3) and differential equations (Chap. 4) fail to yield the solution for the force. Then we must find the function $f(L, V, \rho, \mu)$ experimentally.

Generally speaking, it takes about 10 experimental points to define a curve. To find the effect of body length in Eq. (5.1) we shall have to run the experiment for 10 lengths L. For each L we shall need 10 values of V, 10 values of ρ, 10 values of μ, making a grand total of 10^4, or 10,000, experiments. At \$5 per experiment—well, you see what we are getting into. However, with dimensional analysis, we can immediately reduce Eq. (5.1) to the equivalent form

$$\frac{F}{\rho V^2 L^2} = g\left(\frac{\rho V L}{\mu}\right)$$

or
$$C_F = g(\text{Re}) \qquad (5.2)$$

that is, the dimensionless *force coefficient* $F/\rho V^2 L^2$ is a function only of the dimensionless *Reynolds number* $\rho V L/\mu$. We shall learn exactly how to make this reduction in Secs. 5.2 and 5.3.

The function g is different mathematically from the original function f, but it contains all the same information. Nothing is lost in a dimensional analysis. And think of the saving: we can establish g by running the experiment for only 10 values of the single variable called the Reynolds number. We do not have to vary L, V, ρ, or μ separately but only the *grouping* $\rho V L/\mu$. This we do merely by varying velocity V in say, a wind tunnel or drop test or water channel, and there is no need to build 10 different bodies or find 100 different fluids with 10 densities and 10 viscosities. The cost is now about \$50, maybe less.

A second side benefit of dimensional analysis is that it helps our thinking and planning for an experiment or theory. It suggests dimensionless ways of writing equations before we waste money on computer time to find solutions. It suggests variables which can be discarded; sometimes dimensional analysis will immediately reject variables, and sometimes it groups them off to the side, where a few simple tests will show them to be unimportant. Finally, dimensional analysis will often give a great deal of insight into the form of the physical relationship we are trying to study.

A third benefit is that dimensional analysis provides *scaling laws* which can convert data from a cheap, small *model* into design information for an expensive, large *prototype*. We do not build a million-dollar airplane and see whether it has enough lift force. We measure the lift on a small model and use a scaling law to predict the lift on the full-scale prototype airplane. There are rules we shall explain for finding scaling laws. When the scaling law is valid, we say that a condition of *similarity* exists between model and prototype. In the simple case of Eq. (5.1), similarity is achieved if the Reynolds number is the same for the model and prototype because the function g then requires the force coefficient to be the same also:

$$\text{If } \text{Re}_m = \text{Re}_p \quad \text{then} \quad C_{Fm} = C_{Fp} \qquad (5.3)$$

where subscripts m and p mean model and prototype, respectively. From the definition of force coefficient, this means that

$$\frac{F_p}{F_m} = \frac{\rho_p}{\rho_m}\left(\frac{V_p}{V_m}\right)^2\left(\frac{L_p}{L_m}\right)^2 \qquad (5.4)$$

for data taken where $\rho_p V_p L_p/\mu_p = \rho_m V_m L_m/\mu_m$. Equation (5.4) is a scaling law: if you measure the model force at the model Reynolds number, the prototype force at the same Reynolds number equals the model force times the density ratio times the velocity ratio squared times the length ratio squared. We shall give more examples later.

Do you understand these introductory explanations? Be careful; learning dimensional analysis is like learning to play tennis: there are levels of the game. We can establish some ground rules and do some fairly good work in this brief chapter, but dimensional analysis in the broad view has many subtleties and nuances which only time and practice and maturity enable one to master. Although dimensional analysis has a firm physical and mathematical foundation, considerable art and skill are needed to use it effectively.

EXAMPLE 5.1   A copepod is a water crustacean approximately 1 mm in diameter. We want to know the drag force on the copepod when it moves slowly in fresh water. A scale model 100 times larger is made and tested in glycerin at $V = 30$ cm/s. The measured drag on the model is 1.3 N. For similar conditions, what are the velocity and drag of the actual copepod in water? Assume that Eq. (5.1) applies and the temperature is 20°C.

Solution   From Table 1.3 the fluid properties are:

Water (prototype): $\mu_p = 0.001$ kg/(m·s) $\rho_p = 999$ kg/m³

Glycerin (model): $\mu_m = 1.5$ kg/(m·s) $\rho_p = 1263$ kg/m³

The length scales are $L_m = 100$ mm and $L_p = 1$ mm. We are given enough model data to compute the Reynolds number and force coefficient

$$\text{Re}_m = \frac{\rho_m V_m L_m}{\mu_m} = \frac{(1263 \text{ kg/m}^3)(0.3 \text{ m/s})(0.1 \text{ m})}{1.5 \text{ kg/(m·s)}} = 25.3$$

$$C_{Fm} = \frac{F_m}{\rho_m V_m^2 L_m^2} = \frac{1.3 \text{ N}}{(1263 \text{ kg/m}^3)(0.3 \text{ m/s})^2(0.1 \text{ m})^2} = 1.14$$

Both these numbers are dimensionless, as you can check. For conditions of similarity, the prototype Reynolds number must be the same, and Eq. (5.2) then requires the prototype force coefficient to be the same

$$\text{Re}_p = \text{Re}_m = 25.3 = \frac{999 V_p(0.001)}{0.001}$$

or $V_p = 0.0253$ m/s $= 2.53$ cm/s   *Ans.*

$$C_{Fp} = C_{Fm} = 1.14 = \frac{F_p}{999(0.0253)^2(0.001)^2}$$

or $F_p = 7.31 \times 10^{-7}$ N   *Ans.*

It would obviously be difficult to measure such a tiny drag force.

Historically, the first person to write extensively about units and dimensional reasoning in physical relations was Euler in 1765. Euler's ideas were far ahead of his time, as were those of Joseph Fourier, whose 1822 book, "Analytical Theory of Heat," outlined what is now called the principle of dimensional homogeneity and even developed some similarity rules for heat flow. There were no further significant advances until Lord Rayleigh's book in 1877, "Theory of Sound," which proposed a "method of dimensions" and gave several examples of dimensional analysis. The final breakthrough which established the method as we know it today is generally credited to E. Buckingham in 1914 [24], whose paper outlined what is now called the *Buckingham pi theorem* for describing dimensionless parameters (see Sec. 5.4). However, it is now known that a Frenchman, A. Vaschy in 1892, and a Russian, D. Riabouchinsky in 1911, had independently published papers reporting results equivalent to the pi theorem. Following Buckingham's paper, P. W. Bridgman published a classic book in 1922 [1] outlining the general theory of dimensional analysis. The subject continues to be controversial because there is so much art and subtlety in using dimensional analysis. Thus, since Bridgman there have been at least 20 books published on the subject [1–20]. There will probably be more, but seeing the whole list might make some fledgling authors think twice. Nor is dimensional analysis limited to fluid mechanics or even engineering. Specialized books have been written on the application of dimensional analysis to metrology [21], astrophysics [22], and even economics [23].

5.2 THE PRINCIPLE OF DIMENSIONAL HOMOGENEITY

In making the remarkable jump from the five-variable Eq. (5.1) to the two-variable Eq. (5.2) we were exploiting a rule which is almost a self-evident axiom in physics. This rule, the *principle of dimensional homogeneity*, can be stated as follows:

> If an equation truly expresses a proper relationship between variables in a physical process, it will be *dimensionally homogeneous*; i.e., each of its additive terms will have the same dimensions.

All the equations which are derived from the theory of mechanics are of this form. For example, consider the relation which expresses the displacement of a falling body

$$S = S_0 + V_0 t + \tfrac{1}{2}gt^2 \tag{5.5}$$

Each term in this equation is a displacement, or length, and has dimensions $\{L\}$. The equation is dimensionally homogeneous. Note also that any consistent set of units can be used to calculate a result.

Consider Bernoulli's equation for incompressible flow

$$\frac{p}{\rho} + \tfrac{1}{2}V^2 + gz = \text{const} \tag{5.6}$$

Each term, including the constant, has dimensions of velocity squared, or $\{L^2 T^{-2}\}$. The equation is dimensionally homogeneous and gives proper results for any consistent set of units.

Students count on dimensional homogeneity and use it to check themselves when they cannot quite remember an equation during an exam. For example, which is it:

$$S = \tfrac{1}{2}gt^2? \quad \text{or} \quad S = \tfrac{1}{2}g^2t? \tag{5.7}$$

By checking the dimensions, we reject the second form and back up our faulty memory. We are exploiting the principle of dimensional homogeneity (PDH), and this chapter simply exploits it further.

Equations (5.5) and (5.6) also illustrate some other factors that often enter into a dimensional analysis:

Dimensional variables are the quantities which actually vary during a given case and would be plotted against each other to show the data. In Eq. (5.5), they are S and t, in Eq. (5.6) they are p, V, and z. All have dimensions, and all can be nondimensionalized as a dimensional-analysis technique.

Dimensional constants may vary from case to case but are held constant during a given run. In Eq. (5.5) they are S_0, V_0, and g, and in Eq. (5.6) they are ρ, g, and C. They all have dimensions and conceivably could be nondimensionalized, but they are normally used to help nondimensionalize the variables in the problem.

Pure constants have no dimensions and never did. They arise from mathematical manipulations. In both Eqs. (5.5) and (5.6) they are $\tfrac{1}{2}$ and the exponent 2, both of which came from an integration: $\int t\, dt = \tfrac{1}{2}t^2$, $\int V\, dV = \tfrac{1}{2}V^2$. Other common dimensionless constants are π and e.

Note that integration and differentiation of an equation may change the dimensions but not the homogeneity of the equation. For example, integrate or differentiate Eq. (5.5):

$$\int S\, dt = S_0 t + \tfrac{1}{2}V_0 t^2 + \tfrac{1}{6}gt^3 \tag{5.8a}$$

$$\frac{dS}{dt} = V_0 + gt \tag{5.8b}$$

In the integrated form (5.8a) every term has dimensions of $\{LT\}$, while in the derivative form (5.8b) every term is a velocity $\{LT^{-1}\}$.

Finally, there are some physical variables that are naturally dimensionless by virtue of their definition as ratios of dimensional quantities. Some examples are strain (change in length per unit length), Poisson's ratio (ratio of transverse strain to longitudinal strain), and specific gravity (ratio of density to standard water density). All angles are dimensionless (ratio of arc length to radius) and should be taken in radians for this reason.

The motive behind dimensional analysis is that any dimensionally homogeneous equation can be written in an entirely equivalent nondimensional form which is more compact. The exact details are spelled out in Sec. 5.3 as the pi theorem. For example, Eq. (5.5) is handled by defining dimensionless variables

$$S^* = \frac{S}{S_0} \qquad t^* = \frac{V_0 t}{S_0} \tag{5.9a}$$

or

$$S^{**} = \frac{gS}{V_0^2} \qquad t^{**} = \frac{gt}{V_0} \tag{5.9b}$$

Notice that there were two ways to nondimensionalize the variables. This is quite common; sometimes there are three or more ways. Which way is best? Usually neither, it is a matter of taste, custom, and the user's choice. You must accept the fact that there are several equivalent formulations of most dimensional-analysis problems, all of which are correct.

There are two don'ts involved in operations like Eq. (5.9). First, *don't* nondimensionalize variables upside down:

$$S^* = \frac{S_0}{S} \qquad t^* = \frac{S_0}{V_0 t} \tag{5.10}$$

These are dimensionless, no question about it. But with the constants in the top and the variables in the bottom, there will be singularities where S and $t = 0$, the plots will look funny, users of your data will be confused, and the supervisor will be angry. It is not a good idea. Put your most important variable in the numerator and use parametric constants in the denominator.

Second, don't—repeat, *don't*—mix your variables (S, t) together in one definition:

$$S^* = \frac{V_0 t}{S} \tag{5.11}$$

This is beautiful and intriguing, but you will have mathematical problems and vexing presentation problems also. This idea sometimes works in an advanced technique called *similarity theory* (see, for example, Ref. 11), but it should not be used in dimensional analysis.

Now try our definitions (5.9) in Eq. (5.5):

$$S_0 S^* = S_0 + V_0 \frac{S_0 t^*}{V_0} + \tfrac{1}{2} g \left(\frac{S_0 t^*}{V_0} \right)^2 \tag{5.12a}$$

$$\frac{V_0^2 S^{**}}{g} = S_0 + V_0 \frac{V_0 t^{**}}{g} + \tfrac{1}{2} g \left(\frac{V_0 t^{**}}{g} \right)^2 \tag{5.12b}$$

These still have dimensions of length, but if we divide through and isolate a dimensionless variable, for example, S^* or S^{**}, the PDH guarantees that *all* terms will be dimensionless. Thus divide (5.12a) by S_0 and divide (5.12b) by V_0^2/g

$$S^* = 1 + t^* + \frac{1}{2} \frac{gS_0}{V_0^2} t^{*2} \tag{5.13a}$$

$$S^{**} = \frac{gS_0}{V_0^2} + t^{**} + \tfrac{1}{2} t^{**2} \tag{5.13b}$$

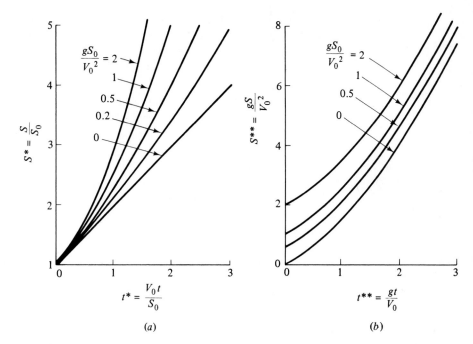

Fig. 5.1 Two entirely equivalent dimensionless forms of the falling-body equation (5.5): (a) Eq. (5.13a) and (b) Eq. (5.13b). Which form is more effective?

These are both dimensionless equations, equivalent to each other and equivalent in every respect to the original Eq. (5.5). They are plotted in Fig. 5.1. Which form do you feel is better and more effective? You are asked to explain your choice in Prob. 5.1.

Whereas Eq. (5.5) was of the form

$$S = f(t, S_0, V_0, g) \tag{5.14}$$

and involved five dimensional quantities, Eqs. (5.13) are each of the form

$$S' = g(t', \alpha) \qquad \alpha = \frac{gS_0}{V_0^2} \tag{5.15}$$

and involve only three dimensionless quantities. The parameter α commonly occurs in processes affected by gravity and is a form of the Froude number (see Table 5.2).

This example checks with our earlier statements about the dimensional-analysis technique. The original function of five variables is reduced to a dimensionless function of three variables. The reduction $5 - 3 = 2$ should equal the number of dimensions ($MLT\Theta$) involved in the problem. Check our variables:

$$\{S, S_0\} = \{L\} \qquad \{t\} = \{T\} \qquad \{V_0\} = \{LT^{-1}\} \qquad \{g\} = \{LT^{-2}\} \tag{5.16}$$

As expected, there are only two dimensions involved, $\{L\}$ and $\{T\}$. This idea culminates in the pi theorem (Sec. 5.3).

Some Peculiar Engineering Equations

The foundation of the dimensional-analysis method rests on two assumptions: (1) that the proposed physical relation is dimensionally homogeneous and (2) that all the relevant variables have been included in the proposed relation.

If a relevant variable is missing, dimensional analysis will fail, giving either algebraic difficulties or, worse, yielding a dimensionless formulation which does not resolve the process. A typical case is Manning's open-channel formula, discussed in Example 1.4:

$$V = \frac{1.49}{n} R^{2/3} S^{1/2} \tag{1.5}$$

Since V is velocity, R is a radius, and n and S are dimensionless, the formula is not dimensionally homogeneous. This should be a warning that (1) the formula changes if the *units* of V and R change and (2) if valid, it represents a very special case. Equation (1.5) predates the dimensional-analysis technique and is valid only for water in rough channels at moderate velocities and large radii in BG units.

Such dimensionally inhomogeneous formulas abound in the hydraulics literature. Another example is the Hazen-Williams formula [25] for volume flow of water through a straight smooth pipe

$$Q = 61.9 D^{2.63} \left(\frac{dp}{dx}\right)^{0.54} \tag{5.17}$$

where D is diameter and dp/dx the pressure gradient. Some of these formulas arise because numbers have been inserted for fluid properties and other physical data into perfectly legitimate homogeneous formulas. We shall not give the units of Eq. (5.17) to avoid encouraging its use.

On the other hand, some formulas are "constructs" which cannot be made dimensionally homogeneous. The "variables" they relate cannot be analyzed by the dimensional-analysis technique. Most of these formulas are raw empiricisms convenient to a small group of specialists. Here are three examples:

$$B = \frac{25,000}{100 - R} \tag{5.18}$$

$$S = \frac{140}{130 + \text{API}} \tag{5.19}$$

$$0.0147 D_E - \frac{3.74}{D_E} = 0.26 t_R - \frac{172}{t_R} \tag{5.20}$$

Equation (5.18) relates the Brinell hardness B of a metal to its Rockwell hardness R. Equation (5.19) relates the specific gravity S of an oil to its density in degrees API. Equation (5.20) relates the viscosity of a liquid in D_E, or degrees Engler, to its viscosity t_R in Saybolt seconds. Such formulas have a certain usefulness when communicated between fellow specialists, but we cannot handle them here. Variables like Brinell hardness and Saybolt viscosity are not suited to an $MLT\Theta$ dimensional system.

5.3 THE PI THEOREM

There are several methods of reducing a number of dimensional variables into a smaller number of dimensionless groups. The scheme given here was proposed in 1914 by Buckingham [24] and is now called the *Buckingham pi theorem*. The name pi comes from the mathematical notation Π, meaning a product of variables. The dimensionless groups found from the theorem are power products denoted by Π_1, Π_2, Π_3, etc. The method allows the pis to be found in sequential order without resorting to free exponents.

The first part of the pi theorem explains what reduction in variables to expect:

If a physical process satisfies the PDH and involves n dimensional variables, it can be reduced to a relation between only k dimensionless variables or Π's. The reduction $j = n - k$ equals the maximum number of variables which do not form a pi among themselves and is always less than or equal to the number of dimensions describing the variables.

Take the specific case of force on an immersed body: Eq. (5.1) contains five variables F, L, U, ρ, and μ described by three dimensions (MLT). Thus $n = 5$ and $j \leq 3$. Therefore it is a good guess that we can reduce the problem to k pis, with $k = n - j \geq 5 - 3 = 2$. And this is exactly what we obtained: two dimensionless variables, $\Pi_1 = C_F$ and $\Pi_2 = \text{Re}$. On rare occasions it may take more pis than this minimum (see Example 5.5).

The second part of the theorem shows how to find the pis one at a time:

Find the reduction j, then select j variables which do not form a pi among themselves.[1] Each desired pi group will be a power product of these j variables plus one additional variable which is assigned any convenient nonzero exponent. Each pi group thus found is independent.

To be specific, suppose that the process involves five variables

$$v_1 = f(v_2, v_3, v_4, v_5)$$

Suppose that there are three dimensions (MLT) and we search around and find that indeed $j = 3$. Then $k = 5 - 3 = 2$ and we expect, from the theorem, two and only two pi groups. Pick out three convenient variables which do *not* form a pi and suppose these turn out to be v_2, v_3, and v_4. Then the two pi groups are formed by power products of these three plus one additional variable

$$\Pi_1 = (v_2)^a (v_3)^b (v_4)^c v_1 = M^0 L^0 T^0 \qquad \Pi_2 = (v_2)^a (v_3)^b (v_4)^c v_5 = M^0 L^0 T^0$$

Here we have arbitrarily chosen v_1 and v_5, the added variables, to have unit exponents. Equating exponents of the various dimensions is guaranteed by the theorem to give unique values of a, b, and c for each pi. And they are independent because only Π_1 contains v_1 and only Π_2 contains v_5. It is a very neat system once you get used to the procedure. We shall illustrate it with several examples.

---

[1] Make a clever choice here because all pis will contain these j variables in various groupings.

Typically, there are six steps involved:

1. List and count the n variables involved in the problem. If any important variables are missing, dimensional analysis will fail.
2. List the dimensions of each variable according to $MLT\Theta$ or $FLT\Theta$. A list is given in Table 5.1.
3. Find j. Initially guess j equal to the number of different dimensions present and look for j variables which do not form a pi product. If no luck, reduce j by 1 and look again. With practice, you will find j rapidly.
4. Select j variables which do not form a pi product. Make sure they please you and have some generality if possible, because they will then appear in every one of your pi groups. Pick density or velocity or length. Do not pick surface tension, for example, or you will form six different independent Weber-number parameters and thoroughly annoy your colleagues.

Table 5.1

DIMENSIONS OF FLUID-MECHANICS PROPERTIES

Quantity	Symbol	Dimensions	
		$\{MLT\Theta\}$	$\{FLT\Theta\}$
Length	L	L	L
Area	A	L^2	L^2
Volume	$\mathcal{V}$	L^3	L^3
Velocity	V	LT^{-1}	LT^{-1}
Speed of sound	a	LT^{-1}	LT^{-1}
Volume flow	Q	L^3T^{-1}	L^3T^{-1}
Mass flow	$\dot{m}$	MT^{-1}	FTL^{-1}
Pressure, stress	p, σ	$ML^{-1}T^{-2}$	FL^{-2}
Strain rate	$\dot{\epsilon}$	T^{-1}	T^{-1}
Angle	θ	None	None
Angular velocity	ω	T^{-1}	T^{-1}
Viscosity	μ	$ML^{-1}T^{-1}$	FTL^{-2}
Kinematic viscosity	ν	L^2T^{-1}	L^2T^{-1}
Surface tension	Υ	MT^{-2}	FL^{-1}
Force	F	MLT^{-2}	F
Moment, torque	M	ML^2T^{-2}	FL
Power	P	ML^2T^{-3}	FLT^{-1}
Work, energy	W, E	ML^2T^{-2}	FL
Density	ρ	ML^{-3}	FT^2L^{-4}
Temperature	T	Θ	Θ
Specific heat	c_p, c_v	$L^2T^{-2}\Theta^{-1}$	$L^2T^{-2}\Theta^{-1}$
Thermal conductivity	k	$MLT^{-3}\Theta^{-1}$	$FT^{-1}\Theta^{-1}$
Expansion coefficient	β	Θ^{-1}	Θ^{-1}

5. Add one additional variable to your j variables and form a power product. Algebraically find the exponents which make the product dimensionless. Try to arrange for your output or *dependent* variables (force, pressure drop, torque, power) to appear in the numerator and your plots will look better. Do this sequentially, adding one new variable each time, and you will find all $n - j = k$ desired pi products.
6. Write the final dimensionless function and check your work to make sure all pi groups are dimensionless.

EXAMPLE 5.2   Repeat the development of Eq. (5.2) from Eq. (5.1), using the pi theorem.

Solution

Step 1   Write the function and count variables:

$$F = f(L, U, \rho, \mu) \qquad \text{there are five variables } (n = 5)$$

Step 2   List dimensions of each variable. From Table 5.1

F	L	U	ρ	μ
$\{MLT^{-2}\}$	$\{L\}$	$\{LT^{-1}\}$	$\{ML^{-3}\}$	$\{ML^{-1}T^{-1}\}$

Step 3   Find j. No variable contains the dimension Θ, and so j is less than or equal to 3 (MLT). We inspect the list and see that L, U, and ρ cannot form a pi group because only ρ contains mass and only U contains time. Therefore j does equal 3, and $n - j = 5 - 3 = 2 = k$. The pi theorem guarantees for this problem that there will be exactly two independent dimensionless groups.

Step 4   Select j variables. The group L, U, ρ we found in Step 3 will do fine.

Step 5   Combine L, U, ρ with one additional variable, in sequence, to find the two pi products.

First add force to find Π_1. You may select *any* exponent on this additional term as you please, to place it in the numerator or denominator to any power. Since F is the output, or dependent, variable, we select it to appear to the first power in the numerator

$$\Pi_1 = L^a U^b \rho^c F = (L)^a (LT^{-1})^b (ML^{-3})^c (MLT^{-2}) = M^0 L^0 T^0$$

Equate exponents:

Length: $a + b - 3c + 1 = 0$

Mass: $c + 1 = 0$

Time: $-b \qquad -2 = 0$

We can solve explicitly for

$$a = -2 \qquad b = -2 \qquad c = -1$$

Therefore

$$\Pi_1 = L^{-2} U^{-2} \rho^{-1} F = \frac{F}{\rho U^2 L^2} = C_F \qquad\qquad Ans.$$

This is exactly the right pi group as in Eq. (5.2). By varying the exponent on F, we could have found other equivalent groups such as $UL\rho^{1/2}/F^{1/2}$.

Finally, add viscosity to L, U, and ρ to find Π_2. Select any power you like for viscosity. By hindsight and custom, we select the power -1 to place it in the denominator:

$$\Pi_2 = L^a U^b \rho^c \mu^{-1} = L^a (LT^{-1})^b (ML^{-3})^c (ML^{-1}T^{-1})^{-1} = M^0 L^0 T^0$$

Equate exponents:

Length: $\qquad\qquad\qquad\qquad\qquad a + b - 3c + 1 = 0$

Mass: $\qquad\qquad\qquad\qquad\qquad\qquad\quad c - 1 = 0$

Time: $\qquad\qquad\qquad\qquad\qquad\quad -b \qquad + 1 = 0$

from which we find

$$a = b = c = 1$$

Therefore $\qquad\qquad\qquad \Pi_2 = L^1 U^1 \rho^1 \mu^{-1} = \frac{\rho U L}{\mu} = \text{Re} \qquad\qquad$ *Ans.*

We know we are finished; this is the second and last pi group. The theorem guarantees that the functional relationship must be of the equivalent form

$$\frac{F}{\rho U^2 L^2} = g\left(\frac{\rho U L}{\mu}\right) \qquad\qquad$$ *Ans.*

which is exactly Eq. (5.2).

EXAMPLE 5.3 Reduce the falling-body relationship, Eq. (5.14), to a function of dimensionless variables. Why are there two different formulations?

Solution Write the function and count variables

$$S = f(t, S_0, V_0, g) \qquad \text{five variables } (n = 5)$$

List the dimensions of each variable, from Table 5.1:

S	t	S_0	V_0	g
$\{L\}$	$\{T\}$	$\{L\}$	$\{LT^{-1}\}$	$\{LT^{-2}\}$

There are only two primary dimensions (L, T), so that $j \leq 2$. By inspection we can easily find two variables which cannot be combined to form a pi, for example, V_0 and g. Then $j = 2$, and we expect $5 - 2 = 3$ pi products. Select j variables among the parameters S_0, V_0, and g. Avoid S and t since they are the dependent variables, which should not be repeated in pi groups. We select V_0 and g. Combine V_0 and g with one additional variable, in sequence, to find the three pi groups.

First add S_0, to the first power, and find Π_1

$$\Pi_1 = V_0^a g^b S_0^1 = (LT^{-1})^a (LT^{-2})^b (L) = L^0 T^0$$

Equate exponents:

Length: $\qquad\qquad\qquad\qquad\qquad a + b + 1 = 0$

Time: $\qquad\qquad\qquad\qquad\qquad -a - 2b \qquad = 0$

Solve for $a = -2, b = 1$; hence $\Pi_1 = V_0^{-2}g^1S_0^1$, or

$$\Pi_1 = \frac{gS_0}{V_0^2} = \alpha \qquad\qquad Ans.$$

Now combine V_0 and g with t to form Π_2

$$\Pi_2 = V_0^a g^b t^1 = (LT^{-1})^a(LT^{-2})^b(T) = L^0 T^0$$

Equate exponents and solve for $a = -1, b = 1$. Then

$$\Pi_2 = \frac{gt}{V_0} = t^{**} \qquad\qquad Ans.$$

Finally, combine V_0 and g with S to form Π_3. Since S and S_0 have the same dimensions, Π_3 will be identical in form to Π_1

$$\Pi_3 = \frac{gS}{V_0^2} = S^{**} \qquad\qquad Ans.$$

Once we have found the three pi groups, the theorem tells us that the functional relationship must be of the form

$$\frac{gS}{V_0^2} = f\left(\frac{gt}{V_0}, \frac{gS_0}{V_0^2}\right) \qquad\qquad Ans.$$

or

$$S^{**} = f(t^{**}, \alpha)$$

This is the relationship we found analytically in Eq. (5.13b). Even if no theory were available, experimentation with falling bodies would produce curves nearly identical to Fig. 5.1b.

Suppose we had chosen our j variables to be V_0 and S_0 rather than V_0 and g. Then by repeating the steps above we would find the following pi groups:

$$\Pi_1 = V_0^a S_0^b g^1 = \frac{gS_0}{V_0^2} = \alpha$$

$$\Pi_2 = V_0^a S_0^b t^1 = \frac{V_0 t}{S_0} = t^*$$

$$\Pi_3 = V_0^a S_0^b S^1 = \frac{S}{S_0} = S^*$$

The functional relationship now would be of the form

$$S^* = f(t^*, \alpha)$$

and experimental data would plot according to Fig. 5.1a. Both formulations are equivalent and contain all the same information. The only difference lies in the choice of the initial j variables. Thus dimensional analysis always allows some arbitrary choices for parameters, giving the user some flexibility in presenting results.

EXAMPLE 5.4 At low velocities (laminar flow), the volume flow Q through a small-bore tube is a function only of the tube radius R, the fluid viscosity μ, and the pressure drop per unit tube length dp/dx. Using the pi theorem, find an appropriate dimensionless relationship.

Solution   Write the given relation and count variables

$$Q = f\left(R, \mu, \frac{dp}{dx}\right) \qquad \text{four variables } (n = 4)$$

Make a list of the dimensions of these variables from Table 5.1:

Q	R	μ	dp/dx
$\{L^3T^{-1}\}$	$\{L\}$	$\{ML^{-1}T^{-1}\}$	$\{ML^{-2}T^{-2}\}$

There are three primary dimensions (M, L, T), hence $j \leq 3$. By trial and error we determine that R, μ, and dp/dx cannot be combined into a pi group. Then $j = 3$, and $n - j = 4 - 3 = 1$. There is only *one* pi group, which we find by combining Q in a power product with the other three

$$\Pi_1 = R^a\mu^b\left(\frac{dp}{dx}\right)^c Q^1 = (L)^a(ML^{-1}T^{-1})^b(ML^{-2}T^{-2})^c(L^3T^{-1})$$
$$= M^0L^0T^0$$

Equate exponents:

Mass: $\qquad\qquad\qquad\qquad\qquad\qquad b + c \quad = 0$

Length: $\qquad\qquad\qquad\qquad\qquad a - b - 2c + 3 = 0$

Time: $\qquad\qquad\qquad\qquad\qquad\quad -b - 2c - 1 = 0$

Solving simultaneously, we obtain $a = -4$, $b = 1$, $c = -1$. Then

$$\Pi_1 = R^{-4}\mu^1\left(\frac{dp}{dx}\right)^{-1} Q$$

or
$$\Pi_1 = \frac{Q\mu}{R^4(dp/dx)} = \text{const} \qquad\qquad\qquad\qquad Ans.$$

Since there is only one pi group, it must equal a dimensionless constant. This is as far as dimensional analysis can take us. The laminar-flow theory of Sec. 6.4 shows that the value of the constant is $\pi/8$.

EXAMPLE 5.5   Assume that the tip deflection δ of a cantilever beam is a function of the tip load P, the beam length L, area moment of inertia I, and the material modulus of elasticity E; that is, $\delta = f(P, L, I, E)$. Rewrite this function in dimensionless form and comment on its complexity and the peculiar value of j.

Solution   List the variables and their dimensions:

δ	P	L	I	E
$\{L\}$	$\{MLT^{-2}\}$	$\{L\}$	$\{L^4\}$	$\{ML^{-1}T^{-2}\}$

There are five variables $(n = 5)$ and three primary dimensions (M, L, T), hence $j \leq 3$. But try as we may, we *cannot* find any combination of three variables which do not form a pi group. This is because $\{M\}$ and $\{T\}$ occur only in P and E and only in the same form, $\{MT^{-2}\}$.

Thus we have encountered a special case of $j = 2$, which is less than the number of dimensions (M, L, T). To gain more insight into this peculiarity, you should rework the problem using the (F, L, T) system of dimensions.

With $j = 2$, we select L and E as two variables which cannot form a pi group and then add other variables to form the three desired pis

$$\Pi_1 = L^a E^b I^1 = (L)^a (ML^{-1}T^{-2})^b (L^4) = M^0 L^0 T^0$$

from which, after equating exponents, we find that $a = -4$, $b = 0$, or $\Pi_1 = I/L^4$. Then

$$\Pi_2 = L^a E^b P^1 = (L)^a (ML^{-1}T^{-2})^b (MLT^{-2}) = M^0 L^0 T^0$$

from which we find $a = -2$, $b = -1$, or $\Pi_2 = P/EL^2$, and

$$\Pi_3 = L^a E^b \delta^1 = (L)^a (ML^{-1}T^{-2})^b (L) = M^0 L^0 T^0$$

from which $a = -1$, $b = 0$, or $\Pi_3 = \delta/L$. The proper dimensionless function is $\Pi_3 = f(\Pi_2, \Pi_1)$, or

$$\frac{\delta}{L} = f\left(\frac{P}{EL^2}, \frac{I}{L^4}\right) \qquad\qquad Ans. \quad (1)$$

This is a complex three-variable function, but dimensional analysis alone can take us no further.

We can "improve" Eq. (1) by taking advantage of some physical reasoning, as Langhaar points out [8, p. 91]. For small elastic deflections, δ is proportional to load P and inversely proportional to moment of inertia I. Since P and I occur separately in Eq. (1), this means that Π_3 must be proportional to Π_2 and inversely proportional to Π_1. Thus, for these conditions,

$$\frac{\delta}{L} = (\text{const}) \frac{P}{EL^2} \frac{L^4}{I}$$

or

$$\delta = (\text{const}) \frac{PL^3}{EI} \qquad\qquad (2)$$

This could not be predicted by a pure dimensional analysis. Strength-of-materials theory predicts that the value of the constant is $\frac{1}{3}$.

5.4 NONDIMENSIONALIZATION OF THE BASIC EQUATIONS

We could use the pi-theorem method of the previous section to analyze problem after problem after problem, finding the dimensionless parameters which govern in each case. Textbooks on dimensional analysis [e.g., 7] do this. An alternate and very powerful technique is to attack the basic equations of flow from Chap. 4. Even though these equations cannot be solved in general, they will reveal basic dimensionless parameters, e.g., Reynolds number, in their proper form and proper position, giving clues to when they are negligible. The boundary conditions must also be nondimensionalized.

Let us briefly apply this technique to the incompressible-flow continuity and momentum equations with constant viscosity:

Continuity:
$$\nabla \cdot \mathbf{V} = 0 \tag{5.21a}$$

Momentum:
$$\rho \frac{d\mathbf{V}}{dt} = \rho\mathbf{g} - \nabla p + \mu \nabla^2 \mathbf{V} \tag{5.21b}$$

Typical boundary conditions for these two equations are

Fixed solid surface:
$$\mathbf{V} = 0$$

Inlet or outlet:
$$\text{Known } \mathbf{V}, p \tag{5.22}$$

Free surface, $z = \eta$:
$$w = \frac{d\eta}{dt} \qquad p = p_a - \Upsilon(R_x^{-1} + R_y^{-1})$$

We omit the energy equation (4.75) and assign its dimensionless form in the problems (Probs. 5.36 and 5.38).

Equations (5.21), and (5.22) contain the three basic dimensions M, L, and T. All variables p, $\mathbf{V}$, x, y, z, and t can be nondimensionalized using density and two reference constants which might be characteristic of the particular fluid flow:

$$\text{Reference velocity} = U \qquad \text{reference length} = L$$

For example, U may be the inlet or upstream velocity and L the diameter of a body immersed in the stream.

Now define all relevant dimensionless variables, denoting them by an asterisk:

$$\mathbf{V}^* = \frac{\mathbf{V}}{U}$$

$$x^* = \frac{x}{L} \qquad y^* = \frac{y}{L} \qquad z^* = \frac{z}{L} \tag{5.23}$$

$$t^* = \frac{tU}{L} \qquad p^* = \frac{p + \rho g z}{\rho U^2}$$

All these are fairly obvious except for p^*, where we have slyly introduced the gravity effect, assuming that z is "up." This is a hindsight idea suggested by Bernoulli's equation (3.63).

Since ρ, U, and L are all constants, the derivatives in Eqs. (5.21) can all be handled in dimensionless form with dimensional coefficients. For example,

$$\frac{\partial u}{\partial x} = \frac{\partial(Uu^*)}{\partial(Lx^*)} = \frac{U}{L}\frac{\partial u^*}{\partial x^*}$$

Substitute the variables from Eqs. (5.23) into Eqs. (5.21) and (5.22) and divide through by the leading dimensional coefficient, in the same way we handled Eq. (5.12). The resulting dimensionless equations of motion are:

Continuity:
$$\nabla^* \cdot \mathbf{V}^* = 0 \tag{5.24a}$$

Momentum:
$$\frac{d\mathbf{V}^*}{dt^*} = -\nabla^* p^* + \frac{\mu}{\rho U L} \nabla^{*2}(\mathbf{V}^*) \tag{5.24b}$$

The dimensionless boundary conditions are:

Fixed solid surface: $\mathbf{V}^* = 0$

Inlet or outlet: Known $\mathbf{V}^*$, p^*

Free surface, $z^* = \eta^*$: $w^* = \dfrac{d\eta^*}{dt^*}$ (5.25)

$$p^* = \frac{p_a}{\rho U^2} + \frac{gL}{U^2} z^* + \frac{\Upsilon}{\rho U^2 L} (R_x^{*-1} + R_y^{*-1})$$

These equations reveal a total of four dimensionless parameters, one in the momentum equation and three in the free-surface-pressure boundary condition.

Dimensionless Parameters

In the continuity equation there are no parameters. The momentum equation contains one, generally accepted as the most important parameter in fluid mechanics:

$$\text{Reynolds number Re} = \frac{\rho U L}{\mu}$$

It is named after Osborne Reynolds (1842–1912), a British engineer who first proposed it in 1883 (Ref. 4 of Chap. 6). The Reynolds number is always important, with or without a free surface, and can be neglected only in flow regions away from high velocity gradients, e.g., away from solid surfaces, jets, or wakes.

The no-slip and inlet-exit boundary conditions contain no parameters. The free-surface-pressure condition contains three:

$$\text{Euler number (pressure coefficient) Eu} = \frac{p_a}{\rho U^2}$$

This is named after Leonhard Euler (1707–1783) and is rarely important unless the pressure drops low enough to cause vapor formation (cavitation) in a liquid. The Euler number is often written in terms of pressure differences, $\text{Eu} = \Delta p / \rho U^2$. If Δp involves vapor pressure p_v, it is called the cavitation number $\text{Ca} = (p_a - p_v)/\rho U^2$.

The second pressure parameter is much more important:

$$\text{Froude number Fr} = \frac{U^2}{gL}$$

It is named after William Froude (1810–1879), a British naval architect who, with his son Robert, developed the ship-model towing-tank concept and proposed similarity rules for free-surface flows (ship resistance, surface waves, open channels). The Froude number is the dominant effect in free-surface flows and is totally unimportant if there is no free surface. Chapter 10 investigates Froude-number effects in detail.

The final free-surface parameter is

$$\text{Weber number We} = \frac{\rho U^2 L}{\Upsilon}$$

It is named after Moritz Weber (1871–1951) of the Polytechnic Institute of Berlin, who developed the laws of similitude in their modern form. It was Weber who named Re and Fr after Reynolds and Froude. The Weber number is important only if it is of order unity or less, which typically occurs when the surface curvature is comparable in size to the liquid depth, e.g., in droplets, capillary flows, ripple waves, and very small hydraulic models. If We is large, its effect may be neglected.

If there is no free surface, Fr, Eu, and We drop out entirely, except for the possibility of cavitation of a liquid at very small Eu. Thus, in low-speed viscous flows with no free surface, the Reynolds number is the only important dimensionless parameter.

Compressibility Parameters

In high-speed flow of a gas there are significant changes in pressure, density, and temperature which must be related by an equation of state such as the perfect-gas law, Eq. (1.29). These thermodynamic changes introduce two additional dimensionless parameters mentioned briefly in earlier chapters:

$$\text{Mach number Ma} = \frac{U}{a} \qquad \text{Specific-heat ratio } \gamma = \frac{c_p}{c_v}$$

The Mach number is named after Ernst Mach (1838–1916), an Austrian physicist. The effect of γ is only slight to moderate, but Ma exerts a strong effect on compressible-flow properties if it is greater than about 0.3. These effects are studied in Chap. 9.

Oscillating Flows

If the flow pattern is oscillating, a seventh parameter enters through the inlet boundary condition. For example, suppose that the inlet stream is of the form

$$u = U \cos \omega t$$

Nondimensionalization of this relation results in

$$\frac{u}{U} = u^* = \cos\left(\frac{\omega L}{U} t^*\right)$$

The argument of the cosine contains the new parameter

$$\text{Strouhal number St} = \frac{\omega L}{U}$$

The dimensionless forces and moments, friction, and heat transfer, etc., of such an oscillating flow would be a function of both Reynolds and Strouhal number. This parameter is named after V. Strouhal, a German physicist who experimented in 1878 with wires singing in the wind.

Some flows which you might guess to be perfectly steady actually have an oscillatory pattern which is dependent on the Reynolds number. An example is the periodic vortex shedding behind a blunt body immersed in a steady stream of velocity U. Figure 5.2a shows an array of alternating vortices shed from a circular

(a)

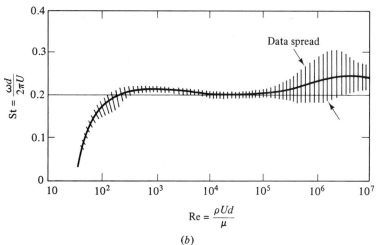

Fig. 5.2 Vortex shedding from a circular cylinder: (*a*) vortex street behind a circular cylinder (*from Ref. 28, courtesy of U.S. Naval Research Laboratory*); (*b*) experimental shedding frequencies (*data from Refs. 26 and 27*).

(b)

cylinder immersed in a steady crossflow. This regular, periodic shedding is called a *Kármán vortex street*, after T. von Kármán, who explained it theoretically in 1912. The shedding occurs in the range $10^2 < \text{Re} < 10^7$, with an average Strouhal number $\omega d/2\pi U \approx 0.21$. Figure 5.2b shows measured shedding frequencies.

Resonance can occur if a vortex shedding frequency is near a body's structural-vibration frequency. Electric transmission wires sing in the wind, undersea mooring lines gallop at certain current speeds, and slender structures flutter at critical wind or vehicle speeds. A striking example is the disastrous failure of the Tacoma Narrows suspension bridge in 1940, when wind-excited vortex shedding caused resonance with the natural torsional oscillations of the bridge.

Other Dimensionless Parameters

We have discussed seven important parameters in fluid mechanics, and there are others. Four additional parameters arise from nondimensionalization of the energy equation (4.75) and its boundary conditions. These four (Prandtl number, Eckert number, Grashof number, and wall-temperature ratio) are listed in Table 5.2 just in case you fail to solve Prob. 5.36. Another important and rather sneaky parameter is the wall-roughness ratio ϵ/L (in Table 5.2).[1] Slight changes in surface roughness have a striking effect in the turbulent flow or high-Reynolds-number range, as we shall see in Chap. 6.

This book is primarily concerned with Reynolds-, Mach-, and Froude-number effects, which dominate most flows. Note that we discovered all these parameters (except ϵ/L) simply by nondimensionalizing the basic equations without actually solving them.

If the reader is not satiated with the 15 parameters given in Table 5.2, Ref. 29 contains a list of over 300 dimensionless parameters in use in engineering.

A Successful Application

Dimensional analysis is fun, but does it work? Yes; if all important variables are included in the proposed function, the dimensionless function found by dimensional analysis will collapse all the data onto a single curve or set of curves.

An example of the success of dimensional analysis is given in Fig. 5.3 for the measured drag on smooth cylinders and spheres. The flow is normal to the axis of the cylinder, which is extremely long, $L/d \to \infty$. The data are from many sources, for both liquids and gases, and include bodies from several meters in diameter down to fine wires and balls less than 1 mm in size. Both curves in Fig. 5.3a are entirely experimental; the analysis of immersed body drag is one of the weakest areas of modern fluid-mechanics theory. Except for some isolated digital-computer calculations, there is no theory for cylinder and sphere drag except *creeping flow*, $\text{Re} < 1$.

---

[1] Roughness is easy to overlook because it is a slight geometric effect which does not appear in the equations of motion.

Table 5.2

DIMENSIONLESS GROUPS IN FLUID MECHANICS

Parameter	Definition	Qualitative ratio of effects	Importance
✓ Reynolds number	$\mathrm{Re} = \dfrac{\rho U L}{\mu}$	$\dfrac{\text{Inertia}}{\text{Viscosity}}$	Always
Mach number	$\mathrm{Ma} = \dfrac{U}{a}$	$\dfrac{\text{Flow speed}}{\text{Sound speed}}$	Compressible flow
✓ Froude number	$\mathrm{Fr} = \dfrac{U^2}{gL}$	$\dfrac{\text{Inertia}}{\text{Gravity}}$	Free-surface flow
Weber number	$\mathrm{We} = \dfrac{\rho U^2 L}{\Upsilon}$	$\dfrac{\text{Inertia}}{\text{Surface tension}}$	Free-surface flow
Cavitation number (Euler number)	$\mathrm{Ca} = \dfrac{p - p_v}{\rho U^2}$	$\dfrac{\text{Pressure}}{\text{Inertia}}$	Cavitation
Prandtl number	$\mathrm{Pr} = \dfrac{\mu c_p}{k}$	$\dfrac{\text{Dissipation}}{\text{Conduction}}$	Heat convection
Eckert number	$\mathrm{Ec} = \dfrac{U^2}{c_p T_0}$	$\dfrac{\text{Kinetic energy}}{\text{Enthalpy}}$	Dissipation
Specific-heat ratio	$\gamma = \dfrac{c_p}{c_v}$	$\dfrac{\text{Enthalpy}}{\text{Internal energy}}$	Compressible flow
Strouhal number	$\mathrm{St} = \dfrac{\omega L}{U}$	$\dfrac{\text{Oscillation}}{\text{Mean speed}}$	Oscillating flow
Roughness ratio	$\dfrac{\epsilon}{L}$	$\dfrac{\text{Wall roughness}}{\text{Body length}}$	Turbulent, rough walls
Grashof number	$\mathrm{Gr} = \dfrac{\beta \,\Delta T\, g L^3 \rho^2}{\mu^2}$	$\dfrac{\text{Buoyancy}}{\text{Viscosity}}$	Natural convection
Temperature ratio	$\dfrac{T_w}{T_0}$	$\dfrac{\text{Wall temperature}}{\text{Stream temperature}}$	Heat transfer
Pressure coefficient	$C_p = \dfrac{p - p_\infty}{\frac{1}{2}\rho U^2}$	$\dfrac{\text{Static pressure}}{\text{Dynamic pressure}}$	Aerodynamics, hydrodynamics
Lift coefficient	$C_L = \dfrac{L}{\frac{1}{2}\rho U^2 A}$	$\dfrac{\text{Lift force}}{\text{Dynamic force}}$	Aerodynamics, hydrodynamics
✓ Drag coefficient	$C_D = \dfrac{D}{\frac{1}{2}\rho U^2 A}$	$\dfrac{\text{Drag force}}{\text{Dynamic force}}$	Aerodynamics, hydrodynamics

Handwritten annotations:

Euler # (next to Pressure coefficient)

$\rightarrow L^2$ (next to Lift coefficient A)

$\rightarrow L^2$ (next to Drag coefficient A)

Power Coefficient $P = F_D \, U$

Torque (Moment) $C_T = \dfrac{T}{\frac{1}{2}\rho U^2 L^3}$

Power $C_P = \dfrac{\text{Power}}{\frac{1}{2}\rho U^3 L^2}$

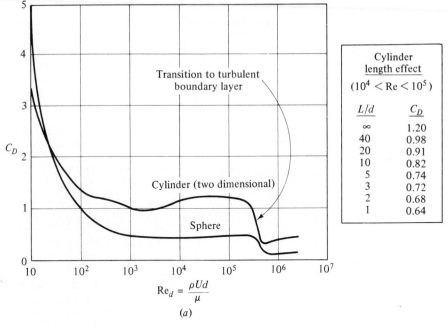

Cylinder length effect $(10^4 < \mathrm{Re} < 10^5)$	
L/d	C_D
∞	1.20
40	0.98
20	0.91
10	0.82
5	0.74
3	0.72
2	0.68
1	0.64

$$\mathrm{Re}_d = \frac{\rho U d}{\mu}$$

(a)

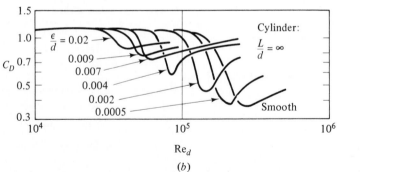

(b)

Fig. 5.3 The proof of practical dimensional analysis: drag coefficients of a cylinder and sphere: (a) drag coefficient of a smooth cylinder and sphere (data from many sources); (b) increased roughness causes earlier transition to a turbulent boundary layer.

The Reynolds number of both bodies is based upon diameter, hence the notation Re_d. But the drag coefficients are defined differently

$$C_D = \begin{cases} \dfrac{\text{drag}}{\frac{1}{2}\rho U^2 L d} & \text{cylinder} \\[2ex] \dfrac{\text{drag}}{\frac{1}{2}\rho U^2 \frac{1}{4}\pi d^2} & \text{sphere} \end{cases} \tag{5.26}$$

They both have a factor $\frac{1}{2}$ as a traditional tribute to Bernoulli and Euler, and both are based on the projected area, i.e., the area one sees when looking toward the body from upstream. The usual definition of C_D is thus

$$C_D = \frac{\text{drag}}{\frac{1}{2}\rho U^2 (\text{projected area})} \tag{5.27}$$

However, one should carefully check the definitions of C_D, Re, etc., before using data in the literature.

Figure 5.3a is for long, smooth cylinders. If wall roughness and cylinder length are included as variables, we obtain from dimensional analysis a complex three-parameter function

$$C_D = f\left(\text{Re}_d, \frac{\epsilon}{d}, \frac{L}{d}\right) \qquad (5.28)$$

To describe this function completely would require 1000 or more experiments. Therefore it is customary to explore the length and roughness effects separately to establish trends.

The table with Fig. 5.3a shows the length effect with zero wall roughness. As length decreases, the drag decreases by up to 50 percent. Physically, the pressure is "relieved" at the ends as the flow is allowed to skirt around the tips instead of deflecting over and under the body.

Figure 5.3b shows the effect of wall roughness for an infinitely long cylinder. The sharp drop in drag occurs at lower Re_d as roughness causes an earlier transition to a turbulent boundary layer on the surface of the body. Roughness has the same effect on sphere drag, a fact which is exploited in sports by deliberate dimpling of golf balls to give them less drag at their flight $\text{Re}_d \approx 10^5$.

Figure 5.3 is a typical experimental study of a fluid-mechanics problem, aided by dimensional analysis. As time and money and demand allow, the complete three-parameter relation (5.28) could be filled out by further experiments.

EXAMPLE 5.6   The capillary rise h of a liquid in a tube varies with tube diameter d, gravity g, fluid density ρ, surface tension Υ, and the contact angle θ. (a) Find a dimensionless statement of this relation. (b) If $h = 3$ cm in a given experiment, what will h be in a similar case if diameter and surface tension are half as much, density is twice as much, and the contact angle is the same?

Solution

Part (a)   *Step 1*   Write down the function and count variables

$$h = f(d, g, \rho, \Upsilon, \theta) \qquad n = 6 \text{ variables}$$

Step 2   List the dimensions (FLT) from Table 5.2:

h	d	g	ρ	Υ	θ
$\{L\}$	$\{L\}$	$\{LT^{-2}\}$	$\{FT^2L^{-4}$	$\{FL^{-1}\}$	None

Step 3   Find j. Several groups of three form no pi: Υ, ρ, and g or ρ, g, and d. Therefore $j = 3$, and we expect $n - j = 6 - 3 = 3$ dimensionless groups. One of these is obviously θ, which is already dimensionless:

$$\Pi_3 = \theta \qquad\qquad Ans. \ (a)$$

If we chose carelessly to search for it using steps 4 and 5, we would still find $\Pi_3 = \theta$.

Step 4   Select j variables which do not form a pi group: ρ, g, d.

Step 5 Add one additional variable in sequence to form the pis:

Add h: $$\Pi_1 = \rho^a g^b d^c h = (FT^2L^{-4})^a(LT^{-2})^b(L)^c(L) = F^0L^0T^0$$

Solve for

$$a = b = 0 \qquad c = -1$$

Therefore $$\Pi_1 = \rho^0 g^0 d^{-1} h = \frac{h}{d}$$ *Ans. (a)*

Finally add Υ, again selecting its exponent to be 1

$$\Pi_2 = \rho^a g^b d^c \Upsilon = (FT^2L^{-4})^a(LT^{-2})^b(L)^c(FL^{-1}) = F^0L^0T^0$$

Solve for

$$a = b = -1 \qquad c = -2$$

Therefore $$\Pi_2 = \rho^{-1}g^{-1}d^{-2}\Upsilon = \frac{\Upsilon}{\rho g d^2}$$ *Ans. (a)*

Step 6 The complete dimensionless relation for this problem is thus

$$\frac{h}{d} = F\left(\frac{\Upsilon}{\rho g d^2}, \theta\right)$$ *Ans. (a)* (1)

This is as far as dimensional analysis goes. Theory, however, establishes that h is proportional to Υ. Since Υ occurs only in the second parameter, we can slip it outside

$$\left(\frac{h}{d}\right)_{actual} = \frac{\Upsilon}{\rho g d^2} F_1(\theta) \quad \text{or} \quad \frac{h\rho g d}{\Upsilon} = F_1(\theta)$$

Example 1.13 showed theoretically that $F_1(\theta). = 4\cos\theta$.

Part (b) We are given h_1 for certain conditions d_1, Υ_1, ρ_1, and θ_1, and θ_1. If $h_1 = 3$ cm, what is h_2 for $d_2 = \frac{1}{2}d$, $\Upsilon_2 = \frac{1}{2}\Upsilon_1$, $\rho_2 = 2\rho_1$, and $\theta_2 = \theta_1$? We know the functional relation, Eq. (1), must still hold at condition 2

$$\frac{h_2}{d_2} = F\left(\frac{\Upsilon_2}{\rho_2 g d_2^2}, \theta_2\right)$$

But

$$\frac{\Upsilon_2}{\rho_2 g d_2^2} = \frac{\frac{1}{2}\Upsilon_1}{2\rho_1 g(\frac{1}{2}d_1)^2} = \frac{\Upsilon_1}{\rho_1 g d_1^2}$$

Therefore, functionally,

$$\frac{h_2}{d_2} = F\left(\frac{\Upsilon_1}{\rho_1 g d_1^2}, \theta_1\right) = \frac{h_1}{d_1}$$

We are given a condition 2 which is exactly similar to condition 1, and therefore a scaling law holds

$$h_2 = h_1 \frac{d_2}{d_1} = (3 \text{ cm})\frac{\frac{1}{2}d_1}{d_1} = 1.5 \text{ cm}$$ *Ans. (b)*

If the pi groups had not been exactly the same for both conditions, we would have to know more about the functional relation F to calculate h_2.

5.5 MODELING AND ITS PITFALLS

So far we have learned about dimensional homogeneity and the pi theorem method, using power products, for converting a homogeneous physical relation into dimensionless form. This is straightforward mathematically, but there are certain engineering difficulties which need to be discussed.

First, we have more or less taken for granted that the variables which affect the process can be listed and analyzed. Actually, selection of the important variables requires considerable judgment and experience. The engineer must decide for example whether viscosity can be neglected. Are there significant temperature effects? Is surface tension important? What about wall roughness? Each pi group which is retained increases the expense and effort required. Judgment in selecting variables will come through practice and maturity; this book should provide some of the necessary experience.

Once the variables are selected and the dimensional analysis performed, the experimenter seeks to achieve *similarity* between the model tested and the prototype to be designed. With sufficient testing, the model data will reveal the desired dimensionless function between variables

$$\Pi_1 = f(\Pi_2, \Pi_3, \ldots, \Pi_k) \tag{5.29}$$

With Eq. (5.29) available in chart, graphical, or analytical form, we are then in a position to ensure complete similarity between model and prototype. A formal statement would be as follows:

> Flow conditions for a model test are completely similar if all relevant dimensionless parameters have the same corresponding values for model and prototype.

This follows mathematically from Eq. (5.29). If $\Pi_{2m} = \Pi_{2p}$, $\Pi_{3m} = \Pi_{3p}$, etc., Eq. (5.29) guarantees that the desired output Π_{1m} will equal Π_{1p}. But this is easier said than done, as we now discuss.

Instead of complete similarity, the engineering literature speaks of particular types of similarity, the most common being geometric, kinematic, dynamic, and thermal. Let us consider each separately.

Geometric Similarity

Geometric similarity concerns the length dimension $\{L\}$ and must be assured before any sensible model testing can proceed. A formal definition is as follows:

> A model and prototype are geometrically similar if and only if all body dimensions in all three coordinates have the same linear-scale ratio.

Note that *all* length scales must be the same. It is as if you took a photograph of the prototype and reduced it or enlarged it until it fitted the size of the model. If the model is to be made one-tenth the prototype size, its length, width, and height must each be one-tenth as large. Not only that, but its entire shape must be one-tenth as large, and technically we speak of *homologous* points, which are points which have the same relative location. For example the nose of the prototype is homologous to the nose of the model. The left wingtip of the prototype is homologous to the left wingtip of the model. Then geometric similarity requires that all homologous points be related by the same linear-scale ratio. This applies to the fluid geometry as well as the model geometry:

> All angles are preserved in geometric similarity. All flow directions are preserved. The orientation of model and prototype with respect to the surroundings must be identical.

Figure 5.4 illustrates a prototype wing and a one-tenth-scale model. The model lengths are all one-tenth as large, but its angle of attack with respect to the free stream is the same: 10° not 1°. All physical details on the model must be scaled, and some of them are rather subtle and sometimes overlooked:

1. The model nose radius must be one-tenth as large.
2. The model surface roughness must be one-tenth as large.
3. If the prototype has a 5-mm boundary-layer trip wire 1.5 m from the leading edge, the model should have a 0.5-mm trip wire 0.15 m from its leading edge.
4. If the prototype is constructed with protruding fasteners, the model should have homologous protruding fasteners one-tenth as large.

And so on. Any departure from these details is a violation of geometric similarity and must be justified by experimental comparison to show that the prototype behavior was not significantly affected by the discrepancy.

Models which appear similar in shape but which clearly violate geometric similarity should not be compared except at your own risk. Figure 5.5 illustrates this point. The spheres in Fig. 5.5a are all geometrically similar and can be tested with a high expectation of success if the Reynolds number or Froude number, etc., are matched. But the ellipsoids in Fig. 5.5b merely *look* similar. They actually have different linear-scale ratios and therefore cannot be compared in a rational manner, even though they may have identical Reynolds and Froude numbers, etc. The data

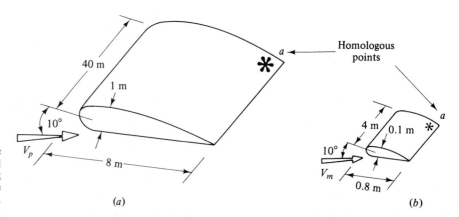

Fig. 5.4 Geometric similarity in model testing: (a) prototype; (b) one-tenth scale model.

(a)

(b)

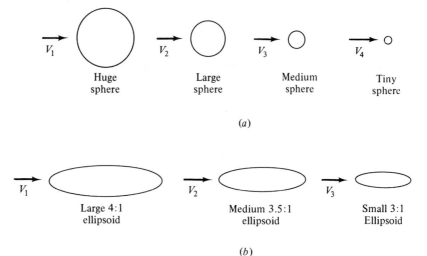

Fig. 5.5 Geometric similarity and dissimilarity of flows: (*a*) similar; (*b*) dissimilar.

will not be the same for these ellipsoids and any attempt to "compare" them is a matter of rough engineering judgment.

Kinematic Similarity

Kinematic similarity requires that the model and prototype have the same length-scale ratio and also the same time-scale ratio. The result is that the velocity-scale ratio will be the same for both. As Langhaar [8] states it:

> "The motions of two systems are kinematically similar if homologous particles lie at homologous points at homologous times."

Length-scale equivalence simply implies geometric similarity, but time-scale equivalence may require additional dynamic considerations such as equivalence of the Reynolds and Mach numbers.

One special case is incompressible frictionless flow with no free surface, as sketched in Fig. 5.6*a*. These perfect-fluid flows are kinematically similar with independent length and time scales, and no additional parameters are necessary (see Chap. 8 for further details).

Frictionless flows with a free surface, as in Fig. 5.6*b*, are kinematically similar if their Froude numbers are equal

$$\text{Fr}_m = \frac{V_m^2}{gL_m} = \frac{V_p^2}{gL_p} = \text{Fr}_p \tag{5.30}$$

Note that Froude number contains only length and time dimensions and hence is a purely kinematic parameter which fixes the relation between length and time. From Eq. (5.30), if the length scale is

$$L_m = \alpha L_p \tag{5.31}$$

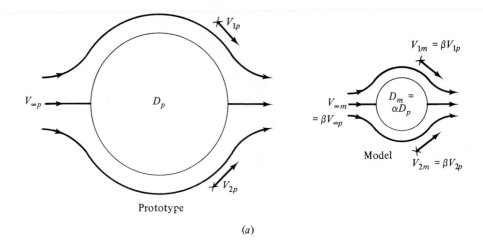

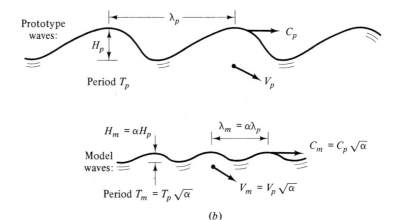

Fig. 5.6 Frictionless low-speed flows are kinematically similar: (a) flows with no free surface are kinematically similar with independent length- and time-scale ratios; (b) free-surface flows are kinematically similar with length and time scales related by the Froude number.

where α is a dimensionless ratio, the velocity scale is

$$\frac{V_m}{V_p} = \left(\frac{L_m}{L_p}\right)^{1/2} = \sqrt{\alpha} \tag{5.32}$$

and the time scale is

$$\frac{T_m}{T_p} = \frac{L_m/V_m}{L_p/V_p} = \sqrt{\alpha} \tag{5.33}$$

These Froude-scaling kinematic relations are illustrated in Fig. 5.6b for wave-motion modeling. If the waves are related by the length scale α, the wave period, propagation speed, and particle velocities are related by $\sqrt{\alpha}$.

If viscosity, surface tension, or compressibility is important, kinematic similarity is dependent upon the achievement of dynamic similarity.

Dynamic Similarity

Dynamic similarity exists when model and prototype have the same length-scale ratio, time-scale ratio, and force-scale (or mass-scale) ratio. Again geometric similarity is a first requirement; otherwise proceed no further. Then dynamic similarity exists, simultaneous with kinematic similarity, if model and prototype force and pressure coefficients are identical. This is assured if:

1. For compressible flow, model and prototype Reynolds number and Mach number and specific-heat ratio are correspondingly equal.
2. For incompressible flow
   a. With no free surface: model and prototype Reynolds number are equal.
   b. With a free surface: model and prototype Reynolds number, Froude number, and (if necessary) Weber number and cavitation number are correspondingly equal.

Mathematically, Newton's law for any fluid particle requires that the sum of the pressure force, gravity force, and friction force equal the acceleration term, or inertia force,

$$\mathbf{F}_p + \mathbf{F}_g + \mathbf{F}_f = \mathbf{F}_i$$

The dynamic-similarity laws listed above ensure that each of these forces will be in the same ratio and have equivalent directions between model and prototype. Figure 5.7 shows an example for flow through a sluice gate. The force polygons at homologous points have exactly the same shape if the Reynolds and Froude numbers are equal (neglecting surface tension and cavitation, of course). Kinematic similarity is also assured by these model laws.

Fig. 5.7 Dynamic similarity in sluice-gate flow. Model and prototype yield identical homologous force polygons if the Reynolds and Froude numbers are the same corresponding values: (*a*) prototype; (*b*) model.

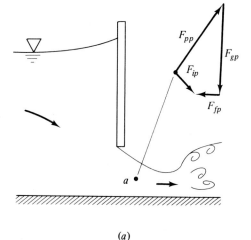

(*a*)

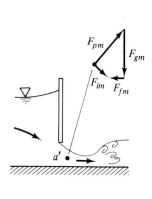

(*b*)

Discrepancies in Water and Air Testing

The perfect dynamic similarity shown in Fig. 5.7 is more of a dream than a reality because true equivalence of Reynolds and Froude numbers can be achieved only by dramatic changes in fluid properties, whereas in fact most model testing is simply done with water or air, the cheapest fluids available.

First consider hydraulic model testing with a free surface. Dynamic similarity requires equivalent Froude numbers, Eq. (5.30), *and* equivalent Reynolds numbers

$$\frac{V_m L_m}{\nu_m} = \frac{V_p L_p}{\nu_p} \tag{5.34}$$

But both velocity and length are constrained by the Froude number, Eqs. (5.31) and (5.32). Therefore, for a given length-scale ratio α, Eq. (5.34) is true only if

$$\frac{\nu_m}{\nu_p} = \frac{L_m}{L_p} \frac{V_m}{V_p} = \alpha \sqrt{\alpha} = \alpha^{3/2} \tag{5.35}$$

For example, for a one-tenth-scale model, $\alpha = 0.1$, and $\alpha^{3/2} = 0.032$. Since ν_p is undoubtedly water, we need a fluid with only 0.032 times the kinematic viscosity of water to achieve dynamic similarity. Referring back to Table 1.3, we see that this is impossible: even mercury has only one-ninth the kinematic viscosity of water, and a mercury hydraulic model would be expensive and bad for your health. In practice, water is used for both model and prototype, and the Reynolds-number similarity (5.34) is unavoidably violated. The Froude number is held constant since it is the dominant parameter in free-surface flows. Typically the Reynolds number of the model flow is too small by a factor of 10 to 1000. As shown in Fig. 5.8, the low-Reynolds-number model data are used to estimate by extrapolation the desired high-Reynolds-number prototype data. As the figure indicates, there is obviously considerable uncertainty in using such an extrapolation, but there is no other practical alternative in hydraulic model testing.

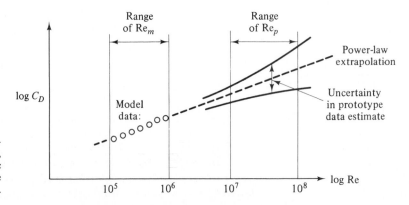

Fig. 5.8 Reynolds-number extrapolation, or scaling, of hydraulic data with equal Froude numbers.

Second, consider aerodynamic model testing in air with no free surface. The important parameters are the Reynolds number and the Mach number. Equation (5.34) should be satisfied, plus the compressibility criterion

$$\frac{V_m}{a_m} = \frac{V_p}{a_p} \tag{5.36}$$

Elimination of V_m/V_p between (5.34) and (5.36) gives

$$\frac{v_m}{v_p} = \frac{L_m}{L_p} \frac{a_p}{a_m} \tag{5.37}$$

Since the prototype is no doubt an air operation, we need a wind-tunnel fluid of low viscosity and high speed of sound. Hydrogen is the only practical example, but clearly it is too expensive and dangerous. Therefore wind tunnels normally operate with air as the working fluid. Cooling and pressurizing the air will bring Eq. (5.37) into better agreement but not enough to satisfy a length-scale reduction of, say, one-tenth. Therefore Reynolds-number scaling is also commonly violated in aerodynamic testing, and an extrapolation like that in Fig. 5.8 is required here also.

In fact, as aerodynamic vehicle speeds and sizes increase, the Reynolds-number gap between prototype and model is actually increasing, as shown in Fig. 5.9.

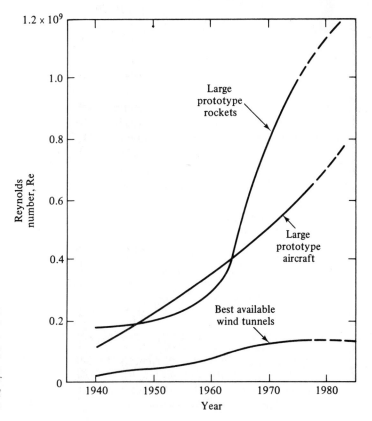

Fig. 5.9 The growing Reynolds-number gap in wind-tunnel testing. (*Adapted from Ref. 30, with additional data, by permission of the American Institute of Aeronautics and Astronautics.*)

Fig. 5.10 Hydraulic model of a barrier-beach inlet at Little River, South Carolina. Such models of necessity violate geometric similarity and do not model the Reynolds number of the prototype inlet. (*Courtesy of U.S. Army Engineer Waterways Experiment Station.*)

Lukasiewicz [30] uses Fig. 5.9 to argue the need for new wind tunnels of higher-Reynolds-number capability.

Finally, a serious discrepancy of another type occurs in hydraulic models of natural flow systems such as rivers, harbors, estuaries, and embayments. Such flows have large horizontal dimensions and small relative vertical dimensions. If we were to scale an estuary model by a uniform linear length ratio of, say, 1:1000, the resulting model would be only a few millimeters deep and dominated by entirely spurious surface-tension or Weber-number effects. Therefore such hydraulic models commonly violate *geometric* similarity by "distorting" the vertical scale by a factor of 10 or more. Figure 5.10 shows a hydraulic model of a barrier-beach inlet in South Carolina. The horizontal scale reduction is 1:300, but the vertical scale is only 1:60. Since a deeper channel flows more efficiently, the model channel bottom is deliberately roughened more than the natural channel to correct for the geometric discrepancy. Thus the friction effect of the discrepancy can be corrected, but its effect on say dispersion of heat and mass is less well known.

5.6   INVENTIVE USE OF THE DATA

The methods of dimensional analysis discussed here allow one to organize both theory and experiment efficiently. The parameters arrived at are customary and traditional: Reynolds number, Froude number, drag coefficient, etc. They are not necessarily the best parameters for a given task, and sometimes they do not give a clear indication of what is happening physically in an experiment. The remedy for this is to regroup the parameters until the particular problem under investigation is most clearly revealed.

As an example of a regrouping procedure, consider Fig. 5.3a for the drag coefficient of a sphere in a uniform stream. This figure is a classic and is reproduced in nearly every textbook on fluid mechanics, but it is a drag-oriented figure. One is supposed to be given the fluid, the diameter, and the velocity, and hence compute the Reynolds number, read the drag coefficient, and compute the sphere drag. Suppose instead that the drag is known but the fluid velocity is not. Then, since V is contained in both C_D and Re, one must iterate back and forth on the chart in Fig. 5.3a until the proper velocity is found. With luck the iteration procedure converges. Consider the following numerical example.

EXAMPLE 5.7   A 0.1-ft-diameter steel sphere ($\rho_s = 15.2$ slugs/ft^3) is dropped in water [$\rho = 1.94$ slugs/ft^3, $\mu = 0.000021$ slug/(ft · s)] until it reaches terminal velocity or zero acceleration. From the sphere data in Fig. 5.3a, compute the terminal velocity of the falling sphere in feet per second.

Solution   At terminal velocity, the net weight of the sphere equals the drag; hence the drag is known in this problem

$$D = W_{\text{net}} = (\rho_s - \rho)g \, \frac{\pi}{6} d^3$$

$$= (15.2 - 1.94 \text{ slugs/ft}^3)(32.2 \text{ ft/s}^2) \frac{\pi}{6} (0.1 \text{ ft})^3 = 0.224 \text{ lbf}$$

We can compute that portion of C_D and Re which excludes the unknown velocity

$$C_D = \frac{D}{\frac{1}{2}\rho(\pi/4)d^2 V^2} = \frac{0.224 \text{ lbf}}{1.94(\pi/8)(0.1 \text{ ft})^2 V^2} = \frac{29.4}{(V \text{ ft/s})^2}$$

$$\text{Re} = \frac{\rho V d}{\mu} = \frac{(1.94 \text{ slugs/ft}^3)(V)(0.1 \text{ ft})}{2.1 \times 10^{-5} \text{ slug/(ft · s)}} = 9240V \text{ ft/s}$$

Now we will just have to guess an initial velocity V to get started on the iteration.

Guess $V = 1.0$ ft/s; then Re $= 9240(1.0) = 9240$. From Fig. 5.3a read $C_D \approx 0.38$; then $V \approx (29.4/C_D)^{1/2} = 8.8$ ft/s. Now try again with this new guess.

Guess $V = 8.8$ ft/s, Re $= 9240(8.8) = 81,000$. From Fig. 5.3a read $C_D \approx 0.52$, $V \approx (29.4/0.52)^{1/2} = 7.5$ ft/s. One more try will give pretty good convergence.

Guess $V = 7.5$ ft/s, Re $= 9240(7.5) = 69,000$. From Fig. 5.3a read $C_D \approx 0.51$, $V \approx (29.4/0.51)^{1/2} = 7.6$ ft/s. To the accuracy of the figure

$$V_{\text{term}} \approx 7.6 \text{ ft/s} \qquad\qquad Ans.$$

The iteration in Example 5.7 converged to a proper terminal velocity. However, near the transition point Re $\approx 3 \times 10^5$, convergence is erratic, and the iteration may oscillate and not settle down. The process of computation is also laborious. The remedy is to regroup the pi products so that only one contains the unknown velocity. It happens that a velocity-free parameter is

$$C_F' = \frac{D\rho}{\mu^2} = \frac{D}{\mu\nu} = \frac{\pi}{8} C_D \text{ Re}^2 \qquad\qquad (5.38)$$

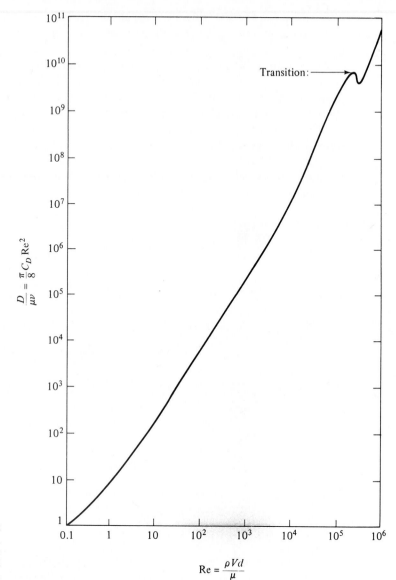

Fig. 5.11 Crossplot of sphere-drag data from Fig. 5.3a to isolate diameter and velocity.

$$\frac{D}{\mu v} = \frac{\pi}{8} C_D \, \mathrm{Re}^2$$

Transition:

$$\mathrm{Re} = \frac{\rho V d}{\mu}$$

This is a perfectly good parameter, if rather uncommon, and a plot of C_F' versus Re is equivalent in every way to a plot of C_D versus Re. Such a regrouped plot is shown in Fig. 5.11. If D, ρ, μ, and d are known, this plot can be read directly for the velocity. It also shows the actual shape of the variation of sphere drag with velocity. Sphere drag increases rapidly with velocity up to transition, where there is a slight drop, after which drag increases faster than ever. This is in contrast to Fig. 5.3a, which might be misread to imply that drag decreases with velocity and drops dramatically at transition.

EXAMPLE 5.8 Repeat Example 5.7, using the regrouped chart, Fig. 5.11.

Solution We must repeat the calculation of the net weight to establish that $D = W_{net} = 0.224$ lbf. But now we can go directly to the new drag coefficient

$$C_F' = \frac{(0.224 \text{ lbf})(1.94 \text{ slugs/ft}^3)}{[2.1 \times 10^{-5} \text{ slug/(ft} \cdot \text{s)}]^2} = 9.8 \times 10^8 = \frac{D\rho}{\mu^2}$$

Now enter Fig. 5.11 and read Re $\approx$ 70,000. Then the desired velocity is

$$V = \frac{\mu \text{ Re}}{\rho d} = \frac{[2.1 \times 10^{-5} \text{ slug/(ft} \cdot \text{s)}](70,000)}{(1.94 \text{ slugs/ft}^3)(0.1 \text{ ft})}$$

or $$V_{term} = 7.6 \text{ ft/s} \qquad\qquad Ans.$$

No iteration is required.

This example should illustrate the power and glory of regrouping dimensionless variables to display certain effects convenient to a particular analysis. For a one-shot calculation, it would take too much time to draw the new regrouped plot (Fig. 5.11), and one might as well hack about with Fig. 5.3a for a single calculation. In general, however, it is up to you, the analyst, to display your dimensionless results in the best manner for the purpose desired. There is no need to constantly mimic the traditional parameters unless they are convenient to the problem at hand.

SUMMARY

Chapters 3 and 4 presented integral and differential methods of mathematical analysis of fluid flow. The present chapter introduces the third and final method: experimentation, as supplemented by the technique of dimensional analysis. Tests and experiments are used both to strengthen existing theories and to provide useful engineering results when theory is inadequate.

The chapter begins with a discussion of some familiar physical relations and how they can be recast in dimensionless form because they satisfy the principle of dimensional homogeneity. A general technique, the pi theorem, is then presented for systematically finding a set of dimensionless parameters by grouping a list of variables which govern any particular physical process. Alternately, direct application of dimensional analysis to the basic equations of fluid mechanics yields the fundamental parameters governing flow patterns: Reynolds number, Froude number, Prandtl number, Mach number, and others.

It is shown that model testing in air and water often leads to scaling difficulties for which compromises must be made. Many model tests do not achieve true dynamic similarity. The chapter ends by pointing out that classic dimensionless charts and data can be manipulated and recast to provide direct solutions to problems that would otherwise be quite cumbersome and laboriously iterative.

RECOMMENDED FILMS

Films numbered 1, 12, 32, and 33 in Appendix C.

REFERENCES

1. P. W. Bridgman, "Dimensional Analysis," Yale University Press, New Haven, Conn., 1922, rev. ed., 1931.
2. A. W. Porter, "The Method of Dimensions," Methuen, London, 1933.
3. F. M. Lanchester, "The Theory of Dimensions and Its Applications for Engineers," Crosby-Lockwood, London, 1940.
4. R. Esnault-Pelterie, "L'Analyse dimensionelle," F. Rouge, Lausanne, 1946.
5. G. W. Stubbings, "Dimensions in Engineering Theory," Crosby-Lockwood, London, 1948.
6. G. Murphy, "Similitude in Engineering," Ronald, New York, 1950.
7. H. E. Huntley, "Dimensional Analysis," Rinehart, New York, 1951.
8. H. L. Langhaar, "Dimensional Analysis and the Theory of Models," Wiley, New York, 1951.
9. W. J. Duncan, "Physical Similarity and Dimensional Analysis," Arnold, London, 1953.
10. C. M. Focken, "Dimensional Methods and Their Applications," Arnold, London, 1953.
11. L. I. Sedov, "Similarity and Dimensional Methods in Mechanics," Academic, New York, 1959.
12. E. C. Ipsen, "Units, Dimensions, and Dimensionless Numbers," McGraw-Hill, New York, 1960.
13. E. E. Jupp, "An Introduction to Dimensional Methods," Cleaver-Hume, London, 1962.
14. R. Pankhurst, "Dimensional Analysis and Scale Factors," Reinhold, New York, 1964.
15. S. J. Kline, "Similitude and Approximation Theory," McGraw-Hill, New York, 1965.
16. B. S. Massey, "Units, Dimensional Analysis, and Physical Similarity," Van Nostrand Reinhold, New York, 1971.
17. J. Zierep, "Similarity Laws and Modeling," Dekker, New York, 1971.
18. W. E. Baker et al., "Similarity Methods in Engineering Dynamics," Spartan, Rochelle Park, N.J., 1973.
19. E. S. Taylor, "Dimensional Analysis for Engineers," Clarendon Press, Oxford, 1974.
20. E. de St. Q. Isaacson and M. de St. Q. Isaacson, "Dimensional Methods in Engineering and Physics," Arnold, London, 1975.
21. R. Esnault-Pelterie, "Dimensional Analysis and Metrology," F. Rouge, Lausanne, 1950.
22. R. Kurth, "Dimensional Analysis and Group Theory in Astrophysics," Pergamon, New York, 1972.
23. F. J. Jong, "Dimensional Analysis for Economists," North Holland, Amsterdam, 1967.
24. E. Buckingham, On Physically Similar Systems: Illustrations of the Use of Dimensional Equations, *Phys. Rev.*, vol. 4, no. 4, pp. 345–376, 1914.
25. Flow of Fluids Through Valves, Fittings, and Pipe, *Crane Co. Tech. Pap.* 410, Chicago, 1957.
26. A. Roshko, On the Development of Turbulent Wakes from Vortex Streets, *NACA Rep.* 1191, 1954.
27. G. W. Jones, Jr., Unsteady Lift Forces Generated by Vortex Shedding about a Large, Stationary, Oscillating Cylinder at High Reynolds Numbers, *ASME Symp. Unsteady Flow, 1968.*
28. O. M. Griffin and S. E. Ramberg, The Vortex Street Wakes of Vibrating Cylinders, *J. Fluid Mech.*, vol. 66, pt. 3, pp. 553–576, 1974.

29. "Encyclopedia of Science and Technology," 3d ed., McGraw-Hill, New York, 1970.
30. J. Lukasiewicz, The Need for Developing a High Reynolds Number Transonic Wind Tunnel in the U.S., *Aeronaut. Astronaut.*, vol. 9, pp. 64–71, April 1971.
31. H. A. Becker, "Dimensionless Parameters," Halstead Press (Wiley), New York, 1976.

Problems

PROBLEM DISTRIBUTION

5.1 Which of the two dimensionless forms of the falling-body relation in Fig. 5.1 is more effective, (*a*) or (*b*)? Explain the reasons for your choice.

5.2 For axial flow inside a circular tube, the Reynolds number which causes transition to turbulence is approximately 2300, based on tube diameter and average flow velocity. If the tube diameter is 4 cm and the fluid is gasoline at 20°C, find the volume flow rate in cubic meters per hour which causes transition.

5.3 In crossflow over a circular tube, the Reynolds number for transition to a turbulent boundary layer is 250,000, based on tube diameter and stream velocity. If the tube diameter is 5 cm and the fluid is argon gas at 20°C and 1 atm, find the stream velocity which causes transition.

5.4 A 6-cm-diameter sphere is tested in water at 20°C and a velocity of 3 m/s and has a measured drag of 6 N. What will be the velocity and drag force of a 2-m-diameter weather balloon moving in air at 20°C and 1 atm under similar conditions?

5.5 An 8-cm-diameter sphere is tested in SAE 30 oil at 20°C at velocities of 1, 2, and 3 m/s and found to have drag forces of 1.92, 6.40, and 13.2 N, respectively. Estimate the drag force if the same sphere is tested in glycerin at 20°C and a velocity of 6 m/s.

5.6 The *Brinkman number*, often used in analysis of organic-liquid flows, is the ratio of viscous dissipation to heat conduction in a fluid. It is a dimensionless combination of viscosity μ, flow velocity V, thermal conductivity k, fluid temperature T. Derive the Brinkman number, using the fact that it is proportional to viscosity.

5.7 For a particle moving in a circle, assume that the centripetal acceleration a is a function of velocity V and radius R. Without using any calculus, i.e., by pure dimensional reasoning, show that the proper form is $a = (\text{const})(V^2)/R$.

5.8 The velocity of sound a of a gas varies with pressure p and density ρ. Show by dimensional reasoning that the proper form must be $a = (\text{const})(p/\rho)^{1/2}$.

5.9 The speed of propagation C of a capillary wave in deep water is known to be a function only of density ρ, wavelength λ, and surface tension Υ. Find the proper functional relationship, completing it with a dimensionless constant. For a given density and wavelength, how does the propagation speed change if surface tension is doubled?

5.10 The excess pressure Δp inside a bubble is known to be a function of the surface tension and the radius. By dimensional reasoning determine how the excess pressure will vary if we double (*a*) the radius and (*b*) the surface tension.

5.11 An airplane has a characteristic length of 125 ft and is designed to fly at 200 mi/h at 10,000 ft standard altitude. The drag coefficient as defined by Eq. (5.2) is measured with a small model and found to be 0.01. What is the horsepower required to drive the prototype airplane?

5.12 It is desired to measure the drag on an airplane whose velocity is 300 mi/h. Is it feasible to test a one-twentieth-scale model of the plane in a wind tunnel at the same pressure and temperature to determine the prototype drag coefficient?

5.13 Convection heat-transfer data are often reported as a *heat-transfer coefficient h*, defined by

$$\dot{Q} = hA\,\Delta T$$

where $\dot{Q}$ = heat flow, J/s
A = surface area, m²
ΔT = temperature difference, K.

The dimensionless form of h, called the *Stanton number*, is a combination of h, fluid density ρ, specific heat c_p, and flow velocity V. Derive the Stanton number if it is proportional to h.

5.14 In some heat-transfer textbooks, e.g., J. P. Holman, "Heat Transfer," 5th ed., McGraw-Hill, 1981, p. 285, simplified formulas are given for the heat-transfer coefficient from Prob. 5.13 for buoyant or *natural* convection over hot surfaces. An example formula is

$$h = 1.42\left(\frac{\Delta T}{L}\right)^{1/4}$$

where L is the length of the hot surface. Comment on the dimensional homogeneity of this formula. What might be the SI units of the constants 1.42 and 1/4? What parameters might be missing or hidden?

5.15 A one-twentieth-scale model of a submarine is tested at 200 ft/s in a wind tunnel using sea-level standard air. What is the prototype speed in 20°C seawater for dynamic similarity? If the model drag is 1 lbf, what is the prototype drag?

5.16 Under laminar conditions, the volume flow Q through a small triangular-section pore of side length b and length L is a function of viscosity μ, pressure drop per unit length $\Delta p/L$, and b. Using the pi theorem, rewrite this relation in dimensionless form. How does the volume flow change if the pore size b is doubled?

5.17 The period of swing T of a simple pendulum is assumed to be a function of its length L, bob mass m, the acceleration of gravity, and the swing angle θ. Use the pi theorem to rewrite this relationship as a dimensionless function. Did anything interesting happen? What happens to the period if the length is doubled and all other parameters remain the same?

5.18 The period of oscillation T of a water surface wave is assumed to be a function of density ρ, wavelength λ, depth h, gravity g, and surface tension Υ. Rewrite this relationship in dimensionless form. What results if Υ is negligible?

5.19 The power input P to a centrifugal pump is assumed to be a function of volume flow Q, impeller diameter D, rotational rate Ω, and the density ρ and viscosity μ of the fluid. Rewrite this as a dimensionless relationship.

5.20 Modify Prob. 5.19 by replacing power P by the pressure rise $\Delta p/\rho$ across the pump as a function of the same five variables: Q, D, Ω, ρ, and μ. Rewrite in dimensionless form. How does the relation simplify if viscosity is negligible?

5.21 In Example 5.1 we used the pi theorem to develop Eq. (5.2) from Eq. (5.1). Instead of merely listing the primary dimensions of each variable, some workers list the *powers* of each primary dimension for each variable in an array:

$$
\begin{array}{c}
\ \\ M \\ L \\ T
\end{array}
\begin{array}{c}
\begin{array}{ccccc} F & L & U & \rho & \mu \end{array} \\
\left[\begin{array}{ccccc}
1 & 0 & 0 & 1 & 1 \\
1 & 1 & 1 & -3 & -1 \\
-2 & 0 & -1 & 0 & -1
\end{array}\right]
\end{array}
$$

This array of exponents is called the *dimensional matrix* for the given function. Show that the *rank* of this matrix (the size of the largest nonzero determinant) is equal to $j = n - k$, the desired reduction between original variables and the pi groups. This is a general property of dimensional matrices, as noted by Buckingham [24].

5.22 The resistance force F of a surface ship is a function of its length L, velocity V, gravity g, and the density ρ and viscosity μ of the water. Rewrite in dimensionless form.

5.23 The period T of vibration of a beam is a function of its length L, area moment of inertia I, modulus of elasticity E, density ρ, and Poisson's ratio σ. Rewrite this relation in dimensionless form. What further reduction can we make if E and I can only occur in the product form EI?

5.24 The lift force F on a missile is a function of its length L, velocity V, diameter D, angle of attack α, density ρ, viscosity μ, and speed of sound a of the air. Write out the dimensional matrix of this function and determine its rank. (See Prob. 5.21 for an explanation of this concept.) Rewrite the function in terms of pi groups.

5.25 When a viscous fluid is confined between two long concentric cylinders as in Fig. P4.34, the torque per unit length T' required to turn the inner cylinder at angular velocity Ω is a function of Ω, cylinder radii a and b, and viscosity μ. Find the equivalent dimensionless function. What happens to the torque if both a and b are doubled?

5.26 The torque M on an axial-flow turbine is a function of fluid density ρ, rotor diameter D, angular rotation rate Ω, and volume flow Q. Rewrite in dimensionless form. If it is known that M is proportional to Q for a particular turbine, how would M vary with Ω and D for that turbine?

5.27 The period of heave oscillation T of a simple spar buoy (see Prob. 2.113) varies with its cross-sectional area A, its mass m, gravity g, and the water density ρ. Rewrite in dimensionless form. What happens to T if the area is doubled? Instrument buoys should have a very long period to avoid wave resonance. Sketch a design which would have long period.

5.28 According to elementary kinetic theory (Ref. 7 of Chap. 1), the thermal conductivity k of a gas is a function of its density ρ, gas constant R, mean free path λ, and absolute temperature T. Rewrite in dimensionless form. If R is doubled with other parameters constant, how will k change?

5.29 In forced convection, the heat transfer coefficient h, as defined in Prob. 5.13, is known to be a function of stream velocity U, body size L, and the fluid properties ρ, μ, c_p, and k. Rewrite this function in dimensionless form and note by name any parameters you recognize.

5.30 The heat-transfer rate per unit area q to a body from a fluid in natural or gravitational convection is a function of temperature difference ΔT, gravity g, body length L, and three fluid properties: kinematic viscosity v, conductivity k, and thermal expansion coefficient β. Rewrite in dimensionless form if it is known that g and β appear only as the product $g\beta$.

5.31 A *weir* is an obstruction in a channel flow which can be calibrated to measure flow rate, as in Fig. P5.31. The volume flow Q varies with gravity g, weir width b into the paper, and upstream water height H above the weir crest. If it is known that Q is proportional to b, use the pi theorem to find a unique functional relationship $Q(g, b, H)$.

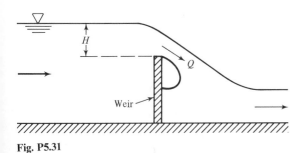

Fig. P5.31

5.32 Experiments show that the fluid velocity u very near a wall in turbulent flow varies only with distance y from the wall, wall shear stress τ_w, and the fluid properties ρ and μ. Rewrite this relation in dimensionless form.

5.33 The flow velocity u very near a rotating disk varies only with disk angular velocity ω, local radius R, distance z from the disk, and kinematic viscosity v. Rewrite this relation in dimensionless form.

5.34 The pressure difference Δp across an explosion or blast wave is a function of distance r from the blast center, time t, speed of sound a of the medium and total energy E in the blast. Rewrite this relation in dimensionless form (see Ref. 18, chap. 4, for further details of blast-wave scaling). How does Δp change if E is doubled?

5.35 The size d of droplets produced by a liquid spray nozzle is thought to depend upon the nozzle diameter D, jet velocity U, and the properties of the liquid ρ, μ, and Υ. Rewrite this relation in dimensionless form.

5.36 Nondimensionalize the energy equation (4.75) and its boundary conditions (4.62), (4.63), and (4.70) by defining $T^* = T/T_0$, where T_0 is the inlet temperature, assumed constant. Use other dimensionless variables as needed from Eqs. (5.23). Isolate all dimensionless parameters you find and relate them to the list given in Table 5.2.

5.37 The differential equation of salt conservation for flowing seawater is

$$\frac{\partial S}{\partial t} + u\frac{\partial S}{\partial x} + v\frac{\partial S}{\partial y} + w\frac{\partial S}{\partial z}$$

$$= \kappa\left(\frac{\partial^2 S}{\partial x^2} + \frac{\partial^2 S}{\partial y^2} + \frac{\partial^2 S}{\partial z^2}\right)$$

where κ is a (constant) coefficient of diffusion, with typical units of square meters per second, and S is the salinity in parts per thousand. Nondimensionalize this equation and discuss any parameters which appear.

5.38 In natural-convection problems the variation of density due to temperature difference ΔT creates an important buoyancy term in the momentum equation (5.30). To first-order accuracy the density variation would be $\rho \approx \rho_0(1 - \beta\,\Delta T)$, where β is the thermal-expansion coefficient. The momentum equation thus becomes

$$\rho_0 \frac{d\mathbf{V}}{dt} = -\nabla(p + \rho_0 gz) + \rho_0\beta\,\Delta T\,g\mathbf{k} + \mu\,\nabla^2\mathbf{V}$$

where we have assumed that z is "up." Nondimensionalize this equation using Eqs. (5.23) and relate the parameters you find to the list in Table 5.2.

5.39 The differential equation for compressible inviscid flow of a gas in the xy plane is

$$\frac{\partial^2 \phi}{\partial t^2} + \frac{\partial}{\partial t}(u^2 + v^2) + (u^2 - a^2)\frac{\partial^2 \phi}{\partial x^2}$$
$$+ (v^2 - a^2)\frac{\partial^2 \phi}{\partial y^2} + 2uv\frac{\partial^2 \phi}{\partial x\, \partial y} = 0$$

where ϕ is the velocity potential and a is the (variable) speed of sound of the gas. Nondimensionalize this relation using a reference length L and the inlet speed of sound a_0 as parameters for defining dimensionless variables.

5.40 The differential equation for small-amplitude vibrations $y(x, t)$ of a simple beam is given by

$$\rho A \frac{\partial^2 y}{\partial t^2} + EI \frac{\partial^4 y}{\partial x^4} = 0$$

where ρ = beam material density
A = cross-sectional area
I = area moment of inertia
E = Young's modulus

Use only the quantities ρ, E, and A to nondimensionalize y, x, and t and rewrite the differential equation in dimensionless form. Do any parameters remain? Could they be removed by further manipulation of the variables?

5.41 A smooth steel (SG = 7.86) sphere is immersed in a stream of ethanol at 20°C moving at 1.5 m/s. Estimate its drag in newtons from Fig. 5.3a. What stream velocity would quadruple its drag? Take $D = 2.5$ cm.

5.42 The sphere in Prob. 5.41 is dropped in water at 20°C. Ignoring its acceleration phase, what will its terminal (constant) fall velocity be, from Fig. 5.3a?

5.43 Repeat Prob. 5.42 if the sphere is dropping in glycerin at 20°C.

5.44 A ship is towing a sonar array which approximates a submerged cylinder 1 ft in diameter and 30 ft long with its axis normal to the direction of tow. If the tow speed is 12 knots (1 knot = 1.69 ft/s), estimate the horse-power required to tow this cylinder. What will be the frequency of vortices shed from the cylinder? Use Figs. 5.2 and 5.3.

5.45 A 1-in-diameter telephone wire is mounted in air at 20°C and has a natural vibration frequency of 12 Hz.

What wind velocity in feet per second will cause the wire to sing? At this condition what will the average drag force per unit wire length be?

5.46 Vortex shedding can be used to design a *vortex flowmeter* (Fig. 6.31). A blunt rod stretched across the pipe sheds vortices whose frequency is read by the sensor downstream. Suppose the pipe diameter is 5 cm and the rod is a cylinder of diameter 8 mm. If the sensor reads 5400 counts per minute, estimate the volume flow rate of water in cubic meters per hour. How might the meter react to other liquids?

5.47 A vertical 18-cm-diameter piling in seawater is 4.5 m deep. If the water-current velocity is 2.2 m/s, estimate the drag in newtons on the piling.

5.48 A flagpole is 12 cm in diameter and 15 m high and stands in sea-level standard air. When the wind velocity is 20 knots (1 knot = 0.5144 m/s), estimate the wind-induced bending moment about the base of the pole.

5.49 We wish to know the drag of a blimp which will move in 20°C air at 6 m/s. If a one-thirtieth scale model is tested in water at 20°C what should the water velocity be? At this velocity, if the measured water drag on the model is 2700 N, what is the drag on the prototype blimp and the power required to propel it?

5.50 A prototype water pump has an impeller diameter of 2 ft and is designed to pump 12 ft³/s at 750 r/min. A 1-ft-diameter model pump is tested in 20°C air at 1800 r/min, and Reynolds-number effects are found to be negligible. For similar conditions, what will the volume flow of the model be in cubic feet per second? If the model pump requires 0.082 hp to drive it, what horsepower is required for the prototype?

5.51 If viscosity is neglected, typical pump-flow results from Probs. 5.19 and 5.20 are shown in Fig. P5.51 for a model pump tested in water. The pressure

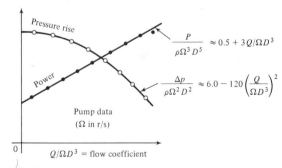

Fig. P5.51

rise decreases and the power required increases with the dimensionless flow coefficient. Curve-fit expressions are given for the data. Suppose a similar pump of 12 cm diameter is built to move gasoline at 20°C and a flow rate of 25 m^3/h. If the pump rotation speed is 30 r/s, find (*a*) the pressure rise and (*b*) the power required.

5.52 Modify Prob. 5.51 so that the rotation speed is unknown but $D = 12$ cm and $Q = 25$ m^3/h. What is the maximum rotation speed for which the power will not exceed 300 W? What will the pressure rise be for this condition?

5.53 The pressure drop per unit length $\Delta p/L$ in smooth pipe flow is known to be a function only of average velocity V, diameter D, and fluid properties ρ and μ. The following data were obtained for flow of water at 20°C in an 8-cm-diameter pipe 50 m long:

Q, m^3/s	0.005	0.01	0.015	0.020
Δp, Pa	5800	20,300	42,100	70,800

Use these data to estimate the pressure drop for flow of kerosine at 20°C in a smooth pipe of diameter 5 cm and length 200 m if the flow rate is 60 m^3/h. *Hint*: A dimensional analysis is needed to begin working with the given data.

5.54 A certain fluid of specific gravity 0.92 in a tube of 3 cm diameter is found to have a capillary rise of 2 mm. What will its capillary rise be in a 5-cm-diameter tube? For what diameter will the capillary rise be 1 cm?

5.55 A torpedo 8 m below the surface in 20°C seawater cavitates at a speed of 21 m/s when atmospheric pressure is 101 kPa. If Reynolds-number and Froude-number effects are negligible, at what speed will it cavitate when running at a depth of 20 m? At what depth should it be to avoid cavitation at 30 m/s?

5.56 A one-fifth-scale model automobile is tested in a wind tunnel in the same air properties as the prototype. The prototype velocity is 50 km/h. For dynamically similar conditions the model drag is 350 N. What are the drag of the prototype automobile and the power in kilowatts required to overcome this drag?

5.57 A rotary mixer is to be designed for stirring ethyl alcohol. Tests with a one-fourth-scale model in SAE 30 oil indicate most efficient mixing at 1770 r/min. What should the speed of the prototype mixer be in revolutions per minute?

5.58 Modify Prob. 5.53 so that the diameter is unknown but the pipe is 200 m long, the flow rate is 60 m^3/h, and the fluid is kerosine at 20°C. What is the minimum pipe diameter for which the total pressure drop will be no more than 200 kPa?

5.59 A one-twelfth-scale model of a weir (see Fig. P5.31) has a measured flow rate of 2.2 ft^3/s when the upstream water height is $H = 6.5$ in. Use the results of Prob. 5.31 to predict the prototype flow rate when $H = 3.6$ ft.

5.60 The equation defining an ellipsoid is

$$\left(\frac{x}{a}\right)^2 + \left(\frac{y}{b}\right)^2 + \left(\frac{z}{c}\right)^2 = 1$$

where *a*, *b*, and *c* are reference lengths. If the prototype ellipsoid has $a = 5$ m, $b = 3$ m, and $c = 1$ m, what should *b* and *c* be for the model ellipsoid if *a* is 1 m?

5.61 An airplane is designed to fly at 260 m/s at 10,000 m U.S. standard altitude. If a one-tenth-scale model is tested in a pressurized wind tunnel at 20°C, what should the tunnel pressure be in pascals to scale both the Reynolds and Mach numbers correctly?

5.62 The pressure drop in a venturi meter (Fig. P3.145) varies only with fluid density, pipe approach velocity, and the diameter ratio of the meter. A model venturi meter tested in water at 20°C shows a 5-kPa drop when the approach velocity is 4 m/s. A geometrically similar prototype meter is used to measure gasoline at 20°C and a flow rate of 9 m^3/min. If the prototype pressure gage is most accurate at 15 kPa, what should the upstream pipe diameter be?

5.63 A one-fifteenth-scale model of a parachute has a drag of 450 lbf when tested at 20 ft/s in a water tunnel. If Reynolds-number effects are negligible, estimate the terminal fall velocity at 5000 ft standard altitude of a parachutist using the prototype if chute and chutist together weigh 160 lbf. Neglect the drag coefficient of the woman.

5.64 The yawing moment on a torpedo control surface is tested on a one-eighth-scale model in a water tunnel at 20 m/s using Reynolds scaling. If the model measured moment is 14 N · m, what will the prototype moment be under similar conditions?

5.65 A one-tenth-scale model of a supersonic wing tested at 700 m/s in air at 20°C and 1 atm shows a pitching moment of 0.25 kN · m. If Reynolds-number effects are negligible, what will the pitching moment of the prototype wing be flying at the same Mach number at 8 km standard altitude?

5.66 An axial compressor is intended to pump helium at 1200 r/min. A one-third-scale model is tested in air at 600 r/min and exhibits a flow rate of 6 ft^3/s, a pressure rise of 145 Pa, and a power input of 1.0 kW. For dynamically similar conditions compute Q, Δp, and the power input for the prototype. Neglect Mach- and Reynolds-number effects and assume sea-level conditions.

5.67 A dam spillway is to be tested using Froude scaling with a one-thirtieth-scale model. The model flow has an average velocity of 0.6 m/s and a volume flow of 0.05 m^3/s. What will the velocity and flow of the prototype be? If the measured force on a certain part of the model is 1.5 N, what will the corresponding force on the prototype be?

5.68 A prototype spillway has a characteristic velocity of 3 m/s and a characteristic length of 10 m. A small model is constructed using Froude scaling. What is the minimum scale ratio of the model which will ensure that its minimum Weber number is 100? Both flows use water at 20°C.

5.69 An East coast estuary has a tidal period of 12.42 h (the semidiurnal lunar tide) and tidal currents of approximately 80 cm/s. If a one-five-hundredth-scale model is constructed with tides driven by a pump and storage apparatus, what should the period of the model tides be and what model current speeds are expected?

5.70 A prototype ship is 35 m long and designed to cruise at 11 m/s (about 21 knots). Its drag is to be simulated by a 1-m-long model pulled in a tow tank. For Froude scaling find (*a*) the tow speed, (*b*) the ratio of prototype to model drag, and (*c*) the ratio of prototype to model power.

5.71 A prototype ship is 400 ft long and has a wetted area of 30,000 ft^2. A one-eightieth-scale model is tested in a tow tank according to Froude scaling at speeds of 1.3, 2.0, and 2.7 knots (1 knot = 1.689 ft/s). The measured friction drag of the model at these speeds is 0.11, 0.24, and 0.41 lbf, respectively. What are the three prototype speeds? What is the estimated prototype friction drag at these speeds if we correct for the Reynolds-number discrepancy by extrapolation?

5.72 A one-fortieth-scale model of a ship's propeller is tested in a tow tank at 1200 r/min and exhibits a power output of 1.4 (ft · lbf)/s. According to Froude scaling laws, what should the revolutions per minute and horsepower output of the prototype propeller be under dynamically similar conditions?

5.73 A prototype ocean-platform piling is expected to encounter currents of 150 cm/s and waves of 12 s period and 3 m height. If a one-fifteenth-scale model is tested in a wave channel, what current speed, wave period, and wave height should be encountered by the model?

5.74 Solve Probs. 5.42 and 5.43 using the modified drag chart of Fig. 5.11. You should be able to obtain both answers in less time than it took to iterate either original problem from Fig. 5.3*a*.

5.75 Let us look again inventively at Prob. 5.52. It was difficult to solve for Ω there because both the power and flow coefficients in Fig. P5.51 contained Ω. Resolve this by defining a new power coefficient nondimensionalized with ρ, D, and Q, not with Ω. Plot the new power grouping versus $Q/\Omega D^3$ using the data of Fig. P5.51. You can enter this chart directly and solve for Ω.

5.76 Modify Prob. 5.51 so that the diameter is unknown but $\Omega = 30$ r/s and $Q = 25$ m^3/h. What is the maximum diameter if the power is not to exceed 300 W? Solve this inventively by defining a new power grouping nondimensionalized by ρ, Ω, and Q but not D. Plot the new pi group versus $Q/\Omega D^3$; this chart can be read directly for D.

5.77 Problem 5.58 was difficult because both of the original pi groups, $\Delta p\, D/L\rho V^2$ and $\rho V D/\mu$, contained the diameter D. Modify the first group by nondimensionalizing $\Delta p/L$ with ρ, Q, and μ, not with D. Then note that $\rho V D/\mu$ is identical to $4\rho Q/\pi D\mu$ and is the Reynolds number. Plot the new pi group versus Reynolds number from the given data. This new chart can be read directly for D.

Viscous Flow in Ducts

6.1 REYNOLDS-NUMBER REGIMES

The remainder of this book is concerned with specific applications of fluid-flow analysis. For example, the present chapter studies viscous flow within confining walls such as pipes or diffusers.

Now that we have derived and studied the basic flow equations in Chap. 4, you would think that we could just whip off a myriad of beautiful solutions illustrating the full range of fluid behavior, of course expressing all these educational results in dimensionless form, using our new tool from Chap. 5, dimensional analysis.

The fact of the matter is that no general analysis of fluid motion yet exists. There are several dozen known particular solutions, there are some rather specific digital-computer solutions, and there are a great many experimental data. There is a lot of theory available if we neglect such important effects as viscosity and compressibility (Chap. 8), but there is no general theory and there may never be. The reason is that a profound and vexing change in fluid behavior occurs at moderate Reynolds numbers. The flow ceases being smooth and steady (*laminar*) and becomes fluctuating and agitated (*turbulent*). The changeover is called *transition* to turbulence. In Fig. 5.3*a* we saw that transition on the cylinder and sphere occurred at about Re $= 3 \times 10^5$, where the sharp drop in drag coefficient appeared. Transition depends upon many effects, e.g., wall roughness (Fig. 5.3*b*) or fluctuations in the inlet stream, but the primary parameter is the Reynolds number. There are a great many data on transition but only a small amount of theory [1–3].

Turbulence can be detected from a measurement by a small, sensitive instrument such as a hot-wire anemometer (Fig. 6.28*e*) or a piezoelectric pressure transducer (Fig. 2.26*d*). The flow will appear steady on the average but will reveal rapid, random fluctuations if turbulence is present, as sketched in Fig. 6.1. If the flow is laminar, there may be occasional natural disturbances which damp out quickly

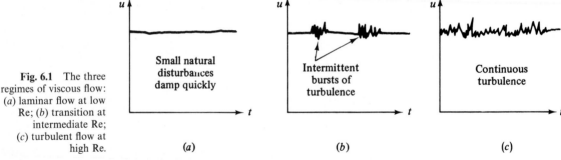

Fig. 6.1 The three regimes of viscous flow: (*a*) laminar flow at low Re; (*b*) transition at intermediate Re; (*c*) turbulent flow at high Re.

(Fig. 6.1*a*). If transition is occurring, there will be sharp bursts of turbulent fluctuation (Fig. 6.1*b*) as the increasing Reynolds number causes a breakdown or instability of laminar motion. At sufficiently large Re, the flow will fluctuate continually (Fig. 6.1*c*) and is termed *fully turbulent*. The fluctuations, typically ranging from 1 to 20 percent of the average velocity, are not strictly periodic but are random and encompass a continuous range, or spectrum, of frequencies. In a typical wind-tunnel flow at high Re the turbulent frequency ranges from 1 to 10,000 Hz, and the wavelength ranges from about 0.01 to 400 cm.

EXAMPLE 6.1   The accepted transition Reynolds number for flow past a smooth sphere is $Re_{crit} = 250,000$ (Fig. 5.3*a*). At what velocity will this occur for airflow at 20°C past a 12-cm-diameter sphere?

Solution   From Table 1.3 read $v = 1.51 \times 10^{-5}$ m²/s for air. The critical Reynolds number is

$$Re_{crit} = 250,000 = \frac{VD}{v} = \frac{V(0.12 \text{ m})}{1.51 \times 10^{-5} \text{ m}^2/\text{s}}$$

Solve for $V = 31.5$ m/s   *Ans.*

This speed, about 70 mi/h, is in the range of many important engineering problems, so that transition and turbulent flow often occur in practical flow studies.

In free-surface flows turbulence can be observed directly. Figure 6.2 shows the water jet issuing from an ordinary faucet. The low-Reynolds-number jet (Fig. 6.2*a*) is smooth and laminar. The higher-Reynolds-number turbulent flow (Fig. 6.2*b*) is unsteady and irregular but steady and predictable in the mean.

Similar fluctuations are visible on the surface of a shallow-water-channel flow, Fig. 6.3. In the transition range (Fig. 6.3*a*) the turbulence is confined to patches or "spots," while in fully turbulent flow (Fig. 6.3*b*) the fluctuations are more or less uniformly distributed.

A complete description of the statistical aspects of turbulence is given in Ref. 1, while theory and data on transition effects are given in Refs. 2 and 3. At this introductory level we merely point out that the primary parameter affecting transition is the Reynolds number. If $Re = UL/v$, where U is the average stream

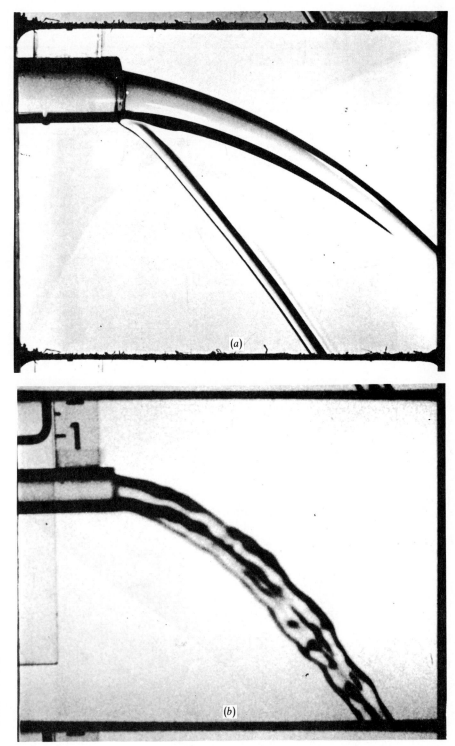

Fig. 6.2 Flow issuing at constant speed from a pipe: (*a*) high viscosity, low Reynolds number, laminar flow; (*b*) low viscosity, high Reynolds number, turbulent flow. [*From Illustrated Experiments in Fluid Mechanics* (*The NCFMF Book of Film Notes*), *National Committee for Fluid Mechanics Films, Education Development Center, Inc., copyright 1972.*]

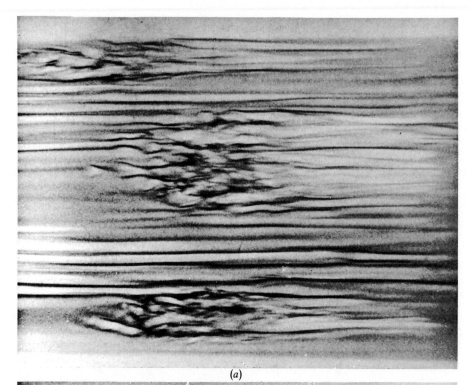

(a)

Fig. 6.3 Visualization of transition in a boundary layer: (a) turbulent bursts appear at transition Re; (b) fully turbulent flow condition at high Re. [*From* Illustrated Experiments in Fluid Mechanics (*The NCFMF Book of Film Notes*), *National Committee for Fluid Mechanics Films, Education Development Center, Inc., copyright 1972.*]

(b)

velocity and L is the "width," or transverse thickness, of the shear layer, the following approximate ranges occur:

$$0 < Re < 1: \text{highly viscous laminar "creeping" motion}$$
$$1 < Re < 100: \text{laminar, strong Reynolds-number dependence}$$
$$100 < Re < 10^3: \text{laminar, boundary-layer theory useful}$$
$$10^3 < Re < 10^4: \text{transition to turbulence}$$
$$10^4 < Re < 10^6: \text{turbulent, moderate Reynolds-number dependence}$$
$$10^6 < Re < \infty : \text{turbulent, slight Reynolds-number dependence}$$

These are representative ranges which vary somewhat with flow geometry, surface roughness, and the level of fluctuations in the inlet stream. The great majority of our analyses are concerned with laminar flow or with turbulent flow, and one should not normally design a flow operation in the transition region.

Historical Outline

Since turbulent flow is more prevalent than laminar flow, experimenters have observed turbulence for centuries without being aware of the details. Before 1930 flow instruments were too insensitive to record rapid fluctuations, and workers simply reported mean values of velocity, pressure, force, etc. But turbulence can change the mean values dramatically, e.g., the sharp drop in drag coefficient in Fig. 5.3. A German engineer, G. H. L. Hagen, first reported in 1839 that there might be *two* regimes of viscous flow. He measured water flow in long brass pipes and deduced a pressure-drop law

$$\Delta p = (\text{const}) \frac{LQ}{R^4} + \text{entrance effect} \tag{6.1}$$

This is exactly our laminar-flow scaling law from Example 5.4, but Hagen did not realize that the constant was proportional to the fluid viscosity.

The formula broke down as Hagen increased Q beyond a certain limit, i.e., past the critical Reynolds number, and he stated in his paper that there must be a second mode of flow characterized by "strong movements of water for which Δp varies as the second power of the discharge ...." He admitted that he could not clarify the reasons for the change.

A typical example of Hagen's data is shown in Fig. 6.4. The pressure drop varies linearly with $V = Q/A$ up to about 1.1 ft/s, where there is a sharp change. Above about $V = 2.2$ ft/s pressure drop is nearly quadratic with V. The actual power $\Delta p \propto V^{1.75}$ seems impossible on dimensional grounds but is easily explained when the dimensionless pipe-flow data or Moody chart (Fig. 6.13) is displayed.

In 1883 Osborne Reynolds, a British engineering professor, showed that the change depended upon the parameter $\rho V d / \mu$, now named in his honor. By introducing a dye streak into a pipe flow, Reynolds could observe transition and turbulence. His sketches of the flow behavior are shown in Fig. 6.5.

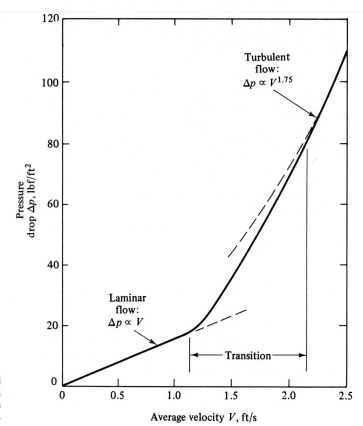

Fig. 6.4 Experimental evidence of transition for water flow in a $\frac{1}{4}$-in smooth pipe 10 ft long.

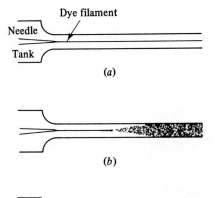

Fig. 6.5 Reynolds' sketches of pipe-flow transition: (a) low-speed, laminar flow; (b) high-speed, turbulent flow; (c) spark photograph of condition (b). (*From Ref. 4.*)

If we examine Hagen's data and compute the Reynolds number at $V = 1.1$ ft/s, we obtain $Re_d = 2100$. The flow became fully turbulent, $V = 2.2$ ft/s, at $Re_d = 4200$. The accepted design value for pipe-flow transition is now taken to be

$$Re_{d, \text{ crit}} \approx 2300 \tag{6.2}$$

This is accurate for commercial pipes (Fig. 6.13), although with special care in providing a rounded entrance, smooth walls, and a steady inlet stream, $Re_{d, \text{ crit}}$ can be delayed until much higher values.

Transition also occurs in external flows around bodies such as the sphere and cylinder in Fig. 5.3. Ludwig Prandtl, a German engineering professor, showed in 1914 that the thin boundary layer surrounding the body was undergoing transition from laminar to turbulent flow. Thereafter the force coefficient of a body was acknowledged to be a function of Reynolds number [Eq. (5.2)].

There are now extensive theories and experiments of laminar-flow instability which explain why a flow changes to turbulence. Reference 5 is an advanced textbook on this subject.

Laminar-flow theory is now well developed, and many solutions are known [2, 3], but there are no analyses, even digital-computer solutions, which can simulate the fine-scale random fluctuations of turbulent flow.[1] Therefore existing turbulent-flow theory is semiempirical, based upon dimensional analysis and physical reasoning; it is concerned with the mean flow properties only and the mean of the fluctuations, not their rapid variations. The turbulent-flow "theory" presented here in Chaps. 6 and 7 is unbelievably crude yet surprisingly effective. We shall attempt a rational approach which places turbulent-flow analysis on a firm physical basis.

6.2 INTERNAL VERSUS EXTERNAL VISCOUS FLOWS

Both laminar and turbulent flow may be either internal, i.e., "bounded" by walls, or external and unbounded. This chapter treats internal flows, and Chap. 7 studies external flows.

An internal flow is constrained by the bounding walls, and the viscous effects will grow and meet and permeate the entire flow. Figure 6.6 shows an internal flow in a long duct. There is an *entrance region* where a nearly inviscid upstream flow converges and enters the tube. Viscous boundary layers grow downstream, retarding the axial flow $u(r, x)$ at the wall and thereby accelerating the center-core flow to maintain the incompressible continuity requirement

$$Q = \int u \, dA = \text{const} \tag{6.3}$$

At a finite distance from the entrance the boundary layers merge and the inviscid core disappears. The tube flow is then entirely viscous, and the axial velocity adjusts slightly further until at $x = L_e$ it no longer changes with x and is said to be *fully developed*, $u \approx u(r)$ only. Downstream of $x = L_e$ the velocity profile is

---

[1] Reference 32 is a computer model of large-scale turbulent fluctuations.

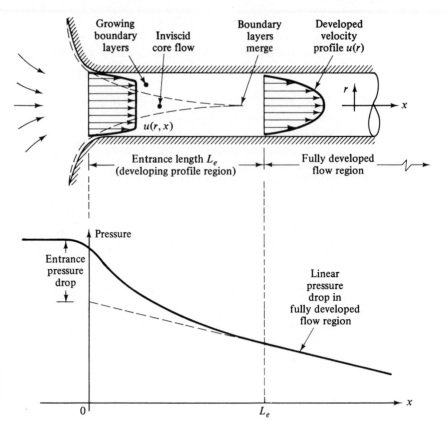

Fig. 6.6 Developing velocity profiles and pressure changes in the entrance of a duct flow.

constant, the wall shear is constant, and the pressure drops linearly with x, for either laminar or turbulent flow. All these details are shown in Fig. 6.6.

Dimensional analysis shows that the Reynolds number is the only parameter affecting entrance length. If

$$L_e = f(d, V, \rho, \mu) \qquad V = \frac{Q}{A}$$

then

$$\frac{L_e}{d} = g\left(\frac{\rho V d}{\mu}\right) = g(\text{Re}) \qquad (6.4)$$

For laminar flow [2, 3], the accepted correlation is

$$\frac{L_e}{d} \approx 0.06\,\text{Re} \qquad \text{laminar} \qquad (6.5)$$

The maximum laminar entrance length, at $\text{Re}_{d,\,\text{crit}} = 2300$, is $L_e = 138d$, which is the longest development length possible.

In turbulent flow the boundary layers grow faster, and L_e is relatively shorter, according to the approximation

$$\frac{L_e}{d} \approx 4.4\,\text{Re}_d^{1/6} \qquad \text{turbulent} \qquad (6.6)$$

Some computed turbulent entrance lengths are thus

R_e	4000	10^4	10^5	10^6	10^7	10^8
L_e/d	18	20	30	44	65	95

Now 44 diameters may seem "long," but typical pipe-flow applications involve L/d of 1000 or more, in which case the entrance effect may be neglected and a simple analysis made for fully developed flow (Sec. 6.4). This is possible for both laminar and turbulent flows, including rough walls and noncircular cross sections.

EXAMPLE 6.2   A $\frac{1}{2}$-in-diameter water pipe is 60 ft long and delivers water at 5 gal/min at 20°C. What fraction of this pipe is taken up by the entrance region?

Solution   Convert

$$Q = (5 \text{ gal/min}) \frac{0.00223 \text{ ft}^3/\text{s}}{1 \text{ gal/min}} = 0.0111 \text{ ft}^3/\text{s}$$

The average velocity is

$$V = \frac{Q}{A} = \frac{0.0111 \text{ ft}^3/\text{s}}{(\pi/4)(\frac{1}{2}/12 \text{ ft})^2} = 8.17 \text{ ft/s}$$

From Table 1.3 read for water $v = 1.01 \times 10^{-6} \text{ m}^2/\text{s} = 1.09 \times 10^{-5} \text{ ft}^2/\text{s}$. Then the pipe Reynolds number is

$$\text{Re}_d = \frac{Vd}{v} = \frac{(8.17 \text{ ft/s})(\frac{1}{2}/12 \text{ ft})}{1.09 \times 10^{-5} \text{ ft}^2/\text{s}} = 31,300$$

This is greater than 4000; hence the flow is fully turbulent and Eq. (6.6) applies for entrance length

$$\frac{L_e}{d} \approx 4.4 \, \text{Re}_d^{1/6} = (4.4)(31,300)^{1/6} = 25$$

The actual pipe has $L/d = (60 \text{ ft})/[(\frac{1}{2}/12) \text{ ft}] = 1440$. Hence the entrance region takes up the fraction

$$\frac{L_e}{L} = \frac{25}{1440} = 0.017 = 1.7\% \qquad\qquad Ans.$$

This is a very small percentage, so that we can reasonably treat this pipe flow as essentially fully developed.

Shortness can be a virtue in duct flow if one wishes to maintain the inviscid core. For example, a "long" wind tunnel would be ridiculous, since the viscous core would invalidate the purpose of simulating free-flight conditions. A typical laboratory low-speed wind-tunnel test section is 1 m in diameter and 5 m long, with $V = 30$ m/s. If we take $v_{air} = 1.51 \times 10^{-5} \text{ m}^2/\text{s}$ from Table 1.3, then $\text{Re}_d = 1.99 \times 10^6$ and, from Eq. (6.6), $L_e/d \approx 49$. The test section has $L/d = 5$, which is much shorter than the development length. At the end of the section the wall boundary layers are only 10 cm thick, leaving 80 cm of inviscid core suitable for model testing.

An external flow has no restraining walls and is free to expand no matter how thick the viscous layers on the immersed body may become. Thus, far from the body the flow is nearly inviscid, and our analytical technique, treated in Chap. 7, is to patch an inviscid flow solution onto a viscous boundary-layer solution computed for the wall region. There is no external equivalent of fully developed internal flow.

6.3 SEMIEMPIRICAL TURBULENT SHEAR CORRELATIONS

Throughout this chapter we assume constant density and viscosity and no thermal interaction, so that only the continuity and momentum equations are to be solved for velocity and pressure

Continuity:
$$\frac{\partial u}{\partial x} + \frac{\partial v}{\partial y} + \frac{\partial w}{\partial z} = 0$$

$$(6.7)$$

Momentum:
$$\rho \frac{d\mathbf{V}}{dt} = -\nabla p + \rho \mathbf{g} + \mu \nabla^2 \mathbf{V}$$

subject to no slip at the walls and known inlet and exit conditions. (We shall save our free-surface solutions for Chap. 10.)

Both laminar and turbulent flows satisfy Eqs. (6.7). For laminar flow, where there are no random fluctuations, we go right to the attack and solve them for a variety of geometries [2, 3], leaving many more of course for problems.

Reynolds' Time-Averaging Concept

For turbulent flow, because of the fluctuations, every velocity and pressure term in Eqs. (6.7) is a rapidly varying random function of time and space. At present our mathematics cannot handle such instantaneous fluctuating variables. No single pair of random functions $\mathbf{V}(x, y, z, t)$ and $p(x, y, z, t)$ is known to be a solution to

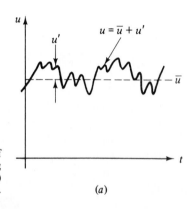

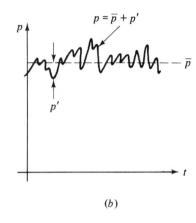

Fig. 6.7 Definition of mean and fluctuating turbulent variables: (*a*) velocity; (*b*) pressure.

(*a*)

(*b*)

Eqs. (6.7). Moreover, our attention as engineers is toward the average or *mean* values of velocity, pressure, shear stress, etc., in a high-Reynolds-number (turbulent) flow. This approach led Osborne Reynolds in 1895 to rewrite Eqs. (6.7) in terms of mean or time-averaged turbulent variables.

The time mean $\bar{u}$ of a turbulent function $u(x, y, z, t)$ is defined by

$$\bar{u} = \frac{1}{T} \int_0^T u \, dt \tag{6.8}$$

where T is an averaging period taken to be longer than any significant period of the fluctuations themselves. The mean values of turbulent velocity and pressure are illustrated in Fig. 6.7. For turbulent gas and water flows an averaging period $T \approx 5$ s is usually quite adequate.

The fluctuation u' is defined as the deviation of u from its average value

$$u' = u - \bar{u} \tag{6.9}$$

as shown also in Fig. 6.7. It follows by definition that a fluctuation has zero mean value

$$\overline{u'} = \frac{1}{T} \int_0^T (u - \bar{u}) \, dt = \bar{u} - \bar{u} = 0 \tag{6.10}$$

However, the mean square of a fluctuation is not zero and is a measure of the *intensity* of the turbulence

$$\overline{u'^2} = \frac{1}{T} \int_0^T u'^2 \, dt \neq 0 \tag{6.11}$$

Nor in general are the mean fluctuation products such as $\overline{u'v'}$ and $\overline{u'p'}$ zero in a typical turbulent flow.

Reynolds' idea was to split each property into mean plus fluctuating variables

$$u = \bar{u} + u' \qquad v = \bar{v} + v' \qquad w = \bar{w} + w' \qquad p = \bar{p} + p' \tag{6.12}$$

Substitute these into Eqs. (6.7) and take the time mean of each equation. The continuity relation reduces to

$$\frac{\partial \bar{u}}{\partial x} + \frac{\partial \bar{v}}{\partial y} + \frac{\partial \bar{w}}{\partial z} = 0 \tag{6.13}$$

which is no different from a laminar continuity relation.

However, each component of the momentum equation (6.7b), after time averaging, will contain mean values plus three mean products, or *correlations*, of fluctuating velocities. The most important of these is the momentum relation in the mainstream, or x, direction, which takes the form

$$\rho \frac{d\bar{u}}{dt} = -\frac{\partial \bar{p}}{\partial x} + \rho g_x + \frac{\partial}{\partial x}\left(\mu \frac{\partial \bar{u}}{\partial x} - \rho \overline{u'^2}\right)$$

$$+ \frac{\partial}{\partial y}\left(\mu \frac{\partial \bar{u}}{\partial y} - \rho \overline{u'v'}\right) + \frac{\partial}{\partial z}\left(\mu \frac{\partial \bar{u}}{\partial z} - \rho \overline{u'w'}\right) \tag{6.14}$$

The three correlation terms $-\rho\overline{u'^2}$, $-\rho\overline{u'v'}$, and $-\rho\overline{u'w'}$ are called *turbulent stresses* because they have the same dimensions and occur right alongside the newtonian (laminar) stress terms $\mu(\partial\bar{u}/\partial x)$, etc. Actually, they are convective acceleration terms (which is why the density appears), not stresses, but they have the mathematical effect of stress and are so termed almost universally in the literature.

The turbulent stresses are unknown a priori and must be related by experiment to geometry and flow conditions, as detailed in Refs. 1 to 3. Fortunately, in duct and boundary-layer flow, the stress $-\rho\overline{u'v'}$ associated with direction y normal to the wall is dominant, and we can approximate with excellent accuracy a simpler streamwise momentum equation

$$\rho\frac{d\bar{u}}{dt} \approx -\frac{\partial\bar{p}}{\partial x} + \rho g_x + \frac{\partial\tau}{\partial y} \tag{6.15}$$

where

$$\tau = \mu\frac{\partial\bar{u}}{\partial y} - \rho\overline{u'v'} = \tau_{\text{lam}} + \tau_{\text{turb}} \tag{6.16}$$

Figure 6.8 shows the distribution of τ_{lam} and τ_{turb} from typical measurements across a turbulent-shear layer near a wall. Laminar shear is dominant near the wall (the *wall layer*), and turbulent shear dominates in the *outer layer*. There is an intermediate region, called the *overlap layer*, where both laminar and turbulent shear are important. These three regions are labeled in Fig. 6.8.

In the outer layer τ_{turb} is two or three orders of magnitude greater than τ_{lam}, and vice versa in the wall layer. These experimental facts enable us to use a crude but very effective model for the velocity distribution $\bar{u}(y)$ across a turbulent wall layer.

The Logarithmic-Overlap Law

We have seen in Fig. 6.8 that there are three regions in turbulent flow near a wall:

1. Wall layer: viscous shear dominates.
2. Outer layer: turbulent shear dominates.
3. Overlap layer: both types of shear important.

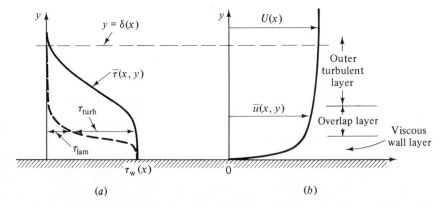

Fig. 6.8 Typical velocity and shear distributions in turbulent flow near a wall: (*a*) shear; (*b*) velocity.

From now on let us agree to drop the overbar from velocity $\bar{u}$. Let τ_w be the wall shear stress and let δ and U represent the thickness and velocity at the edge of the outer layer, $y = \delta$.

For the wall layer, Prandtl deduced in 1930 that u must be independent of the shear-layer thickness

$$u = f(\mu, \tau_w, \rho, y) \tag{6.17}$$

By dimensional analysis, this is equivalent to

$$u^+ = \frac{u}{u^*} = F\left(\frac{yu^*}{\nu}\right) \qquad u^* = \left(\frac{\tau_w}{\rho}\right)^{1/2} \tag{6.18}$$

Equation (6.18) is called the *law of the wall*, and the quantity u^* is termed the *friction velocity* because it has dimensions $\{LT^{-1}\}$, although it is not actually a flow velocity.

Subsequently, Kármán in 1933 deduced that u in the outer layer is independent of molecular viscosity but its deviation from the stream velocity U must depend on layer thickness δ and the other properties

$$(U - u)_{\text{outer}} = g(\delta, \tau_w, \rho, y) \tag{6.19}$$

Again, by dimensional analysis we rewrite this as

$$\frac{U - u}{u^*} = G\left(\frac{y}{\delta}\right) \tag{6.20}$$

where u^* has the same meaning as in Eq. (6.18). Equation (6.20) is called the *velocity-defect law* for the outer layer.

Both the wall law (6.18) and the defect law (6.20) are found to be accurate for a wide variety of experimental turbulent duct and boundary-layer flows [1–3]. They are different in form yet they must overlap smoothly in the intermediate layer. In 1937 C. B. Millikan showed that this can be true only if the overlap-layer velocity varies logarithmically with y:

$$\frac{u}{u^*} = \frac{1}{\kappa} \ln \frac{yu^*}{\nu} + B \qquad \text{overlap layer} \tag{6.21}$$

Over the full range of turbulent wall flows the dimensionless constants κ and B are found to have the approximate values $\kappa \approx 0.41$ and $B \approx 5.0$. Equation (6.21) is called the *logarithmic-overlap layer*.

Thus by dimensional reasoning and physical insight we infer that a plot of u versus $\ln y$ in a turbulent-shear layer will show a curved wall region, a curved outer region, and a straight-line logarithmic overlap. Figure 6.9 shows that this is exactly the case. The four outer-law profiles shown all merge smoothly with the logarithmic-overlap law but have different magnitudes because they vary in external pressure gradient. The wall law is unique and follows the linear viscous relation

$$u^+ = \frac{u}{u^*} = \frac{yu^*}{\nu} = y^+ \tag{6.22}$$

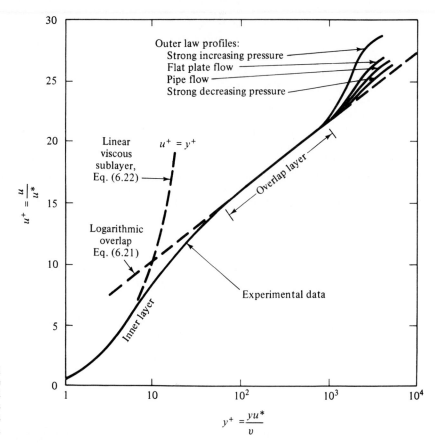

Fig. 6.9 Experimental verification of the inner-, outer-, and overlap-layer laws relating velocity profiles in turbulent wall flow.

from the wall to about $y^+ = 5$, thereafter curving over to merge with the logarithmic law at about $y^+ = 30$.

Believe it or not, Fig. 6.9, which is nothing more than a shrewd correlation of velocity profiles, is the basis for all existing "theory" of turbulent-shear flows. Notice that we have not solved any equations at all but have merely expressed the streamwise velocity in a neat form.

There is serendipity in Fig. 6.9: the logarithmic law (6.21), instead of just being a short overlapping link, actually approximates nearly the entire velocity profile, except for the outer law when the pressure is increasing strongly downstream (as in a diffuser). The inner-wall law typically extends over less than 2 percent of the profile and can be neglected. Thus we can use Eq. (6.21) as an excellent approximation to solve nearly every turbulent-flow problem presented in this and the next chapter. Many additional applications are given in Refs. 2 and 3.

EXAMPLE 6.3 Air at 20°C flows through a 14-cm-diameter tube under fully developed conditions. The centerline velocity is $u_0 = 5$ m/s. Estimate from Fig. 6.9 (a) the friction velocity u^*, (b) the wall shear stress τ_w, and (c) the average velocity $V = Q/A$.

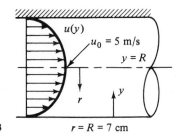

Fig. E6.3 $r = R = 7$ cm

Solution
Part (a)
For pipe flow Fig. 6.9 shows that the logarithmic law, Eq. (6.21), is accurate all the way to the center of the tube. From Fig. E6.3 $y = R - r$ should go from the wall to the centerline as shown. At the center $u = u_0$, $y = R$, and Eq. (6.21) becomes

$$\frac{u_0}{u^*} = \frac{1}{0.41} \ln \frac{Ru^*}{v} + 5.0 \tag{1}$$

Since we know that $u_0 = 5$ m/s and $R = 0.07$ m, u^* is the only unknown in Eq. (1). Find the solution by trial and error

$$u^* = 0.228 \text{ m/s} = 22.8 \text{ cm/s} \qquad\qquad Ans. (a)$$

where we have taken $v = 1.51 \times 10^{-5}$ m²/s for air from Table 1.3.

Part (b)
Assuming a pressure of 1 atm, we have $\rho = p/RT = 1.205$ kg/m³. Since by definition $u^* = (\tau_w/\rho)^{1/2}$, we compute

$$\tau_w = \rho u^{*2} = (1.205 \text{ kg/m}^3)(0.228 \text{ m/s})^2 = 0.062 \text{ kg/(m}\cdot\text{s}^2) = 0.062 \text{ Pa} \quad Ans. (b)$$

This is a very small shear stress, but it will cause a large pressure drop in a long pipe (170 Pa for every 100 m of pipe).

Part (c)
The average velocity V is found by integrating the logarithmic-law velocity distribution

$$V = \frac{Q}{A} = \frac{1}{\pi R^2} \int_0^R u 2\pi r \, dr \tag{2}$$

Introducing $u = u^*[1/\kappa \ln (yu^*/v) + B]$ from Eq. (6.21) and noting that $y = R - r$, we can carry out the integration of Eq. (2), which is rather laborious. The final result is

$$V = 0.835 u_0 = 4.17 \text{ m/s} \qquad\qquad Ans. (c)$$

We shall not bother showing the integration here because it is all worked out and a very neat formula is given in Eqs. (6.49) and (6.59).

Notice that we started from almost nothing (the pipe diameter and the centerline velocity) and found the answers without solving the differential equations of continuity and momentum. We just used the logarithmic-law, Eq. (6.21), which makes the differential equations unnecessary for pipe flow. This is a powerful technique, but you should remember that all we are doing is using an experimental velocity correlation to approximate the actual solution to the problem.

We should check the Reynolds number to ensure turbulent flow

$$\text{Re}_d = \frac{Vd}{v} = \frac{(4.17 \text{ m/s})(0.14 \text{ m})}{1.51 \times 10^{-5} \text{ m}^2/\text{s}} = 38,700$$

Since this is greater than 4000, the flow is definitely turbulent.

6.4 FLOW IN A CIRCULAR PIPE

As our first example of a specific viscous-flow analysis, we take the classic problem of flow in a full pipe, driven either by pressure or by gravity or both. Figure 6.10 shows the geometry of the pipe of radius R. The x axis is taken in the flow direction and is inclined to the horizontal at an angle ϕ.

Before proceeding with a solution to the equations of motion, we can learn a lot by making a control-volume analysis of the flow between sections 1 and 2 in Fig. 6.10. The continuity relation, Eq. (3.23), reduces to

$$Q_1 = Q_2 = \text{const}$$

or

$$V_1 = \frac{Q_1}{A_1} = V_2 = \frac{Q_2}{A_2} \tag{6.23}$$

since the pipe is of constant area. The steady-flow energy equation (3.92) reduces to

$$\frac{p_1}{\rho} + \tfrac{1}{2}\alpha_1 V_1^2 + gz_1 = \frac{p_2}{\rho} + \tfrac{1}{2}\alpha_2 V_2^2 + gz_2 + gh_f \tag{6.24}$$

since there are no shaft-work or heat-transfer effects. Now assume that the flow is fully developed (Fig. 6.6) and correct later for entrance effects. Then the kinetic-

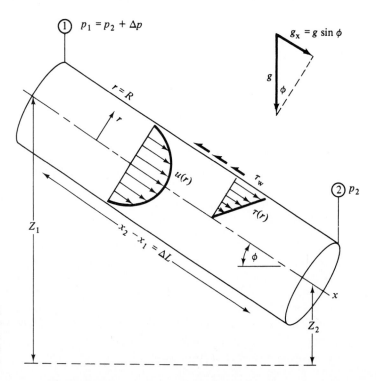

Fig. 6.10 Control volume of steady fully developed flow between two sections in an inclined pipe.

energy correction factor $\alpha_1 = \alpha_2$ and, since $V_1 = V_2$ from (6.23), Eq. (6.24) now reduces to a simple expression for the friction-head loss h_f

$$h_f = \left(z_1 + \frac{p_1}{\rho g}\right) - \left(z_2 + \frac{p_2}{\rho g}\right) = \Delta\left(z + \frac{p}{\rho g}\right) = \Delta z + \frac{\Delta p}{\rho g} \qquad (6.25)$$

The pipe-head loss equals the change in the sum of pressure and gravity head, i.e., the change in height of the HGL. Since the velocity head is constant through the pipe, h_f also equals the height change of the EGL. Recall from Fig. 3.18 that the EGL decreases downstream in a flow with losses unless it passes through an energy source, e.g., as a pump or heat exchanger.

Finally apply the momentum relation (3.40) to the control volume in Fig. 6.10, accounting for applied forces due to pressure, gravity, and shear

$$\Delta p\, \pi R^2 + \rho g(\pi R^2)\, \Delta L \sin \phi - \tau_w(2\pi R)\, \Delta L = \dot{m}(V_1 - V_2) = 0 \qquad (6.26)$$

This equation relates h_f to the wall shear stress

$$\Delta z + \frac{\Delta p}{\rho g} = h_f = \frac{2\tau_w}{\rho g}\frac{\Delta L}{R} \qquad (6.27)$$

where we have substituted $\Delta z = \Delta L \sin \phi$ from Fig. 6.10.

So far we have not assumed either laminar or turbulent flow. If we can correlate τ_w with flow conditions, we have resolved the problem of head loss in pipe flow. Functionally, we can assume that

$$\tau_w = F(\rho, V, \mu, d, \epsilon) \qquad (6.28)$$

where ϵ is the wall-roughness height. Then dimensional analysis tells us that

$$\frac{8\tau_w}{\rho V^2} = f = F\left(\text{Re}_d, \frac{\epsilon}{d}\right) \qquad (6.29)$$

The dimensionless parameter f is called the *Darcy friction factor*, after Henry Darcy (1803–1858), a French engineer whose pipe-flow experiments in 1857 first established the effect of roughness on pipe resistance.

Combining Eqs. (6.27) and (6.29), we obtain the desired expression for finding pipe-head loss

$$h_f = f\frac{L}{d}\frac{V^2}{2g} \qquad (6.30)$$

This is the Darcy-Weisbach equation, valid for duct flows of any cross section and for laminar and turbulent flow. It was proposed by Julius Weisbach, a German professor who in 1850 published the first modern textbook on hydrodynamics.

Our only remaining problem is to find the form of the function F in Eq. (6.29) and plot it in the Moody chart of Fig. 6.13.

Equations of Motion

For either laminar or turbulent flow the continuity equation in cylindrical coordinates is given by (Appendix E)

$$\frac{1}{r}\frac{\partial}{\partial r}(rv_r) + \frac{1}{r}\frac{\partial}{\partial\theta}(v_\theta) + \frac{\partial u}{\partial z} = 0 \tag{6.31}$$

We assume that there is no swirl or circumferential variation, $v_\theta = \partial/\partial\theta = 0$, and fully developed flow: $u = u(r)$ only. Then Eq. (6.31) reduces to

$$\frac{1}{r}\frac{\partial}{\partial r}(rv_r) = 0$$

or

$$rv_r = \text{const} \tag{6.32}$$

But at the wall, $r = R$, $v_r = 0$ (no slip); therefore (6.32) implies that $v_r = 0$ everywhere. Thus in fully developed flow there is only one velocity component, $u = u(r)$.

The momentum differential equation in cylindrical coordinates now reduces to

$$\rho u \frac{\partial u}{\partial x} = -\frac{dp}{dx} + \rho g_x + \frac{1}{r}\frac{\partial}{\partial r}(r\tau) \tag{6.33}$$

where τ can represent either laminar or turbulent shear. But the left-hand side vanishes because $u = u(r)$ only. Rearrange, noting from Fig. 6.10 that $g_x = g\sin\phi$

$$\frac{1}{r}\frac{\partial}{\partial r}(r\tau) = \frac{d}{dx}(p - \rho gx\sin\phi) = \frac{d}{dx}(p + \rho gz) \tag{6.34}$$

Since the left-hand side varies only with r and the right-hand side varies only with x, it follows that both sides must be equal to the same constant.[1] Therefore we can integrate Eq. (6.34) to find the shear distribution across the pipe, utilizing the fact that $\tau = 0$ at $r = 0$

$$\tau = \tfrac{1}{2}r\frac{d}{dx}(p + \rho gz) = (\text{const})(r) \tag{6.35}$$

Thus the shear varies linearly from the centerline to the wall, for either laminar or turbulent flow. This is also shown in Fig. 6.10. At $r = R$, we have the wall shear

$$\tau_w = \tfrac{1}{2}R\frac{\Delta p + \rho g\,\Delta z}{\Delta L} \tag{6.36}$$

which is identical with our momentum relation (6.27). We can now complete our study of pipe flow by applying either laminar or turbulent assumptions to fill out Eq. (6.35).

---

[1] Ask your instructor to explain this to you if necessary.

Laminar-Flow Solution

Note in Eq. (6.35) that the HGL slope $d(p + \rho gz)/dx$ is *negative* because both pressure and height drop with x. For laminar flow, $\tau = \mu \, du/dr$, which we substitute in Eq. (6.35)

$$\mu \frac{du}{dr} = \tfrac{1}{2}rK \qquad K = \frac{d}{dx}(p + \rho gz) \tag{6.37}$$

Integrate once

$$u = \tfrac{1}{4}r^2 \frac{K}{\mu} + C_1 \tag{6.38}$$

The constant C_1 is evaluated from the no-slip condition at the wall: $u = 0$ at $r = R$

$$0 = \tfrac{1}{4}R^2 \frac{K}{\mu} + C_1 \tag{6.39}$$

or $C_1 = -\tfrac{1}{4}R^2 K/\mu$. Introduce into Eq. (6.38) to obtain the exact solution for laminar fully developed pipe flow

$$u = \frac{1}{4\mu}\left[-\frac{d}{dx}(p + \rho gz) \right](R^2 - r^2) \tag{6.40}$$

The laminar-flow profile is thus a paraboloid falling to zero at the wall and reaching a maximum at the axis

$$u_{max} = \frac{R^2}{4\mu}\left[-\frac{d}{dx}(p + \rho gz) \right] \tag{6.41}$$

It resembles the sketch of $u(r)$ given in Fig. 6.10.

The laminar distribution (6.40) is called *Hagen-Poiseuille flow* to commemorate the experimental work of G. Hagen in 1839 and J. L. Poiseuille in 1840, both of whom established the pressure-drop law, Eq. (6.1). The first theoretical derivation of Eq. (6.40) was given independently by E. Hagenbach and by F. Neumann around 1859.

Other pipe-flow results follow immediately from Eq. (6.40). The volume flow is

$$Q = \int u \, dA = \int_0^R u_{max}\left(1 - \frac{r^2}{R^2} \right) 2\pi r \, dr$$

$$= \tfrac{1}{2}u_{max} \pi R^2 = \frac{\pi R^4}{8\mu}\left[-\frac{d}{dx}(p + \rho gz) \right] \tag{6.42}$$

Thus the average velocity in laminar flow is one-half the maximum velocity

$$V = \frac{Q}{A} = \frac{Q}{\pi R^2} = \tfrac{1}{2}u_{max} \tag{6.43}$$

For a horizontal tube ($\Delta z = 0$), Eq. (6.42) is of the form predicted by Hagen's experiment, Eq. (6.1)

$$\Delta p = \frac{8\mu L Q}{\pi R^4} \tag{6.44}$$

The wall shear is computed from the wall velocity gradient

$$\tau_w = \left| \mu \frac{du}{dr} \right|_{r=R} = \frac{2\mu u_{max}}{R} = \tfrac{1}{2} R \left| \frac{d}{dx}(p + \rho g z) \right| \tag{6.45}$$

This gives an exact theory for laminar Darcy friction factor

$$f = \frac{8\tau_w}{\rho V^2} = \frac{8(8\mu V/d)}{\rho V^2} = \frac{64\mu}{\rho V d}$$

or

$$f_{lam} = \frac{64}{\mathrm{Re}_d} \tag{6.46}$$

This is plotted in the Moody chart, Fig. 6.13. The fact that f drops off with increasing Re_d should not mislead us into thinking that shear decreases with velocity: Eq. (6.45) clearly shows that τ_w is proportional to u_{max}; it is interesting to note that τ_w is independent of density because the fluid acceleration is zero.

The laminar head loss follows from Eq. (6.30)

$$h_{f,\,lam} = \frac{64\mu}{\rho V d}\frac{L}{d}\frac{V^2}{2g} = \frac{32\mu L V}{\rho g d^2} = \frac{128\mu L Q}{\pi \rho g d^4} \tag{6.47}$$

We see that laminar head loss is proportional to V.

EXAMPLE 6.4 An oil with $\rho = 900\ \mathrm{kg/m^3}$ and $\nu = 0.0002\ \mathrm{m^2/s}$ flows upward through an inclined pipe as shown in Fig. E6.4. The pressure and elevation are known at sections 1 and 2, 10 m apart. Assuming steady laminar flow, (a) verify that the flow is up, (b) compute h_f between 1 and 2, and compute (c) Q, (d) V, and (e) Re_d. Is the flow really laminar?

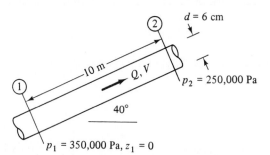

Fig. E6.4

Solution

Part (a) For later use, calculate

$$\mu = \rho \nu = (900\ \mathrm{kg/m^3})(0.0002\ \mathrm{m^2/s}) = 0.18\ \mathrm{kg/(m \cdot s)}$$

$$z_2 = \Delta L \sin 40° = (10\ \mathrm{m})(0.643) = 6.43\ \mathrm{m}$$

The flow goes in the direction of falling HGL; therefore compute the grade-line height at each section

$$HGL_1 = z_1 + \frac{p_1}{\rho g} = 0 + \frac{350{,}000}{900(9.807)} = 39.65 \text{ m}$$

$$HGL_2 = z_2 + \frac{p_2}{\rho g} = 6.43 + \frac{250{,}000}{900(9.807)} = 34.75 \text{ m}$$

The HGL is lower at section 2; hence the flow is from 1 to 2 as assumed.　　　*Ans. (a)*

Part (b)　　The head loss is the change in HGL height

$$h_f = HGL_1 - HGL_2 = 39.65 \text{ m} - 34.75 \text{ m} = 4.9 \text{ m}$$　　　*Ans. (b)*

Half the length of the pipe is quite a large head loss.

Part (c)　　We can compute Q from various laminar-flow formulas, notably Eq. (6.47)

$$Q = \frac{\pi \rho g d^4 h_f}{128 \mu L} = \frac{\pi(900)(9.807)(0.06)^4(4.9)}{128(0.18)(10)} = 0.0076 \text{ m}^3/\text{s}$$　　　*Ans. (c)*

Part (d)　　Divide Q by the pipe area to get the average velocity

$$V = \frac{Q}{\pi R^2} = \frac{0.0076}{\pi(0.03)^2} = 2.7 \text{ m/s}$$　　　*Ans. (d)*

Part (e)　　With V known, the Reynolds number is

$$Re_d = \frac{Vd}{\nu} = \frac{2.7(0.06)}{0.0002} = 810$$　　　*Ans. (e)*

This is well below the transition value $Re_d = 2300$, and so we are fairly certain the flow is laminar.

Notice that by sticking entirely to consistent SI units (meters, seconds, kilograms, newtons) for all variables we avoid the need for any conversion factors in the calculations.

EXAMPLE 6.5　　A liquid of specific weight $\rho g = 58$ lb/ft^3 flows by gravity through a 1-ft tank and a 1-ft capillary tube at a rate of 0.15 ft^3/h, as shown in Fig. E6.5. Sections 1 and 2 are at

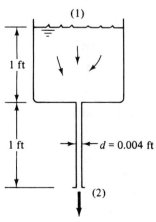

Fig. E6.5　　　$Q = 0.15 \text{ ft}^3/\text{h}$

atmospheric pressure. Neglecting entrance effects, compute the viscosity of the liquid in slugs per foot-second.

Solution Apply the steady-flow energy equation (3.86) with no heat transfer or shaft work

$$\frac{p_1}{\rho g} + \frac{V_1^2}{2g} + z_1 = \left(\frac{p_2}{\rho g} + \frac{V_2^2}{2g} + z_2\right) + h_f$$

But $p_1 = p_2 = p_a$, and V_1 is negligible. Therefore, approximately,

$$h_f = z_1 - z_2 - \frac{V_2^2}{2g} = 2\text{ ft} - \frac{V_2^2}{2g} \tag{1}$$

But V_2 can be computed from the known volume flow and pipe diameter

$$V_2 = \frac{Q}{\pi R^2} = \frac{0.15/3600\text{ ft}^3/\text{s}}{\pi(0.002\text{ ft})^2} = 3.32\text{ ft/s}$$

Substitution into Eq. (1) gives the net head loss

$$h_f = 2.0 - \frac{(3.32)^2}{2(32.2)} = 1.83\text{ ft} \tag{2}$$

Note that h_f includes the entire 2-ft drop through the system and not just the 1 ft of capillary pipe length.

Up to this point we have not specified laminar or turbulent flow. For laminar flow with negligible entrance loss, the head loss is given by Eq. (6.47)

$$h_f = 1.83\text{ ft} = \frac{32\mu L V}{\rho g d^2} = \frac{32\mu(1.0\text{ ft})(3.32\text{ ft/s})}{(58\text{ lb/ft}^3)(0.004\text{ ft})^2} = 114{,}500\mu$$

or

$$\mu = \frac{1.83}{114{,}500} = 1.60 \times 10^{-5}\text{ slug/(ft}\cdot\text{s)} \qquad \qquad Ans.$$

Note that L in this formula is the *pipe* length of 1 ft. Check the Reynolds number to see whether it is really laminar flow

$$\rho = \frac{\rho g}{g} = \frac{58.0}{32.2} = 1.80\text{ slugs/ft}^3$$

$$\text{Re}_d = \frac{\rho V d}{\mu} = \frac{(1.80)(3.32)(0.004)}{1.60 \times 10^{-5}} = 1500 \qquad \text{laminar}$$

Since this is less than 2300, we seem to verify that the flow is laminar. Actually, we may be quite wrong, as Example 6.8 will show.

Turbulent-Flow Solution

For turbulent pipe flow we need not solve a differential equation but instead proceed with the logarithmic law, as in Example 6.3. Assume that Eq. (6.21) correlates the local mean velocity $u(r)$ all the way across the pipe

$$\frac{u(r)}{u^*} \approx \frac{1}{\kappa}\ln\frac{(R-r)u^*}{\nu} + B \tag{6.48}$$

where we have replaced y by $R - r$. Compute the average velocity from this profile

$$V = \frac{Q}{A} = \frac{1}{\pi R^2} \int_0^R u^* \left[\frac{1}{\kappa} \ln \frac{(R - r)u^*}{\nu} + B \right] 2\pi r \, dr$$

$$= \frac{1}{2} u^* \left(\frac{2}{\kappa} \ln \frac{Ru^*}{\nu} + 2B - \frac{3}{\kappa} \right) \tag{6.49}$$

Introducing $\kappa = 0.41$ and $B = 5.0$, we obtain, numerically,

$$\frac{V}{u^*} \approx 2.44 \ln \frac{Ru^*}{\nu} + 1.34 \tag{6.50}$$

This looks only marginally interesting until we realize that V/u^* is directly related to the Darcy friction factor

$$\frac{V}{u^*} = \left(\frac{\rho V^2}{\tau_w} \right)^{1/2} = \left(\frac{8}{f} \right)^{1/2} \tag{6.51}$$

Moreover, the argument of the logarithm in (6.50) is equivalent to

$$\frac{Ru^*}{\nu} = \frac{\frac{1}{2} V d}{\nu} \frac{u^*}{V} = \frac{1}{2} \operatorname{Re}_d \left(\frac{f}{8} \right)^{1/2} \tag{6.52}$$

Introducing (6.52) and (6.51) into Eq. (6.50), changing to a base-10 logarithm and rearranging, we obtain

$$\frac{1}{f^{1/2}} \approx 1.99 \log \left(\operatorname{Re}_d f^{1/2} \right) - 1.02 \tag{6.53}$$

In other words, by simply computing the mean velocity from the logarithmic-law correlation we obtain a relation between friction factor and Reynolds number for turbulent pipe flow. Prandtl derived Eq. (6.53) in 1935 and then adjusted the constants slightly to fit friction data better

$$\frac{1}{f^{1/2}} = 2.0 \log \left(\operatorname{Re}_d f^{1/2} \right) - 0.8 \tag{6.54}$$

This is the accepted formula for a smooth-walled pipe. Some numerical values may be listed as follows:

Re_d	4000	10^4	10^5	10^6	10^7	10^8
f	0.0399	0.0309	0.0180	0.0116	0.0081	0.0059

Thus f drops by only a factor of 5 over a 10,000-fold increase in Reynolds number. Equation (6.54) is cumbersome to solve if Re_d is known and f is wanted. There are many alternate approximations in the literature from which f can be computed explicitly from Re_d

$$f = \begin{cases} 0.316\,Re_d^{-1/4} & 4000 < Re_d < 10^5 \qquad \text{H. Blasius (1911)} \\ \left(1.8 \log \dfrac{Re_d}{6.9}\right)^{-2} & \qquad\qquad\qquad\qquad \text{Ref. 9} \end{cases} \qquad (6.55)$$

Blasius, a student of Prandtl, presented his formula in the first correlation ever made of pipe friction versus Reynolds number. Although his formula has a limited range, it illustrates what was happening to Hagen's 1839 pressure-drop data. For a horizontal pipe, from Eq. (6.55),

$$h_f = \frac{\Delta p}{\rho g} = f \frac{L}{d} \frac{V^2}{2g} \approx 0.316 \left(\frac{\mu}{\rho V d}\right)^{1/4} \frac{L}{d} \frac{V^2}{2g}$$

or

$$\Delta p \approx 0.158 L \rho^{3/4} \mu^{1/4} d^{-5/4} V^{7/4} \qquad (6.56)$$

at low turbulent Reynolds numbers. This explains why Hagen's data for pressure drop begin to increase as the 1.75 power of the velocity, in Fig. 6.4. Note that Δp varies only slightly with viscosity, which is characteristic of turbulent flow. Introducing $Q = \frac{1}{4}\pi d^2 V$ into Eq. (6.56), we obtain the alternate form

$$\Delta p \approx 0.241 L \rho^{3/4} \mu^{1/4} d^{-4.75} Q^{1.75} \qquad (6.57)$$

For a given flow rate Q, the turbulent pressure drop decreases with diameter even more sharply than the laminar formula (6.47). Thus the quickest way to reduce required pumping pressure is to increase the pipe size, although of course the larger pipe is more expensive. Doubling the pipe size decreases Δp by a factor of about 27 for a given Q.

The maximum velocity in turbulent pipe flow is given by Eq. (6.48) evaluated at $r = 0$

$$\frac{u_{max}}{u^*} \approx \frac{1}{\kappa} \ln \frac{Ru^*}{\nu} + B \qquad (6.58)$$

Combining this with Eq. (6.49), we obtain the formula relating mean velocity to maximum velocity

$$\frac{V}{u_{max}} \approx (1 + 1.33\sqrt{f})^{-1} \qquad (6.59)$$

Some numerical values are

Re_d	4000	10^4	10^5	10^6	10^7	10^8
V/u_{max}	0.790	0.811	0.849	0.875	0.893	0.907

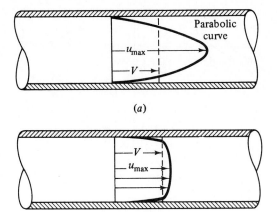

Fig. 6.11 Comparison of laminar and turbulent pipe-flow velocity profiles for the same volume flow: (*a*) laminar flow; (*b*) turbulent flow.

(*a*)

(*b*)

The ratio varies with Reynolds number and is much larger than the value of 0.5 predicted for all laminar pipe flow in Eq. (6.43). Thus a turbulent velocity profile, as shown in Fig. 6.11, is very flat in the center and drops off sharply to zero at the wall.

Effect of Rough Walls

It was not known until experiments in 1800 by Coulomb [6] that surface roughness has an effect on friction resistance. It turns out that the effect is negligible for laminar pipe flow, and all the laminar formulas derived in this section are valid for rough walls also. But turbulent flow is strongly affected by roughness. In Fig. 6.9 the linear viscous sublayer only extends out to $y^+ = yu^*/v = 5$. Thus, compared with the diameter, the sublayer thickness y_s is only

$$\frac{y_s}{d} = \frac{5v/u^*}{d} = \frac{14.1}{\text{Re}_d f^{1/2}} \tag{6.60}$$

For example, at $\text{Re}_d = 10^5$, $f = 0.0180$, $y_s/d = 0.001$. A wall roughness of about $0.001d$ will break up the sublayer and profoundly change the wall law in Fig. 6.9.

Measurements of $u(y)$ in turbulent rough-wall flow by Prandtl's student Nikuradse [7] show, as in Fig. 6.12*a*, that a roughness height ϵ will force the logarithm-law profile outward on the abscissa by an amount approximately equal to $\ln \epsilon^+$, where $\epsilon^+ = \epsilon u^*/v$. The slope of the logarithm law remains the same, $1/\kappa$, but the shift outward causes the constant B to be less by an amount $\Delta B \approx 1/\kappa \ln \epsilon^+$. The actual measured values of ΔB are shown in Fig. 6.12*b* for sand-grain roughness [7] and commercially rough pipes [8].

Fig. 6.12*b* reveals three regimes of rough walls:

$\epsilon u^*/v < 5$: *hydraulically smooth* walls, no effect of roughness on friction
$5 < \epsilon u^*/v < 70$: *transitional* roughness, moderate Reynolds-number effect
$\epsilon u^*/v > 70$: *fully rough* flow, sublayer totally broken up and friction independent of Reynolds number

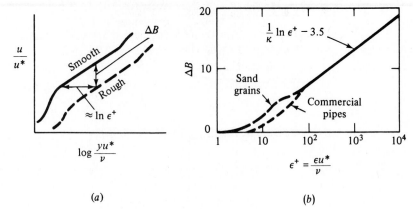

Fig. 6.12 Effect of wall roughness on turbulent pipe-flow velocity profiles: (*a*) logarithm-law downshift; (*b*) correlation with roughness.

For fully rough flow, $\epsilon^+ > 70$, the data in Fig. 6.12*b* follow the straight line

$$\Delta B \approx \frac{1}{\kappa} \ln \epsilon^+ - 3.5 \tag{6.61}$$

and the logarithm law modified for roughness becomes

$$u^+ = \frac{1}{\kappa} \ln y^+ + B - \Delta B = \frac{1}{\kappa} \ln \frac{y}{\epsilon} + 8.5 \tag{6.62}$$

The viscosity vanishes, and hence fully rough flow is independent of Reynolds number. If we integrate Eq. (6.62) to obtain the average velocity in the pipe, we obtain

$$\frac{V}{u^*} = 2.44 \ln \frac{d}{\epsilon} + 3.2$$

or

$$\frac{1}{f^{1/2}} = -2.0 \log \frac{\epsilon/d}{3.7} \qquad \text{fully rough flow} \tag{6.63}$$

There is no Reynolds-number effect; hence the head loss varies exactly as the square of the velocity in this case. Some numerical values of friction factor may be listed:

ϵ/d	0.00001	0.0001	0.001	0.01	0.05
f	0.00806	0.0120	0.0196	0.0379	0.0716

The friction factor increases by 9 times as the roughness increases by a factor of 5000.

The Moody Chart

In 1939 to cover the transitionally rough range Colebrook [9] combined the smooth wall [Eq. (6.54)] and fully rough [Eq.(6.63)] relations into a clever interpolation formula

$$\frac{1}{f^{1/2}} = -2.0 \log\left(\frac{\epsilon/d}{3.7} + \frac{2.51}{\mathrm{Re}_d f^{1/2}}\right) \tag{6.64}$$

This is the accepted design formula for turbulent friction. It was plotted in 1944 by Moody [8] into what is now called the *Moody chart* for pipe friction (Fig. 6.13). The Moody chart is probably the most famous and useful figure in fluid mechanics. It is accurate to ± 15 percent for design calculations over the full range shown in Fig. 6.13. It can be used for circular and noncircular (Sec. 6.6) pipe flows and also for open-channel flows (Chap. 10). The data can even be adapted as an approximation to boundary-layer flows (Chap. 7).

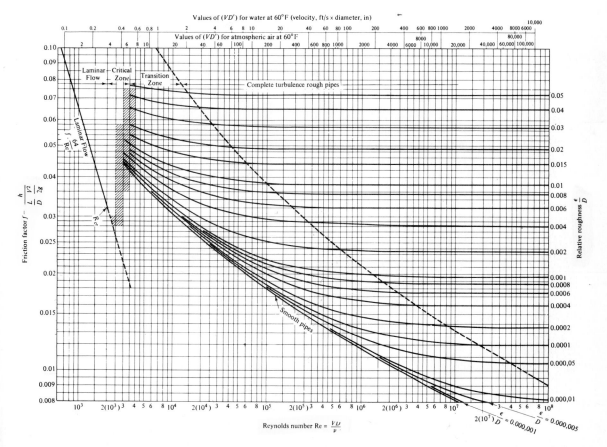

Fig. 6.13 The Moody chart for pipe friction with smooth and rough walls. This chart is identical to Eq. (6.64) for turbulent flow. (*From Ref. 8, by permission of the ASME.*)

Table 6.1

AVERAGE ROUGHNESS OF COMMERCIAL PIPES

	ϵ	
Material (new)	ft	mm
Riveted steel	0.003–0.03	0.9–9.0
Concrete	0.001–0.01	0.3–3.0
Wood stave	0.0006–0.003	0.18–0.9
Cast iron	0.00085	0.26
Galvanized iron	0.0005	0.15
Asphalted cast iron	0.0004	0.12
Commercial steel or wrought iron	0.00015	0.046
Drawn tubing	0.000005	0.0015
Glass	"Smooth"	"Smooth"

Equation (6.64) is cumbersome to evaluate for f if Re_d is known. An alternate explicit formula given by Haaland [33] as

$$\frac{1}{f^{1/2}} \approx -1.8 \log\left[\frac{6.9}{\mathrm{Re}_d} + \left(\frac{\epsilon/d}{3.7}\right)^{1.11}\right] \tag{6.64a}$$

varies less than 2 percent from Eq. (6.64).

The shaded area in the Moody chart indicates the range where transition from laminar to turbulent flow occurs. There are no reliable friction factors in this range, $2000 < \mathrm{Re}_d < 4000$. Notice that the roughness curves are horizontal in the fully rough regime to the right of the dashed line.

From tests with commercial pipes Moody gave the values for average pipe roughness listed in Table 6.1.

EXAMPLE 6.6[1] Compute the loss of head and pressure drop in 200 ft of horizontal 6-in-diameter asphalted cast-iron pipe carrying water with a mean velocity of 6 ft/s.

Solution One can estimate the Reynolds number of water and air from the Moody chart. Look across the top of the chart to $V(\text{ft/s}) \times d(\text{in}) = 36$ and then look directly down to the bottom abscissa to find that $\mathrm{Re}_d(\text{water}) \approx 2.7 \times 10^5$. The roughness ratio for asphalted cast iron ($\epsilon = 0.0004$ ft) is

$$\frac{\epsilon}{d} = \frac{0.0004}{6/12} = 0.0008$$

Find the line on the right side for $\epsilon/d = 0.0008$ and follow it to the left until it intersects the vertical line for $\mathrm{Re} = 2.7 \times 10^5$. Read, approximately, $f = 0.02$ [or compute $f = 0.0197$ from Eq. (6.64a)]. Then the head loss is

$$h_f = f\,\frac{L}{d}\,\frac{V^2}{2g} = (0.02)\,\frac{200}{0.5}\,\frac{(6\ \text{ft/s})^2}{2(32.2\ \text{ft/s}^2)} = 4.5\ \text{ft} \qquad\qquad Ans.$$

[1] This example was given by Moody in his 1944 paper [8].

The pressure drop for a horizontal pipe ($z_1 = z_2$) is

$$\Delta p = \rho g h_f = (62.4 \text{ lbf/ft}^3)(4.5 \text{ ft}) = 280 \text{ lbf/ft}^2 \qquad Ans.$$

Moody points out that this computation, even for clean new pipe, can be considered accurate only to about ± 10 percent.

EXAMPLE 6.7   Oil, with $\rho = 900 \text{ kg/m}^3$ and $\nu = 0.00001 \text{ m}^2/\text{s}$, flows at $0.2 \text{ m}^3/\text{s}$ through 500 m of 200-mm-diameter cast-iron pipe. Determine (a) the head loss and (b) the pressure drop if the pipe slopes down at $10°$ in the flow direction.

Solution   First compute the velocity from the known flow rate

$$V = \frac{Q}{\pi R^2} = \frac{0.2 \text{ m}^3/\text{s}}{\pi (0.1 \text{ m})^2} = 6.4 \text{ m/s}$$

Then the Reynolds number is

$$\text{Re}_d = \frac{Vd}{\nu} = \frac{(6.4 \text{ m/s})(0.2 \text{ m})}{0.00001 \text{ m}^2/\text{s}} = 128,000$$

From Table 6.1, $\epsilon = 0.26 \text{ mm}$ for cast-iron pipe. Then

$$\frac{\epsilon}{d} = \frac{0.26 \text{ mm}}{200 \text{ mm}} = 0.0013$$

Enter the Moody chart on the right at $\epsilon/d = 0.0013$ (you will have to interpolate) and move to the left to intersect with $\text{Re} = 128,000$. Read $f \approx 0.0225$ [from Eq. (6.64) for these values we could compute $f = 0.0227$]. Then the head loss is

$$h_f = f \frac{L}{d} \frac{V^2}{2g} = (0.0225) \frac{500 \text{ m}}{0.2 \text{ m}} \frac{(6.4 \text{ m/s})^2}{2(9.81 \text{ m/s}^2)} = 117 \text{ m} \qquad Ans. (a)$$

From Eq. (6.25) for the inclined pipe,

$$h_f = \frac{\Delta p}{\rho g} + z_1 - z_2 = \frac{\Delta p}{\rho g} + L \sin 10°$$

or

$$\Delta p = \rho g [h_f - (500 \text{ m}) \sin 10°] = \rho g (117 \text{ m} - 87 \text{ m})$$
$$= (900 \text{ kg/m}^3)(9.81 \text{ m/s}^2)(30 \text{ m}) = 265,000 \text{ kg/(m} \cdot \text{s}^2) = 265,000 \text{ Pa} \quad Ans. (b)$$

EXAMPLE 6.8   Repeat Example 6.5 to see whether there is any possible turbulent-flow solution for a smooth-walled pipe.

Solution   As mentioned in that example, we determined the head loss $h_f = 1.83 \text{ ft}$ independent of any assumption of laminar or turbulent flow. Then the friction factor is

$$f = h_f \frac{d}{L} \frac{2g}{V^2} = (1.83 \text{ ft}) \frac{0.004 \text{ ft}}{1.0 \text{ ft}} \frac{2(32.2 \text{ ft/s}^2)}{(3.32 \text{ ft/s})^2} = 0.0428$$

Assuming laminar flow, $f = 64/\text{Re}_d$, we could then predict $\text{Re}_d = 64/0.0428 = 1500$, which is what we did in Example 6.5. However, the Moody chart shows that a *turbulent* flow can also have $f = 0.0428$. For a smooth wall, we read $f = 0.0428$ at about $\text{Re}_d = 3300$, very near the shaded transition area. We can also compute Re from our formulas. If $f = 0.0428$,

$$\text{Re}_d = \begin{cases} 3300 & \text{Eq. (6.55}b) \\ 3200 & \text{Eq. (6.64)} \end{cases}$$

So the flow *might* have been turbulent, in which case the viscosity of the fluid would have been

$$\mu = \frac{\rho V d}{\text{Re}_d} = \frac{1.80(3.32)(0.004)}{3300} = 7.2 \times 10^{-6} \text{ slug/(ft·s)} \qquad \qquad Ans.$$

This is about 55 percent less than our laminar estimate in Example 6.5. The moral is to keep the capillary-flow Reynolds number below about 1000 to avoid such duplicate solutions.

6.5 ALTERNATE FORMS OF THE MOODY CHART[1]

The Moody chart (Fig. 6.13) can be used to solve almost any problem involving friction losses in long pipe flows. However, many such problems involve considerable iteration and repeated calculations using the chart because the standard Moody chart is essentially a *head-loss chart*. One is supposed to know all other variables, compute Re_d, enter the chart, find f, and hence compute h_f. This is one of three fundamental problems which are commonly encountered in pipe flow calculations:

1. Given d, L, and V or Q, ρ, μ, and g, compute the head loss h_f (head-loss problem).
2. Given d, L, h_f, ρ, μ, and g, compute the velocity V or flow rate Q (flow-rate problem).
3. Given Q, L, h_f, ρ, μ, and g, compute the diameter d of the pipe (sizing problem).

Only problem 1 is well suited to the Moody chart. We have to iterate to compute velocity or diameter because both d and V are contained in the ordinate *and* the abscissa of the chart.

For a one-shot velocity or sizing problem, we might as well iterate back and forth on the Moody chart. But if many flow-rate problems or sizing problems are contemplated, we should rearrange the chart to isolate the unknown. This is in the spirit of Sec. 5.6, where we rearranged the sphere-drag problem to isolate the velocity of a falling sphere.

A Modified Moody Chart for Flow Rate

Suppose that problem 2 is specified, so that the velocity is unknown. Then we must eliminate V from either f or Re_d and replot the Moody chart. The product of f and Re_d^2 eliminates velocity

$$\alpha = \tfrac{1}{2} f \, \text{Re}_d^2 = \frac{gd^3 h_f}{Lv^2} \qquad (6.65)$$

---

[1] This section may be omitted without loss of continuity.

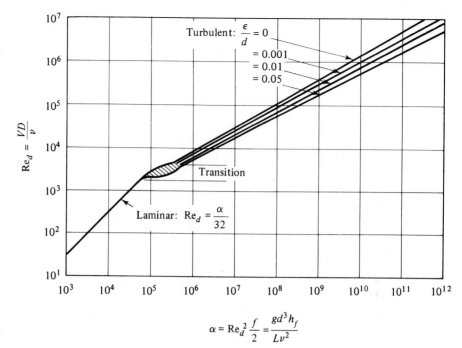

$Re_d = \dfrac{VD}{\nu}$

Turbulent: $\dfrac{\epsilon}{d} = 0$

= 0.001
= 0.01
= 0.05

Transition

Laminar: $Re_d = \dfrac{\alpha}{32}$

Fig. 6.14 Modified Moody chart for finding pipe-flow velocity.

$$\alpha = Re_d^2 \frac{f}{2} = \frac{gd^3 h_f}{L\nu^2}$$

We can keep the parameter ϵ/d because d is known. We now replot Re_d versus this new parameter[1] α for various values of ϵ/d in Fig. 6.14. With the given data we can compute α and ϵ/d, enter the chart, find Re_d, hence compute the velocity V and the flow rate $Q = \pi V d^2/4$. No iteration whatever is required.

Not only that, but the chart is not really necessary, because when α is introduced into Colebrook's formula, Eq. (6.64), we can solve explicitly for Re_d

$$Re_d = -\sqrt{8\alpha}\, \log\!\left(\frac{\epsilon/d}{3.7} + \frac{2.51}{\sqrt{2\alpha}}\right) \tag{6.66}$$

So the computation can be made immediately and the chart serves mostly as visual reassurance. For laminar flow, the Hagen-Poiseuille formula $f = 64/Re_d$ rearranges to

$$Re_d = \frac{\alpha}{32} \tag{6.67}$$

which is also plotted on Fig. 6.14.

EXAMPLE 6.9 Oil, with $\rho = 950 \text{ kg/m}^3$ and $\nu = 0.00002 \text{ m}^2/\text{s}$, flows through a 30-cm-diameter pipe 100 m long with a head loss of 8 m. The roughness ratio is $\epsilon/d = 0.0002$. Find the average velocity and flow rate.

---

[1] The parameter α was suggested by H. Rouse in 1942.

Solution   From Eq. (6.65) calculate

$$\alpha = \frac{gd^3 h_f}{Lv^2} = \frac{(9.807 \text{ m/s}^2)(0.3 \text{ m})^3(8 \text{ m})}{(100 \text{ m})(0.00002 \text{ m}^2/\text{s})^2} = 5.3 \times 10^7$$

Enter Fig. 6.14 for $\epsilon/d = 0.0002$ and read $\text{Re}_d \approx 70{,}000$. For more accuracy, compute $\text{Re}_d = 72{,}600$ from Eq. (6.66). Then

$$V = \frac{v \, \text{Re}_d}{d} = \frac{0.00002(72{,}600)}{0.3} = 4.84 \text{ m/s}$$

$$Q = \tfrac{1}{4}\pi(0.3 \text{ m})^2(4.84 \text{ m/s}) = 0.342 \text{ m}^3/\text{s} \qquad\qquad \textit{Ans.}$$

There is no iteration.

EXAMPLE 6.10   Work Moody's example, Example 6.6, backward, assuming that the head loss is known and the velocity unknown.

Solution   Compute α from the given information

$$\alpha = \frac{(32.2 \text{ ft/s}^2)(0.5 \text{ ft})^3(4.5 \text{ ft})}{(200 \text{ ft})(0.000011 \text{ ft}^2/\text{s})^2} = 7.48 \times 10^8$$

where we have taken $v = 0.000011 \text{ ft}^2/\text{s}$ for water at 60°F. Enter Fig. 6.14 for $\epsilon/d = 0.0008$ (given in Example 6.6) and read $\text{Re}_d \approx 250{,}000$. For more accuracy, compute from Eq. (6.66) that $\text{Re} = 275{,}000$. Then

$$V = \frac{v \, \text{Re}_d}{d} = \frac{0.000011(275{,}000)}{0.5} = 6.04 \text{ ft/s} \qquad\qquad \textit{Ans.}$$

We did not get $V = 6.0$ ft/s exactly because the head loss $h_f = 4.5$ ft in Example 6.6 was rounded off.

A Modified Moody Chart for Pipe Sizing

If the pipe diameter is unknown, considerable manipulation is needed to rearrange the Moody chart since d occurs in all three parameters, Re_d, ϵ/d, and f. Further, it depends upon whether we know the velocity V or the flow rate Q; we cannot know both, or else we could compute $d = (4Q/\pi V)^{1/2}$.

Let us assume that we know the volume flow rate Q. Then the Reynolds number takes the form

$$\text{Re}_d = \frac{Vd}{v} = \frac{4Q}{\pi \, dv} \qquad\qquad (6.68)$$

Similarly, in terms of flow rate, the friction factor becomes

$$f = h_f \frac{d}{L} \frac{2g}{V^2} = \frac{\pi^2}{8} \frac{g h_f d^5}{L Q^2} \qquad\qquad (6.69)$$

Therefore diameter is eliminated by the grouping

$$\beta = (f \text{Re}_d^5)^{1/2} = \left(\frac{128 g h_f Q^3}{\pi^3 L v^5} \right)^{1/2} \qquad\qquad (6.70)$$

Meanwhile, we can eliminate d from roughness ratio through the grouping

$$\frac{\epsilon v}{Q} = \frac{4}{\pi} \frac{\epsilon/d}{\text{Re}_d} \tag{6.71}$$

Thus, if we plot the functional relation

$$\text{Re}_d = f\left(\beta, \frac{\epsilon v}{Q}\right) \tag{6.72}$$

we solve directly for the diameter. This plot is shown in Fig. 6.15 and is simply a rearrangement of the Moody chart (Fig. 6.13). It happens that the data, even for extremely rough pipes, collapse to a narrow range about the smooth-wall curve, which has the approximate power-law form

Smooth walls: $\qquad\qquad\qquad \text{Re}_d \approx 1.43\beta^{0.416} \tag{6.73}$

This is a good first approximation for any problem. The equivalent rearrangement of the Colebrook formula, Eq. (6.64), is not so successful since Re_d remains both inside and outside the logarithm

$$\text{Re}_d^{2.5} = -2.0\beta \log\left[\frac{\pi(\epsilon v/Q)\,\text{Re}_d}{14.8} + \frac{2.51}{\beta}\,\text{Re}_d^{1.5}\right] \tag{6.74}$$

Starting from scratch with known β and $\epsilon v/Q$, this formula would require iteration. However, if one first reads Fig. 6.15 for Re_d and substitutes this value inside the

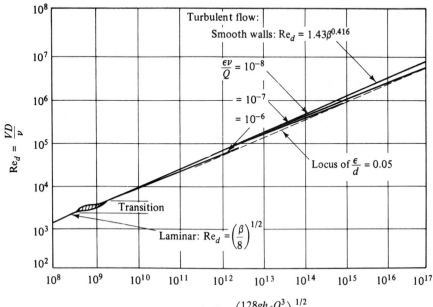

Fig 6.15 Modified Moody chart for finding pipe diameter.

$$\beta = (f\,\text{Re}_d^5)^{1/2} = \left(\frac{128 g h_f Q^3}{\pi^3 L v^5}\right)^{1/2}$$

logarithm, the value of $\mathrm{Re}_d^{2.5}$ is then very accurate and no iteration is needed. Alternately, the initial guess for Re_d could be computed from Eq. (6.73). In Fig. 6.15 the laminar formula is $\mathrm{Re}_d = (\beta/8)^{1/2}$.

EXAMPLE 6.11 Work Example 6.9 backward, assuming that $Q = 0.342\,\mathrm{m^3/s}$ and $\epsilon = 0.06\,\mathrm{mm}$ are known but d is unknown.

Solution Compute $\epsilon v/Q$ and β

$$\frac{\epsilon v}{Q} = \frac{(0.00006\ \mathrm{m})(0.00002\ \mathrm{m^2/s})}{0.342\ \mathrm{m^3/s}} = 3.51 \times 10^{-9}$$

$$\beta = \left[\frac{128(9.81)(8.0)(0.342)^3}{\pi^3(100)(0.00002)^5}\right]^{1/2} = 2.012 \times 10^{11}$$

From Fig. 6.15, this point obviously falls almost exactly on the smooth-wall curve. Therefore, from Eq. (6.73),

$$\mathrm{Re}_d \approx 1.43(2.012 \times 10^{11})^{0.416} = 72{,}100$$

and $\qquad d = \dfrac{4Q}{\pi v\ \mathrm{Re}_d} = \dfrac{4(0.342)}{\pi(0.00002)(72{,}100)} = 0.302\ \mathrm{m}$ $\qquad$ *Ans.*

This is within 1 percent of the originally given diameter of 30 cm.

EXAMPLE 6.12 Work Moody's problem, Example 6.6, backward, assuming that the diameter is the only unknown. $Q = 1.178\ \mathrm{ft^3/s}$.

Solution Evaluate $\epsilon v/Q$ and β

$$\frac{\epsilon v}{Q} = \frac{(0.0004\ \mathrm{ft})(0.000011\ \mathrm{ft^2/s})}{1.178\ \mathrm{ft^3/s}} = 3.74 \times 10^{-9}$$

$$\beta = \left[\frac{128(32.2)(4.5)(1.178)^3}{\pi^3(200)(0.000011)^5}\right]^{1/2} = 5.51 \times 10^{12}$$

Again, from Fig. 6.15 we see that $\epsilon v/Q$ is so small that we are effectively on the smooth-walled curve. Therefore, from Eq. (6.73),

$$\mathrm{Re}_d \approx 1.43(5.51 \times 10^{12})^{0.416} = 285{,}000$$

Thus, approximately,

$$d = \frac{4Q}{\pi v\ \mathrm{Re}_d} = \frac{4(1.178)}{\pi(0.000011)(285{,}000)} = 0.478\ \mathrm{ft} = 5.74\ \mathrm{in} \qquad \textit{Ans.}$$

This is 4 percent low from the true value of 6 in. Most of the discrepancy is due to roughness. Substituting $\mathrm{Re}_d \approx 285{,}000$ into Eq. (6.74), we could, if desired, obtain a better estimate

$$\mathrm{Re}_d^{2.5} = -2.0(5.51 \times 10^{12})\log\left[\frac{\pi(3.74 \times 10^{-9})(285{,}000)}{14.8} + \frac{2.51(285{,}000)^{1.5}}{5.51 \times 10^{12}}\right]$$

$$= 3.89 \times 10^{13}$$

or $\qquad \mathrm{Re}_d = (3.89 \times 10^{13})^{0.4} = 273{,}000$

Table 6.2

**NOMINAL AND ACTUAL SIZES OF
SCHEDULE 40 WROUGHT-STEEL PIPE†**

Nominal size, in	Actual ID, in
$\frac{1}{8}$	0.269
$\frac{1}{4}$	0.364
$\frac{3}{8}$	0.493
$\frac{1}{2}$	0.622
$\frac{3}{4}$	0.824
1	1.049
$1\frac{1}{2}$	1.610
2	2.067
$2\frac{1}{2}$	2.469
3	3.068

† Nominal size within 1 % for 4 in or larger.

Then the improved estimate is

$$d = \frac{4Q}{\pi v \, \text{Re}_d} = 0.499 \text{ ft} = 5.99 \text{ in}$$

The only discrepancy now results from rounding off the head loss in Example 6.6 to 4.5 ft.

In discussing pipe-sizing problems, we should remark that commercial pipes are made only in certain specific sizes. Table 6.2 lists standard water-pipe sizes in the United States. If the sizing calculation gives an intermediate diameter, the next largest pipe size should be selected.

6.6   FLOW IN NONCIRCULAR DUCTS[1]

If the duct is noncircular, the analysis of fully developed flow follows that of the circular pipe but is more complicated algebraically. For laminar flow, one can solve the exact equations of continuity and momentum. For turbulent flow, the logarithm-law velocity profile can be used, or (better and simpler) the hydraulic diameter is an excellent approximation.

The Hydraulic Diameter

For a noncircular duct, the control-volume concept of Fig. 6.10 is still valid, but the cross-sectional area A does not equal πR^2 and the cross-sectional perimeter wetted by the shear stress $\mathscr{P}$ does not equal $2\pi R$. The momentum equation (6.26) thus becomes

$$\Delta p \, A + \rho g A \, \Delta L \sin \phi - \bar{\tau}_w \mathscr{P} \, \Delta L = 0$$

or

$$h_f = \frac{\Delta p}{\rho g} + \Delta z = \frac{\bar{\tau}_w}{\rho g} \frac{\Delta L}{A/\mathscr{P}} \tag{6.75}$$

[1] This section may be omitted without loss of continuity.

This is identical to Eq. (6.27) except that (1) the shear stress is an average value integrated around the perimeter and (2) the length scale $A/\mathcal{P}$ takes the place of the pipe radius R. For this reason a noncircular duct is said to have a *hydraulic radius* R_h defined by

$$R_h = \frac{A}{\mathcal{P}} = \frac{\text{cross-sectional area}}{\text{wetted perimeter}} \qquad (6.76)$$

This concept receives constant use in open-channel flow (Chap. 10), where the channel cross section is almost never circular. If, by comparison to Eq. (6.29) for pipe flow, we define the friction factor in terms of average shear

$$f_{\text{NCD}} = \frac{8\bar{\tau}_w}{\rho V^2} \qquad (6.77)$$

where NCD stands for noncircular duct and $V = Q/A$ as usual, Eq. (6.75) becomes

$$h_f = f \frac{L}{4R_h} \frac{V^2}{2g} \qquad (6.78)$$

This is equivalent to Eq. (6.30) for pipe flow except that d is replaced by $4R_h$. Therefore we customarily define the *hydraulic diameter* as

$$D_h = \frac{4A}{\mathcal{P}} = \frac{4 \times \text{area}}{\text{wetted perimeter}} = 4R_h \qquad (6.79)$$

We should stress that wetted perimeter includes all surfaces acted upon by the shear stress. For example, in a circular annulus, both the outer and inner perimeter should be added. The fact that D_h equals $4R_h$ is just one of those things: chalk it up to an engineer's sense of humor. Note that for the degenerate case of a circular pipe, $D_h = 4\pi R^2/2\pi R = 2R$, as expected.

We would therefore expect by dimensional analysis that this friction factor f based upon hydraulic diameter as in Eq. (6.78) would correlate with Reynolds number and roughness ratio based upon hydraulic diameter

$$f = F\left(\frac{VD_h}{\nu}, \frac{\epsilon}{D_h}\right) \qquad (6.80)$$

and this is the way the data are correlated. But we should not necessarily expect the Moody chart (Fig. 6.13) to hold exactly in terms of this new length scale; and it does not, but it is surprisingly accurate:

$$f \approx \begin{cases} \dfrac{64}{\text{Re}_{D_h}} & \pm 40\% & \text{laminar flow} \\[12pt] f_{\text{Moody}}\left(\text{Re}_{D_h}, \dfrac{\epsilon}{D_h}\right) & \pm 15\% & \text{turbulent flow} \end{cases} \qquad (6.81)$$

Now let us look at some particular cases.

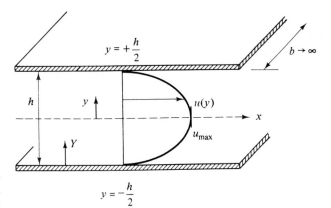

Fig. 6.16 Fully developed flow between parallel plates.

Flow between Parallel Plates

As shown in Fig. 6.16, flow between parallel plates a distance h apart is the limiting case of flow through a very wide rectangular channel. For fully developed flow, $u = u(y)$ only, which satisfies continuity identically. The momentum equation in cartesian coordinates reduces to

$$0 = -\frac{dp}{dx} + \rho g_x + \frac{d\tau}{dy} \qquad \tau_{\text{lam}} = \mu \frac{du}{dy} \qquad (6.82)$$

subject to no slip: $u = 0$ at $y = +h/2$ and $-h/2$. Introducing the laminar-flow shear stress and rearranging, we obtain

$$\mu \frac{d^2u}{dy^2} = \frac{d}{dx}(p + \rho g z) = \text{const} = -K \qquad (6.83)$$

When we integrate twice, the velocity distribution is

$$u = \frac{-K}{2\mu} y^2 + C_1 y + C_2 \qquad (6.84)$$

The constants are found from the two no-slip conditions

$$u\left(\frac{+h}{2}\right) = 0 = \frac{-K}{2\mu}\frac{h^2}{4} + \frac{C_1 h}{2} + C_2$$

$$u\left(\frac{-h}{2}\right) = 0 = \frac{-K}{2\mu} - \frac{C_1 h}{2} + C_2 \qquad (6.85)$$

Solving, we obtain $C_1 = 0$ and $C_2 = Kh^2/8\mu$. The desired solution for laminar parallel-plate flow is thus

$$u = \frac{1}{8\mu}\left[-\frac{d}{dx}(p + \rho g z) \right](h^2 - 4y^2) \qquad (6.86)$$

This is a parabolic distribution very similar to Hagen-Poiseuille pipe flow, Eq. (6.40). The volume flow rate, assuming a very large width b, is

$$Q = \int_{-h/2}^{+h/2} u(b\,dy) = \frac{bh^3}{12\mu}\left[-\frac{d}{dx}(p + \rho gz) \right] \tag{6.87}$$

The average velocity between the plates is given by

$$V = \frac{Q}{bh} = \frac{h^2}{12\mu}\left[-\frac{d}{dx}(p + \rho gz) \right] = \tfrac{2}{3}u_{max} \tag{6.88}$$

The wall shear stress is computed from

$$\tau_w = \left| \mu \frac{du}{dy}\right|_{y=\pm h/2} = \frac{h}{2}\left[-\frac{d}{dx}(p + \rho gz) \right] \tag{6.89}$$

Wall shear is a constant in this particular problem. The friction factor is

$$f = \frac{8\tau_w}{\rho V^2} = \frac{48\mu}{\rho Vh} = \frac{48}{Re_h} \tag{6.90}$$

These are exact analytic results, and there is no reason to resort to the hydraulic-diameter concept. However, if we did use D_h, a discrepancy would arise. For parallel plates

$$D_h = \lim_{b\to\infty} \frac{4bh}{2b + 2h} = 2h \tag{6.91}$$

or twice the distance between the plates. Substitution into Eq. (6.90), using the subscript PP for parallel plates, gives

$$f_{PP} = \frac{96\mu}{2h\rho V} = \frac{96}{Re_{D_h}} \tag{6.92}$$

If we could not work out the laminar theory and chose to use the approximation $f \approx 64/Re_{D_h}$, we would be 33 percent low. Thus the hydraulic-diameter approximation is a crude one in laminar calculations or noncircular ducts, as Eq. (6.81) states.

Just as in circular-pipe flow, the laminar solution above becomes unstable at about $Re_{D_h} = 2300$; transition occurs, and turbulent duct flow results.

For turbulent flow between parallel plates, we can again use the logarithm law, Eq. (6.21), as an approximation across the entire flow field, using not y but a wall coordinate Y, as shown in Fig. 6.16

$$\frac{u(y)}{u^*} \approx \frac{1}{\kappa}\ln\frac{Yu^*}{\nu} + B \qquad 0 < Y < \frac{h}{2} \tag{6.93}$$

This distribution looks very much like the flat turbulent profile for pipe flow in Fig. 6.11b, and the mean velocity is

$$V = \frac{2}{h}\int_0^{h/2} u\,dY = u^*\left(\frac{1}{\kappa}\ln\frac{hu^*}{2\nu} + B - \frac{1}{\kappa} \right) \tag{6.94}$$

Recalling that $V/u^* = (8/f)^{1/2}$, we see that Eq. (6.94) is equivalent to a parallel-plate friction law. Rearranging and cleaning up the constant terms, we obtain

$$\frac{1}{f^{1/2}} \approx 2.0 \log (\text{Re}_{D_h} f^{1/2}) - 1.19 \qquad (6.95)$$

where we have introduced the hydraulic diameter $D_h = 2h$. This is remarkably close to the pipe-friction law, Eq. (6.54). Therefore we conclude that the use of the hydraulic diameter in this turbulent case is quite successful. That turns out to be true for other noncircular turbulent flows also.

Equation (6.95) can be brought into exact agreement with the pipe law by rewriting it in the form

$$\frac{1}{f^{1/2}} = 2.0 \log (0.64 \, \text{Re}_{D_h} f^{1/2}) - 0.8 \qquad (6.96)$$

Thus the turbulent friction is predicted most accurately when we use an effective diameter D_{eff} equal to 0.64 times the hydraulic diameter. The effect on f itself is much less, about 10 percent at most. We can compare with Eq. (6.92) for laminar flow, which predicted

Parallel plates: $\qquad\qquad D_{\text{eff}} = \frac{64}{96} D_h = \frac{2}{3} D_h \qquad (6.97)$

This close resemblance ($0.64 D_h$ versus $0.667 D_h$) occurs so often in noncircular duct flow that we may take it to be a general rule for computing turbulent friction in ducts:

$$D_{\text{eff}} = D_h = \frac{4A}{\mathscr{P}} \qquad \text{reasonable accuracy}$$

$$D_{\text{eff}}(\text{laminar theory}) \qquad \text{extreme accuracy} \qquad (6.98)$$

Jones [10] shows that the effective-laminar-diameter idea collapses all data for rectangular ducts of arbitrary height-to-width ratio onto the Moody chart for pipe flow. We recommend this idea for all noncircular ducts.

EXAMPLE 6.13

Fluid flows at an average velocity of 6 ft/s between horizontal parallel plates a distance of 2.4 in apart. Find the head loss and pressure drop for each 100 ft of length for $\rho = 1.9$ slugs/ft^3 and (a) $v = 0.00002$ ft^3/s and (b) $v = 0.002$ ft^3/s. Assume smooth walls.

Solution

Part (a)

The viscosity $\mu = \rho v = 3.8 \times 10^{-5}$ slug/(ft · s). The spacing is $h = 2.4$ in $= 0.2$ ft, and $D_h = 2h = 0.4$ ft. The Reynolds number is

$$\text{Re}_{D_h} = \frac{V D_h}{v} = \frac{(6.0 \text{ ft/s})(0.4 \text{ ft})}{0.00002 \text{ ft}^2/\text{s}} = 120{,}000$$

The flow is therefore turbulent. For reasonable accuracy, simply look on the Moody chart (Fig. 6.13) for smooth walls

$$f \approx 0.0173 \qquad h_f \approx f \frac{L}{D_h} \frac{V^2}{2g} = 0.0173 \frac{100}{0.4} \frac{(6.0)^2}{2(32.2)} \approx 2.42 \text{ ft} \qquad \textit{Ans. (a)}$$

Since there is no change in elevation,

$$\Delta p = \rho g h_f = 1.9(32.2)(2.42) = 148 \text{ lbf/ft}^2 \qquad \qquad \textit{Ans. (a)}$$

This is the head loss and pressure drop per 100 ft of channel. For more accuracy, take $D_{\text{eff}} = \frac{2}{3}D_h$ from laminar theory; then

$$\text{Re}_{\text{eff}} = \tfrac{2}{3}(120{,}000) = 80{,}000$$

and from the Moody chart read $f \approx 0.0189$ for smooth walls. Thus a better estimate is

$$h_f = 0.0189 \frac{100}{0.4} \frac{(6.0)^2}{2(32.2)} = 2.64 \text{ ft}$$

and

$$\Delta p = 1.9(32.2)(2.64) = 161 \text{ lbf/ft}^2 \qquad \qquad \textit{Better ans. (a)}$$

The more accurate formula predicts friction about 9 percent higher.

Part (b) Compute $\mu = \rho v = 0.0038$ slug/(ft·s). The Reynolds number is $6.0(0.4)/0.002 = 1200$; therefore the flow is laminar, since Re is less than 2300.

You could use the laminar-flow friction factor, Eq. (6.92)

$$f_{\text{lam}} = \frac{96}{\text{Re}_{D_h}} = \frac{96}{1200} = 0.08$$

from which

$$h_f = 0.08 \frac{100}{0.4} \frac{(6.0)^2}{2(32.2)} = 11.2 \text{ ft}$$

and

$$\Delta p = 1.9(32.2)(11.2) = 684 \text{ lbf/ft}^2 \qquad \qquad \textit{Ans. (b)}$$

Alternatively, you can finesse the Reynolds number and go directly to the appropriate laminar-flow formula, Eq. (6.88)

$$V = \frac{h^2}{12\mu} \frac{\Delta p}{L}$$

or

$$\Delta p = \frac{12(6.0 \text{ ft/s})[0.0038 \text{ slug/(ft·s)}](100 \text{ ft})}{(0.2 \text{ ft})^2} = 684 \text{ slugs/(ft·s}^2) = 684 \text{ lbf/ft}^2$$

and

$$h_f = \frac{\Delta p}{\rho g} = \frac{684}{1.9(32.2)} = 11.2 \text{ ft}$$

This is one of those—perhaps unexpected—problems where the laminar friction is greater than the turbulent friction.

Flow through a Concentric Annulus

Consider steady axial laminar flow in the annular space between two concentric cylinders, as in Fig. 6.17. There is no slip at the inner $(r = b)$ and outer radius $(r = a)$. For $u = u(r)$ only, the governing relation is Eq. (6.34)

$$\frac{d}{dr}\left(r\mu \frac{du}{dr}\right) = Kr \qquad K = \frac{d}{dx}(p + \rho g z) \qquad (6.99)$$

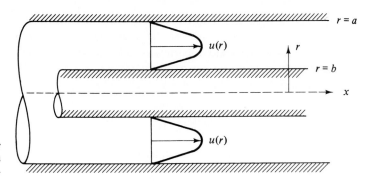

Fig. 6.17 Fully developed flow through a concentric annulus.

Integrate this twice

$$u = \tfrac{1}{4}r^2 \frac{K}{\mu} + C_1 \ln r + C_2 \tag{6.100}$$

The constants are found from the two no-slip conditions

$$u(r = a) = 0 = \tfrac{1}{4}a^2 \frac{K}{\mu} + C_1 \ln a + C_2$$

$$u(r = b) = 0 = \tfrac{1}{4}b^2 \frac{K}{\mu} + C_1 \ln b + C_2 \tag{6.101}$$

The final solution for the velocity profile is

$$u = \frac{1}{4\mu} \left[-\frac{d}{dx}(p + \rho g z) \right] \left[a^2 - r^2 + \frac{a^2 - b^2}{\ln (b/a)} \ln \frac{a}{r} \right] \tag{6.102}$$

The volume flow is given by

$$Q = \int_b^a u 2\pi r \, dr = \frac{\pi}{8\mu} \left[-\frac{d}{dx}(p + \rho g z) \right] \left[a^4 - b^4 - \frac{(a^2 - b^2)^2}{\ln (a/b)} \right] \tag{6.103}$$

The velocity profile $u(r)$ resembles a parabola wrapped around in a circle to form a split doughnut, as in Fig. 6.17. The maximum velocity occurs at the radius

$$r' = \left[\frac{a^2 - b^2}{2 \ln (a/b)} \right]^{1/2} \qquad u = u_{max} \tag{6.104}$$

This maximum is closer to the inner radius but approaches the midpoint between cylinders as the clearance $a - b$ becomes small. Some numerical values are as follows:

$\dfrac{b}{a}$	0.01	0.1	0.2	0.5	0.8	0.9	0.99
$\dfrac{r' - b}{a - b}$	0.323	0.404	0.433	0.471	0.491	0.496	0.499

Also, as the clearance becomes small, the profile approaches a parabolic distribution, as if the flow were between two parallel plates [Eq. (6.86)].

It is confusing to base the friction factor on the wall shear because there are two shear stresses, the inner stress being greater than the outer. It is better to define f with respect to the head loss, as in Eq. (6.88),

$$f = h_f \frac{D_h}{L} \frac{2g}{V^2} \quad \text{where } V = \frac{Q}{\pi(a^2 - b^2)} \tag{6.105}$$

The hydraulic diameter for an annulus is

$$D_h = \frac{4\pi(a^2 - b^2)}{2\pi(a + b)} = 2(a - b) \tag{6.106}$$

It is twice the clearance, rather like the parallel-plate result of twice the distance between plates [Eq. (6.91)].

Substituting h_f, D_h, and V into Eq. (6.105), we find that the friction factor for laminar flow in a concentric annulus is of the form

$$f = \frac{64\zeta}{\mathrm{Re}_{D_h}} \quad \zeta = \frac{(a - b)^2(a^2 - b^2)}{a^4 - b^4 - (a^2 - b^2)^2/[\ln(a/b)]} \tag{6.107}$$

The dimensionless term ζ is a sort of correction factor for the hydraulic diameter. We could rewrite Eq. (6.107) as

Concentric annulus: $\qquad f = \dfrac{64}{\mathrm{Re}_{\mathrm{eff}}} \qquad \mathrm{Re}_{\mathrm{eff}} = \dfrac{1}{\zeta} \, \mathrm{Re}_{D_h}$ \hfill (6.108)

Some numerical values of $f\,\mathrm{Re}_{D_h}$ and $D_{\mathrm{eff}}/D_h = 1/\zeta$ are given in Table 6.3.

Table 6.3

LAMINAR FRICTION FACTORS FOR A CONCENTRIC ANNULUS

b/a	$f\,\mathrm{Re}_{D_h}$	D_{eff}/D_h
0.0	64.0	1.000
0.00001	70.09	0.913
0.0001	71.78	0.892
0.001	74.68	0.857
0.01	80.11	0.799
0.05	86.27	0.742
0.1	89.37	0.716
0.2	92.35	0.693
0.4	94.71	0.676
0.6	95.59	0.670
0.8	95.92	0.667
1.0	96.0	0.667

For turbulent flow through a concentric annulus, the analysis might proceed by patching together two logarithmic-law profiles, one going out from the inner wall to meet the other coming in from the outer wall. We omit such a scheme here and proceed directly to the friction factor. According to the general rule proposed in Eq. (6.98), turbulent friction is predicted with excellent accuracy by replacing d in the Moody chart by $D_{\text{eff}} = 2(a - b)/\zeta$, with values listed in Table 6.3.[1] This idea includes roughness, also (replace ϵ/d in the chart by ϵ/D_{eff}). For a quick design number with about 10 percent accuracy, one can simply use the hydraulic diameter $D_h = 2(a - b)$.

EXAMPLE 6.14 What should the reservoir level h be to maintain a flow of 0.01 m³/s through the commercial steel annulus 30 m long shown in Fig. E6.14? Neglect entrance effects and take $\rho = 1000 \text{ kg/m}^3$ and $\nu = 1.02 \times 10^{-6} \text{ m}^2/\text{s}$ for water.

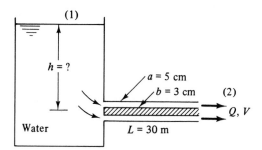

Fig. E6.14

Solution Compute the average velocity and hydraulic diameter

$$V = \frac{Q}{A} = \frac{0.01 \text{ m}^3/\text{s}}{\pi[(0.05 \text{ m})^2 - (0.03 \text{ m})^2]} = 1.99 \text{m/s}$$

$$D_h = 2(a - b) = 2(0.05 - 0.03) \text{ m} = 0.04 \text{ m}$$

Apply the steady-flow energy equation between (1) and (2)

$$\frac{p_1}{\rho} + \tfrac{1}{2}V_1^2 + gz_1 = \left(\frac{p_2}{\rho} + \tfrac{1}{2}V_2^2 + gz_2\right) + gh_f$$

But $p_1 = p_2 = p_a$, $V_1 \approx 0$, and $V_2 = V$ in the pipe. Therefore solve for

$$h_f = f\frac{L}{D_h}\frac{V^2}{2g} = z_1 - z_2 - \frac{V^2}{2g}$$

But $z_1 - z_2 = h$, the desired reservoir height. Thus, finally,

$$h = \frac{V^2}{2g}\left(1 + f\frac{L}{D_h}\right) \tag{1}$$

[1] Jones shows [44] that data for annular flow also satisfy the effective-laminar-diameter idea.

Since V, L, and D_h are known, our only remaining problem is to compute the annulus friction factor f. For a quick approximation, take $D_{eff} = D_h = 0.04$ m. Then

$$\text{Re}_{D_h} = \frac{VD_h}{\nu} = \frac{1.99(0.04)}{1.02 \times 10^{-6}} = 78{,}000$$

$$\frac{\epsilon}{D_h} = \frac{0.046 \text{ mm}}{40 \text{ mm}} = 0.00115$$

where $\epsilon = 0.046$ mm has been read from Table 6.1 for commercial steel surfaces. From the Moody chart, read $f = 0.0232$. Then, from Eq. (1) above,

$$h \approx \frac{(1.99 \text{ m/s})^2}{2(9.81 \text{ m/s}^2)}\left(1 + 0.0232\frac{30 \text{ m}}{0.04 \text{ m}}\right) = 3.71 \text{ m} \qquad \textit{Crude ans.}$$

For better accuracy, take $D_{eff} = D_h/\zeta = 0.670 D_h = 2.68$ cm, where the correction factor 0.670 has been read from Table 6.3 for $b/a = 3/5 = 0.6$. Then the corrected Reynolds number and roughness ratio are

$$\text{Re}_{eff} = \frac{VD_{eff}}{\nu} = 52{,}300 \qquad \frac{\epsilon}{D_{eff}} = 0.00172$$

From the Moody chart, read $f = 0.0257$. Then the improved computation for reservoir height is

$$h = \frac{(1.99 \text{ m/s})^2}{2(9.81 \text{ m/s}^2)}\left(1 + 0.0257\frac{30 \text{ m}}{0.04 \text{ m}}\right) = 4.09 \text{ m} \qquad \textit{Better ans.}$$

The uncorrected hydraulic diameter estimate is about 9 percent low. Note that we do *not* replace D_h by D_{eff} in the ratio L/D_h in Eq. (1) since this is implicit in the definition of friction factor.

Other Noncircular Cross Sections

In principle, any duct cross section can be solved analytically for the laminar-flow velocity distribution, volume flow, and friction factor. This is because any cross section can be mapped onto a circle by the methods of complex variable, and other powerful analytical techniques are also available. Many examples are given by White [3, pp. 123–128], Berker [11], and Olson [12, pp. 224–225]. Some new results for cross sections formed by overlapping rectangles and circles are given by Zarling [13]. Reference 34 is devoted entirely to laminar duct flow.

In general, however, most unusual duct sections have strictly academic and not commercial value. We list here only the rectangular and isosceles-triangular sections, in Table 6.4, leaving other cross sections for you to find in the references.

For turbulent flow in a duct of unusual cross section one should replace d by D_h on the Moody chart if no laminar theory is available. If laminar results are known, such as Table 6.4, replace d by $D_{eff} = [64/(f \text{ Re})]D_h$ for the particular geometry of the duct.

Table 6.4

LAMINAR FRICTION CONSTANTS f Re FOR RECTANGULAR AND TRIANGULAR DUCTS

Rectangular		Isosceles triangle	
b/a	f Re_{D_h}	θ, deg	f Re_{D_h}
0.0	96.00	0	48.0
0.05	89.91	10	51.6
0.1	84.68	20	52.9
0.125	82.34	30	53.3
0.167	78.81	40	52.9
0.25	72.93	50	52.0
0.4	65.47	60	51.1
0.5	62.19	70	49.5
0.75	57.89	80	48.3
1.0	56.91	90	48.0

For laminar flow in rectangles and triangles, the wall friction varies greatly, being largest near the midpoints of the sides and zero in the corners. In turbulent flow through the same sections, the shear is nearly constant along the sides, dropping off sharply to zero in the corners. This is because of the phenomenon of turbulent *secondary flow*, in which there are nonzero mean velocities v and w in the plane of the cross section. Some measurements of axial velocity and secondary-flow patterns are shown in Fig. 6.18, as sketched by Nikuradse in his 1926 dissertation.

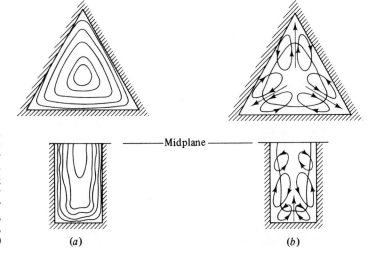

Fig. 6.18 Illustration of secondary turbulent flow in noncircular ducts: (*a*) axial mean-velocity contours; (*b*) secondary-flow cellular motions. (*After J. Nikuradse, dissertation, Göttingen, 1926.*)

Midplane

(*a*)   (*b*)

The secondary-flow "cells" drive the mean flow toward the corners, so that the axial-velocity contours are similar to the cross section and the wall shear nearly constant. This is why the hydraulic-diameter concept is so successful for turbulent flow. Laminar flow in a straight noncircular duct has no secondary flow. An accurate theoretical prediction of turbulent secondary flow has yet to be achieved, although numerical models are improving [36].

EXAMPLE 6.15   Air, with $\rho = 0.00237$ slug/ft^3 and $v = 0.000157$ ft^2/s, is forced through a horizontal square 9- by 9-in duct 100 ft long at 25 ft^3/s. Find the pressure drop if $\epsilon = 0.0003$ ft.

Solution   Compute the mean velocity and hydraulic diameter

$$V = \frac{25 \text{ ft}^3/\text{s}}{(0.75 \text{ ft})^2} = 44.4 \text{ ft/s}$$

$$D_h = \frac{4A}{\mathscr{P}} = \frac{4(81 \text{ in}^2)}{36 \text{ in}} = 9 \text{ in} = 0.75 \text{ ft}$$

From Table 6.4, for $b/a = 1.0$, the effective diameter is

$$D_{\text{eff}} = \frac{64}{56.91} D_h = 0.843 \text{ ft}$$

whence

$$\text{Re}_{\text{eff}} = \frac{VD_{\text{eff}}}{v} = \frac{44.4(0.843)}{0.000157} = 239{,}000$$

$$\frac{\epsilon}{D_{\text{eff}}} = \frac{0.0003}{0.843} = 0.000356$$

From the Moody chart, read $f = 0.0177$. Then the pressure drop is

$$\Delta p = \rho g h_f = \rho g \left(f \frac{L}{D_h} \frac{V^2}{2g} \right) = 0.00237(32.2) \left[0.0177 \frac{100}{0.75} \frac{44.4^2}{2(32.2)} \right]$$

or
$$\Delta p = 5.5 \text{ lbf/ft}^2 \qquad\qquad \textit{Ans.}$$

Pressure drop in air ducts is usually small because of the low density.

6.7   MINOR LOSSES IN PIPE SYSTEMS[1]

For any pipe system, in addition to the Moody-type friction loss computed for the length of pipe, there are additional so-called *minor losses* due to

1. Pipe entrance or exit
2. Sudden expansion or contraction
3. Bends, elbows, tees, and other fittings
4. Valves, open or partially closed
5. Gradual expansions or contractions

---

[1] This section may be omitted without loss of continuity.

The losses may not be so minor; e.g., a partially closed valve can cause a greater pressure drop than a long pipe.

Since the flow pattern in fittings and valves is quite complex, the theory is very weak. The losses are commonly measured experimentally and correlated with the pipe-flow parameters. The data, especially for valves, are somewhat dependent upon the particular manufacturer's design, so that the values listed here must be taken as average design estimates [15, 16, 35, 43].

The measured minor loss is usually given as a ratio of the head loss $h_m = \Delta p/\rho g$ through the device to the velocity head $V^2/2g$ of the associated piping system

$$\text{Loss coefficient } K = \frac{h_m}{V^2/2g} \tag{6.109}$$

Although K is dimensionless, it unfortunately is not correlated in the literature with Reynolds number and roughness ratio but rather simply with the raw size of the pipe in, say, inches. Almost all data are reported for turbulent-flow conditions.

An alternate, and less desirable, procedure is to report the minor loss as if it were an *equivalent length* L_{eq} of pipe, satisfying the Darcy friction-factor relation

$$h_m = f \frac{L_{eq}}{d} \frac{V^2}{2g} = K \frac{V^2}{2g}$$

or

$$L_{eq} = \frac{Kd}{f} \tag{6.110}$$

Although the equivalent length should take some of the variability out of the loss data, it is an artificial concept and will not be pursued here.

A single pipe system may have many minor losses. Since all of them are correlated with $V^2/2g$, they can be summed into a single total system loss if the pipe has constant diameter

$$\Delta h_{tot} = h_f + \sum h_m = \frac{V^2}{2g}\left(\frac{fL}{d} + \sum K\right) \tag{6.111}$$

Note, however, that we must sum the losses separately if the pipe size changes so that V^2 changes. The length L in Eq. (6.111) is the total length of the pipe axis, including any bends.

Table 6.5 lists loss coefficients K for four common types of valves, three standard elbows, and a tee fitting. Small fittings normally use internal screw connections, and large fittings are connected by flanges; hence the two listings. We see that K generally decreases with the size of the connected pipe, which is consistent with the increased Reynolds number and decreased roughness ratio associated with large pipes.

The valve-loss coefficients are for the fully open condition. Enormous losses occur for a partially open valve. Table 6.6 lists the approximate increase in loss for partially open conditions of gate and globe valves. Of all the fittings, valves, because of their complex geometry, are most sensitive to the manufacturer's design

Table 6.5

RESISTANCE COEFFICIENTS $K = h_m/V^2/2g$ **FOR OPEN VALVES, ELBOWS, AND TEES**

Nominal diameter, in	Screwed				Flanged				
	$\frac{1}{2}$	1	2	4	1	2	4	8	20
Valves (fully open):									
Globe	14	8.2	6.9	5.7	13	8.5	6.0	5.8	5.5
Gate	0.30	0.24	0.16	0.11	0.80	0.35	0.16	0.07	0.03
Swing check	5.1	2.9	2.1	2.0	2.0	2.0	2.0	2.0	2.0
Angle	9.0	4.7	2.0	1.0	4.5	2.4	2.0	2.0	2.0
Elbows:									
45° regular	0.39	0.32	0.30	0.29					
45° long radius					0.21	0.20	0.19	0.16	0.14
90° regular	2.0	1.5	0.95	0.64	0.50	0.39	0.30	0.26	0.21
90° long radius	1.0	0.72	0.41	0.23	0.40	0.30	0.19	0.15	0.10
180° regular	2.0	1.5	0.95	0.64	0.41	0.35	0.30	0.25	0.20
180° long radius					0.40	0.30	0.21	0.15	0.10
Tees:									
Line flow	0.90	0.90	0.90	0.90	0.24	0.19	0.14	0.10	0.07
Branch flow	2.4	1.8	1.4	1.1	1.0	0.80	0.64	0.58	0.41

details. If accurate losses are desired, the manufacturer's data should always be consulted. Lyons' handbook [35] gives design data.

A bend or curve in a pipe, as in Fig. 6.19, always induces a loss larger than the simple Moody friction loss, due to flow separation at the walls and a swirling secondary flow arising from the centripetal acceleration. The loss coefficients K in Fig. 6.19 are for this additional bend loss. The Moody loss due to the axial length of the bend must be computed separately; i.e., the bend length should be added to the pipe length.

As shown in Fig. 6.20, entrance losses are highly dependent upon entrance geometry, but exit losses are not. Sharp edges or protrusions in the entrance cause

Table 6.6

INCREASED LOSSES OF PARTIALLY OPEN VALVES

Condition	Ratio K/K(open condition)	
	Gate value	Globe value
Open	1.0	1.0
Closed, 25%	3.0–5.0	1.5–2.0
50%	12–22	2.0–3.0
75%	70–120	6.0–8.0

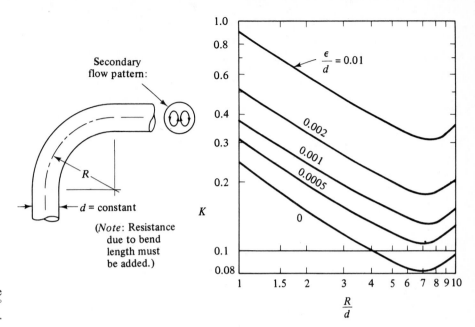

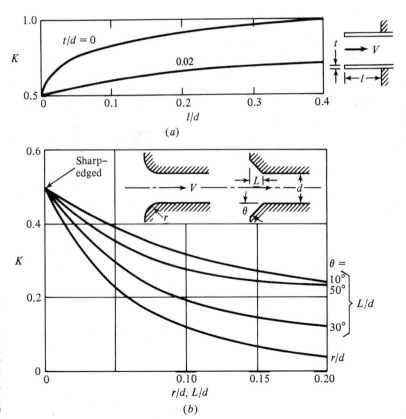

Fig. 6.19 Resistance coefficients for 90° bends.

(Note: Resistance due to bend length must be added.)

Fig. 6.20 Entrance and exit loss coefficients: (a) reentrant inlets; (b) rounded and beveled inlets. Exit losses are $K \approx 1.0$ for all shapes of exit (reentrant, sharp, beveled, or rounded). (From Ref. 37.)

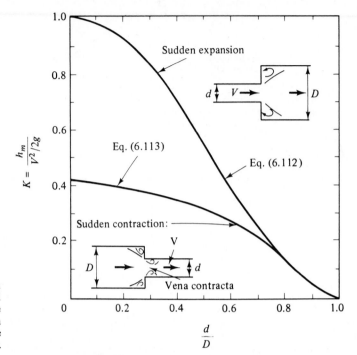

Fig. 6.21 Sudden expansion and contraction losses. Note that the loss is based on velocity head in the small pipe.

large zones of flow separation and large losses. A little rounding goes a long way, and a well-rounded entrance ($r = 0.2d$) has a nearly negligible loss $K = 0.05$. At an exit, on the other hand, the flow simply passes out of the pipe into the large downstream reservoir and loses all its velocity head due to viscous dissipation. Therefore $K = 1.0$ for all submerged exits, no matter how well rounded.

If the entrance is from a finite reservoir, it is termed a *sudden contraction* (SC) between two sizes of pipe. If the exit is to a finite-sized pipe, it is termed a *sudden expansion* (SE). The losses for both are graphed in Fig. 6.21. For the sudden expansion, the shear stress in the corner separated flow, or deadwater region, is negligible, so that a control-volume analysis between the expansion section and the end of the separation zone gives a theoretical loss

$$K_{SE} = \left(1 - \frac{d^2}{D^2}\right)^2 = \frac{h_m}{V^2/2g} \tag{6.112}$$

Note that K is based on the velocity head in the small pipe. Equation (6.112) is in excellent agreement with experiment.

For the sudden contraction, however, flow separation in the downstream pipe causes the main stream to contract through a minimum diameter d_{min} called the *vena contracta*, as sketched in Fig. 6.21. Because the theory of the vena contracta is not well developed, the loss coefficient in the figure for sudden contraction is experimental. It fits the empirical formula

$$K_{SC} \approx 0.42\left(1 - \frac{d^2}{D^2}\right) \tag{6.113}$$

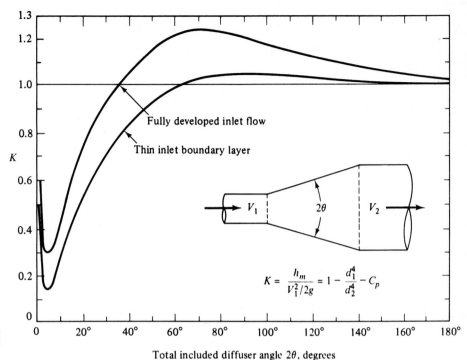

Fig. 6.22 Flow losses in a gradual conical expansion region.

up to the value $d/D = 0.76$, above which it merges into the sudden-expansion prediction, Eq. (6.112).

If the expansion or contraction is gradual, the losses are quite different. Figure 6.22 shows the loss through a gradual conical expansion, usually called a *diffuser* [14]. There is a spread in the data, depending upon the boundary-layer conditions in the upstream pipe. A thinner entrance boundary layer, like the entrance profile in Fig. 6.6, gives a smaller loss. Since a diffuser is intended to raise the static pressure of the flow, diffuser data list the pressure-recovery coefficient of the flow

$$C_p = \frac{p_2 - p_1}{\frac{1}{2}\rho V_1^2} \tag{6.114}$$

The loss coefficient is related to this parameter by

$$K = \frac{h_m}{V^2/2g} = 1 - \frac{d_1^4}{d_2^4} - C_p \tag{6.115}$$

For a given area ratio, the higher the pressure recovery, the lower the loss; hence large C_p means a successful diffuser. From Fig. 6.22 the minimum loss (maximum recovery) occurs for a cone angle 2θ equal to about $5°$. Angles smaller than this give a large Moody-type loss because of their excessive length. For cone angles greater than 40 to $60°$, the loss is so excessive that it would actually be better to use a sudden expansion. This unexpected effect is due to gross flow separation in a wide-angle diffuser, as we shall see soon when we study boundary layers. Reference 14 has extensive data on diffusers.

For a gradual contraction, the loss is very small, as seen from the following experimental values [15]:

Contraction cone angle 2θ, deg	30	45	60
K for gradual contraction	0.02	0.04	0.07

References 15, 16, and 43 contain additional data on minor losses.

EXAMPLE 6.16   Water, $\rho = 1.94$ slugs/ft^3 and $v = 0.000011$ ft^2/s, is pumped between two reservoirs at 0.2 ft^3/s through 400 ft of 2-in-diameter pipe and several minor losses, as shown in Fig. E6.16. The roughness ratio is $\epsilon/d = 0.001$. Compute the pump horsepower required.

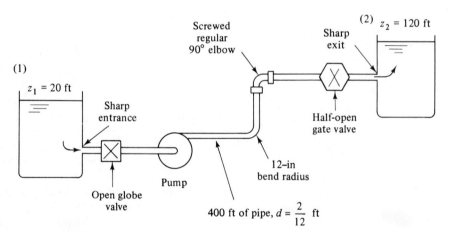

Fig. E6.16

Solution   Write the steady-flow energy equation between sections 1 and 2, the two reservoir surfaces

$$\frac{p_1}{\rho g} + \frac{V_1^2}{2g} + z_1 = \left(\frac{p_2}{\rho g} + \frac{V_2^2}{2g} + z_2\right) + h_f + \sum h_m - h_p$$

where h_p is the head increase across the pump. But since $p_1 = p_2$ and $V_1 = V_2 \approx 0$, solve for the pump head

$$h_p = z_2 - z_1 + h_f + \sum h_m = 120 \text{ ft} - 20 \text{ ft} + \frac{V^2}{2g}\left(\frac{fL}{d} + \sum K\right) \qquad (1)$$

Now with the flow rate known, calculate

$$V = \frac{Q}{A} = \frac{0.2 \text{ ft}^3/\text{s}}{\frac{1}{4}\pi(\frac{2}{12} \text{ ft})^2} = 9.17 \text{ ft/s}$$

Now list and sum the minor loss coefficients:

Loss	K
Sharp entrance (Fig. 6.20)	0.5
Open globe valve (2 in, Table 6.5)	6.9
12-in bend (Fig. 6.19)	0.15
Regular 90° elbow (Table 6.5)	0.95
Half-closed gate valve (0.16 from Table 6.5 times 17 from Table 6.6)	2.7
Sharp exit (Fig. 6.20)	1.0
	$\sum K = 12.2$

Calculate the Reynolds number and pipe-friction factor

$$\text{Re}_d = \frac{Vd}{\nu} = \frac{9.17(\frac{2}{12})}{0.000011} = 139{,}000$$

For $\epsilon/d = 0.001$, from the Moody chart read $f = 0.0216$. Substitute into Eq. (1)

$$h_p = 100 \text{ ft} + \frac{(9.17 \text{ ft/s})^2}{2(32.2 \text{ ft/s}^2)} \left[\frac{0.0216(400)}{\frac{2}{12}} + 12.2 \right]$$

$$= 100 \text{ ft} + 84 \text{ ft} = 184 \text{ ft} \qquad \text{pump head}$$

The pump must provide a power to the water of

$$P = \rho g Q h_p = [1.94(32.2) \text{ lbf/ft}^3](0.2 \text{ ft}^3/\text{s})(184 \text{ ft}) = 2300 \text{ (ft} \cdot \text{lbf)/s}$$

The conversion factor is 1 hp = 550 (ft · lbf)/s. Therefore

$$P = \frac{2300}{550} = 4.2 \text{ hp} \qquad\qquad Ans.$$

Allowing for an efficiency of 70 to 80 percent, a pump is needed with an input of about 6 hp.

6.8 MULTIPLE-PIPE SYSTEMS[1]

If you can solve one pipe, you can solve them all; but when systems contain two or more pipes, certain basic rules make the calculations very smooth. Any resemblance between these rules and the rules for handling electric circuits is not coincidental.

Figure 6.23 shows three examples of multiple-pipe systems. The first is a set of three (or more) pipes in series. Rule 1 is that the flow rate is the same in all pipes

$$Q_1 = Q_2 = Q_3 = \text{const}$$

or

$$V_1 d_1^2 = V_2 d_2^2 = V_3 d_3^2 \qquad\qquad (6.116)$$

---

[1] This section may be omitted without loss of continuity.

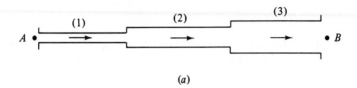

(a)

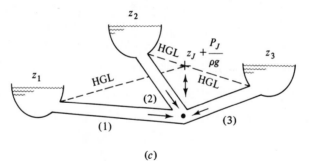

(b)

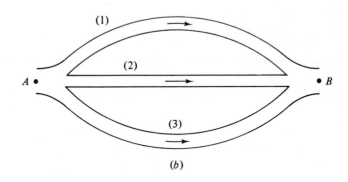

Fig. 6.23 Examples of multiple-pipe systems: (a) pipes in series; (b) pipes in parallel; (c) the three-reservoir junction problem.

(c)

Rule 2 is that the total head loss through the system equals the sum of the head loss in each pipe

$$\Delta h_{A \to B} = \Delta h_1 + \Delta h_2 + \Delta h_3 \tag{6.117}$$

In terms of the friction and minor losses in each pipe, we could rewrite this as

$$\Delta h_{A \to B} = \frac{V_1^2}{2g}\left(\frac{f_1 L_1}{d_1} + \sum K_1\right) + \frac{V_2^2}{2g}\left(\frac{f_2 L_2}{d_2} + \sum K_2\right)$$
$$+ \frac{V_3^2}{2g}\left(\frac{f_3 L_3}{d_3} + \sum K_3\right) \tag{6.118}$$

and so on for any number of pipes in the series. Since V_2 and V_3 are proportional to V_1 from Eq. (6.116), Eq. (6.118) is of the form

$$\Delta h_{A \to B} = \frac{V_1^2}{2g}(\alpha_0 + \alpha_1 f_1 + \alpha_2 f_2 + \alpha_3 f_3) \tag{6.119}$$

where the α_i are dimensionless constants. If the flow rate is given, we can evaluate the right-hand side and hence the total head loss. If the head loss is given, a little iteration is needed, since $f_{1,2,3}$ all depend upon V_1 through the Reynolds number. Begin by calculating $f_{1,2,3}$ assuming fully rough flow, and the solution for V_1 will converge with one or two iterations.

EXAMPLE 6.17 Given a three-pipe series system, as in Fig. 6.23a. The total pressure drop is $p_A - p_B = 150,000$ Pa, and the elevation drop is $z_A - z_B = 5$ m. The pipe data are

Pipe	L, m	d, cm	ϵ, mm	ϵ/d
1	100	8	0.24	0.003
2	150	6	0.12	0.002
3	80	4	0.20	0.005

The fluid is water, $\rho = 1000$ kg/m^3, and $\nu = 1.02 \times 10^{-6}$ m^2/s. Calculate the flow rate Q in cubic meters per hour through the system.

Solution The total head loss across the system is

$$\Delta h_{A \rightarrow B} = \frac{p_A - p_B}{\rho g} + z_A - z_B = \frac{150,000}{1000(9.81)} + 5 \text{ m} = 20.3 \text{ m}$$

From the continuity relation (6.116) the velocities are

$$V_2 = \frac{d_1^2}{d_2^2} V_1 = \tfrac{16}{9} V_1 \qquad V_3 = \frac{d_1^2}{d_3^2} V_1 = 4V_1$$

and
$$\text{Re}_2 = \frac{V_2 d_2}{V_1 d_1} \text{Re}_1 = \tfrac{4}{3} \text{Re}_1 \qquad \text{Re}_3 = 2 \text{ Re}_1$$

Neglecting minor losses and substituting into Eq. (6.118), we obtain

$$\Delta h_{A \rightarrow B} = \frac{V_1^2}{2g} [1250 f_1 + 2500(\tfrac{16}{9})^2 f_2 + 2000(4)^2 f_3]$$

or
$$20.3 \text{ m} = \frac{V_1^2}{2g} (1250 f_1 + 7900 f_2 + 32,000 f_3) \tag{1}$$

This is the form which was hinted at in Eq. (6.119). It seems to be dominated by the third pipe loss $32,000 f_3$. Begin by estimating $f_{1,2,3}$ from the Moody-chart fully rough regime

$$f_1 = 0.0262 \qquad f_2 = 0.0234 \qquad f_3 = 0.0304$$

Substitute in Eq. (1) to find $V_1^2 \approx 2g(20.3)/(33 + 185 + 973)$. The first estimate thus is $V_1 = 0.58$ m/s, from which

$$\text{Re}_1 \approx 45,400 \qquad \text{Re}_2 = 60,500 \qquad \text{Re}_3 = 90,800$$

Hence, from the Moody chart,

$$f_1 = 0.0288 \qquad f_2 = 0.0260 \qquad f_3 = 0.0314$$

Substitution into the Eq. (1) gives the better estimate

$$V_1 = 0.565 \text{ m/s} \qquad Q = \tfrac{1}{4}\pi d_1^2 V_1 = 2.84 \times 10^{-3} \text{ m}^3/\text{s}$$

or
$$Q_1 = 10.2 \text{ m}^3/\text{h} \qquad\qquad\qquad\qquad Ans.$$

A second iteration gives $Q = 10.22 \text{ m}^3/\text{h}$, a negligible change.

The second multiple-pipe system is the *parallel*-flow case shown in Fig. 6.23b. Here the loss is the same in each pipe, and the total flow is the sum of the individual flows

$$\Delta h_{A \to B} = \Delta h_1 = \Delta h_2 = \Delta h_3 \qquad (6.120a)$$

$$Q = Q_1 + Q_2 + Q_3 \qquad (6.120b)$$

If the total head loss is known, it is relatively simple to solve for each Q_i and add them up. The reverse problem of known total flow rate Q requires quite a bit of iteration to determine how this flow splits up among the various pipes. The normal procedure is to guess, say, $Q_1 = Q/3$, and compute the head loss and hence Q_2 and Q_3 from Eq. (6.120a). Then, if the sum is incorrect, such as $Q_1 + Q_2 + Q_3 = 1.14Q$, scale down the original guess to $Q_{1,\text{new}} = Q_{1,\text{old}}/1.14$ and recompute Q_2 and Q_3 and test the sum again. Scale Q_1 up or down again if necessary. The process converges.

EXAMPLE 6.18

Assume that the same three pipes in Example 6.17 are now in parallel with the same total head loss of 20.3 m. Compute the total flow rate Q, neglecting minor losses.

Solution

From Eq. (6.120a) we can solve for each V separately

$$20.3 \text{ m} = \frac{V_1^2}{2g} 1250 f_1 = \frac{V_2^2}{2g} 2500 f_2 = \frac{V_3^2}{2g} 2000 f_3 \qquad (1)$$

Guess fully rough flow in pipe 1: $f_1 = 0.0262$, $V_1 = 3.49$ m/s; hence $\text{Re}_1 = V_1 d_1/\nu = 273{,}000$. From the Moody chart read $f_1 = 0.0267$; recompute $V_1 = 3.46$ m/s, $Q_1 = 62.5 \text{ m}^3/\text{h}$. (This problem can also be solved from Fig. 6.14.)

Next guess pipe 2: $f_2 \approx 0.0234$, $V_2 \approx 2.61$ m/s; then $\text{Re}_2 = 153{,}000$, and hence $f_2 = 0.0246$, $V_2 = 2.55$ m/s, $Q_2 = 25.9 \text{ m}^3/\text{h}$.

Finally guess pipe 3: $f_3 \approx 0.0304$, $V_3 \approx 2.56$ m/s; then $\text{Re}_3 = 100{,}000$, and hence $f_3 = 0.0313$, $V_3 = 2.52$ m/s, $Q_3 = 11.4 \text{ m}^3/\text{h}$.

This is satisfactory convergence. The total flow rate is

$$Q = Q_1 + Q_2 + Q_3 = 62.5 + 25.9 + 11.4 = 99.8 \text{ m}^3/\text{h} \qquad Ans.$$

These three pipes carry 10 times more flow in parallel than in series.

Consider the third example of a *three-reservoir pipe junction*, as in Fig. 6.23c. If all flows are considered positive toward the junction, then

$$Q_1 + Q_2 + Q_3 = 0 \qquad (6.121)$$

which obviously implies that one or two of the flows must be away from the junction. The pressure must change through each pipe so as to give the same static pressure p_J at the junction. In other words, let the HGL at the junction have the elevation

$$h_J = z_J + \frac{p_J}{\rho g} \qquad (6.122)$$

where p_J is in gage pressure for simplicity. Then the head loss through each, assuming $p_1 = p_2 = p_3 = 0$ (gage) at each reservoir surface, must be such that

$$\Delta h_1 = \frac{V_1^2}{2g} \frac{f_1 L_1}{d_1} = z_1 - h_J$$

$$\Delta h_2 = \frac{V_2^2}{2g} \frac{f_2 L_2}{d_2} = z_2 - h_J \qquad (6.123)$$

$$\Delta h_3 = \frac{V_3^2}{2g} \frac{f_3 L_3}{d_3} = z_3 - h_J$$

We guess the position h_J and solve Eqs. (6.123) for $V_{1,2,3}$ and hence $Q_{1,2,3}$, iterating until the flow rates balance at the junction according to Eq. (6.121). If we guess h_J too *high*, the sum $Q_1 + Q_2 + Q_3$ will be *negative* and the remedy is to reduce h_J, and vice versa.

EXAMPLE 6.19   Take the same three pipes as in Example 6.17 and assume that they connect three reservoirs at these surface elevations

$$z_1 = 20 \text{ m} \qquad z_2 = 100 \text{ m} \qquad z_3 = 40 \text{ m}$$

Find the resulting flow rates in each pipe, neglecting minor losses.

Solution   As a first guess, take h_J equal to the middle reservoir height, $z_3 = h_J = 40$ m. This saves one calculation ($Q_3 = 0$) and enables us to get the lay of the land:

Reservoir	h_J, m	$z_i - h_J$, m	f_i	V_i, m/s	Q_i, m³/h	L_i/d_i
1	40	− 20	0.0267	− 3.43	− 62.1	1250
2	40	60	0.0241	4.42	45.0	2500
3	40	0		0	0	2000
					$\sum Q = -17.1$	

Since the sum of the flow rates toward the junction is negative, we guessed h_J too high. Reduce h_J to 30 m and repeat:

Reservoir	h_J, m	$z_i - h_J$, m	f_i	V_i, m/s	Q_i, m³/h
1	30	− 10	0.0269	− 2.42	− 43.7
2	30	70	0.0241	4.78	48.6
3	30	10	0.0317	1.76	8.0
				$\sum Q =$	12.9

This is positive $\sum Q$, and so we can linearly interpolate to get an accurate guess: $h_J \approx 34.3$ m. Make one final list:

Reservoir	h_J, m	$z_i - h_J$, m	f_i	V_i, m/s	Q_i, m³/h
1	34.3	−14.3	0.0268	−2.90	−52.4
2	34.3	65.7	0.0241	4.63	47.1
3	34.3	5.7	0.0321	1.32	6.0
				$\sum Q =$	0.7

This is close enough; hence we calculate that the flow rate is 52.4 m³/h toward reservoir 3, balanced by 47.1 m³/h away from reservoir 1 and 6.0 m³/h away from reservoir 3.

One further iteration with this problem would give $h_J = 34.53$ m, resulting in $Q_1 = -52.8$, $Q_2 = 47.0$, and $Q_3 = 5.8$ m³/h, so that $\sum Q = 0$ to three-place accuracy. Pedagogically speaking, we would then be exhausted.

The ultimate case of a multipipe system is the *piping network* illustrated in Fig. 6.24. This might represent a water-supply system for an apartment or subdivision or even a city. This network is quite complex algebraically but follows the same basic rules:

1. The net flow into any junction must be zero.
2. The net head loss around any closed loop must be zero. In other words, the HGL at each junction must have one and only one elevation.
3. All head losses must satisfy the Moody and minor-loss friction correlations.

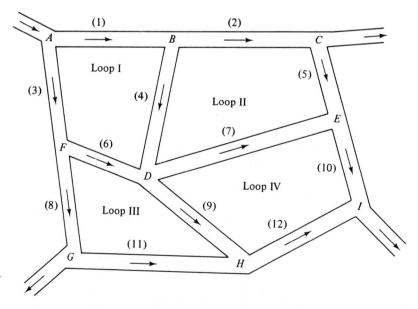

Fig. 6.24 Schematic of a piping network.

By applying these rules to each junction and independent loop in the network one obtains a set of simultaneous algebraic equations for the flow rates in each pipe leg and the HGL at each junction. Since the equations are not linear, the solution is obtained by numerical iteration, as in Example 6.19. The earliest iteration technique was developed by Cross in 1936 [17]. The method is explained in detail with a computer program for its use in Ref. 18, pp. 565–581. Several other, fast and improved, methods have been developed for network solution on a digital computer [e.g., 19]. This work is beyond the scope and intent of the present text.

6.9 EXPERIMENTAL DUCT FLOWS; DIFFUSER PERFORMANCE

The Moody chart is such a great correlation for tubes of any cross section with any roughness or flow rate that we may be deluded into thinking that the world of internal-flow prediction is at our feet. Not so. The theory is reliable only for ducts of constant cross section. As soon as the section varies, we must rely principally upon experiment to determine the flow properties. As mentioned many times before, experiment is a vital part of fluid mechanics.

Literally thousands of papers in the literature report experimental data for specific internal and external viscous flows. We have already seen several examples:

1. Vortex shedding from a cylinder (Fig. 5.2)
2. Drag of a sphere and a cylinder (Fig. 5.3)
3. Hydraulic model of an estuary (Fig. 5.10)
4. Rough-wall pipe flows (Fig. 6.12)
5. Secondary flow in ducts (Fig. 6.18)
6. Minor duct-loss coefficients (Sec. 6.7)

Chapter 7 will treat a great many more external-flow experiments, especially in Sec. 7.5. Here we shall show data for one type of internal flow, the diffuser.

Diffuser Performance

A diffuser, shown in Fig. 6.25a and b, is an expansion or area increase intended to reduce velocity in order to recover the pressure head of the flow. Rouse and Ince [6] relate that it may have been invented by customers of the early Roman (about A.D. 100) water-supply system, where water flowed continuously and was billed according to pipe size. The ingenious customers discovered that they could increase the flow rate at no extra cost by flaring the outlet section of the pipe.

Engineers have always designed diffusers to increase pressure and reduce kinetic energy of ducted flows, but until about 1950, diffuser design was a combination of art, luck, and vast amounts of empiricism. Small changes in design parameters caused large changes in performance. The Bernoulli equation seemed highly suspect as a useful tool.

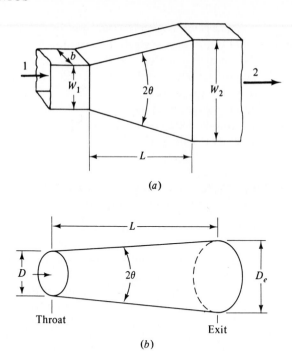

(a)

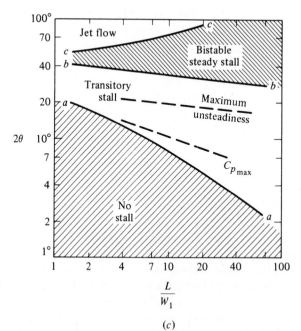

Fig. 6.25 Diffuser geometry and typical flow regimes: (a) geometry of a flat-walled diffuser; (b) geometry of a conical diffuser; (c) flat-diffuser stability map. (*From Ref. 14, by permission of Creare, Inc.*)

Neglecting losses and gravity effects, the incompressible Bernoulli equation predicts that

$$p + \tfrac{1}{2}\rho V^2 = p_0 = \text{const} \tag{6.124}$$

where p_0 is the stagnation pressure which the fluid would achieve if the fluid were slowed to rest ($V = 0$) without losses.

The basic output of a diffuser is the pressure-recovery coefficient C_p, defined as

$$C_p = \frac{p_e - p_t}{p_{0t} - p_t} \tag{6.125}$$

where subscripts e and t mean the exit and the throat (or inlet), respectively. Higher C_p means better performance.

Consider the flat-walled diffuser in Fig. 6.25a, where section 1 is the inlet and section 2 the exit. Application of Bernoulli's equation (6.124) to this diffuser predicts that

$$p_{01} = p_1 + \tfrac{1}{2}\rho V_1^2 = p_2 + \tfrac{1}{2}\rho V_2^2 = p_{02}$$

or
$$C_{p,\text{frictionless}} = 1 - \left(\frac{V_2}{V_1}\right)^2 \tag{6.126}$$

Meanwhile, steady one-dimensional continuity would require that

$$Q = V_1 A_1 = V_2 A_2 \tag{6.127}$$

Combining (6.126) and (6.127), we can write the performance in terms of the area ratio $\text{AR} = A_2/A_1$, which is a basic parameter in diffuser design:

$$C_{p,\text{frictionless}} = 1 - (\text{AR})^{-2} \tag{6.128}$$

A typical design would have $\text{AR} = 5:1$, for which Eq. (6.128) predicts $C_p = 0.96$, or nearly full recovery. But in fact measured values of C_p for this area ratio [14] are only as high as 0.86 and can be as low as 0.24.

The basic reason for the discrepancy is flow separation, as sketched in Fig. 6.26. The increasing pressure in the diffuser is an unfavorable gradient (Sec. 7.4), which causes the viscous boundary layers to break away from the walls and greatly reduces the performance. We are just beginning to develop theories which predict this behavior (see, for example, Ref. 20).

As an added complication to boundary-layer separation, the flow patterns in a diffuser are highly variable and were considered mysterious and erratic until 1955, when Kline revealed the structure of these patterns with flow-visualization techniques in a simple water channel.

A complete *stability map* of diffuser flow patterns was published in 1962 by Fox and Kline [21], as shown in Fig. 6.25c. There are four basic regions. Below the line *aa* there is steady viscous flow, no separation, and moderately good performance. Note that even a very short diffuser will separate, or stall, if its half-angle is greater than 10°.

Between lines *aa* and *bb* is a transitory stall pattern with strongly unsteady flow. Best performance, i.e., highest C_p, occurs in this region. The third pattern, between

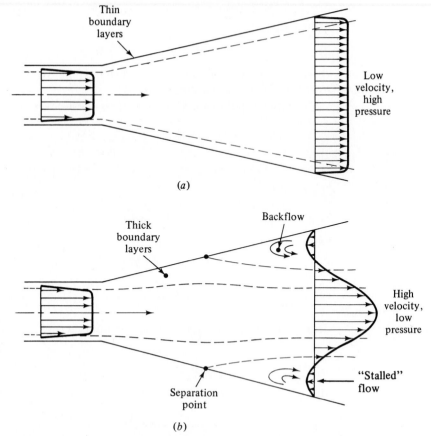

Thin
boundary
layers

Low
velocity,
high
pressure

(a)

Thick
boundary
layers

Backflow

High
velocity,
low
pressure

"Stalled"
flow

Separation
point

(b)

Fig. 6.26 Diffuser performance: (a) ideal pattern with good performance; (b) actual measured pattern with boundary-layer separation and resultant poor performance.

bb and *cc*, is steady bistable stall from one wall only. The stall pattern may flip-flop from one wall to the other, and performance is poor.

The fourth pattern, above line *cc*, is *jet flow*, where the wall separation is so gross and pervasive that the main stream ignores the walls and simply passes on through at nearly constant area. Performance is extremely poor in this region.

Dimensional analysis of a flat-walled or conical diffuser shows that C_p should depend upon the following parameters:

1. Any two of the following geometric parameters:
   a. Area ratio $AR = A_2/A_1$ or $(D_e/D)^2$
   b. Divergence angle 2θ
   c. Slenderness L/W_1 or L/D
2. Inlet Reynolds number $Re_t = V_1 W_1/\nu$ or $V_1 D/\nu$
3. Inlet Mach number $Ma_t = V_1/a_1$
4. Inlet boundary-layer *blockage factor* $B_t = A_{BL}/A_1$ where A_{BL} is the wall area blocked, or displaced, by the retarded boundary-layer flow in the inlet (typically B_t varies from 0.03 to 0.12)

A flat-walled diffuser would require an additional shape parameter to describe its cross section:

5. Aspect ratio $AS = b/W_1$

Even with this formidable list, we have omitted five possible important effects: inlet turbulence, inlet swirl, inlet profile vorticity, superimposed pulsations, and downstream obstructions, all of which occur in practical machinery applications.

The three most important parameters are AR, θ, and B_t. Typical performance maps for diffusers are shown in Fig. 6.27. For this case of 8 to 9 percent blockage, both the flat-walled and conical types give about the same maximum performance, $C_p = 0.70$, but at different divergence angles (9° flat versus 4.5° conical). Both types fall far short of the Bernoulli estimates of $C_p = 0.93$ (flat) and 0.99 (conical), primarily because of the blockage effect.

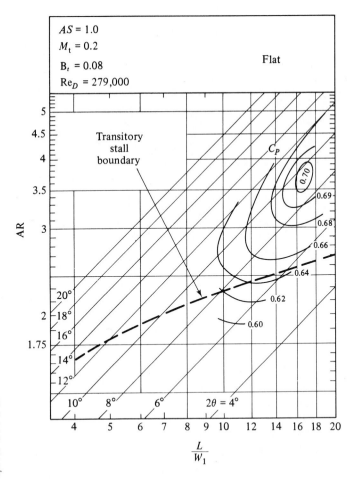

Fig. 6.27 Typical performance maps for flat-wall and conical diffusers at similar operating conditions: (a) flat wall. (*From Ref. 14, by permission of Creare, Inc.*)

(a)

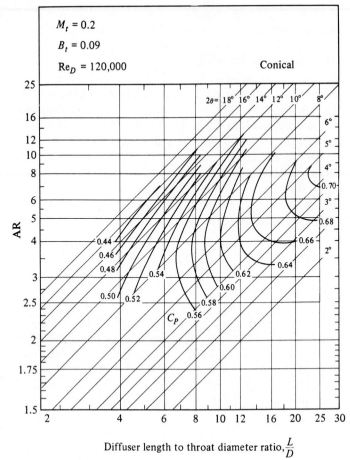

$M_t = 0.2$

$B_t = 0.09$

$Re_D = 120,000$

Conical

Fig. 6.27 Typical
performance maps for
flat-wall and conical
diffusers at similar
operating conditions:
(*b*) conical wall. (*From
Ref. 14, by permission of
Creare, Inc.*)

Diffuser length to throat diameter ratio, $\dfrac{L}{D}$

(*b*)

Table 6.7

MAXIMUM DIFFUSER-PERFORMANCE DATA [14]

Inlet blockage B_t	Flat-walled		Conical	
	$C_{p,\,\text{max}}$	L/W_1	$C_{p,\,\text{max}}$	L/D
0.02	0.86	18	0.83	20
0.04	0.80·	18	0.78	22
0.06	0.75	19	0.74	24
0.08	0.70	20	0.71	26
0.10	0.66	18	0.68	28
0.12	0.63	16	0.65	30

From the data of Ref. 14 we can determine that, in general, performance decreases with blockage and is approximately the same for both flat-walled and conical diffusers, as shown in Table 6.7. In all cases, the best conical diffuser is 10 to 80 percent longer than the best flat-walled design. Therefore, if length is limited in the design, the flat-walled design will show the better performance.

The experimental design of a diffuser is an excellent example of a successful attempt to minimize the undesirable effects of adverse pressure gradient and flow separation.

6.10 FLUID METERS

Almost all practical fluids engineering problems are associated with the need for an accurate flow measurement. There is a need to measure *local* properties (velocity, pressure, temperature, density, viscosity, turbulent intensity), *integrated* properties (mass flow and volume flow), and *global* properties (visualization of the entire flow field). We shall concentrate in this section on velocity and volume flow measurements.

We have discussed pressure measurement in Sec. 2.10. Measurement of other thermodynamic properties, such as density, temperature, and viscosity, are beyond the scope of this text and are treated in specialized books such as Refs. 22 and 23. Global visualization techniques were discussed in Sec. 1.7 for low-speed flows, and the special optical techniques used in high-speed flows are treated in Ref. 18 of Chap. 1. Flow-measurement schemes suitable for open-channel and other free-surface flows are treated in Chap. 10.

Local-Velocity Measurements

Velocity averaged over a small region, or point, can be measured by several different physical principles, listed in order of increasing complexity and sophistication:

1. Trajectory of floats or neutrally buoyant particles
2. Rotating mechanical devices
   a. Cup anemometer
   b. Savonius rotor
   c. Propeller meter
   d. Turbine meter
3. Pitot tube (Fig. 6.29)
4. Electromagnetic current meter
5. Hot wires and hot films
6. Laser-doppler anemometer

Some of these meters are sketched in Fig. 6.28

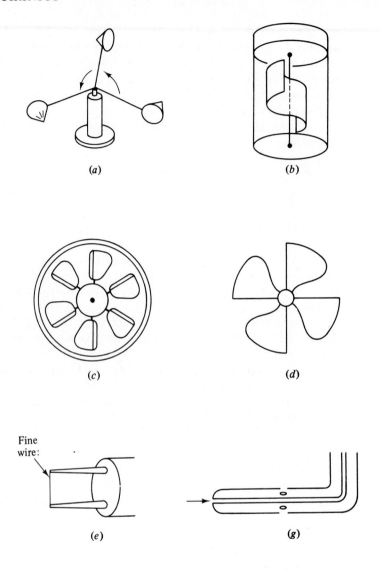

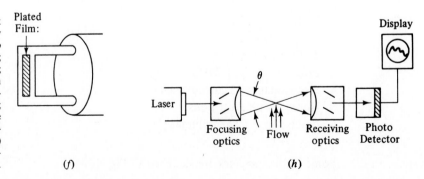

Fig. 6.28 Eight common velocity meters: (*a*) three-cup anemometer; (*b*) Savonius rotor; (*c*) turbine mounted in a duct; (*d*) free-propeller meter; (*e*) hot-wire anemometer; (*f*) hot-film anemometer; (*g*) pitot tube; (*h*) laser-doppler anemometer.

Floats or buoyant particles A simple but effective estimate of flow velocity can be found from visible particles entrained in the flow. Examples would be flakes on the surface of a channel flow, small neutrally buoyant spheres mixed with a liquid, or hydrogen bubbles. Sometimes gas flows can be estimated from the motion of entrained dust particles. One must establish whether the particle motion truly simulates the fluid motion. Floats are commonly used to track the movement of ocean waters and can be designed to move at the surface, along the bottom, or at any given depth [24]. Many official tidal-current charts [25] were obtained by releasing and timing a floating spar attached to a length of string. One can release whole groups of spars to determine a flow pattern.

Rotating sensors The rotating devices of Fig. 6.28a to d can be used in either gases or liquids, and their rotation rate is approximately proportional to the flow velocity. The cup anemometer (Fig. 6.28a) and Savonius rotor (Fig. 6.28b) always rotate the same way, regardless of flow direction. They are popular in atmospheric and oceanographic applications and can be fitted with a direction vane to align themselves with the flow. The ducted-propeller (Fig. 6.28c) and free-propeller (Fig. 6.28d) meters must be aligned with the flow parallel to their axis of rotation. They can sense reverse flow because they will then rotate in the opposite direction. All these rotating sensors can be attached to counters or sensed by electromagnetic or slip-ring devices for either a continuous or digital reading of flow velocity. All have the disadvantage of being relatively large and thus not representing a "point."

Pitot tube A slender tube aligned with the flow (Figs. 6.28g and 6.29) can measure local velocity by means of a pressure difference. It has side-wall holes to measure the static pressure p_s in the moving stream and a hole in the front to measure the *stagnation* pressure p_0, where the stream is decelerated to zero velocity. Instead of measuring p_0 or p_s separately, it is customary to measure their difference with, say, a manometer, as in Fig. 6.29.

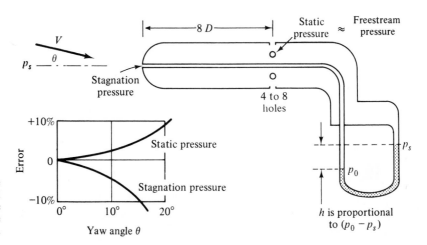

Fig. 6.29 Pitot tube for combined measurement of static and stagnation pressure in a moving stream.

If $Re_D > 1000$, where D is the probe diameter, the flow around the probe is nearly frictionless and Bernoulli's relation, Eq. (3.63), applies with good accuracy. For incompressible flow

$$p_s + \tfrac{1}{2}\rho V^2 + \rho g z_s \approx p_0 + \tfrac{1}{2}\rho(0)^2 + \rho g z_0$$

Assuming that the elevation pressure difference $\rho g(z_s - z_0)$ is negligible, this reduces to

$$V \approx \left[\frac{2(p_0 - p_s)}{\rho}\right]^{1/2} \tag{6.129}$$

This is the *Pitot formula*, named after a French engineer who designed the device in 1732.

The primary disadvantage of the pitot tube is that it must be aligned with the flow direction, which may be unknown. For yaw angles greater than 5°, there are substantial errors in both the p_0 and p_s measurements, as shown in Fig. 6.29. The pitot tube is useful in liquids and gases; for gases a compressibility correction is necessary if the stream Mach number is high (Chap. 9). Because of the slow response of the fluid-filled tubes leading to the pressure sensors, it is not useful for unsteady-flow measurements. It does resemble a point and can be made small enough to measure, for example, blood flow in arteries and veins. It is not suitable for low-velocity measurement in gases because of the small pressure differences developed. For example, if $V = 1$ ft/s in standard air, from Eq. (6.129) we compute $p_0 - p$ equal to only 0.001 lbf/ft² (0.048 Pa). This is beyond the resolution of most pressure gages.

Electromagnetic meter   If a magnetic field is applied across a conducting fluid, the fluid motion will induce a voltage across two electrodes placed in or near the flow. The electrodes can be streamlined or built into the wall and cause little or no flow resistance. The output is very strong for highly conducting fluids such as liquid metals. Seawater also gives good output, and electromagnetic current meters are in common use in oceanography. Even low-conductivity fresh water can be measured by amplifying the output and insulating the electrodes. Commercial instruments are available for most liquid flows but are relatively costly. Electromagnetic flow meters are treated in Ref. 26.

Hot-wire anemometer   A very fine wire ($d = 0.01$ mm or less) heated between two small probes, as in Fig. 6.28e, is ideally suited to measure rapidly fluctuating flows such as the turbulent boundary layer. The idea dates back to work by L. V. King in 1914 on heat loss from long thin cylinders. If electric power is supplied to heat the cylinder, the loss varies with flow velocity across the cylinder according to *King's law*

$$q = I^2 R \approx a + b(\rho V)^n \tag{6.130}$$

where $n \approx \tfrac{1}{3}$ at very low Reynolds numbers and equals $\tfrac{1}{2}$ at high Reynolds numbers. The hot wire normally operates in the high-Reynolds-number range but should be calibrated in each situation to find the best fit a, b, and n. The wire can be operated either at constant current I, so that the resistance R is a measure of V, or at constant resistance R (constant temperature), with I a measure of velocity. In either case, the output is a nonlinear function of V, and the equipment should contain a *linearizer* to produce convenient velocity data. Many varieties of commercial hot-wire equipment are available, as are do-it-yourself designs [27]. Excellent detailed discussions of the hot wire are given in Refs. 1 and 28.

Because of its frailty, the hot wire is not suited to liquid flows, whose high density and entrained sediment will knock the wire right off. A more stable yet quite sensitive alternative for liquid-flow measurement is the hot-film anemometer, Fig. 6.28*f*. A thin metallic film, usually platinum, is plated onto a relatively thick support which can be a wedge, a cone, or a cylinder. The operation is similar to the hot wire. The cone gives best reponse but is liable to error when the flow is yawed to its axis.

Hot wires can easily be arranged in groups to measure two- and three-dimensional velocity components.

Laser-doppler anemometer   In the LDA a laser beam provides highly focused, coherent monochromatic light which is passed through the flow. When this light is scattered from a moving particle in the flow, a stationary observer can detect a change, or *doppler shift*, in the frequency of the scattered light. The shift Δf is proportional to the velocity of the particle. There is essentially zero disturbance of the flow by the laser.

Figure 6.28*h* shows the popular dual-beam mode of LDA. A focusing device splits the laser into two beams, which cross the flow at an angle θ. Their intersection, which is the measuring volume or resolution of the measurement, resembles an ellipsoid about 0.5 mm wide and 0.1 mm in diameter. Particles passing through this measuring volume scatter the beams; they then pass through receiving optics to a photodetector which converts the light into an electric signal. A signal processor then converts electric frequency into a voltage which can either be displayed or stored. If λ is the wavelength of the laser light, the measured velocity is given by

$$V = \frac{\lambda \, \Delta f}{2 \sin (\theta/2)} \tag{6.131}$$

Multiple components of velocity can be detected by using more than one photodetector and other operating modes. Either liquids or gases can be measured as long as scattering particles are present. In liquids normal impurities serve as scatterers, but gases may have to be seeded. The particles may be as small as the wavelength of the light. Although the measuring volume is not as small as with a hot wire, the LDA is capable of measuring turbulent fluctuations.

The advantages of the LDA are as follows:

1. No disturbance of the flow
2. High spatial resolution of the flow field
3. Velocity data that are independent of the fluid thermodynamic properties
4. An output voltage that is linear with velocity
5. No need for calibration

The disadvantages are that both the apparatus and the fluid must be transparent to light and the cost is high (a basic system shown in Fig. 6.28h begins at about $20,000).

Once installed, an LDA can map the entire flow field in minutest detail. To truly appreciate the power of the LDA one should examine for example the amazingly detailed three-dimensional flow profiles measured by Eckardt [29] in a high-speed centrifugal compressor impeller. Extensive discussions of laser velocimetry are given in Refs. 38 and 39.

EXAMPLE 6.20 The pitot tube of Fig. 6.29 uses mercury as a manometer fluid. When placed in a water flow, the manometer-height reading is $h = 8.4$ in. Neglecting yaw and other errors, what is the flow velocity V in feet per second?

Solution From the two-fluid manometer relation (2.33), with $z_A = z_2$, the pressure difference is related to h by

$$p_0 - p_s = (\rho_M g - \rho_w g)h$$

Taking the specific weights of mercury and water from Table 2.1, we have

$$p_0 - p_s = (846 - 62.4 \text{ lbf/ft}^3)\, \frac{8.4}{12} \text{ ft} = 549 \text{ lbf/ft}^2$$

The density of water is $62.4/32.2 = 1.94$ slugs/ft^3. Introducing these values into the pitot formula (2.75), we obtain

$$V = \left[\frac{2(549 \text{ lbf/ft}^2)}{1.94 \text{ slugs/ft}^3}\right]^{1/2} = 23.8 \text{ ft/s} \qquad\qquad Ans.$$

Since this is a low-speed flow, no compressibility correction is needed.

Volume-Flow Measurements

It is often desirable to measure the integrated mass, or volume flow, passing through a duct. Accurate measurement of flow is vital when billing customers for a given amount of liquid or gas passing through a duct. The different devices available to make these measurements are discussed in great detail in the ASME text on fluid meters [30]. These devices split into two classes, mechanical instruments and head-loss instruments.

The mechanical instruments measure actual mass or volume of fluid by trapping it and counting it. The various types of measurement are

1. Mass measurement
   a. Weighing tanks
   b. Tilting traps
2. Volume measurement
   a. Volume tanks
   b. Reciprocating pistons
   c. Rotating slotted rings

d. Nutating disk

e. Sliding vanes

f. Gear or lobed impellers

g. Reciprocating bellows

h. Sealed-drum compartments

The last three of these are suitable for gas-flow measurement.

The head-loss devices obstruct the flow and cause a pressure drop which is a measure of flux:

1. Bernoulli-type devices

  a. Thin-plate orifice

  b. Flow nozzle

  c. Venturi tube

2. Friction-loss devices

  a. Capillary tube

  b. Porous plug

The friction-loss meters cause a large nonrecoverable head loss and obstruct the flow too much to be generally useful.

Three other widely used meters operate on different physical principles:

1. Turbine meter

2. Vortex meter

3. Ultrasonic flowmeter

Turbine meter   The turbine meter, sometimes called a propeller meter, is a freely rotating propeller which can be installed in a pipeline. A typical design is shown in Fig. 6.30a. There are flow straighteners upstream and downstream of the rotor, and the rotation is measured by electric or magnetic pickup of pulses caused by passage of a point on the rotor. The rotor rotation is approximately proportional to the volume flow in the pipe.

A major advantage of the turbine meter is that each pulse corresponds to a finite incremental volume of fluid, and the pulses are digital and can easily be summed. Liquid-flow turbine meters have as few as two blades and produce a constant number of pulses per unit fluid volume over a 5:1 flow-rate range with ± 0.25 percent accuracy. Gas meters need many blades to produce sufficient torque and are accurate to ± 1 percent.

Since turbine meters are very individualistic, flow calibration is an absolute necessity. A typical liquid-meter calibration curve is shown in Fig. 6.30b. Researchers attempting to establish universal calibration curves have met with little practical success as a result of manufacturing variabilities.

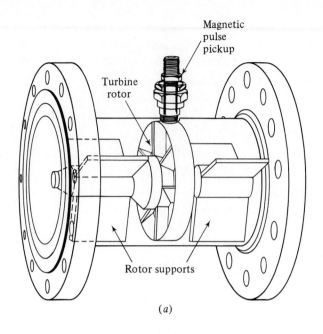

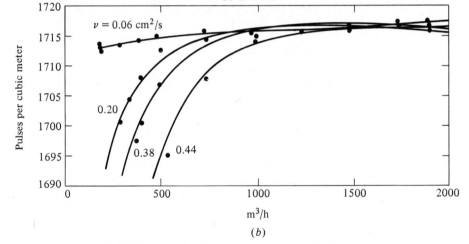

Fig. 6.30 The turbine meter widely used in the oil, gas, and water-supply industries: (*a*) basic design; (*b*) typical calibration curve for a range of crude oils. (*Daniel Industries, Inc., Flow Products Division.*)

Vortex flowmeters Recall from Fig. 5.2 that a bluff body placed in a uniform crossflow sheds alternating vortices at a nearly uniform Strouhal number, St = fL/U, where U is the approach velocity and L a characteristic body width. Since L and St are constant, this means that the shedding frequency is proportional to velocity

$$f = (\text{const})(LU) \tag{6.132}$$

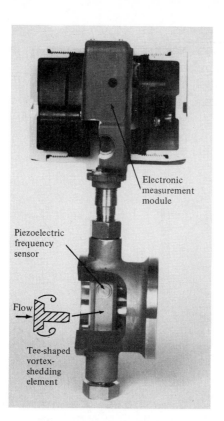

Electronic
measurement
module

Piezoelectric
frequency
sensor

Flow

Tee-shaped
vortex-
shedding
element

Fig. 6.31 A vortex
flowmeter. (*The
Foxboro Company.*)

The vortex meter introduces a shedding element across a pipe flow and picks up the
shedding frequency downstream with a pressure, ultrasonic, or heat-transfer type
of sensor. A typical design is shown in Fig. 6.31.

The advantages of a vortex meter are as follows:

1. Absence of moving parts
2. Accuracy to ± 1 percent over a wide flow-rate range (up to $100:1$)
3. Ability to handle very hot or very cold fluids
4. Requirement of only a short pipe length
5. Calibration insensitive to fluid density or viscosity

For further details see Ref. 40.

Ultrasonic flowmeters　The sound-wave analog of the laser velocimeter of Fig.
6.28*h* is the ultrasonic flowmeter. Two examples are shown in Fig. 6.32. The pulse-
type flowmeter is shown in Fig. 6.32*a*. Upstream piezoelectric transducer A is
excited with a short sonic pulse which propagates across the flow to downstream
transducer B. The arrival at B triggers another pulse to be created at A, resulting in
a regular pulse frequency f_A. The same process is duplicated in the reverse
direction from B to A, creating frequency f_B. The difference $f_A - f_B$ is proportional

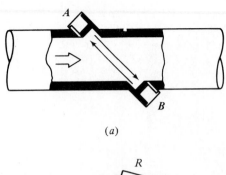

(a)

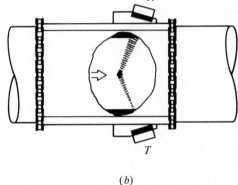

Fig. 6.32 Ultrasonic flowmeters: (*a*) pulse type; (*b*) doppler shift type. (*From Ref. 41.*)

(b)

to the flow rate. Figure 6.32*b* shows a doppler-type arrangement, where sound waves from transmitter *T* are scattered by particles or contaminants in the flow to receiver *R*. Comparison of the two signals reveals a doppler frequency shift which is proportional to flow rate. Ultrasonic meters are nonintrusive and can be directly attached to pipe flows in the field. Their quoted uncertainty of ± 1 to 2 percent can rise to ± 5 percent or more due to irregularities in velocity profile, fluid temperature, or Reynolds number. For further details see Ref. 41.

Bernoulli obstruction theory   Consider the generalized flow obstruction shown in Fig. 6.33. The flow in the basic duct of diameter *D* is forced through an obstruction of diameter *d*; the β ratio of the device is a key parameter

$$\beta = \frac{d}{D} \tag{6.133}$$

After leaving the obstruction the flow may neck down even more through a vena contracta of diameter $D_2 < d$, as shown. Apply the Bernoulli and continuity equations for incompressible steady frictionless flow to estimate the pressure change:

Continuity:
$$Q = \frac{\pi}{4} D^2 V_1 = \frac{\pi}{4} D_2^2 V_2$$

Bernoulli:
$$p_0 = p_1 + \tfrac{1}{2}\rho V_1^2 = p_2 + \tfrac{1}{2}\rho V_2^2$$

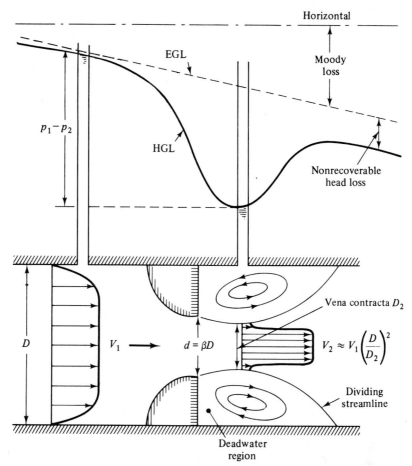

Fig. 6.33 Velocity and pressure change through a generalized Bernoulli obstruction meter.

Eliminating V_1, we solve these for V_2 or Q in terms of the pressure change $p_1 - p_2$

$$\frac{Q}{A_2} = V_2 \approx \left[\frac{2(p_1 - p_2)}{\rho(1 - D_2^4/D^4)} \right]^{1/2} \tag{6.134}$$

But this is surely inaccurate because we have neglected friction in a duct flow, where we know friction will be very important. Nor do we want to get into the business of measuring vena contracta ratios D_2/d for use in (6.134). Therefore we assume that $D_2/D \approx \beta$ and then calibrate the device to fit the relation

$$Q = A_t V_t = C_d A_t \left[\frac{2(p_1 - p_2)/\rho}{1 - \beta^4} \right]^{1/2} \tag{6.135}$$

where subscript t denotes the throat of the obstruction. The dimensionless *discharge coefficient* C_d accounts for the discrepancies in the approximate analysis. By dimensional analysis for a given design we expect

$$C_d = f(\beta, \mathrm{Re}_D) \qquad \text{where } \mathrm{Re}_D = \frac{V_1 D}{\nu} \tag{6.136}$$

The geometric factor involving β in (6.135) is called the *velocity-of-approach factor*

$$E = (1 - \beta^4)^{-1/2} \tag{6.137}$$

One can also group C_d and E in Eq. (6.135) to form the dimensionless *flow coefficient* α

$$\alpha = C_d E = \frac{C_d}{(1 - \beta^4)^{1/2}} \tag{6.138}$$

Thus Eq. (6.135) can be written in the equivalent form

$$Q = \alpha A_t \left[\frac{2(p_1 - p_2)}{\rho} \right]^{1/2} \tag{6.139}$$

Obviously the flow coefficient is correlated in the same manner:

$$\alpha = f(\beta, \text{Re}_D) \tag{6.140}$$

Occasionally one uses the throat Reynolds number instead of the approach Reynolds number

$$\text{Re}_d = \frac{V_t d}{\nu} = \frac{\text{Re}_D}{\beta} \tag{6.141}$$

Since the design parameters are assumed known, the correlation of α from Eq. (6.140) or of C_d from Eq. (6.136) is the desired solution to the fluid-metering problem.

The mass flow is related to Q by

$$\dot{m} = \rho Q \tag{6.142}$$

and is thus correlated by exactly the same formulas.

Figure 6.34 shows the three basic devices recommended for use by the International Organization for Standardization (ISO) [31], the orifice, nozzle, and venturi tube.

Thin-plate orifice The thin-plate orifice, Fig. 6.34b, can be made with β in the range of 0.2 to 0.8, except that the hole diameter d should not be less than 12.5 mm. To measure p_1 and p_2 three types of tappings are commonly used:

1. Corner taps where the plate meets the pipe wall
2. $D : \frac{1}{2}D$ taps: pipe-wall taps at D upstream and $\frac{1}{2}D$ downstream
3. Flange taps: 1 in (25 mm) upstream and 1 in (25 mm) downstream of the plate, regardless of the size D

Types 1 and 2 approximate geometric similarity, but since the flange taps 3 do not, they must be correlated separately for every single size of pipe in which a flange-tap plate is used [30, 31].

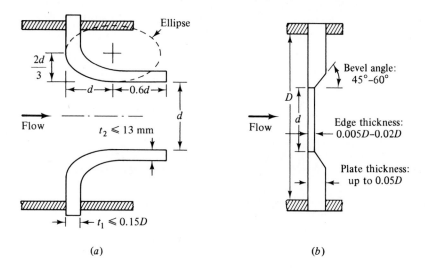

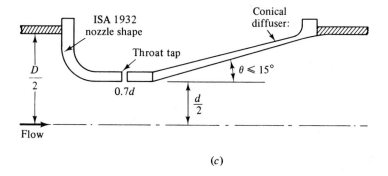

Fig. 6.34 International standard shapes for the three primary Bernoulli obstruction-type meters: (*a*) long-radius nozzle; (*b*) thin-plate orifice; (*c*) venturi nozzle. (*From Ref. 31 by permission of the International Organization for Standardization.*)

Figure 6.35 shows the discharge coefficient of an orifice with $D:\frac{1}{2}D$ or type 2 taps in the Reynolds number range $Re_D = 10^4$ to 10^7 of normal use. Although detailed charts such as Fig. 6.35 are available for designers [30], the ASME recommends use of the curve-fit formulas developed by the ISO [31]. The basic form of the curve fit is [42]

$$C_d = f(\beta) + 91.71\beta^{2.5}Re_D^{-0.75} + \frac{0.09\beta^4}{1 - \beta^4}F_1 - 0.0337\beta^3 F_2 \qquad (6.143)$$

where $$f(\beta) = 0.5959 + 0.0312\beta^{2.1} - 0.184\beta^8$$

The correction factors F_1 and F_2 vary with tap position:

Corner taps: $\qquad\qquad\qquad\qquad F_1 = 0 \qquad F_2 = 0 \qquad\qquad\qquad (6.144a)$

$D:\frac{1}{2}D$ taps: $\qquad\qquad\qquad F_1 = 0.4333 \qquad F_2 = 0.47 \qquad\qquad (6.144b)$

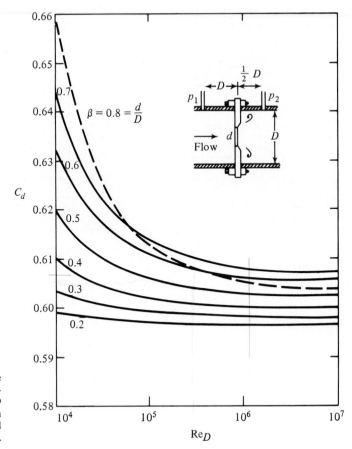

Fig. 6.35 Discharge coefficient for a thin-plate orifice with $D: \frac{1}{2}D$ tappings, plotted from Eqs. (6.143) and (6.144b).

Flange taps:

$$F_1 = \begin{cases} \dfrac{1}{D\,(\text{in})} & D > 2.3\,\text{in} \\[2mm] 0.4333 & 2.0 \le D \le 2.3\,\text{in} \end{cases} \qquad F_2 = \dfrac{1}{D\,(\text{in})} \qquad\qquad (6.144c)$$

Note that the flange taps (6.144c), not being geometrically similar, use raw diameter in inches in the formula. The constants will change if other diameter units are used. We cautioned against such dimensional formulas in Example 1.4 and Eq. (5.17) and give Eq. (6.144c) only because flange taps are widely used in the United States.

Flow nozzle   The flow nozzle comes in two types, a long-radius type shown in Fig. 6.34a and a short-radius type (not shown) called the ISA 1932 nozzle [30, 31]. The flow nozzle, with its smooth rounded entrance convergence, practically eliminates the vena contracta and gives discharge coefficients near unity. The nonrecoverable loss is still large because there is no diffuser provided for gradual expansion.

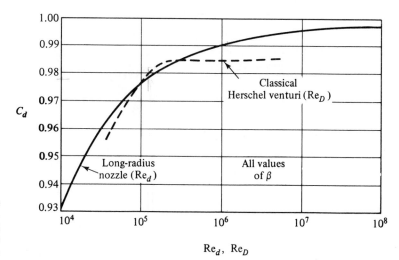

Fig. 6.36 Discharge
coefficient for long-
radius nozzle and
classical Herschel-type
venturi.

The ISO recommended correlation for long-radius-nozzle discharge coefficient is

$$C_d \approx 0.9965 - 0.00653\beta^{1/2}\left(\frac{10^6}{\mathrm{Re}_D}\right)^{1/2} = 0.9965 - 0.00653\left(\frac{10^6}{\mathrm{Re}_d}\right)^{1/2} \quad (6.145)$$

The second form is independent of β ratio and is plotted in Fig. 6.36. A similar ISO correlation is recommended for the short-radius ISA 1932 flow nozzle

$$C_d \approx 0.9900 - 0.2262\beta^{4.1}$$

$$+ (0.000215 - 0.001125\beta + 0.00249\beta^{4.7})\left(\frac{10^6}{\mathrm{Re}_D}\right)^{1.15} \quad (6.146)$$

Flow nozzles may have β between 0.2 and 0.8.

Venturi meter The third and final type of obstruction meter is the venturi, named in honor of Giovanni Venturi (1746–1822), an Italian physicist who first tested conical expansions and contractions. The original, or *classical*, venturi was invented by an American engineer, Clemens Herschel, in 1898. It consisted of a 21° conical contraction, a straight throat of diameter d and length d, then a 7 to 15° conical expansion. The discharge coefficient is near unity and the nonrecoverable loss very small. Herschel venturis are seldom used now.

The modern venturi nozzle, Fig. 6.34c, consists of an ISA 1932 nozzle entrance and a conical expansion of half-angle no greater than 15°. It is intended to be operated in a narrow Reynolds-number range 1.5×10^5 to 2×10^6. Its discharge coefficient, shown in Fig. 6.37, is given by the ISO correlation formula

$$C_d \approx 0.9858 - 0.196\beta^{4.5} \quad (6.147)$$

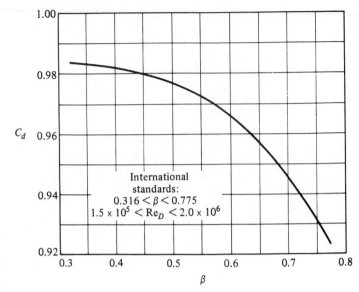

Fig. 6.37 Discharge coefficient for a venturi nozzle.

It is independent of Re_D within the given range. The Herschel venturi discharge varies with Re_D but not with β, as shown in Fig. 6.36. Both have very low net losses.

The choice of meter depends upon the loss and the cost and can be illustrated by the following table:

Type of meter	Net head loss	Cost
Orifice	Large	Small
Nozzle	Medium	Medium
Venturi	Small	Large

As so often happens, the product of inefficiency and initial cost is approximately constant.

The average nonrecoverable head losses for the three types of meters, expressed as a fraction of the throat velocity head $V_t^2/2g$, are shown in Fig. 6.38. The orifice has the greatest loss and the venturi the least, as discussed. The orifice and nozzle simulate partially closed valves as in Table 6.6, while the venturi is a very minor loss. When the loss is given as a fraction of the measured *pressure drop*, the orifice and nozzle have nearly equal losses, as Example 6.20 will illustrate.

The other types of instruments discussed earlier in this section can also serve as flow meters if properly constructed. For example, a hot wire mounted in a tube can be calibrated to read volume flow rather than point velocity. Such hot-wire meters are commercially available, as are other meters modified to use velocity instruments. For further details see Ref. 30.

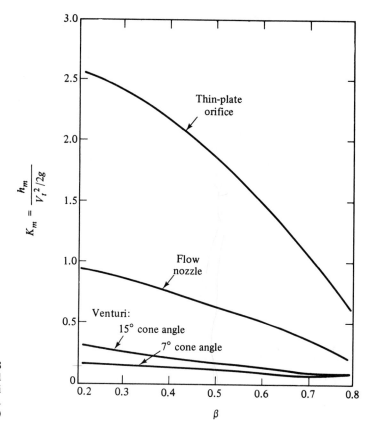

$$K_m = \frac{h_m}{V_t^2/2g}$$

Thin-plate orifice

Flow nozzle

Venturi:

15° cone angle

7° cone angle

0.2　0.3　0.4　0.5　0.6　0.7　0.8

β

Fig. 6.38
Nonrecoverable head loss in Bernoulli obstruction meters. (*Adapted from Ref. 30.*)

EXAMPLE 6.21　We want to meter the volume flow of water ($\rho = 1000 \text{ kg/m}^3$, $v = 1.02 \times 10^{-6} \text{ m}^2/\text{s}$) moving through a 200-mm-diameter pipe at an average velocity of 2.0 m/s. If the differential pressure gage selected reads accurately at $p_1 - p_2 = 50{,}000$ Pa, what size meter should be selected for installing (a) an orifice with $D:\frac{1}{2}D$ taps, (b) a long-radius flow nozzle, or (c) a venturi nozzle? What would be the nonrecoverable head loss for each design?

Solution　Here the unknown is the β ratio of the meter. Since the discharge coefficient is a complicated function of β, iteration will be necessary or else we shall have to devise an ingenious new chart like the modified Moody charts in Figs. 6.14 and 6.15. Let's stick with iteration for now. We are given $D = 0.2$ m and $V_1 = 2.0$ m/s. The pipe-approach Reynolds number is thus

$$\text{Re}_D = \frac{V_1 D}{v} = \frac{(2.0)(0.2)}{1.02 \times 10^{-6}} = 392{,}000$$

The generalized formula (6.139) gives the throat velocity as

$$V_t = \frac{V_1}{\beta^2} = \alpha \left[\frac{2(p_1 - p_2)}{\rho} \right]^{1/2}$$

where everything is known except α and β. Solve for β^2

$$\beta^2 = \frac{1}{\alpha} \left(\frac{\rho V_1^2}{2\Delta p} \right)^{1/2} \tag{1}$$

With $V_1 = 2.0$ m/s and $\Delta p = 50,000$ Pa, this should be valid for all three meters. Substitute in these numbers:

$$\beta^2 = \frac{1}{\alpha}\left[\frac{(1000)(2.0)^2}{2(50,000)}\right]^{1/2} = \frac{0.2}{\alpha}$$

or

$$\beta = \frac{0.447}{\alpha^{1/2}} \qquad (2)$$

The solution depends only on getting the proper flow coefficient.

Part (a) A good initial guess for an orifice is $\alpha \approx 0.62$. From Eq. (2) compute $\beta \approx 0.447/(0.62)^{1/2} = 0.568$. From Eq. (6.143) or Fig. 6.35 compute the discharge coefficient

$$C_d = 0.6064$$

for $\qquad \beta = 0.568 \qquad$ and $\qquad \text{Re}_D = 392,000$

Then $\qquad E = [1 - (0.568)^4]^{-1/2} = 1.0565 \qquad \alpha = C_d E = 0.6407$

Iterate Eq. (2) again

$$\beta = \frac{0.447}{(0.6407)^{1/2}} = 0.558 \qquad C_d = 0.6061 \qquad \alpha = 0.6378$$

Stop. We have converged satisfactorily to a design value

$$\beta \approx 0.56 \qquad d = \beta D = 112 \text{ mm} \qquad \qquad Ans. (a)$$

The throat velocity is $V_t = V_1/\beta^2 = 2.0/(0.56)^2 = 6.38$ m/s. The throat head is

$$\frac{V_t^2}{2g} = \frac{(6.38)^2}{2(9.81)} = 2.07 \text{ m}$$

From Fig. 6.38 for the orifice at $\beta = 0.56$, estimate $K_m \approx 1.7$. Then the nonrecoverable loss of the orifice will be

$$h_m = K_m\frac{V_t^2}{2g} \approx 1.7(2.07) = 3.5 \text{ m} \qquad \qquad Ans. (a)$$

Part (b) A good guess for flow-nozzle design is $\alpha = 1.0$. Iterate Eq. (2) and list the results:

α	β, Eq. (2)	C_d, Fig. 6.36	E	$\alpha = EC_d$
1.0	0.447	0.9895	1.0206	1.0099
1.0099	0.445	0.9895	1.0202	1.0095
1.0095	0.445			

Convergence is rapid to

$$\beta = 0.445 \qquad d = \beta D = 89 \text{ mm} \qquad \qquad Ans. (b)$$

The throat velocity is $2.0/(0.445)^2 = 10.1$ m/s; the throat head is $(10.1)^2/[2(9.81)] = 5.2$ m. From Fig. 6.38 for the nozzle read $K_m \approx 0.7$. Then the nozzle loss is

$$h_m = 0.7(5.2 \text{ m}) = 3.6 \text{ m} \qquad \qquad Ans. (b)$$

This is the same as the orifice because of the higher throat head.

Part (c)   For the venturi guess $\alpha \approx 1.0$ and iterate Eq. (2):

$$\alpha = 1.0 \qquad \beta[\text{Eq. (2)}] = 0.447 \qquad C_d = (\text{Fig. 6.33}) = 0.9806$$

$$= 1.0007 \qquad\qquad = 0.4468 \qquad\qquad\qquad = 0.9806$$

This rapidly converges to give

$$\beta = 0.4468 \qquad d = \beta D = 89 \text{ mm} \qquad\qquad\qquad Ans. \ (c)$$

The throat velocity is $2.0/(0.4468)^2 = 10.0 \text{ m/s}$, and the throat head is $(10.0)^2/[2(9.81)] = 5.12 \text{ m}$. From Fig. 6.38 for the venturi estimate $K_m \approx 0.15$; hence

$$h_m = 0.15(5.12 \text{ m}) = 0.8 \text{ m} \qquad\qquad\qquad Ans. \ (c)$$

This is only 22 percent of the orifice and nozzle losses.

SUMMARY

This chapter is concerned with internal pipe and duct flows, which are probably the most common problems encountered in engineering fluid mechanics. Such flows are very sensitive to the Reynolds number and change from laminar to transitional to turbulent flow as the Reynolds number increases.

The various Reynolds-number regimes are outlined, and a semiempirical approach to turbulent-flow modeling is presented. The chapter then makes a detailed analysis of flow through a straight circular pipe, leading to the famous Moody chart (Fig. 6.13) for the friction factor. Possible modifications to the Moody chart are discussed for flow rate and sizing problems, as well as the application of the Moody chart to noncircular ducts using an equivalent duct "diameter." The addition of minor losses due to valves, elbows, fittings, and other devices is presented in the form of loss coefficients to be incorporated along with Moody-type friction losses. Multiple-pipe systems are discussed briefly and are seen to be quite complex algebraically and appropriate for computer solution.

Diffusers are added to ducts to increase pressure recovery at the exit of a system. Their behavior is presented as experimental data, since the theory of real diffusers is still not well developed. The chapter ends with a discussion of flow meters, especially the pitot tube and the Bernoulli-obstruction type of meter. Flow meters also require careful experimental calibration.

REFERENCES

1. J. O. Hinze, "Turbulence," 2d ed., McGraw-Hill, New York, 1975.
2. H. Schlichting, "Boundary Layer Theory," 7th ed., McGraw-Hill, New York, 1979.
3. F. M. White, "Viscous Fluid Flow," McGraw-Hill, New York, 1974.
4. O. Reynolds, An Experimental Investigation of the Circumstances Which Determine Whether the Motion of Water Shall Be Direct or Sinuous and of the Law of Resistance in Parallel Channels, *Phil. Trans. R. Soc.*, vol. 174, pp. 935–982, 1883.
5. R. Betchov and W. O. Criminale, "Stability of Parallel Flows," Academic, New York, 1967.
6. H. Rouse and S. Ince, "History of Hydraulics," Iowa Institute of Hydraulic Research, State University of Iowa, Iowa City, 1957.

7. J. Nikuradse, Strömungsgesetze in Rauhen Rohren, *VDI Forschungsh.* 361, 1933; English trans., *NACA Tech. Mem.* 1292.

8. L. F. Moody, Friction Factors for Pipe Flow, *ASME Trans.*, vol. 66; pp. 671–684, 1944.

9. C. F. Colebrook, Turbulent Flow in Pipes, with Particular Reference to the Transition between the Smooth and Rough Pipe Laws, *J. Inst. Civ. Eng. Lond.*, vol. 11, pp. 133–156, 1938–1939.

10. O. C. Jones, Jr., An Improvement in the Calculations of Turbulent Friction in Rectangular Ducts, *J. Fluids Eng.*, June 1976, pp. 173–181.

11. R. Berker, "Handbuch der Physik," vol. VII, no. 2, pp. 1–384, Springer-Verlag, Berlin, 1963.

12. R. M. Olson, "Essentials of Engineering Fluid Mechanics," 4th ed., Harper, New York, 1980.

13. J. P. Zarling, An Analysis of Laminar Flow and Pressure Drop in Complex Shaped Ducts, *J. Fluids Eng.*, December 1976, pp. 702–706.

14. P. W. Runstadler, Jr., et al., Diffuser Data Book, *Creare Inc. Tech. Note* 186, Hanover, N.H., 1975.

15. Flow of Fluids through Valves, Fittings, and Pipe, *Crane Co. Tech. Pap.* 410, Chicago, 1957.

16. "Pipe Friction Manual," 3d ed., The Hydraulic Institute, New York, 1961.

17. Hardy Cross, Analysis of Flow in Networks of Conduits or Conductors, *Univ. Ill. Bull.* 286, November 1936.

18. V. L. Streeter and E. B. Wylie, "Fluid Mechanics," 8th ed., McGraw-Hill, New York, 1985.

19. T. A. Marlow et al., Improved Design of Fluid Networks with Computers, *ASCE J. Hydraul. Div.*, July 1966.

20. Y. Senoo and M. Nishi, Prediction of Flow Separation in a Diffuser by a Boundary Layer Calculation, *J. Fluids Eng.*, June 1977, pp. 379–389.

21. R. W. Fox and S. J. Kline, Flow Regime Data and Design Methods for Curved Subsonic Diffusers, *J. Basic Eng.*, vol. 84, pp. 303–312, 1962.

22. J. P. Holman, "Experimental Methods for Engineers," 4th ed., McGraw-Hill, New York, 1983.

23. R. P. Benedict, "Fundamentals of Temperature, Pressure, and Flow Measurement," Wiley, New York, 1969.

24. G. Neumann and W. J. Pierson Jr., "Principles of Physical Oceanography," Prentice-Hall, Englewood Cliffs, N.J., 1966.

25. U.S. Department of Commerce, "Tidal Current Tables," National Oceanographic and Atmospheric Administration, Washington, 1971.

26. J. A. Shercliff, "Electromagnetic Flow Measurement," Cambridge University Press, New York, 1962.

27. J. A. Miller, A Simple Linearized Hot-Wire Anemometer, *J. Fluids Eng.*, December 1976, pp. 749–752.

28. V. A. Sandborn, "Class Notes for Experimental Methods in Fluid Mechanics," Civil Engineering Department, Colorado State University, Fort Collins, Colo., 1972.

29. D. Eckardt, Detailed Flow Investigations within a High Speed Centrifugal Compressor Impeller, *J. Fluids Eng.*, September 1976, pp. 390–402.

30. H. S. Bean (ed.), "Fluid Meters: Their Theory and Application," 6th ed., American Society of Mechanical Engineers, New York, 1971.

31. Measurement of Fluid Flow by Means of Orifice Plates, Nozzles, and Venturi Tubes Inserted in Circular Cross Section Conduits Running Full, *Int. Organ. Stand. Rep.* DIS-5167, Geneva, April 1976.

32. P. Moin and J. Kim, Numerical Investigation of Turbulent Channel Flow, *J. Fluid Mechanics*, vol. 118, pp. 341–377, 1982.
33. S. E. Haaland, Simple and Explicit Formulas for the Friction Factor in Turbulent Pipe Flow, *J. Fluids Eng.*, March 1983, pp. 89–90.
34. R. K. Shah and A. L. London, "Laminar Flow Forced Convection in Ducts," Academic, New York, 1979.
35. J. L. Lyons, "Lyons' Valve Designers Handbook," Van Nostrand Reinhold, New York, 1982.
36. A. O. Demuren and W. Rodi, Calculations of Turbulence-Driven Secondary Motion in Non-circular Ducts, *J. Fluid Mech.*, vol. 140, pp. 189–222, 1984.
37. "ASHRAE Handbook of Fundamentals," chap. 33, ASHRAE, Atlanta, 1981.
38. F. Durst, A. Melling, and J. H. Whitelaw, "Principles and Practice of Laser-Doppler Anemometry," 2d ed., Academic, New York, 1981.
39. H. W. Coleman and P. A. Pfund (eds.), Engineering Applications of Laser Velocimetry, *ASME Symp. Proc.*, vol. H00230, November 1982.
40. J. G. Kopp, Vortex Flowmeters, *Meas. Control*, June 1983, pp. 280–284.
41. J. C. Graber, Jr., Ultrasonic Flow, *Meas. Control*, October 1983, pp. 258–266.
42. ASME Fluid Meters Research Committee, The ISO-ASME Orifice Coefficient Equation, *Mech. Eng.* July 1981, pp. 44–45.
43. R. D. Blevins, "Applied Fluid Dynamics Handbook," Van Nostrand Reinhold, New York, 1984.
44. O. C. Jones, Jr. and J. C. M. Leung, An Improvement in the Calculation of Turbulent Friction in Smooth Concentric Annuli, *J. Fluids Eng.*, December 1981, pp. 615–623.

RECOMMENDED FILMS

Films numbered 3, 4, 11, 12, 15, 23, 24, 26, and 34 in Appendix C.

Problems

PROBLEM DISTRIBUTION

Section	Topic	Problems
6.1	Reynolds-number regimes	6.1–6.7
6.2	Internal and external flows	None
6.3	Laminar and turbulent flows	6.8–6.25
6.4	Flow in a circular pipe: laminar	6.26–6.37
6.4	Flow in a circular pipe: turbulent	6.38–6.71
6.5	Alternate forms of the Moody chart	6.72–6.77
6.6	Flow in noncircular ducts	6.78–6.85
6.7	Minor losses in pipe systems	6.86–6.96
6.8	Multiple pipes: series and parallel	6.97–6.103
6.8	Multiple pipes: networks	6.104–6.111
6.9	Experimental diffuser performance	6.112–6.115
6.10	The pitot tube	6.116–6.122
6.10	Fluid meters: orifice plates	6.123–6.130
6.10	Fluid meters: the flow nozzle	6.131–6.134
6.10	Fluid meters: venturi meter	6.135–6.138

6.1 Repeat Example 6.1 if the fluid is (a) water at 20°C; (b) hydrogen at 20°C. What practical examples can you give of sphere motions where turbulence might be encountered?

6.2 For a thin wing moving parallel to its chord line, transition to a turbulent boundary layer normally occurs at $Re_x = 2.8 \times 10^6$, where x is the distance from the leading edge [2, 3]. If the wing is moving at 20 m/s, at what point on the wing will transition occur at 20°C for (a) air; (b) water?

6.3 If the wing of Prob. 6.2 is tested in a wind or water tunnel it may undergo transition earlier than $Re_x = 2.8 \times 10^6$ if the test stream itself contains turbulent fluctuations. A semiempirical correlation for this case [3, p. 434] is

$$Re_{x_{crit}}^{1/2} \approx \frac{-1 + (1 + 13.25\zeta^2)^{1/2}}{0.00392\zeta^2}$$

where ζ is the tunnel-turbulence intensity in percent. If $V = 20$ m/s in air at 20°C, use this formula to plot the transition position on the wing versus stream turbulence for ζ between 0 and 2 percent. At what value of ζ is x_{crit} decreased 50 percent from its value at $\zeta = 0$?

6.4 For flow of SAE 10 oil through a 5-cm-diameter pipe, from Fig. A-1 in Appendix A, for what flow rate in cubic meters per hour would we expect transition to turbulence at (a) 20°C; (b) 100°C?

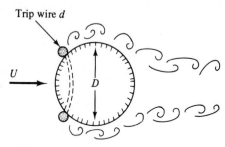

Trip wire d

U

D

Fig. P6.5

6.5 In flow past a body or wall early transition to turbulence can be induced by placing a trip wire on the wall across the flow, as in Fig. P6.5. If the trip wire in Fig. P6.5 is placed where the local velocity is U, it will trigger turbulence if $Ud/\nu = 826$, where d is the wire diameter [3, p. 437]. If the sphere diameter is 20 cm and transition is observed at $Re_D = 90,000$, what is the diameter of the trip wire in millimeters?

6.6 A fluid at 20°C flows at 700 cm³/s through an 8-cm-diameter pipe. Determine whether the flow is laminar or turbulent if the fluid is (a) hydrogen, (b) air, (c) gasoline, (d) water, (e) mercury, or (f) glycerin.

6.7 Oil (SG = 0.9, $\nu = 0.0003$ m²/s) enters a 4-cm-diameter tube. Estimate the entrance length in meters if the flow rate is (a) 0.001 m³/s; (b) 1 m³/s.

6.8 Derive the time-averaged x-momentum equation (6.14) by direct substitution of Eqs. (6.12) into the momentum equation (6.7). It is convenient to write the convective acceleration as

$$\frac{du}{dt} = \frac{\partial}{\partial x}(u^2) + \frac{\partial}{\partial y}(uv) + \frac{\partial}{\partial z}(uw)$$

which is valid because of the continuity relation, Eq. (6.7).

6.9 By analogy with Eq. (6.14) write the turbulent mean-momentum differential equation for (a) the y direction and (b) the z direction. How many turbulent stress terms appear in each equation? How many unique turbulent stresses are there for the total of three directions?

6.10 Two infinite plates a distance h apart are parallel to the xz plane with the upper plate moving at speed V, as in Fig. P6.10. There is a fluid of viscosity μ and

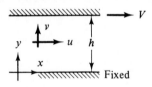

V

v

y u h

x

Fixed

Fig. P6.10

constant pressure between the plates. Neglecting gravity and assuming an incompressible laminar flow pattern $u = u(y)$, $v = w = 0$, use Eqs. (6.7) and appropriate boundary conditions to derive the velocity profile $u(y)$.

6.11 Repeat Prob. 6.10 for turbulent flow, using the logarithm law, Eq. (6.21), to formulate the velocity $u(y)$ between the plates. How does the shear stress vary with y between plates?

6.12 For the flow of Fig. P6.10, if $V = 4$ m/s and $h = 1.8$ cm, compute the shear stress at the upper and lower walls if the fluid is SAE 30 oil at 20°C.

6.13 For the flow of Fig. P6.10, if $V = 6$ m/s and $h = 1.5$ cm, compute the shear stress at the upper and lower walls if the fluid is water at 20°C. *Hint*: The flow is turbulent.

6.14 Just as laminar shear equals $\mu\, du/dy$, Boussinesq in 1877 postulated that turbulent shear could also be related to velocity gradient $\tau_{turb} = \epsilon \, du/dy$, where ϵ is called the *eddy viscosity* and is much larger than μ. For the Couette flow of Fig. P6.10, use the log law, Eq. (6.21), to show that the eddy viscosity between the plates is given by $\epsilon = \kappa \rho u^* y$, for $y \leq h/2$. Compute the ratio ϵ/μ at $y = h/2$ if $\text{Re}_h = Vh/\nu = 10^5$.

6.15 Theodore von Kármán in 1930 theorized that turbulent shear could be represented by $\tau_{turb} = \epsilon \, du/dy$, where $\epsilon = \rho \kappa^2 y^2 \, |du/dy|$ is called the *mixing-length eddy viscosity* and $\kappa \approx 0.41$ is Kármán's dimensionless *mixing-length constant* [2, 3]. Assuming that $\tau_{turb} \approx \tau_w$ near the wall, show that this expression can be integrated to yield the logarithmic-overlap law, Eq. (6.21).

6.16 Water at 20°C flows in a 9-cm-diameter pipe under fully developed conditions. The centerline velocity is 10 m/s. Compute (a) Q, (b) V, (c) τ_w, and (d) Δp for a 100-m pipe length.

6.17 Repeat Prob. 6.16 for SAE 30 oil at 20°C.

6.18 A viscous fluid flows between two fixed parallel plates, as shown in Fig. P6.18. The velocity pattern is $u = u(y)$, $v = w = 0$, and the pressure drops linearly, $p = -Cx + D$. Reduce the continuity and momentum equations (6.7) to fit this case, assuming laminar flow, and solve for $u(y)$ and the shear stress at each wall.

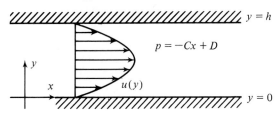

Fig. P6.18

6.19 A viscous fluid between two long cylinders as in Fig. P6.19 is set into motion by rotation counterclockwise of the inner cylinder at angular velocity Ω. The

outer cylinder is fixed. Assuming steady laminar flow, derive an expression for the tangential velocity $v_\theta(r)$ between cylinders.

6.20 Repeat Prob. 6.19 for the case where the outer cylinder is rotating and the inner cylinder is fixed. Can this solution be added linearly to the solution of Prob. 6.19 when both cylinders are rotating at different rates, say Ω_1 and Ω_2?

6.21 Show how you would analyze Prob. 6.19 for the case of steady *turbulent* flow, using the logarithmic law (6.21), some physics, some math, some sketches, and some talk.

6.22 Show how you would analyze Prob. 6.18 for the case of steady *turbulent* flow, using the same tools listed in Prob. 6.21.

6.23 The space between the two long concentric cylinders in Fig. P6.19 is filled with a viscous fluid. The inner cylinder moves axially at speed V, and the outer cylinder is fixed. Assuming steady laminar flow, derive an expression for the axial-velocity distribution $v_z(r)$ in the fluid.

6.24 For Prob. 6.23, if $a = 6$ cm, $b = 5$ cm, and $V = 4$ m/s, compute the shear stress at each cylinder wall for glycerin at 20°C.

6.25 For steady laminar flow, determine whether the solutions to Probs. 6.19 and 6.23 can be added linearly for the case of the inner cylinder which is both rotating and moving axially.

In Probs. 6.26 to 6.85 neglect minor losses

6.26 The 6-cm-diameter pipe in Fig. P6.26 contains glycerin at 20°C flowing at a rate of 8 m³/h. Verify that the flow is laminar. For the pressure measurements shown, is the flow up or down? What is the indicated head loss for these pressures?

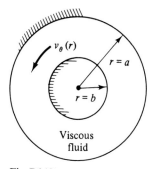

Fig. P6.19

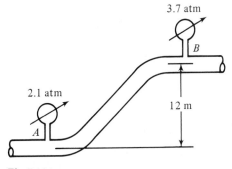

Fig. P6.26

6.27 In Prob. 6.26 compute the theoretical head loss if the pipe length is 30 m between A and B. Compare with the indicated pressure drop $p_B - p_A$.

6.28 In Example 6.5 suppose the flow rate is unknown but the liquid viscosity is 2×10^{-5} slug/(ft·s). What will be the flow rate in cubic feet per hour? Is the flow still laminar?

6.29 A steady push on the piston in Fig. P6.29 causes a flow rate $Q = 0.4$ cm³/s through the needle. The fluid has $\rho = 900$ kg/m³ and $\mu = 0.002$ kg/(m·s). What force F is required to maintain the flow?

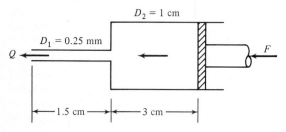

Fig. P6.29

6.30 An oil (SG = 0.9) issues from the pipe in Fig. P6.30 at $Q = 45$ ft³/h. What is the kinematic viscosity of the oil in square feet per second? Is the flow laminar?

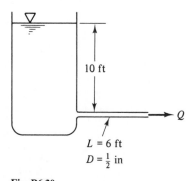

Fig. P6.30

6.31 In Prob. 6.30 what will the flow rate Q be if the fluid is SAE 10 oil at 20°C?

6.32 For the tank and capillary system of Fig. E6.5, suppose the fluid is kerosine at 20°C and the flow rate Q is unknown. Set up a differential equation and integrate to find the time required to drain the entire system. For simplicity, neglect the kinetic energy at the exit.

6.33 For the configuration shown in Fig. P6.33, the fluid is ethyl alcohol at 20°C and the tanks are very wide. Find the flow rate which occurs in cubic meters per hour. Is the flow laminar?

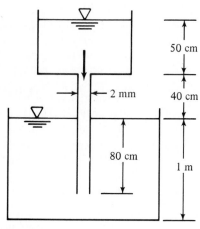

Fig. P6.33

6.34 For the system in Fig. P6.33, if the fluid has density of 920 kg/m³ and the flow rate is unknown, for what value of viscosity will the capillary Reynolds number exactly equal the critical value of 2300?

6.35 Let us attack Prob. 6.33 in symbolic fashion, using Fig. P6.35. All parameters are constant except the upper tank depth $Z(t)$. Find an expression for the flow

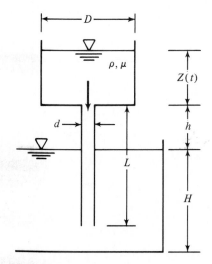

Fig. P6.35

rate $Q(t)$ as a function of $Z(t)$. Set up a differential equation and solve for the time t_0 to drain the upper tank completely. Assume quasi-steady laminar flow.

6.36 SAE 30 oil at 20°C flows in the 3-cm-diameter pipe in Fig. P6.36, which slopes at 37°. For the pressure measurements shown, determine (a) whether the flow is up or down, and (b) the flow rate in cubic meters per hour.

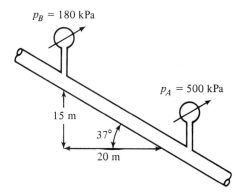

$p_B = 180$ kPa

$p_A = 500$ kPa

15 m

37°

20 m

Fig. P6.36

6.37 Modify and repeat Prob. 6.36 if the pressures are the same but there is a pump between A and B which adds a 10-m head rise in the flow direction. Verify that the flow is laminar.

6.38 Show that if Eq. (6.49) is accurate, the point in a turbulent pipe flow where local velocity u equals average velocity V occurs exactly at $r = 0.777R$ regardless of Reynolds number.

6.39 Derive Eq. (6.59), showing all steps. The constant 1.33 dates back to Prandtl's work in 1935 and may change slightly to 1.30 in your own analysis.

6.40 For the system in Fig. P6.30 find the flow rate Q if the fluid is water at 20°C.

6.41 If 1 mi (5280 ft) of 3-in-diameter wrought-iron pipe carries water at 20°C and $V = 7$ m/s, compute the head loss in feet and the pressure drop in pounds force per square inch.

6.42 Mercury at 20°C flows through 3 m of 6-mm-diameter glass tubing at an average velocity of 2.5 m/s. Compute the head loss in meters and the pressure drop in kilopascals.

6.43 Gasoline at 20°C is pumped at 0.2 m³/s through 15 km of 18-cm-diameter cast-iron pipe. Compute the

power in kilowatts required if the pumps are 80 percent efficient.

6.44 Oil (SG = 0.9, $v = 0.00003$ ft²/s) flows at 1 ft³/s through a 6-in asphalted cast-iron pipe. The pipe is 2000 ft long and slopes upward at 5° in the flow direction. Compute the head loss in feet and the pressure change.

6.45 The viscous sublayer (Fig. 6.9) is normally less than 1 percent of the pipe diameter and therefore very difficult to probe with a finite-sized instrument. In an effort to generate a thick sublayer for probing, Pennsylvania State University in 1964 built a pipe with a flow of glycerin. Assume a smooth 12-in-diameter pipe with $V = 60$ ft/s and glycerin at 20°C. Compute the sublayer thickness in inches and the pumping horsepower required at 75 percent efficiency if $L = 40$ ft.

6.46 The pipe flow in Fig. P6.46 is driven by pressurized air in the tank. What gage pressure p_1 is needed to provide a water flow rate $Q = 50$ m³/h?

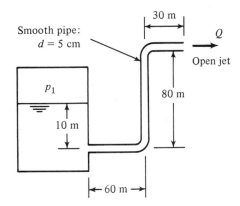

Smooth pipe: $d = 5$ cm

30 m

Q

Open jet

p_1

80 m

10 m

60 m

Fig. P6.46

6.47 In Fig. P6.46 suppose the fluid is methanol at 20°C and $p_1 = 900$ kPa gage. What flow rate Q results, in cubic meters per hour? What makes the use of Fig. 6.13 a cumbersome process?

6.48 In Fig. P6.46 suppose the fluid is carbon tetrachloride at 20°C and $p_1 = 1300$ kPa. What pipe diameter in centimeters is needed to provide a flow rate of 20 m²/h? What makes this problem cumbersome?

6.49 The reservoirs in Fig. P6.49 contain water at 20°C. If the pipe is smooth with $L = 7000$ m and $D = 5$

cm, what will the flow rate in cubic meters per hour be for $\Delta z = 100$ m?

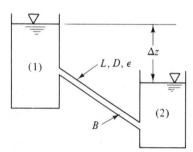

Fig. P6.49

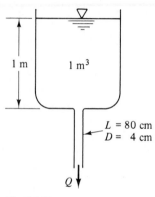

Fig. P6.56

6.50 Repeat Prob. 6.49 to find Q in gallons per minute if $L = 2500$ ft, $D = 3$ in, and $\Delta z = 80$ ft. What will the percentage decrease in flow rate be if the pipe has a roughness of 0.2 mm?

6.51 In Fig. P6.49 assume that the pipe is cast iron with $L = 120$ m and $D = 8$ cm. If a 75 percent efficient pump is placed at point B, what input power in kilowatts is required to pump the water upward from reservoir 2 to 1 at a rate of 0.05 m³/s?

6.52 Ethyl alcohol at 20°C flows at $V = 300$ cm/s through 10-cm-diameter drawn tubing. Compute (a) the head loss per 100 m of tube, (b) the wall shear stress, and (c) the local velocity u at $r = 2$ cm. By what percentage is the head loss increased due to the roughness of the tube?

6.53 Water at 20°C flows through a 600-m pipe 15 cm in diameter at a flow rate of 0.06 m³/s. If the head loss is 50 m, estimate the pipe roughness in millimeters. If this roughness is doubled, what will the percentage increase in head loss be?

6.54 A long 4-in commercial steel pipe is to be laid on a slope so that 200 gal/min of water at 20°C flows through it due to gravity only. What should the angle of slope of the pipe be?

6.55 In Prob. 6.54 how many gallons per minute of water will flow by gravity if the pipe is laid at a slope angle of 4°?

6.56 A tank contains 1 m³ of water at 20°C and has a drawn-capillary outlet tube at the bottom, as in Fig. P6.56. Find the outlet volume flux Q in cubic meters per hour at this instant.

6.57 For the system in Fig. P6.56 solve for the flow rate in cubic meters per hour if the fluid is SAE 30 oil at 20°C. Is the flow laminar or turbulent?

6.58 In Prob. 6.56 the initial flow is turbulent. As the water drains out of the tank, will the flow revert to laminar motion as the tank becomes nearly empty? If so, at what tank depth? Estimate the time in hours to drain the tank completely.

6.59 What level h must be maintained in the tank in Fig. P6.59 to deliver a flow rate of 0.02 ft³/s through the $\frac{1}{2}$-in commercial-steel exit pipe?

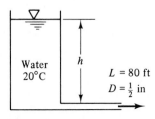

Fig. P6.59

6.60 In Fig. P6.59 suppose the fluid is benzene at 20°C and $h = 100$ ft. What pipe diameter is required for the flow rate to be 0.02 ft³/s?

6.61 You wish to water your garden with 100 ft of $\frac{1}{2}$-in-diameter garden hose whose roughness is 0.01 in. What will the delivery in cubic feet per second be if the pressure at the faucet is 60 lbf/in² gage? If the hose cannot quite reach the garden, what is the maximum distance the stream of water at 20°C will carry?

6.62 Water at 20°C is to be pumped through 2000 ft of pipe from reservoir 1 to 2 at a rate of 3 ft³/s, as shown in

Fig. P6.62. If the pipe is cast iron of diameter 6 in and the pump is 75 percent efficient, what horsepower pump is needed?

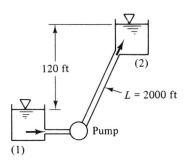

Fig. P6.62

6.63 For Prob. 6.62 suppose the only pump available can deliver 80 hp to the fluid. What is the proper pipe size in inches to maintain the 3 ft³/s flow rate?

6.64 In Fig. P6.62 suppose the pipe is 6-in-diameter cast iron and the pump delivers 75 hp to the flow. What flow rate Q results in cubic feet per second?

6.65 It is desired to solve Prob. 6.62 for the most economical pump and cast-iron pipe system. If the pump costs $125 per horse-power delivered to the fluid and the pipe costs $7000 per inch of diameter, what is the minimum cost and the pipe and pump size to maintain 3 ft³/s flow rate? Make some simplifying assumptions.

6.66 The small turbine in Fig. P6.66 extracts 400 W of power from the water flow. Both pipes are wrought iron. Compute the flow rate Q in cubic meters per hour. Sketch the EGL and HGL accurately.

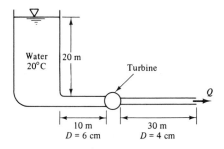

Fig. P6.66

6.67 In Fig. P6.67 the connecting pipe is commercial steel 6 cm in diameter. Compute the flow rate in cubic

meters per hour if the fluid is SAE 30 oil at 20°C. Which way is the flow?

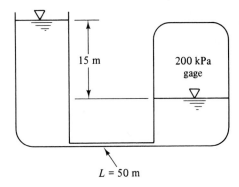

Fig. P6.67

6.68 In Figure P6.67, if the fluid is SAE 30 oil at 20°C, what should the pipe diameter in centimeters be to maintain a flow rate of 25 m³/h?

6.69 The head-versus-flow-rate characteristics of a centrifugal pump are shown in Fig. P6.69. If this pump drives water at 20°C through 120 m of 30-cm diameter cast-iron pipe, what will the resulting flow rate in cubic meters per second be?

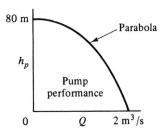

Fig. P6.69

6.70 The pump in Fig. P6.69 is used to drive gasoline at 20°C through 500 m of 25-cm-diameter galvanized-iron pipe. What flow rate will result in cubic meters per second? (Note that the pump head in Fig. P6.69 is in meters of gasoline for this problem.)

6.71 The pump in Fig. P6.69 has its maximum efficiency at a head of 51 m. If it is used to pump ethyl alcohol at 20°C through 150 m of commercial steel pipe, what is the proper pipe diameter to operate the pump at maximum efficiency?

6.72   Solve Prob. 6.49 using the modified Moody chart of Fig. 6.14.

6.73   For the system of Fig. P6.33, if the fluid is mercury at 20°C, use the modified Moody chart (Fig. 6.14) to find the resulting flow rate. Is the flow laminar or turbulent?

6.74   Solve Prob. 6.67 using the modified Moody chart of Fig. 6.14.

6.75   For the system of Fig. P6.49, suppose $\Delta z = 90$ m and the cast-iron pipe is 200 m long. If the resulting flow rate is 5 m³/h, use the modified Moody chart (Fig. 6.15) to determine the proper pipe diameter.

6.76   If gasoline at 20°C is to be pumped through 45 m of cast-iron pipe at 90 m³/h with a head loss no greater than 20 m, use the modified Moody chart (Fig. 6.15) to determine the minimum possible pipe diameter.

6.77   The pump in Fig. P6.69 is used to drive methanol at 20°C through 80 m of wrought-iron pipe at 0.5 m³/s. Using the modified Moody chart (Fig. 6.15), find the proper pipe diameter.

6.78   A commercial steel annulus 40 ft long, with $a = 1$ in and $b = \frac{1}{2}$ in, connects two reservoirs which differ in surface height by 20 ft. Compute the flow rate in cubic feed per second through the annulus if the fluid is water at 20°C.

6.79   Show that for laminar flow through an annulus of very small clearance the flow rate Q is approximately proportional to the cube of the clearance $a - b$.

6.80   An annulus of narrow clearance causes a very large pressure drop and is useful as an accurate measurement of viscosity. If a smooth annulus 1 m long with $a = 50$ mm and $b = 49$ carries an oil flow at 0.001 m³/s, what is the oil viscosity if the pressure drop is 250 kPa?

6.81   A sheet-steel ventilation duct carries air at 20°C and 1 atm (approximately). The duct section is an equilateral triangle 12 in on a side, and its length is 120 ft. If a blower can deliver 1 hp to the air, what flow rate in cubic feet per second can occur?

6.82   SAE 30 oil at 20°C flows between two smooth parallel plates 4 cm apart at an average velocity of 3m/s. Compute the pressure drop, centerline velocity, and head loss for each 100 m of length.

6.83   Air at sea-level standard conditions is blown through a 30-cm-square steel duct 150 m long at $V = 25$ m/s. Compute the head loss, the pressure drop, and the power required in kilowatts if the blower efficiency is 60 percent.

6.84   A wind tunnel has a wooden rectangular section 40 cm by 1 m by 50 m long. The average flow velocity is 45 m/s for air at sea-level standard conditions. Compute the pressure drop, assuming fully developed conditions, and the power required if the fan has 65 percent efficiency.

6.85   It is desired to pump hydrogen at 20°C and 1 atm through a smooth rectangular duct 80 m long of aspect ratio 6:1. If the flow rate is 0.6 m³/s and the pressure drop 75 Pa, what should the width and height of the duct cross section be?

6.86   Repeat Prob. 6.49 by including losses due to a sharp-edged entrance, the exit, and a fully open flanged globe valve. By what percentage is the flow rate decreased?

6.87   Repeat Prob. 6.59 by including losses due to a sharp entrance and a fully open screwed swing-check valve. By what percentage is the required tank level h increased?

6.88   A 70 percent efficent pump delivers water at 20°C from one reservoir to another 20 ft higher, as in Fig. P6.88. The piping system consists of 60 ft of galvanized-iron 2-in pipe, a reentrant entrance, two screwed 90° long-radius elbows, a screwed open gate valve, and a sharp exit. What is the input power required in horsepower with and without a 6° well-designed conical expansion added to the exit? The flow rate is 0.4 ft³/s.

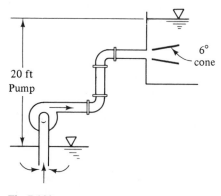

Fig. P6.88

6.89   The two reservoirs in Fig. P6.89 are connected by 20-ft-long wrought-iron pipes joined abruptly. The entrance and exit are sharp-edged. The fluid is water at 20°C. Including minor losses, compute the flow rate in cubic feet per second if reservoir 1 is 60 ft higher than reservoir 2.

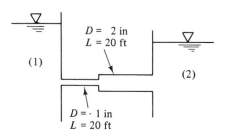

Fig. P6.89

6.90 Two reservoirs containing water at 20°C are connected by 700 m of 18-cm cast-iron pipe, including a sharp entrance, a submerged exit, a gate valve 75 percent open, two 1-m radius bends, and six regular 90° elbows. If the flow rate is 0.15 m³/s, what is the difference in reservoir elevations?

6.91 The system in Fig. P6.91 consists of 1200 m of 5 cm cast-iron pipe, two 45° and four 90° flanged long-radius elbows, a fully open flanged globe valve, and a sharp exit into a reservoir. If the elevation at point 1 is 400 m, what gage pressure is required at point 1 to deliver 0.005 m³/s of water at 20°C into the reservoir?

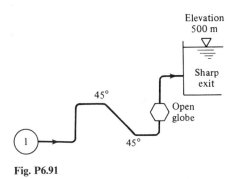

Fig. P6.91

6.92 The water pipe in Fig. P6.92 slopes upward at 30°. The pipe is 1-in-diameter and smooth. The flanged globe valve is fully open. If the mercury manometer shows a 7-in deflection, what is the flow rate in cubic feet per second?

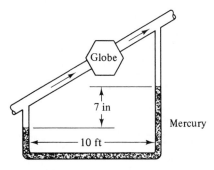

Fig. P6.92

6.93 Despite the fact that the figures are expressed in different systems of units, suppose the pump of Fig. P6.69 is used to drive the flow system in Fig. P6.88, including the 6° cone diffuser. What flow rate will result, in cubic feet per second? Are the pump and system well matched?

6.94 In Fig. P6.94 there are 150 ft of 2-in pipe, 90 ft of 6-in pipe, and 180 ft of 3-in pipe, all wrought iron. There are two 90° elbows and an open globe valve, all screwed. If the exit elevation is zero, what power in horsepower is extracted by the turbine from the 20°C water when the flow rate is 0.15 ft³/s?

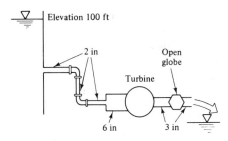

Fig. P6.94

6.95 A pump delivers water at 20°C from one reservoir to another 20 m higher through 80 m of 10 cm wrought-iron pipe. Minor losses are a sharp entrance and exit and screwed connections: two 45° elbows, one 90° elbow, and an open angle valve. The pump head versus flow rate approximates

$$h_p \approx 50 - 30{,}000Q^4$$

with h_p in meters and Q in cubic meters per second. For this system what will Q be?

6.96 In Fig. P6.96 the pipe entrance is sharp-edged. If the flow rate is 0.004 m³/s, what power in watts is extracted by the turbine?

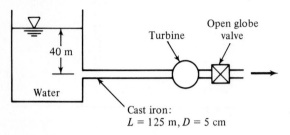

Fig. P6.96

6.97 For the parallel pipe system of Fig. P6.97, each pipe is cast iron, and the pressure drop $p_1 - p_2 = 3$ lbf/in². Compute the total flow rate between 1 and 2 if the fluid is SAE 30 oil at 20°C.

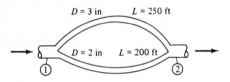

Fig. P6.97

6.98 If the two pipes in Fig. P6.97 are instead laid in series with the same total pressure drop of 3 lbf/in², what will the flow rate be? The fluid is gasoline at 20°C.

6.99 In the parallel pipe system of Fig. P6.99, pipe 2 is assisted by a pump which delivers 38 kW to the flow. If the water flow rate is 0.036 m³/s, compute (a) the flow rate in each pipe and (b) the total pressure drop. Neglect minor losses. Both pipes are galvanized iron.

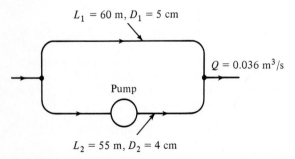

Fig. P6.99

6.100 For the parallel-series pipe system of Fig. P6.100 all pipes are 6-cm-diameter wrought iron containing water at 20°C. If the total flow rate from 1 to 2 is 0.01 m³/s, compute the total pressure drop $p_1 - p_2$. How does Q divide between the two parallel pipes? Neglect minor losses.

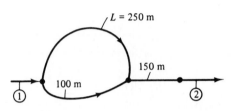

Fig. P6.100

6.101 Modify Prob. 6.100 so that the flow rate is unknown but the total pressure drop is 900 kPa. Find the resulting flow rate Q.

6.102 For the piping system of Fig. P6.102, all pipes are concrete with roughness of 0.03 in. If the flow rate is 18 ft³/s of water at 20°C, compute (a) the pressure drop $p_1 - p_2$ and (b) how the flow divides among the three parallel pipes.

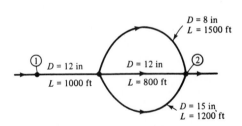

Fig. P6.102

6.103 For the system of Fig. P6.102 with gasoline at 20°C, compute the flow rate in all pipes if the pressure drop $p_1 - p_2$ is 35 lbf/in². Neglect minor losses.

6.104 For the three-reservoir system of Fig. P6.104, $z_1 = 30$ m, $L_1 = 80$ m, $z_2 = 130$ m, $L_2 = 150$ m, $z_3 = 70$ m, and $L_3 = 110$ m. All pipes are 25-cm-diameter concrete with roughness height of 0.5 mm. Compute the flow rate in all pipes for water at 20°C.

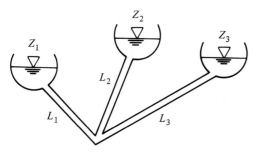

Fig. P6.104

6.105 The three pipes in Fig. P6.105 are cast iron: $D_1 = 7$ in, $L_1 = 2000$ ft; $D_2 = 5$ in, $L_2 = 1000$ ft; $D_3 = 8$ in, $L_3 = 1500$ ft. Compute the flow rate in all pipes for water at 20°C.

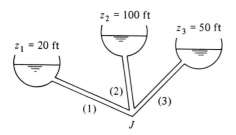

Fig. P6.105

6.106 In Fig. P6.106 all pipes are cast iron and $p_1 - p_2 = 35$ lbf/in². Compute the total flow rate of water at 20°C.

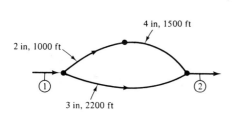

Fig. P6.106

6.107 In Fig. P6.106 all pipes are cast iron, and the total flow rate from 1 to 2 is 0.5 ft³/s of water at 20°C. Compute the pressure drop $p_1 - p_2$ and the division of flow between the two paths.

6.108 In the five-pipe network of Fig. P6.108 assume that all pipes have a friction factor of 0.02. The pressure

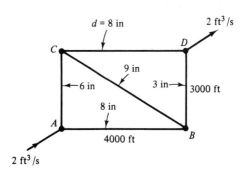

Fig. P6.108

at A is 100 lbf/in² gage. Find the flow rate in all pipes and the pressures at B, C, and D. The fluid is water at 20°C. *Hint:* The flow directions may surprise you.

6.109 In Fig. P6.109 four equal-sized pipes of 30 m long and 6 cm in diameter are joined at junction a. Pressures are known at each of the pipe ends: $p_1 = 900$ kPa, $p_2 = 300$ kPa, $p_3 = 700$ kPa, $p_4 = 100$ kPa. All pipes are cast iron. Neglecting gravity and minor losses, compute the water flow rate in all pipes.

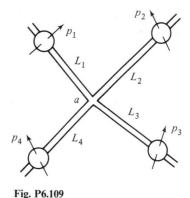

Fig. P6.109

6.110 The pipe dimensions in Fig. P6.110 are the same as in Fig. P6.108. The pressure $p_A = 100$ lbf/in² gage, and the fluid is water at 20°C. Assume that $f = 0.02$ in all pipes and compute pressure at B, C, and D and the flow rate in all pipes.

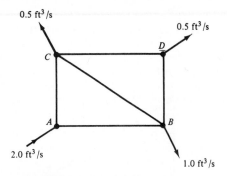

0.5 ft³/s

0.5 ft³/s

D

C

A

B

2.0 ft³/s

1.0 ft³/s

Fig. P6.110

6.111   In Fig. P6.111 all three pipes are cast iron. Neglecting minor losses, compute the flow rate in each pipe for water at 20°C.

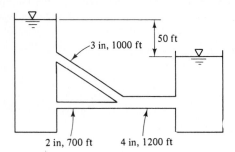

3 in, 1000 ft

50 ft

2 in, 700 ft   4 in, 1200 ft

Fig. P6.111

6.112   A water-tunnel test section has a 1-m diameter and flow properties $V = 20$ m/s, $p = 100$ kPa, and $T = 20$°C. The boundary-layer blockage at the end of the section is 9 percent. If a conical diffuser is to be added at the end of the section to achieve maximum pressure recovery, what should its angle, length, exit diameter, and exit pressure be?

6.113   For Prob. 6.112 suppose we are limited by space to a total diffuser length of 10 m. What should the diffuser angle, exit diameter, and exit pressure be for maximum recovery?

6.114   A wind-tunnel test section is 3 ft square with flow properties $V = 280$ ft/s, $p = 15$ lbf/in² absolute, and $T = 68$°F. Boundary-layer blockage at the end of the test section is 8 percent. Find the angle, length, exit height, and exit pressure of a flat-walled diffuser added onto the section to achieve maximum pressure recovery.

6.115   For Prob. 6.114 suppose we are limited by space to a total diffuser length of 30 ft. What should the diffuser angle, exit height, and exit pressure be for maximum recovery?

6.116   A pitot tube (Fig. 6.29) is inserted into a flowing airstream at 15°C and 110 kPa. The differential pressure reads 9 mm on an air-mercury manometer. What is the indicated airspeed?

6.117   For the pitot-static pressure arrangement of Fig. P6.117, compute (*a*) the centerline velocity, (*b*) the pipe volume flow, and (*c*) the wall shear stress. The manometer fluid is Meriam red oil (SG = 0.827).

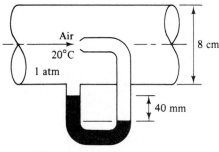

Air
20°C
1 atm

8 cm

40 mm

Fig. P6.117

6.118   For the 20°C water flow of Fig. P6.118 use the manometer measurement to estimate (*a*) the centerline velocity and (*b*) the volume flow in the 6-in-diameter pipe. Assume a smooth wall.

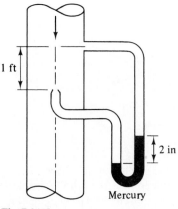

1 ft

2 in

Mercury

Fig. P6.118

6.119 A pitot tube placed in a flow of helium at 20°C and 1 atm shows a helium-water differential manometer reading of 11 mm. What is the helium velocity? What will the manometer reading be if the helium velocity is 25 m/s?

6.120 A small airplane flying at 8000 m altitude uses a pitot stagnation probe without a static tube. The measured stagnation pressure is 39 kPa. What is the indicated airplane speed in miles per hour and its probable uncertainty? Is a compressibility correction needed?

6.121 An engineer who took college fluid mechanics on a pass-fail basis has placed the static pressure hole far upstream of the stagnation probe, as in Fig. P6.121, thus contaminating the pitot measurement ridiculously with pipe friction losses. If the pipe flow is air at 20°C and 1 atm and the manometer fluid is Meriam red oil (SG = 0.827), estimate the air centerline velocity for the given manometer reading of 16 cm. Assume a smooth-walled tube.

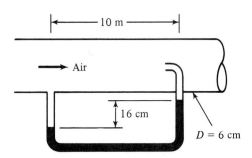

Fig. P6.121

6.122 A pitot-tube traverse has resulted in the following local-velocity measurements for airflow in a 6-in-radius pipe:

r, in	0	1	2	3	4	5	6
V, ft/s	4.56	4.45	4.31	4.15	3.93	3.57	0.0

Simply using these raw data without introducing friction laws or other theory, compute the volume flux in cubic feet per second.

6.123 Water at 20°C in a 10-cm-diameter pipe flows through a 5-cm-diameter thin-plate orifice with $D:\frac{1}{2}D$ taps. If the measured pressure drop is 65 kPa, what is the flow rate in cubic meters per hour? What is the nonrecoverable head loss?

6.124 To solve Prob. 6.123 using Fig. 6.35 requires iteration because both ordinate and abscissa contain the unknown Q. In the spirit of Sec. 6.5 suggest an alternate abscissa which does not contain Q so that the solution can be read directly from the ordinate. How is your new abscissa variable related to C_d and Re_D?

6.125 Gasoline at 20°C flows in a 5-cm-diameter pipe at 25 m³/h. If a 2-cm-diameter thin-plate orifice with corner taps is installed in the pipe, what will the measured pressure drop be in kilopascals?

6.126 We are to meter 125 m³/h of ethyl alcohol at 20°C flowing in a 9-cm-diameter pipe using a thin-plate orifice with corner taps. If the differential pressure transducer reads best at about 35 kPa, what should the β ratio of the orifice be?

6.127 A 1-m-diameter cylindrical tank in Fig. P6.127 is initially filled with kerosine at 20°C to a depth of 2 m. There is a 3-cm-diameter thin-plate orifice in the bottom similar to Fig. 6.34b. When the orifice is opened, what will the initial flow rate be in cubic meters per hour? How long will it take $h(t)$ to drop from 2 to 1.4 m?

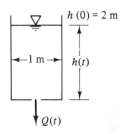

Fig. P6.127

6.128 In Fig. P6.128 the fluid is water at 20°C. The smooth 5-cm-diameter pipe has a sharp entrance and

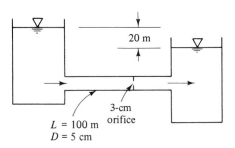

Fig. P6.128

exit. Compute the volume flow rate through the pipe and the pressure drop across the orifice for flange taps.

6.129   A 2.4-cm-diameter orifice plate is installed with no taps in a very long 8-cm-diameter commercial steel pipe. There are pressure taps 20 m upstream and 20 m downstream of the orifice. When carbon tetrachloride at 20°C flows through the pipe, the measured pressure difference is 1.27 MPa. Estimate the volume flow rate and comment on its accuracy.

6.130   Air flows through a 6-cm-diameter smooth pipe which has a 2-m long perforated section containing 500 holes (diameter 1 mm), as in Fig. P6.130. Pressure outside the pipe is sea-level standard air. If $p_1 = 105$ kPa and $Q_1 = 110 \text{ m}^3/\text{h}$, estimate p_2 and Q_2 assuming that the holes are approximated by thin-plate orifices. *Hint*: A momentum control volume may be very useful.

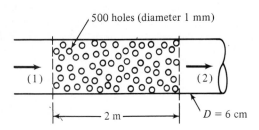

Fig. P6.130

6.131   Gasoline at 20°C flows at 0.06 m³/s through a 15-cm pipe and is metered by a 9-cm long-radius flow nozzle (Fig. 6.34a). What is the expected pressure drop across the nozzle?

6.132   Ethyl alcohol at 20°C flowing in a 6-cm-diameter pipe is metered through a 3-cm long-radius flow nozzle. If the measured pressure drop is 45 kPa, what is the estimated volume flow in cubic meters per hour?

6.133   Kerosine at 20°C flows at 20 m³/h in an 8-cm-diameter pipe. The flow is to be metered by an ISA 1932 flow nozzle so that the pressure drop is 7000 Pa. What is the proper nozzle diameter?

6.134   Two water tanks, each with base area of 1 ft², are connected by a 0.5-in-diameter long-radius nozzle as in Fig. P6.134. If $h = 1$ ft as shown for $t = 0$, estimate the time for $h(t)$ to drop to 0.25 ft.

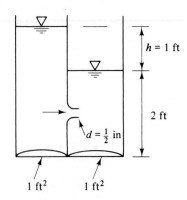

Fig. P6.134

6.135   It is planned to equip a 15-cm-diameter pipe carrying water at 20°C with a modern venturi nozzle (Fig. 6.34c). In order for international standards to be valid (Fig. 6.37), what is the permissible range of (a) flow rates, (b) nozzle diameters, and (c) pressure drops? Would compressibility be a problem for the highest pressure-drop condition?

6.136   Light oil (SG = 0.92, $v = 10^{-5}\text{ft}^2/\text{s}$) flows down a 6-in vertical pipe through a 3-in venturi nozzle as in Fig. P6.136. If the mercury manometer reads a deflection of 4 in, what is the estimated flow rate in the pipe in cubic feet per second?

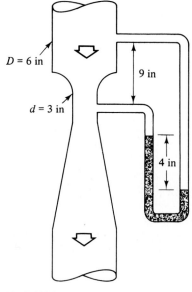

Fig. P6.136

6.137 Modify Prob. 6.128 by replacing the orifice by a 3-cm-diameter Herschel venturi nozzle.

6.138 A modern venturi nozzle is tested in a laboratory flow with water at 20°C. The pipe diameter is 5.5 cm and the venturi throat diameter is 3.5 cm. The flow rate is measured by a weigh tank and the pressure drop by a water-mercury manometer. The mass flow rate and manometer readings are as follows:

$\dot{m}$, kg/s	0.95	1.98	2.99	5.06	8.15
h, mm	3.7	15.9	36.2	102.4	264.4

Use these data to plot a calibration curve of venturi discharge coefficient versus Reynolds number. Compare with the accepted correlation, Eq. (6.145).

Boundary-Layer Flows

7.1 INTRODUCTION

The previous chapter considered the effect of viscosity on "internal" flows, as in pipes, ducts, and diffusers. In that case the viscous boundary layers which begin to grow outward from the walls at the entrance soon fill the entire duct, and viscous stresses dominate the flow. For example, the Moody chart of Fig. 6.13 is essentially a wall shear-stress correlation for ducts of constant cross section.

An external or unbounded flow moves around a solid surface and is free to expand no matter how thick the viscous layers grow. Thus, far from the body the flow is nearly inviscid, and a very important analysis technique, called *boundary-layer theory* (Sec. 7.3), is to compute the viscous-layer motion near the walls and "patch" it onto the outer inviscid flow. The patching is more successful as the Reynolds number becomes larger, as Fig. 7.1 illustrates.

In Fig 7.1 a uniform stream U moves parallel to a sharp flat plate of length L. If the Reynolds number UL/ν is low (Fig. 7.1a), the viscous region is very broad and extends far ahead and to the sides of the plate. The plate retards the oncoming stream greatly, and small changes in flow parameters cause large changes in the pressure distribution along the plate. Thus, although it should in principle be possible to patch the viscous and inviscid layers in a mathematical analysis, their interaction is strong and nonlinear [1–3]. There is no existing simple theory for external-flow analysis at Reynolds numbers from 1 to about 1000. Such thick-shear-layer flows are typically studied by experiment or by numerical modeling of the flow field on a digital computer [4].

A high-Reynolds-number flow (Fig. 7.1b) is much more amenable to boundary-layer patching, as first pointed out by Prandtl in 1904. The viscous layers, either laminar or turbulent, are very thin, thinner even than the drawing shows. We define the boundary-layer thickness δ as the locus of points where the velocity u parallel

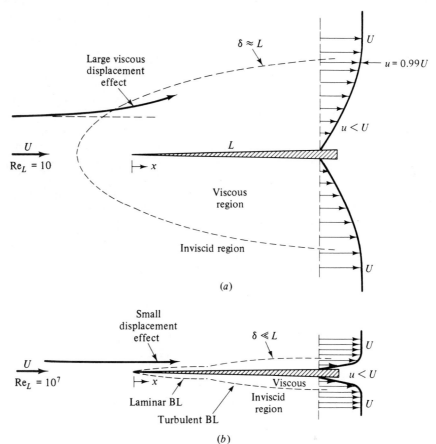

to the plate reaches 99 percent of the external velocity U. As we shall see in Sec. 7.4, the accepted formulas for flat-plate flow are

$$\frac{\delta}{x} \approx \begin{cases} \dfrac{5.0}{\text{Re}_x^{1/2}} & \text{laminar} \qquad\qquad (7.1a) \\[2ex] \dfrac{0.16}{\text{Re}_x^{1/7}} & \text{turbulent} \qquad\quad (7.1b) \end{cases}$$

where $\text{Re}_x = Ux/\nu$ is called the *local Reynolds number* of the flow along the plate surface. The turbulent-flow formula applies for Re_x greater than approximately 10^6.

Some computed values from Eq. (7.1) are

Re_x	10^4	10^5	10^6	10^7	10^8
$(\delta/x)_{\text{lam}}$	0.050	0.016	0.005		
$(\delta/x)_{\text{turb}}$			0.022	0.016	0.011

The blanks indicate that the formula is not applicable. In all cases these boundary layers are so thin that their displacement effect on the outer inviscid layer is negligible. Thus the pressure distribution along the plate can be computed from inviscid theory as if the boundary layer were not even there. This external pressure field then "drives" the boundary-layer flow, acting as a forcing function in the momentum equation along the surface. We shall explain this boundary-layer theory in Secs. 7.4 and 7.5.

For slender bodies, such as plates and airfoils parallel to the oncoming stream, we conclude that this assumption of negligible interaction between the boundary layer and the outer pressure distribution is an excellent approximation.

For a blunt-body flow, however, even at very high Reynolds numbers, there is a discrepancy in the viscous-inviscid patching concept. Figure 7.2 shows two sketches of flow past a two- or three-dimensional blunt body. In the idealized sketch (7.2a) there is a thin film of boundary layer about the body and a narrow sheet of viscous wake in the rear. The patching theory would be glorious for this picture, but it is false. In the actual flow (Fig. 7.2b) the boundary layer is thin on the front, or windward, side of the body, where the pressure decreases along the surface (*favorable* pressure gradient). But in the rear the boundary layer encounters increasing pressure (*adverse* pressure gradient) and breaks off, or separates, into a broad, pulsating wake. (See Fig. 5.2a for a photograph of a specific example.) The main stream is deflected by this wake, so that the external flow is quite different from the prediction from inviscid theory with the addition of a thin boundary layer.

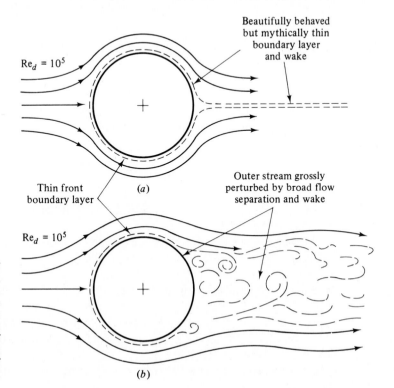

Fig. 7.2 Illustration of the strong interaction between viscous and inviscid regions in the rear of blunt-body flow: (a) idealized and definitely false picture of blunt-body flow; (b) actual picture of blunt-body flow.

The theory of strong interaction between blunt-body viscous and inviscid layers is not well developed. Flows like Fig. 7.2b are normally studied experimentally. Reference 5 is an example of recent intensive effort to improve the theory of separated-boundary-layer flows. Reference 6 is a textbook devoted to separated flow.

EXAMPLE 7.1 A long thin flat plate is placed parallel to a 20 ft/s stream of water at 20°C. At what distance x from the leading edge will the boundary-layer thickness be 1 in?

Solution Since we do not know the Reynolds number, we must guess which of Eqs. (7.1) applies. From Table 1.3 for water $v = 1.09 \times 10^{-5}$ ft^2/s; hence

$$\frac{U}{v} = \frac{20 \text{ ft/s}}{1.09 \times 10^{-5} \text{ ft}^2/\text{s}} = 1.84 \times 10^6 \text{ ft}^{-1}$$

With $\delta = 1$ in $= \frac{1}{12}$ ft, try Eq. (7.1a):

Laminar flow:
$$\frac{\delta}{x} = \frac{5}{(Ux/v)^{1/2}}$$

or
$$x = \frac{\delta^2(U/v)}{5^2} = \frac{(\frac{1}{12} \text{ ft})^2(1.84 \times 10^6 \text{ ft}^{-1})}{25} = 511 \text{ ft}$$

Now we can test the Reynolds number to see whether the formula applied:

$$\text{Re}_x = \frac{U}{v} x = \frac{(20 \text{ ft/s})(511 \text{ ft})}{1.09 \times 10^{-5} \text{ ft}^2/\text{s}} = 9.4 \times 10^8$$

This is impossible since the maximum Re_x for laminar flow past a flat plate is 3×10^6. So we try again with Eq. (7.1b):

Turbulent flow:
$$\frac{\delta}{x} = \frac{0.16}{(Ux/v)^{1/7}}$$

or
$$x = \left[\frac{\delta(U/v)^{1/7}}{0.16}\right]^{7/6} = \left[\frac{(\frac{1}{12} \text{ ft})(1.84 \times 10^6 \text{ ft}^{-1})^{1/7}}{0.16}\right]^{7/6} = (4.09)^{7/6} = 5.17 \text{ ft} \qquad Ans.$$

Test
$$\text{Re}_x = \frac{(20 \text{ ft/s})(5.17 \text{ ft})}{1.09 \times 10^{-5} \text{ ft}^2/\text{s}} = 9.5 \times 10^6$$

This is a perfectly proper turbulent-flow condition; hence we have found the correct position x on our second try.

7.2 MOMENTUM-INTEGRAL ESTIMATES

When we derived the momentum-integral relation, Eq. (3.37), and applied it to a flat-plate boundary layer in Example 3.11, we promised to consider it further in Chap. 7. Well, here we are! Let us review the problem using Fig. 7.3.

A shear layer of unknown thickness grows along the sharp flat plate in Fig. 7.3. The no-slip wall condition retards the flow, making it into a rounded profile $u(y)$, which merges into the external velocity $U = $ constant at a "thickness" $y = \delta(x)$. By

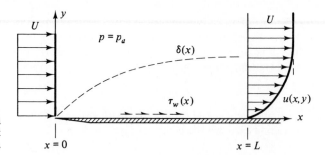

Fig. 7.3 Growth of a boundary layer on a flat plate.

utilizing the control volume of Fig. 3.11 we found (without making any assumptions about laminar versus turbulent flow) in Example 3.11 that the drag force on the plate is given by the following momentum integral across the exit plane

$$D(x) = \rho b \int_0^{\delta(x)} u(U - u) \, dy \qquad (7.2)$$

where b is the plate width into the paper and the integration is carried out along a vertical plane $x = $ constant. You should review the momentum-integral relation (3.37) and its use in Example 3.11.

Kármán's Analysis of the Flat Plate

Equation (7.2) was derived in 1921 by Kármán [7], who wrote it in the convenient form of the *momentum thickness* θ

$$D(x) = \rho b U^2 \theta \qquad \theta = \int_0^{\delta} \frac{u}{U}\left(1 - \frac{u}{U}\right) dy \qquad (7.3)$$

Momentum thickness is thus a measure of total plate drag. Kármán then noted that the drag also equals the integrated wall shear stress along the plate

$$D(x) = b \int_0^x \tau_w(x) \, dx$$

or

$$\frac{dD}{dx} = b\tau_w \qquad (7.4)$$

Meanwhile, the derivative of Eq. (7.3), with $U = $ constant, is

$$\frac{dD}{dx} = \rho b U^2 \frac{d\theta}{dx}$$

By comparing this with Eq. (7.4) Kármán arrived at what is now called the *momentum-integral relation* for flat-plate boundary-layer flow

$$\tau_w = \rho U^2 \frac{d\theta}{dx} \qquad (7.5)$$

It is valid for either laminar or turbulent flow.

To get a numerical result for laminar flow, Kármán assumed that the velocity profiles had an approximately parabolic shape

$$u(x, y) \approx U\left(\frac{2y}{\delta} - \frac{y^2}{\delta^2}\right) \qquad 0 \le y \le \delta(x) \tag{7.6}$$

making it possible to estimate both momentum thickness and wall shear

$$\theta = \int_0^\delta \left(\frac{2y}{\delta} - \frac{y^2}{\delta^2}\right)\left(1 - \frac{2y}{\delta} + \frac{y^2}{\delta^2}\right) dy \approx \tfrac{2}{15}\delta$$

$$\tau_w = \mu \frac{\partial u}{\partial y}\bigg|_{y=0} \approx \frac{2\mu U}{\delta} \tag{7.7}$$

By substituting (7.7) into (7.5) and rearranging we obtain

$$\delta \, d\delta \approx 15\frac{v}{U}\, dx \tag{7.8}$$

where $v = \mu/\rho$. We can integrate from 0 to x, assuming that $\delta = 0$ at $x = 0$, the leading edge

$$\tfrac{1}{2}\delta^2 = \frac{15vx}{U}$$

or
$$\frac{\delta}{x} \approx 5.5\left(\frac{v}{Ux}\right)^{1/2} = \frac{5.5}{\mathrm{Re}_x^{1/2}} \tag{7.9}$$

This is the desired thickness estimate. It is all approximate, of course, part of Kármán's *momentum-integral theory* [7], but it is startlingly accurate, being only 10 percent higher than the known exact solution for laminar flat-plate flow, which we gave as Eq. (7.1a).

By combining Eqs. (7.9) and (7.7) we also obtain a shear-stress estimate along the plate

$$c_f = \frac{2\tau_w}{\rho U^2} \approx \left(\frac{\frac{8}{15}}{\mathrm{Re}_x}\right)^{1/2} = \frac{0.73}{\mathrm{Re}_x^{1/2}} \tag{7.10}$$

Again this estimate, in spite of the crudeness of the profile assumption (7.3), is only 10 percent higher than the known exact laminar-plate-flow solution $c_f = 0.664/\mathrm{Re}_x^{1/2}$, to be treated in Sec. 7.4. The dimensionless quantity c_f, called the *skin-friction coefficient*, is analogous to the friction factor f in ducts.

A boundary layer can be judged as "thin" if, say, the ratio δ/x is less than about 0.1. This occurs at $\delta/x = 0.1 = 5.0/\mathrm{Re}_x^{1/2}$ or at $\mathrm{Re}_x = 2500$. For Re_x less than 2500 we can estimate that boundary-layer theory fails because the thick layer has a significant effect on the outer inviscid flow. The upper limit on Re_x for laminar flow is about 3×10^6, where measurements on a smooth flat plate [8] show that the flow undergoes transition to a turbulent boundary layer. From 3×10^6 upward the turbulent Reynolds number may be arbitrarily large, and a practical limit at present is 5×10^9 for oil supertankers.

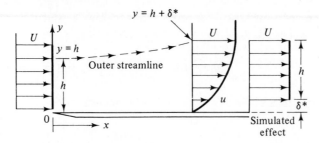

Fig. 7.4 Displacement effect of a boundary layer.

Displacement Thickness

Another interesting effect of a boundary layer is its small but finite displacement of the outer streamlines. As shown in Fig. 7.4, outer streamlines must deflect outward a distance $\delta^*(x)$ to satisfy conservation of mass between the inlet and outlet

$$\int_0^h \rho U b \, dy = \int_0^\delta \rho u b \, dy \qquad \delta = h + \delta^* \tag{7.11}$$

The quantity δ^* is called the *displacement thickness* of the boundary layer. To relate it to $u(y)$, cancel ρ and b from Eq. (7.11), evaluate the left integral, and slyly add and subtract U from the right integrand

$$Uh = \int_0^\delta (U + u - U) \, dy = U(h + \delta^*) + \int_0^\delta (u - U) \, dy$$

or

$$\delta^* = \int_0^\delta \left(1 - \frac{u}{U}\right) dy \tag{7.12}$$

Thus the ratio of δ^*/δ varies only with the dimensionless velocity-profile shape u/U.

Introducing our profile approximation (7.6) into (7.12), we obtain by integration the approximate result

$$\delta^* \approx \tfrac{1}{3}\delta \qquad \frac{\delta^*}{x} \approx \frac{1.83}{\mathrm{Re}_x^{1/2}} \tag{7.13}$$

These estimates are only 6 percent away from the exact solutions for laminar flat-plate flow given in Sec. 7.4: $\delta^* = 0.344\delta = 1.721x/\mathrm{Re}_x^{1/2}$. Since δ^* is much smaller than x for large Re_x and the outer streamline slope V/U is proportional to δ^*, we conclude that the velocity normal to the wall is much smaller than the velocity parallel to the wall. This is a key assumption in boundary-layer theory (Sec. 7.3).

We also conclude from the success of these simple parabolic estimates that Kármán's momentum-integral theory is effective and useful. Many details of this theory are given in Refs. 1 to 3.

EXAMPLE 7.2 Are low-speed, small-scale air and water boundary layers really thin? Consider flow at $U = 1$ ft/s past a flat plate 1 ft long. Compute the boundary-layer thickness at the trailing edge for (*a*) air and (*b*) water at 20°C.

Solution
Part (a)   From Table 1.3 $v_{air} = 1.63 \times 10^{-4}$ ft^2/s. The trailing-edge Reynolds number thus is

$$\text{Re}_L = \frac{UL}{v} = \frac{(1 \text{ ft/s})(1 \text{ ft})}{1.63 \times 10^{-4} \text{ ft}^2/\text{s}} = 6150$$

Since this is less than 10^6, the flow is laminar, and since it is greater than 2500, we expect a thin boundary layer. The actual thickness from Eq. (7.1a) is

$$\frac{\delta}{x} = \frac{5.0}{(6150)^{1/2}} = 0.0637$$

or at $x = 1$ ft

$$\delta = 0.0637 \text{ ft} = 0.765 \text{ in} \hspace{3cm} \textit{Ans. (a)}$$

Part (b)   From Table 1.3, $v_{water} = 1.09 \times 10^{-5}$ ft^2/s. The Reynolds number is

$$\text{Re}_L = \frac{1(1)}{1.09 \times 10^{-5}} = 92,000$$

This again satisfies the laminar and thinness conditions. The trailing-edge thickness is

$$\frac{\delta}{x} = \frac{5.0}{(92,000)^{1/2}} = 0.0165$$

or
$$\delta(\text{at } x = 1 \text{ ft}) = 0.0165 \text{ ft} = 0.20 \text{ in} \hspace{2cm} \textit{Ans. (b)}$$

Thus, even at such low velocities and short lengths, both air and water satisfy the boundary-layer approximations.

7.3   THE BOUNDARY-LAYER EQUATIONS

In Chap. 6 we learned that there are several dozen known laminar-flow solutions [1–3]. Almost none are for external flows around immersed bodies although this is one of the primary applications of fluid mechanics. And no exact turbulent-flow solutions are known whatever.

There are presently three techniques used to treat external flows: (1) numerical digital-computer solutions, (2) experimentation, and (3) boundary-layer theory.

Numerical fluid mechanics is a bright new tool described in the advanced text by Roache [4]. Hundreds of interesting computer solutions have been published, and computer speeds and graphical presentations are improving each year. To date almost all are laminar-flow solutions, but there is some progress toward computer analysis of turbulent flows [9]. Accurate numerical modeling of turbulence is a frontier research topic, but the time is not yet ripe for discussion of numerical fluid dynamics in an introductory textbook.

Experimentation is the most common method of studying external flows. Chapter 5 outlined the dimensional-analysis technique for presenting data, and several experimental results were given in Chaps. 5 and 6. We shall give a great many experimental results on external flows in Sec. 7.6.

The third tool is boundary-layer theory, first formulated by Ludwig Prandtl in 1904. We shall follow Prandtl's ideas here and make certain order-of-magnitude assumptions to greatly simplify the Navier-Stokes equations (4.38) into boundary-layer equations which are solved relatively easily and patched onto the outer inviscid-flow field.

One of the great achievements of boundary-layer theory is its ability to predict the flow separation illustrated in Fig. 7.2b. Before 1904 no one realized that such thin shear layers could cause such a gross effect as flow separation. Unfortunately, even today the theory cannot accurately predict the behavior of the separated flow region and its interaction with the outer layer. This is the weakness of boundary-layer theory, which we hope will be resolved by intensive research into the dynamics of separated flows [6].

Derivation for Two-Dimensional Flow

We consider only steady two-dimensional incompressible viscous flow with the x direction along the wall and y normal to the wall, as in Fig. 7.3.[1] We neglect gravity, which is important only in boundary layers where fluid buoyancy is dominant [2, sec. 4.12]. From Chap. 4, the complete equations of motion consist of continuity and the x- and y-momentum relations

$$\frac{\partial u}{\partial x} + \frac{\partial v}{\partial y} = 0 \tag{7.14a}$$

$$\rho\left(u \frac{\partial u}{\partial x} + v \frac{\partial u}{\partial y}\right) = -\frac{\partial p}{\partial x} + \mu\left(\frac{\partial^2 u}{\partial x^2} + \frac{\partial^2 u}{\partial y^2}\right) \tag{7.14b}$$

$$\rho\left(u \frac{\partial v}{\partial x} + v \frac{\partial v}{\partial y}\right) = -\frac{\partial p}{\partial y} + \mu\left(\frac{\partial^2 v}{\partial x^2} + \frac{\partial^2 v}{\partial y^2}\right) \tag{7.14c}$$

These should be solved for u, v, and p subject to typical no-slip, inlet, and exit boundary conditions, but in fact they are too difficult to handle for most external flows.

In 1904 Prandtl correctly deduced that a shear layer must be very thin if the Reynolds number is large, so that the following approximations apply:

Velocities: $$v \ll u \tag{7.15a}$$

Rates of change: $$\frac{\partial}{\partial x} \ll \frac{\partial}{\partial y} \tag{7.15b}$$

Our discussion of displacement thickness in the previous section was intended to justify these assumptions.

Applying these approximations to Eq. (7.14c) results in a powerful simplification

$$\frac{\partial p}{\partial y} \approx 0 \quad\text{or}\quad p \approx p(x) \text{ only} \tag{7.16}$$

---

[1] For a curved wall, x can represent arc length along the wall and y can be everywhere normal to x with negligible change in the boundary-layer equations as long as the radius of curvature of the wall is large compared with boundary-layer thickness [1–3].

In other words, the y-momentum equation can be neglected entirely, and the pressure varies only *along* the boundary layer, not through it. The pressure-gradient term in Eq. (7.14b) is assumed to be known in advance from Bernoulli's equation applied to the outer inviscid flow

$$\frac{\partial p}{\partial x} = \frac{dp}{dx} = -\rho U \frac{dU}{dx} \tag{7.17}$$

Presumably we have already made the inviscid analysis and know the distribution of $U(x)$ along the wall (Chap. 8).

Meanwhile, one term in Eq. (7.14b) is negligible due to Eqs. (7.15)

$$\frac{\partial^2 u}{\partial x^2} \ll \frac{\partial^2 u}{\partial y^2} \tag{7.18}$$

However, neither term in the continuity relation (7.14a) can be neglected—another warning that continuity is always a vital part of any fluid-flow analysis.

The net result is that the three full equations of motion (7.14) are reduced to Prandtl's two boundary-layer equations

Continuity:
$$\frac{\partial u}{\partial x} + \frac{\partial v}{\partial y} = 0 \tag{7.19a}$$

Momentum along wall:
$$u \frac{\partial u}{\partial x} + v \frac{\partial u}{\partial y} \approx U \frac{dU}{dx} + \frac{1}{\rho} \frac{\partial \tau}{\partial y} \tag{7.19b}$$

where
$$\tau = \begin{cases} \mu \dfrac{\partial u}{\partial y} & \text{laminar flow} \\[2ex] \mu \dfrac{\partial u}{\partial y} - \rho \overline{u'v'} & \text{turbulent flow} \end{cases}$$

These are to be solved for $u(x, y)$ and $v(x, y)$, with $U(x)$ assumed to be a known function from the outer inviscid-flow analysis. There are two boundary conditions on u and one on v

At $y = 0$ (wall): $\qquad\qquad u = v = 0 \qquad$ (no slip) $\qquad\qquad$ (7.20a)

At $y = \delta(x)$ (outer stream): $\quad u = U(x) \qquad$ (patching) $\qquad\qquad$ (7.20b)

Unlike the Navier-Stokes equations (7.14), which are mathematically elliptic and must be solved simultaneously over the entire flow field, the boundary-layer equations (7.19) are mathematically parabolic and are solved by beginning at the leading edge and marching downstream as far as you like, stopping at the separation point or earlier if you prefer.[1]

The boundary-layer equations have been solved for scores of interesting cases of internal and external flow for both laminar and turbulent flow, utilizing the inviscid

---

[1] For further mathematical details, see Ref. 2, sec. 2.8.

distribution $U(x)$ appropriate to each flow. Full details of boundary-layer theory and results and comparison with experiment are given in Refs. 1 to 3 and 9. Here we shall confine ourselves primarily to flat-plate solutions (See. 7.4).

7.4 THE FLAT-PLATE BOUNDARY LAYER

The classic and most often used solution of boundary-layer theory is for flat-plate flow, as in Fig. 7.3, which can represent either laminar or turbulent flow.

Laminar Flow

For laminar flow past the plate, the boundary-layer equations (7.19) can be solved exactly for u and v, assuming that the free-stream velocity U is constant ($dU/dx = 0$). The solution was given by Prandtl's student, Blasius, in his 1908 dissertation from Göttingen. With an ingenious coordinate transformation, Blasius showed that the dimensionless velocity profile u/U is a function only of the single composite dimensionless variable $(y)(U/vx)^{1/2}$:

$$\frac{u}{U} = f'(\eta) \qquad \eta = y\left(\frac{U}{vx}\right)^{1/2} \tag{7.21}$$

where the prime denotes differentiation with respect to η. Substitution of (7.21) into the boundary-layer equations (7.19) reduces the problem, after much algebra, to a single third-order nonlinear ordinary differential equation for f

$$f''' + \tfrac{1}{2}ff'' = 0 \tag{7.22}$$

The boundary conditions (7.20) become

At $y = 0$: $\qquad\qquad\qquad f(0) = f'(0) = 0 \tag{7.23a}$

As $y \to \infty$: $\qquad\qquad\qquad f'(\infty) \to 1.0 \tag{7.23b}$

This is the *Blasius equation*, for which accurate solutions have been obtained only by numerical integration. Some tabulated values of the velocity-profile shape $f'(\eta) = u/U$ are given in Table 7.1.

Since u/U approaches 1.0 only as $y \to \infty$, it is customary to select the boundary-layer thickness δ as that point where $u/U = 0.99$. From the table, this occurs at $\eta \approx 5.0$:

$$\delta_{99\%}\left(\frac{U}{vx}\right)^{1/2} \approx 5.0$$

or $\qquad\qquad \dfrac{\delta}{x} \approx \dfrac{5.0}{\mathrm{Re}_x^{1/2}} \qquad$ Blasius (1908) $\tag{7.24}$

With the profile known, Blasius of course could also compute the wall shear and displacement thickness

$$c_f = \frac{0.664}{\mathrm{Re}_x^{1/2}} \qquad \frac{\delta^*}{x} = \frac{1.721}{\mathrm{Re}_x^{1/2}} \tag{7.25}$$

Table 7.1

THE BLASIUS VELOCITY PROFILE [1-3]

$y(U/vx)^{1/2}$	u/U	$y(U/vx)^{1/2}$	u/U
0.0	0.0	2.8	0.81152
0.2	0.06641	3.0	0.84605
0.4	0.13277	3.2	0.87609
0.6	0.19894	3.4	0.90177
0.8	0.26471	3.6	0.92333
1.0	0.32979	3.8	0.94112
1.2	0.39378	4.0	0.95552
1.4	0.45627	4.2	0.96696
1.6	0.51676	4.4	0.97587
1.8	0.57477	4.6	0.98269
2.0	0.62977	4.8	0.98779
2.2	0.68132	5.0	0.99155
2.4	0.72899	∞	1.00000
2.6	0.77246		

Notice how close these are to our integral estimates, Eqs. (7.9), (7.10), and (7.13). When c_f is converted into dimensional form, we have

$$\tau_w(x) = \frac{0.332\rho^{1/2}\mu^{1/2}U^{1.5}}{x^{1/2}}$$

The wall shear drops off with $x^{1/2}$ because of boundary-layer growth and varies as velocity to the 1.5 power. This is in contrast to laminar pipe flow, where $\tau_w \propto U$ and is independent of x.

If $\tau_w(x)$ is substituted into Eq. (7.4), we compute the total drag force

$$D(x) = b\int_0^x \tau_w(x)\,dx = 0.664b\rho^{1/2}\mu^{1/2}U^{1.5}x^{1/2} \tag{7.26}$$

The drag increases only as the square root of the plate length. The nondimensional *drag coefficient* is defined as

$$C_D = \frac{2D(L)}{\rho U^2 bL} = \frac{1.328}{\text{Re}_L^{1/2}} = 2c_f(L) \tag{7.27}$$

Thus, for laminar plate flow, C_D equals twice the value of the skin-friction coefficient at the trailing edge. This is the drag on one side of the plate.

Kármán pointed out that the drag could also be computed from the momentum relation (7.2). In dimensionless form, Eq. (7.2) becomes

$$C_D = \frac{2}{L}\int_0^\delta \frac{u}{U}\left(1 - \frac{u}{U}\right)dy \tag{7.28}$$

This can be rewritten in terms of the momentum thickness at the trailing edge

$$C_D = \frac{2\theta(L)}{L} \tag{7.29}$$

Computation of θ from the profile u/U or from C_D gives

$$\frac{\theta}{x} = \frac{0.664}{\text{Re}_x^{1/2}} \qquad \text{laminar flat plate} \tag{7.30}$$

Since δ is so ill defined, the momentum thickness, being definite, is often used to correlate data taken for a variety of boundary layers under differing conditions. The ratio of displacement to momentum thickness, called the dimensionless-profile *shape factor*, is also useful in integral theories. For laminar flat-plate flow

$$H = \frac{\delta^*}{\theta} = \frac{1.721}{0.664} = 2.59 \tag{7.31}$$

As we shall see, a large shape factor implies that boundary-layer separation is about to occur.

If we plot the Blasius velocity profile from Table 7.1 in the form of u/U versus y/δ, we can see why the simple integral-theory guess, Eq. (7.6), was such a great success. This is done in Fig. 7.5. The simple parabolic approximation is not far from

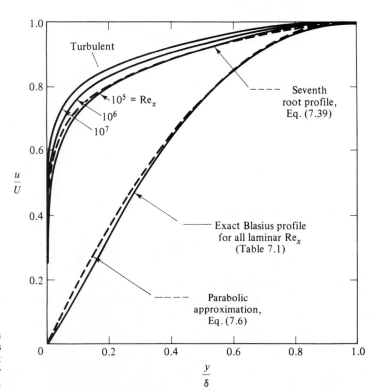

Fig. 7.5 Comparison of dimensionless laminar and turbulent flat-plate velocity profiles.

the true Blasius profile; hence its momentum thickness is within 10 percent of the true value. Also shown in Fig. 7.5 are three typical turbulent flat-plate velocity profiles. Notice how strikingly different in shape they are from the laminar profiles. Instead of decreasing monotonically to zero, the turbulent profiles are very flat and then drop off sharply at the wall. As you might guess, they follow the logarithmic-law shape and thus can be analyzed by momentum-integral theory if this shape is properly represented.

EXAMPLE 7.3   A sharp flat plate with $L = 1$ m and $b = 3$ m is immersed parallel to a stream of velocity 2 m/s. Find the drag on one side of the plate, and at the trailing edge find the thicknesses δ, δ^*, and θ for (a) air, $\rho = 1.23$ kg/m^3 and $v = 1.46 \times 10^{-5}$ m^2/s and (b) water, $\rho = 1000$ kg/m^3 and $v = 1.02 \times 10^{-6}$ m^2/s.

Solution
Part (a)   The airflow Reynolds number is

$$\frac{VL}{v} = \frac{(2.0 \text{ m/s})(1.0 \text{ m})}{1.46 \times 10^{-5} \text{ m}^2/\text{s}} = 137{,}000$$

Since this is less than 3×10^6, we assume that the boundary layer is laminar. From Eq. (7.27), the drag coefficient is

$$C_D = \frac{1.328}{(137{,}000)^{1/2}} = 0.00359$$

Thus the drag on one side in the airflow is

$$D = C_D \tfrac{1}{2}\rho U^2 bL = 0.00359(\tfrac{1}{2})(1.23)(2.0)^2(3.0)(1.0) = 0.0265 \text{ N} \qquad \textit{Ans. (a)}$$

The boundary-layer thickness at the end of the plate is

$$\frac{\delta}{L} = \frac{5.0}{\text{Re}_L^{1/2}} = \frac{5.0}{(137{,}000)^{1/2}} = 0.0135$$

or $\qquad\qquad\qquad \delta = 0.0135(1.0) = 0.0135 \text{ m} = 13.5 \text{ mm} \qquad\qquad\qquad \textit{Ans. (a)}$

We find the other two thicknesses simply by ratios:

$$\delta^* = \frac{1.721}{5.0}\delta = 4.65 \text{ mm} \qquad \theta = \frac{\delta^*}{2.59} = 1.79 \text{ mm} \qquad \textit{Ans. (a)}$$

Notice that no conversion factors are needed with SI units.

Part (b)   The water Reynolds number is

$$\text{Re}_L = \frac{2.0(1.0)}{1.02 \times 10^{-6}} = 1.96 \times 10^6$$

This is rather close to the critical value of 3×10^6, so that a rough surface or noisy free stream might trigger transition to turbulence; but let us assume that the flow is laminar. The water drag coefficient is

$$C_D = \frac{1.328}{(1.96 \times 10^6)^{1/2}} = 0.000949$$

and $\qquad\qquad D = 0.000949(\tfrac{1}{2})(1000)(2.0)^2(3.0)(1.0) = 5.70 \text{ N} \qquad\qquad \textit{Ans. (b)}$

The drag is 215 times more for water in spite of the higher Reynolds number and lower drag coefficient because water is 57 times more viscous and 813 times denser than air. From Eq. (7.26), in laminar flow, it should have $(57)^{1/2}(813)^{1/2} = 7.53(28.5) = 215$ times more drag.

The boundary-layer thickness is given by

$$\frac{\delta}{L} = \frac{5.0}{(1.96 \times 10^6)^{1/2}} = 0.00357$$

or
$$\delta = 0.00357(1000 \text{ mm}) = 3.57 \text{ mm} \qquad \qquad \textit{Ans.} (b)$$

By ratioing down we have

$$\delta^* = \frac{1.721}{5.0} \delta = 1.23 \text{ mm} \qquad \theta = \frac{\delta^*}{2.59} = 0.48 \text{ mm} \qquad \textit{Ans.} (b)$$

The water layer is 3.8 times thinner than the air layer, which reflects the square root of the 14.3 ratio of air to water kinematic viscosity.

Turbulent Flow

There is no exact theory for turbulent flat-plate flow, although there are many elegant computer solutions of the boundary-layer equations using various empirical models for the turbulent eddy viscosity [9]. The most widely accepted result is simply an integral analysis similar to our study of the laminar-profile approximation (7.6).

We begin with Eq. (7.5), which is valid for laminar or turbulent flow. Write it here for convenient reference:

$$\tau_w(x) = \rho U^2 \frac{d\theta}{dx} \tag{7.32}$$

From the definition of c_f, Eq. (7.10), this can be rewritten as

$$c_f = 2\frac{d\theta}{dx} \tag{7.33}$$

Now recall from Fig. 7.5 that the turbulent profiles are nowhere near parabolic. Going back to Fig. 6.9, we see that flat-plate flow is very nearly logarithmic, with a slight outer wake and a thin viscous sublayer. Therefore, just as in turbulent pipe flow, we assume that the logarithmic law (6.21) holds all the way across the boundary layer

$$\frac{u}{u^*} \approx \frac{1}{\kappa} \ln \frac{yu^*}{v} + B \qquad u^* = \left(\frac{\tau_w}{\rho}\right)^{1/2} \tag{7.34}$$

with, as usual, $\kappa = 0.41$ and $B = 5.0$. At the outer edge of the boundary layer, $y = \delta$ and $u = U$, and Eq. (7.34) becomes

$$\frac{U}{u^*} = \frac{1}{\kappa} \ln\left(\frac{\delta u^*}{v}\right) + B \tag{7.35}$$

But the definition of skin-friction coefficient, Eq. (7.10), is such that the following identities hold:

$$\frac{U}{u^*} \equiv \left(\frac{2}{c_f}\right)^{1/2} \qquad \frac{\delta u^*}{v} \equiv \mathrm{Re}_\delta \left(\frac{c_f}{2}\right)^{1/2} \qquad (7.36)$$

Therefore Eq. (7.35) is a *skin-friction law* for turbulent flat-plate flow

$$\left(\frac{2}{c_f}\right)^{1/2} \approx 2.44 \ln\left[\mathrm{Re}_\delta \left(\frac{c_f}{2}\right)^{1/2}\right] + 5.0 \qquad (7.37)$$

It is a complicated law, but we can at least solve for a few values and list them:

Re_δ	10^4	10^5	10^6	10^7
c_f	0.00493	0.00315	0.00217	0.00158

Following a suggestion of Prandtl, we can forget the complex log-friction law (7.37) and simply fit the numbers in the table to a power-law approximation

$$c_f \approx 0.02 \, \mathrm{Re}_\delta^{-1/6} \qquad (7.38)$$

This we shall use as the left-hand side of Eq. (7.33). For the right-hand side, we need an estimate for $\theta(x)$ in terms of $\delta(x)$. If we use the logarithmic-law profile (7.34), we shall be up to our hips in logarithmic integrations for the momentum thickness. Instead we follow another suggestion of Prandtl's, who pointed out that the turbulent profiles in Fig. 7.5 can be approximated by a one-seventh-power law

$$\left(\frac{u}{U}\right)_{\mathrm{turb}} \approx \left(\frac{y}{\delta}\right)^{1/7} \qquad (7.39)$$

This is shown as a dashed line in Fig. 7.5. It is an excellent fit to the low-Reynolds-number turbulent data, which were all that were available to Prandtl at the time. With this simple approximation, the momentum thickness (7.28) can easily be evaluated

$$\theta \approx \int_0^\delta \left(\frac{y}{\delta}\right)^{1/7}\left[1 - \left(\frac{y}{\delta}\right)^{1/7}\right] dy = \tfrac{7}{72}\delta \qquad (7.40)$$

We accept this result and substitute Eqs. (7.38) and (7.40) into Kármán's momentum law (7.33)

$$c_f = 0.02 \, \mathrm{Re}_\delta^{-1/6} = 2\frac{d}{dx}(\tfrac{7}{72}\delta)$$

or

$$\mathrm{Re}_\delta^{-1/6} = 9.72\frac{d\delta}{dx} = 9.72\frac{d(\mathrm{Re}_\delta)}{d(\mathrm{Re}_x)} \qquad (7.41)$$

Separate the variables and integrate, assuming $\delta = 0$ at $x = 0$:

$$\mathrm{Re}_\delta \approx 0.16 \, \mathrm{Re}_x^{6/7} \qquad \text{or} \qquad \frac{\delta}{x} \approx \frac{0.16}{\mathrm{Re}_x^{1/7}} \qquad (7.42)$$

Thus the thickness of a turbulent boundary layer increases as $x^{6/7}$, far faster than the laminar increase $x^{1/2}$. Equation (7.42) is the solution to the problem, because all other parameters are now available. For example, combining Eqs. (7.42) and (7.38), we obtain the friction variation

$$c_f \approx \frac{0.027}{\text{Re}_x^{1/7}} \tag{7.43}$$

Writing this out in dimensional form, we have

$$\tau_{w,\text{turb}} \approx \frac{0.0135 \mu^{1/7} \rho^{6/7} U^{13/7}}{x^{1/7}} \tag{7.44}$$

Turbulent plate friction drops slowly with x, increases nearly as ρ and U^2, and is rather insensitive to viscosity.

We can evaluate the drag coefficient from Eq. (7.29)

$$C_D = \frac{0.031}{\text{Re}_L^{1/7}} = \tfrac{7}{6} c_f(L) \tag{7.45}$$

Then C_D is only 16 percent greater than the trailing-edge skin friction [compare with Eq. (7.27) for laminar flow].

The displacement thickness can be estimated from the one-seventh power law and Eq. (7.12)

$$\delta^* \approx \int_0^\delta \left[1 - \left(\frac{y}{\delta}\right)^{1/7} \right] dy = \tfrac{1}{8} \delta \tag{7.46}$$

The turbulent flat-plate shape factor is approximately

$$H = \frac{\delta^*}{\theta} = \frac{1/8}{7/72} = 1.3 \tag{7.47}$$

These are the basic results of turbulent flat-plate theory.

Figure 7.6 shows flat-plate drag coefficients for both laminar and turbulent flow conditions. The smooth-wall relations (7.27) and (7.45) are shown, along with the effect of wall roughness, which is quite strong. The proper roughness parameter here is x/ϵ or L/ϵ, by analogy with the pipe parameter ϵ/d. In the fully rough regime, C_D is independent of Reynolds number, so that the drag varies exactly as U^2 and is independent of μ. Reference 2 presents a theory of rough flat-plate flow, and Ref. 1 gives a curve fit for skin friction and drag in the fully rough regime:

$$c_f \approx \left(2.87 + 1.58 \log \frac{x}{\epsilon} \right)^{-2.5} \tag{7.48a}$$

$$C_D \approx \left(1.89 + 1.62 \log \frac{L}{\epsilon} \right)^{-2.5} \tag{7.48b}$$

Equation (7.48b) is plotted to the right of the dashed line in Fig. 7.6. The figure also shows the behavior of the drag coefficient in the transition region, $5 \times 10^5 < \text{Re}_L < 8 \times 10^7$, where the laminar drag at the leading edge is an appreciable

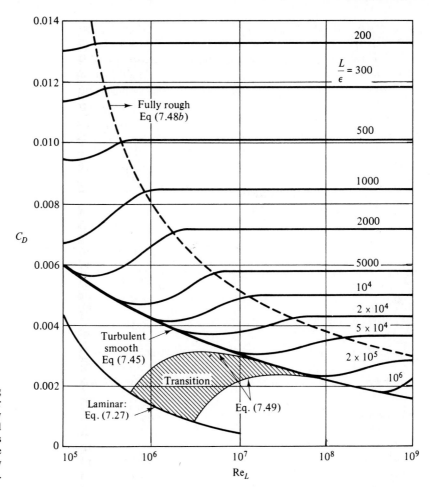

Fig. 7.6 Drag coefficient of laminar and turbulent boundary layer on smooth and rough flat plates. This chart is the flat-plate analog of the Moody diagram of Fig. 6.13.

fraction of the total drag. Schlichting [1] suggests the following curve fits for these transition drag curves depending upon the Reynolds number Re_{trans} where transition begins:

$$C_D \approx \begin{cases} \dfrac{0.031}{Re_L^{1/7}} - \dfrac{1440}{Re_L} & Re_{trans} = 5 \times 10^5 \qquad (7.49a) \\[4mm] \dfrac{0.031}{Re_L^{1/7}} - \dfrac{8700}{Re_L} & Re_{trans} = 3 \times 10^6 \qquad (7.49b) \end{cases}$$

EXAMPLE 7.4 A hydrofoil 1.2 ft long and 6 ft wide is placed in a water flow of 40 ft/s, with $\rho = 1.99$ slug/ft^3 and $v = 0.000011$ ft^2/s. Estimate (a) the boundary-layer thickness at the end of the plate. Estimate the friction drag for (b) turbulent smooth-wall flow from the leading edge, (c) laminar-turbulent flow with $Re_{trans} = 5 \times 10^5$, and (d) turbulent rough-wall flow with $\epsilon = 0.0004$ ft.

Solution

Part (a)   The Reynolds number is

$$\mathrm{Re}_L = \frac{UL}{\nu} = \frac{(40\,\text{ft/s})(1.2\,\text{ft})}{0.000011\,\text{ft}^2/\text{s}} = 4.36 \times 10^6$$

Thus the trailing-edge flow is certainly turbulent. The maximum boundary-layer thickness would occur for turbulent flow starting at the leading edge. From Eq. (7.42),

$$\frac{\delta(L)}{L} = \frac{0.16}{(4.36 \times 10^6)^{1/7}} = 0.018$$

or
$$\delta = 0.018(1.2\,\text{ft}) = 0.0216\,\text{ft} \qquad\qquad Ans.\,(a)$$

This is 7.5 times thicker than a fully laminar boundary layer at the same Reynolds number.

Part (b)   For fully turbulent smooth-wall flow, the drag coefficient on one side of the plate is, from Eq. (7.45),

$$C_D = \frac{0.031}{(4.36 \times 10^6)^{1/7}} = 0.00349$$

Then the drag on both sides of the foil is approximately

$$D = 2C_D(\tfrac{1}{2}\rho U^2)bL = 2(0.00349)(\tfrac{1}{2})(1.99)(40)^2(6.0)(1.2) = 80\,\text{lb} \qquad Ans.\,(b)$$

Part (c)   With a laminar leading edge and $\mathrm{Re}_{\text{trans}} = 5 \times 10^5$, Eq. (7.49a) applies:

$$C_D = 0.00349 - \frac{1440}{4.36 \times 10^6} = 0.00316$$

The drag can be recomputed for this lower drag coefficient:

$$D = 2C_D(\tfrac{1}{2}\rho U^2)bL = 72\,\text{lbf} \qquad\qquad Ans.\,(c)$$

Part (d)   Finally, for the rough wall, we calculate

$$\frac{L}{\epsilon} = \frac{(1.2\,\text{ft})}{0.0004\,\text{ft}} = 3000$$

From Fig. 7.6 at $\mathrm{Re}_L = 4.36 \times 10^6$, this condition is just inside the fully rough regime. Equation (7.48b) applies:

$$C_D = (1.89 + 1.62 \log 3000)^{-2.5} = 0.00644$$

and the drag estimate is

$$D = 2C_D(\tfrac{1}{2}\rho U^2)bL = 148\,\text{lb} \qquad\qquad Ans.\,(d)$$

This small roughness nearly doubles the drag. It is probable that the total hydrofoil drag is still another factor of 2 larger because of trailing-edge flow-separation effects.

7.5 BOUNDARY LAYERS WITH PRESSURE GRADIENT[1]

The flat-plate analysis of the previous section should give us a good feeling for the behavior of both laminar and turbulent boundary layers, except for one important effect: flow separation. Prandtl showed that separation like that in Fig. 7.2b is

[1] This section may be omitted without loss of continuity.

caused by excessive momentum loss near the wall in a boundary layer trying to move downstream against increasing pressure, $dp/dx > 0$, which is called an adverse pressure gradient. The opposite case of decreasing pressure, $dp/dx < 0$, is called a favorable gradient, where flow separation can never occur. In a typical immersed body flow, e.g., Fig. 7.2b, the favorable gradient is on the front of the body and the adverse gradient in the rear, as will be discussed in detail in Chap. 8.

We can explain flow separation with a geometrical argument about the second derivative of velocity u at the wall. From the momentum equation (7.19b) at the wall, where $u = v = 0$, we obtain

$$\left.\frac{\partial \tau}{\partial y}\right|_{\text{wall}} = \left.\mu \frac{\partial^2 u}{\partial y^2}\right|_{\text{wall}} = -\rho U \frac{dU}{dx} = \frac{dp}{dx}$$

or

$$\left.\frac{\partial^2 u}{\partial y^2}\right|_{\text{wall}} = \frac{1}{\mu}\frac{dp}{dx} \qquad (7.50)$$

for either laminar or turbulent flow. Thus in an adverse gradient the second derivative of velocity is positive at the wall, yet it must be negative at the outer layer $(y = \delta)$ to merge smoothly with the mainstream flow $U(x)$. It follows that the second derivative must pass through zero somewhere in between, at a point of inflection, and any boundary-layer profile in an adverse gradient must exhibit a characteristic S shape.

Figure 7.7 illustrates the general case. In a favorable gradient (Fig. 7.7a) the profile is very rounded, there is no point of inflection, there can be no separation, and laminar profiles of this type are very resistant to a transition to turbulence [1–3].

In a zero pressure gradient (Fig. 7.7b), e.g., flat-plate flow, the point of inflection is at the wall itself. There can be no separation, and the flow will undergo transition at Re_x no greater than about 3×10^6, as discussed earlier.

In an adverse gradient (Fig. 7.7c to e), a point of inflection (PI) occurs in the boundary layer, its distance from the wall increasing with the strength of the adverse gradient. For a weak gradient (Fig. 7.7c) the flow does not actually separate, but it is vulnerable to transition to turbulence at Re_x as low as 10^5 [1, 2]. At a moderate gradient, a critical condition (Fig. 7.7d) is reached where the wall shear is exactly zero $(\partial u/\partial y = 0)$. This is defined as the *separation point* $(\tau_w = 0)$, because any stronger gradient will actually cause backflow at the wall (Fig. 7.7e): the boundary layer thickens greatly, and the main flow breaks away, or separates, from the wall (Fig. 7.2b).

The flow profiles of Fig. 7.7 usually occur in sequence as the boundary layer progresses along the wall of a body. For example, in Fig. 7.2a, a favorable gradient occurs on the front of the body, zero pressure gradient occurs just upstream of the shoulder, and an adverse gradient occurs successively as we move around the rear of the body.

A second practical example is the flow in a duct consisting of a nozzle, throat, and diffuser, as in Fig. 7.8. The nozzle flow is a favorable gradient and never

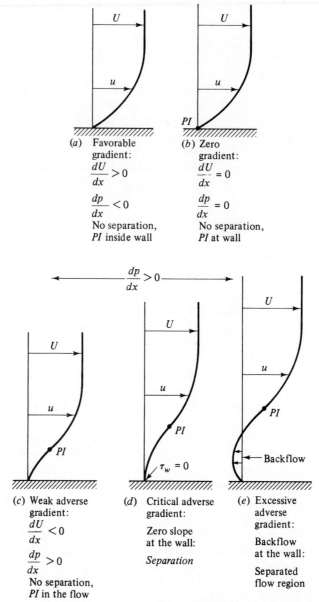

Fig. 7.7 Effect of pressure gradient on boundary-layer profiles; PI = profile point of inflection.

(a) Favorable gradient:

$$\frac{dU}{dx} > 0$$

$$\frac{dp}{dx} < 0$$

No separation, PI inside wall

(b) Zero gradient:

$$\frac{dU}{dx} = 0$$

$$\frac{dp}{dx} = 0$$

No separation, PI at wall

$$\frac{dp}{dx} > 0$$

(c) Weak adverse gradient:

$$\frac{dU}{dx} < 0$$

$$\frac{dp}{dx} > 0$$

No separation, PI in the flow

(d) Critical adverse gradient:

Zero slope at the wall:

Separation

$\tau_w = 0$

(e) Excessive adverse gradient:

Backflow at the wall:

Separated flow region

Backflow

separates nor does the throat flow where the pressure gradient is approximately zero. But the expanding-area diffuser produces low velocity and increasing pressure, an adverse gradient. If the diffuser angle is too large, the adverse gradient is excessive and the boundary layer will separate at one or both walls, with backflow, increased losses, and poor pressure recovery. In the diffuser literature [10] this condition is called *diffuser stall*, a term used also in airfoil aerodynamics (Sec. 7.6) to denote airfoil boundary-layer separation. Thus the boundary-layer

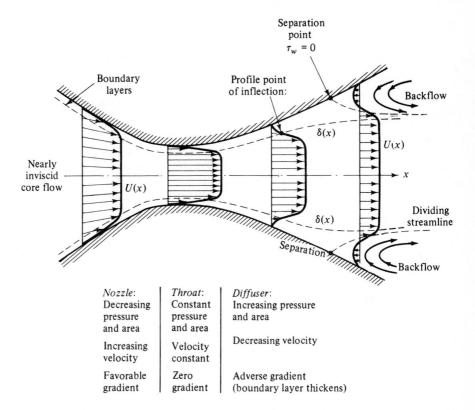

Fig. 7.8 Boundary-layer separation in a diffuser.

Nozzle:	Throat:	Diffuser:
Decreasing pressure and area	Constant pressure and area	Increasing pressure and area
Increasing velocity	Velocity constant	Decreasing velocity
Favorable gradient	Zero gradient	Adverse gradient (boundary layer thickens)

behavior explains why a large-angle diffuser has heavy flow losses (Fig. 6.22) and poor performance (Fig. 6.27).

Presently boundary-layer theory can only compute up to the separation point, after which it is invalid. New techniques are now being developed for analyzing the strong interaction effects caused by separated flows [5, 6]. Only when they prove successful will boundary-layer theory be an unqualified success.

Laminar Integral Theory

Both laminar and turbulent theories can be developed from Kármán's general two-dimensional boundary-layer integral relation [7], which extends Eq. (7.33) to variable $U(x)$

$$\frac{\tau_w}{\rho U^2} = \tfrac{1}{2}c_f = \frac{d\theta}{dx} + (2 + H)\frac{\theta}{U}\frac{dU}{dx} \qquad (7.51)$$

where $\theta(x)$ is the momentum thickness and $H(x) = \delta^*(x)/\theta(x)$ is the shape factor. From Eq. (7.17) negative dU/dx is equivalent to positive dp/dx, that is, an adverse gradient.

We can integrate Eq. (7.51) to determine $\theta(x)$ for a given $U(x)$ if we correlate c_f and H with momentum thickness. This has been done by examining typical velocity profiles of laminar and turbulent boundary-layer flows for various pressure gradients. Some examples are given in Fig. 7.9, showing that the shape factor H is a good indicator of pressure gradient. The higher the H the stronger the adverse gradient, and separation occurs approximately at

$$H \approx \begin{cases} 3.5 & \text{laminar flow} \\ 2.4 & \text{turbulent flow} \end{cases} \tag{7.52}$$

The laminar profiles (Fig. 7.9a) clearly exhibit the S shape and a point of inflection with an adverse gradient. But in the turbulent profiles (Fig. 7.9b) the points of inflection are typically buried deep within the thin viscous sublayer, which can hardly be seen on the scale of the figure.

There are scores of turbulent theories in the literature, but they are all complicated algebraically and will be omitted here. The reader is referred to advanced texts [1, 2, 9].

For laminar flow, a simple and effective method was developed by Thwaites [11], who found that Eq. (7.51) can be correlated by a single dimensionless momentum-thickness variable λ, defined as

$$\lambda = \frac{\theta^2}{\nu} \frac{dU}{dx} \tag{7.53}$$

Using a straight-line fit to his correlation, Thwaites was able to integrate Eq. (7.51) in closed form, with the result

$$\theta^2 = \theta_0^2 + \frac{0.45\nu}{U^6} \int_0^x U^5 \, dx \tag{7.54}$$

where θ_0 is the momentum thickness at $x = 0$ (usually taken to be zero). Separation ($c_f = 0$) was found to occur at a particular value of λ

Separation: $\qquad\qquad\qquad\qquad\qquad \lambda = -0.09 \qquad\qquad\qquad\qquad\qquad (7.55)$

Finally, Thwaites correlated values of the dimensionless shear stress $S = \tau_w \theta / \mu U$ with λ, and his graphed result can be curve-fit as follows:

$$S(\lambda) = \frac{\tau_w \theta}{\mu U} \approx (\lambda + 0.09)^{0.62} \tag{7.56}$$

This parameter is related to skin friction by the identity

$$S \equiv \tfrac{1}{2} c_f \, \text{Re}_\theta \tag{7.57}$$

Equations (7.54) to (7.56) constitute a complete theory for the laminar boundary layer with variable $U(x)$, with an accuracy of ± 10 percent compared with exact digital-computer solutions of the laminar-boundary-layer equations (7.19). Complete details of Thwaites' and other laminar theories are given in Refs. 2 and 3.

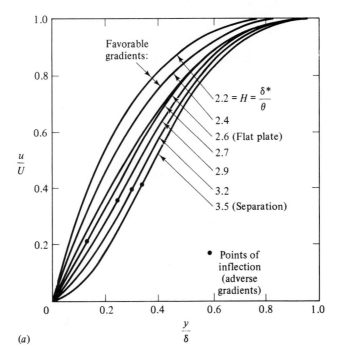

(a)

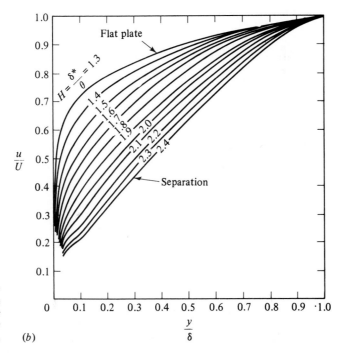

Fig. 7.9 Velocity profiles with pressure gradient: (*a*) laminar flow; (*b*) turbulent flow with adverse gradients.

(b)

As a demonstration of Thwaites' method, take a flat plate, where $U = $ constant, $\lambda = 0$, and $\theta_0 = 0$. Equation (7.54) integrates to

$$\theta^2 = \frac{0.45 \nu x}{U}$$

or

$$\frac{\theta}{x} = \frac{0.671}{\mathrm{Re}_x^{1/2}} \qquad (7.58)$$

This is within 1 percent of Blasius' exact solution, Eq. (7.30).

With $\lambda = 0$, Eq. (7.56) predicts the flat-plate shear to be

$$\frac{\tau_w \theta}{\mu U} = (0.09)^{0.62} = 0.225$$

or

$$c_f = \frac{2\tau_w}{\rho U^2} = \frac{0.671}{\mathrm{Re}_x^{1/2}} \qquad (7.59)$$

This is also within 1 percent of the Blasius result, Eq. (7.25). However, the general accuracy of this method is poorer than 1 percent because Thwaites actually "tuned" his correlation constants to make them agree with exact flat-plate theory.

We shall not compute any more boundary-layer details here, but as we go along investigating various immersed-body flows, especially in Chap. 8, we shall use Thwaites' method to make qualitative assessments of the boundary-layer behavior.

EXAMPLE 7.5   In 1938 Howarth proposed a linearly decelerating external-velocity distribution

$$U(x) = U_0\left(1 - \frac{x}{L}\right) \qquad (1)$$

as a theoretical model for laminar-boundary-layer study. Use Thwaites' method to compute (a) the separation point x_{sep} for $\theta_0 = 0$ and compare with the exact digital-computer solution $x_{\text{sep}}/L = 0.119863$ given by H. Wipperman in 1966. Also compute (b) the value of $c_f = 2\tau_w/\rho U^2$ at $x/L = 0.1$.

Solution

Part (a)   First note that $dU/dx = -U_0/L = $ constant: velocity decreases, pressure increases, and the pressure gradient is adverse throughout. Now integrate Eq. (7.54)

$$\theta^2 = \frac{0.45 \nu}{U_0^6(1 - x/L)^6} \int_0^x U_0^5\left(1 - \frac{x}{L}\right)^5 dx = 0.075 \frac{\nu L}{U_0}\left[\left(1 - \frac{x}{L}\right)^{-6} - 1\right] \qquad (2)$$

Then the dimensionless factor λ is given by

$$\lambda = \frac{\theta^2}{\nu}\frac{dU}{dx} = -\frac{\theta^2 U_0}{\nu L} = -0.075\left[\left(1 - \frac{x}{L}\right)^{-6} - 1\right] \qquad (3)$$

From Eq. (7.55) we set this equal to -0.09 for separation

$$\lambda_{sep} = -0.09 = -0.075\left[\left(1 - \frac{x_{sep}}{L}\right)^{-6} - 1\right]$$

or

$$\frac{x_{sep}}{L} = 1 - (2.2)^{-1/6} = 0.123 \qquad \textit{Ans. (a)}$$

This is less than 3 percent higher than Wipperman's exact solution, and the computational effort is very modest.

Part (b) To compute c_f at $x/L = 0.1$ (just before separation), we first compute λ at this point, using Eq. (3)

$$\lambda(x = 0.1L) = -0.075[(1 - 0.1)^{-6} - 1] = -0.0661$$

Then from Eq. (7.56) the shear parameter is

$$S(x = 0.1L) = (-0.0661 + 0.09)^{0.62} = 0.099 = \tfrac{1}{2}c_f\,\mathrm{Re}_\theta \qquad (4)$$

We can compute Re_θ in terms of Re_L from Eq. (2) or (3)

$$\frac{\theta^2}{L^2} = \frac{0.0661}{UL/\nu} = \frac{0.0661}{\mathrm{Re}_L}$$

or

$$\mathrm{Re}_\theta = 0.257\,\mathrm{Re}_L^{1/2} \qquad \text{at } \frac{x}{L} = 0.1$$

Substitute into Eq. (4):

$$0.099 = \tfrac{1}{2}c_f(0.257\,\mathrm{Re}_L^{1/2})$$

or

$$c_f = \frac{0.77}{\mathrm{Re}_L^{1/2}} \qquad \mathrm{Re}_L = \frac{UL}{\nu} \qquad \textit{Ans. (b)}$$

We cannot actually compute c_f without the value of, say, $U_0 L/\nu$.

7.6 EXPERIMENTAL EXTERNAL FLOWS

Boundary-layer theory is very interesting and illuminating and gives us a great qualitative grasp of viscous-flow behavior, but, because of flow separation, the theory does not generally allow a quantitative computation of the complete flow field. In particular, there is at present no satisfactory theory for the forces on an arbitrary body immersed in a stream flowing at an arbitrary Reynolds number. Therefore experimentation is the key to treating external flows.

Literally thousands of papers in the literature report experimental data on specific external viscous flows. This section will give a brief description of the following external-flow problems:

1. Drag of two- and three-dimensional bodies
   a. Blunt bodies
   b. Streamlined shapes

2. Performance of lifting bodies
  a. Airfoils and aircraft
  b. Projectiles and finned bodies
  c. Birds and insects

For further reading see the goldmine of data compiled in Hoerner [12]. In later chapters we shall study data on supersonic airfoils (Chap. 9), open-channel friction (Chap. 10), and turbomachinery performance (Chap. 11).

Drag of Immersed Bodies

Any body of any shape when immersed in a fluid stream will experience forces and moments from the flow. If the body has arbitrary shape and orientation, the flow will exert forces and moments about all three coordinate axes, as shown in Fig. 7.10. It is customary to choose one axis parallel to the free stream and positive downstream. The force on the body along this axis is called *drag*, and the moment about that axis the *rolling moment*. The drag is essentially a flow loss and must be overcome if the body is to move against the stream.

A second and very important force is perpendicular to the drag and usually performs a useful job, such as bearing the weight of the body. It is called the *lift*. The moment about the lift axis is called *yaw*.

The third component, neither a loss nor a gain, is the *side force*, and about this axis is the *pitching* moment. To deal with this three-dimensional force-moment situation is more properly the role of a textbook on aerodynamics [e.g. 13]. We shall limit the discussion here to lift and drag.

When the body has symmetry about the lift-drag axis, e.g., airplanes, ships, and cars moving directly into a stream, side force, yaw, and roll vanish, and the problem reduces to a two-dimensional case: two forces, lift and drag, and one moment, pitch.

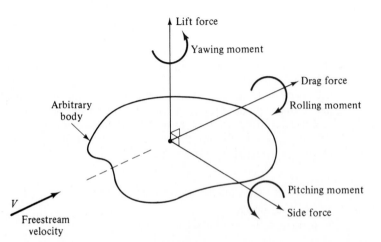

Fig. 7.10 Definition of forces and moments on a body immersed in a uniform flow.

Lift force

Yawing moment

Drag force

Rolling moment

Arbitrary body

Pitching moment

Side force

V

Freestream velocity

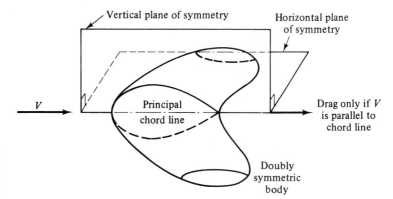

Fig. 7.11 Only the drag force occurs if the flow is parallel to both planes of symmetry.

A final simplification often occurs when the body has two planes of symmetry, as in Fig. 7.11. A wide variety of shapes such as cylinders, wings, and all bodies of revolution satisfy this requirement. If the free stream is parallel to the intersection of these two planes, called the *principal chord line of the body*, the body experiences drag only, with no lift, side force, or moments.[1] This type of degenerate one-force drag data is what is most commonly reported in the literature, but if the free stream is not parallel to the chord line, the body will have an unsymmetric orientation and all three forces and three moments can theoretically arise.

In low-speed flow past geometrically similar bodies with identical orientation and relative roughness, the drag coefficient should be a function only of the body Reynolds number

$$C_D = f(\text{Re}) \tag{7.60}$$

The Reynolds number is based upon free-stream velocity V and a characteristic length L of the body, usually the chord or body length parallel to the stream

$$\text{Re} = \frac{VL}{\nu} \tag{7.61}$$

The drag coefficient could be based upon L^2 but is by custom based upon a characteristic body area A

$$C_D = \frac{\text{drag}}{\frac{1}{2}\rho V^2 A} \tag{7.62}$$

The factor $\frac{1}{2}$ is our traditional tribute to Euler and Bernoulli. The area A is usually one of three types:

1. *Frontal area*, the body as seen from the stream; suitable for thick stubby bodies, such as spheres, cylinders, cars, missiles, projectiles, and torpedoes

---

[1] In bodies with shed vortices, such as the cylinder in Fig. 5.2, there may be *oscillating* lift, side force, and moments, but their mean value is zero.

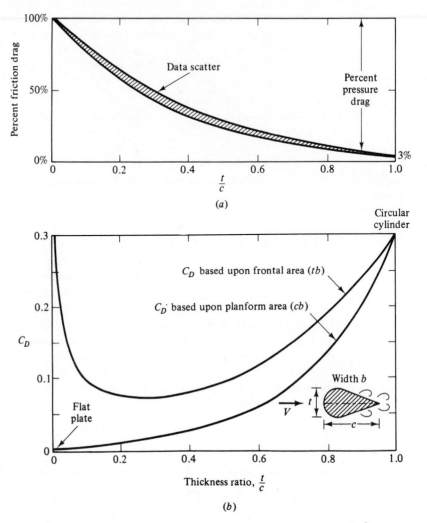

Fig. 7.12 Drag of a streamlined two dimensional cylinder at $Re_c = 10^6$: (a) effect of thickness ratio on percentage friction drag; (b) total drag versus thickness when based upon two different areas.

2. *Planform area*, the body area as seen from above; suitable for wide flat bodies such as wings and hydrofoils

3. *Wetted area*, customary for surface ships and barges

When using drag or other fluid-force data, it is important to note what length and area are being used to scale the measured coefficients.

As we have mentioned, the theory of drag is weak and inadequate, except for the flat plate. This is because of flow separation. Boundary-layer theory can predict the separation point but cannot accurately estimate the (usually low) pressure distribution in the separated region. The difference between the high pressure in the front stagnation region and the low pressure in the rear separated region causes a large drag contribution called *pressure drag*. This is added to the integrated shear stress or *friction drag* of the body, which it often exceeds:

$$C_D = C_{D,\text{press}} + C_{D,\text{fric}} \tag{7.63}$$

The relative contribution of friction and pressure drag depends upon the body shape, especially its thickness. Figure 7.12 shows drag data for a streamlined cylinder of very large depth into the paper. At zero thickness the body is a flat plate and exhibits 100 percent friction drag. At thickness equal to the chord length, simulating a circular cylinder, the friction drag is only about 3 percent. Friction and pressure drag are about equal at thickness $t/c = 0.25$. Note that C_D in Fig. 7.12b looks quite different when based upon frontal area instead of planform area, planform being the usual choice for this body shape. The two curves in Fig. 7.12b represent exactly the same drag data.

Figure 7.13 illustrates the dramatic effect of separated flow and the subsequent failure of boundary-layer theory. The theoretical inviscid pressure distribution on a circular cylinder (Chap. 8) is shown as the dashed line in Fig. 7.13c:

$$C_p = \frac{p - p_\infty}{\frac{1}{2}\rho V^2} = 1 - 4\sin^2\theta \tag{7.64}$$

where p_∞ and V are the pressure and velocity in the free stream. The actual laminar and turbulent boundary-layer pressure distributions in Fig. 7.13c are startlingly

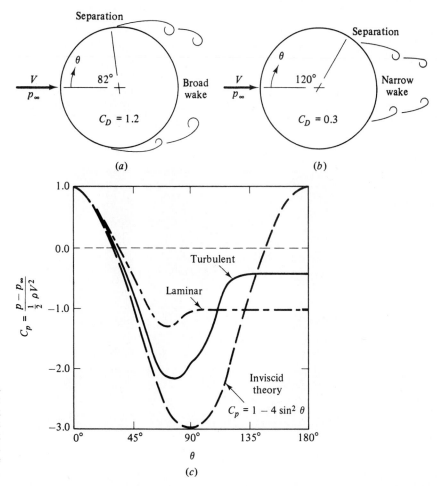

Fig. 7.13 Flow past a circular cylinder: (a) laminar separation; (b) turbulent separation; (c) theoretical and actual surface-pressure distributions.

different from theory. Laminar flow is very vulnerable to the adverse gradient on the rear of the cylinder, and separation occurs at $\theta = 82°$, which certainly could not have been predicted from inviscid theory. The broad wake and very low pressure in the separated laminar region causes the large drag $C_D = 1.2$.

The turbulent boundary layer in Fig. 7.13b is more resistant, and separation is delayed until $\theta = 120°$, with a resulting smaller wake, higher pressure on the rear, and 75 percent less drag, $C_D = 0.3$. This explains the sharp drop in drag at transition in Fig. 5.3.

The same sharp difference between vulnerable laminar separation and resistant turbulent separation can be seen for a sphere in Fig. 7.14. The laminar flow (Fig. 7.14a) separates at about 80°, $C_D = 0.5$, while the turbulent flow (Fig. 7.14b) separates at 120°, $C_D = 0.2$. Here the Reynolds numbers are exactly the same, and the turbulent boundary layer is induced by a patch of sand roughness at the nose of the ball. Golf balls fly in this range of Reynolds numbers, which is why they are deliberately dimpled to induce a turbulent boundary layer and lower drag. Again we would find the actual pressure distribution on the sphere to be quite different from that predicted by inviscid theory.

In general, we cannot overstress the importance of body streamlining to reduce drag at Reynolds numbers above about 100. This is illustrated in Fig. 7.15. The rectangular cylinder (Fig. 7.15a) has rampant separation at all sharp corners and very high drag. Rounding its nose (Fig. 7.15b) reduces drag by about 45 percent, but C_D is still high. Streamlining its rear to a sharp trailing edge (Fig. 7.15c) reduces its drag another 85 percent to a practical minimum for the given thickness. As a dramatic contrast, the circular cylinder (Fig. 7.15d) has one-eighth the thickness and one-three-hundredth the cross section of case (Fig. 7.15c), yet it has the same

Fig. 7.14 Strong differences in laminar and turbulent separation on an 8.5-in bowling ball entering water at 25 ft/s: (a) smooth ball, laminar boundary layer; (b) same entry, turbulent flow induced by patch of nose-sand roughness. (*U.S. Navy photograph, Ordnance Test Station, Pasadena Annex.*)

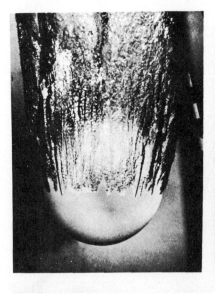

(a)

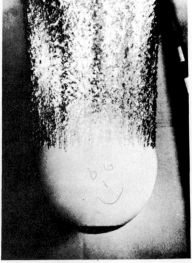

(b)

Fig. 7.15 The importance of streamlining in reducing drag of a body (C_D based on frontal area): (a) rectangular cylinder; (b) rounded nose; (c) rounded nose and streamlined sharp trailing edge; (d) circular cylinder with the same drag as case (c).

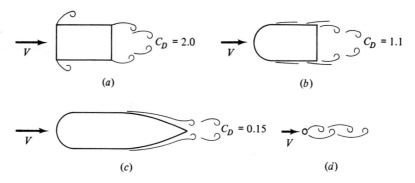

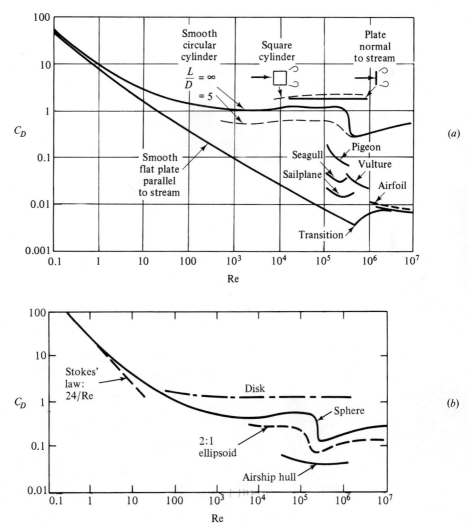

Fig. 7.16 Drag coefficients of smooth bodies at low Mach numbers: (a) two-dimensional bodies; (b) three-dimensional bodies. Note the Reynolds-number independence of blunt bodies at high Re.

Table 7.2

DRAG OF TWO-DIMENSIONAL BODIES AT Re $\geq 10^4$

Shape	C_D based on frontal area	Shape	C_D based on frontal area
Plate:		**Half-cylinder:**	
⟶ |	2.0	⟶ ◗	1.2
Square cylinder:			
⟶ ☐	2.1	⟶ ◖	1.7
⟶ ◇	1.6	**Equilateral triangle:**	
		⟶ ◁	1.6
Half tube:			
⟶ (	1.2	⟶ ▷	2.0
⟶)	2.3		

Elliptical cylinder:		Laminar	Turbulent
1:1 ⟶ ◯		1.2	0.3
2:1 ⟶ ⬭		0.6	0.2
4:1 ⟶ ⬬		0.35	0.15
8:1 ⟶ ⬬		0.25	0.1

drag. For high-performance vehicles and other moving bodies, the name of the game is drag reduction, for which intense research is now in progress for both aerodynamic and hydrodynamic applications [20].

The drag of some representative wide-span (nearly two-dimensional) bodies is shown versus Reynolds number in Fig. 7.16a. All bodies have high C_D at very low (*creeping flow*) Re $\leq$ 1.0, while they spread apart at high Reynolds number according to their degree of streamlining. All values of C_D are based on the planform area except the plate normal to the flow. The birds and the sailplane are of course not very two-dimensional, having only modest span length. Note that birds are not nearly as efficient as modern sailplanes or airfoils [14].

Table 7.2 gives a few data on drag, based on frontal area, of two-dimensional bodies of various cross section, at Re $\geq 10^4$. The sharp-edged bodies, which tend to cause flow separation regardless of the character of the boundary layer, are insensitive to Reynolds number. The elliptic cylinders, being smoothly rounded, have the laminar to turbulent transition effect of Figs. 7.13 and 7.14 and are therefore quite sensitive to the character of the boundary layer.

EXAMPLE 7.6   A square 6-in piling is acted on by a water flow of 5 ft/s 20 ft deep, as shown in Fig. E7.6. Estimate the maximum bending exerted by the flow on the bottom of the piling.

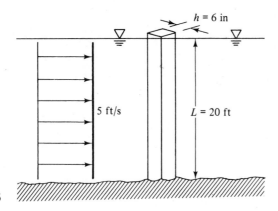

Fig. E7.6

Solution   Assume seawater with $\rho = 1.99$ slugs/ft^3 and kinematic viscosity $v = 0.000011$ ft^2/s. With piling width of 0.5 ft, we have

$$\mathrm{Re}_h = \frac{(5 \text{ ft/s})(0.5 \text{ ft})}{0.000011 \text{ ft}^2/\text{s}} = 2.3 \times 10^5$$

This is the range where Table 7.2 applies. The worst case occurs when the flow strikes the flat side of the piling, $C_D \approx 2.1$. The frontal area is $A = Lh = (20 \text{ ft})(0.5 \text{ ft}) = 10 \text{ ft}^2$. The drag is estimated by

$$F = C_D(\tfrac{1}{2}\rho V^2 A) \approx 2.1(\tfrac{1}{2})(1.99 \text{ slugs/ft}^3)(5 \text{ ft/s})^2(10 \text{ ft}^2) = 522 \text{ lbf}$$

If the flow is uniform, the center of this force should be at approximately middepth. Therefore the bottom bending moment is

$$M_0 \approx \frac{FL}{2} = 522(10) = 5220 \text{ ft} \cdot \text{lbf} \qquad\qquad Ans.$$

According to the flexure formula from strength of materials, the bending stress at the bottom would be

$$S = \frac{M_0 y}{I} = \frac{(5220 \text{ ft} \cdot \text{lb})(0.25 \text{ ft})}{\tfrac{1}{12}(0.5 \text{ ft})^4} = 251{,}000 \text{ lbf/ft}^2 = 1740 \text{ lbf/in}^2$$

to be multiplied, of course, by the stress-concentration factor due to the built-in end conditions.

Table 7.3

DRAG OF THREE-DIMENSIONAL BODIES AT $Re \geq 10^4$

Body	Ratio	C_D based on frontal area

Cube:

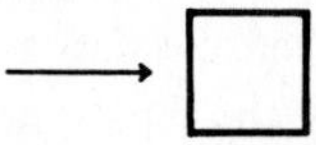

1.07

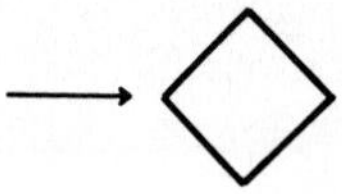

0.81

60° cone:

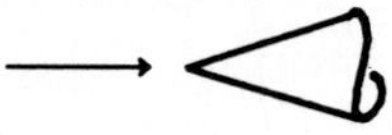

0.5

Disk:

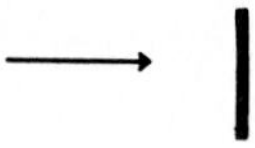

1.17

Cup:

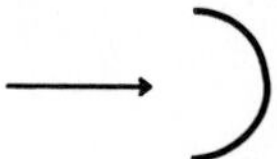

1.4

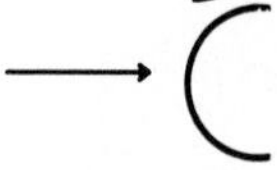

0.4

Parachute (low porosity):

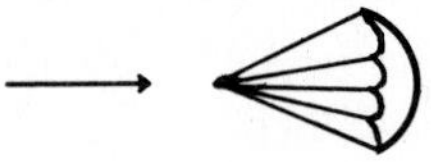

1.2

Rectangular plate:

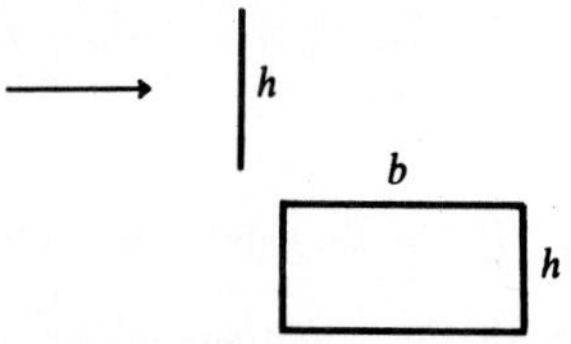

	b/h	
	1	1.18
	5	1.2
	10	1.3
	20	1.5
	∞	2.0

Flat-faced cylinder:

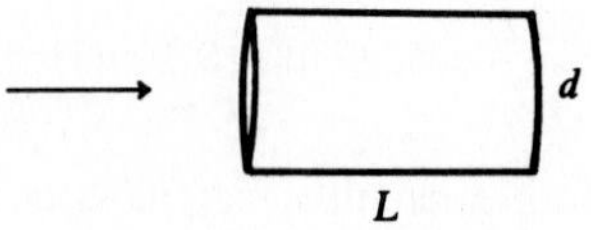

	L/d	
	0.5	1.15
	1	0.90
	2	0.85
	4	0.87
	8	0.99

Ellipsoid:

	L/d	Laminar	Turbulent
	0.75	0.5	0.2
	1	0.47	0.2
	2	0.27	0.13
	4	0.25	0.1
	8	0.2	0.08

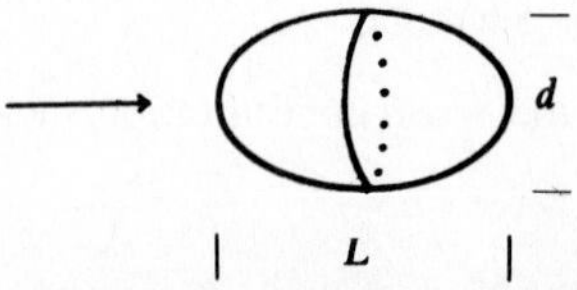

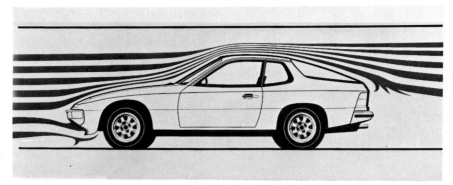

Fig. 7.17 Wind-tunnel test of a modern high-performance automobile. Streamlines made visible by smoke injection upstream. (*Porsche/Audi Division of Volkswagen of America, Inc.*)

Some drag coefficients of three-dimensional bodies are listed in Table 7.3 and Fig. 7.16b. Again we can conclude that sharp edges always cause flow separation and high drag which is insensitive to Reynolds number. Rounded bodies like the ellipsoid have drag which depends upon the point of separation, so that both Reynolds number and the character of the boundary layer are important. Body length will generally decrease pressure drag by making the body relatively more slender, but sooner or later the friction drag will catch up. For the flat-faced cylinder in Table 7.3, pressure drag decreases with L/d but friction increases, so that minimum drag occurs at about $L/d = 2$.

Modern vehicles are being designed with reduced drag achieved by wind-tunnel tests. In the United States the idea before 1974 was to increase speed for a given horsepower, while in the rest of the world it was to reduce fuel consumption. Figure 7.17 shows a wind-tunnel test of a modern automobile. The streamlines are made visible by smoke introduced into the flow. This vehicle has a drag coefficient of about 0.3 based on frontal area, while antique cars, with their flat faces and numerous appendages, have $C_D \approx 0.9$. For details see Ref. 21.

Figure 7.18 shows the horsepower required to drive a typical tractor-trailer truck at speeds up to 80 mi/h (117 ft/s or 36 m/s). The rolling resistance increases

Fig. 7.18 Drag reduction of a tractor-trailer truck: (a) horsepower required to overcome resistance; (b) deflector added to cab reduces air drag by 20 percent. (*Uniroyal Inc.*)

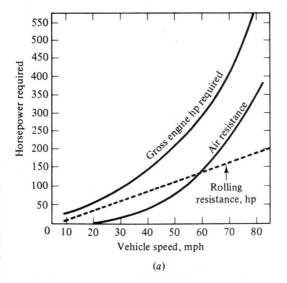

(a)

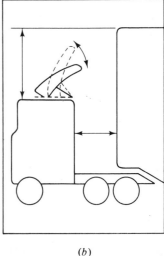

(b)

linearly and the air drag quadratically with speed ($C_D \approx 1.0$). The two are about equally important at 55 mi/h, which is the national speed limit in the United States. As shown in Fig. 7.18b, air drag can be reduced by attaching a deflector to the top of the tractor. If the angle of the deflector is adjusted to carry the flow smoothly over the top and around the sides of the trailer, the reduction in C_D is about 20 percent. Thus, at 55 mi/h total resistance is reduced 10 percent, with corresponding reduction in fuel costs and/or trip time for the trucker. This type of applied fluids engineering can be a large factor in many of the conservation-oriented transportation problems of the future.

EXAMPLE 7.7

A high-speed car with $m = 2000$ kg, $C_D = 0.3$, and $A = 1$ m², deploys a 2-m parachute to slow down from an initial velocity of 100 m/s (Fig. E7.7). Assuming constant C_D, brakes free, and no rolling resistance, calculate the distance and velocity of the car after 1, 10, 100, and 1000 s. For air, assume $\rho = 1.2$ kg/m³ and neglect interference between the wake of the car and the parachute.

$d_p = 2$ m $\qquad\qquad\qquad V_0 = 100$ m/s

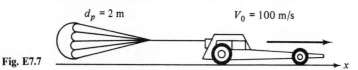

Fig. E7.7

Solution

Newton's law applied in the direction of motion gives

$$F_x = m\frac{dV}{dt} = -F_c - F_p = -\tfrac{1}{2}\rho V^2(C_{Dc}A_c + C_{Dp}A_p)$$

where subscript c is the car and subscript p the parachute. This is of the form

$$\frac{dV}{dt} = -\frac{K}{m}V^2 \qquad K = \sum C_D A\frac{\rho}{2}$$

Separate the variables and integrate

$$\int_{V_0}^{V}\frac{dV}{V^2} = -\frac{K}{m}\int_0^t dt$$

or

$$V_0^{-1} - V^{-1} = -\frac{K}{m}t$$

Rearrange and solve for the velocity V:

$$V = \frac{V_0}{1 + (K/m)V_0 t} \qquad K = \frac{(C_{Dc}A_c + C_{Dp}A_p)\rho}{2} \tag{1}$$

We can integrate this to find the distance traveled;

$$S = \frac{V_0}{\alpha}\ln(1 + \alpha t) \qquad \alpha = \frac{K}{m}V_0 \tag{2}$$

Now work out some numbers. From Table 7.3, $C_{Dp} \approx 1.2$; hence

$$C_{Dc}A_c + C_{Dp}A_p = 0.3(1 \text{ m}^2) + 1.2\frac{\pi}{4}(2 \text{ m})^2 = 4.07 \text{ m}^2$$

Then $$\frac{K}{m}V_0 = \frac{\tfrac{1}{2}(4.07 \text{ m}^2)(1.2 \text{ kg/m}^3)(100 \text{ m/s})}{2000 \text{ kg}} = 0.122 \text{ s}^{-1} = \alpha$$

Now make a table of the results for V and S from Eqs. (1, 2):

t, s	1	10	100	1000
V, m/s	89	45	7.6	0.8
S, m	94	654	2110	3940

Air resistance alone will not stop a body completely. If you don't apply the brakes, you'll be halfway to the Yukon Territory and still going.

Forces on Lifting Bodies

Lifting bodies (airfoil, hydrofoils, or vanes) are intended to provide a large force normal to the free stream and as little drag as possible. Conventional design practice has evolved a shape not unlike a bird's wing, i.e., relatively thin ($t/c \leq 0.18$) with a rounded leading edge and a sharp trailing edge. A typical shape is sketched in Fig. 7.19.

For our purposes we consider the body to be symmetric, as in Fig. 7.11, with the free-stream velocity in the vertical plane. If the chord line between leading and trailing edge is not a line of symmetry, the airfoil is said to be *cambered*. The camber line is the line midway between the upper and lower surfaces of the vane.

The angle between the free stream and the chord line is called the *angle of attack* α. The lift L and the drag D vary with this angle. The dimensionless forces are defined with respect to the planform area $A_p = bc$:

Lift coefficient: $$C_L = \frac{L}{\tfrac{1}{2}\rho V^2 A_p} \tag{7.65a}$$

Drag coefficient: $$C_D = \frac{D}{\tfrac{1}{2}\rho V^2 A_p} \tag{7.65b}$$

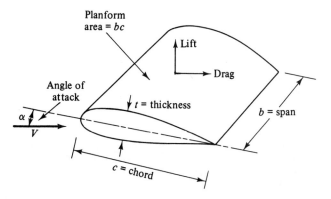

Fig. 7.19 Definition sketch for a lifting vane.

If chord length is not constant, as in the tapered wings of modern aircraft, $A_p = \int c\, db$.

For low-speed flow with a given roughness ratio, C_L and C_D should vary with α and the chord Reynolds number

$$C_L \quad \text{or} \quad C_D = f(\alpha, \text{Re}_c) \tag{7.66}$$

where $\text{Re}_c = Vc/\nu$. The Reynolds numbers are commonly in the turbulent-boundary-layer range and have a modest effect.

The rounded leading edge prevents flow separation there, but the sharp trailing edge causes a separation which generates the lift. Figure 7.20 shows what happens when a flow starts up past a lifting vane or airfoil.

Just after start-up in Fig. 7.20a the streamline motion is irrotational and inviscid. The rear stagnation point, assuming a positive angle of attack, is on the upper surface, and there is no lift; but the flow cannot long negotiate the sharp turn at the trailing edge: it separates, and a *starting vortex* forms in Fig. 7.20b. This starting vortex is shed downstream in Fig. 7.20c and d and a smooth streamline flow develops over the wing, leaving the foil in a direction approximately parallel to the chord line. Lift at this time is fully developed, and the starting vortex is gone. Should the flow now cease, a *stopping vortex* of opposite (clockwise) sense will form and be shed. During flight, increases or decreases in lift will cause incremental starting or stopping vortices, always with the effect of maintaining a smooth parallel flow at the trailing edge. We pursue this idea mathematically in Chap. 8. The details of start-up and stopping are clearly shown in the film on Vorticity, listed as number 2 in Appendix C.

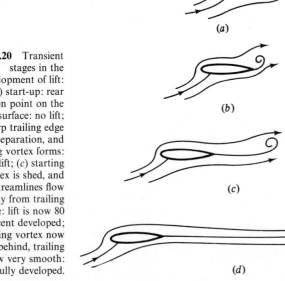

(a)

(b)

(c)

(d)

Fig. 7.20 Transient stages in the development of lift: (a) start-up: rear stagnation point on the upper surface: no lift; (b) sharp trailing edge induces separation, and a starting vortex forms: slight lift; (c) starting vortex is shed, and streamlines flow smoothly from trailing edge: lift is now 80 percent developed; (d) starting vortex now shed far behind, trailing edge now very smooth: lift fully developed.

Fig. 7.21 At high angle of attack smoke-flow visualization shows stalled flow on the upper surface of a lifting vane. [*From Ref. 19, Illustrated Experiments in Fluid Mechanics* (*The NCFMF Book of Film Notes*), *National Committee for Fluid Mechanics Films, Education Development Center, Inc., copyright 1972.*]

At a low angle of attack, the rear surfaces have an adverse pressure gradient but not enough to cause significant boundary-layer separation. The flow pattern is smooth, as in Fig. 7.20d, and drag is small and lift excellent. As the angle of attack is increased, the upper-surface adverse gradient becomes stronger, and generally a *separation bubble* begins to creep forward on the upper surface.[1] At a certain angle $\alpha = 15$ to $20°$, the flow is separated completely from the upper surface, as in Fig. 7.21. The airfoil is said to be *stalled*: lift drops off markedly, drag increases markedly, and the foil is no longer flyable.

Early airfoils were thin, modeled after birds' wings. The German engineer Otto Lilienthal (1848–1896) experimented with flat and cambered plates on a rotating arm. He and his brother Gustav flew the world's first glider in 1891. Horatio Frederick Phillips (1845–1912) built the first wind tunnel in 1884 and measured lift and drag of cambered vanes. The first theory of lift was proposed by Frederick W. Lanchester shortly afterward. Modern airfoil theory dates from 1905, when the Russian hydrodynamicist N. E. Joukowsky (1847–1921) developed a circulation theorem (Chap. 8) for computing airfoil lift for arbitrary camber and thickness. With this basic theory, as extended and developed by Prandtl and Kármán and their students, it is now possible to design a low-speed airfoil to satisfy particular surface-pressure distributions and boundary-layer characteristics. There are whole families of airfoil designs, notably those developed in the United States under the

---

[1] For some airfoils the bubble leaps, not creeps, forward, and stall occurs rapidly and dangerously.

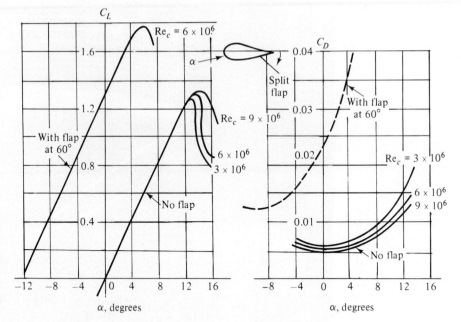

sponsorship of the NACA (now NASA). Extensive theory and data on these airfoils are contained in Ref. 16. We shall discuss this further in Chap. 8.

Figure 7.22 shows the lift and drag on a symmetrical airfoil denoted as the NACA 0009 foil, the last digit indicating the thickness of 9 percent. With no flap extended, this airfoil, as expected, has zero lift at zero angle of attack. Up to about 12° the lift coefficient increases linearly with a slope of 0.1 per degree or 6.0 per radian. This is in agreement with the theory outlined in Chap. 8:

$$C_{L, \text{theory}} \approx 2\pi \sin\left(\alpha + \frac{2h}{c}\right) \tag{7.67}$$

where h/c is the maximum camber expressed as a fraction of the chord. The NACA 0009 has zero camber; hence $C_L = 2\pi \sin \alpha \approx 0.11\alpha$, where α is in degrees. This is excellent agreement.

The drag coefficient of the smooth model airfoils in Fig. 7.22 is as low as 0.005, which is actually lower than both sides of a flat plate in turbulent flow. This is misleading inasmuch as a commercial foil will have roughness effects; e.g., a paint job will double the drag coefficient.

The effect of increasing Reynolds number in Fig. 7.22 is to increase maximum lift and stall angle (without changing the slope appreciably) and to reduce drag coefficient. This is a salutary effect, since the prototype will probably be at a higher Reynolds number than the model (10^7 or more).

For takeoff and landing, the lift is greatly increased by deflecting a split flap, as shown in Fig. 7.22. This makes the airfoil unsymmetric (or effectively cambered) and changes the zero-lift point to $\alpha = -12°$. The drag is also greatly increased by the flap, but the reduction in takeoff and landing distance is worth the extra power needed.

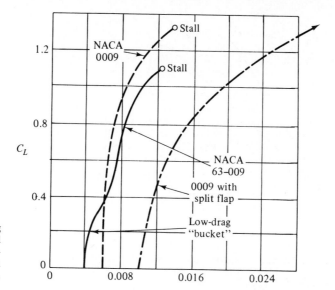

Fig. 7.23 Lift-drag polar plot for standard (0009) and a laminar-flow (63-009) NACA airfoil.

A lifting craft cruises at low angle of attack, where the lift is much larger than the drag. Maximum lift-to-drag ratios for the common airfoils lie between 20 and 50.

Some airfoils, such as the NACA 6 series, are shaped to provide favorable gradients over much of the upper surface at low angles. Thus separation is small, and transition to turbulence is delayed; the airfoil retains a good length of laminar flow even at high Reynolds numbers. The lift-drag polar plot in Fig. 7.23 shows the NACA 0009 data from Fig. 7.22 and also a laminar-flow airfoil, NACA 63–009, of the same thickness. The laminar-flow airfoil has a low drag bucket at small angles but also suffers lower stall angle and lower maximum lift coefficient. The drag is 30 percent less in the bucket, but the bucket disappears if there is significant surface roughness.

All the data in Figs. 7.22 and 7.23 are for infinite span, i.e., a two-dimensional flow pattern about wings without tips. The effect of finite span can be correlated with the dimensionless slenderness, or *aspect ratio*

$$AR = \frac{b^2}{A_p} = \frac{b}{\bar{c}} \tag{7.68}$$

where $\bar{c}$ is the average chord length. Finite-span effects are shown in Fig. 7.24. The lift slope decreases, but the zero-lift angle is the same, and the drag increases, but the zero-lift drag is the same. The theory of finite-span airfoils [16] predicts that the effective angle of attack increases, as in Fig. 7.24, by the amount

$$\Delta\alpha \approx \frac{C_L}{\pi AR} \tag{7.69}$$

When applied to Eq. (7.67), the finite-span lift becomes

$$C_L \approx \frac{2\pi \sin{(\alpha + 2h/c)}}{1 + 2/AR} \tag{7.70}$$

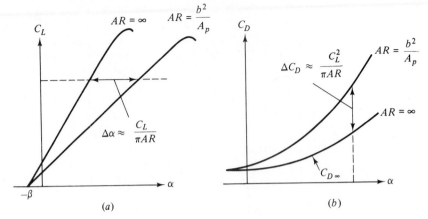

Fig. 7.24 Effect of finite aspect ratio on lift and drag of an airfoil: (a) effective angle increase; (b) induced drag increase.

The associated drag increase is $\Delta C_D \approx C_L \sin \Delta\alpha \approx C_L \Delta\alpha$, or

$$C_D \approx C_{D\infty} + \frac{C_L^2}{\pi AR} \tag{7.71}$$

where $C_{D\infty}$ is the drag of the infinite-span airfoil, as sketched in Fig. 7.22. These correlations are in good agreement with experiments on finite-span wings [16].

The existence of a maximum lift coefficient implies the existence of a minimum speed, or *stall speed*, for a craft whose lift supports its weight

$$L = W = C_{L,\max}(\tfrac{1}{2}\rho V_s^2 A_p)$$

or

$$V_s = \left(\frac{2W}{C_{L,\max}\rho A_p}\right)^{1/2} \tag{7.72}$$

The stall speed of typical aircraft varies between 60 and 200 ft/s, depending upon the weight and value of $C_{L,\max}$. The pilot must hold speed greater than about $1.2V_s$ in order to avoid the instability associated with complete stall.

The split flap in Fig. 7.22 is only one of many devices used to secure high lift at low speeds. Figure 7.25a shows six such devices whose lift performance is given in 7.25b along with a standard (A) and laminar-flow (B) airfoil. The double-slotted flap achieves $C_{L,\max} \approx 3.4$, and a combination of this plus a leading-edge slat can achieve $C_{L,\max} \approx 4.0$. These are not scientific curiosities; e.g., the Boeing 727 commercial jet aircraft uses a triple-slotted flap plus a leading-edge slat during landing.

Also shown as C in Figure 7.25b is the Kline-Fogleman airfoil [17], not yet a reality. The designers are amateur model-plane enthusiasts who did not know that conventional aerodynamic wisdom forbids a sharp leading edge and a step cutout from the trailing edge. The Kline-Fogleman airfoil has relatively high drag but shows an amazing continual increase in lift out to $\alpha = 45°$. In fact, we may fairly say that this airfoil does not stall and provides smooth performance over a tremendous range of flight conditions. No explanation for this behavior has yet been given by

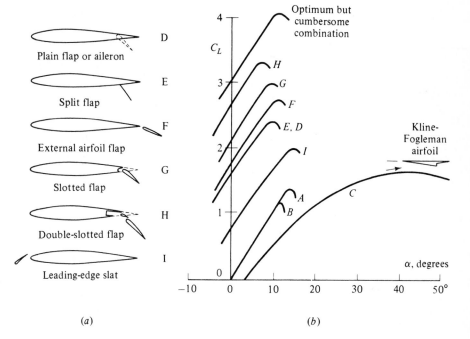

Fig. 7.25 Performance of airfoils with and without high-lift devices: A = NACA 0009; B = NACA 63-009; C= Kline-Fogleman Airfoil (Ref. 17); D to I shown in (a): (a) types of high-lift devices; (b) lift coefficients for various devices.

Plain flap or aileron   D

Split flap   E

External airfoil flap   F

Slotted flap   G

Double-slotted flap   H

Leading-edge slat   I

(a)

Optimum but cumbersome combination

Kline-Fogleman airfoil

(b)

any aerodynamicist. This airfoil is presently under study by NASA and may or may not have any commercial value.

Further information on the performance of lifting craft can be found in Refs. 12, 13, and 16. We discuss this matter again briefly in Chap. 8.

EXAMPLE 7.8   An aircraft weights 75,000 lb, has a planform area of 2500 ft^2, and can deliver constant thrust of 12,000 lb. It has an aspect ratio of 7, and $C_D \approx 0.02$. Neglecting rolling resistance, estimate the takeoff distance at sea level if takeoff speed equals 1.2 times stall speed. Take $C_{L,\,\mathrm{max}} = 2.0$.

Solution   The stall speed from Eq. (7.72), with sea-level density $\rho = 0.00237$ slug/ft^3, is

$$V_s = \left(\frac{2W}{C_{L,\,\mathrm{max}}\rho A_p}\right)^{1/2} = \left[\frac{2(75,000)}{2.0(0.00237)(2500)}\right]^{1/2} = 112.5 \text{ ft/s}$$

Hence takeoff speed $V_0 = 1.2 V_s = 135$ ft/s. The drag is estimated from Eq. (7.71) for AR = 7 as

$$C_D \approx 0.02 + \frac{C_L^2}{7\pi} = 0.02 + 0.0455 C_L^2$$

A force balance in the direction of takeoff gives

$$F_s = m\frac{dV}{dt} = \text{thrust} - \text{drag} = T - kV^2 \qquad k = \tfrac{1}{2}C_D\rho A_p \qquad (1)$$

Since we are looking for distance, not time, introduce $dV/dt = V(dV/ds)$ into Eq. (1), separate variables, and integrate

$$\int_0^{S_0} dS = \frac{m}{2} \int_0^{V_0} \frac{d(V^2)}{T - kV^2} \qquad k \approx \text{const}$$

or

$$S_0 = \frac{m}{2k} \ln \frac{T}{T - kV_0^2} = \frac{m}{2k} \ln \frac{T}{T - D_0} \qquad (2)$$

where $D_0 = kV_0^2$ is the takeoff drag. Equation (2) is the desired theoretical relation for takeoff distance. For the particular numerical values, take

$$m = \frac{75,000}{32.2} = 2329 \text{ slugs}$$

$$C_{L_0} = \frac{W}{\frac{1}{2}\rho V_0^2 A_p} = \frac{75,000}{\frac{1}{2}(0.00237)(135)^2(2500)} = 1.39$$

$$C_{D_0} = 0.02 + 0.0455(C_{L_0})^2 = 0.108$$

$$k \approx \tfrac{1}{2} C_{D_0} \rho A_p = (\tfrac{1}{2})(0.108)(0.00237)(2500) = 0.319 \text{ slug/ft}$$

$$D_0 = kV_0^2 = 5820 \text{ lb}$$

Then Eq. (2) predicts that

$$S_0 = \frac{2329 \text{ slugs}}{2(0.319 \text{ slug/ft})} \ln \frac{12,000}{12,000 - 5820} = 3650 \ln 1.94 = 2420 \text{ ft} \qquad \text{Ans.}$$

A more exact analysis accounting for variable k [13] gives the same result to within 1 percent.

SUMMARY

This chapter has dealt with viscous effects in external flow past bodies immersed in a stream. When the Reynolds number is large, viscous forces are confined to a thin boundary layer and wake in the vicinity of the body. Flow outside these "shear layers" is essentially inviscid and can be predicted by potential theory and Bernoulli's equation.

The chapter begins with a discussion of the flat-plate boundary layer and the use of momentum-integral estimates to predict wall shear, friction drag, and thickness of such layers. These approximations suggest how to eliminate certain small terms in the Navier-Stokes equations, resulting in Prandtl's boundary-layer equations for laminar and turbulent flow. Section 7.4 then solves the boundary-layer equations to give very accurate formulas for flat-plate flow at high Reynolds numbers. Rough wall effects are included, and Sec. 7.5 gives a brief introduction to pressure-gradient effects. An adverse (decelerating) gradient is seen to cause flow separation, where the boundary layer breaks away from the surface and forms a broad, low-pressure wake.

Boundary-layer theory fails in separated flows, which are commonly studied by experiment. Section 7.6 gives data on drag coefficients of various two- and three-dimensional body shapes. The chapter ends with a brief discussion of lift forces generated by lifting bodies such as airfoils and hydrofoils. Airfoils also suffer flow separation or *stall* at high angles of incidence.

REFERENCES

1. H. Schlichting, "Boundary Layer Theory," 7th ed., McGraw-Hill, New York, 1979.
2. F. M. White, "Viscous Fluid Flow," McGraw-Hill, New York, 1974.
3. L. Rosenhead (ed.), "Laminar Boundary Layers," Oxford University Press, London, 1963.
4. P. J. Roache, "Computational Fluid Dynamics," Hermosa, Albuquerque, N.M., 1976.
5. R. H. Pletcher and C. L. Dancey, "A Direct Method of Calculating through Separated Regions in Boundary Layer Flow," *J. Fluids Eng.*, September 1976, pp. 568–572.
6. P. K. Chang, "Control of Flow Separation," McGraw-Hill, New York, 1976.
7. T. von Kármán, On Laminar and Turbulent Friction, *Z. Angew. Math. Mech.*, vol. 1, 1921, pp. 235–236.
8. G. B. Schubauer and H. K. Skramstad, Laminar Boundary Layer Oscillations and Stability of Laminar Flow, *Natl. Bur. Stand. Res. Pap.* 1772, April 1943 (see also *J. Aero. Sci.*, vol. 14, 1947, pp. 69–78, and *NACA Rep.* 909, 1947).
9. T. Cebeci and A. M. O. Smith, "Analysis of Turbulent Boundary Layers," Academic, New York, 1974.
10. P. W. Runstadler, Jr., et al., Diffuser Data Book, Creare Inc., *Tech. Note* 186, Hanover, N. H., May 1975.
11. B. Thwaites, Approximate Calculation of the Laminar Boundary Layer, *Aeronaut. Q.*, vol. 1, 1949, pp. 245–280.
12. S. F. Hoerner, "Fluid Dynamic Drag," published by the author, Midland Park, N.J., 1965.
13. Dommasch, D. O., S. S. Sherby, and T. F. Connally, "Airplane Aerodynamics," 4th ed., Pitman, New York, 1967.
14. V. Tucker and G. C. Parrott, Aerodynamics of Gliding Flight of Falcons and Other Birds, *J. Exp. Biol.*, vol. 52, 1970, pp. 345–368.
15. J. P. Comstock (ed.), "Principles of Naval Architecture," Society of Naval Architects and Marine Engineers, New York, 1967.
16. I. H. Abbott and A. E. von Doenhoff, "Theory of Wing Sections," Dover, New York, 1959.
17. R. L. Kline and F. F. Fogleman, Airfoil for Aircraft, U.S. Patent 3,706,430, Dec. 19, 1972.
18. T. Y. Yu et al. (eds.), "Swimming and Flying in Nature: Proceedings of a Symposium," 2 vol., Plenum, New York, 1975.
19. National Committee for Fluid Mechanics Films, "Illustrated Experiments in Fluid Mechanics," M.I.T. Press, Cambridge, Mass., 1972.
20. E. M. Uram and H. E. Weber (eds.), Laminar and Turbulent Boundary Layers, *ASME Symp. Proc.*, vol. I00167, February, 1984.
21. G. Sovran, T. Morel, and W. T. Mason Jr. (eds.), "Aerodynamic Drag Mechanisms of Bluff Bodies and Road Vehicles," Plenum, New York, 1978.

RECOMMENDED FILMS

Films numbered 1, 2, 8, 9, 14, 15, 26, 34, and 35 in Appendix C.

Problems

PROBLEM DISTRIBUTION

Section	Topic	Problems
7.1	Introduction	7.1–7.6
7.2	Momentum-integral estimates	7.7–7.11
7.3	The boundary-layer equations	7.12–7.14
7.4	The flat plate: laminar flow	7.15, 7.23
7.4	The flat plate: turbulent flow	7.16–7.22, 7.24–7.36
7.5	Boundary layers with pressure gradients	7.37–7.43
7.6	External flows: drag forces	7.44–7.82
7.6	External flows: lift forces	7.83–7.96

7.1 Air at 20°C flows at 25 m/s past a thin flat plate. Estimate the distance x from the leading edge at which boundary-layer thickness will be (a) 10 cm; (b) 1 mm.

7.2 Mercury at 20°C flows past a thin flat plate at 7 ft/s. Estimate the distance x from the leading edge at which boundary-layer thickness will be (a) 1 in; (b) 0.01 in.

7.3 Equation (7.1b) assumes that the boundary layer on the plate is turbulent from the leading edge onward. Devise a scheme for determining the boundary-layer thickness more accurately when the flow is laminar up to a point $\mathrm{Re}_{x,\,\mathrm{crit}}$ and turbulent thereafter. Apply this scheme to computation of the boundary-layer thickness at $x = 5$ ft in airflow at 120 ft/s and 20°C past a flat plate. Compare with Eq. (7.1b). Assume $\mathrm{Re}_{x,\,\mathrm{crit}} = 10^6$.

7.4 Sea-level standard air flows at 18 m/s past a smooth flat plate. At what point x will the boundary-layer thickness be 6 mm? Why do Eqs. (7.1) seem to fail? Make a sketch showing the discrepancy. Assuming that $\mathrm{Re}_{x,\,\mathrm{crit}} = 8 \times 10^5$, use the ideas in Prob. 7.3 to complete this problem correctly.

7.5 Helium at 20°C and 1 atm flows into the region between two parallel plates 12 cm apart at an average velocity of 22 m/s. Assuming the boundary layers grow from the entrance, estimate the distance downstream at which the upper and lower plate layers meet.

7.6 SAE 30 oil at 20°C flows at 1.8 ft³/s from a reservoir into a 6-in diameter pipe. Use flat-plate theory to estimate the position x where the pipe-wall boundary layers meet in the center. Compare with Eq. (6.5) and give some explanations for the discrepancy.

7.7 For the laminar parabolic boundary-layer profile of Eq. (7.6), compute the shape factor H and compare with the exact Blasius result, Eq. (7.31).

7.8 Repeat the flat-plate momentum-integral analysis of Sec. 7.2 by replacing the parabolic profile, Eq. (7.6), with the linear profile $u/U \approx y/\delta$. Compute c_f, θ/x, δ^*/x, and H.

7.9 Repeat Prob. 7.8 but use instead the (extremely accurate) sinusoidal-velocity-profile assumption

$$\frac{u}{U} \approx \sin\frac{\pi y}{2\delta}$$

Compare this profile shape with the Blasius profile in Fig. 7.5.

7.10 Repeat Prob. 7.8 using the polynomial profile suggested by K. Pohlhausen in 1921:

$$\frac{u}{U} \approx 2\frac{y}{\delta} - 2\frac{y^3}{\delta^3} + \frac{y^4}{\delta^4}$$

Does this profile satisfy the boundary conditions of laminar flat-plate flow?

7.11 Find the correct form for a cubic velocity-profile polynomial

$$u = A + By + Cy^2 + Dy^3$$

to replace Eq. (7.6) in a flat-plate momentum analysis. Find the value of θ/δ for this profile but do not pursue the analysis further.

7.12 Derive modified forms of the laminar boundary-layer equations (7.19) for the case of axisymmetric flow along the outside of a circular cylinder of constant

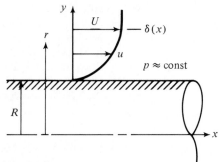

Fig. P7.12

radius R, as in Fig. P7.12. Consider the two special cases (a) $\delta \ll R$; and (b) $\delta \approx R$. What are the proper boundary conditions?

7.13 Show that the two-dimensional laminar flow pattern with $dp/dx = 0$,

$$u = U_0(1 - e^{Cy}) \qquad v = v_0 < 0$$

is an exact solution to the boundary-layer equations (7.19). Find the value of the constant C in terms of flow parameters. Are the boundary conditions satisfied? What might this flow represent?

7.14 Discuss whether the laminar flow between parallel plates, Eq. (6.86), is an exact solution to the boundary-layer equations (7.19) and the boundary conditions (7.20). In what sense are duct flows boundary layers?

7.15 A thin flat plate 45 by 90 cm is immersed in a stream of glycerin at 20°C and a velocity of 7 m/s. Compute the viscous drag if the plate side parallel to the stream is (a) the short side and (b) the long side.

7.16 Repeat Prob. 7.15 if the fluid is water at 20°C. Assume that transition occurs at $Re_x = 10^6$ on a smooth plate.

7.17 Repeat Prob. 7.16 if the plate has a roughness height of 2 mm and a transition Reynolds number of 5×10^5.

7.18 A plate 15 ft wide and 35 ft long is towed at 14 ft/s through seawater at 20°C. Estimate the drag and the tow horsepower required if the plate is (a) smooth; (b) rough with $\epsilon = 0.01$ in.

7.19 A ship is 200 m long and has a wetted area of 8500 m². Estimate the power required to overcome friction for a smooth surface when the ship moves at 15 knots in seawater at 20°C.

7.20 A blimp approximates an ellipsoid 250 ft long and 50 ft in diameter. Estimate its skin-friction drag in pounds and the horse-power required to overcome it when the blimp moves at 60 mi/h (88 ft/s) through still air at 68°F and 12 lbf/in² absolute.

7.21 A wind tunnel has a test section 1 m square and 6 m long with air at 20°C moving at an average velocity of 30 m/s. It is planned to slant the walls outward slightly to account for the growing boundary-layer displacement thickness on the four walls, thus keeping the test-section velocity constant. At what angle should they be slanted at keep V constant between $x = 2$ m and $x = 4$ m?

7.22 A hydrofoil 60 cm long and 3 m wide moves in water at 20°C at a speed of 14 m/s. Using flat-plate theory, estimate its drag in newtons if $Re_{tr} = 5 \times 10^5$ for (a) a smooth wall and (b) a rough wall, $\epsilon = 0.1$ mm.

7.23 A thin smooth disk of diameter D is immersed parallel to a uniform stream at velocity U_0. Assuming laminar flow and using flat-plate theory as a guide, develop an expression for the friction drag of the disk.

7.24 A thin equilateral triangle plate is immersed parallel to a 12 m/s stream of water at 20°C, as in Fig. P7.24. Assuming $Re_{tr} = 5 \times 10^5$, estimate the drag of this plate.

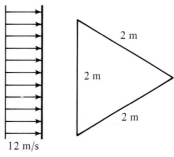

12 m/s

Fig. P7.24

7.25 A four-bladed helicopter rotor turns in still air at 150 r/min. Each blade is 3 m long and 30 cm wide. Estimate the power required to overcome friction at sea level.

7.26 The transition Reynolds number in pipe flow is approximately 2300 from Eq. (6.2). How does this value relate to flat-plate transition $Re_{x, \text{crit}}$ if U in the boundary layer is analogous to $u_{\max}$ in the pipe and δ is analogous to pipe radius? Can you interpret this result concerning ability to resist transition?

7.27 Since C_D decreases with Re_L in Fig. 7.6 for both laminar and turbulent flow over smooth plates, does this mean that drag decreases with velocity? Define a better dimensionless parameter ζ which we can plot versus Re_L and show directly the variation of drag with velocity. Write an expression for ζ versus Re_L for turbulent smooth-plate flow.

7.28 Atmospheric boundary layers are very thick but follow formulas very similar to those of flat-plate theory. Consider wind blowing at 10 m/s at a height of 80 m above a smooth beach. Estimate the wind shear stress in pascals on the beach if the air is standard sea-level conditions. What will the wind velocity striking your nose be if (a) you are standing up and your nose is 170 cm off the ground; (b) you are lying on the beach and your nose is 17 cm off the ground?

7.29 A jet airliner travels at 250 m/s at 10,000 m standard altitude. It has a smooth wing 7 m long and 55 m wide. Estimate the power required to overcome friction drag. If the wing is rough and requires 13 MW to overcome friction, estimate the wing roughness in millimeters.

7.30 Flow straighteners are arrays of narrow ducts placed in wind tunnels to remove swirl and other in-plane secondary velocities. They can be idealized as square boxes constructed by vertical and horizontal plates, as in Fig. P7.30. The cross section is a by a and the box length is L. Assuming laminar flat-plate flow and an array of N by N boxes, derive a formula for (a) the total drag on the bundle of boxes and (b) the effective pressure drop across the bundle.

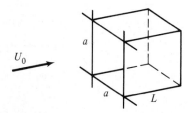

Fig. P7.30

7.31 Suppose the flow straighteners in Fig. P7.30 form an array of 10 by 10 boxes of side length $a = 10$ cm and length $L = 30$ cm. If the approach velocity is $U_0 = 15$ m/s and the fluid is sea-level standard air, compute (a) the total drag of the array (b) the pressure drop across the array. Use flat-plate theory and compare with duct-flow estimates from Sec. 6.6.

7.32 A flat plate of length L and height δ is placed at a wall and is parallel to an approaching boundary layer, as in Fig. P7.32. Assume that the flow over the plate is

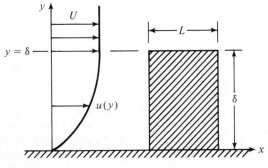

Fig. P7.32

fully turbulent and that the approaching flow is a one-seventh power law

$$u(y) = U_0 \left(\frac{y}{\delta}\right)^{1/7}$$

Using strip theory, derive a formula for the drag coefficient of this plate. Compare this result with the drag of the same plate immersed in a uniform stream U_0.

7.33 An alternate analysis of turbulent flat-plate flow was given by Prandtl in 1927, using a wall shear-stress formula from pipe flow

$$\tau_w = 0.0225 \rho U^2 \left(\frac{\nu}{U\delta}\right)^{1/4}$$

Show that this formula can be combined with Eqs. (7.33) and (7.40) to derive the following relations for turbulent flat-plate flow:

$$\frac{\delta}{x} = \frac{0.37}{\mathrm{Re}_x^{1/5}}$$

$$c_f = \frac{0.0577}{\mathrm{Re}_x^{1/5}}$$

$$C_D = \frac{0.072}{\mathrm{Re}_L^{1/5}}$$

These formulas are limited to Re_x between 5×10^5 and 10^7.

7.34 A flat barge 14 m wide and 40 m long moves at 3 knots in seawater at 20°C. Estimate the friction drag on the bottom of the barge for (a) a smooth wall and (b) wall roughness height of 3 mm.

7.35 A torpedo 55 cm in diameter and 5 m long moves at 45 knots in seawater at 20°C. Estimate the power required to overcome friction drag if $\mathrm{Re}_{crit} = 5 \times 10^5$ and $\epsilon = 0.5$ mm.

7.36 A ship is 150 m long and has a wetted area of 5000 m². If it is encrusted with barnacles, the ship requires 7000 hp to overcome friction drag when moving in seawater at 15 knots and 20°C. What is the average roughness of the barnacles? How fast would the ship move with the same power if the surface were smooth? Neglect wave drag.

7.37 As a case similar to Example 7.5. Howarth also proposed the adverse-gradient velocity distribution $U = U_0(1 - x^2/L^2)$ and computed separation at $x_{sep}/L = 0.271$ by a series-expansion method. Compute separation by Thwaites' method and compare.

7.38 In 1957 H. Görtler proposed the adverse-gradient test cases

$$U = \frac{U_0}{(1 + x/L)^n}$$

and computed separation for laminar flow at $n = 1$ to be $x_{sep}/L = 0.159$. Compare with Thwaites' method, assuming $\theta_0 = 0$.

7.39 Repeat Prob. 7.38 for $n = 2$ and compare with Görtler's series-expansion computation $x_{sep}/L = 0.078$.

7.40 For flow past a cylinder of radius R as in Fig. P7.40 the theoretical inviscid velocity distribution along the surface is $U = 2U_0 \sin(x/R)$, where U_0 is the oncoming stream velocity and x is arc length measured

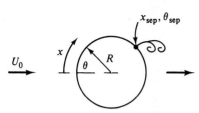

Fig. P7.40

from the nose (Chap. 8). Compute the laminar separation point x_{sep} and θ_{sep} by Thwaites' method and compare with the digital-computer solution $x_{sep}/R = 1.823$ ($\theta_{sep} = 104.5°$) given by R. M. Terrill in 1960.

7.41 The velocity-profile shape

$$u = U(1 - e^{-4.6y/\delta})$$

is a smooth curve with $u = 0$ at $y = 0$ and $u = 0.99U$ at $y = \delta$ and would therefore seem to be a reasonable substitute for the parabolic profile of Eq. (7.3). Yet when this new shape is used in the flat-plate integral analysis of Sec. 7.2, we get the lousy results $\delta/x = 9.2/Re_x^{1/2}$ and $c_f = 1.00/Re_x^{1/2}$, which are 80 and 50 percent high respectively. What is the reason for the discrepancy?

7.42 Consider the flat-walled diffuser in Fig. P7.42, which is similar to that of Fig. 6.25a, with constant width b. If x is measured from the inlet and the wall boundary layers are thin, show that the core velocity $U(x)$ in the diffuser is given approximately by

$$U = \frac{U_0}{1 + (2x \tan \theta)/W}$$

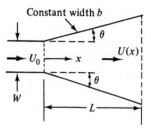

Fig. P7.42

where W is the inlet height. Use this velocity distribution with Thwaites' method to compute the wall angle θ for which laminar separation will occur in the exit plane when diffuser length $L = 2W$. Note that the result is independent of Reynolds number.

7.43 In Fig. P7.43 a slanted upper wall creates a favorable pressure gradient on the upper surface of the flat plate. Use Thwaites' theory to estimate

$$C_D = \frac{F}{\frac{1}{2}\rho U_0^2 bL}$$

on the upper plate surface if $U_0 L/\nu = 10^5$. Compare with Eq. (7.27).

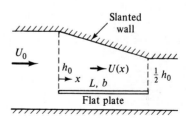

Fig. P7.43

7.44 From Table 7.2, C_D for a plate normal to the stream is 2.0. If the stream has velocity U_∞ and pressure p_∞, and if the average pressure on the front of the plate is the stagnation value p_0 (actually it is slightly less), show that the base pressure on the rear of the plate is $p_b \approx p_\infty - \frac{1}{2}\rho U_\infty^2$.

7.45 A very wide flat plate 1 m long is immersed in a stream of sea-level standard air moving at 10 m/s. Compute the drag per unit width if the 1-m-long edge of the plate is (a) parallel and (b) normal to the stream.

7.46 A chimney 2.5 m in diameter and 45 m high is exposed to sea-level storm winds at 22 m/s. Estimate the drag force and the bending moment about the bottom of the chimney.

7.47 A logging boat tows a log 2.5 m in diameter and 25 m long at 4 m/s in fresh water at 20°C. Estimate the power required if the axis of the log is (a) parallel and (b) normal to the tow direction.

7.48 A parachutist jumps from a plane at 8000 ft standard altitude with a 28-ft-diameter chute. The total weight of chutist and chute is 185 lbf. Estimate (a) the terminal fall velocity at 8000 ft and (b) the time to fall from 8000 to 4000 ft.

7.49 The drag of a sphere at very low Reynolds numbers $Re_D \ll 1$ was given analytically by G. G. Stokes in 1851: $F = 3\pi\mu VD$ [2, pp. 204–208]. This formula is an example of Stokes' *law of creeping motion*, where inertia is negligible. Show that the drag coefficient in this region is $C_D = 24/Re_D$. A 1-mm-diameter sphere falls in 20°C glycerin at 2.5 mm/s. Is this a creeping motion? Compute (a) the Reynolds number, (b) the drag, and (c) the specific gravity of the sphere.

7.50 A football weighs 0.91 lbf and approximates an ellipsoid 6 in in diameter and 12 in long (Table 7.3). If the football is thrown at 45° with an initial velocity of 70 ft/s, estimate (a) the initial deceleration and (b) the horizontal travel distance, assuming sea-level standard air. Neglect lift and spin.

7.51 A baseball weighs 145 g and is 7.35 cm in diameter. If it is driven upward at a 45° angle and initial velocity of 48 m/s, estimate the initial drag and distance traveled. *U.S. students*: Would this be a home run down the first-base line at Fenway Park?

7.52 Two baseballs from Prob. 7.51 are connected to a rod 6 mm in diameter and 56 cm long spinning about the center at 250 r/min, as in Fig. P7.52. What power in watts is required to maintain this rotation? Include the drag of the rod and assume sea-level standard air.

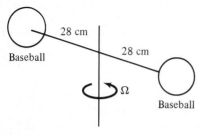

Fig. P7.52

7.53 An anemometer has two 5-cm-diameter cups on a rod of radius 15 cm, as in Fig. P7.53. The drag of the rod may be neglected. If the resisting torque of the

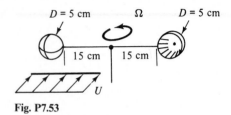

Fig. P7.53

bearing is $T_0 = 0.01$ N·m, compute the rotation speed Ω of the anemometer in revolutions per minute if $U =$ (a) 5 m/s; (b) 10 m/s. Assume sea-level standard air.

7.54 A settling tank for a municipal water supply is 3 m deep, and water flows through continuously at 30 cm/s. What should the minimum length of the tank be to ensure that sediment particles of 1 mm diameter (SG = 2.6) will all fall to the bottom before the water leaves the tank?

7.55 Repeat Prob. 7.54 for sediment particles of 0.1 mm diameter.

7.56 A fishnet consists of 1-mm-diameter strings overlapped and knotted to form 1- by 1-cm squares. Estimate the drag of 1 m² of such a net when towed normal to its plane at 3 m/s in 20°C seawater. What horsepower is required to tow 500 ft² of this net?

7.57 A filter may be idealized as an array of cylindrical fibers normal to the flow, as in Fig. P7.57. Assuming that the fibers are uniformly distributed and have drag coefficients given by Fig. 7.16a, derive an approximate expression for the pressure drop Δp through a filter of thickness L.

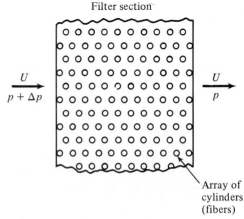

Fig. P7.57

7.58 Apply Prob. 7.57 to a filter consisting of 0.2-mm-diameter fibers packed 1000 per square centimeter of filter section in the plane of Fig. P7.57. For sea-level standard air flowing at $U = 1.2$ m/s, estimate the pressure drop through a filter 3 cm thick.

7.59 Consider a heavy solid sphere of diameter D and density ρ_s suddenly dropped from rest in a fluid of density ρ and viscosity μ. Set up a differential equation and solve for (a) the time history $V(t)$; (b) the final or *terminal* velocity V_f; and (c) the time required for V to reach 99 percent of V_f. Assume that C_D is constant.

7.60 A solid 6-in-cannonball (SG = 7.9) is dropped into the ocean at 20°C. What will its terminal velocity be? How long will it take to fall to the bottom 12,000 ft deep?

7.61 A helium-filled balloon is designed to have a sea-level terminal ascent velocity of 5 m/s. The helium pressure is 125 kPa. The balloon and payload weigh 300 N, not including the helium weight. Neglecting payload drag, estimate the proper balloon diameter.

7.62 In Prob. 7.61 what is the minimum diameter for which the balloon will rise? What will be the terminal ascent velocity if $D = 4.5$ m?

7.63 Which has a higher terminal velocity, a 1-mm air bubble rising in fresh water or a 1-mm water droplet falling in air? What is the ratio of their terminal velocities?

7.64 Assume that a radioactive dust particle approximates a sphere with a specific weight of 26 kN/m³. Approximately how long in days will it take such a particle to settle to earth from an altitude of 10 km if the particle diameter is (a) 1 μm; (b) 10 μm?

7.65 A ping-pong ball weighs 2.6 g and has a diameter of 3.8 cm. It can be supported by an air jet from a vacuum-cleaner outlet, as in Fig. P7.65. For sea-level standard air, what jet velocity is required?

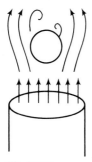

Fig. P7.65

7.66 A parachutist is to land at sea level at a vertical velocity of 9 mi/h. Jumper and pack weigh 200 lbf. What should the diameter of the chute be?

7.67 The average skydiver with parachute unopened weighs 175 lbf and has a drag area $C_D A = 9$ ft² spread-eagled and 1.2 ft² falling feet first [12, p. 313]. What are the minimum and maximum terminal speeds that can be achieved by a skydiver at 5000 ft standard altitude?

7.68 Make a dynamic analysis of Prob. 7.67 to determine the acceleration of the skydiver starting from rest, assuming constant drag-area $C_D A$. Show that the diver reaches 99 percent of terminal velocity at a time $t \approx 2.65(2W/\rho g^2 C_D A)^{1/2}$. Compute this acceleration time in seconds for the two drag cases given in Prob. 7.67.

7.69 A heavy sphere attached to a string should hang at an angle θ when immersed in a stream of velocity U, as in Fig. P7.69. Derive an expression for θ as a function of sphere and flow properties. What is θ if the sphere is steel (SG = 7.86) of diameter 3 cm and the flow is sea-level standard air at $U = 40$ m/s? Neglect the string drag.

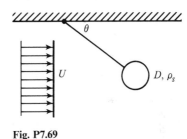

Fig. P7.69

7.70 A high-speed car has a drag coefficient of 0.3 and a frontal area of 1 m². A parachute is to be used to slow this car down from 80 to 40 m/s in 8 s. What should the chute diameter be? What distance will be traveled during this deceleration? Take $m = 2000$ kg.

7.71 A tractor-trailer truck has a drag area $C_D A = 8$ m² bare and 6.7 m² with an aerodynamic deflector (Fig. 7.18b). Its rolling resistance is 50 N for each mile per hour of speed. Calculate the total horsepower required at sea level with and without the deflector if the truck moves at (a) 55 mi/h; (b) 75 mi/h.

7.72 A pickup truck has a clean drag area $C_D A$ of 35 ft². Estimate the horsepower required to drive the truck at 55 mi/h (a) clean; and (b) with the 3- by 6-ft sign in Fig. P7.72 installed if rolling resistance is 150 lbf at sea level.

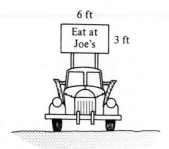

Fig. P7.72

7.73 A water tower is approximated by a 15-m-diameter sphere mounted on a 1-m-diameter rod 20 m long. Estimate the bending moment at the root of the rod due to aerodynamic forces during hurricane winds of 40 m/s.

7.74 In the great hurricane of 1938, winds of 85 mi/h blew over a boxcar in Providence, Rhode Island. The boxcar was 10 ft high, 40 ft long, and 6 ft wide, with a 3-ft clearance above tracks 4.8 ft apart. What windspeed would topple a boxcar weighing 40,000 lbf?

7.75 A hot-film probe on a 60° cone is mounted on a 5-mm-diameter rod in a sea-level airstream at 45 m/s, as in Fig. P7.75. Estimate the total drag on the instrument, the bending moment at the root (point B), and the rod stress at the root in newtons per square centimeter.

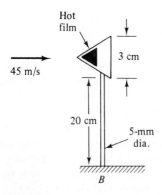

Fig. P7.75

7.76 Assume that an automobile engine can deliver 40 hp to the wheels. The frontal area is 20 ft², and rolling resistance is 1 lbf for each mile per hour of speed. Compute the maximum speed in miles per hour for (*a*) a modern car, $C_D = 0.3$; (*b*) an antique, $C_D = 0.9$.

7.77 A rotary mixer consists of two 1-m-long half-tubes rotating around a central arm, as in Fig. P7.77. Using the drag from Table 7.2, derive an expression for the torque T required to drive the mixer at angular velocity Ω in a fluid of density ρ. Suppose the fluid is water at 20°C and the maximum driving power available is 20 kW. What is the maximum rotation speed Ω in revolutions per minute?

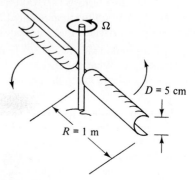

Fig. P7.77

7.78 A 50,000-lb airplane with a drag area $C_D A = 180$ ft² lands at sea level at 180 ft/s and deploys a drag chute 28 ft in diameter. How long will it take the plane to slow down to 100 ft/s if no other brakes are applied? How far will it travel in this time?

7.79 A buoyant ball of specific gravity $s < 1$ dropped into water at inlet velocity V_0 will penetrate a distance h and then pop out again, as in Fig. P7.79. Make a dynamic analysis of this problem, assuming constant drag coefficient, and derive an expression for h as a function of the system properties. How far will a 5-cm-diameter ball with $s = 0.5$ and $C_D \approx 0.47$ penetrate if it enters at 10 m/s?

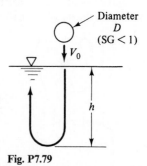

Fig. P7.79

7.80 Icebergs can be driven at substantial speeds by the wind. Let the iceberg be idealized as a large flat cylinder, $D \gg L$, with one-eighth of its bulk exposed, as in Fig. P7.80. Let the seawater be at rest. If the upper and lower drag forces depend upon relative velocities between berg and fluid, derive an approximate expression for the steady iceberg speed V when driven by wind velocity U.

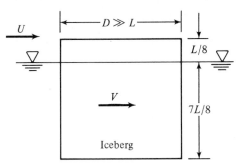

Fig. P7.80

7.81 Solve Prob. 7.80 for V for the special case $D = 800$ m, $L = 100$ m, and $U = 15$ m/s.

7.82 When immersed in a uniform stream V, a heavy rod hinged at A will hang at *Pode's angle* θ, after an analysis by L. Pode in 1951 (Fig. P7.82). Assume that the cylinder has normal drag coefficient C_{DN} and tangential coefficient C_{DT} which relate the drag forces to V_N and V_T, respectively. Derive an expression for Pode's angle as a function of flow and rod parameters. Find θ for a steel rod, $L = 40$ cm, $D = 1$ cm, hanging in sea-level air at $V = 35$ m/s.

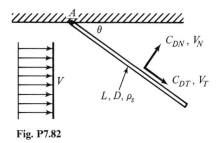

Fig. P7.82

7.83 An airplane weighs 180 kN and has a wing area of 160 m^2 and a mean chord of 4 m. The airfoil properties are given by Fig. 7.22. If the airplane cruises at 250 mi/h at 3000 m standard altitude, what propulsive power is required to overcome wing drag?

7.84 The airplane of Prob. 7.83 is designed to land at $V_0 = 1.2V_{\text{stall}}$, using a split flap set at 60°. What is the

proper landing speed in miles per hour? What power is required for takeoff at the same speed?

7.85 Suppose that the airplane of Prob. 7.83 takes off at sea level without benefit of flaps, with C_L constant so that takeoff speed is 100 mi/h. Estimate the takeoff distance if the thrust is 10 kN. How much thrust is needed to make the takeoff distance 1250 m?

7.86 Suppose that the airplane of Prob. 7.83 is fitted with all the best high-lift devices of Fig. 7.25. What is its minimum stall speed in miles per hour? Estimate the stopping distance if the plane lands at $V_0 = 1.25V_{\text{stall}}$ with constant $C_L = 3.0$ and $C_D = 0.2$ and the braking force is 20 percent of the weight on the wheels.

7.87 For finite wings below stall Eqs. (7.70) and (7.71) are excellent approximations. Use them to show that the maximum lift-to-drag ratio occurs when $C_D = 2C_{D\infty}$. What are $(L/D)_{\text{max}}$ and α for a symmetric wing ($h = 0$) when AR $= 7$ and $C_{D\infty} = 0.006$?

7.88 In gliding flight an airplane or bird is moving downward at velocity V and gliding angle θ (measured down from the horizontal) such that its lift and drag are in equilibrium with its weight. Show that for this condition the sinking, or downward velocity component, u, is such that

$$\frac{u}{V} \approx \tan \theta \approx \frac{\text{drag}}{\text{lift}}$$

Minimum gliding angle thus occurs at the maximum L/D for which the maximum gliding range can be attained. What would the minimum gliding angle be for the wing of Prob. 7.87?

7.89 A sailplane weighs 2 kN and has a wing area of 10 m^2 and an aspect ratio of 9, with an NACA 0009 airfoil. What is its stall speed in meters per second? What is its minimum gliding angle? What is the maximum distance it can glide in still air when it is 1500 m above level ground?

7.90 A typical falcon has a wingspan of 80 cm and a wing area of 0.11 m^2, with a minimum glide angle of 5.5° at $C_L = 1.1$ [14]. Estimate the minimum drag $C_{D\infty}$ of the falcon and compare with Fig. 7.16.

7.91 The minimum, or "parasite," drag of a pigeon is given in Fig. 7.16. A typical pigeon weighs 1.2 lbf, with a wingspan of 2 ft, a wing area of 0.85 ft^2, and $C_{L,\text{max}} = 1.5$. What is the stall speed of a pigeon? What is its minimum glide angle? In what speed range can it glide in equilibrium at 10°?

7.92 In prewar days there was a controversy, perhaps apocryphal, about whether the bumblebee has a legitimate aerodynamic right to fly. The average bumblebee (*Bombus terrestris*) weighs 0.88 g, with a wing span of 1.73 cm and a wing area of 1.26 cm². It can indeed fly at 10 m/s. Using fixed-wing theory, what is the lift coefficient of the bee at this speed? Is this reasonable for typical airfoils?

7.93 The bumblebee can hover at zero speed by flapping its wings. Using the data of Prob. 7.92 devise a theory for flapping wings where the downstroke approximates a short flat plate normal to the flow (Table 7.3) and the upstroke is feathered at nearly zero drag. How many flaps per second of such a model wing are needed to support the bee's weight? (Actual measurements of bees show a flapping rate of 194 Hz.)

7.94 A 6000-lb boat is outfitted with a hydrofoil of 1 ft chord and 6 ft span, with $C_{L,\max} = 1.4$ and $C_{D\infty} = 0.05$. The engine can deliver 130 hp to the water. What is the minimum speed in seawater for which the foil supports the boat weight? What is the maximum speed attainable?

7.95 A pitcher can throw a curve ball at about 60 mi/h with a spin of 1500 r/min about, say, a vertical axis. Experiments with spinning spheres indicate a lift coefficient of about 0.3 for this condition. How far will such a curve ball have deviated from its straight-line path when it reaches home plate 60 ft away? Take a baseball to weigh 0.32 lbf with a diameter of 2.9 in.

7.96 A symmetrical wing has a uniform rectangular planform such that its width $b = 8c$, where c is the chord length. It is designed to develop 35 kN of lift at sea level at an angle of attack of 5° and velocity of 125 m/s. Compute (*a*) the proper chord length c and (*b*) the wing drag for this condition.

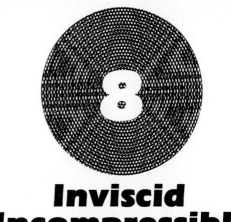

Inviscid Incompressible Flow

8.1 INTRODUCTION

This chapter treats inviscid flow in detail, using the definitions given in Secs. 4.8 and 4.9 and the additional insight gained from Chaps. 6 and 7.

We found that in high-Reynolds-number flows the viscous effects (especially in external flows) could be confined to thin boundary layers near solid surfaces plus separated flow and wake regions which occur in adverse gradients. Flow outside the boundary layer is essentially inviscid, and we further assume here that it is incompressible. We now develop techniques for detailed analysis of inviscid flow, which is then patched onto the boundary layer, as explained in Chap. 7.

Figure 8.1 reminds us of the problem to be faced. The front of almost any body is a favorable-gradient region, and the boundary layer will be attached and thin: inviscid theory will give excellent results for the outer flow. For the internal flow the wall boundary layers grow and meet, and the inviscid core vanishes. But inviscid theory applies in a "short" duct such as the nozzle of a wind tunnel. Even in a "long" internal flow ($L/D > 10$) we are tempted to use inviscid theory as a crude approximation because it is a relatively easy theory, but we must keep in mind that such calculations are only a rough approximation of the real internal flow.

Inviscid theory should work well for the external flow in Fig. 8.1, especially near the front of the body. If there is boundary-layer separation at the rear, the separated flow will deflect and modify the inviscid streamlines. Unless the interaction is properly modeled (presently a research area), the theory will only be qualitative there.

As we saw in Sec. 4.9, the equations of motion for a frictionless incompressible flow are continuity and momentum

$$\mathbf{V} \cdot \mathbf{V} = 0 \qquad \rho \frac{d\mathbf{V}}{dt} = -\nabla p + \rho \mathbf{g} \tag{8.1}$$

441

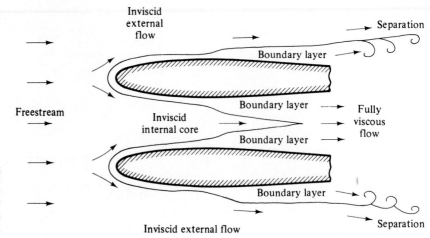

Fig. 8.1 Patching viscous and inviscid flow regions. The theory in the present chapter does not apply to the boundary-layer regions.

We would solve these equations for velocity and pressure subject to given boundary conditions; but in fact we do not have to. If we neglect (1) viscous effects, (2) entropy gradients, (3) stratification, and (4) noninertial effects, the flow will be irrotational, so that

$$\mathbf{V} = \nabla\phi$$

or

$$u = \frac{\partial\phi}{\partial x} \qquad v = \frac{\partial\phi}{\partial y} \qquad w = \frac{\partial\phi}{\partial z} \qquad (8.2)$$

where ϕ is the scalar velocity-potential function. The continuity equation becomes Laplace's equation

$$\nabla^2\phi = 0 \qquad (8.3)$$

and the momentum equation becomes Bernoulli's equation

$$\frac{\partial\phi}{\partial t} + \frac{p}{\rho} + \tfrac{1}{2}V^2 + gz = \text{const} \qquad (8.4)$$

Most of our examples will be steady flows, where $\partial\phi/\partial t \equiv 0$. The boundary conditions on Eq. (8.3) are (1) a known velocity in the free stream or other open-stream boundary

$$\text{Known } \frac{\partial\phi}{\partial x}, \frac{\partial\phi}{\partial y}, \text{ and } \frac{\partial\phi}{\partial z} \qquad (8.5a)$$

and (2) no velocity normal to the boundary at solid surfaces

$$\frac{\partial\phi}{\partial n} = 0 \qquad (8.5b)$$

There is no condition on tangential velocity $= \partial\phi/\partial s$ at a solid surface, where s is the coordinate along the surface, and this becomes part of the solution to the problem.

In plane polar coordinates (r, θ) the velocity components are related to velocity potential as follows:

$$v_r = \frac{\partial\phi}{\partial r} \qquad v_\theta = \frac{1}{r}\frac{\partial\phi}{\partial\theta} \tag{8.6}$$

These are convenient in the source, sink, and vortex analyses of the next section.

Occasionally the problem involves a free surface, for which the boundary pressure is known and equal to p_a, usually constant. The Bernoulli equation (8.4) then gives a boundary relation between $V = |\nabla\phi|$ and the location z of the free surface. For steady flow, for example,

$$V^2 = |\nabla\phi|^2 = \text{const} - 2gz \tag{8.7}$$

at the free surface.

Is this a better procedure than a direct attack on Eqs. (8.1)? Yes, definitely. The analysis of Laplace's equation (8.3) is very well developed and is termed *potential theory*, with whole books written about it [1]. There are many techniques for finding so-called potential functions which satisfy Laplace's equation, including (1) superposition of elementary functions, (2) numerical analysis, (3) conformal mapping, (4) electrical analogs, and (5) mechanical analogs. Having found $\phi(x, y, z, t)$ from such an analysis, we then compute $\mathbf{V}$ by direct differentiation in Eq. (8.2), after which we can compute p from Eq. (8.4). The procedure is quite straightforward, and many interesting albeit idealized results can be obtained. Whole books have also been written on potential flow of inviscid fluids [e.g., 2, 3].

Recall from Sec. 4.7 that if the flow is described by only two coordinates, the stream function ψ also exists as an alternate approach. In plane irrotational flow ψ also satisfies Laplace's equation. For example, in xy coordinates

$$u = \frac{\partial\psi}{\partial y} \qquad v = -\frac{\partial\psi}{\partial x} \tag{8.8}$$

The condition of irrotationality reduces to

$$2\omega_z = 0 = \frac{\partial v}{\partial x} - \frac{\partial u}{\partial y} = \frac{\partial}{\partial x}\left(-\frac{\partial\psi}{\partial x}\right) - \frac{\partial}{\partial y}\left(\frac{\partial\psi}{\partial y}\right) = -\nabla^2\psi(x, y) \tag{8.9}$$

Therefore irrotational inviscid plane flow requires that

$$\frac{\partial^2\psi}{\partial x^2} + \frac{\partial^2\psi}{\partial y^2} = 0 \tag{8.10}$$

In like manner, the stream function in polar coordinates is

$$v_r = \frac{1}{r}\frac{\partial\psi}{\partial\theta} \qquad v_\theta = -\frac{\partial\psi}{\partial r} \tag{8.11}$$

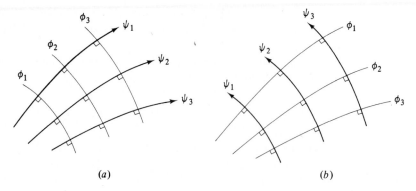

Fig. 8.2 Streamlines and potential lines are orthogonal and may reverse roles if results are useful: (*a*) typical inviscid-flow pattern; (*b*) same as (*a*) with roles reversed.

(*a*)   (*b*)

The condition of irrotationality is

$$\omega_z = 0 = \frac{1}{r}\frac{\partial}{\partial r}(rv_\theta) - \frac{1}{r}\frac{\partial}{\partial \theta}(v_r)$$

$$= \frac{1}{r}\frac{\partial}{\partial r}\left(-r\frac{\partial \psi}{\partial r}\right) - \frac{1}{r}\frac{\partial}{\partial \theta}\left(\frac{1}{r}\frac{\partial \psi}{\partial \theta}\right) = -\nabla^2\psi(r, \theta) \qquad (8.12)$$

This is also Laplace's equation in polar coordinates

$$\frac{1}{r}\frac{\partial}{\partial r}\left(r\frac{\partial \psi}{\partial r}\right) + \frac{1}{r^2}\frac{\partial^2 \psi}{\partial \theta^2} = 0 \qquad (8.13)$$

The boundary conditions again are known velocity in the stream and no flow through any solid surface:

Free stream: Known $\dfrac{\partial \psi}{\partial x}, \dfrac{\partial \psi}{\partial y}$ or $\dfrac{\partial \psi}{\partial r}, \dfrac{\partial \psi}{\partial \theta}$ $\qquad (8.14a)$

Solid surface: $\psi_{\text{body}} = \text{const}$ $\qquad (8.14b)$

Equation (8.14*b*) is particularly interesting because any line of constant ψ in a flow can therefore be interpreted as a body shape and may lead to interesting results.

We can compute either ϕ or ψ or both, and the solution will be an orthogonal *flow net* like Fig. 8.2. Once found, either set of lines may be considered the ϕ lines, and the other will be the ψ lines. Both sets of lines are laplacian solutions and can serve either role.

An intriguing facet of potential flow is that the governing equations (8.3) and (8.10) contain no parameters, nor do the boundary conditions. Therefore flow-net solutions are purely geometric, depending only on the body shape, free-stream orientation, and—surprisingly—the position of the rear stagnation point.[1] There is no Reynolds, Froude, or Mach number to complicate the dynamic similarity. Inviscid potential flows are kinematically similar with no additional parameters needed.

---

[1] The rear stagnation condition establishes the net amount of circulation about the body, giving rise to a lift force. Otherwise the solution is not unique. See Sec. 8.4, Eq. (8.67) and Fig. 8.22.

In three-dimensional potential flow, the stream function ψ does not generally exist, but the ϕ function does and we solve Eq. (8.3) for $\phi(x, y, z, t)$. For the special case of three-dimensional axisymmetric flow a stream function $\psi(r, z)$ does exist, but it does not satisfy Laplace's equation. Even in axisymmetric flow it is more common to solve for ϕ than for ψ.

8.2 ELEMENTARY PLANE-FLOW SOLUTIONS

Several potential-flow problems of interest can be constructed from three types of elementary solutions:

1. Uniform stream
2. Source or sink
3. Vortex

The potential and stream functions for these three can be added to produce various useful results. In doing this we rely on the principle of superposition, which is valid because Laplace's equation is linear. That is, if ϕ_1 and ϕ_2 are separate solutions, their sum is also a solution. If

$$\nabla^2 \phi_1 = 0 \quad \text{and} \quad \nabla^2 \phi_2 = 0$$

then
$$\nabla^2(\phi_1 + \phi_2) = 0 \tag{8.15}$$

And of course the same is true for two solutions ψ_1 and ψ_2. Let us now derive the elementary solutions.

Uniform Stream

A stream of constant velocity U_∞ has zero spatial derivatives and hence satisfies continuity and irrotationality identically. First let the stream be in the x direction and solve for the resulting ϕ and ψ functions:

$$u = U_\infty = \frac{\partial \psi}{\partial y} = \frac{\partial \phi}{\partial x} = \text{const}$$

$$v = 0 = -\frac{\partial \psi}{\partial x} = \frac{\partial \phi}{\partial y} \tag{8.16}$$

Integrating, we obtain
$$\psi = U_\infty y + C_1 \qquad \phi = U_\infty x + C_2 \tag{8.17}$$

Now the integration constants C_1 and C_2 have no effect whatever on velocities or pressures in the flow. Therefore we shall consistently ignore such irrelevant constants and for a uniform stream in the x direction

$$\psi = U_\infty y \qquad \phi = U_\infty x \tag{8.18}$$

These are plotted in Fig. 8.3a and consist of a rectangular mesh of straight stream-lines normal to straight potential lines. It is customary to put arrows on the streamlines showing the direction of the flow.

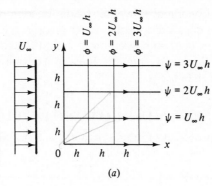

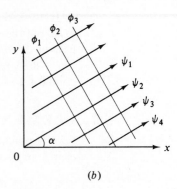

Fig. 8.3 Flow net for a uniform stream: (*a*) stream in the *x* direction; (*b*) stream at angle α, Eq. (8.21).

(*a*) (*b*)

In terms of plane polar coordinates (r, θ) Eq. (8.18) becomes

Uniform stream: $\psi = U_\infty r \sin \theta$ $\phi = U_\infty r \cos \theta$ (8.19)

The plot of course is exactly the same as Fig. 8.3*a*.

If we generalize to a uniform stream at an angle α to the *x* axis, as in Fig. 8.3*b*, we require that

$$u = U_\infty \cos \alpha = \frac{\partial \psi}{\partial y} = \frac{\partial \phi}{\partial x} \qquad v = U_\infty \sin \alpha = -\frac{\partial \psi}{\partial x} = \frac{\partial \phi}{\partial y} \qquad (8.20)$$

Integrating, we obtain for a uniform stream at angle α

$$\psi = U_\infty(y \cos \alpha - x \sin \alpha) \qquad \phi = U_\infty(x \cos \alpha + y \sin \alpha) \qquad (8.21)$$

These are useful in airfoil angle-of-attack problems.

Line Source or Sink

Suppose that the *z* axis were a sort of thin pipe manifold through which fluid issued at a uniform rate along its length. Looking end on at the *z* axis, we would see a cylindrical radial outflow, as sketched in Fig. 8.4*a*. In steady flow the amount of fluid crossing any given cylindrical surface of radius *r* and length *b* is constant

$$Q = v_r(2\pi rb) = \text{const} = 2\pi bm \qquad (8.22)$$

or $$v_{r,\,\text{source}} = \frac{m}{r}$$

where *m* is a convenient constant. This is called a *line source* if *m* is positive and a *line sink* if *m* is negative. Obviously the source streamlines flow outward, as in Fig. 8.4*a*, and the tangential velocity v_θ is zero. We can solve for the plane polar version of ψ and ϕ

$$v_r = \frac{m}{r} = \frac{1}{r}\frac{\partial \psi}{\partial \theta} = \frac{\partial \phi}{\partial r} \qquad v_\theta = 0 = -\frac{\partial \psi}{\partial r} = \frac{1}{r}\frac{\partial \phi}{\partial \theta} \qquad (8.23)$$

Integrating, we obtain for a line source or sink

$$\psi = m\theta \qquad \phi = m \ln r \qquad (8.24)$$

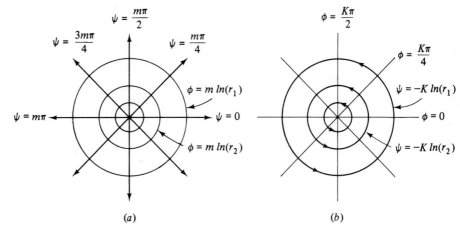

Fig. 8.4 Flow net for a line source and line vortex: (*a*) line source, Eq. (8.24); (*b*) roles are reversed to form a line vortex, Eq. (8.26).

(*a*)   (*b*)

These are sketched in Fig. 8.4*a* and are quite convenient compared with the equivalent cartesian forms

$$\psi = m \tan^{-1} \frac{y}{x} \qquad \phi = m \ln (x^2 + y^2)^{1/2} \tag{8.25}$$

We can verify by direct substitution that both ψ and ϕ satisfy Laplace's equation in either coordinate system. We emphasize that this is a line, not a point, source. The entire z axis is a line of singularity where the radial velocity is infinite and ψ and ϕ are not defined. When using the line source in a practical computation we keep this singularity hidden within a boundary or body-shape streamline.

Line Vortex

Now suppose we reverse the roles of ψ and ϕ in Eq. (8.24), yielding

$$\psi = -K \ln r \qquad \phi = K\theta \tag{8.26}$$

By direct differentiation of either one we obtain the velocity pattern

$$v_r = 0 \qquad v_\theta = \frac{K}{r} \tag{8.27}$$

This is a purely circulating flow with tangential velocity dropping off as $1/r$. It is sketched in Fig. 8.4*b* and again has a singularity at the origin, where velocity is infinite and ψ and ϕ are not defined. Again we should keep the vortex center hidden within a body streamline. The vortex strength K has the same dimensions as the source strength m, namely, velocity times length.

Circulation

The line-vortex flow is irrotational everywhere except at the origin, where the vorticity $\nabla \times \mathbf{V}$ is infinite. This means that a certain line integral called the *fluid circulation* Γ does not vanish when taken around a vortex center.

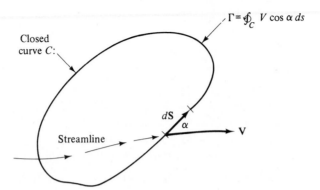

$\Gamma = \displaystyle\oint_C V \cos \alpha \, ds$

Closed curve C:

Streamline

$d\mathbf{S}$

α

$\mathbf{V}$

Fig. 8.5 Definition of the fluid circulation Γ.

With reference to Fig. 8.5, the circulation is defined as the counterclockwise line integral, around a closed curve C, of the arc length ds times the velocity component tangent to the curve

$$\Gamma = \oint_C V \cos \alpha \, ds = \int_C \mathbf{V} \cdot d\mathbf{s} = \int_C (u \, dx + v \, dy + w \, dz) \tag{8.28}$$

From the definition of ϕ, $\mathbf{V} \cdot d\mathbf{s} = \nabla\phi \cdot d\mathbf{s} = d\phi$ for an irrotational flow; hence normally Γ in an irrotational flow would equal the final value of ϕ minus the initial value of ϕ. Since we start and end at the same point, we would compute $\Gamma = 0$; but not for vortex flow: with $\phi = K\theta$ from Eq. (8.26) there is a change in ϕ of amount $2\pi K$ as we make one complete circle

Path enclosing a vortex: $\qquad\qquad \Gamma = 2\pi K \tag{8.29}$

Alternately the calculation can be made by defining a circular path of radius r around the vortex center, from Eq. (8.28)

$$\Gamma = \int_C v_\theta \, ds = \int_0^{2\pi} \frac{K}{r} r \, d\phi = 2\pi K \tag{8.30}$$

In general, Γ denotes the net algebraic strength of all the vortex filaments contained within the closed curve. In the next section we shall see that a region of finite circulation within a flowing stream will be subjected to a lift force proportional to both U_∞ and Γ.

It is easy to show, using Eq. (8.28), that a source or sink creates no circulation. If there are no vortices present, the circulation will be zero for any path enclosing any number of sources and sinks.

8.3 SUPERPOSITION OF PLANE-FLOW SOLUTIONS

We can now form a variety of interesting potential flows by summing the velocity-potential and stream functions of a uniform stream, source or sink, and vortex. Most of the results are classic, of course, needing only a brief treatment here.

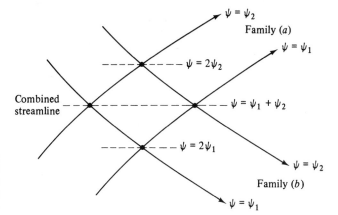

$\psi = \psi_2$
Family (*a*)

$\psi = \psi_1$

$---\bullet--- \psi = 2\psi_2$

Combined
streamline $--\bullet--------\bullet-- \psi = \psi_1 + \psi_2$

$----\bullet--- \psi = 2\psi_1$

$\psi = \psi_2$

Family (*b*)

$\psi = \psi_1$

Fig. 8.6 Intersections of elementary streamlines can be joined to form a combined streamline.

To plot the streamlines of the superposition we can form the combined function $\psi_{\text{tot}} = \sum \psi_i$ and set it equal to various constants, plotting x versus y to satisfy this constant. But a neat trick is first to plot a lot of the individual streamlines ψ_i, as sketched in Fig. 8.6. The intersections of these are a guide to the form of the combined streamlines. In Fig. 8.6 the intersection of ψ_1 from family *a* and ψ_2 from family *b* is a point on the combined streamline $\psi_1 + \psi_2$, as is the intersection of ψ_2 (family *a*) and ψ_1 (family *b*). This graphical superposition is often the fastest way to solve the problem.

Source plus an Equal Sink

Consider a source of strength m at $(x, y) = (-a, 0)$ combined with a sink of strength $-m$ at $(+a, 0)$, as in Fig. 8.7. From Fig. 8.5*a*, each component has

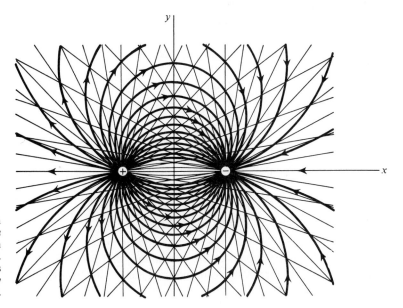

Fig. 8.7 Superposition of a source at $x = -a$ with an equal-strength sink at $x = +a$. Streamlines are circles with centers on the y axis.

individual streamlines consisting of radial lines extending from their respective centers. These are sketched lightly in Fig. 8.7. For the source the ψ values increase counterclockwise while the sink ψ values increase clockwise. The combined streamlines pass through intersections whose sum has a constant value. These are sketched as heavy lines with arrows in Fig. 8.7 and turn out to be circles passing from source to sink.

We could have found the same result analytically by combining the streamline functions for the source and sink, mindful of the different center for each. From Eq. (8.25)

$$\psi = \psi_{\text{source}} + \psi_{\text{sink}} = m \tan^{-1} \frac{y}{x+a} - m \tan^{-1} \frac{y}{x-a} \qquad (8.31)$$

But if you set this equal to various constants and try to plot the streamlines without any hints, you will find it is algebraically very cumbersome compared with the graphical solution.

Once the pattern is revealed by the graphical solution, we can manipulate Eq. (8.31) until we get what we expect to find. Use the trigonometric identity

$$\tan^{-1} \alpha - \tan^{-1} \beta \equiv \tan^{-1} \frac{\alpha - \beta}{1 + \alpha\beta} \qquad (8.32)$$

to rewrite Eq. (8.31) as

$$\psi = -m \tan^{-1} \frac{2ay}{x^2 + y^2 - a^2} \qquad (8.33)$$

This can be rearranged to exhibit the functional form

$$x^2 + \left(y + a \cot \frac{\psi}{m}\right)^2 = a^2 \csc^2 \frac{\psi}{m} \qquad (8.34)$$

Thus the streamlines (constant ψ) are circles of radius $a \csc (\psi/m)$ with centers on the y axis at $y = -a \cot (\psi/m)$. All streamlines pass through the source-sink centers $(\pm a, 0)$.

From Eq. (8.25) the velocity potential of the source-sink pair is a similar superposition:

$$\phi = \tfrac{1}{2} m \ln \frac{(x+a)^2 + y^2}{(x-a)^2 + y^2} \qquad (8.35)$$

After considerable manipulation this can be rewritten as

$$\left(x - a \coth \frac{\phi}{m}\right)^2 + y^2 = a^2 \csch^2 \frac{\phi}{m} \qquad (8.36)$$

Thus the potential lines are a family of circles with centers on the x axis. Each ϕ circle is orthogonal to the streamlines and encloses either the source or the sink center. In Eq. (8.36) coth and csch denote the hyperbolic cotangent and cosecant respectively.

EXAMPLE 8.1   As shown in Fig. E8.1a, a flow is caused by a uniform stream plus a source of volume flow 100π m^3/s per meter depth and a sink of volume flow 80π m^3/s per meter depth. What is the net velocity induced at the point P, where $(x, y) = (3 \text{ m}, 4 \text{ m})$?

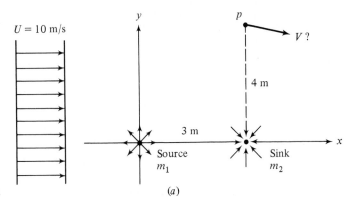

Fig. E8.1a   (a)

Solution   Since the source and sink are not of equal strength, Eq. (8.33) cannot be used. From Eq. (8.22), the strengths are

Source:
$$m_1 = \frac{Q_1}{2\pi b} = \frac{100\pi \text{ m}^3/\text{s}}{2\pi(1 \text{ m})} = +50 \text{ m}^2/\text{s}$$

Sink:
$$m_2 = \frac{Q_2}{2\pi b} = \frac{-80\pi \text{ m}^2/\text{s}}{2\pi(1 \text{ m})} = -40 \text{ m}^2/\text{s}$$

The problem could be done graphically by computing the radial velocities induced at $(3, 4)$ by the source and sink and adding these vectorially to U. Or we could write the composite stream function for the flow

$$\psi_{\text{net}} = Uy + m_1 \tan^{-1}\frac{y}{x} + m_2 \tan^{-1}\frac{y}{x - 3}$$

and differentiate it for the x and y components

$$u = \frac{\partial \psi}{\partial y} = U + \frac{m_1 x}{x^2 + y^2} + \frac{m_2(x - 3)}{(x - 3)^2 + y^2} \tag{1}$$

$$v = -\frac{\partial \psi}{\partial x} = 0 + \frac{m_1 y}{x^2 + y^2} + \frac{m_2 y}{(x - 3)^2 + y^2} \tag{2}$$

Considerable algebra has been omitted in giving these results. Alternately, polar coordinates could have been used.

Introduce numerical values at the point $(3, 4)$ into Eqs. (1) and (2):

$$u(3, 4) = 10 + \frac{50(3)}{3^2 + 4^2} + \frac{-40(3 - 3)}{(3 - 3)^2 + 4^2} = 16 \text{ m/s}$$

$$v(3, 4) = 0 + \frac{50(4)}{3^2 + 4^2} + \frac{-40(4)}{(3 - 3)^2 + 4^2} = -2 \text{ m/s}$$

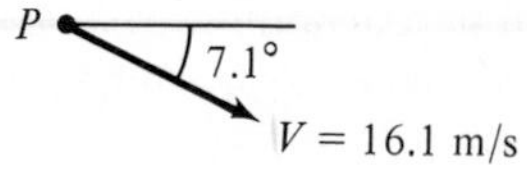

P

7.1°

$V = 16.1$ m/s

Fig. E8.1b   (b)

Thus the net velocity at point P is (Fig. E8.1b)

$$V = 16\mathbf{i} - 2\mathbf{j} \text{ m/s} = 16.12 \text{ m/s at } \theta = -7.1° \qquad \textit{Ans.}$$

You should repeat this example using the combined velocity potential as an alternate approach.

The Doublet

As we move far away from the source-sink pair of Fig. 8.7, the flow pattern begins to resemble a family of circles tangent to the origin, as in Fig. 8.8. This limit of vanishingly small distance a is called a *doublet*. To keep the flow strength large enough to exhibit decent velocities as a becomes small, we specify that the product $2am$ remain constant. Let us call this constant λ. Then the stream function of a doublet is

$$\psi = \lim_{\substack{a \to 0 \\ 2am = \lambda}} \left(-m \tan^{-1} \frac{2ay}{x^2 + y^2 - a^2} \right) = - \frac{2amy}{x^2 + y^2} = - \frac{\lambda y}{x^2 + y^2} \qquad (8.37)$$

We have used the fact that $\tan^{-1} \alpha \approx \alpha$ as α becomes small. The quantity λ is called the *strength* of the doublet.

Equation (8.37) can be rearranged to yield

$$x^2 + \left(y + \frac{\lambda}{2\psi} \right)^2 = \left(\frac{\lambda}{2\psi} \right)^2 \qquad (8.38)$$

so that, as advertised, the streamlines are circles tangent to the origin with centers on the y axis. This pattern is sketched in Fig. 8.8.

In a similar manner the velocity potential of a doublet is found by taking the limit of Eq. (8.35) as $a \to 0$ and $2am = \lambda$

$$\phi_{\text{doublet}} = \frac{\lambda x}{x^2 + y^2}$$

or

$$\left(x - \frac{\lambda}{2\phi} \right)^2 + y^2 = \left(\frac{\lambda}{2\phi} \right)^2 \qquad (8.39)$$

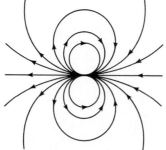

Fig. 8.8 A doublet, or source-sink pair, is the limiting case of Fig. 8.7 viewed from afar. Streamlines are circles tangent to the x axis at the origin.

The potential lines are circles tangent to the origin with centers on the x axis. Simply turn Fig. 8.8 clockwise 90° to visualize the ϕ lines, which are everywhere normal to the streamlines.

The doublet functions can also be written in polar coordinates

$$\psi = -\frac{\lambda \sin \theta}{r} \qquad \phi = \frac{\lambda \cos \theta}{r} \tag{8.40}$$

These forms are convenient for the cylinder flows of the next section.

Sink plus a Vortex

An interesting flow pattern, which you may have solved earlier as Prob. 4.22, occurs when a vortex is added to a source or a sink with the same center. The vortex plus sink is the superposition of Eqs. (8.24) and (8.26)

$$\psi = m\theta - K \ln r \qquad \phi = m \ln r + K\theta \tag{8.41}$$

The streamlines are plotted by the graphical method in Fig. 8.9. Equally spaced radial lines for the sink are combined with circles for the vortex whose diameter increases in the same ratio (here taken as 1.65). The streamlines cross the intersections of these two families and are seen to be spirals inward toward the origin, or common center. For a vortex plus source, reverse the arrows.

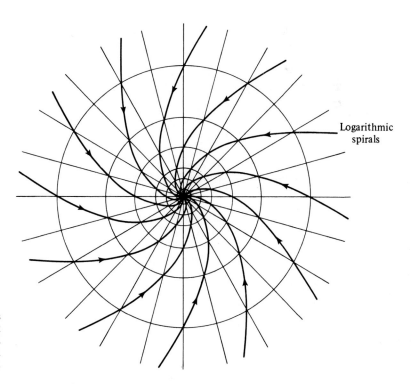

Logarithmic spirals

Fig. 8.9 Streamlines for a vortex plus sink [Eq. (8.41)] constructed by the graphical method.

The streamlines in Eq. (8.41) are of the form

$$r = C_1 e^{m\theta/K} \qquad C_1 = e^{\psi/K} = \text{const} \tag{8.42}$$

These are logarithmic, or *equiangular*, spirals which you may recall from the study of analytic geometry. For inward flow (vortex to sink) they simulate the flow toward a drain hole in a tank, and for outward flow (vortex to source) they resemble the streamline pattern in a vaneless diffuser (Prob. 4.22).

The potential lines for the vortex to sink are also logarithmic spirals with opposite curvature. To see them turn to the back of Fig. 8.9 and hold the page up to the light.

The Rankine Half-Body

An interesting body shape appears if we let a uniform stream flow against an isolated source or sink. If the stream is U_∞ in the x direction and the source is at the origin, the combined stream function in polar coordinates is

$$\psi = U_\infty r \sin\theta + m\theta \tag{8.43}$$

We can set this equal to various constants and plot or use the graphical method to intersect the horizontal lines of the stream with the radial lines of the source. In any case the results are shown in Fig. 8.10a. A curved roughly elliptical half-body shape

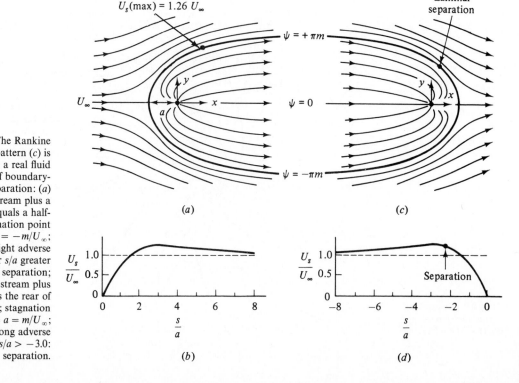

Fig. 8.10 The Rankine half-body; pattern (*c*) is not found in a real fluid because of boundary-layer separation: (*a*) uniform stream plus a source equals a half-body; stagnation point at $x = -a = -m/U_\infty$; (*b*) slight adverse gradient for s/a greater than 3.0: no separation; (*c*) uniform stream plus a sink equals the rear of a half-body; stagnation point at $x = a = m/U_\infty$; (*d*) strong adverse gradient for $s/a > -3.0$: separation.

appears, which separates the source flow from the stream flow. The upper part of the body shape is the line $\psi = \pi m$, or

$$r = \frac{m(\pi - \theta)}{U_\infty \sin \theta} \tag{8.44}$$

from which we can plot r versus θ. It is not a true ellipse. The lower part of the body shape is $\psi = -\pi m$. They meet at a stagnation point $(V = 0)$ at $x = -a = -m/U_\infty$, where the streamline $\psi = 0$ also crosses. Recall that a stagnation point is the only place where streamlines can cross.

The cartesian velocity components are

$$u = \frac{\partial \psi}{\partial y} = U_\infty + \frac{m}{r} \cos \theta \qquad v = -\frac{\partial \psi}{\partial x} = \frac{m}{r} \sin \theta \tag{8.45}$$

Setting $u = v = 0$, we find a stagnation point at $\theta = 180°$ and $r = m/U_\infty$, or $(x, y) = (-m/U_\infty, 0)$ as stated.

The resultant velocity at any point is given by

$$V^2 = u^2 + v^2 = U_\infty^2 \left(1 + \frac{a^2}{r^2} + 2\frac{a}{r}\cos\theta\right) \tag{8.46}$$

where we have substituted $m = U_\infty a$. By substituting values of r from Eq. (8.44) we can plot the velocity U_s along the half-body surface as a function of arc length s measured from the stagnation point. This is shown in Fig. 8.10b. There is a favorable pressure gradient from the stagnation point up to $s \approx 3a$ ($\theta = 63°$), where $U_{s,\max} = 1.26 U_\infty$, after which there is a mild adverse gradient, where $U_s \to U_\infty$ as $s \to \infty$. With Fig. 8.10b we can apply boundary-layer theory from Chap. 7 to see whether separation occurs. By Thwaites' method, Eqs. (7.54) and (7.55), no separation is predicted. Therefore we conclude that Fig. 8.10a is a very realistic and useful flow pattern simulating the front of a cylindrical body immersed in a stream.

As $x \to \infty$ in Fig. 8.10a the half-body streamlines approach the straight lines $y = \pm \pi a$; that is, far downstream of the source the half-body has a uniform thickness $2\pi a$.

The flow of a uniform stream past a sink is given by

$$\psi = U_\infty r \sin \theta - m\theta \tag{8.47}$$

The streamlines are plotted in Fig. 8.10c and are the exact mirror images of those in Fig. 8.10a. The stagnation point is at $x = +a = m/U_\infty$. The mirror-image surface-velocity distribution is shown in Fig. 8.10d. Application of Thwaites' method shows that separation will occur at $s \approx 2.2a$ or $\theta \approx 110°$. Therefore if the half-body shape is a solid surface, Fig. 8.10c is not realistic and separation and a broad wake will actually occur. However, as in Example 8.2, it correctly simulates a real fluid approaching a real sink, so that the half-body shape is a fluid line rather than a solid surface.

EXAMPLE 8.2 An offshore power-plant cooling-water intake sucks in 1500 ft³/s in water 30 ft deep, as in Fig. E8.2. If the tidal velocity approaching the intake is 0.7 ft/s, (a) how far downstream does the intake effect extend and (b) how much width L of tidal flow is entrained into the intake?

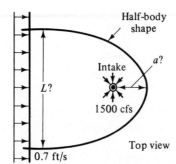

Fig. E8.2

Solution Recall from Eq. (8.22) that the sink strength m is related to the volume flow Q and the depth b into the paper

$$m = \frac{Q}{2\pi b} = \frac{1500\ \text{ft}^3/\text{s}}{2\pi(30\ \text{ft})} = 7.96\ \text{ft}^2/\text{s}$$

Therefore from Fig. 8.10 the desired lengths a and L are

$$a = \frac{m}{U_\infty} = \frac{7.96\ \text{ft}^2/\text{s}}{0.7\ \text{ft/s}} = 11.4\ \text{ft} \qquad\qquad Ans.\ (a)$$

$$L = 2\pi a = 2\pi(11.4\ \text{ft}) = 71\ \text{ft} \qquad\qquad Ans.\ (b)$$

Flow Past a Vortex

Consider a uniform stream U_∞ in the x direction flowing past a vortex of strength K with center at the origin. By superposition the combined stream function is

$$\psi = \psi_{\text{stream}} + \psi_{\text{vortex}} = U_\infty r \sin\theta - K \ln r \qquad (8.48)$$

The velocity components are given by

$$v_r = \frac{1}{r}\frac{\partial\psi}{\partial\theta} = U_\infty \cos\theta \qquad v_\theta = -\frac{\partial\psi}{\partial r} = -U_\infty \sin\theta + \frac{K}{r} \qquad (8.49)$$

The streamlines are plotted in Fig. 8.11 by the graphical method, intersecting the circular streamlines of the vortex with the horizontal lines of the uniform stream.

By setting $v_r = v_\theta = 0$ from (8.49) we find a stagnation point at $\theta = 90°$, $r = a = K/U_\infty$, or $(x, y) = (0, a)$. This is where the counterclockwise vortex velocity K/r exactly cancels the stream velocity U_∞.

Probably the most interesting thing about this example is that there is a nonzero lift force normal to the stream on the surface of any region enclosing the vortex, but we postpone this discussion until the next section.

An Infinite Row of Vortices

Consider an infinite row of vortices of equal strength K and equal spacing a, as in Fig. 8.12a. This case is included here to illustrate the interesting concept of a *vortex sheet*.

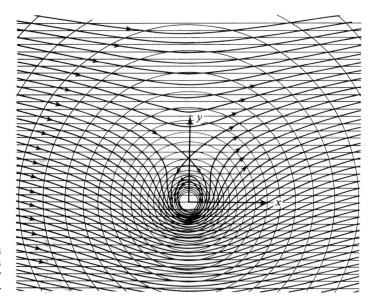

Fig. 8.11 Flow of a uniform stream past a vortex constructed by the graphical method.

From Eq. (8.26), the ith vortex in Fig. 8.12a has a stream function $\psi_i = -K \ln r_i$, so that the total infinite row has a combined stream function

$$\psi = -K \sum_{i=1}^{\infty} \ln r_i \tag{8.50}$$

It can be shown [2, sec. 4.51] that this infinite sum of logarithms is equivalent to a closed-form function

$$\psi = -\tfrac{1}{2}K \ln \left[\tfrac{1}{2}\left(\cosh \frac{2\pi y}{a} - \cos \frac{2\pi x}{a} \right) \right] \tag{8.51}$$

Since the proof uses the complex variable $z = x + iy$, $i = (-1)^{1/2}$, we are not going to show the details here.

The streamlines from Eq. (8.51) are plotted in Fig. 8.12b, showing what is called a *cat's-eye* pattern of enclosed flow cells surrounding the individual vortices. Above the cat's-eyes the flow is entirely to the left, and below the cat's-eyes the flow is to the right. Moreover, these left and right flows are uniform if $|y| \gg a$, which follows by differentiating Eq. (8.51)

$$u = \frac{\partial \psi}{\partial y}\bigg|_{|y| \gg a} = \pm \frac{\pi K}{a} \tag{8.52}$$

where the plus sign applies below the row and the minus sign above the row. This uniform left and right streaming is sketched in Fig. 8.12c. We stress that this effect is induced by the row of vortices: there is no uniform stream approaching the row in this example.

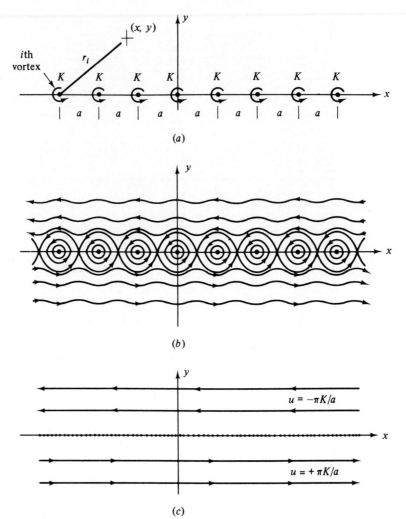

Fig. 8.12 Superposition of vortices: (a) an infinite row of equal strength; (b) streamline pattern for part (a); (c) vortex sheet: part (b) viewed from afar.

The Vortex Sheet

When Fig. 8.12b is viewed from afar, the streaming motion is uniform left above and uniform right below, as in Fig. 8.12c, and the vortices are packed so closely together that they are smudged into a continuous *vortex sheet*. The strength of the sheet is defined as

$$\gamma = \frac{2\pi K}{a} \tag{8.53}$$

and in the general case γ can vary with x. The circulation about any closed curve which encloses a short length dx of the sheet would be, from Eqs. (8.28) and (8.52),

$$d\Gamma = u_l \, dx - u_u \, dx = (u_l - u_u) \, dx = \frac{2\pi K}{a} \, dx = \gamma \, dx \tag{8.54}$$

where the subscripts l and u show for lower and upper, respectively. Thus the sheet strength $\gamma = d\Gamma/dx$ is the circulation per unit length of the sheet. Thus when a vortex sheet is immersed in a uniform stream, γ is proportional to the lift per unit length of any surface enclosing the sheet.

Note that there is no velocity normal to the sheet at the sheet surface. Therefore a vortex sheet can simulate a thin-body shape, e.g., plate or thin airfoil. This is the basis of the thin-airfoil theory mentioned in Sec. 8.7.

8.4 PLANE FLOW PAST CLOSED BODY SHAPES

A variety of closed-body external flows can be constructed by superimposing a uniform stream with sources, sinks, and vortices. The body shape will be closed only if the net source outflow equals the net sink inflow.

The Rankine Oval

A cylindrical shape called a *Rankine oval*, which is long compared with its height, is formed by a source-sink pair aligned parallel to a uniform stream, as in Fig. 8.13a.

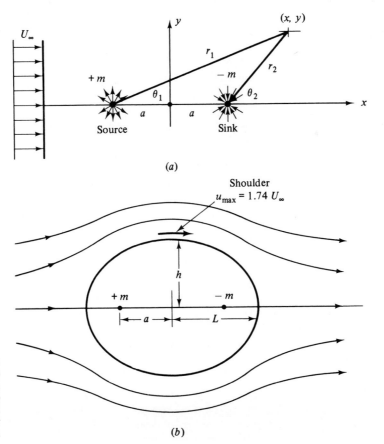

Fig. 8.13 Flow past a Rankine oval: (*a*) uniform stream plus a source-sink pair; (*b*) oval shape and streamlines for $m/U_\infty a = 1.0$.

From Eqs. (8.18) and (8.33) the combined stream function is

$$\psi = U_\infty y - m \tan^{-1} \frac{2ay}{x^2 + y^2 - a^2} = U_\infty r \sin\theta + m(\theta_1 - \theta_2) \quad (8.55)$$

When streamlines of constant ψ are plotted from Eq. (8.55), an oval body shape appears, as in Fig. 8.13b. The half-length L and half-height h of the oval depend upon the relative strength of source and stream, i.e., the ratio $m/U_\infty a$, which equals 1.0 in Fig. 8.13b. The circulating streamlines inside the oval are uninteresting and not usually shown. The oval is the line $\psi = 0$.

There are stagnation points at the front and rear, $x = \pm L$, and points of maximum velocity and minimum pressure at the shoulders, $y = \pm h$, of the oval. All these parameters are a function of the basic dimensionless parameter $m/U_\infty a$, which we can determine from Eq. (8.55):

$$\frac{h}{a} = \cot \frac{h/a}{2m/U_\infty a} \qquad \frac{L}{a} = \left(1 + \frac{2m}{U_\infty a}\right)^{1/2}$$

$$\frac{u_{max}}{U_\infty} = 1 + \frac{2m/U_\infty a}{1 + h^2/a^2} \qquad (8.56)$$

As we increase $m/U_\infty a$ from zero to large values, the oval shape increases in size and thickness from a flat plate of length $2a$ to a huge, nearly circular cylinder. This is shown in Table 8.1. In the limit as $m/U_\infty a \to \infty$, $L/h \to 1.0$ and $u_{max}/U_\infty \to 2.0$, which is equivalent to flow past a circular cylinder.

All the Rankine ovals except very thin ones have a large adverse pressure gradient on their leeward surface. Thus boundary-layer separation will occur in the rear with a broad wake flow, and the inviscid pattern is unrealistic in that region.

Flow Past a Circular Cylinder with Circulation

From Table 8.1 at large source strength the Rankine oval becomes a large circle, much greater in diameter than the source-sink spacing $2a$. Viewed on the scale of

Table 8.1

RANKINE-OVAL PARAMETERS FROM EQ. (8.56)

$m/U_\infty a$	h/a	L/a	L/h	u_{max}/U_∞
0.0	0.0	1.0	∞	1.0
0.01	0.031	1.010	32.79	1.020
0.1	0.263	1.095	4.169	1.187
1.0	1.307	1.732	1.326	1.739
10.0	4.435	4.583	1.033	1.968
100.0	14.130	14.177	1.003	1.997
∞	∞	∞	1.000	2.000

the cylinder, this is equivalent to a uniform stream plus a doublet. We also throw in a vortex at the doublet center, which does not change the shape of the cylinder.

Thus the stream function for flow past a circular cylinder with circulation, centered at the origin, is a uniform stream plus a doublet plus a vortex

$$\psi = U_\infty r \sin\theta - \frac{\lambda \sin\theta}{r} - K \ln r + \text{const} \qquad (8.57)$$

The doublet strength λ has units of velocity times length squared. For convenience, let $\lambda = U_\infty a^2$, where a is a length, and let the arbitrary constant in Eq. (8.57) equal $K \ln a$. Then the stream function becomes

$$\psi = U_\infty \sin\theta\left(r - \frac{a^2}{r}\right) - K \ln \frac{r}{a} \qquad (8.58)$$

The streamlines are plotted in Fig. 8.14 for four different values of the dimensionless vortex strength $K/U_\infty a$. For all cases the line $\psi = 0$ corresponds to the circle $r = a$, that is, the shape of the cylindrical body. As circulation $\Gamma = 2\pi K$ increases, the velocity becomes faster and faster below the cylinder and slower and slower above it.

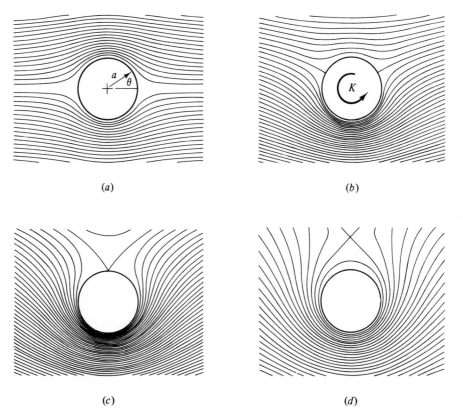

(a)   (b)

Fig. 8.14 Flow past a circular cylinder with circulation for values of $K/U_\infty a$ of (a) 0, (b) 1.0, (c) 2.0, and (d) 3.0.

(c)   (d)

The velocity components in the flow are given by

$$v_r = \frac{1}{r}\frac{\partial \psi}{\partial \theta} = U_\infty \cos\theta \left(1 - \frac{a^2}{r^2}\right)$$

$$v_\theta = -\frac{\partial \psi}{\partial r} = -U_\infty \sin\theta \left(1 + \frac{a^2}{r^2}\right) + \frac{K}{r} \tag{8.59}$$

The velocity at the cylinder surface $r = a$ is purely tangential, as expected

$$v_r(r = a) = 0 \qquad v_\theta(r = a) = -2U_\infty \sin\theta + \frac{K}{a} \tag{8.60}$$

For small K, two stagnation points appear on the surface at angles θ_s where $v_\theta = 0$, or, from Eq. (8.60),

$$\sin\theta_s = \frac{K}{2U_\infty a} \tag{8.61}$$

Figure 8.14a is for $K = 0$, $\theta_s = 0$ and 180°, or doubly symmetric inviscid flow past a cylinder with no circulation. Figure 8.14b is for $K/U_\infty a = 1$, $\theta_s = 30$ and 150°, and Fig. 8.14c is the limiting case where the two stagnation points meet at the top, $K/U_\infty a = 2$, $\theta_s = 90°$.

For $K > 2U_\infty a$, Eq. (8.61) is invalid, and the single stagnation point is above the cylinder, as in Fig. 8.14d, at a point $y = h$ given by

$$\frac{h}{a} = \tfrac{1}{2}[\beta + (\beta^2 - 4)^{1/2}] \qquad \beta = \frac{K}{U_\infty a} > 2 \tag{8.62}$$

In Fig. 8.14d, $K/U_\infty a = 3.0$, and $h/a = 2.6$.

The Kutta-Joukowski Lift Theorem

For the cylinder flows of Fig. 8.14b to d there is a downward force, or negative lift, called the *Magnus effect*, which is proportional to stream velocity and vortex strength. We can see from the streamline pattern that the velocity on the top of the cylinder is less and therefore the pressure higher from Bernoulli's equation; this explains the force. There is no viscous force of course because our theory is inviscid.

The surface velocity is given by Eq. (8.60). From Bernoulli's equation (8.4), neglecting gravity, the surface pressure p_s is given by

$$p_\infty + \tfrac{1}{2}\rho U_\infty^2 = p_s + \tfrac{1}{2}\rho\left(-2U_\infty \sin\theta + \frac{K}{a}\right)^2$$

or

$$p_s - p_\infty = \tfrac{1}{2}\rho U_\infty^2 (1 - 4\sin^2\theta + 4\beta\sin\theta - \beta^2) \tag{8.63}$$

where $\beta = K/U_\infty a$ and p_∞ is the free-stream pressure. If b is the cylinder depth into the paper, the drag D is the integral over the surface of the horizontal component of pressure force

$$D = -\int_0^{2\pi} (p_s - p_\infty) \cos\theta \, ba \, d\theta \tag{8.64}$$

where $p_s - p_\infty$ is substituted from Eq. (8.63). But the integral of $\cos \theta$ times any power of $\sin \theta$ over a full cycle 2π is identically zero. Thus we obtain the (perhaps surprising) result

$$D(\text{cylinder with circulation}) = 0 \qquad (8.65)$$

This is a special case of d'Alembert's paradox mentioned in Sec. 1.10:

> According to inviscid theory the drag of any body of any shape immersed in a uniform stream is identically zero.

D'Alembert published this result in 1752 and pointed out himself that it did not square with the facts for real fluid flows. This unfortunate paradox caused everyone to overreact and reject all inviscid theory until 1904, when Prandtl first pointed out the profound effect of the thin viscous boundary layer on the flow pattern in the rear, as in Fig. 7.2b, for example.

The lift force L normal to the stream, taken positive upward, is given by summation of vertical pressure forces

$$L = -\int_0^{2\pi} (p_s - p_\infty) \sin \theta \, ba \, d\theta \qquad (8.66)$$

Since the integral over 2π of any odd power of $\sin \theta$ is zero, only the third term in the parentheses in Eq. (8.63) contributes to the lift:

$$L = -\tfrac{1}{2}\rho U_\infty^2 \frac{4K}{aU_\infty} ba \int_0^{2\pi} \sin^2 \theta \, d\theta = -\rho U_\infty (2\pi K) b$$

or
$$\frac{L}{b} = -\rho U_\infty \Gamma \qquad (8.67)$$

Notice that the lift is independent of the radius a of the cylinder. Actually, though, as we shall see in Sec. 8.7, the circulation Γ depends upon the body size and orientation through a physical requirement.

Equation (8.67) was generalized by W. M. Kutta in 1902 and independently by N. Joukowski in 1906 as follows:

> According to inviscid theory, the lift per unit depth of any cylinder of any shape immersed in a uniform stream equals $\rho U_\infty \Gamma$, where Γ is the total net circulation contained within the body shape. The direction of the lift is 90° from the stream direction, rotating opposite to the circulation.

The problem in airfoil analysis, Sec. 8.7, is thus to determine the circulation Γ as a function of airfoil shape and orientation.

Experimental Lift and Drag of Rotating Cylinders

It is nearly impossible to test Fig. 8.14 by constructing a doublet and vortex with the same center and then letting a stream flow past them. But one physical realization would be a rotating cylinder in a stream. The viscous no-slip condition

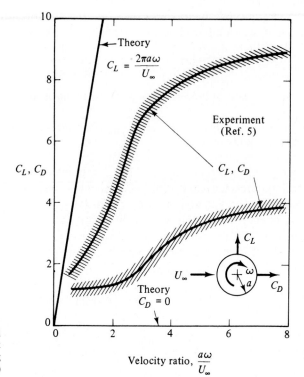

Fig. 8.15 Theoretical and experimental lift and drag of a rotating cylinder. (*From Ref. 5.*)

would cause the fluid in contact with the cylinder to move tangentially at the cylinder peripheral speed $v_\theta = a\omega$. A net circulation Γ would be set up by this no-slip mechanism, but it turns out to be less than 50 percent of the value expected from inviscid theory, primarily because of flow separation behind the cylinder.

Figure 8.15 shows experimental lift and drag coefficients, based on planform area $2ba$, of rotating cylinders. From Eq. (8.65) the theoretical drag is zero, but the actual C_D is quite large, more even than the stationary cylinder of Fig. 5.3. The theoretical lift follows from Eq. (8.67)

$$C_L = \frac{L}{\frac{1}{2}\rho U_\infty^2 (2ba)} = \frac{2\pi\rho U_\infty K b}{\rho U_\infty^2 ba} = \frac{2\pi v_{\theta s}}{U_\infty} \tag{8.68}$$

where $v_{\theta s} = K/a$ is the peripheral speed of the cylinder.

Figure 8.15 shows that the theoretical lift from Eq. (8.68) is much too high, but the measured lift is quite respectable, much larger in fact than a typical airfoil of the same chord length, e.g., Fig. 7.22. Thus rotating cylinders have practical possibilities. The Flettner rotor ship built in Germany in 1924 employed rotating vertical cylinders which developed a thrust normal to any winds blowing past the ship. The Flettner design did not gain any popularity, but such inventions may be more attractive in this era of high energy costs.

EXAMPLE 8.3   The experimental Flettner rotor sailboat at the University of Rhode Island is shown in Fig. E8.3. The rotor is 2.5 ft in diameter and 10 ft long and rotates at 220 r/min. It is driven by a small lawnmower engine. If the wind is a steady 10 knots and boat relative motion is neglected, what is the maximum thrust expected for the rotor? Assume standard air and water density.

Fig. E8.3

Solution   Convert the rotation rate to $\omega = 2\pi(220)/60 = 23.04$ rad/s. The wind velocity is 10 knots = 16.88 ft/s, so the velocity ratio is

$$\frac{a\omega}{U_\infty} = \frac{(1.25 \text{ ft})(23.04 \text{ rad/s})}{16.88 \text{ ft/s}} = 1.71$$

Entering Fig. 8.15, we read $C_L \approx 3.3$ and $C_D \approx 1.2$. From Appendix Table A.6, standard air density is 0.002377 slug/ft^3. Then the estimated rotor lift and drag are

$$L = C_L \tfrac{1}{2}\rho U_\infty^2 2ba = 3.3(\tfrac{1}{2})(0.002377)(16.88)^2(2)(10)(1.25)$$
$$= 27.9 \text{ lbf}$$

$$D = C_D \tfrac{1}{2}\rho U_\infty^2 2ba = L\frac{C_D}{C_L} = 27.9\,\frac{1.2}{3.3} = 10.2 \text{ lbf}$$

The maximum thrust available is the resultant of these two

$$F = [(27.9)^2 + (10.2)^2]^{1/2} = 29 \text{ lbf} \qquad\qquad Ans.$$

If aligned along the boat's keel, this thrust will drive the boat at a speed of about 5 knots through the water.

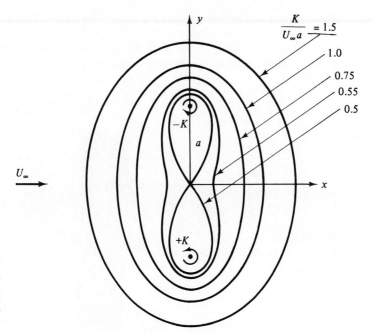

Fig. 8.16 Kelvin-oval body shapes as a function of the vortex-strength parameter $K/U_\infty a$; outer streamlines not shown.

The Kelvin Oval

A family of body shapes taller than they are wide can be formed by letting a uniform stream flow normal to a vortex pair. If U_∞ is to the right, the negative vortex $-K$ is placed at $y = +a$ and the counterclockwise vortex $+K$ placed at $y = -a$, as in Fig. 8.16. The combined stream function is

$$\psi = U_\infty y - \tfrac{1}{2}K \ln \frac{x^2 + (y + a)^2}{x^2 + (y - a)^2} \tag{8.69}$$

The body shape is the line $\psi = 0$, and some of these shapes are shown in Fig. 8.16. For $K/U_\infty a > 10$ the shape is within 1 percent of a Rankine oval (Fig. 8.13) turned 90°, but for small $K/U_\infty a$ the waist becomes pinched in, and a figure-eight shape occurs at 0.5. For $K/U_\infty a < 0.5$ the stream blasts right between the vortices and isolates two more or less circular body shapes, one surrounding each vortex.

A closed body of practically any shape can be constructed by proper superposition of sources, sinks, and vortices. See the advanced work in Refs. 2 to 4 for further details.

Potential-Flow Analogs

For complicated potential-flow geometries, one can resort to other methods than superposition of sources, sinks, and vortices. There are a variety of devices which simulate solutions to Laplace's equation.

In 1897–1899 Hele-Shaw [6] developed a technique whereby laminar viscous flow between very closely spaced parallel plates simulated potential flow when viewed from above the plates. When obstructions are placed between the plates, the

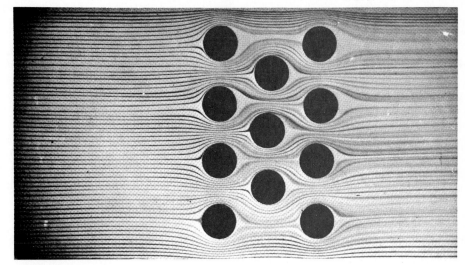

Fig. 8.17 Potential flow past an array of circular cylinders constrained between plane walls, visualized by the Hele-Shaw method. (*Tecquipment Ltd., Nottingham, England.*)

streamlines as visualized by dye streaks are the same as for potential flow about the obstruction. Figure 8.17 illustrates a Hele-Shaw flow experiment past an array of cylinders constrained by upper and lower walls. This pattern might be used to approximate the flow through rod bundles in a heat exchanger. The Hele-Shaw apparatus makes an excellent laboratory demonstration of fluid-flow patterns.

Other fluid-flow mapping techniques are discussed in Ref. 7. Electromagnetic fields satisfy Laplace's equation also, with voltage analogous to velocity potential and current lines analogous to streamlines. Commercial analog field plotters are available [8] using thin conducting paper which can be cut to the shape of the flow geometry. Lines of constant voltage are plotted by probing the paper with a potentiometer pointer.

There are also many numerical finite-difference algorithms for solving Laplace's equation [summarized in 2, chap. 10]. A hand-sketching technique is also given [9, sec. 7.7].

EXAMPLE 8.4   A Kelvin oval from Fig. 8.16 has $K/U_\infty a = 1.0$. Compute the velocity at the top shoulder of the oval in terms of U_∞.

Solution   We must locate the shoulder $y = h$ from Eq. (8.69) for $\psi = 0$ and then compute the velocity by differentiation. At $\psi = 0$ and $y = h$ and $x = 0$ Eq. (8.69) becomes

$$\frac{h}{a} = \frac{K}{U_\infty a} \ln \frac{1 + h/a}{1 - h/a}$$

With $K/U_\infty a = 1.0$ and the initial guess $h/a \approx 1.5$ from Fig. 8.16, we iterate and find the location $h/a = 1.5434$.

By inspection $v = 0$ at the shoulder because the streamline is horizontal. Therefore the shoulder velocity is, from Eq. (8.69),

$$u\bigg|_{y=h} = \frac{\partial \psi}{\partial y}\bigg|_{y=h} = U_\infty + \frac{K}{h-a} - \frac{K}{h+a}$$

Introducing $K = U_\infty a$ and $h = 1.5434a$, we obtain

$$u_{\text{shoulder}} = U_\infty(1.0 + 1.84 - 0.39) = 2.45U_\infty \qquad \textit{Ans.}$$

Because they are short-waisted compared with a circular cylinder, all the Kelvin ovals have shoulder velocity greater than the cylinder result $2.0U_\infty$ from Eq. (8.60).

8.5 OTHER PLANE POTENTIAL FLOWS[1]

References 2 to 4 treat many other potential flows of interest in addition to the cases presented in Secs. 8.3 and 8.4. In principle, any plane potential flow can be solved by the method of *conformal mapping*, using the complex variable

$$z = x + iy \qquad i = (-1)^{1/2} \qquad (8.70)$$

It turns out that any arbitrary analytic function of this complex variable z has the remarkable property that both its real and imaginary parts are solutions of Laplace's equation. If

$$f(z) = f(x + iy) = f_1(x, y) + if_2(x, y)$$

then

$$\frac{\partial^2 f_1}{\partial x^2} + \frac{\partial^2 f_1}{\partial y^2} = 0 = \frac{\partial^2 f_2}{\partial x^2} + \frac{\partial^2 f_2}{\partial y^2} \qquad (8.71)$$

We shall assign the proof of this as a problem. Even more remarkable if you have never seen it before is that lines of constant f_1 will be everywhere perpendicular to lines of constant f_2:

$$\left(\frac{dy}{dx}\right)_{f_1 = C} = -\frac{1}{(dy/dx)_{f_2 = C}} \qquad (8.72)$$

We also leave this proof as a problem exercise. This is true for totally arbitrary $f(z)$ as long as this function is analytic; i.e., it must have a unique derivative df/dz at every point in the region.

The net result of Eqs. (8.71) and (8.72) is that the functions f_1 and f_2 can be interpreted to be the potential lines and streamlines of an inviscid flow. By long custom we let the real part of $f(z)$ be velocity potential and the imaginary part be stream function

$$f(z) = \phi(x, y) + i\psi(x, y) \qquad (8.73)$$

We try various functions $f(z)$ and see whether any interesting flow pattern results. Of course, most of them have already been found, and we simply report on them here.

We shall not go into the details, but there are excellent treatments of this complex-variable technique both on an introductory [4, chap. 5; 10, chap. 5] and more advanced [2, 3] level.

---

[1] This section may be omitted without loss of continuity.

As a simple example, consider the linear function

$$f(z) = U_\infty z = U_\infty x + iU_\infty y$$

It follows from Eq. (8.73) that $\phi = U_\infty x$ and $\psi = U_\infty y$, which, we recall from Eq. (8.18), represents a uniform stream in the x direction. Once you get used to the complex variable, the solution practically falls in your lap.

To find the velocities, you may either separate ϕ and ψ from $f(z)$ and differentiate or differentiate f directly

$$\frac{df}{dz} = \frac{\partial \phi}{\partial x} + i\frac{\partial \psi}{\partial x} = -i\frac{\partial \phi}{\partial y} + \frac{\partial \psi}{\partial y} = u - iv \tag{8.74}$$

Thus the real part of df/dz equals $u(x, y)$, and the imaginary part equals $-v(x, y)$. To get a practical result the derivative df/dz must exist and be unique; hence the requirement that f be an analytic function. For Eq. (8.74) $df/dz = U_\infty = u$, since it is real, and $v = 0$, as expected.

Sometimes it is convenient to use the polar coordinate form of the complex variable

$$z = x + iy = re^{i\theta} = r\cos\theta + ir\sin\theta$$

where

$$r = (x^2 + y^2)^{1/2} \qquad \theta = \tan^{-1}\frac{y}{x}$$

This form is especially convenient when powers of z occur.

Uniform Stream at an Angle of Attack

All the elementary plane flows of Sec. 8.2 have a complex-variable formulation. The uniform stream U_∞ at an angle of attack α (see Fig. 8.3b) has the complex potential

$$f(z) = U_\infty ze^{-i\alpha} \tag{8.75}$$

Compare this form with Eq. (8.21).

Line Source at a Point z_0

Consider a line source of strength m placed off the origin at a point $z_0 = x_0 + iy_0$. Its complex potential is

$$f(z) = m \ln (z - z_0) \tag{8.76}$$

This can be compared with Eq. (8.24), which is valid only for the source at the origin. For a line sink, the strength m is negative.

Line Vortex at a Point z_0

If a line vortex of strength K is placed at point z_0, its complex potential is

$$f(z) = -iK \ln (z - z_0) \tag{8.77}$$

to be compared with Eq. (8.26). Also compare to Eq. (8.76) to see that we reverse the meaning of ϕ and ψ simply by multiplying the complex potential by $-i$.

Flow around a Corner of Arbitrary Angle

Corner flow is an example of a pattern that cannot be conveniently produced by superimposing sources, sinks, and vortices. It has a strikingly simple complex representation

$$f(z) = Az^n = Ar^n e^{in\theta} = Ar^n \cos n\theta + iAr^n \sin n\theta$$

where A and n are constants.

It follows from Eq. (8.73) that for this pattern

$$\phi = Ar^n \cos n\theta \qquad \psi = Ar^n \sin n\theta \qquad (8.78)$$

Streamlines from Eq. (8.78) are plotted in Fig. 8.18 for five different values of n. The flow is seen to represent a stream turning through an angle $\beta = \pi/n$. Patterns in Fig. 8.18d and e are not realistic on the downstream side of the corner, where separation will occur due to the adverse pressure gradient and sudden change of direction. In general, separation always occurs downstream of salient, or protruding, corners, except in creeping flows at low Reynolds number Re < 1.

Since $360° = 2\pi$ is the largest possible corner, the patterns for $n < \frac{1}{2}$ do not represent corner flow. They are peculiar-looking and we ask you to plot one as a problem.

If we expand the plot of Fig. 8.18a to c to double size, we can represent stagnation flow toward a corner of angle $2\beta = 2\pi/n$. This is done in Fig. 8.19 for $n = 3, 2$, and 1.5. These are very realistic flows; although they slip at the wall, they can be patched to boundary-layer theories very successfully. We took a brief look at corner flows before, in Examples 4.6 and 4.8 and in Probs. 4.52 to 4.54.

Flow Normal to a Flat Plate

We treat this case separately because the Kelvin ovals of Fig. 8.16 failed to degenerate into a flat plate as K became small. The flat plate normal to a uniform stream is an extreme case worthy of our attention.

Although the result is quite simple, the derivation is very complicated and is given, for example, in Ref. 2, sec. 9.3. There are three changes of complex variable, or *mappings*, beginning with the basic cylinder-flow solution of Fig. 8.14a. First the uniform stream is rotated to be vertical upward, then the cylinder is squeezed down into a plate shape, and finally the free stream is rotated back to the horizontal. The final result for complex potential is

$$f(z) = \phi + i\psi = U_\infty (z^2 + a^2)^{1/2} \qquad (8.79)$$

where $2a$ is the height of the plate. To isolate ϕ or ψ, square both sides and separate real and imaginary parts

$$\phi^2 - \psi^2 = U_\infty^2 (x^2 - y^2 + a^2) \qquad \phi\psi = U_\infty^2 xy \qquad (8.80)$$

We can solve for ψ to determine the streamlines

$$\psi^4 + \psi^2 U_\infty^2 (x^2 - y^2 + a^2) = U_\infty^4 x^2 y^2 \qquad (8.81)$$

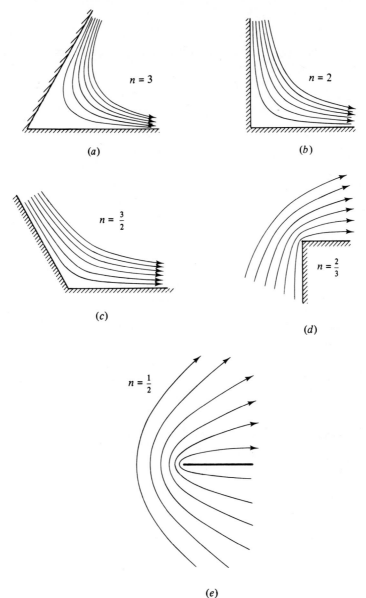

Fig. 8.18 Streamlines for corner flow, Eq. (8.78) for corner angle β of (a) 60°, (b) 90°, (c) 120°, (d) 270°, and (e) 360°.

Equation (8.81) is plotted in Fig. 8.20a, revealing a doubly symmetric pattern of streamlines which approach very closely to the plate and then deflect up and over, with very high velocities and low pressures near the plate tips.

The velocity v_s along the plate surface is found by computing df/dz from Eq. (8.79) and isolating the imaginary part

$$\frac{v_s}{U_\infty}\bigg|_{\text{plate surface}} = \frac{y/a}{(1 - y^2/a^2)^{1/2}} \qquad (8.82)$$

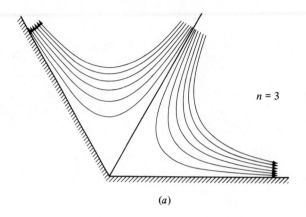

$n = 3$

(a)

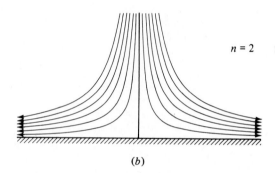

$n = 2$

(b)

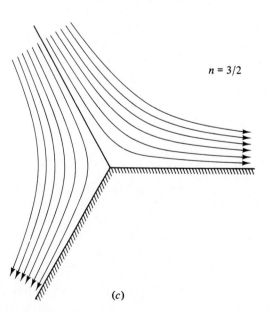

$n = 3/2$

Fig. 8.19 Streamlines for stagnation flow from Eq. (8.78) for corner angle 2β of (a) 120°, (b) 180°, and (c) 240°.

(c)

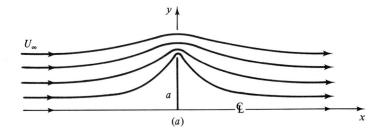

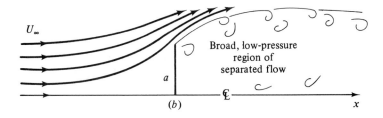

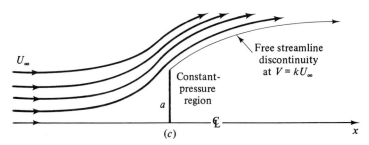

Fig. 8.20 Streamlines in upper half-plane for flow normal to a flat plate of height $2a$: (a) continuous potential-flow theory, Eq. (8.81); (b) actual measured flow pattern; (c) discontinuous potential theory with $k \approx 1.5$.

Some values of surface velocity can be tabulated as follows:

y/a	0.0	0.2	0.4	0.6	0.71	0.8	0.9	1.0
v_s/U_∞	0.0	0.204	0.436	0.750	1.00	1.33	2.07	∞

The origin is a stagnation point; then the velocity grows linearly at first and very rapidly near the tip, with both velocity and acceleration being infinite at the tip.

As you might guess, Fig. 8.20a is not realistic. In a real flow the sharp salient edge causes separation and a broad, low-pressure wake forms in the lee, as in Fig. 8.20b. Instead of being zero, the drag coefficient is very large, $C_D \approx 2.0$ from Table 7.2.

A discontinuous-potential-flow theory which accounts for flow separation was devised by Helmholtz in 1868 and Kirchhoff in 1869. This free-streamline solution is shown in Fig. 8.20c, with the streamline which breaks away from the tip having a constant velocity $V = kU_\infty$. From Bernoulli's equation the pressure in the dead-water region behind the plate will equal $p_r = p_\infty + \frac{1}{2}\rho U_\infty^2(1 - k^2)$ to match the pressure along the free streamline. For $k = 1.5$ this Helmholtz-Kirchhoff theory predicts $p_r = p_\infty - 0.625\rho U_\infty^2$ and an average pressure on the front $p_f = p_\infty + 0.375\rho U_\infty^2$, giving an overall drag coefficient of 2.0, in agreement with experiment. However, the coefficient k is a priori unknown and must be tuned to experimental data, so that free-streamline theory can be considered only a qualified success. For further details see Ref. 2, sec. 11.2.

8.6   IMAGES[1]

The previous solutions have all been for unbounded flows, such as a circular cylinder immersed in a broad expanse of uniformly streaming fluid, Fig. 8.14a. However, many practical problems involve a nearby rigid boundary constraining the flow, e.g., (1) groundwater flow near the bottom of a dam, (2) an airfoil near the ground, simulating landing or takeoff, or (3) a cylinder mounted in a wind tunnel with narrow walls. In such cases the basic unbounded-potential-flow solutions can be modified for wall effects by the method of *images*.

Consider a line source placed a distance a from a wall, as in Fig. 8.21a. To create the desired wall, an image source of identical strength is placed the same distance below the wall. By symmetry the two sources create a plane-surface streamline between them, which is taken to be the wall.

In Fig. 8.21b a vortex near a wall requires an image vortex the same distance below but of *opposite* rotation. We have shaded in the wall, but of course the pattern could also be interpreted as the flow near a vortex pair in an unbounded fluid.

In Fig. 8.21c an airfoil in a uniform stream near the ground is created by an image airfoil below the ground of opposite circulation and lift. This looks easy, but actually it is not because the airfoils are so close together that they interact and distort each other's shapes. A rule of thumb is that nonnegligible shape distortion occurs if the body shape is within two chord lengths of the wall. To eliminate distortion, a whole series of "corrective" images must be added to the flow to recapture the shape of the original isolated airfoil. Reference 2, sec. 7.75, has a good discussion of this procedure, which usually requires digital-computer summation of the multiple images needed.

Figure 8.21d shows a source constrained between two walls. One wall required only one image in Fig. 8.21a, but *two* walls require an infinite array of image sources above and below the desired pattern, as shown. Usually computer summation is necessary, but sometimes a closed-form summation can be achieved, as in the infinite vortex row of Eq. (8.51).

EXAMPLE 8.5   For the source near a wall as in Fig. 8.21a, the wall velocity is zero between the sources, rises to a maximum moving out along the wall, and then drops to zero far from the sources. If the source strength is $8 \text{ m}^2/\text{s}$, how far from the wall should the source be to ensure that the maximum velocity along the wall will be 5 m/s?

---

[1] This section may be omitted without loss of continuity.

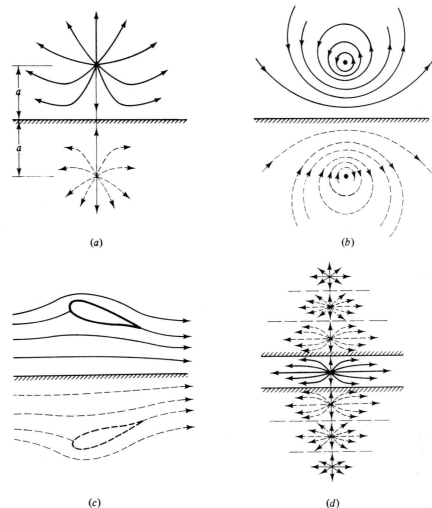

(a)

(b)

Fig. 8.21 Constraining walls can be created by image flows: (a) source near a wall with identical image source; (b) vortex near a wall with image vortex of opposite sense; (c) airfoil in ground effect with image airfoil of opposite circulation; (d) source between two walls requiring an infinite row of images.

(c)

(d)

Solution At any point x along the wall, as in Fig. E8.5, each source induces a radial outward velocity $v_r = m/r$, which has a component $v_r \cos \theta$ along the wall. The total wall velocity is thus

$$u_{\text{wall}} = 2v_r \cos \theta$$

From the geometry of Fig. E8.5, $r = (x^2 + a^2)^{1/2}$ and $\cos \theta = x/r$. Then the total wall velocity can be expressed as

$$u = \frac{2mx}{x^2 + a^2}$$

This is zero at $x = 0$ and also at $x \to \infty$. To find the maximum velocity, differentiate and set equal to zero

$$\frac{du}{dx} = 0 \quad \text{at } x = a \qquad \text{and} \qquad u_{\text{max}} = \frac{m}{a}$$

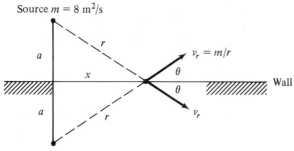

Fig. E8.5   Source m

We have omitted a bit of algebra in giving these results. For the given source strength and maximum velocity, the proper distance a is

$$a = \frac{m}{u_{max}} = \frac{8\ \text{m}^2/\text{s}}{5\ \text{m/s}} = 1.625\ \text{m} \qquad\qquad Ans.$$

For $x > a$, there is an adverse pressure gradient along the wall and boundary-layer theory should be used to predict separation.

8.7   AIRFOIL THEORY[1]

As mentioned in conjunction with the Kutta-Joukowski lift theorem, Eq. (8.67), the problem in airfoil theory is to determine the net circulation Γ as a function of airfoil shape and free-stream angle of attack α.

The Kutta Condition

Even if the airfoil shape and free-stream angle of attack are specified, the potential-theory solution is nonunique: an infinite family of solutions can be found corresponding to different values of circulation Γ. Four examples of this nonuniqueness were shown for the cylinder flows in Fig. 8.14. The same is true of the airfoil, and Fig. 8.22 shows three mathematically acceptable "solutions" to a given airfoil flow for small (Fig. 8.22a), large (Fig. 8.22b), and medium (Fig. 8.22c) net circulation. You can guess which case best simulates a real airfoil from the earlier discussion of transient-lift development in Fig. 7.20. It is the case (Fig. 8.22c) where the upper and lower flows meet and leave the trailing edge smoothly. If the trailing edge is rounded slightly, there will be a stagnation point there. If the trailing edge is sharp, approximating most airfoil designs, the upper- and lower-surface flow velocity will be equal as they meet and leave the airfoil.

This statement of the physically proper value of Γ is generally attributed to W. M. Kutta, hence the name *Kutta condition*, although some texts give credit to Joukowski and/or Chaplygin. All airfoil theories use the Kutta condition, which is

---

[1] This section may be omitted without loss of continuity.

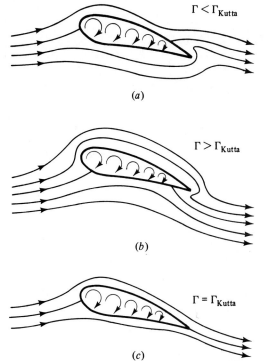

$\Gamma < \Gamma_{\text{Kutta}}$

(a)

$\Gamma > \Gamma_{\text{Kutta}}$

(b)

$\Gamma = \Gamma_{\text{Kutta}}$

(c)

Fig. 8.22 The Kutta condition properly simulates the flow about an airfoil: (a) too little circulation, stagnation point on rear upper surface; (b) too much, stagnation point on rear lower surface; (c) just right, Kutta condition requires smooth flow at trailing edge.

in good agreement with experiment. It turns out that the correct circulation Γ_{Kutta} depends upon flow velocity, angle of attack, and airfoil shape.

Flat-Plate Airfoil Vortex-Sheet Theory

The flat plate is the simplest airfoil, having no thickness or "shape," but even its theory is not so simple. The problem can be solved by a complex-variable mapping [2, p. 480], but here we shall use a vortex-sheet approach. Figure 8.23a shows a flat plate of length C simulated by a vortex sheet of variable strength $\gamma(x)$. The free-stream U_∞ is at an angle of attack α with respect to the plate chord line.

To make the lift "up" with flow from left to right as shown, we specify here that the circulation is positive clockwise. Recall from Fig. 8.12c that there is a jump in tangential velocity across a sheet equal to the local strength

$$u_u - u_l = \gamma(x) \tag{8.83}$$

If we omit the free stream, the sheet should cause a rightward flow $\delta u = +\frac{1}{2}\gamma$ on the upper surface and an equal and opposite leftward flow on the lower surface, as shown in Fig. 8.23a. The Kutta condition for this sharp trailing edge requires that this velocity difference vanish at the trailing edge to keep the exit flow smooth and parallel

$$\gamma(C) = 0 \tag{8.84}$$

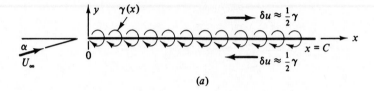

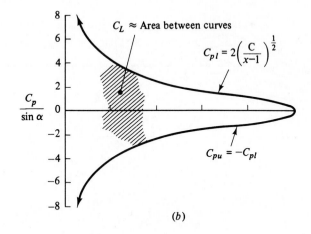

Fig. 8.23 Vortex-sheet solution for the flat-plate airfoil: (*a*) sheet geometry; (*b*) theoretical pressure coefficient on upper and lower surfaces; (*c*) upper-surface velocity with laminar separation points *S*.

The proper solution must satisfy this condition, after which the total lift can be computed by summing the sheet strength over the whole airfoil. From Eq. (8.67) for a foil of depth b

$$L = \rho U_\infty b \Gamma \qquad \Gamma = \int_0^C \gamma(x)\, dx \tag{8.85}$$

An alternate way to compute lift is from the dimensionless pressure coefficient C_p on the upper and lower surfaces

$$C_{p_{u,l}} = \frac{p_{u,l} - p_\infty}{\frac{1}{2}\rho U_\infty^2} = 1 - \frac{U_{u,l}^2}{U_\infty^2} \tag{8.86}$$

where the last expression follows from Bernoulli's equation. The surface velocity squared is given by combining the uniform stream and the vortex-sheet velocity components from Fig. 8.23a:

$$U^2_{u,l} = (U_\infty \cos \alpha \pm \delta u)^2 + (U_\infty \sin \alpha)^2$$

$$= U^2_\infty \pm 2U_\infty \, \delta u \cos \alpha + \delta u^2 \approx U^2_\infty \left(1 \pm \frac{2\delta u}{U_\infty} \right) \tag{8.87}$$

where we have made the approximations $\delta u \ll U_\infty$ and $\cos \alpha \approx 1$ in the last expression, assuming a small angle of attack. Equations (8.86) and (8.87) combine to the first-order approximation

$$C_{p_{u,l}} = \mp \frac{2\delta u}{U_\infty} = \mp \frac{\gamma}{U_\infty} \tag{8.88}$$

The lift force is the integral of the pressure difference over the length of the airfoil, assuming depth b

$$L = \int_0^C (p_l - p_u)b \, dx$$

or

$$C_L = \frac{L}{\frac{1}{2}\rho U^2_\infty bC} = \int_0^1 (C_{p_l} - C_{p_u}) \frac{dx}{C} = 2 \int_0^1 \frac{\gamma}{U_\infty} \, d\left(\frac{x}{C} \right) \tag{8.89}$$

Equations (8.85) and (8.89) are entirely equivalent within the small-angle approximations.

The sheet strength $\gamma(x)$ is computed from the requirement that the net normal velocity $v(x)$ be zero at the sheet ($y = 0$), since the sheet represents a solid plate or stream surface. Consider a small piece of sheet $\gamma \, dx$ located at position x_0. The velocity v at point x on the sheet is that of an infinitesimal line vortex of strength $d\Gamma = -\gamma \, dx$

$$dv \bigg|_x = \frac{d\Gamma}{2\pi r}\bigg|_{x_0 \to x} = \frac{-\gamma \, dx}{2\pi(x_0 - x)} \tag{8.90}$$

The total normal velocity induced by the entire sheet at point x is thus

$$v_{\text{sheet}} = -\int_0^C \frac{\gamma \, dx}{2\pi(x_0 - x)} \tag{8.91}$$

Meanwhile, from Fig. 8.23a, the uniform stream induces a constant normal velocity at every point on the sheet given by

$$v_{\text{stream}} = U_\infty \sin \alpha \tag{8.92}$$

Setting the sum of v_{sheet} and v_{stream} equal to zero gives the integral equation

$$\int_0^C \frac{\gamma \, dx}{x_0 - x} = 2\pi U_\infty \sin \alpha \tag{8.93}$$

to be solved for $\gamma(x)$ subject to the Kutta condition $\gamma(C) = 0$ from Eq. (8.84).

Although Eq. (8.93) is quite formidable (and not only for beginners), in fact it was solved long ago using integral formulas developed by Poisson in the nineteenth century. The sheet strength which satisfies Eq. (8.93) is

$$\gamma(x) = 2U_\infty \sin \alpha \left(\frac{C}{x} - 1\right)^{1/2} \tag{8.94}$$

From Eq. (8.88) the surface-pressure coefficients are thus

$$c_{p_{u,l}} = \mp 2 \sin \alpha \left(\frac{C}{x} - 1\right)^{1/2} \tag{8.95}$$

Details of the calculations are given in advanced texts [e.g., 11, chap. 17].

The pressure coefficients from Eq. (8.95) are plotted in Fig. 8.23b, showing that the upper surface has pressure continually increasing with x, that is, an adverse gradient. The upper-surface velocity $U_u \approx U_\infty + \delta u = U_\infty + \frac{1}{2}\gamma$ is plotted in Fig. 8.23c for various angles of attack. Above $\alpha = 5°$ the sheet contribution δu is about 20 percent of U_∞ so that the small-disturbance assumption is violated. Figure 8.23c also shows separation points computed by Thwaites' laminar-boundary-layer method, Eqs. (7.54) and (7.55). The prediction is that a flat plate would be extensively stalled on the upper surface for $\alpha > 6°$, which is approximately correct.

The lift coefficient of the airfoil is proportional to the area between c_{p_l} and c_{p_u} in Fig. 8.23b, from Eq. (8.89):

$$C_L = 2 \int_0^1 \frac{\gamma}{U} d\frac{x}{C} = 4 \sin \alpha \int_0^1 \left(\frac{C}{x} - 1\right)^{1/2} d\frac{x}{C} = 2\pi \sin \alpha \approx 2\pi\alpha \tag{8.96}$$

This is a classic result which was alluded to earlier in Eq. (7.70) without proof.

Also of interest is the moment coefficient about the leading edge of the airfoil, taken as positive counterclockwise

$$C_{M_{LE}} = \frac{M_{LE}}{\frac{1}{2}\rho U_\infty^2 bC^2} = \int_0^1 (C_{p_l} - C_{p_u})\frac{x}{C} d\frac{x}{C} = \frac{\pi}{2} \sin \alpha = \frac{1}{4}C_L \tag{8.97}$$

Thus the *center of pressure*, or position of the resultant lift force, is at the one-quarter-chord point

$$\left(\frac{x}{C}\right)_{CP} = \frac{1}{4} \tag{8.98}$$

This theoretical result is independent of the angle of attack.

These results can be compared with experimental results for NACA airfoils in Fig. 8.24. The thinnest NACA airfoil is $t/C = 0.06$, and the thickest is 24 percent or $t/C = 0.24$. The lift-curve slope $dC_L/d\alpha$ is within 9 percent of the theoretical value of 2π for all of the various airfoil families at all thicknesses. Increasing thickness tends to increase both $C_{L,max}$ and the stall angle. The stall angle at $t/C = 0.06$ is about 8° and would be even less for a flat plate, verifying the boundary-layer separation estimates in Fig. 8.23c. Best performance is usually at about the 12 percent thickness point for any airfoil.

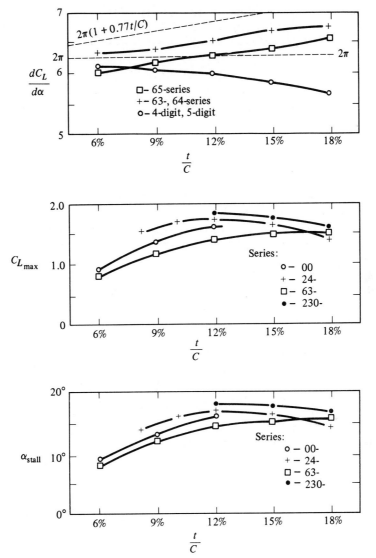

Fig. 8.24 Lift characteristics of smooth NACA airfoils as a function of thickness ratio, for infinite aspect ratio. (*From Ref. 12.*)

Potential Theory for Thick Cambered Airfoils

The theory of thick cambered airfoils is covered in advanced texts [e.g., 2 to 4]; Ref. 13 has a thorough and comprehensive review of both inviscid and viscous aspects of airfoil behavior.

Basically the theory uses a complex-variable mapping which transforms the flow about a cylinder with circulation in Fig. 8.14 into flow about a foil shape with circulation. The circulation is then adjusted to match the Kutta condition of smooth exit flow from the trailing edge.

Regardless of the exact airfoil shape, the inviscid mapping theory predicts that the correct circulation for any thick cambered airfoil is

$$\Gamma_{\text{Kutta}} = \pi b C U_\infty \left(1 + 0.77 \frac{t}{C} \right) \sin (\alpha + \beta) \tag{8.99}$$

where $\beta = \tan^{-1}(2h/C)$ and h is the maximum camber, or maximum deviation of the airfoil midline from its chord line, as in Fig. 8.25a.

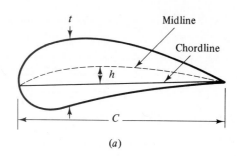

(a)

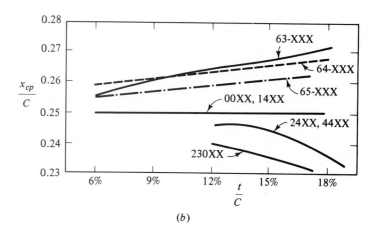

(b)

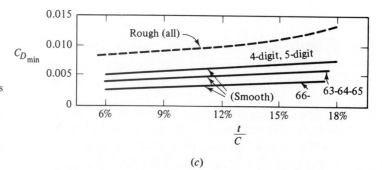

(c)

Fig. 8.25 Characteristics of NACA airfoils: (a) typical thick cambered airfoil, (b) center-of-pressure data, and (c) minimum drag coefficient.

The lift coefficient of the infinite-span airfoil is thus

$$C_L = \frac{\rho U_\infty \Gamma}{\frac{1}{2}\rho U_\infty^2 bC} = 2\pi\left(1 + 0.77\frac{t}{C}\right)\sin(\alpha + \beta) \qquad (8.100)$$

This reduces to Eq. (8.96) when the thickness and camber are zero. Figure 8.24 shows that the thickness effect $[1 + 0.77(t/C)]$ is not verified by experiment. Some airfoils increase lift with thickness, some decrease, and none approach the theory very closely, the primary reason being the boundary-layer growth on the upper surface affecting the airfoil "shape." Thus it is customary to drop the thickness effect from the theory

$$C_L \approx 2\pi \sin(\alpha + \beta) \qquad (8.101)$$

The theory correctly predicts that a cambered airfoil will have finite lift at zero angle of attack and zero lift at an angle

$$\alpha_{ZL} = -\beta = -\tan^{-1}\frac{2h}{C} \qquad (8.102)$$

Equation (8.102) overpredicts the measured zero-lift angle by $1°$ or so, as shown in Table 8.2. The measured values are essentially independent of thickness. The designation XX in the NACA series indicates the thickness in percent, and the other digits refer to camber and other details. For example the 2415 airfoil has 2 percent maximum camber (the first digit) occurring at 40 percent chord (the second digit) with 15 percent maximum thickness (the last two digits). The maximum thickness need not occur at the same position as maximum camber.

Figure 8.25b shows the measured position of the center of pressure of the various NACA airfoils, both symmetric and cambered. In all cases x_{CP} is within 0.02 chord length of the theoretical quarter-chord point predicted by Eq. (8.98). The standard cambered airfoils (24, 44, and 230 series) lie slightly forward of $x/C = 0.25$ and the low-drag (60 series) foils slightly aft. The symmetric airfoils are at 0.25.

Figure 8.25c shows minimum drag coefficient of NACA airfoils as a function of thickness. As mentioned earlier in conjunction with Fig. 7.22, these foils when smooth actually have less drag than turbulent flow parallel to a flat plate, especially the low-drag 60 series. However, for standard surface roughness all foils have about the same minimum drag, roughly 30 percent greater than a smooth flat plate.

Table 8.2
ZERO-LIFT ANGLE OF NACA AIRFOILS

Airfoil series	Camber h/C, %	Measured α_{ZL}, deg	Theory $-\beta$, deg
24XX	2.0	−2.1	−2.3
44XX	4.0	−4.0	−4.6
230XX	1.8	−1.3	−2.1
63-2XX	2.2	−1.8	−2.5
63-4XX	4.4	−3.1	−5.0
64-1XX	1.1	−0.8	−1.2

Wings of Finite Span

The results of airfoil theory and experiment in the previous subsection were for two-dimensional, or infinite-span, wings. But all real wings have tips and are therefore of finite span or finite aspect ratio AR, defined by

$$AR = \frac{b^2}{A_p} = \frac{b}{\bar{C}} \tag{8.103}$$

where b is the span length from tip to tip and A_p is the planform area of the wing as seen from above. The lift and drag coefficients of a finite-aspect-ratio wing depend strongly upon aspect ratio and slightly upon the planform shape of the wing.

Vortices cannot end in a fluid; they must either extend to the boundary or form a closed loop. Figure 8.26a shows how the vortices which provide the wing circulation bend downstream at finite wing tips and extend far behind the wing to join the starting vortex (Fig. 7.20) downstream. The strongest vortices are shed from the tips, but some are shed from the body of the wing, as sketched schematically in Fig. 8.26b. The effective circulation $\Gamma(y)$ of these trailing shed vortices is zero at the tips and usually a maximum at the center plane, or root, of the wing. In 1918 Prandtl successfully modeled this flow by replacing the wing by a single lifting line and a continuous sheet of semi-infinite trailing vortices of strength $\gamma(y) = d\Gamma/dy$, as in Fig. 8.26c. Each elemental piece of trailing sheet $\gamma(\eta)\,d\eta$ induces a downwash, or downward velocity, $dw(y)$, given by

$$dw(y) = \frac{\gamma(\eta)\,d\eta}{4\pi(y - \eta)} \tag{8.104}$$

at position y on the lifting line. Note the denominator term 4π rather than 2π because the trailing vortex extends only from 0 to ∞ rather than from $-\infty$ to $+\infty$.

The total downwash $w(y)$ induced by the entire trailing vortex system is thus

$$w(y) = \frac{1}{4\pi} \int_{-(1/2)b}^{(1/2)b} \frac{\gamma(\eta)\,d\eta}{y - \eta} \tag{8.105}$$

When this downwash is vectorially added to the approaching free-stream U_∞, the effective angle of attack at this section of the wing is reduced to

$$\alpha_{\text{eff}} = \alpha - \alpha_i \qquad \alpha_i = \tan^{-1}\frac{w}{U_\infty} \approx \frac{w}{U_\infty} \tag{8.106}$$

where we have used a small-amplitude approximation $w \ll U_\infty$.

The final step is to assume that the local circulation $\Gamma(y)$ is equal to that of a two-dimensional wing of the same shape and same effective angle of attack. From thin-airfoil theory, Eqs. (8.85) and (8.96), we have the estimate

$$C_L = \frac{\rho U_\infty \Gamma b}{\frac{1}{2}\rho U_\infty^2 bC} \approx 2\pi\alpha_{\text{eff}}$$

or

$$\Gamma \approx \pi C U_\infty \alpha_{\text{eff}} \tag{8.107}$$

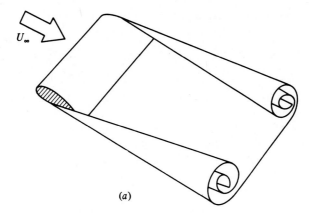

(a)

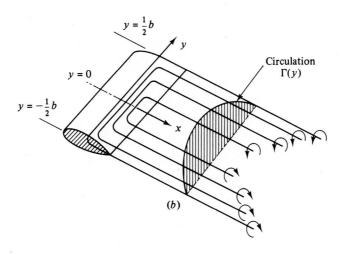

$y = \frac{1}{2}b$

y

$y = 0$

$y = -\frac{1}{2}b$

x

Circulation
$\Gamma(y)$

(b)

Wing
replaced by
"lifting line"

y, η

x

$\gamma(\eta)\, d\eta$ = vortex sheet element

dw = downwash due to $\gamma\, d\eta$

(c)

Fig. 8.26 Lifting-line
theory for a finite wing:
(a) actual trailing-
vortex system behind a
wing; (b) simulation by
vortex system "bound"
to the wing; (c)
downwash on the wing
due to an element of the
trailing-vortex system.

Combining Eqs. (8.105) to (8.107), we obtain Prandtl's lifting-line theory for a finite-span wing

$$\Gamma(y) = \pi C(y) U_\infty \left[\alpha(y) - \frac{1}{4\pi U_\infty} \int_{-(1/2)b}^{(1/2)b} \frac{(d\Gamma/d\eta)\, d\eta}{y - \eta} \right] \tag{8.108}$$

This is an integrodifferential equation to be solved for $\Gamma(y)$ subject to the conditions $\Gamma(\frac{1}{2}b) = \Gamma(-\frac{1}{2}b) = 0$. It is similar to the thin-airfoil integral equation (8.93) and even more formidable. Once it is solved, the total wing lift and induced drag are given by

$$L = \rho U_\infty \int_{-(1/2)b}^{(1/2)b} \Gamma(y)\, dy \qquad D_i = \rho U_\infty \int_{-(1/2)b}^{(1/2)b} \Gamma(y)\alpha_i(y)\, dy \tag{8.109}$$

Here is a case where the drag is not zero in a frictionless theory because the downwash causes the lift to slant backward by the angle α_i so that it has a drag component parallel to the free-stream direction, $dD_i = dL \sin \alpha_i \approx dL\alpha_i$.

The complete solution to Eq. (8.108) for arbitrary wing planform $C(y)$ and arbitrary twist $\alpha(y)$ is treated in advanced texts [e.g., 11, chap. 19]. It turns out that there is a simple representative solution for an untwisted wing of elliptical planform

$$C(y) = C_0 \left[1 - \left(\frac{2y}{b} \right)^2 \right]^{1/2} \tag{8.110}$$

The area and aspect ratio of this wing are

$$A_p = \int_{-(1/2)b}^{(1/2)b} C\, dy = \tfrac{1}{4}\pi b C_0 \qquad AR = \frac{2b}{\pi C_0} \tag{8.111}$$

The solution to Eq. (8.108) for this $C(y)$ is an elliptical circulation distribution of exactly similar shape

$$\Gamma(y) = \Gamma_0 \left[1 - \left(\frac{2y}{b} \right)^2 \right]^{1/2} \tag{8.112}$$

Substituting into Eq. (8.108) and integrating gives a relation between Γ_0 and C_0

$$\Gamma_0 = \frac{\pi C_0 U_\infty \alpha}{1 + 2/AR} \tag{8.113}$$

where α is assumed constant across the untwisted wing.

Substitution into Eq. (8.109) gives the elliptical-wing lift

$$L = \tfrac{1}{4}\pi^2 b C_0 \rho U_\infty^2 \alpha$$

or

$$C_L = \frac{2\pi\alpha}{1 + 2/AR} \tag{8.114}$$

If we generalize this to a thick cambered finite wing of approximately elliptical planform, we obtain

$$C_L = \frac{2\pi \sin(\alpha + \beta)}{1 + 2/AR} \tag{8.115}$$

This result was given without proof as Eq. (7.70). From Eq. (8.105) the computed downwash for the elliptical wing is constant

$$w(y) = \frac{2U_\infty \alpha}{2 + \text{AR}} = \text{const} \tag{8.116}$$

Finally, the induced drag coefficient from Eq. (8.109) is

$$C_{D_i} = C_L \frac{w}{U_\infty} = \frac{C_L^2}{\pi \text{AR}} \tag{8.117}$$

This was given without proof as Eq. (7.71).

Figure 8.27 shows the effectiveness of this theory when tested against a nonelliptical cambered wing by Prandtl in 1921 [14]. Figure 8.27a and b shows the measured lift curves and drag polars for five different aspect ratios. Note the

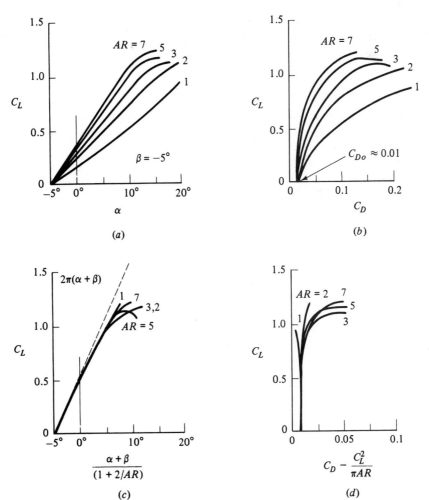

Fig. 8.27 Comparison of theory and experiment for a finite wing: (a) measured lift [14]; (b) measured drag polar [14]; (c) lift reduced to infinite aspect ratio; (d) drag polar reduced to infinite aspect ratio.

increase in stall angle and drag and decrease in lift slope as the aspect ratio decreases.

Figure 8.27c shows the lift data replotted against effective angle of attack $\alpha_{\text{eff}} = (\alpha + \beta)/(1 + 2/\text{AR})$, as predicted by Eq. (8.115). These curves should be equivalent to an infinite-aspect-ratio wing, and they do collapse together except near stall. Their common slope $dC_L/d\alpha$ is about 10 percent less than the theoretical value 2π, but this is consistent with the thickness and shape effects noted in Fig. 8.24.

Figure 8.27d shows the drag data replotted with the theoretical induced drag $C_{Di} = C_L^2/(\pi\text{AR})$ subtracted out. Again, except near stall, the data collapse onto a single line of nearly constant infinite-aspect-ratio drag $C_{D0} \approx 0.01$. We conclude that the finite-wing theory is very effective and may be used for design calculations.

8.8 AXISYMMETRIC POTENTIAL FLOW[1]

The same superposition technique which worked so well for plane flow in Sec. 8.3 is also successful for axisymmetric potential flow. We give some brief examples here.

Most of the basic results carry over from plane to axisymmetric flow with only slight changes owing to the geometric differences. Consider the following related flows:

Basic plane flow	Counterpart axisymmetric flow
Uniform stream	Uniform stream
Line source or sink	Point source or sink
Line doublet	Point doublet
Line vortex	No counterpart
Rankine half-body cylinder	Rankine half-body of revolution
Rankine oval cylinder	Rankine oval of revolution
Circular cylinder	Sphere
Symmetric airfoil	Tear-shaped body

Since there is no such thing as a point vortex, we must forgo the pleasure of studying circulation effects in axisymmetric flow. However, as any smoker knows, there is an axisymmetric ring vortex, and there are also ring sources and ring sinks, which we leave to advanced texts [e.g., 3].

Spherical Polar Coordinates

Axisymmetric potential flows are conveniently treated in the spherical polar coordinates of Fig. 8.28. There are only two coordinates (r, θ), and flow properties are constant on a circle of radius $r \sin \theta$ about the x axis.

---

[1] This section may be omitted without loss of continuity.

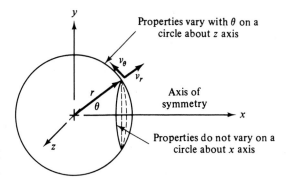

Fig. 8.28 Spherical polar coordinates for axisymmetric flow.

The equation of continuity for incompressible flow in these coordinates is

$$\frac{\partial}{\partial r}(r^2 v_r \sin \theta) + \frac{\partial}{\partial \theta}(r v_\theta \sin \theta) = 0 \qquad (8.118)$$

where v_r and v_θ are radial and tangential velocity as shown. Thus a spherical polar stream function[1] exists such that

$$v_r = -\frac{1}{r^2 \sin \theta}\frac{\partial \psi}{\partial \theta} \qquad v_\theta = \frac{1}{r \sin \theta}\frac{\partial \psi}{\partial r} \qquad (8.119)$$

In like manner a velocity potential $\phi(r, \theta)$ exists such that

$$v_r = \frac{\partial \phi}{\partial r} \qquad v_\theta = \frac{1}{r}\frac{\partial \phi}{\partial \theta} \qquad (8.120)$$

These formulas serve to deduce the ψ and ϕ functions for various elementary axisymmetric potential flows.

Uniform Stream in the x Direction

A stream U_∞ in the x direction has components

$$v_r = U_\infty \cos \theta \qquad v_\theta = -U_\infty \sin \theta \qquad (8.121)$$

Substitution into Eqs. (8.119) and (8.120) and integrating gives

Uniform stream: $\qquad \psi = -\tfrac{1}{2}U_\infty r^2 \sin^2 \theta \qquad \phi = U_\infty r \cos \theta \qquad (8.122)$

As usual, arbitrary constants of integration have been neglected.

Point Source or Sink

Consider a volume flux Q issuing from a point source. The flow will spread out radially and at radius r will equal Q divided by the area $4\pi r^2$ of a sphere. Thus

$$v_r = \frac{Q}{4\pi r^2} = \frac{m}{r^2} \qquad v_\theta = 0 \qquad (8.123)$$

---

[1] It is often called *Stokes' stream function*, having been used in a paper Stokes wrote in 1851 on viscous sphere flow.

with $m = Q/4\pi$ for convenience. Integrating (8.119) and (8.120) gives

Point source: $\qquad\qquad\qquad \psi = m \cos\theta \qquad \phi = -\dfrac{m}{r}$ $\qquad\qquad$ (8.124)

For a point sink change m to $-m$ in Eq. (8.124).

Point Doublet

Exactly as in Figs. 8.7 and 8.8, place a source at $(x, y) = (-a, 0)$ and an equal sink at $(+a, 0)$, taking the limit as a becomes small with the product $2am = \lambda$ held constant

$$\psi_{\text{doublet}} = \lim_{\substack{a \to 0 \\ 2am = \lambda}} (m \cos\theta_{\text{source}} - m \cos\theta_{\text{sink}}) = \frac{\lambda \sin^2\theta}{r} \qquad (8.125)$$

We leave the proof of this limit as a problem. The point-doublet velocity potential becomes

$$\phi_{\text{doublet}} = \lim_{\substack{a \to 0 \\ 2am = \lambda}} \left(-\frac{m}{r_{\text{source}}} + \frac{m}{r_{\text{sink}}} \right) = \frac{\lambda \cos\theta}{r^2} \qquad (8.126)$$

The streamlines and potential lines are shown in Fig. 8.29. Unlike the plane doublet flow of Fig. 8.8, neither set of lines represents perfect circles.

Uniform Stream plus a Point Source

By combining Eqs. (8.122) and (8.124) we obtain the stream function for a uniform stream plus a point source at the origin

$$\psi = -\tfrac{1}{2}U_\infty r^2 \sin^2\theta + m \cos\theta \qquad (8.127)$$

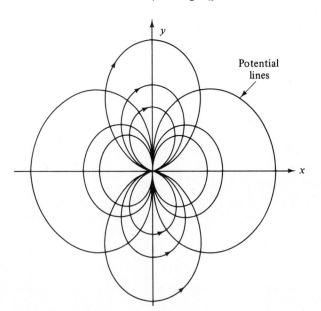

Fig. 8.29 Streamlines and potential lines due to a point doublet at the origin, from Eqs. (8.125) and (8.126).

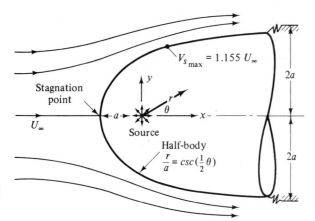

$V_{s_{max}} = 1.155\, U_\infty$

$2a$

Stagnation point

U_∞

a

Source

Half-body

$\dfrac{r}{a} = csc(\tfrac{1}{2}\theta)$

$2a$

Fig. 8.30 Streamlines for a Rankine half-body of revolution.

From Eq. (8.119) the velocity components are, by differentiation,

$$v_r = U_\infty \cos\theta + \frac{m}{r^2} \qquad v_\theta = -U_\infty \sin\theta \tag{8.128}$$

Setting these equal to zero reveals a stagnation point at $\theta = 180°$ and $r = a = (m/U_\infty)^{1/2}$, as shown in Fig. 8.30. If we let $m = U_\infty a^2$, the stream function can be rewritten as

$$\frac{\psi}{U_\infty a^2} = \cos\theta - \frac{1}{2}\left(\frac{r}{a}\right)^2 \sin^2\theta \tag{8.129}$$

The stream surface which passes through the stagnation point $(r, \theta) = (a, \pi)$ has the value $\psi = -U_\infty a^2$ and forms a half-body of revolution enclosing the point source, as shown in Fig. 8.30. This half-body can be used to simulate a pitot tube. Far downstream the half-body approaches the constant radius $R = 2a$ about the x axis. The maximum velocity and minimum pressure along the half-body surface occur at $\theta = 70.5°$, $r = a\sqrt{3}$, $V_s = 1.155U_\infty$. Downstream of this point there is an adverse gradient as V_s slowly decelerates to U_∞, but boundary-layer theory indicates no flow separation. Thus Eq. (8.129) is a very realistic simulation of a real half-body flow. But when the uniform stream is added to a sink to form a half-body rear surface, e.g., similar to Fig. 8.10c, separation is predicted and the inviscid pattern is not realistic.

Uniform Stream plus a Point Doublet

From Eqs. (8.122) and (8.125), combination of a uniform stream and a point doublet at the origin gives

$$\psi = -\tfrac{1}{2}U_\infty r^2 \sin^2\theta + \frac{\lambda}{r}\sin^2\theta \tag{8.130}$$

Examination of this relation reveals that the stream surface $\psi = 0$ corresponds to the sphere of radius

$$r = a = \left(\frac{2\lambda}{U_\infty}\right)^{1/3} \tag{8.131}$$

This is exactly analogous to the cylinder flow of Fig. 8.14a formed by combining a uniform stream and a line doublet.

Letting $\lambda = \frac{1}{2}U_\infty a^3$ for convenience, we rewrite Eq. (8.130) as

$$\frac{\psi}{\frac{1}{2}U_\infty a^2} = -\sin^2\theta\left(\frac{r^2}{a^2} - \frac{a}{r}\right) \tag{8.132}$$

The streamlines for this sphere flow are plotted in Fig. 8.31. By differentiation from Eq. (8.119) the velocity components are

$$v_r = U_\infty \cos\theta\left(1 - \frac{a^3}{r^3}\right) \qquad v_\theta = -\frac{1}{2}U_\infty \sin\theta\left(2 + \frac{a^3}{r^3}\right) \tag{8.133}$$

We see that the radial velocity vanishes at the sphere surface $r = a$, as expected. There is a stagnation point at the front (a, π) and the rear $(a, 0)$ of the sphere. The maximum velocity occurs at the shoulder $(a, \pm\frac{1}{2}\pi)$, where $v_r = 0$ and $v_\theta = -1.5U_\infty$. The surface-velocity distribution is

$$V_s = -v_\theta|_{r=a} = \frac{3}{2}U_\infty \sin\theta \tag{8.134}$$

Note the similarity to the cylinder surface velocity equal to $2U_\infty \sin\theta$ from Eq. (8.60) with zero circulation.

Equation (8.134) predicts, as expected, an adverse pressure gradient on the rear $(\theta < 90°)$ of the sphere. If we use this distribution with laminar-boundary-layer theory [e.g., 15, p. 346], separation is computed to occur at about $\theta = 76°$, so that in the actual flow pattern of Fig. 7.14 a broad wake forms in the rear. This wake interacts with the free stream and causes Eq. (8.134) to be inaccurate even in the

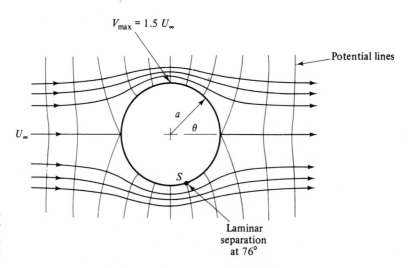

Fig. 8.31 Streamlines and potential lines for inviscid flow past a sphere.

front of the sphere. The measured maximum surface velocity is equal only to about $1.3U_\infty$ and occurs at about $\theta = 107°$ (see Ref. 15, sec. 4.9.5, for further details).

The Concept of Hydrodynamic Mass

When a body moves through a fluid, it must push a finite mass of fluid out of the way. If the body is accelerated, the surrounding fluid must also be accelerated. The body behaves as if it were heavier by an amount called the *hydrodynamic mass* (also called the added or virtual mass) of the fluid. If the instantaneous body velocity is $U(t)$, the summation of forces must include this effect

$$\sum \mathbf{F} = (m + m_h) \frac{d\mathbf{U}}{dt} \tag{8.135}$$

where m_h, the hydrodynamic mass, is a function of body shape, the direction of motion, and (to a lesser extent) flow parameters such as Reynolds number.

According to potential theory [2, sec. 6.4; 3, sec. 9.22], m_h depends only on shape and direction of motion and can be computed by summing the total kinetic energy of the fluid relative to the body and setting this equal to an equivalent body energy

$$\mathrm{KE}_{\mathrm{fluid}} = \int \tfrac{1}{2} \, dm \, V_{\mathrm{rel}}^2 = \tfrac{1}{2} m_h U^2 \tag{8.136}$$

The integration of fluid kinetic energy can also be accomplished by a body-surface integral involving the velocity potential [16, sec. 11].

Consider the previous example of a sphere immersed in a uniform stream. By subtracting out the stream velocity we can replot the flow as in Fig. 8.32, showing

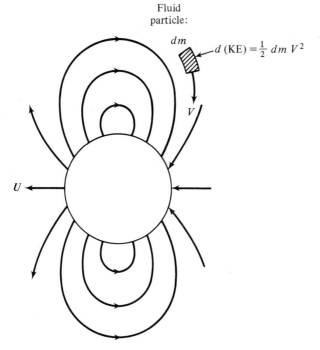

Fig. 8.32 Potential-flow streamlines relative to a moving sphere. Compare with Figs. 8.29 and 8.31.

the streamlines relative to the moving sphere. Note the similarity to the doublet flow in Fig. 8.29. The relative-velocity components are found by substracting U from Eqs. (8.133)

$$v_r = -\frac{Ua^3 \cos \theta}{r^3} \qquad v_\theta = -\frac{Ua^3 \sin \theta}{2r^3}$$

The element of fluid mass, in spherical polar coordinates, is

$$dm = \rho(2\pi r \sin \theta)r \, dr \, d\theta$$

When dm and $V_{\text{rel}}^2 = v_r^2 + v_\theta^2$ are substituted into Eq. (8.136), the integral can be evaluated

$$\text{KE}_{\text{fluid}} = \tfrac{1}{3}\rho\pi a^3 U^2$$

or
$$m_h(\text{sphere}) = \tfrac{2}{3}\rho\pi a^3 \qquad (8.137)$$

Thus, according to potential theory, the hydrodynamic mass of a sphere equals one-half of its displaced mass, independent of the direction of motion.

A similar result for a cylinder moving normal to its axis can be computed from Eqs. (8.59) after subtracting out the stream velocity. The result is

$$m_h(\text{cylinder}) = \rho\pi a^2 L \qquad (8.138)$$

for a cylinder of length L, assuming two-dimensional motion. The cylinder's hydrodynamic mass equals its displaced mass.

Tables of hydrodynamic mass for various body shapes and directions of motion are given by Patton [17].

8.9 NUMERICAL ANALYSIS

When potential flow involves irregular boundaries or unusual stream conditions, the classical superposition technique of the previous sections becomes unattractive. Numerical analysis is appropriate for complex flow fields, and there are at least three different approaches:

1. Integral methods with distributed singularities [18]
2. The finite-element method [19]
3. The finite-difference method [20]

The integral methods create arbitrary body shapes and flow fields by summing a large number of distributed singularities such as sources, sinks, and vortices. These may be distributed either inside the body or along its surface. The strengths of the singularities are determined by an integral equation expressing the fact that the body shape must be a streamline of the flow. To be valid this technique requires superposition and thus is not applicable to nonlinear flow fields such as boundary layers or arbitrary viscous flows.

The finite-element method is applicable to all types of linear and nonlinear partial differential equations in physics and engineering. It approximates the unknown field variable (stream function, velocity potential, pressure, temperature, etc.) by algebraic expressions valid over small flow regions called *finite elements*. The coefficients of the algebraic formulas are then found by minimizing the residual errors over the entire flow field. Finite-element methods are a relatively new and extremely active part of numerical-analysis research.

The finite-difference method approximates the partial derivatives in a physical equation by "differences" between nodal values at points a finite distance apart. The partial differential equation is thus replaced by a set of algebraic equations for the nodal values. These algebraic relations are linear for potential flow but generally nonlinear for viscous flows. The solution is usually obtained by iteration.

Chow [21] explains integral methods and finite-difference methods in fluid mechanics, while Huebner [19] applies the finite-element method to many different problems, including inviscid and viscous flows.

Potential Flow by Finite Differences

Since this is not a textbook on numerical analysis of fluid flow, like Chow [21] or Roache [22], we illustrate only one popular method, the finite-difference analysis of two-dimensional potential flow, to give the reader the flavor of numerical flow analysis.

The appropriate dependent variable is the stream function, which satisfies Laplace's equation

$$\frac{\partial^2 \psi}{\partial x^2} + \frac{\partial^2 \psi}{\partial y^2} = 0 \tag{8.139}$$

subject to known values of ψ along any body shape and known values of $\partial \psi / \partial x$ and $\partial \psi / \partial y$ in the free stream.

The finite-difference technique divides the flow field into equally spaced nodes, as shown in Fig. 8.33. To economize on the use of parentheses or functional

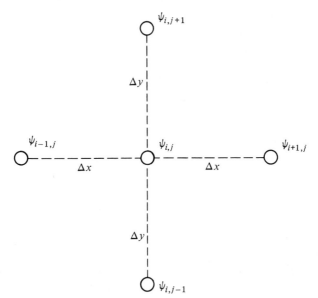

Fig. 8.33 Definition sketch for a two-dimensional rectangular finite-difference grid.

notation, the subscripts i and j denote the position of an arbitrary, equally spaced node, and $\psi_{i,j}$ denotes the value of the stream function at that node

$$\psi_{i,j} = \psi(x_0 + i\,\Delta x, y_0 + j\,\Delta y)$$

Thus, $\psi_{i+1,j}$ is just to the right of $\psi_{i,j}$, and $\psi_{i,j+1}$ is just above.

An algebraic approximation for the derivative $\partial\psi/\partial x$ is

$$\frac{\partial\psi}{\partial x} \approx \frac{\psi(x+\Delta x, y) - \psi(x, y)}{\Delta x}$$

A similar approximation for the second derivative is

$$\frac{\partial^2\psi}{\partial x^2} \approx \frac{1}{\Delta x}\left[\frac{\psi(x+\Delta x, y) - \psi(x, y)}{\Delta x} - \frac{\psi(x, y) - \psi(x-\Delta x, y)}{\Delta x}\right]$$

The subscript notation makes these expressions more compact

$$\frac{\partial\psi}{\partial x} \approx \frac{1}{\Delta x}(\psi_{i+1,j} - \psi_{i,j})$$

$$\frac{\partial^2\psi}{\partial x^2} \approx \frac{1}{\Delta x^2}(\psi_{i+1,j} - 2\psi_{i,j} + \psi_{i-1,j})$$

(8.140)

These formulas are exact in the calculus limit as $\Delta x \to 0$, but in numerical analysis we keep Δx and Δy finite, hence the term *finite differences*.

In an exactly similar manner we can derive the equivalent difference expressions for the y direction

$$\frac{\partial\psi}{\partial y} \approx \frac{1}{\Delta y}(\psi_{i,j+1} - \psi_{i,j})$$

$$\frac{\partial^2\psi}{\partial y^2} \approx \frac{1}{\Delta y^2}(\psi_{i,j+1} - 2\psi_{i,j} + \psi_{i,j-1})$$

(8.141)

The use of subscript notation allows these expressions to be programmed directly into a scientific computer language such as BASIC or FORTRAN.

When (8.140) and (8.141) are substituted into Laplace's equation (8.139), the result is the algebraic formula

$$2(1+\beta)\psi_{i,j} \approx \psi_{i,j+1} + \psi_{i+1,j} + \beta(\psi_{i,j-1} + \psi_{i-1,j}) \tag{8.142}$$

where $\beta = (\Delta x/\Delta y)^2$ depends upon the mesh size selected. This finite-difference model of Laplace's equation states that every nodal stream-function value $\psi_{i,j}$ is a linear combination of its four nearest neighbors.

The most commonly programmed case is a square mesh ($\beta = 1$), for which Eq. (8.143) reduces to

$$\psi_{i,j} \approx \tfrac{1}{4}(\psi_{i,j+1} + \psi_{i,j-1} + \psi_{i+1,j} + \psi_{i-1,j}) \tag{8.143}$$

Thus, for a square mesh, each nodal value equals the arithmetic average of the four neighbors shown in Fig. 8.33. The formula is easily remembered and easily programmed. If P(I, J) is a subscripted variable stream function, the BASIC or FORTRAN statement of (8.143) is

$$P(I, J) = 0.25 * (P(I, J + 1) + P(I, J - 1) + P(I + 1, J) + P(I - 1, J)) \quad (8.144)$$

This is applied in iterative fashion sweeping over each of the internal nodes (I, J), with known values of P specified at each of the surrounding boundary nodes. Any initial guesses can be specified for the internal nodes P(I, J) and the iteration process will converge to the final algebraic solution in a finite number of sweeps. The numerical error, compared with the exact solution of Laplace's equation, is proportional to the square of the mesh size.

Convergence can be speeded up by the successive overrelaxation (SOR) method, discussed by Chow [21]. The modified SOR form of the iteration is

$$\begin{aligned} P(I, J) = P(I, J) + 0.25 * A * (P(I, J + 1) + P(I, J - 1) \\ + P(I + 1, J) + P(I - 1, J) - 4 * P(I, J)) \end{aligned} \quad (8.145)$$

The recommended value of the SOR convergence factor A is about 1.7. Note that the value $A = 1.0$ reduces Eq. (8.145) to (8.144).

Let us illustrate the finite-difference method with an example.

EXAMPLE 8.6 Make a numerical analysis, using $\Delta x = \Delta y = 0.2$ m, of potential flow in the duct expansion shown in Fig. 8.34. The flow enters at a uniform 10 m/s, where the duct width is 1 m, and is assumed to leave at a uniform velocity of 5 m/s, where the duct width is 2 m. There is a straight section 1 m long, a 45° expansion section, and a final straight section 1 m long.

Solution Using the mesh shown in Fig. 8.34 results in 45 boundary nodes and 91 internal nodes, with i varying from 1 to 16 and j varying from 1 to 11. The internal points are modeled by Eq. (8.144). For convenience, let the stream function be zero along the lower wall. Then, since the

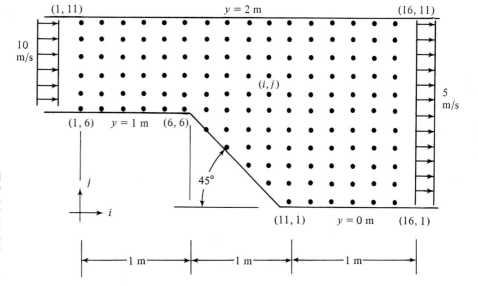

Fig. 8.34 Numerical model of potential flow through a two-dimensional 45° expansion. The nodal points shown are 20 cm apart. There are 45 boundary nodes and 91 internal nodes.

$\psi=$ 10.00	10.00	10.00	10.00	10.00	10.00	10.00	10.00	10.00	10.00	10.00	10.00	10.00	10.00	10.00	10.00
8.00	8.02	8.04	8.07	8.12	8.20	8.30	8.41	8.52	8.62	8.71	8.79	8.85	8.91	8.95	9.00
6.00	6.03	6.06	6.12	6.22	6.37	6.58	6.82	7.05	7.26	7.44	7.59	7.71	7.82	7.91	8.00
4.00	4.03	4.07	4.13	4.26	4.48	4.84	5.24	5.61	5.93	6.19	6.41	6.59	6.74	6.88	7.00
2.00	2.02	2.05	2.09	2.20	2.44	3.08	3.69	4.22	4.65	5.00	5.28	5.50	5.69	5.85	6.00
$\psi=$ 0.00	0.00	0.00	0.00	0.00	0.00	1.33	2.22	2.92	3.45	3.87	4.19	4.45	4.66	4.84	5.00
						0.00	1.00	1.77	2.37	2.83	3.18	3.45	3.66	3.84	4.00
							0.00	0.80	1.42	1.90	2.24	2.50	2.70	2.86	3.00
								0.00	0.63	1.09	1.40	1.61	1.77	1.89	2.00
									0.00	0.44	0.66	0.79	0.87	0.94	1.00
										0.00	0.00	0.00	0.00	0.00	0.00

Fig. 8.35 Stream-function nodal values for the potential flow of Fig. 8.34. Boundary values are known inputs. Internal nodes are solutions to Eq. (8.44).

volume flow is $(10\,\text{m/s})(1\,\text{m}) = 10\,\text{m}^2/\text{s}$ per unit depth, the stream function must equal $10\,\text{m}^2/\text{s}$ along the upper wall. Over the entrance and exit planes, the stream function must vary linearly to give uniform velocities:

Inlet: $\qquad\qquad\qquad \psi(1, J) = 2 * (J - 6) \qquad$ for $J = 7$ to 10

Exit: $\qquad\qquad\qquad\quad \psi(16, J) = J - 1 \qquad$ for $J = 2$ to 10

All these boundary values must be input to the program and are shown printed in Fig. 8.35.

Initial guesses are stored for the internal points, say, zero or an average value of $5.0\,\text{m}^2/\text{s}$. The program then starts at any convenient point, such as the upper left (2, 10), and evaluates Eq. (8.144) at every internal point, repeating this sweep iteratively until there are no further changes (within some selected maximum change) in the nodal values. The results are the finite-difference simulation of this potential flow for this mesh size; they are shown printed in Fig. 8.35 to three-digit accuracy. The reader should test a few nodes in Fig. 8.35 to verify that Eq. (8.144) is satisfied everywhere. The numerical accuracy of these printed values is difficult to estimate, since there is no known exact solution to this problem. In practice, one would keep decreasing the mesh size to see whether there were any significant changes in nodal values.

This problem is well within the capability of a small personal computer. The values shown in Fig. 8.35 were obtained after 100 iterations, or 6 min of execution time, on a TRS-80 Model 4 personal computer, using BASIC.

Although Fig. 8.35 is the computer solution to the problem, these numbers must be manipulated to yield practical engineering results. For example, one can interpolate these numbers to sketch various streamlines of the flow. This is done in Fig. 8.36a. We see that the streamlines are curved both upstream and downstream of the corner regions, especially near the lower wall. This indicates that the flow is not one-dimensional.

The velocities at any point in the flow can be computed from finite-difference formulas such as Eqs. (8.140) and (8.141). For example, at the point $(I, J) = (3, 6)$, from Eq. (8.141), the horizontal velocity is approximately

$$u(3, 6) \approx \frac{\psi(3, 7) - \psi(3, 6)}{\Delta y} = \frac{2.09 - 0.00}{0.2} = 10.45\,\text{m/s}$$

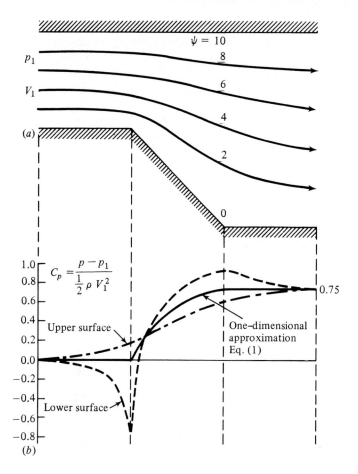

$\psi = 10$

p_1

V_1

(a)

$$C_p = \frac{p - p_1}{\frac{1}{2}\rho V_1^2}$$

Upper surface

One–dimensional approximation Eq. (1)

0.75

Lower surface

(b)

Fig. 8.36 Useful results computed from Fig. 8.35: (*a*) streamlines of the flow; (*b*) pressure-coefficient distribution along each wall.

and the vertical velocity is zero from Eq. (8.140). Directly above this on the upper wall, we estimate

$$u(3, 11) \approx \frac{\psi(3, 11) - \psi(3, 10)}{\Delta y} = \frac{10.00 - 8.07}{0.2} = 9.65 \text{ m/s}$$

The flow is not truly one-dimensional in the entrance duct. The lower wall, which contains the diverging section, accelerates the fluid, while the flat upper wall is actually decelerating the fluid.

Another output function, useful in making boundary-layer analyses of the wall regions, is the pressure distribution along the walls. If p_1 and V_1 are the pressure and velocity at the entrance (I = 1), conditions at any other point are computed from Bernoulli's equation (8.4), neglecting gravity

$$p + \tfrac{1}{2}\rho V^2 = p_1 + \tfrac{1}{2}\rho V_1^2$$

which can be rewritten as a dimensionless pressure coefficient

$$C_p = \frac{p - p_1}{\frac{1}{2}\rho V_1^2} = 1 - \left(\frac{V}{V_1}\right)^2$$

This determines p after V is computed from the stream function differences in Fig. 8.35.

Figure 8.36b shows the computed wall-pressure distributions as compared with the one-dimensional continuity approximation $V_1 A_1 \approx V(x)A(x)$, or

$$C_p(\text{one-dim}) \approx 1 - \left(\frac{A_1}{A}\right)^2 \qquad (1)$$

The one-dimensional approximation, which is rather crude for this large (45°) expansion, lies between the upper and lower wall pressures. One-dimensional theory would be much more accurate for a 10° expansion.

Analyzing Fig. 8.36b, we predict that boundary-layer separation will probably occur on the lower wall between the corners, where pressure is strongly rising (highly adverse gradient). Therefore potential theory is probably not too realistic for this flow, where viscous effects are strong. (Recall Figs. 6.26 and 7.8.)

Potential theory is *reversible*; i.e., when we reverse the flow arrows in Fig. 8.36a, Fig. 8.36b is still valid and would represent a 45° *contraction* flow. The pressure would fall on both walls (no separation) from $x = 3$ m to $x = 1$ m. Between $x = 1$ m and $x = 0$, the pressure rises on the lower surface, indicating possible separation, probably just downstream of the corner.

This example should give the reader an idea of the usefulness and generality of numerical analysis of fluid flows.

SUMMARY

This chapter has analyzed a highly idealized but very useful type of flow: inviscid, incompressible, irrotational flow, for which Laplace's equation holds for the velocity potential (8.3) and for the plane stream function (8.10). The mathematics is well developed, and solutions of potential flows can be obtained for practically any body shape.

Some solution techniques outlined here are (1) superposition of elementary line or point solutions in both plane and axisymmetric flow; (2) the analytic functions of a complex variable; (3) use of variable-strength vortex sheets; and (4) numerical analysis on a digital computer. Potential theory is especially useful and accurate for thin bodies such as airfoils. The only requirement is that the boundary layer be thin, i.e., that the Reynolds number be large.

For blunt bodies or highly divergent flows, potential theory serves as a first approximation, to be used as input to a boundary-layer analysis. The reader should consult the advanced texts [e.g., 2-4, 10-13] for further applications of potential theory.

REFERENCES

1. O. D. Kellogg, "Foundations of Potential Theory," Dover, New York, 1953.
2. J. M. Robertson, "Hydrodynamics in Theory and Application," Prentice-Hall, Englewood Cliffs, N.J., 1965.
3. L. M. Milne-Thomson, "Theoretical Hydrodynamics," 4th ed., Macmillan, New York, 1960.
4. H. R. Vallentine, "Applied Hydrodynamics," 2d ed., Plenum, New York, 1967.
5. H. Rouse, "Elementary Mechanics of Fluids," Wiley, New York, 1946.

6. H. J. S. Hele-Shaw, Investigation of the Nature of the Surface Resistance of Water and of Streamline Motion under Certain Experimental Conditions, *Trans. Inst. Nav. Archit.*, vol. 40, p. 25, 1898.
7. A. D. Moore, Fields from Fluid Flow Mappers, *J. Appl. Phys.*, vol. 20, pp. 790–804, 1949.
8. Instructions for Analog Field Plotter, General Electric Co., Schenectady, N.Y., catalogs 112L152-G1,G2.
9. V. L. Streeter and E. B. Wylie, "Fluid Mechanics," 8th ed., McGraw-Hill, New York, 1985.
10. R. H. F. Pao, "Fluid Dynamics," Merrill, Columbus, Ohio, 1967.
11. K. Karamcheti, "Principles of Ideal-Fluid Aerodynamics," Wiley, New York, 1966.
12. I. H. Abbott and A. E. von Doenhoff, "Theory of Wing Sections," Dover, New York, 1959.
13. B. Thwaites (ed.), "Incompressible Aerodynamics," Clarendon Press, Oxford, 1960.
14. L. Prandtl, Applications of Modern Hydrodynamics to Aeronautics, *NACA Rep.* 116, 1921.
15. F. M. White, "Viscous Fluid Flow," McGraw-Hill, New York, 1974.
16. C. S. Yih, "Fluid Mechanics," McGraw-Hill, New York, 1969.
17. K. T. Patton, Tables of Hydrodynamic Mass Factors for Translational Motion, *ASME Winter Ann. Meet.*, Pap. 65-WA/UNT-2, 1965.
18. J. L. Hess, Review of Integral-Equation Techniques for Solving Potential Flow Problems, with Emphasis on the Surface-Source Method, *Comput. Methods Appl. Mech. Eng.*, vol. 5, 1975, pp. 145–196.
19. K. H. Huebner and E. A. Thornton, "The Finite Element Method for Engineers," 2d ed., Wiley, New York, 1983.
20. R. D. Richtmeyer and K. W. Morton, "Difference Methods for Initial Value Problems," 2d ed., Wiley-Interscience, New York, 1976.
21. C. Y. Chow, "An Introduction to Computational Fluid Mechanics," Wiley, New York, 1979.
22. P. J. Roache, "Computational Fluid Dynamics," Hermosa, Albuquerque, N.M., 1976.
23. R. L. Panton, "Incompressible Flow," Wiley, New York, 1984.

Problems

PROBLEM DISTRIBUTION

Section	Topic	Problems	Section	Topic	Problems
8.1	Introduction	8.1–8.7	8.6	Images	8.60–8.66
8.2	Elementary plane-flow solutions	8.8–8.12	8.7	Airfoil theory: two-dimensional	8.67–8.68
8.3	Superposition of plane flows	8.13–8.29	8.7	Airfoil theory: finite-span wings	8.69–8.72
8.4	Plane flow past closed body shapes	8.30–8.48	8.8	Axisymmetric potential flows	8.73–8.82
8.5	Other plane flows; the complex potential	8.49–8.59	8.8	Hydrodynamic mass	8.83–8.84
			8.9	Numerical analysis of potential flow	8.85–8.88

8.1   Prove that the streamlines $\psi(r, \theta)$ in polar coordinates from Eqs. (8.11) are orthogonal to the potential lines $\phi(r, \theta)$ from Eqs. (8.6).

8.2   A uniform stream flows downward, $u = 0$, $v = -U_\infty$. Write the stream function and velocity potential for this flow in (a) cartesian coordinates and (b) polar coordinates.

8.3   Using cartesian coordinates, show that each velocity component (u, v, w) of a potential flow satisfies Laplace's equation separately.

8.4   Is the function $1/r$ a legitimate velocity potential in plane polar coordinates? If so, what is the associated stream function $\psi(r, \theta)$?

8.5   Show that the velocity potential as defined by Eq. (8.3) is an exact solution to the full Navier-Stokes relation with constant density and viscosity, Eq. (4.74). If this is so, why do we back away from Eq. (4.74) and use Eq. (8.3) instead?

8.6   Use integration to find the stream function and velocity potential of the incompressible flow field $u = Kx$, $v = -Ky$, $w = 0$, where K is a constant.

8.7   A mathematical relation sometimes used in fluid mechanics is the theorem of Stokes [1]

$$\oint_C \mathbf{V} \cdot d\mathbf{s} = \iint_A (\nabla \times \mathbf{V}) \cdot \mathbf{n}\, dA$$

where A is any surface and C is the curve enclosing that surface. The vector $d\mathbf{s}$ is the differential arc length along C, and $\mathbf{n}$ is the unit outward normal vector to A. How does this relation simplify for irrotational flow, and how does the resulting line integral relate to velocity potential?

8.8   A power plant discharges cooling water through the manifold in Fig. P8.8, which is 55 cm in diameter,

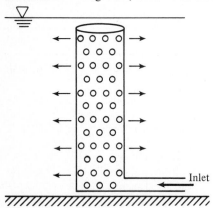

Fig. P8.8

8 m high, and perforated with 25,000 holes 1 cm in diameter. Does this manifold simulate a line source? If so, what is the equivalent source strength m?

8.9   Consider the flow due to a vortex of strength K at the origin. Evaluate the circulation from Eq. (8.28) about the clockwise path from $(r, \theta) = (a, 0)$ to $(2a, 0)$ to $(2a, 3\pi/2)$ to $(a, 3\pi/2)$ and back to $(a, 0)$. Interpret the result.

8.10   A tornado may be modeled as the circulating flow shown in Fig. P8.10, with $v_r = v_z = 0$ and $v_\theta(r)$ such that

$$v_\theta = \begin{cases} \omega r & r \le R \\ \dfrac{\omega R^2}{r} & r > R \end{cases}$$

Determine whether this flow pattern is irrotational in either the inner or outer region. Using the r-momentum equation (E.5) of Appendix E, determine the pressure distribution $p(r)$ in the tornado, assuming $p = p_\infty$ as $r \to \infty$. Find the location and magnitude of the lowest pressure.

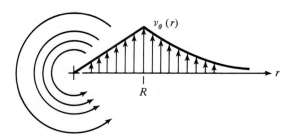

Fig. P8.10

8.11   Solve Prob. 8.10 for the particular case $v_\theta(\text{max}) = 55$ m/s at $R = 100$ m and plot $p(r)$ for r between 0 and 500 m if the far-field pressure is 101 kPa.

8.12   Examine the flow of Fig. 8.34 as an *analytical* (not a numerical) problem. Give the appropriate differential equation and the complete boundary conditions for both the stream function and the velocity potential. Is a Fourier-series solution possible?

8.13   Using the graphical method of Fig. 8.7, plot the streamlines and potential lines of the flow due to a line source of strength m at $(a, 0)$ plus a source $2m$ at $(-a, 0)$. What is the flow pattern viewed from afar?

8.14   Plot the streamlines and potential lines of the flow due to a line source of strength $2m$ at $(a, 0)$ plus a sink $-m$ at $(-a, 0)$. What is the pattern viewed from afar?

8.15 Find the resultant velocity vector induced at point A in Fig. P8.15 by the uniform stream, vortex, and line source.

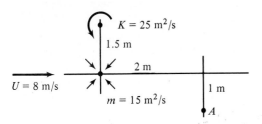

Fig. P8.15

8.16 Line sources of equal strength $m = Ua$, where U is a reference velocity, are placed at $(x, y) = (0, a)$ and $(0, -a)$. Sketch the stream and potential lines in the upper half plane. Is $(y = 0)$ a "wall"? If so, sketch the pressure coefficient,

$$C_p = \frac{p - p_0}{\frac{1}{2}\rho U^2}$$

along the wall, where p_0 is the pressure at $(0, 0)$. Find the minimum pressure point and indicate where flow separation might occur in the boundary layer.

8.17 Find the resultant velocity vector induced at point A in Fig. P8.17 by the uniform stream, line source, line sink, and vortex.

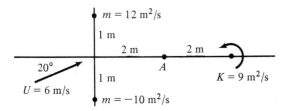

Fig. P8.17

8.18 A counterclockwise line vortex of strength $2K$ at $(x, y) = (0, a)$ is combined with a clockwise vortex K at $(0, -a)$. Plot the streamline and potential-line pattern and find the point of minimum velocity between the two vortices.

8.19 The vortex-sink pattern of Fig. 8.9 can simulate the flow due to a stationary hurricane as in Fig. P8.19, except that viscous effects can be neglected only outside a finite radius of about 40 m, as shown. Suppose that the pressure at $r = 40$ m is 1500 Pa less than the pressure far from the center and the air density is 1.2 kg/m³.

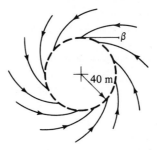

Fig. P8.19

Suppose further that the influx of air across the position $r = 40$ m is 5000 m³/s per meter of depth into the paper. Compute (a) the total flow velocity V at $r = 40$ m; (b) the sink strength $-m$ and vortex strength K in square meters per second; (c) the pressure at $r = 100$ m compared with the pressure at infinity; and (d) the angle β at which the streamlines cross the circle $r = 40$ m.

8.20 Sources of equal strength m are placed at the four symmetric positions $(x, y) = (a, a)$, $(-a, a)$, $(-a, -a)$, and $(a, -a)$. Sketch the streamline and potential line patterns. Do any plane "walls" appear?

8.21 A source of strength $m = 1.5$ m³/s at the origin is combined with a uniform stream $U = 9$ m/s in the x direction. For the half-body which results, find (a) the stagnation point, (b) the body height as it crosses the y axis, (c) the body height at large x, and (d) the maximum surface velocity and its position (x, y).

8.22 A power-plant cooling-water outlet is a vertical manifold which discharges 90 m³/s in tidal waters 7 m deep. If the approaching tidal velocity is 50 cm/s, how far upstream will a stagnation point form? How far to the left and right will the tidal flow be deflected by the manifold flow? What will the maximum flow velocity be at the interface between cooling water and tidal flow?

8.23 A Rankine half-body is formed as shown in Fig. P8.23. For the stream velocity and body dimension

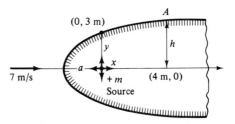

Fig. P8.23

shown, compute (*a*) the source strength *m* in square meters per second, (*b*) the distance *a*, (*c*) the distance *h*, and (*d*) the total velocity at point *A*.

8.24 Sketch the streamlines of the flow due to a line source *m* at $(x, y) = (0, a)$, an equal source *m* at $(0, -a)$, and a uniform stream $U_\infty = m/a$, where *a* is a characteristic length.

8.25 Sketch the streamlines of a uniform stream U_∞ past a line source-sink pair aligned vertically with the source at $+a$ and the sink at $-a$ on the *y* axis. Does a closed body shape form?

8.26 Sketch the streamlines of a uniform stream U_∞ past a line sink $-2m$ at the origin, a source *m* at $(a, 0)$, and a source *m* at $(-a, 0)$. Does a closed body shape appear?

8.27 Consider three equally spaced sources of strength *m* placed at $(x, y) = (-a, 0)$, $(0, 0)$, and $(a, 0)$. Sketch the resulting streamline pattern, noting any stagnation points.

8.28 A uniform stream *U* in the *x* direction combines with a source *m* at $(a, 0)$ and a sink $(-m)$ at $(-a, 0)$. Plot the resulting streamlines and note any stagnation points.

8.29 A stream $U = 10$ m/s combines with a source $m = 5$ m²/s at the origin to form a half-body. Let *x* denote distance along the half-body surface from the front stagnation point. Use the Thwaites laminar-boundary-layer method, Eqs. (7.54) and (7.56), to predict $\theta(x)$ and $\tau_w(x)$ along the surface if the fluid is air at sea-level conditions. Where is the point of minimum shear stress? Does the boundary layer separate?

8.30 When a line source-sink pair with $m = 2$ m²/s is combined with a uniform stream, it forms a Rankine oval whose minimum dimension is 40 cm. If $a = 15$ cm, what is the stream velocity and the velocity at the shoulder? What is the maximum dimension?

8.31 A Rankine oval is formed by a line source-sink pair with $m = 5$ m²/s, $a = 2$ m, and $U_\infty = 8$ m/s. Find the length and width of the body and the flow velocity at the shoulder.

8.32 A Kelvin oval is formed by a line-vortex pair with $K = 12$ m²/s, $a = 1$ m, and $U = 10$ m/s. What are the height, width, and shoulder velocity of this oval?

8.33 For what value of $K/U_\infty a$ does the velocity at the shoulder of a Kelvin oval equal $5U_\infty$? What is the height h/a of this oval?

8.34 Find the doublet strength λ in square meters per second which simulates flow without circulation at a

stream velocity of 4 m/s past a 1-m-diameter cylinder. If the fluid is water at 20°C with a stream pressure of 100 kPa, compute the surface pressure at (*a*) the stagnation point, (*b*) $\theta = 135°$, (*c*) the shoulder, $\theta = 90°$.

8.35 Suppose that circulation is added to the cylinder flow of Prob. 8.34 sufficient to place the stagnation points at $\theta = 50°$ and at 130°. What is the required vortex strength *K* in square meters per second? Compute the resulting pressure and surface velocity at (*a*) the stagnation points and (*b*) the upper and lower shoulders. What will the lift per meter of cylinder width be?

8.36 What circulation *K* must be added to the cylinder flow in Prob. 8.34 to place the stagnation point exactly at the upper shoulder? What will the velocity and pressure at the lower shoulder be then? What value of *K* causes the lower shoulder pressure to be 10 kPa?

8.37 A cylinder is formed by bolting two semicylindrical channels together on the inside, as shown in Fig. P8.37. There are 10 bolts per meter of width on each side, and the inside pressure is 50 kPa (gage). Using potential theory for the outside pressure, compute the tension force in each bolt if the fluid outside is sea-level air.

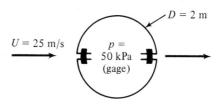

Fig. P8.37

8.38 It is desired to simulate flow past a two-dimensional ridge or bump by using a streamline which passes above the flow over a cylinder, as in Fig. P8.38.

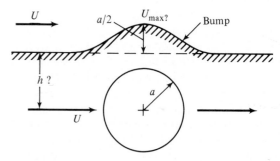

Fig. P8.38

The bump is to be $a/2$ high, where a is the cylinder radius. What is the elevation h of this streamline? What is U_{max} on the bump compared with stream velocity U?

8.39 A 2-m-diameter cylinder is rotating at 1800 r/min in an airstream flowing at 25 m/s. Compute, per unit depth of the cylinder, (a) the theoretical lift and drag forces and (b) the actual lift and drag from experimental data.

8.40 The Flettner rotor sailboat in Fig. E8.3 has a water drag coefficient of 0.006 based on a wetted area of 45 ft². If the rotor spins at 220 r/min, find the maximum boat velocity that can be achieved in a 15 mi/h wind. What is the optimum angle between the boat and the wind?

8.41 The original Flettner rotor-ship was approximately 100 ft long, displaced 800 tons, and had a wetted area of 3500 ft². As sketched in Fig. P8.41, it had two rotors 50 ft high and 9 ft in diameter rotating at 750 r/min, which is far outside the range of Fig. 8.15. The measured lift and drag coefficients for each rotor were about 10 and 4, respectively. If the ship is moored and subjected to a crosswind of 25 ft/s, as in Fig. P8.31, what will the wind force parallel and normal to the ship centerline be? Estimate the power required to drive the rotors.

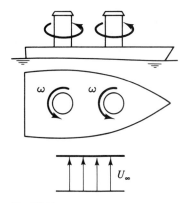

Fig. P8.41

8.42 Assume that the Flettner rotorship of Fig. P8.41 has a water resistance coefficient of 0.005. How fast will the ship sail in seawater at 20°C in a 20 ft/s wind if the keel aligns itself with the resultant force on the rotors? *Hint*: This is a problem in relative velocities.

8.43 Wind at U_∞ and p_∞ flows past a Quonset hut which is a half-cylinder of radius a and length L (Fig.

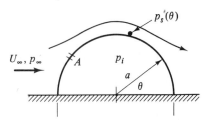

Fig. P8.43

P8.43). The internal pressure is p_i. Using inviscid theory, derive an expression for the upward force on the hut due to the difference between p_i and p_s.

8.44 In strong winds the force in Prob. 8.43 can be quite large. Suppose that a hole is introduced in the hut roof at point A to make p_i equal to the surface pressure there. At what angle θ should hole A be placed to make the net wind force zero?

8.45 In principle it is possible to use rotating cylinders as aircraft wings. Consider a cylinder 30 cm in diameter, rotating at 2400 r/min. It is to lift a 55-kN airplane cruising at 100 m/s. What should the cylinder length be? How much power is required to maintain this speed? Neglect end effects on the rotating wing.

8.46 When a doublet is added to a uniform stream so that the source part of the doublet faces the stream, the cylinder flow of Fig. 8.14a results. Plot the streamlines when the doublet is reversed so that the sink faces the stream. Are there any stagnation points?

8.47 Plot the streamlines due to the combined flow of a line sink $-m$ at the origin plus line sources $+m$ at $(a, 0)$ and $(4a, 0)$. *Hint*: A cylinder of radius $2a$ will appear.

8.48 By analogy with Prob. 8.47 plot the streamlines due to counterclockwise line vortices $+K$ at $(0, 0)$ and $(4a, 0)$ plus a clockwise vortex $-K$ at $(a, 0)$. Again a cylinder appears.

8.49 One of the corner-flow patterns of Fig. 8.19 is given by the cartesian stream function $\psi = A(3yx^2 - y^3)$. Which one? Can the correspondence be proved from Eq. (8.78)?

8.50 Plot the streamlines of Eq. (8.78) in the upper right quadrant for $n = 4$. How does the velocity increase with x outward along the x axis from the origin? For what corner angle and value of n would this increase be linear in x? For what corner angle and n would the increase be as x^5?

8.51 Determine qualitatively from boundary-layer theory (Chap. 7) whether any of the three stagnation-flow patterns of Fig. 8.19 can suffer flow separation along the walls.

8.52 Potential flow past a wedge of half-angle θ leads to an important application of laminar boundary layer theory called the *Falkner-Skan flows* [15, pp. 273–284]. Let x denote distance along the wedge wall, as in Fig. P8.52, and let $\theta = 10°$. Use Eq. (8.78) to find the variation of surface velocity $U(x)$ along the wall. Is the pressure gradient adverse or favorable?

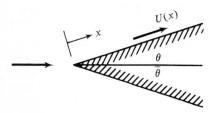

Fig. P8.52

8.53 In the flow normal to a plate of Fig. 8.20a, find the point where the surface velocity (a) equals the stream velocity and (b) is a maximum. (c) Find also the point of maximum surface deceleration. Use of Bernoulli's equation to compute the net pressure force on the front side of the plate gives a negative infinite result, i.e., an infinite thrust. Explain this anomaly.

8.54 Figure P8.54 shows the streamlines and potential lines of flow over a thin-plate weir as computed by the complex potential method. Compare qualitatively with Fig. 10.12a. State the proper boundary conditions at all boundaries. The velocity potential has equally spaced values. Why do the flow-net "squares" become smaller in the overflow jet?

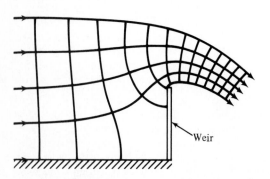

Weir

Fig. P8.54

8.55 Investigate the complex potential function $f(z) = U_\infty(z + a^2/z)$ and interpret the flow pattern.

8.56 Investigate the complex potential function $f(z) = U_\infty z + m \ln[(z + a)/(z - a)]$ and interpret the flow pattern.

8.57 Investigate the complex potential $f(z) = A \cosh[\pi(z/a)]$ and plot the streamlines inside the region shown in Fig. P8.57. What hyphenated word (originally French) might describe such a flow pattern?

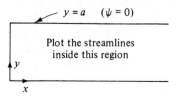

Fig. P8.57

8.58 Show that the complex potential $f = U_\infty\{z + \frac{1}{4}a \coth[\pi(z/a)]\}$ represents flow past an oval shape placed midway between two parallel walls $y = \pm\frac{1}{2}a$. What is a practical application?

8.59 Repeat Prob. 8.38 but let the bump be a streamline above the flow normal to a plate, as in Fig. 8.20a. Again let the bump height be $a/2$ and find h and U_{max}.

8.60 The flow past a cylinder very near a wall might be simulated by doublet images, as in Fig. P8.60. Explain why the result is not very successful and the cylinder shape becomes badly distorted.

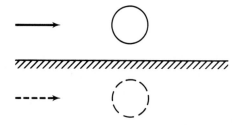

Fig. P8.60

8.61 Use the method of images to approximate the flow pattern past a cylinder a distance $4a$ from a single wall, as in Fig. P8.61. To illustrate the effect of the wall compute the velocities at corresponding points A, B and C, D, comparing with a cylinder flow in an infinite expanse of fluid.

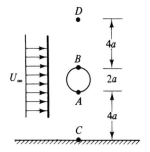

Fig. P8.61

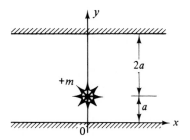

Fig. P8.64

8.62 Use the method of images to construct the flow pattern for a source $+m$ near two walls, as shown in Fig. P8.62. Sketch the velocity distribution along the lower wall ($y = 0$). Is there any danger of flow separation along this wall?

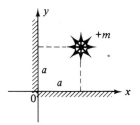

Fig. P8.62

8.63 Set up an image system to compute the flow of a source at unequal distances from two walls, as in Fig. P8.63. Find the point of maximum velocity on the y axis.

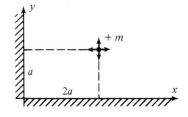

Fig. P8.63

8.64 Explain the system of images needed to simulate the flow of a line source placed unsymmetrically between two parallel walls as in Fig. P8.64. Compute the velocity on the lower wall at $x = a$. How many images are needed to estimate this velocity within 1 percent?

8.65 Discuss how the flow pattern of Prob. 8.47 might be interpreted to be an image-system construction for circular walls. Why are there two images instead of one?

8.66 Indicate the system of images needed to construct the flow of a uniform stream past a Rankine half-body constrained between two parallel walls, as in Fig. P8.66. For the particular dimensions shown in this figure estimate the position of the nose of the resulting half-body.

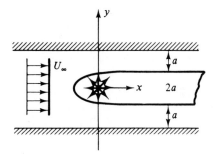

Fig. P8.66

8.67 The NACA 4412 airfoil has 4 percent camber and 12 percent thickness. For two-dimensional flow at an angle of attack of 5°, estimate its theoretical lift coefficient and its moment coefficient about the leading edge.

8.68 A two-dimensional airfoil has 2 percent camber and 10 percent thickness. If $C = 1.75$ m, estimate its lift per meter when immersed in 20°C water at $\alpha = 6°$ and $U = 18$ m/s.

8.69 A boat has a mass of 4500 kg and is supported by a rectangular hydrofoil of aspect ratio 6, 2 percent camber, and 10 percent thickness. If the boat moves at 12 m/s and $\alpha = 4°$, what should the chord length and span of the foil be?

8.70 In steady, level flight an airplane's weight equals its lift. Suppose the plane has a mass of 45,000 kg and has a rectangular wing of 4 m chord length, with 12 percent thickness and zero camber. If the airplane cruises at 6000 m standard altitude with $\alpha = 3°$ and a speed of 150 m/s, estimate (a) the wing span b, (b) the aspect ratio, and (c) the induced drag.

8.71 A wing of 2 percent camber, 5-in chord, and 30-in span is tested at a certain angle of attack in a wind tunnel with sea-level standard air at 200 ft/s and found to have lift of 30 lbf and drag of 1.5 lbf. Estimate from wing theory (a) the angle of attack, (b) the minimum drag of the wing and the angle of attack at which it occurs, and (c) the maximum lift-to-drag ratio.

8.72 A symmetric hydrofoil has a planform area of 4 m^2 and is tested in seawater at 14 m/s. It is found to have maximum lift-to-drag ratio of 16:1 when the lift is 150 kN. Estimate from wing theory the aspect ratio and optimum angle of attack.

8.73 Determine whether the Stokes streamlines from Eq. (8.119) are everywhere normal to the Stokes potential lines from Eq. (8.120), as is the case for cartesian and polar coordinate flows.

8.74 Is the function $\phi = 1/r$ a valid Stokes axisymmetric velocity potential as in Eq. (8.120)? If so, find the corresponding Stokes stream function.

8.75 Show that the axisymmetric potential flow formed by superposition of a point source $+m$ at $(x, y) = (-a, 0)$, a point sink $-m$ at $(+a, 0)$, and a stream U_∞ in the x direction forms a Rankine body of revolution as in Fig. P8.75. Find analytic expressions for determining the length $2L$ and maximum diameter $2R$ of the body in terms of m, U_∞, and a.

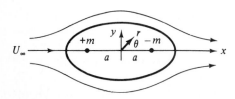

Fig. P8.75

8.76 For the Rankine body of Fig. P8.75, suppose that $a = 1$ m, $U_\infty = 12$ m/s, and $m = 30$ m^3/s. Compute (a) the length, (b) the maximum diameter, and (c) the maximum surface velocity for the body.

8.77 A point source with volume flow $Q = 1000$ ft^3/s is immersed in a uniform stream of speed 10 ft/s. For the resulting Rankine half-body of revolution, compute (a) the distance from the source to the stagnation point, (b) the body diameter far downstream, and (c) the position (r, θ) on the surface where the local velocity equals 11 ft/s.

8.78 The Rankine body of revolution in Fig. P8.78 is 60 cm long and 30 cm in diameter. When immersed in the low-pressure water tunnel as shown, cavitation may appear at point A. Compute the stream velocity U, neglecting surface wave formation, for which cavitation occurs.

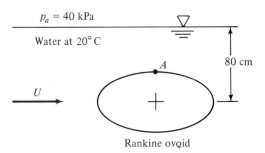

Fig. P8.78

8.79 Consider air flowing past a hemisphere resting on a flat surface, as in Fig. P8.79. If the internal pressure is p_i, find an expression for the pressure force on the hemisphere. By analogy with Prob. 8.44, at what point A on the hemisphere should a hole be cut so that the pressure force will be zero according to inviscid theory?

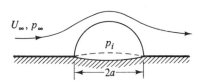

Fig. P8.79

8.80 We have studied the point source (sink) and the line source (sink) of infinite depth into the paper. Does it make any sense to define a finite-length line sink (source) as in Fig. P8.80? If so, how would you establish the mathematical properties of such a finite line sink? When combined with a uniform stream and a point source of equivalent strength as in Fig. P8.80, should a closed body shape be formed? Make a guess and sketch some of these possible shapes for various values of the dimensionless parameter $m/U_\infty L^2$.

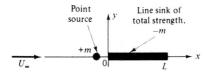

Fig. P8.80

8.81 Normally by its very nature inviscid theory is incapable of predicting body drag, but by analogy with Fig. 8.20c we can analyze flow approaching a hemisphere, as in Fig. P8.81. Assume that the flow on the front follows inviscid sphere theory, Eq. (8.132), and the pressure in the rear equals the shoulder pressure. Compute the drag coefficient and compare with experiment (Table 7.3). What are the defects and limitations of this analysis?

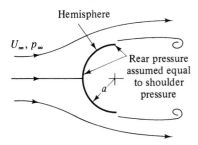

Fig. P8.81

8.82 A 1-m-diameter sphere is being towed at speed V in fresh water at 20°C as shown in Fig. P8.82. Assuming inviscid theory with an undistorted free surface, estimate the speed V in meters per second at which cavitation will first appear on the sphere surface. Where will cavitation appear? For this condition, what will be the pressure at point A on the sphere which is 45° up from the direction of travel?

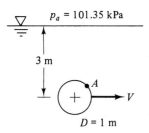

Fig. P8.82

8.83 Consider a cylinder of radius a moving at speed U_∞ through a still fluid, as in Fig. P8.83. Plot the streamlines relative to the cylinder by modifying Eq. (8.58) to give the relative flow with $K = 0$. Integrate to find the total relative kinetic energy and verify the hydrodynamic mass of a cylinder from Eq. (8.138).

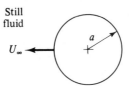

Fig. P8.83

8.84 In Table 7.2 the drag coefficient of a 4:1 elliptical cylinder in laminar boundary layer flow is 0.35. According to Patton [17], the hydrodynamic mass of this cylinder is $\pi\rho hb/4$, where b is width into the paper and h is the maximum thickness. Use these results to derive a formula for the time history $U(t)$ of the cylinder if it is accelerated from rest in a still fluid by the sudden application of a constant force F.

8.85 Laplace's equation in plane polar coordinates, Eq. (8.13), is complicated by the variable radius. Consider the finite-difference mesh in Fig. P8.85, with nodes (i,j) equally spaced $\Delta\theta$ and Δr apart. Derive a finite-difference model for Eq. (8.13) similar to the cartesian expression (8.143).

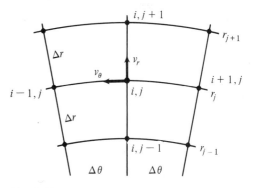

Fig. P8.85

8.86 Set up the numerical problem of Fig. 8.34 for an expansion of 30°. A new grid system and a non-square mesh may be needed. Give the proper nodal equation and boundary conditions. If possible, program this 30° expansion and solve on a digital computer.

8.87 Recently an exact analytical (nonnumerical) solution to the duct expansion problem of Fig. 8.34 was given by L. N. Goenka in an M.S. thesis at the University of Texas. The solution uses a number of complex-variable mappings and is described by Panton [23]. Study this solution as an outside research project and report on the methods and solutions to your class. Compare the exact results with the numerical results in Example 8.6.

8.88 Consider two-dimensional potential flow into a step contraction as in Fig. P8.88. The inlet velocity $U_1 = 7$ m/s and the outlet velocity U_2 is uniform. The nodes (i, j) are labeled in the figure. Set up the complete finite-difference algebraic relations for all nodes. Solve, if possible, on a digital computer and plot the streamlines in the flow.

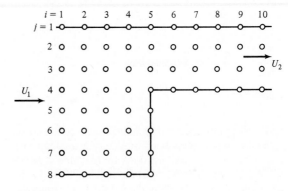

Fig. P8.88

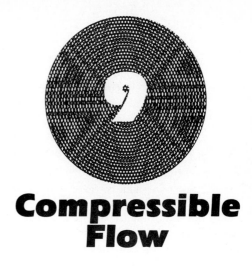

Compressible Flow

9.1 INTRODUCTION

Up to this point we have barely touched on the notion of a compressible flow. For example, one of the key properties of a compressible fluid, its *speed of sound*, was not defined in Chap. 1.

We took a brief look in Chap. 4 [Eqs. (4.13) to (4.17)] to see when we might safely neglect the compressibility inherent in every real fluid. We found that the proper criterion for a nearly incompressible flow was a small Mach number

$$\text{Ma} = \frac{V}{a} \ll 1 \tag{9.1}$$

where V is the flow velocity and a the speed of sound of the fluid. Under small-Mach-number conditions, changes in fluid density are everywhere small in the flow field. The energy equation becomes uncoupled, and temperature effects can either be ignored or put aside for later study. The equation of state degenerates into the simple statement that density is nearly constant. This means that an incompressible flow requires only a momentum and continuity analysis, as we showed with many examples in Chaps. 7 and 8.

The present chapter treats compressible flows, which have Mach numbers greater than about 0.3 and thus exhibit nonnegligible density changes. If the density change is significant, it follows from the equation of state that the temperature and pressure changes are also substantial. Large temperature changes imply that the energy equation can no longer be neglected. Therefore the work is doubled from two basic equations to four

1. Continuity equation
2. Momentum equation
3. Energy equation
4. Equation of state

to be solved simultaneously for four unknowns: pressure, density, temperature, and flow velocity (p, ρ, T, V). Thus the general theory of compressible flow is quite complicated, and we try here to make further simplification, especially by assuming a reversible adiabatic or *isentropic* flow.

The Mach number is the dominant parameter in compressible-flow analysis, with different effects depending upon its magnitude. Aerodynamicists especially make a distinction between the various ranges of Mach number, and the following rough classifications are commonly used:

> Ma < 0.3: *incompressible flow*, where density effects are negligible.
>
> 0.3 < Ma < 0.8: *subsonic flow*, where density effects are important but no shock waves appear.
>
> 0.8 < Ma < 1.2: *transonic flow*, where shock waves first appear, dividing subsonic and supersonic regions of the flow. Powered flight in the transonic region is difficult because of the mixed character of the flow field.
>
> 1.2 < Ma < 3.0: *supersonic flow*, where shock waves are present but there are no subsonic regions.
>
> 3.0 < Ma: *hypersonic flow* [13], where shock waves and other flow changes are especially strong.

The numerical values listed above are only rough guides. These five categories of flow are appropriate to external high-speed aerodynamics. For internal (duct) flows, the most important question is simply whether the flow is subsonic (Ma < 1) or supersonic (Ma > 1), because the effect of area changes reverses, as we show in Sec. 9.4. Since supersonic flow effects may go against one's intuition, you should study these differences carefully.

This text contains only a single chapter on compressible flow, but, as usual, whole books have been written on the subject. References 1 to 6 and 26 are introductory, fairly elementary treatments, while Refs. 7 to 14 and 27 are advanced. From time to time we shall defer some specialized topic to these texts.

We note in passing that there are at least two flow patterns which depend strongly upon very small density differences, acoustics, and natural convection. Acoustics [9, 14] is the study of sound-wave propagation, which is accompanied by extremely small changes in density, pressure, and temperature. Natural convection is the gentle circulating pattern set up by buoyancy forces in a fluid stratified by uneven heating or uneven concentration of dissolved materials. Here we are concerned only with steady compressible flow where the fluid velocity is of magnitude comparable to that of the speed of sound.

The Perfect Gas

In principle, compressible-flow calculations can be made for any fluid equation of state, and we shall assign problems involving the steam tables [15], the gas tables

[16], and liquids [Eq. (1.38)]. But in fact most elementary treatments are confined to the perfect gas with constant specific heats

$$p = \rho R T \qquad R = c_p - c_v = \text{const} \qquad \gamma = \frac{c_p}{c_v} = \text{const} \qquad (9.2)$$

For all real gases, c_p, c_v, and γ vary with temperature but only moderately; for example, c_p of air increases 30 percent as temperature increases from 0 to 5000°F (see Ref. 17, fig. B.17). Since we rarely deal with such large temperature changes, it is quite reasonable to assume constant specific heats.

Recall from Sec. 1.6 that the gas constant is related to a universal constant Λ divided by the gas molecular weight

$$R_{\text{gas}} = \frac{\Lambda}{M_{\text{gas}}} \qquad (9.3)$$

where
$$\Lambda = 49{,}720 \ \text{ft}^2/(\text{s}^2 \cdot {}^\circ\text{R}) = 8314 \ \text{m}^2/(\text{s}^2 \cdot \text{K})$$

For air, $M = 28.97$, and we shall adopt the following property values for air throughout this chapter:

$$R = 1717 \ \text{ft}^2/(\text{s}^2 \cdot {}^\circ\text{R}) = 287 \ \text{m}^2/(\text{s}^2 \cdot \text{K}) \qquad \gamma = 1.400$$

$$c_v = \frac{R}{\gamma - 1} = 4293 \ \text{ft}^2/(\text{s}^2 \cdot {}^\circ\text{R}) = 718 \ \text{m}^2/(\text{s}^2 \cdot \text{K})$$

$$c_p = \frac{\gamma R}{\gamma - 1} = 6010 \ \text{ft}^2/(\text{s}^2 \cdot {}^\circ\text{R}) = 1005 \ \text{m}^2/(\text{s}^2 \cdot \text{K}) \qquad (9.4)$$

Experimental values of γ for eight common gases were shown in Fig. 1.4. From this figure and the molecular weight the other properties can be computed, as in Eq. (9.4).

The changes in internal energy $\hat{u}$ and enthalpy h of a perfect gas are computed for constant specific heats as

$$\hat{u}_2 - \hat{u}_1 = c_v(T_2 - T_1) \qquad h_2 - h_1 = c_p(T_2 - T_1) \qquad (9.5)$$

For variable specific heats one must integrate $\hat{u} = \int c_v \, dT$ and $h = \int c_p \, dT$ or use the gas tables [16].

Isentropic Process

The isentropic approximation is common in compressible-flow theory. We compute the entropy change from the first and second law of thermodynamics for a pure substance [17 or 18]

$$T \, ds = dh - \frac{dp}{\rho} \qquad (9.6)$$

Introducing $dh = c_p \, dT$ for a perfect gas and solving for ds, we substitute $\rho T = p/R$ from the perfect-gas law and obtain

$$\int_1^2 ds = \int_1^2 c_p \frac{dT}{T} - R \int_1^2 \frac{dp}{p} \tag{9.7}$$

If c_p is variable, the gas tables will be needed, but for constant c_p we obtain the analytic results

$$s_2 - s_1 = c_p \ln \frac{T_2}{T_1} - R \ln \frac{p_2}{p_1} = c_v \ln \frac{T_2}{T_1} - R \ln \frac{\rho_2}{\rho_1} \tag{9.8}$$

Equations (9.8) are used to compute the entropy change across a shock wave (Sec. 9.5), which is an irreversible process.

For isentropic flow, we set $s_2 = s_1$ and obtain the interesting power-law relations for an isentropic perfect gas

$$\frac{p_2}{p_1} = \left(\frac{T_2}{T_1}\right)^{\gamma/(\gamma-1)} = \left(\frac{\rho_2}{\rho_1}\right)^{\gamma} \tag{9.9}$$

These relations are used in Sec. 9.3.

EXAMPLE 9.1   Argon flows through a tube such that its initial condition is $p_1 = 250 \, \text{lbf/in}^2$ absolute and $\rho_1 = 1.16 \, \text{lbm/ft}^3$ and its final condition is $p_2 = 30 \, \text{lbf/in}^2$ absolute and $T_2 = 265°\text{F}$. Estimate (a) the initial temperature, (b) the final density, (c) the change in enthalpy, and (d) the change in entropy.

Solution   First get immediately into proper BG units:

$$p_1 = 250(144) = 36,000 \, \text{lbf/ft}^2 \qquad p_2 = 30(144) = 4320 \, \text{lbf/ft}^2$$

$$\rho_1 = \frac{1.16}{32.2} = 0.0360 \, \text{slug/ft}^3 \qquad T_2 = 265 + 460 = 725°\text{R}$$

From Appendix A the molecular weight of argon is 39.944 and its specific-heat ratio from Fig. 1.4 is about 1.67. Therefore its gas constant and specific heats are approximately

$$R = \frac{49,720}{39.944} = 1245 \, \text{ft}^2/(\text{s}^2 \cdot °\text{R}) \qquad c_p = \frac{1.67(1245)}{1.67 - 1} = 3103 \, \text{ft}^2/(\text{s}^2 \cdot °\text{R})$$

Then the initial temperature and final density can be estimated from the perfect-gas law, Eq. (9.2),

$$T_1 = \frac{p_1}{R\rho_1} = \frac{36,000}{1245(0.0360)} = 803°\text{R} \qquad\qquad Ans. (a)$$

$$\rho_2 = \frac{p_2}{RT_2} = \frac{4320}{1245(725)} = 0.00479 \, \text{slug/ft}^3 \qquad\qquad Ans. (b)$$

From Eq. (9.5) the enthalpy change is

$$h_2 - h_1 = c_p(T_2 - T_1) = 3103(725 - 803) = -242,000 \; (\text{ft} \cdot \text{lbf})/\text{slug (or ft}^2/\text{s}^2) \quad Ans. (c)$$

The argon enthalpy decreases as we move down the tube. Actually there may not be any cooling; i.e., the fluid enthalpy may be converted by friction into increased kinetic energy. Finally, the entropy change is computed from Eq. (9.8)

$$s_2 - s_1 = c_p \ln \frac{T_2}{T_1} - R \ln \frac{p_2}{p_1} = 3103 \ln \frac{725}{803} - 1245 \ln \frac{4320}{36,000}$$

$$= -317 + 2640 = 2320 \text{ ft}^2/(\text{s}^2 \cdot {}^\circ\text{R}) \qquad\qquad Ans. \ (d)$$

The fluid entropy has increased. If there is no heat transfer, this indicates an irreversible process. Note that entropy has the same units as the gas constant and specific heat.

This problem is not just arbitrary numbers. It correctly simulates the behavior of argon moving subsonically through a tube with large frictional effects (Sec. 9.7).

9.2 THE SPEED OF SOUND

The so-called sound speed is the rate of propagation of a pressure pulse of infinitesimal strength through a still fluid. It is a thermodynamic property of a fluid. Let us analyze it by first considering a pulse of finite strength, as in Fig. 9.1. In Fig. 9.1a the pulse, or pressure wave, moves at speed C toward the still fluid $(p, \rho, T, V = 0)$ at the left, leaving behind at the right a fluid of increased properties $(p + \Delta p, \rho + \Delta \rho, T + \Delta T)$ and a fluid velocity ΔV toward the left following the wave but much slower. We can determine these effects by making a control-volume analysis across the wave. To avoid the unsteady terms which would be necessary in Fig. 9.1a, we adopt instead the control volume of Fig. 9.1b, which moves at wave

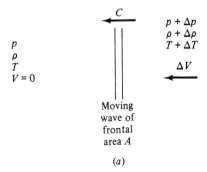

(a)

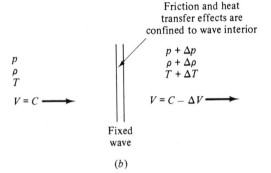

Fig. 9.1 Control-volume analysis of a finite-strength pressure wave: (a) control volume fixed to still fluid at left; (b) control volume moving left at the wave speed C.

(b)

speed C to the left. The wave appears fixed from this viewpoint, and the fluid appears to have velocity C on the left and $C - \Delta V$ on the right. The thermodynamic properties p, ρ, and T are not affected by this change of viewpoint.

The flow in Fig. 9.1b is steady and one-dimensional across the wave. The continuity equation is thus, from Eq. (3.24),

$$\rho A C = (\rho + \Delta\rho)(A)(C - \Delta V)$$

or

$$\Delta V = C \frac{\Delta\rho}{\rho + \Delta\rho} \tag{9.10}$$

This proves our contention that the induced fluid velocity on the right is much smaller than the wave speed C. In the limit of infinitesimal wave strength (sound wave) this speed is itself infinitesimal.

Notice that there are no velocity gradients on either side of the wave. Therefore, even if fluid viscosity is large, frictional effects are confined to the interior of the wave. Advanced texts [e.g., 19, sec. 3-9.4] show that the thickness of pressure waves in gases is of order 10^{-6}-ft·at atmospheric pressure. Thus we can safely neglect friction and apply the one-dimensional momentum equation (3.40) across the wave

$$\sum F_{\text{right}} = \dot{m}(V_{\text{out}} - V_{\text{in}})$$

or

$$pA - (p + \Delta p)A = (\rho A C)(C - \Delta V - C) \tag{9.11}$$

Again the area cancels, and we can solve for the pressure change

$$\Delta p = \rho C \, \Delta V \tag{9.12}$$

If the wave strength is very small, the pressure change is small.

Finally combine Eqs. (9.10) and (9.12) to give an expression for the wave speed

$$C^2 = \frac{\Delta p}{\Delta\rho}\left(1 + \frac{\Delta\rho}{\rho}\right) \tag{9.13}$$

The larger the strength $\Delta\rho/\rho$ of the wave, the faster the wave speed; i.e., powerful explosion waves move much faster than sound waves. In the limit of infinitesimal strength, $\Delta\rho \to 0$, we have what is defined to be the speed of sound a of a fluid:

$$a^2 = \frac{\partial p}{\partial \rho} \tag{9.14}$$

But the evaluation of the derivative requires knowledge of the thermodynamic process undergone by the fluid as the wave passes. Sir Isaac Newton in 1686 made a famous error by deriving a formula for sound speed which was equivalent to assuming an isothermal process, the result being 20 percent low for air, for example. He rationalized the discrepancy as being due to the "crassitude" (dust particles, etc.) in the air; the error is certainly understandable when we reflect that it was made 180 years before the proper basis was laid for the second law of thermodynamics.

We now see that the correct process must be *adiabatic* because there are no temperature gradients except inside the wave itself. For vanishing-strength sound waves we therefore have an infinitesimal adiabatic or isentropic process. The correct expression for the sound speed is

$$a = \left(\frac{\partial p}{\partial \rho}\Big|_s\right)^{1/2} = \left(\gamma \frac{\partial p}{\partial \rho}\Big|_T\right)^{1/2} \tag{9.15}$$

for any fluid, gas or liquid. Even a solid has a sound speed.

For a perfect gas, from Eq. (9.2) or (9.9), we deduce that the speed of sound is

$$a = \left(\frac{\gamma p}{\rho}\right)^{1/2} = (\gamma R T)^{1/2} \tag{9.16}$$

The speed of sound increases as the square root of the absolute temperature. For air, with $\gamma = 1.4$ and $R = 1717$, an easily memorized dimensional formula is

$$a(\text{ft/s}) \approx 49[T(°\text{R})]^{1/2}$$
$$a(\text{m/s}) \approx 20[T(\text{K})]^{1/2} \tag{9.17}$$

At sea-level standard temperature, $60°\text{F} = 520°\text{R}$, $a = 1117$ ft/s. This decreases in the upper atmosphere, which is cooler; at 50,000 ft standard altitude, $T = -69.7°\text{F} = 389.9°\text{R}$ and $a = 49(389.9)^{1/2} = 968$ ft/s, or 13 percent less.

Table 9.1

SOUND SPEED OF VARIOUS MATERIALS AT 60°F (15.5°C) AND 1 atm

Material	a, ft/s	a, m/s
Gases:		
H_2	4,246	1294
He	3,281	1000
Air	1,117	340
Ar	1,040	317
CO_2	873	266
CH_4	607	185
$^{238}UF_6$	297	91
Liquids:		
Glycerin	6,100	1860
Water	4,890	1490
Mercury	4,760	1450
Ethyl alcohol	3,940	1200
Solids:†		
Aluminum	16,900	5150
Steel	16,600	5060
Hickory	13,200	4020
Ice	10,500	3200

† Plane waves. Solids also have a *shear-wave speed.*

Some representative values of sound speed in various materials are given in Table 9.1. For liquids and solids it is common to define the *bulk modulus K* of the material

$$K = -\mathfrak{v} \frac{\partial p}{\partial \mathfrak{v}}\Big|_s = \rho \frac{\partial p}{\partial \rho}\Big|_s \qquad (9.18)$$

For example, at standard conditions, the bulk modulus of carbon tetrachloride is $163{,}000$ lbf/in^2 absolute, and its density is 3.09 slugs/ft^3. Its speed of sound is therefore $[163{,}000(144)/3.09]^{1/2} = 2756$ ft/s, or 840 m/s. Steel has a bulk modulus of about 29×10^6 lbf/in^2 absolute and water about 320×10^3 lbf/in^2 absolute, or 90 times less.

For solids, it is sometimes assumed that bulk modulus is approximately equivalent to Young's modulus of elasticity E, but in fact their ratio depends upon Poisson's ratio σ

$$\frac{E}{K} = 3(1 - 2\sigma) \qquad (9.19)$$

The two are equal for $\sigma = \frac{1}{3}$, which is approximately the case for many common metals such as steel and aluminum.

EXAMPLE 9.2 Estimate the speed of sound of carbon monoxide at 200 kPa pressure and 300°C in meters per second.

Solution From Appendix Table A.4, for CO, the molecular weight is 28.01 and $\gamma \approx 1.40$. Thus from Eq. (9.3) $R_{CO} = 8314/28.01 = 297$ m^2/(s$^2 \cdot$ K), and the given temperature is 300°C + 273 = 573 K. Thus from Eq. (9.16) we estimate

$$a_{CO} = (\gamma R T)^{1/2} = [1.40(297)(573)]^{1/2} = 488 \text{ m/s} \qquad Ans.$$

9.3 ADIABATIC AND ISENTROPIC STEADY FLOW

As mentioned in Sec. 9.1, the isentropic approximation greatly simplifies a compressible-flow calculation. So does the assumption of adiabatic flow, even if nonisentropic.

Consider high-speed flow of a gas past an insulated wall, as in Fig. 9.2. There is no shaft work delivered to any part of the fluid. Therefore every streamtube in the flow satisfies the steady-flow energy equation in the form of Eq. (3.85)

$$h_1 + \tfrac{1}{2}V_1^2 + gz_1 = h_2 + \tfrac{1}{2}V_2^2 + gz_2 - q + w_v \qquad (9.20)$$

where point 1 is upstream of point 2. You may wish to review the details of Eq. (3.85) and its development. We saw in Example 3.20 that potential-energy changes of a gas are extremely small compared with kinetic-energy and enthalpy terms. We shall neglect the terms gz_1 and gz_2 in all gas-dynamic analyses.

Inside the thermal and velocity boundary layers in Fig. 9.2 the heat-transfer and viscous-work terms q and w_v are not zero. But outside the boundary layer q and w_v are zero by definition, so that the outer flow satisfies the simple relation

$$h_1 + \tfrac{1}{2}V_1^2 = h_2 + \tfrac{1}{2}V_2^2 = \text{const} \qquad (9.21)$$

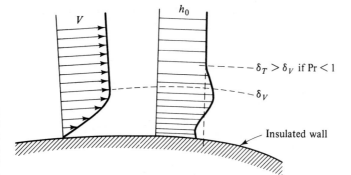

Fig. 9.2 Velocity and stagnation-enthalpy distributions near an insulated wall in a typical high-speed gas flow.

The constant in Eq. (9.21) is equal to the maximum enthalpy which the fluid would achieve if brought to rest adiabatically. We call this value h_0, the *stagnation enthalpy* of the flow. Thus we rewrite Eq. (9.21) in the form

$$h + \tfrac{1}{2}V^2 = h_0 = \text{const} \tag{9.22}$$

This should hold for steady adiabatic flow of any compressible fluid outside the boundary layer. The wall in Fig. 9.2 could be either the surface of an immersed body or the wall of a duct. We have shown the details of Fig. 9.2; typically the thermal-layer thickness δ_T is greater than the velocity-layer thickness δ_V because most gases have a dimensionless Prandtl number $\text{Pr} = \mu c_p/k$ less than unity (see, for example, Ref. 19, sec. 4-3.3). Note that the stagnation enthalpy varies inside the thermal boundary layer, but its average value is the same as the outer layer due to the insulated wall.

For nonperfect gases we may have to use the steam tables [15] or the gas tables [16] to implement Eq. (9.22). But for a perfect gas $h = c_p T$ and Eq. (9.22) becomes

$$c_p T + \tfrac{1}{2}V^2 = c_p T_0 \tag{9.23}$$

This establishes the stagnation temperature T_0 of an adiabatic perfect-gas flow, i.e., the temperature it achieves when decelerated to rest adiabatically.

An alternate interpretation of Eq. (9.22) occurs when the enthalpy and temperature drop to (absolute) zero, so that the velocity achieves a maximum value

$$V_{\max} = (2h_0)^{1/2} = (2c_p T_0)^{1/2} \tag{9.24}$$

No higher flow velocity can occur unless additional energy is added to the fluid through shaft work or heat transfer (Sec. 9.8).

Mach-Number Relations

The dimensionless form of Eq. (9.23) brings in the Mach number Ma as a parameter, using Eq. (9.16) for the speed of sound of a perfect gas. Divide through by $c_p T$ to obtain

$$1 + \frac{V^2}{2c_p T} = \frac{T_0}{T} \tag{9.25}$$

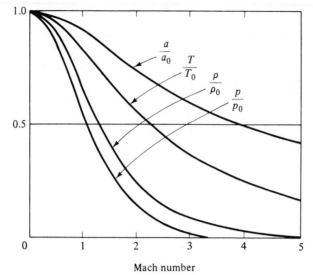

But, from the perfect-gas law, $c_p T = [\gamma R/(\gamma - 1)]T = a^2/(\gamma - 1)$, so that Eq. (9.25) becomes

$$1 + \frac{(\gamma - 1)V^2}{2a^2} = \frac{T_0}{T}$$

or

$$\frac{T_0}{T} = 1 + \frac{\gamma - 1}{2} \, \text{Ma}^2 \qquad \text{Ma} = \frac{V}{a} \tag{9.26}$$

This relation is plotted in Fig. 9.3 versus the Mach number for $\gamma = 1.4$. At Ma = 5 the temperature has dropped to $\frac{1}{6}T_0$.

Since $a \propto T^{1/2}$, the ratio a_0/a is the square root of (9.26)

$$\frac{a_0}{a} = \left(\frac{T_0}{T}\right)^{1/2} = [1 + \tfrac{1}{2}(\gamma - 1)\text{Ma}^2]^{1/2} \tag{9.27}$$

Equation (9.27) is also plotted in Fig. 9.3. At Ma = 5 the speed of sound has dropped to 41 percent of the stagnation value.

Isentropic Pressure and Density Relations

Note that Eqs. (9.26) and (9.27) require only adiabatic flow and hold even in the presence of irreversibilities such as friction losses or shock waves.

If the flow is also *isentropic*, then for a perfect gas the pressure and density ratios can be computed from Eq. (9.9) as a power of the temperature ratio

$$\frac{p_0}{p} = \left(\frac{T_0}{T}\right)^{\gamma/(\gamma - 1)} = [1 + \tfrac{1}{2}(\gamma - 1) \, \text{Ma}^2]^{\gamma/(\gamma - 1)} \tag{9.28a}$$

$$\frac{\rho_0}{\rho} = \left(\frac{T_0}{T}\right)^{1/(\gamma - 1)} = [1 + \tfrac{1}{2}(\gamma - 1) \, \text{Ma}^2]^{1/(\gamma - 1)} \tag{9.28b}$$

These relations are also plotted in Fig. 9.3; at Ma = 5 the density is 1.13 percent of its stagnation value, and the pressure is only 0.19 percent of stagnation pressure.

The quantities p_0 and ρ_0 are the isentropic stagnation pressure and density, respectively, i.e., the pressure and density which the flow would achieve if brought isentropically to rest. In an adiabatic nonisentropic flow p_0 and ρ_0 retain their local meaning, but they vary throughout the flow as the entropy changes due to friction or shock waves. The quantities h_0, T_0, and a_0 are constant in an adiabatic nonisentropic flow (see Sec. 9.7 for further details).

Relationship to Bernoulli's Equation

The isentropic assumptions (9.28) are effective, but are they realistic? Yes; to see why differentiate Eq. (9.22)

Adiabatic: $$dh + V\,dV = 0 \tag{9.29}$$

Meanwhile, from Eq. (9.6), if $ds = 0$ (isentropic process),

$$dh = \frac{dp}{\rho} \tag{9.30}$$

Combining (9.29) and (9.30), we find that an isentropic streamtube flow must be such that

$$\frac{dp}{\rho} + V\,dV = 0 \tag{9.31}$$

But this is exactly the Bernoulli relation, Eq. (3.61), for steady frictionless flow with negligible gravity terms. Thus we see that the isentropic-flow assumption is equivalent to use of the Bernoulli or streamline form of the frictionless momentum equation.

Critical Values at the Sonic Point

The stagnation values (a_0, T_0, p_0, ρ_0) are useful reference conditions in a compressible flow, but of comparable usefulness are the conditions where the flow is sonic, Ma = 1.0. These sonic, or *critical*, properties are denoted by asterisks: p^*, ρ^*, a^*, and T^*. They are certain ratios of the stagnation properties as given by Eqs. (9.26) to (9.28) when Ma = 1.0; for $\gamma = 1.4$

$$\frac{p^*}{p_0} = \left(\frac{2}{\gamma + 1}\right)^{\gamma/(\gamma - 1)} = 0.5283 \qquad \frac{\rho^*}{\rho_0} = \left(\frac{2}{\gamma + 1}\right)^{1/(\gamma - 1)} = 0.6339$$

$$\frac{T^*}{T_0} = \frac{2}{\gamma + 1} = 0.8333 \qquad \frac{a^*}{a_0} = \left(\frac{2}{\gamma + 1}\right)^{1/2} = 0.9129 \tag{9.32}$$

In all isentropic flow all critical properties are constant; in adiabatic nonisentropic flow a^* and T^* are constant, but p^* and ρ^* may vary.

The critical velocity V^* equals the sonic sound speed a^* by definition and is often used as a reference velocity in isentropic or adiabatic flow

$$V^* = a^* = (\gamma RT^*)^{1/2} = \left(\frac{2\gamma}{\gamma + 1} RT_0\right)^{1/2} \tag{9.33}$$

The usefulness of these critical values will become clearer as we study compressible duct flow with friction or heat transfer later in this chapter.

Some Useful Numbers for Air

Since the great bulk of our practical calculations are for air, $\gamma = 1.4$, the stagnation-property ratios p/p_0, etc., from Eqs. (9.26) to (9.28) are tabulated for this value in Appendix Table B.1. The increments in Mach number are rather coarse in this table because the values are only meant as a guide: these equations are now a trivial matter to manipulate on a hand calculator. Twenty-five years ago every text had extensive compressible-flow tables with Mach-number spacings of about 0.01, so that accurate values could be interpolated.

For $\gamma = 1.4$, the following numerical versions of the isentropic and adiabatic flow formulas are obtained

$$\frac{T_0}{T} = 1 + 0.2\,\text{Ma}^2 \qquad \frac{\rho_0}{\rho} = (1 + 0.2\,\text{Ma}^2)^{2.5}$$

$$\tag{9.34}$$

$$\frac{p_0}{p} = (1 + 0.2\,\text{Ma}^2)^{3.5}$$

Or, if we are given the properties, it is equally easy to solve for the Mach number (again with $\gamma = 1.4$)

$$\text{Ma}^2 = 5\left(\frac{T_0}{T} - 1\right) = 5\left[\left(\frac{\rho_0}{\rho}\right)^{2/5} - 1\right] = 5\left[\left(\frac{p_0}{p}\right)^{2/7} - 1\right] \tag{9.35}$$

Note that these isentropic-flow formulas serve as the equivalent of the frictionless adiabatic momentum and energy equations. They relate velocity to physical properties for a perfect gas, but they are *not* the "solution" to a gas-dynamics problem. The complete solution is not obtained until the continuity equation has also been satisfied, for either one-dimensional (Sec. 9.4) or multidimensional (Sec. 9.9) flow.

One final note: these isentropic-ratio–versus–Mach-number formulas are seductive, tempting one to solve all problems by jumping right into the tables. Actually, many problems involving (dimensional) velocity and temperature can be solved more easily from the original raw dimensional energy equation (9.23) plus the perfect-gas law (9.2), as the next example will illustrate.

EXAMPLE 9.3 Air flows adiabatically through a duct. At point 1 the velocity is 800 ft/s, $T_1 = 500°$R, and $p_1 = 25$ lbf/in^2 absolute $= 3600$ lbf/ft^2. Compute (a) T_0; (b) p_{01}; (c) ρ_0; (d) Ma; (e) V_{max}; (f) V^*. At point 2 further downstream $V_2 = 962$ ft/s, and $p_2 = 2850$ lbf/ft^2. (g) What is the stagnation pressure p_{02}?

Solution For air take $\gamma = 1.4$, $c_p = 6010$, and $R = 1717$ (BG units). With V_1 and T_1 known, we can compute T_0 from Eq. (9.23) without using ratios

$$T_{01} = T_1 + \frac{\frac{1}{2}(V_1)^2}{c_p} = 500 + \frac{\frac{1}{2}(800)^2}{6010} = 500 + 53 = 553°\text{R} \qquad Ans.\ (a)$$

Then compute Ma from the known ratio T/T_0, using Eq. (9.35)

$$\text{Ma}_1^2 = 5\left(\frac{553}{500} - 1\right) = 0.53 \qquad \text{Ma}_1 = 0.73 \qquad Ans.\ (d)$$

Alternately compute $a_1 = (\gamma R T_1)^{1/2} \approx 49(500)^{1/2} = 1096\ \text{ft/s}$, whence $\text{Ma}_1 = V_1/a_1 = 800/1096 = 0.73$. The stagnation pressure follows from Eq. (9.34):

$$p_{01} = p_1(1 + 0.2\ \text{Ma}_1^2)^{3.5} = 3600[1 + 0.2(0.73)^2]^{3.5} = 5130\ \text{lbf/ft}^2 \qquad Ans.\ (b)$$

We need the density before we can compute stagnation density:

$$\rho_1 = \frac{p_1}{RT_1} = \frac{3600}{1717(500)} = 0.00419\ \text{slug/ft}^3$$

Then $\rho_0 = \rho_1(1 + 0.2\ \text{Ma}_1^2)^{2.5} = 0.00419[1 + 0.2(0.73)^2]^{2.5} = 0.00540\ \text{slug/ft}^3$ *Ans. (c)*

However, if we were clever, we could compute ρ_0 directly from p_0 and T_0:

$$\rho_0 = \frac{p_0}{RT_0} = \frac{5130}{1717(553)} = 0.00540\ \text{slug/ft}^3$$

The maximum velocity follows from Eq. (9.24)

$$V_{\text{max}} = (2c_p T_0)^{1/2} = [2(6010)(553)]^{1/2} = 2580\ \text{ft/s} \qquad Ans.\ (e)$$

and the sonic velocity from Eq. (9.33)

$$V^* = \left[\frac{2(1.4)}{1.4 + 1}(1717)(553)\right]^{1/2} = 1050\ \text{ft/s} \qquad Ans.\ (f)$$

At point 2, the temperature is not given, but since we know it is adiabatic flow, $T_{02} = T_{01} = 553°\text{R}$ from solution (a). Thus we can compute T_2 from Eq. (9.23)

$$T_2 = T_{02} - \frac{\frac{1}{2}V_2^2}{c_p} = 553 - \frac{\frac{1}{2}(962)^2}{6010} = 476°\text{R}$$

(Trying to find T_2 from the Mach-number relations is a frustratingly laborious procedure.) The speed of sound a_2 thus equals $(\gamma R T_2)^{1/2} \approx 49(476)^{1/2} = 1069\ \text{ft/s}$, whence the Mach number $\text{Ma}_2 = V_2/a_2 = 962/1069 = 0.90$. Finally compute

$$p_{02} = p_2(1 + 0.2\ \text{Ma}_2^2)^{3.5} = 2850[1 + 0.2(0.9)^2]^{3.5}$$
$$= 4820\ \text{lbf/ft}^2 \qquad (6\ \%\ \text{less than }p_{01}) \qquad Ans.\ (g)$$

9.4 ISENTROPIC FLOW WITH AREA CHANGES

By combining the isentropic- and/or adiabatic-flow relations with the equation of continuity we can study practical compressible-flow problems. This section treats the one-dimensional flow approximation.

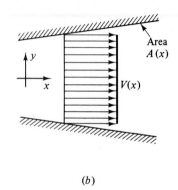

Fig. 9.4 Compressible flow through a duct: (*a*) real-fluid velocity profile; (*b*) one-dimensional approximation.

Figure 9.4 illustrates the one-dimensional flow assumption. A real flow, Fig. 9.4*a*, has no slip at the walls and a velocity profile $V(x, y)$ which varies across the duct section (compare with Fig. 7.8). If, however, the area change is small and wall radius of curvature large

$$\frac{dh}{dx} \ll 1 \qquad h(x) \ll R(x) \tag{9.36}$$

the flow is approximately one-dimensional, as in Fig. 9.4*b*, with $V \approx V(x)$ reacting to area change $A(x)$. Compressible-flow nozzles and diffusers do not always satisfy conditions (9.36), but we use the one-dimensional theory anyway because of its simplicity.

For steady one-dimensional flow the equation of continuity is, from Eq. (3.24),

$$\rho(x)V(x)A(x) = \dot{m} = \text{const} \tag{9.37}$$

Before applying this to duct theory we can learn a lot from the differential form of Eq. (9.37)

$$\frac{d\rho}{\rho} + \frac{dV}{V} + \frac{dA}{A} = 0 \tag{9.38}$$

The differential forms of the frictionless momentum equation (9.31) and the sound-speed relation (9.15) are recalled here for convenience:

Momentum $$\frac{dp}{\rho} + V\,dV = 0$$
$$\tag{9.39}$$

Sound speed: $$dp = a^2\,d\rho$$

Now eliminate dp and $d\rho$ between Eqs. (9.38) and (9.39) to obtain the following relation between velocity change and area change in isentropic duct flow:

$$\frac{dV}{V} = \frac{dA}{A} \frac{1}{\text{Ma}^2 - 1} = -\frac{dp}{\rho V^2} \tag{9.40}$$

Inspection of this equation, without actually solving it, reveals a fascinating aspect of compressible flow: property changes are of opposite sign for subsonic and

Duct geometry	Subsonic Ma < 1	Supersonic Ma > 1
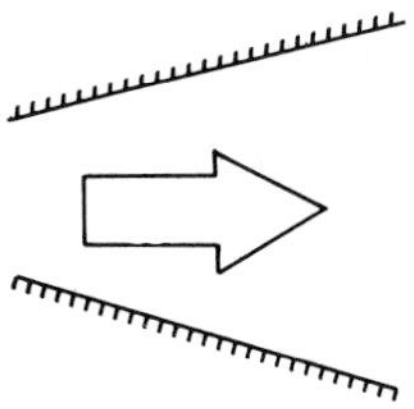 $dA > 0$	$dV < 0$ $dp > 0$ Subsonic diffuser	$dV > 0$ $dp < 0$ Supersonic nozzle
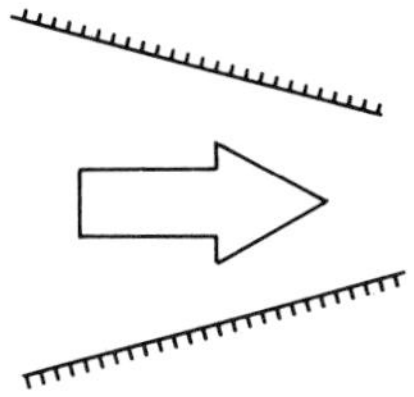 $dA < 0$	$dV > 0$ $dp < 0$ Subsonic nozzle	$dV < 0$ $dp > 0$ Supersonic diffuser

Fig. 9.5 Effect of Mach number on property changes with area change in duct flow.

supersonic flow because of the term $Ma^2 - 1$. There are four combinations of area change and Mach number, summarized in Fig. 9.5.

From earlier chapters we are used to subsonic behavior (Ma < 1): when area increases, velocity decreases and pressure increases, which is denoted a subsonic diffuser. But in supersonic flow (Ma > 1) velocity actually increases when area increases, a supersonic nozzle. The same opposing behavior occurs for an area decrease, which speeds up a subsonic flow and slows down a supersonic flow.

What about the sonic point, Ma = 1? Since infinite acceleration is physically impossible, Eq. (9.40) indicates that dV can be finite only when $dA = 0$, that is, a minimum area (throat) or a maximum area (bulge). In Fig. 9.6 we patch together a throat section and a bulge section using the rules from Fig. 9.5. The throat or converging-diverging section can smoothly accelerate a subsonic flow through sonic to supersonic flow, as in Fig. 9.6a. This is the only way a supersonic flow can be created by expanding the gas from a stagnant reservoir. The bulge section fails; the bulge Mach number moves away from a sonic condition rather than toward it.

Although supersonic flow downstream of a nozzle requires a sonic throat, the opposite is not true: a compressible gas can pass through a throat section without becoming sonic.

Fig. 9.6 From Eq. (9.40), in flow through a throat (a) the fluid can accelerate smoothly through sonic and supersonic flow. In flow through the bulge (b) the flow at the bulge cannot be sonic on physical grounds.

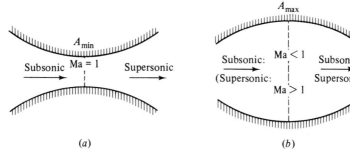

(a) $\qquad$ (b)

Perfect-Gas Relations

We can use the perfect-gas and isentropic-flow relations to convert the continuity relation (9.37) into an algebraic expression involving only area and Mach number, as follows. Equate the mass flow at any section to the mass flow under sonic conditions (which may not actually occur in the duct)

$$\rho V A = \rho^* V^* A^*$$

or

$$\frac{A}{A^*} = \frac{\rho^*}{\rho} \frac{V^*}{V} \qquad (9.41)$$

Both the terms on the right are functions only of Mach number for isentropic flow. From Eqs. (9.28) and (9.32)

$$\frac{\rho^*}{\rho} = \frac{\rho^*}{\rho_0} \frac{\rho_0}{\rho} = \left\{ \frac{2}{\gamma + 1} [1 + \tfrac{1}{2}(\gamma - 1)\, \mathrm{Ma}^2] \right\}^{1/(\gamma - 1)} \qquad (9.42)$$

From Eqs. (9.26) and (9.32) we obtain

$$\frac{V^*}{V} = \frac{(\gamma R T^*)^{1/2}}{V} = \frac{(\gamma R T)^{1/2}}{V} \left(\frac{T^*}{T_0}\right)^{1/2} \left(\frac{T_0}{T}\right)^{1/2}$$

$$= \frac{1}{\mathrm{Ma}} \left\{ \frac{2}{\gamma + 1} [1 + \tfrac{1}{2}(\gamma - 1)\, \mathrm{Ma}^2] \right\}^{1/2} \qquad (9.43)$$

Combining Eqs. (9.41) to (9.43) we get the desired result

$$\frac{A}{A^*} = \frac{1}{\mathrm{Ma}} \left[\frac{1 + \tfrac{1}{2}(\gamma - 1)\, \mathrm{Ma}^2}{\tfrac{1}{2}(\gamma + 1)} \right]^{(1/2)(\gamma + 1)/(\gamma - 1)} \qquad (9.44)$$

For $\gamma = 1.4$ Eq. (9.44) takes the numerical form

$$\frac{A}{A^*} = \frac{1}{\mathrm{Ma}} \frac{(1 + 0.2\, \mathrm{Ma}^2)^3}{1.728} \qquad (9.45)$$

which is plotted in Fig. 9.7. Equations (9.45) and (9.34) enable us to solve any one-dimensional isentropic airflow problem given, say, the shape of the duct $A(x)$ and the stagnation conditions and assuming that there are no shock waves in the duct.

Figure 9.7 shows that the minimum area which can occur in a given isentropic duct flow is the sonic, or critical, throat area. All other duct sections must have A greater than A^*. In many flows a critical sonic throat is not actually present, and the flow in the duct is either entirely subsonic or, more rarely, entirely supersonic.

Choking

From Eq. (9.41) the inverse ratio A^*/A equals $\rho V/\rho^* V^*$, the mass flow per unit area at any section compared with the critical mass flow per unit area. From Fig. 9.7 this inverse ratio rises from zero at $\mathrm{Ma} = 0$ to unity at $\mathrm{Ma} = 1$ and back down to zero at large Ma. Thus, for given stagnation conditions, the maximum possible mass flow passes through a duct when its throat is at the critical or sonic condition. The duct is then said to be *choked* and can carry no additional mass flow unless the throat is widened. If the throat is constricted further, the mass flow through the duct must decrease.

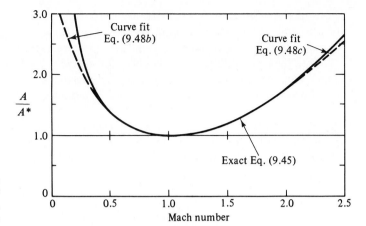

Fig. 9.7 Area ratio versus Mach number for isentropic flow of a perfect gas with $\gamma = 1.4$.

From Eqs. (9.32) and (9.33) the maximum mass flow is

$$\dot{m}_{max} = \rho^* A^* V^* = \rho_0 \left(\frac{2}{\gamma + 1}\right)^{1/(\gamma - 1)} A^* \left(\frac{2\gamma}{\gamma + 1} R T_0\right)^{1/2}$$

$$= \gamma^{1/2} \left(\frac{2}{\gamma + 1}\right)^{(1/2)(\gamma + 1)/(\gamma - 1)} A^* \rho_0 (R T_0)^{1/2} \tag{9.46}$$

For $\gamma = 1.4$ this reduces to

$$\dot{m}_{max} = 0.6847 A^* \rho_0 (R T_0)^{1/2} = \frac{0.6847 p_0 A^*}{(R T_0)^{1/2}} \tag{9.47}$$

For isentropic flow through a duct, the maximum mass flow possible is proportional to the throat area and stagnation pressure and inversely proportional to the square root of the stagnation temperature. These are somewhat abstract facts, so let us illustrate with some examples.

The only cumbersome algebra in these problems is the inversion of Eq. (9.45) to compute the Mach number when A/A^* is known. The writer cannot do it without iteration, but maybe you can think of a way. Meanwhile the following curve-fit formulas are suggested; they estimate the Mach number from A/A^* with ± 2 percent for $\gamma = 1.4$ if you stay within the ranges listed for each formula:

$$\text{Ma} \approx \begin{cases} \dfrac{1 + 0.27(A/A^*)^{-2}}{1.728(A/A^*)} & 1.34 < \dfrac{A}{A^*} < \infty \qquad (9.48a) \\[2pt] & \text{subsonic flow} \\[6pt] 1 - 0.88\left(\ln \dfrac{A}{A^*}\right)^{0.45} & 1.0 < \dfrac{A}{A^*} < 1.34 \qquad (9.48b) \\[10pt] 1 + 1.2\left(\dfrac{A}{A^*} - 1\right)^{1/2} & 1.0 < \dfrac{A}{A^*} < 2.9 \qquad (9.48c) \\[4pt] & \text{supersonic flow} \\[6pt] \left[216 \dfrac{A}{A^*} - 254\left(\dfrac{A}{A^*}\right)^{2/3}\right]^{1/5} & 2.9 < \dfrac{A}{A^*} < \infty \qquad (9.48d) \end{cases}$$

Formulas (9.48a) and (9.48d) are asymptotically correct as $A/A^* \to \infty$, while (9.48b) and (9.48c) are just curve fits. However, formulas (9.48b) and (9.48c) are seen in Fig. 9.7 to be accurate within their recommended ranges.

Note that two solutions are possible for a given A/A^*, one subsonic and one supersonic. The proper solution cannot be selected without further information, e.g., known pressure or temperature at the given duct section.

EXAMPLE 9.4 Air flows isentropically through a duct. At section 1 the area is 1 ft^2 and $V_1 = 600\text{ ft/s}$, $p_1 = 12,000\text{ lbf/ft}^2$, and $T_1 = 850°\text{R}$. Compute (a) T_0, (b) Ma_1, (c) p_0, and (d) A^*. If $A_2 = 0.75\text{ ft}^2$, compute Ma_2 and p_2 if V_2 is (e) subsonic or (f) supersonic.

Solution With V_1 and T_1 known, the energy equation (9.23) gives

$$T_0 = T_1 + \frac{V_1^2}{2c_p} = 850 + \frac{(600)^2}{2(6010)} = 850 + 30 = 880°\text{R} \qquad Ans.\ (a)$$

The sound speed $a_1 \approx 49(850)^{1/2} = 1429\text{ ft/s}$; hence

$$\text{Ma}_1 = \frac{V_1}{a_1} = \frac{600}{1429} = 0.42 \qquad Ans.\ (b)$$

With Ma_1 known

$$\frac{p_0}{p_1} = (1 + 0.2\,\text{Ma}_1^2)^{3.5} = 1.129$$

Hence $\qquad\qquad p_0 = 1.129 p_1 = 1.129(12,000) = 13,550\text{ lbf/ft}^2 \qquad Ans.\ (c)$

Similarly, from Eq. (9.45),

$$\frac{A_1}{A^*} = \frac{[1 + 0.2(0.42)^2]^3}{1.728(0.42)} = 1.529$$

Hence $\qquad\qquad A^* = \frac{A_1}{1.529} = \frac{1\text{ ft}^2}{1.529} = 0.654\text{ ft}^2 \qquad Ans.\ (d)$

This throat must actually be present in the duct to expand the subsonic Ma_1 to supersonic flow downstream.

Given $A_2 = 0.75\text{ ft}^2$, we can compute $A_2/A^* = 0.75/0.654 = 1.147$. For subsonic flow, use Eq. (9.48b) to estimate

$$\text{Ma}_2 \approx 1 - 0.88(\ln 1.147)^{0.45} = 0.64 \qquad Ans.\ (e)$$

whence $\qquad\qquad p_2 = \frac{p_0}{[1 + 0.2(0.64)^2]^{3.5}} = \frac{13,550}{1.317} = 10,300\text{ lbf/ft}^2 \qquad Ans.\ (e)$

This is quite accurate: an exact iteration of Eq. (9.45) would yield $\text{Ma}_2 = 0.6381$ and $p_2 = 10,302\text{ lbf/ft}^2$.

For case (e) there is *no* sonic throat; i.e., the duct area has not decreased sufficiently to create supersonic flow. It may possibly do this further downstream, as illustrated in Fig. E9.4. If, on the other hand, the flow at section 2 is supersonic, we use Eq. (9.48c) to estimate

$$\text{Ma}_2 \approx 1 + 1.2(1.147 - 1)^{1/2} = 1.46 \qquad Ans.\ (f)$$

whence $\qquad\qquad p_2 = \frac{13,550}{[1 + 0.2(1.46)^2]^{3.5}} = \frac{13,550}{3.46} = 3910\text{ lbf/ft}^2 \qquad Ans.\ (f)$

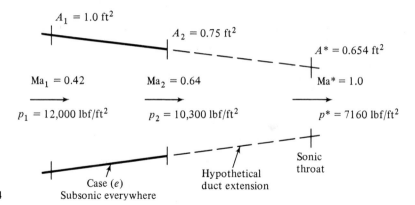

$A_1 = 1.0 \text{ ft}^2$

$A_2 = 0.75 \text{ ft}^2$

$A^* = 0.654 \text{ ft}^2$

$\text{Ma}_1 = 0.42$ $\text{Ma}_2 = 0.64$ $\text{Ma}^* = 1.0$

$p_1 = 12,000 \text{ lbf/ft}^2$ $p_2 = 10,300 \text{ lbf/ft}^2$ $p^* = 7160 \text{ lbf/ft}^2$

Sonic throat

Hypothetical duct extension

Case (e)
Subsonic everywhere

Fig. E9.4

These answers are accurate within less than 1 percent. Note that the supersonic-flow pressure level is much less than the subsonic-flow condition at section 2 for the same duct area, and a sonic throat ($A^* = 0.654 \text{ ft}^2$) *must* have occurred between sections 1 and 2.

EXAMPLE 9.5

It is desired to expand air from $p_0 = 200 \text{ kPa}$ and $T_0 = 500 \text{ K}$ through a throat to an exit Mach number of 2.5. If the desired mass flow is 3 kg/s, compute (a) the throat area and the exit (b) pressure, (c) temperature, (d) velocity, and (e) area, assuming isentropic flow, with $\gamma = 1.4$.

Solution

The throat area follows from Eq. (9.47), because the throat flow must be sonic to produce a supersonic exit:

$$A^* = \frac{\dot{m}(RT_0)^{1/2}}{0.6847 p_0} = \frac{3.0[287(500)]^{1/2}}{0.6847(200,000)} = 0.00830 \text{ m}^2 = \tfrac{1}{4}\pi D^{*2}$$

or
$$D_{\text{throat}} = 10.3 \text{ cm} \qquad \qquad Ans. \ (a)$$

With the exit Mach number known, the isentropic-flow relations give pressure and temperature:

$$p_e = \frac{p_0}{[1 + 0.2(2.5)^2]^{3.5}} = \frac{200,000}{17.08} = 11,700 \text{ Pa} \qquad Ans. \ (b)$$

$$T_e = \frac{T_0}{[1 + 0.2(2.5)^2]} = \frac{500}{2.25} = 222 \text{ K} \qquad Ans. \ (c)$$

The exit velocity follows from the known Mach number and temperature

$$V_e = \text{Ma}_e(\gamma R T_e)^{1/2} = 2.5[1.4(287)(222)]^{1/2} = 2.5(299 \text{ m/s}) = 747 \text{ m/s} \qquad Ans. \ (d)$$

The exit area follows from the known throat area and exit Mach number and Eq. (9.45):

$$\frac{A_e}{A^*} = \frac{[1 + 0.2(2.5)^2]^3}{1.728(2.5)} = 2.64$$

or
$$A_e = 2.64 A^* = 2.64(0.0083 \text{ m}^2) = 0.0219 \text{ m}^2 = \tfrac{1}{4}\pi D_e^2$$

or
$$D_e = 16.7 \text{ cm} \qquad \qquad Ans. \ (e)$$

One point might be noted: the computation of throat area A^* did not depend in any way on the numerical value of the exit Mach number. The exit was supersonic; therefore the throat is sonic and choked, and no further information is needed.

9.5 THE NORMAL-SHOCK WAVE

A common irreversibility occurring in supersonic internal or external flows is the normal-shock wave sketched in Fig. 9.8. Except at near-vacuum pressures such shock waves are very thin (a few micrometers thick) and approximate a discontinuous change in flow properties. We select a control volume just before and after the wave, as in Fig. 9.8.

The analysis is identical to that of Fig. 9.1; that is, a shock wave is a fixed strong pressure wave. To compute all property changes rather than just the wave speed, we use all our basic one-dimensional steady-flow relations, letting section 1 be upstream and section 2 be downstream:

Continuity:
$$\rho_1 V_1 = \rho_2 V_2 = G = \text{const} \tag{9.49a}$$

x momentum:
$$p_1 - p_2 = \rho_2 V_2^2 - \rho_1 V_1^2 \tag{9.49b}$$

Energy:
$$h_1 + \tfrac{1}{2}V_1^2 = h_2 + \tfrac{1}{2}V_2^2 = h_0 = \text{const} \tag{9.49c}$$

Perfect gas:
$$\frac{p_1}{\rho_1 T_1} = \frac{p_2}{\rho_2 T_2} \tag{9.49d}$$

Constant c_p:
$$h = c_p T \qquad \gamma = \text{const} \tag{9.49e}$$

Note that we have canceled out the areas $A_1 \approx A_2$, which is justified even in a variable duct section because of the thinness of the wave. The first successful analyses of these normal shock relations are credited to W. J. M. Rankine (1870) and to A. Hugoniot (1887), hence the modern term *Rankine-Hugoniot relations*. If

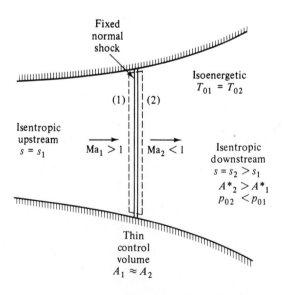

Fig. 9.8 Flow through a fixed normal-shock wave.

we assume that the upstream conditions (p_1, V_1, ρ_1, h_1, T_1) are known, Eqs. (9.49) are five algebraic relations in the five unknowns (p_2, V_2, ρ_2, h_2, T_2). Because of the velocity-squared term, two solutions are found and the correct one is determined from the second law of thermodynamics, which requires that $s_2 > s_1$.

The velocities V_1 and V_2 can be eliminated from Eqs. (9.49a) to (9.49c) to obtain the Rankine-Hugoniot relation

$$h_2 - h_1 = \tfrac{1}{2}(p_2 - p_1)\left(\frac{1}{\rho_2} + \frac{1}{\rho_1}\right) \tag{9.50}$$

This contains only thermodynamic properties and is independent of the equation of state. Introducing the perfect-gas law $h = c_p T = \gamma p/(\gamma - 1)\rho$, we can rewrite this as

$$\frac{\rho_2}{\rho_1} = \frac{1 + \beta p_2/p_1}{\beta + p_2/p_1} \qquad \beta = \frac{\gamma + 1}{\gamma - 1} \tag{9.51}$$

We can compare this with the isentropic-flow relation for a very weak pressure wave in a perfect gas

$$\frac{\rho_2}{\rho_1} = \left(\frac{p_2}{p_1}\right)^{1/\gamma} \tag{9.52}$$

Also, the actual change in entropy across the shock can be computed from the perfect-gas relation

$$\frac{s_2 - s_1}{c_v} = \ln\left[\frac{p_2}{p_1}\left(\frac{\rho_1}{\rho_2}\right)^{\gamma}\right] \tag{9.53}$$

Assuming a given wave strength p_2/p_1, we can compute the density ratio and the entropy change and list them as follows for $\gamma = 1.4$:

$\dfrac{p_2}{p_1}$	ρ_2/ρ_1		$\dfrac{s_2 - s_1}{c_v}$
	Eq. (9.51)	Isentropic	
0.5	0.6154	0.6095	−0.0134
0.9	0.9275	0.9275	−0.00005
1.0	1.0	1.0	0.0
1.1	1.00704	1.00705	0.00004
1.5	1.3333	1.3359	0.0027
2.0	1.6250	1.6407	0.0134

We see that the entropy change is negative if the pressure decreases across the shock, which violates the second law. Thus a rarefaction shock is impossible in a perfect gas.[1] We see also that weak-shock waves ($p_2/p_1 \leq 2.0$) are very nearly isentropic.

---

[1] This is true also for most real gases; see Ref. 14, sec. 7.3.

Mach-Number Relations

For a perfect gas all the property ratios across the normal shock are unique functions of γ and the upstream Mach number Ma_1. For example, if we eliminate ρ_2 and V_2 from Eqs. (9.49a) to (9.49c) and introduce $h = \gamma p/(\gamma - 1)\rho$, we obtain

$$\frac{p_2}{p_1} = \frac{1}{\gamma + 1}\left[\frac{2\rho_1 V_1^2}{p_1} - (\gamma - 1)\right] \tag{9.54}$$

But for a perfect gas $\rho_1 V_1^2/p_1 = \gamma V_1^2/\gamma RT_1 = \gamma\,Ma_1^2$, so that Eq. (9.54) is equivalent to

$$\frac{p_2}{p_1} = \frac{1}{\gamma + 1}[2\gamma\,Ma_1^2 - (\gamma - 1)] \tag{9.55}$$

From this equation we see that, for any γ, $p_2 > p_1$ only if $Ma_1 > 1.0$. Thus for flow through a normal-shock wave the upstream Mach number must be supersonic to satisfy the second law of thermodynamics.

What about the downstream Mach number? From the perfect-gas identity $\rho V^2 = \gamma p\,Ma^2$, we can rewrite Eq. (9.49b) as

$$\frac{p_2}{p_1} = \frac{1 + \gamma\,Ma_1^2}{1 + \gamma\,Ma_2^2} \tag{9.56}$$

which relates the pressure ratio to both Mach numbers. By equating Eqs. (9.55) and (9.56) we can solve for

$$Ma_2^2 = \frac{(\gamma - 1)\,Ma_1^2 + 2}{2\gamma\,Ma_1^2 - (\gamma - 1)} \tag{9.57}$$

Since Ma_1 must be supersonic, this equation predicts for all $\gamma > 1$ that Ma_2 must be subsonic. Thus a normal-shock wave decelerates a flow almost discontinuously from supersonic to subsonic conditions.

Further manipulation of the basic relations (9.49) for a perfect gas gives additional equations relating the change in properties across a normal-shock wave in a perfect gas

$$\frac{\rho_2}{\rho_1} = \frac{(\gamma + 1)\,Ma_1^2}{(\gamma - 1)\,Ma_1^2 + 2} = \frac{V_1}{V_2}$$

$$\frac{T_2}{T_1} = [2 + (\gamma - 1)\,Ma_1^2]\frac{2\gamma\,Ma_1^2 - (\gamma - 1)}{(\gamma + 1)^2\,Ma_1^2} \tag{9.58}$$

$$T_{02} = T_{01}$$

$$\frac{p_{02}}{p_{01}} = \frac{\rho_{02}}{\rho_{01}} = \left[\frac{(\gamma + 1)\,Ma_1^2}{2 + (\gamma - 1)\,Ma_1^2}\right]^{\gamma/(\gamma-1)}\left[\frac{\gamma + 1}{2\gamma\,Ma_1^2 - (\gamma - 1)}\right]^{1/(\gamma-1)}$$

Of additional interest is the fact that the critical, or sonic, throat area A^* in a duct increases across a normal shock

$$\frac{A_2^*}{A_1^*} = \frac{Ma_2}{Ma_1}\left[\frac{2 + (\gamma - 1)\,Ma_1^2}{2 + (\gamma - 1)\,Ma_2^2}\right]^{(1/2)(\gamma+1)/(\gamma-1)} \tag{9.59}$$

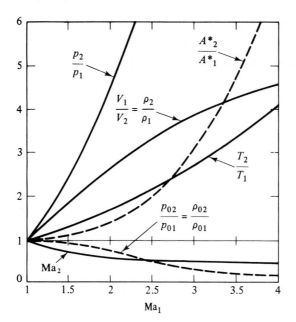

Fig. 9.9 Change in flow properties across a normal-shock wave for $\gamma = 1.4$

All these relations are given in Table B.2 and plotted versus upstream Mach number Ma_1 in Fig. 9.9 for $\gamma = 1.4$. We see that pressure increases greatly while temperature and density increase moderately. The effective throat area A^* increases slowly at first and then rapidly. The failure of students to account for this change in A^* is a common source of error in shock calculations.

The stagnation temperature remains the same, but the stagnation pressure and density decrease in the same ratio; i.e., the flow across the shock is adiabatic but nonisentropic. Other basic principles governing the behavior of shock waves can be summarized as follows:

1. The upstream flow is supersonic and the downstream flow subsonic.
2. For perfect gases (and also for real fluids except under bizarre thermodynamic conditions) rarefaction shocks are impossible, and only a compression shock can exist.
3. The entropy increases across a shock with consequent decreases in stagnation pressure and stagnation density and an increase in the effective sonic-throat area.
4. Weak shock waves are very nearly isentropic.

Normal-shock waves form in ducts under transient conditions, e.g., shock tubes, and also in steady flow for certain ranges of the downstream pressure. Figure 9.10a shows a normal shock in a supersonic nozzle. Flow is from left to right. The oblique wave pattern to the left is formed by roughness elements on the nozzle walls and indicates that the upstream flow is supersonic. Note the absence of these Mach waves (see Sec. 9.10) in the subsonic flow downstream.

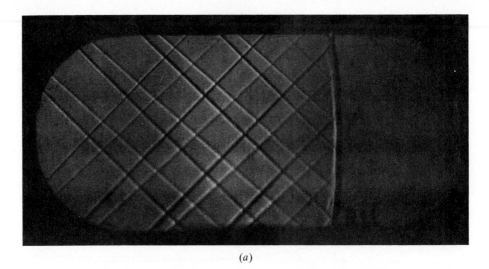

(a)

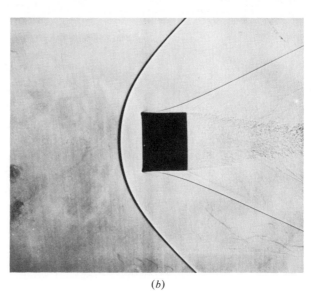

(b)

Fig. 9.10 Normal shocks form in both internal and external flows: (a) normal shock in a duct; note the Mach-wave pattern to the left (upstream), indicating supersonic flow (*courtesy of U.S. Air Force Arnold Engineering Development Center*). (b) Supersonic flow past a blunt body creates a normal shock at the nose; the apparent shock thickness and body-corner curvature are optical distortions (*courtesy of U.S. army Ballistic Research Laboratory, Aberdeen Proving Ground*).

Normal-shock waves occur not only in supersonic duct flows but also in a variety of supersonic external flows. An example is the supersonic flow past a blunt body shown in Fig. 9.10b. The bow shock is curved, with a portion in front of the body which is essentially normal to the oncoming flow. This normal portion of the bow shock satisfies the property-change conditions just as outlined in this section. The flow inside the shock near the body nose is thus subsonic and at relatively high temperature $T_2 > T_1$, so that convective heat transfer is especially high in this region.

Each nonnormal portion of the bow shock in Fig. 9.10b satisfies the oblique-shock relations to be outlined in Sec. 9.9. Note also the oblique recompression shock on the sides of the body. What has happened is that the subsonic nose flow has accelerated around the corners back to supersonic flow at low pressure which must then pass through the second shock to match the higher downstream pressure conditions.

Note the fine-grained turbulent wake structure in the rear of the body in Fig. 9.10b. The turbulent boundary layer along the sides of the body is also clearly visible.

The analysis of a complex multidimensional supersonic flow such as in Fig. 9.10 is beyond the scope of this book. For further information see, for example, Ref. 14, chap. 9; or Ref. 8, chap. 16.

Moving Normal Shocks

The preceding analysis of the fixed shock applies equally well to the moving shock if we reverse the transformation used in Fig. 9.1. To make the upstream conditions simulate a still fluid, we move the shock of Fig. 9.8 to the left at speed V_1; that is, we fix our coordinates to a control volume moving with the shock. The downstream flow then appears to move to the left at a slower speed $V_1 - V_2$ following the shock. The thermodynamic properties are not changed by this transformation, so that all our Eqs. (9.50) to (9.59) are still valid.

EXAMPLE 9.6 Air flows from a reservoir where $p = 300$ kPa and $T = 500$ K through a throat to section 1 in Fig. E9.6, where there is a normal-shock wave. Compute (a) p_1, (b) p_2, (c) p_{02}, (d) A_2^*, (e) p_{03}, (f) A_3^*, (g) p_3, (h) T_{03}, and (i) T_3.

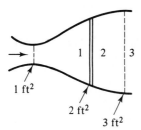

Fig. E9.6

Solution The reservoir conditions are the stagnation properties, which, for assumed one-dimensional adiabatic frictionless flow, hold through the throat up to section 1

$$p_{01} = 300 \text{ kPa} \qquad T_{01} = 500 \text{ K}$$

A shock wave cannot exist unless Ma_1 is supersonic; therefore the flow must have accelerated through a throat which is sonic

$$A_t = A_1^* = 1 \text{ ft}^2$$

We can now find the Mach number Ma_1 from the known isentropic area ratio

$$\frac{A_1}{A_1^*} = \frac{2 \text{ ft}^2}{1 \text{ ft}^2} = 2.0$$

From Eq. (9.48c)

$$\text{Ma}_1 \approx 1 + 1.2(2.0 - 1)^{1/2} = 2.20$$

Further iteration with Eq. (9.45) would give $\text{Ma}_1 = 2.1972$, showing that Eq. (9.48c) gives satisfactory accuracy. The pressure p_1 follows from the isentropic relation (9.28) (or Table B.1)

$$\frac{p_{01}}{p_1} = [1 + 0.2(2.20)^2]^{3.5} = 10.7$$

or

$$p_1 = \frac{300 \text{ kPa}}{10.7} = 28.06 \text{ kPa} \qquad\qquad Ans. (a)$$

The pressure p_2 is now obtained from Ma_1 and the normal-shock relation (9.55) or Table B.2

$$\frac{p_2}{p_1} = \frac{1}{2.4}[2.8(2.20)^2 - 0.4] = 5.48$$

or

$$p_2 = 5.48(28.06) = 154 \text{ kPa} \qquad\qquad Ans. (b)$$

In similar manner, for $\text{Ma}_1 = 2.20$, $p_{02}/p_{01} = 0.628$ from Eq. (9.58) and $A_2^*/A_1^* = 1.592$ from Eq. (9.59) or we can read Table B.2 for these values. Thus

$$p_{02} = 0.628(300 \text{ kPa}) = 188 \text{ kPa} \qquad\qquad Ans. (c)$$
$$A_2^* = 1.592(1 \text{ ft}^2) = 1.592 \text{ ft}^2 \qquad\qquad Ans. (d)$$

The flow from section 2 to 3 is isentropic (but at higher entropy than the flow upstream of the shock). Thus

$$p_{03} = p_{02} = 188 \text{ kPa} \qquad\qquad Ans. (e)$$
$$A_3^* = A_2^* = 1.592 \text{ ft}^2 \qquad\qquad Ans. (f)$$

Knowing A_3^*, we can now compute p_3 by finding Ma_3 and without bothering to find Ma_2 (which happens to equal 0.547). The area ratio at section 3 is

$$\frac{A_3}{A_3^*} = \frac{3 \text{ ft}^2}{1.592 \text{ ft}^2} = 1.884$$

Then, since Ma_3 is known to be subsonic because it is downstream of a normal shock, we use Eq. (9.48a) to estimate

$$\text{Ma}_3 \approx \frac{1 + 0.27/(1.884)^2}{1.728(1.884)} = 0.330$$

The pressure p_3 then follows from the isentropic relation (9.28) or Table B.1

$$\frac{p_{03}}{p_3} = [1 + 0.2(0.330)^2]^{3.5} = 1.078$$

or

$$p_3 = \frac{188 \text{ kPa}}{1.078} = 174 \text{ kPa} \qquad\qquad Ans. (g)$$

Meanwhile, the flow is adiabatic throughout the duct; thus

$$T_{01} = T_{02} = T_{03} = 500 \text{ K} \qquad\qquad Ans. (h)$$

Therefore, finally, from the adiabatic relation (9.26)

$$\frac{T_{03}}{T_3} = 1 + 0.2(0.330)^2 = 1.022$$

or $$T_3 = \frac{500 \text{ K}}{1.022} = 489 \text{ K} \qquad \textit{Ans. (i)}$$

Notice that this type of duct-flow problem, with or without a shock wave, requires straight-forward application of algebraic perfect-gas relations coupled with a little thought given to which formula is appropriate for the particular situation.

EXAMPLE 9.7 An explosion in air, $\gamma = 1.4$, creates a spherical shock wave propagating radially into still air at standard conditions. At the instant shown in Fig. E9.7 the pressure just inside the shock is 200 lbf/in^2 absolute. Estimate (a) the shock speed C and (b) the air velocity V just inside the shock.

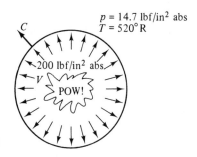

Fig. E9.7

Solution

Part (a) In spite of the spherical geometry the flow across the shock moves normal to the spherical wavefront; hence the normal-shock relations (9.50) to (9.59) apply. Fixing our control volume to the moving shock, we find that the proper conditions to use in Fig. 9.8 are

$$C = V_1 \qquad p_1 = 14.7 \text{ lbf/in}^2 \text{ absolute} \qquad T_1 = 520°\text{R}$$
$$V = V_1 - V_2 \qquad p_2 = 200 \text{ lbf/in}^2 \text{ absolute}$$

The speed of sound outside the shock is $a_1 \approx 49T_1^{1/2} = 1117$ ft/s. We can find Ma_1 from the known pressure ratio across the shock

$$\frac{p_2}{p_1} = \frac{200 \text{ lbf/in}^2 \text{ absolute}}{14.7 \text{ lbf/in}^2 \text{ absolute}} = 13.61$$

From Eq. (9.55) or Table B.2

$$13.61 = \frac{1}{2.4}(2.8 \text{ Ma}_1^2 - 0.4) \qquad \text{or} \qquad \text{Ma}_1 = 3.436$$

Then, by definition of the Mach number,

$$C = V_1 = \text{Ma}_1 a_1 = 3.436(1117 \text{ ft/s}) = 3840 \text{ ft/s} \qquad \textit{Ans. (a)}$$

Part (b) To find V_2, we need the temperature or sound speed inside the shock. Since Ma_1 is known, from Eq. (9.58) or Table B.2 for $Ma_1 = 3.436$ we compute $T_2/T_1 = 3.228$. Then

$$T_2 = 3.228T_1 = 3.228(520°R) = 1679°R$$

At such a high temperature we should account for non-perfect-gas effects or at least use the gas tables [16], but we won't. Here just estimate from the perfect-gas energy equation (9.23) that

$$V_2^2 = 2c_p(T_1 - T_2) + V_1^2 = 2(6010)(520 - 1679) + (3840)^2 = 815,000$$

or

$$V_2 \approx 903 \text{ ft/s}$$

Notice that we did this without bothering to compute Ma_2, which equals 0.454, or $a_2 \approx 49T_2^{1/2} = 2000$ ft/s.

Finally, the air velocity behind the shock is

$$V = V_1 - V_2 = 3840 - 903 \approx 2940 \text{ ft/s} \qquad \qquad Ans. (b)$$

Thus a powerful explosion creates a brief but intense blast wind as it passes.[1]

9.6 OPERATION OF CONVERGING AND DIVERGING NOZZLES

By combining the isentropic-flow and normal-shock relations plus the concept of sonic throat choking, we can outline the characteristics of converging and diverging nozzles.

Converging Nozzle

First consider the converging nozzle sketched in Fig. 9.11a. There is an upstream reservoir at stagnation pressure p_0. The flow is induced by lowering the downstream outside, or *back*, pressure p_b below p_0, resulting in the sequence of states a to e shown in Fig. 9.11b and c.

For a moderate drop in p_b to states a and b, the throat pressure is higher than the critical value p^* which would make the throat sonic. The flow in the nozzle is subsonic throughout, and the jet exit pressure p_e equals the back pressure p_b. The mass flow is predicted by subsonic isentropic theory and is less than the critical value $\dot{m}_{max}$, as shown in Fig. 9.11c.

For condition c, the back pressure exactly equals the critical pressure p^* of the throat. The throat becomes sonic, the jet exit flow is sonic, $p_e = p_b$, and the mass flow equals its maximum value from Eq. (9.46). The flow upstream of the throat is subsonic everywhere and predicted by isentropic theory based on the local area ratio $A(x)/A^*$ and Table B.1.

Finally, if p_b is lowered further to conditions d or e below p^*, the nozzle cannot respond further because it is choked at its maximum throat mass flow. The throat remains sonic with $p_e = p^*$, and the nozzle-pressure distribution is the same as in

---

[1] This is the principle of the *shock-tube wind tunnel*, in which a controlled explosion creates a brief flow at very high Mach number, with data taken by fast-response instruments. See, for example, Ref. 5, sec. 4.5.

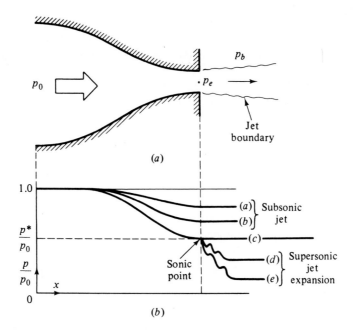

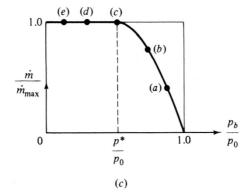

Fig. 9.11   Operation of a converging nozzle: (*a*) nozzle geometry showing characteristic pressures; (*b*) pressure distribution caused by various back pressures; (*c*) mass flow versus back pressure.

state *c*, as sketched in Fig. 9.11*b*. The exit jet expands supersonically so that the jet pressure can be reduced from p^* down to p_b. The jet structure is complex and multidimensional and not shown here. Being supersonic, the jet cannot send any signal upstream to influence the choked flow conditions in the nozzle.

If the stagnation plenum chamber is large or supplemented by a compressor, and if the discharge chamber is large or supplemented by a vacuum pump, the converging-nozzle flow will be steady or nearly so. Otherwise the nozzle will be blowing down, with p_0 decreasing and p_b increasing, and the flow states will be changing from, say, state *e* backward to state *a*. Blowdown calculations are usually made by a quasi-steady analysis based on isentropic steady-flow theory for the instantaneous pressures $p_0(t)$ and $p_b(t)$.

EXAMPLE 9.8   A converging nozzle has a throat area of 1 in² and stagnation air conditions of 120 lbf/in² absolute and 600°R. Compute the exit pressure and mass flow if the back pressure is (a) 90 lbf/in² absolute and (b) 45 lbf/in² absolute. Assume $\gamma = 1.4$.

Solution   We are given

$$p_0 = 120 \text{ lbf/in}^2 \text{ absolute} = 17{,}280 \text{ lbf/ft}^2 \qquad T_0 = 600°R$$

Hence
$$\rho_0 = \frac{p_0}{RT_0} = \frac{17{,}280}{1717(600)} = 0.0168 \text{ slug/ft}^3$$

From Eq. (9.32) the critical throat pressure is

$$\frac{p^*}{p_0} = 0.5283 \qquad \text{or} \qquad p^* = 63.4 \text{ lbf/in}^2 \text{ absolute} = 9130 \text{ lbf/ft}^2$$

Part (a)   Since $p_b = 90$ lbf/in² absolute $> p^*$, the flow is subsonic throughout and similar to condition *b* in Fig. 9.11*b*. The throat Mach number is found from $p_e = p_b$ and Eq. (9.35) or Table B.1

$$\text{Ma}_e^2 = 5[(\tfrac{120}{90})^{2/7} - 1] = 0.4283 \qquad \text{Ma}_e = 0.6545$$

From Ma_e and the isentropic relations (9.34) we can compute the throat temperature and density

$$\frac{T_0}{T_e} = 1 + 0.2(0.6545)^2 = 1.0857 \qquad \frac{\rho_0}{\rho_e} = (1.0857)^{2.5} = 1.2281$$

Hence $T_e = 600/1.0857 = 553°R$, so that $a_e \approx 49(553)^{1/2} = 1152$ ft/s. $\rho_e = 0.0168/1.2281 = 0.0137$ slug/ft³. The mass flow is thus given by

$$\dot{m} = \rho_e A_e V_e = \rho_e A_e (\text{Ma}_e a_e) = 0.0137(\tfrac{1}{144} \text{ ft}^2)[0.6545(1152)] = 0.0715 \text{ slug/s} \quad \textit{Ans. (a)}$$

The exit pressure equals the back pressure:

$$p_e = p_b = 90 \text{ lbf/in}^2 \text{ absolute} \qquad\qquad \textit{Ans. (a)}$$

Part (b)   Since $p_b = 45$ lbf/in² absolute $< p^*$, the throat is choked similar to condition *d* in Fig. 9.11*b*. The exit pressure is sonic

$$p_e = p^* = 63.4 \text{ lbf/in}^2 \text{ absolute} \qquad\qquad \textit{Ans. (b)}$$

The mass flux is a maximum from Eq. (9.47)

$$\dot{m} = \dot{m}_{max} = \frac{0.6847 p_0 A_t}{(RT_0)^{1/2}} = \frac{0.6847(17{,}280 \text{ lbf/ft}^2)(\tfrac{1}{144} \text{ ft}^2)}{[1717(600)]^{1/2}} = 0.0810 \text{ slug/s} \quad \textit{Ans. (b)}$$

Any back pressure less than or equal to $p^* = 63.4$ lbf/in² absolute would cause this same choked condition. Notice that the 50 percent increase in throat Mach number from 0.6545 to 1.0 increases the mass flux only 13 percent, from 0.0715 to 0.0810 slug/s.

Converging-Diverging Nozzle

Now consider the converging-diverging nozzle sketched in Fig. 9.12*a*. If the back pressure p_b is low enough, there will be supersonic flow in the diverging portion and a variety of shock-wave conditions may occur, which are sketched in Fig. 9.12*b*. Let the back pressure be gradually decreased.

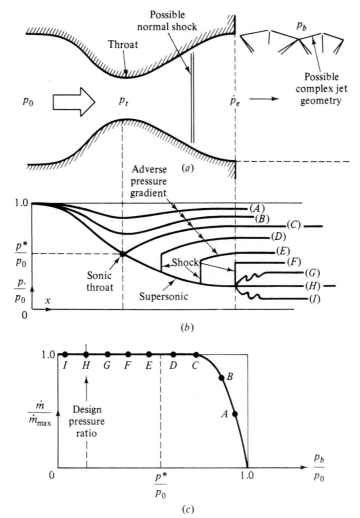

Fig. 9.12 Operation of a converging-diverging nozzle: (*a*) nozzle geometry with possible flow configurations; (*b*) pressure distribution caused by various back pressures; (*c*) mass flow versus back pressure.

For curves A and B in Fig. 9.12b the back pressure is not low enough to induce sonic flow in the throat, and the flow in the nozzle is subsonic throughout. The pressure distribution is computed from subsonic isentropic area-change relations, e.g., Table B.1. The exit pressure $p_e = p_b$, and the jet is subsonic.

For curve C the area ratio A_e/A_t exactly equals the critical ratio A_e/A^* for a subsonic Ma_e in Table B.1. The throat becomes sonic, and mass flux reaches a maximum in Fig. 9.12c. The remainder of the nozzle flow is subsonic, including the exit jet, and $p_e = p_b$.

Now jump for a moment to curve H. here p_b is such that p_b/p_0 exactly corresponds to the critical area ratio A_e/A^* for a *supersonic* Ma_e in Table B.1. The diverging flow is entirely supersonic, including the jet flow, and $p_e = p_b$. This is called the *design pressure ratio* of the nozzle and is the back pressure suitable for operating a supersonic wind tunnel or an efficient rocket exhaust.

Now back up and suppose that p_b lies between curves C and H, which is impossible according to purely isentropic-flow calculations. Then back pressures D to F occur in Fig. 9.12b. The throat remains choked at the sonic value, and we can match $p_e = p_b$ by placing a normal shock at just the right place in the diverging section to cause a *subsonic-diffuser* flow back to the back-pressure condition. The mass flow remains at maximum in Fig. 9.12c. At back pressure F the required normal shock stands in the duct exit. At back pressure G no single normal shock can do the job, and so the flow compresses outside the exit in a complex series of oblique shocks until it matches p_b.

Finally, at back pressure I, p_b is lower than the design pressure H but the nozzle is choked and cannot respond. The exit flow expands in a complex series of supersonic wave motions until it matches the low back pressure. See, for example, Ref. 9, sec. 5.4, for further details of these off-design jet-flow configurations.

Note that for p_b less than back pressure C, there is supersonic flow in the nozzle and the throat can receive no signal from the exit behavior. The flow remains choked, and the throat has no idea what the exit conditions are.

Note also that the normal shock-patching idea is idealized. Downstream of the shock the nozzle flow has an adverse pressure gradient, usually leading to wall boundary-layer separation. Blockage by the greatly thickened separated layer interacts strongly with the core flow (recall Fig. 6.26) and usually induces a series of weak two-dimensional compression shocks rather than a single one-dimensional normal shock (see, for example, Ref. 14, pp. 292–293, for further details).

EXAMPLE 9.9 A converging-diverging nozzle (Fig. 9.12a) has a throat area of 0.002 m^2 and an exit area of 0.008 m^2. Air stagnation conditions are $p_0 = 1000\,\text{kPa}$ and $T_0 = 500\,\text{K}$. Compute the exit pressure and mass flow for (a) design condition and the exit pressure and mass flow if (b) $p_b \approx 300\,\text{kPa}$ and (c) $p_b \approx 900\,\text{kPa}$. Assume $\gamma = 1.4$.

Solution

Part (a) The design condition corresponds to supersonic isentropic flow at the given area ratio $A_e/A_t = 0.008/0.002 = 4.0$. We can find the design Mach number either by iteration of the area-ratio formula (9.45) or by the curve fit (9.48d)

$$\text{Ma}_{e,\text{design}} \approx [216(4.0) - 254(4.0)^{2/3}]^{1/5} \approx 2.95 \qquad (\text{exact} = 2.9402)$$

The accuracy of the curve fit is seen to be satisfactory. The design pressure ratio follows from Eq. (9.34)

$$\frac{p_0}{p_e} = [1 + 0.2(2.95)^2]^{3.5} = 34.1$$

or

$$p_{e,\text{design}} = \frac{1000\,\text{kPa}}{34.1} = 29.3\,\text{kPa} \qquad\qquad\qquad\qquad Ans.\ (a)$$

Since the throat is clearly sonic at design conditions, Eq. (9.47) applies

$$\dot{m}_{\text{design}} = \dot{m}_{\text{max}} = \frac{0.6847 p_0 A_t}{(RT_0)^{1/2}} = \frac{0.6847(10^6\,\text{Pa})(0.002\,\text{m}^2)}{[287(500)]^{1/2}}$$

$$= 3.61\,\text{kg/s} \qquad\qquad\qquad\qquad Ans.\ (a)$$

Part (b)   For $p_b = 300$ kPa we are definitely far below the subsonic isentropic condition C in Fig. 9.12b but we may even be below condition F with a normal shock in the exit, i.e., in condition G, where oblique shocks occur outside the exit plane. If it is condition G, $p_e = p_{e,\text{design}} = 29.3$ kPa because no shock has yet occurred. To find out, compute condition F by assuming an exit normal shock with $\text{Ma}_1 = 2.95$, that is, the design Mach number just upstream of the shock. From Eq. (9.55)

$$\frac{p_2}{p_1} = \frac{1}{2.4}\left[2.8(2.95)^2 - 0.4\right] = 9.99$$

or $$p_2 = 9.99p_1 = 9.99p_{e,\text{design}} = 293 \text{ kPa}$$

Since this is less than the given $p_b = 300$ kPa, there is a normal shock just upstream of the exit plane (condition E). The exit flow is subsonic and equals the back pressure

$$p_e = p_b = 300 \text{ kPa} \qquad\qquad\qquad\qquad \textit{Ans. (b)}$$

Also $$\dot{m} = \dot{m}_{\max} = 3.61 \text{ kg/s} \qquad\qquad\qquad \textit{Ans. (b)}$$

The throat is still sonic and choked at its maximum mass flow.

Part (c)   Finally, for $p_b = 900$ kPa, which is up near condition C, we compute Ma_e and p_e for condition C as a comparison. Again $A_e/A_t = 4.0$ for this condition, with a subsonic Ma_e estimated from the curve-fit Eq. (9.48a):

$$\text{Ma}_e(C) \approx \frac{1 + 0.27/(4.0)^2}{1.728(4.0)} = 0.147 \qquad (\text{exact} = 0.14655)$$

Then the isentropic exit-pressure ratio for this condition is

$$\frac{p_0}{p_e} = \left[1 + 0.2(0.147)^2\right]^{3.5} = 1.0152$$

or $$p_e = \frac{1000}{1.0152} = 985 \text{ kPa}$$

The given back pressure of 900 kPa is less than this value, corresponding roughly to condition D in Fig. 9.12b. Thus for this case there is a normal shock just downstream of the throat and the throat is choked

$$p_e = p_b = 900 \text{ kPa} \qquad \dot{m} = \dot{m}_{\max} = 3.61 \text{ kg/s} \qquad \textit{Ans. (c)}$$

For this large exit-area ratio the exit pressure would have to be larger than 985 kPa to cause a subsonic flow in the throat and a mass flow less than maximum.

9.7   COMPRESSIBLE DUCT FLOW WITH FRICTION[1]

Section 9.4 showed the effect of area change on a compressible flow while neglecting friction and heat transfer. We could now add friction and heat transfer to the area change and consider coupled effects, which is done in advanced texts [e.g., 8, chap. 8]. Instead, as an elementary introduction, this section treats only the

---

[1] This section may be omitted without loss of continuity.

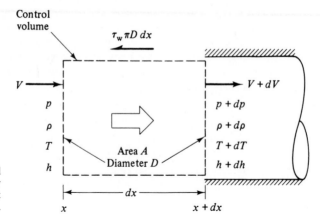

Control volume

$\tau_w \pi D\, dx$

$V \longrightarrow$ $\longrightarrow V + dV$

p $p + dp$

ρ $\rho + d\rho$

T Area A $T + dT$

h Diameter D $h + dh$

$\longleftarrow dx \longrightarrow$

x $x + dx$

Fig. 9.13 Elemental control volume for flow in a constant-area duct with friction.

effect of friction, neglecting area change and heat transfer. The basic assumptions are

1. Steady one-dimensional adiabatic flow
2. Perfect gas with constant specific heats
3. Constant-area straight duct
4. Negligible shaft-work and potential-energy changes
5. Wall shear stress correlated by a Darcy friction factor

In effect, we are studying a Moody-type pipe-friction problem but with large changes in kinetic energy, enthalpy, and pressure in the flow.

Consider the elemental duct control volume of area A and length dx in Fig. 9.13. The area is constant, but other flow properties (p, ρ, T, h, V) may vary with x. Application of the three conservation laws to this control volume gives three differential equations

Continuity:
$$\rho V = \frac{\dot{m}}{A} = G = \text{const}$$

or
$$\frac{d\rho}{\rho} + \frac{dV}{V} = 0 \tag{9.60a}$$

x momentum:
$$pA - (p + dp)A - \tau_w \pi D\, dx = \dot{m}(V + dV - V)$$

or
$$dp + \frac{4\tau_w\, dx}{D} + \rho V\, dV = 0 \tag{9.60b}$$

Energy:
$$h + \tfrac{1}{2}V^2 = h_0 = c_p T_0 = c_p T + \tfrac{1}{2}V^2$$

or
$$c_p\, dT + V\, dV = 0 \tag{9.60c}$$

Since these three equations have five unknowns, p, ρ, T, V, and τ_w, we need two additional relations. One is the perfect-gas law

$$p = \rho R T \qquad \text{or} \qquad \frac{dp}{p} = \frac{d\rho}{\rho} + \frac{dT}{T} \tag{9.61}$$

To eliminate τ_w as an unknown, it is assumed that wall shear is correlated by a local Darcy friction factor f

$$\tau_w = \tfrac{1}{8} f \rho V^2 = \tfrac{1}{8} f \gamma p \, \mathrm{Ma}^2 \tag{9.62}$$

where the last form follows from the perfect-gas speed-of-sound expression $a^2 = \gamma p / \rho$. In practice f can be related to the local Reynolds number and wall roughness from, say, the Moody chart, Fig. 6.13.

Equations (9.60) and (9.61) are first-order differential equations and can be integrated, using friction-factor data, from any inlet section (1), where p_1, T_1, V_1, etc., are known, to determine $p(x)$, $T(x)$, etc., along the duct. It is practically impossible to eliminate all but one variable to give, say, a single differential equation for $p(x)$, but all equations can be written in terms of the Mach number $\mathrm{Ma}(x)$ and the friction factor, using the definition of Mach number

$$V^2 = \mathrm{Ma}^2 \, \gamma R T$$

or

$$\frac{2 \, dV}{V} = \frac{2 \, d\mathrm{Ma}}{\mathrm{Ma}} + \frac{dT}{T} \tag{9.63}$$

Adiabatic Flow

By eliminating variables between Eqs. (9.60) to (9.63), we obtain the working relations

$$\frac{dp}{p} = -\gamma \, \mathrm{Ma}^2 \, \frac{1 + \tfrac{1}{2}(\gamma - 1) \, \mathrm{Ma}^2}{2(1 - \mathrm{Ma}^2)} f \, \frac{dx}{D} \tag{9.64a}$$

$$\frac{d\rho}{\rho} = -\frac{\gamma \, \mathrm{Ma}^2}{2(1 - \mathrm{Ma}^2)} f \, \frac{dx}{D} = -\frac{dV}{V} \tag{9.64b}$$

$$\frac{dp_0}{p_0} = \frac{d\rho_0}{\rho_0} = -\tfrac{1}{2}\gamma \, \mathrm{Ma}^2 f \, \frac{dx}{D} \tag{9.64c}$$

$$\frac{dT}{T} = -\frac{\gamma(\gamma - 1) \, \mathrm{Ma}^4}{2(1 - \mathrm{Ma}^2)} f \, \frac{dx}{D} \tag{9.64d}$$

$$\frac{d \, \mathrm{Ma}^2}{\mathrm{Ma}^2} = \gamma \, \mathrm{Ma}^2 \, \frac{1 + \tfrac{1}{2}(\gamma - 1) \, \mathrm{Ma}^2}{1 - \mathrm{Ma}^2} f \, \frac{dx}{D} \tag{9.64e}$$

All these except dp_0/p_0 have the factor $1 - \mathrm{Ma}^2$ in the denominator, so that, like the area-change formulas in Fig. 9.5, subsonic and supersonic flow have opposite effects:

Property	Subsonic	Supersonic
p	Decreases	Increases
ρ	Decreases	Increases
V	Increases	Decreases
p_0, ρ_0	Decreases	Decreases
T	Decreases	Increases
Ma	Increases	Decreases
Entropy	Increases	Increases

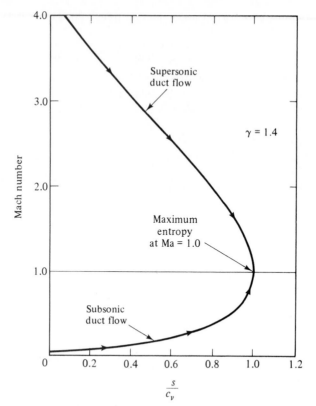

We have added to the list above that entropy must increase along the duct for either subsonic or supersonic flow as a consequence of the second law for adiabatic flow. For the same reason, stagnation pressure and density must both decrease.

The key parameter above is the Mach number. Whether the inlet flow is subsonic or supersonic, the duct Mach number always tends downstream toward Ma = 1 because this is the path along which the entropy increases. If the pressure and density are computed from Eqs. (9.64a) and (9.64b) and the entropy from Eq. (9.53), the result can be plotted in Fig. 9.14 versus Mach number for $\gamma = 1.4$. The maximum entropy occurs at Ma = 1, so that the second law requires that the duct-flow properties continually approach the sonic point. Since p_0 and ρ_0 continually decrease along the duct due to the frictional (nonisentropic) losses, they are not useful as reference properties. Instead, the sonic properties p^*, ρ^*, T^*, p_0^*, and ρ_0^* are the appropriate constant reference quantities in adiabatic duct flow. The theory then computes the ratios p/p^*, T/T^*, etc., as a function of local Mach number and the integrated friction effect.

To derive working formulas, we first attack Eq. (9.64e), which relates Mach number to friction. Separate the variables and integrate,

$$\int_0^{L^*} f \frac{dx}{D} = \int_{\mathrm{Ma}^2}^{1.0} \frac{1 - \mathrm{Ma}^2}{\gamma \, \mathrm{Ma}^4 [1 + \frac{1}{2}(\gamma - 1) \, \mathrm{Ma}^2]} \, d \, \mathrm{Ma}^2 \qquad (9.65)$$

The upper limit is the sonic point, whether or not it is actually reached in the duct flow. The lower limit is arbitrarily placed at the position $x = 0$, where the Mach number is Ma. The result of the integration is

$$\frac{\bar{f}L^*}{D} = \frac{1 - \text{Ma}^2}{\gamma\,\text{Ma}^2} + \frac{\gamma + 1}{2\gamma}\ln\frac{(\gamma + 1)\,\text{Ma}^2}{2 + (\gamma - 1)\,\text{Ma}^2} \tag{9.66}$$

where $\bar{f}$ is the average friction factor between 0 and L^*. In practice, an average f is always assumed, and no attempt is made to account for the slight changes in Reynolds number along the duct. For noncircular ducts D is replaced by the hydraulic diameter $D_h = (4 \times \text{area})/\text{perimeter}$ as in Eq. (6.79).

Equation (9.66) is tabulated versus Mach number in Appendix Table B.3. The length L^* is the length of duct required to develop a duct flow from Mach number Ma to the sonic point. Many problems involve short ducts which never become sonic, for which the solution uses the differences in the tabulated "maximum," or sonic, length. For example, the length ΔL required to develop from Ma_1 to Ma_2 is given by

$$\bar{f}\frac{\Delta L}{D} = \left(\frac{\bar{f}L^*}{D}\right)_1 - \left(\frac{\bar{f}L^*}{D}\right)_2 \tag{9.67}$$

This avoids the need for separate tabulations for short ducts.

It is recommended that the friction factor $\bar{f}$ be estimated from the Moody chart (Fig. 6.13) for the average Reynolds number and wall-roughness ratio of the duct. Available data [20] on duct friction for compressible flow show good agreement with the Moody chart for subsonic flow, but the measured data in supersonic duct flow are up to 50 percent less than the equivalent Moody friction factor.

EXAMPLE 9.10

Air flows subsonically in an adiabatic 1-in-diameter duct. The average friction factor is 0.024. What length of duct is necessary to accelerate the flow from $\text{Ma}_1 = 0.1$ to $\text{Ma}_2 = 0.5$? What additional length will accelerate it to $\text{Ma}_3 = 1.0$? Assume $\gamma = 1.4$.

Solution

Equation (9.67) applies, with values of $\bar{f}L^*/D$ computed from Eq. (9.66) or read from Table B.3:

$$\bar{f}\frac{\Delta L}{D} = \frac{0.024\,\Delta L}{\frac{1}{12}\,\text{ft}} = \left(\frac{\bar{f}L^*}{D}\right)_{\text{Ma}\,=\,0.1} - \left(\frac{\bar{f}L^*}{D}\right)_{\text{Ma}\,=\,0.5}$$

$$= 66.9216 - 1.0691 = 65.8525$$

Thus

$$\Delta L = \frac{65.8525(\frac{1}{12})}{0.024} = 229\,\text{ft} \qquad\qquad Ans.\ (a)$$

The additional length $\Delta L'$ to go from $\text{Ma} = 0.5$ to 1.0 is taken directly from Appendix Table B.3

$$f\frac{\Delta L'}{D} = \left(\frac{fL^*}{D}\right)_{\text{Ma}\,=\,0.5} = 1.0691$$

or

$$\Delta L' = L^*_{\text{Ma}\,=\,0.5} = \frac{1.0691(\frac{1}{12}\,\text{ft})}{0.024} = 3.7\,\text{ft} \qquad\qquad Ans.\ (b)$$

This is typical of these calculations: it takes 229 ft to accelerate up to Ma = 0.5 and then only 3.7 ft more to get all the way up to the sonic point.

Formulas for other flow properties along the duct can be derived from Eqs. (9.64). Equation (9.64e) can be used to eliminate $f\,dx/D$ from each of the other relations, giving, for example, dp/p as a function only of Ma and $(d\,\mathrm{Ma}^2)/\mathrm{Ma}^2$. For convenience in tabulating the results, each expression is then integrated all the way from (p, Ma) to the sonic point $(p^*, 1.0)$. The integrated results are

$$\frac{p}{p^*} = \frac{1}{\mathrm{Ma}}\left[\frac{\gamma + 1}{2 + (\gamma - 1)\,\mathrm{Ma}^2}\right]^{1/2} \tag{9.68a}$$

$$\frac{\rho}{\rho^*} = \frac{V^*}{V} = \frac{1}{\mathrm{Ma}}\left[\frac{2 + (\gamma - 1)\,\mathrm{Ma}^2}{\gamma + 1}\right]^{1/2} \tag{9.68b}$$

$$\frac{T}{T^*} = \frac{a^2}{a^{*2}} = \frac{\gamma + 1}{2 + (\gamma - 1)\,\mathrm{Ma}^2} \tag{9.68c}$$

$$\frac{p_0}{p_0^*} = \frac{\rho_0}{\rho_0^*} = \frac{1}{\mathrm{Ma}}\left[\frac{2 + (\gamma - 1)\,\mathrm{Ma}^2}{\gamma + 1}\right]^{(1/2)(\gamma + 1)/(\gamma - 1)} \tag{9.68d}$$

All these ratios are also tabulated in Table B.3. For finding changes between points Ma_1 and Ma_2 which are not sonic, products of these ratios are used. For example,

$$\frac{p_2}{p_1} = \frac{p_2}{p^*}\frac{p^*}{p_1} \tag{9.69}$$

since p^* is a constant reference value for the flow.

EXAMPLE 9.11   For the duct flow of Example 9.10 assume that at $\mathrm{Ma}_1 = 0.1$ we have $p_1 = 100\,\mathrm{lbf/in}^2$ absolute and $T_1 = 600°\mathrm{R}$. Compute at section 2 farther downstream ($\mathrm{Ma}_2 = 0.5$) (a) p_2; (b) T_2; (c) V_2; and (d) p_{02}.

Solution   As preliminary information we can compute V_1 and p_{01} from the given information:

$$V_1 = \mathrm{Ma}_1 a_1 = 0.1[49(600°\mathrm{R})^{1/2}] = 120\,\mathrm{ft/s}$$

$$p_{01} = p_1[1 + \tfrac{1}{2}(\gamma - 1)\,\mathrm{Ma}_1^2]^{3.5} = 100[1 + 0.2(0.1)^2]^{3.5}$$

$$= 100.7\,\mathrm{lbf/in}^2\ \mathrm{absolute}$$

Now enter Table B.3 or Eqs. (9.68), to find the following ratios:

Section	Ma	p/p^*	T/T^*	V/V^*	p_0/p_0^*
1	0.1	10.9435	1.1976	0.1094	5.8218
2	0.5	2.1381	1.1429	0.5345	1.3399

Use these ratios to compute all properties downstream:

$$p_2 = p_1 \frac{p_2}{p^*} \frac{p^*}{p_1} = \frac{100(2.1381)}{10.9435} = 19.5 \text{ lbf/in}^2 \text{ absolute} \qquad \textit{Ans. (a)}$$

$$T_2 = T_1 \frac{T_2}{T^*} \frac{T^*}{T_1} = \frac{600(1.1429)}{1.1976} = 573°\text{R} \qquad \textit{Ans. (b)}$$

$$V_2 = V_1 \frac{V_2}{V^*} \frac{V^*}{V_1} = \frac{120(0.5345)}{0.1094} = 586 \text{ ft/s} \qquad \textit{Ans. (c)}$$

$$p_{02} = p_{01} \frac{p_{02}}{p_0^*} \frac{p_0^*}{p_{01}} = \frac{100.7(1.3399)}{5.8218} = 23.2 \text{ lbf/in}^2 \text{ absolute} \qquad \textit{Ans. (d)}$$

Note the 77 percent reduction in stagnation pressure due to friction. The formulas are seductive, so check your work by other means.

Choking Due to Friction

The theory here predicts that for adiabatic frictional flow in a constant-area duct, no matter what the inlet Mach number Ma_1 is, the flow downstream tends toward the sonic point. There is a certain duct length $L^*(Ma_1)$ for which the exit Mach number will be exactly unity.

But what if the actual duct length L is greater than the predicted "maximum" length L^*? Then the flow conditions must change, and there are two classifications.

Subsonic inlet If $L > L^*(Ma_1)$, the flow slows down until an inlet Mach number Ma_2 is reached such that $L = L^*(Ma_2)$. The exit flow is sonic, and the mass flow has been reduced by *frictional choking*. Further increases in duct length will continue to decrease the inlet Ma and mass flow.

Supersonic inlet From Table B.3 we see that friction has a very large effect on supersonic duct flow. Even an infinite inlet Mach number will be reduced to sonic conditions in only 41 diameters for $\bar{f} = 0.02$. Some typical numerical values are shown in Fig. 9.15, assuming an inlet Ma = 3.0 and $\bar{f} = 0.02$. For this condition $L^* = 26$ diameters. If L is increased beyond $26D$, the flow will not choke but a normal shock will form at just the right place for the subsequent subsonic frictional flow to become sonic exactly at the exit. Figure 9.15 shows two examples, for $L/D = 40$ and 53. As length increases, the required normal shock moves upstream until, for Fig. 9.15, the shock is at the inlet for $L/D = 63$. Further increase in L causes the shock to move upstream of the inlet into the supersonic nozzle feeding the duct. Yet the mass flow is still the same as for the very short duct, because presumably the feed nozzle still has a sonic throat. Eventually, a very long duct will cause the feed-nozzle throat to become choked, thus reducing the duct mass flow. Thus supersonic friction changes the flow pattern if $L > L^*$ but does not choke the flow until L is much larger than L^*.

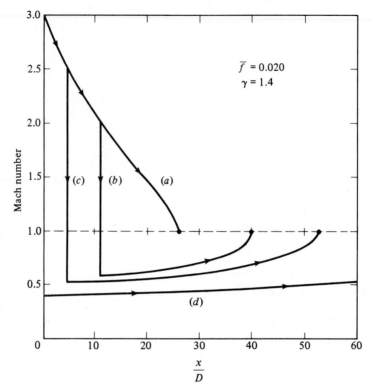

Fig. 9.15 Behavior of duct flow with a nominal supersonic inlet condition Ma = 3.0: (a) $L/D = \leq 26$, flow is supersonic throughout duct; (b) $L/D = 40 > L^*/D$, normal shock at Ma = 2.0 with subsonic flow then accelerating to sonic exit point; (c) $L/D = 53$, shock must now occur at Ma = 2.5; (d) $L/D > 63$, flow must be entirely subsonic and choked at exit.

EXAMPLE 9.12   Air enters a 3-cm-diameter duct at $p_0 = 200$ kPa, $T_0 = 500$ K, and $V_1 = 100$ m/s. The friction factor is 0.02. Compute (a) the maximum duct length for these conditions, (b) the mass flow if the duct length is 15 m, and (c) the reduced mass flow if $L = 30$ m.

Solution

Part (a)   First compute

$$T_1 = T_0 - \frac{\frac{1}{2}V_1^2}{c_p} = 500 - \frac{\frac{1}{2}(100 \text{ m/s})^2}{1005 \text{ m}^2/(\text{s}^2 \cdot \text{K})} = 500 - 5 = 495 \text{ K}$$

$$a_1 = (\gamma R T_1)^{1/2} \approx 20(495 \text{ K})^{1/2} = 445 \text{ m/s}$$

Thus
$$\text{Ma}_1 = \frac{V_1}{a_1} = \frac{100}{445} = 0.225$$

For this Ma_1, from Eq. (9.66) or interpolation in Table B.3,

$$\frac{\bar{f}L^*}{D} = 11.0$$

The maximum duct length possible for these inlet conditions is

$$L^* = \frac{(\bar{f}L^*/D)D}{\bar{f}} = \frac{11.0(0.03 \text{ m})}{0.02} = 16.5 \text{ m} \qquad \qquad Ans. (a)$$

Part (b) The given $L = 15$ m is less than L^*, and so the duct is not choked and the mass flow follows from inlet conditions

$$\rho_{01} = \frac{p_0}{RT_0} = \frac{200{,}000 \text{ Pa}}{287(500 \text{ K})} = 1.394 \text{ kg/m}^3$$

$$\rho_1 = \frac{\rho_{01}}{[1 + 0.2(0.225)^2]^{2.5}} = \frac{1.394}{1.0255} = 1.359 \text{ kg/m}^3$$

whence

$$\dot{m} = \rho_1 A V_1 = (1.359 \text{ kg/m}^3)\left[\frac{\pi}{4}(0.03 \text{ m})^2\right](100 \text{ m/s})$$

$$= 0.0961 \text{ kg/s} \hspace{4cm} Ans. (b)$$

Part (c) Since $L = 30$ m is greater than L^*, the duct must choke back until $L = L^*$, corresponding to a lower inlet Ma_1:

$$L^* = L = 30 \text{ m}$$

$$\frac{\bar{f}L^*}{D} = \frac{0.02(30 \text{ m})}{0.03 \text{ m}} = 20.0$$

From Eq. (9.66) or interpolation in Table B.3 we find that this value of 20.0 corresponds to

$$Ma_{1,\text{choked}} = 0.174 \hspace{1cm} (23\% \text{ less})$$

$$T_{1,\text{new}} = \frac{T_0}{1 + 0.2(0.174)^2} = 497 \text{ K}$$

$$a_{1,\text{new}} \approx 20(497 \text{ K})^{1/2} = 446 \text{ m/s}$$

$$V_{1,\text{new}} = Ma_1 \, a_1 = 0.174(446) = 77.6 \text{ m/s}$$

$$\rho_{1,\text{new}} = \frac{\rho_{01}}{[1 + 0.2(0.174)^2]^{2.5}} = 1.373 \text{ kg/m}^3$$

$$\dot{m}_{\text{new}} = \rho_1 A V_1 = 1.373\left[\frac{\pi}{4}(0.03)^2\right](77.6)$$

$$= 0.0753 \text{ kg/s} \hspace{1cm} (22\% \text{ less}) \hspace{2cm} Ans. (c)$$

Isothermal Flow with Friction

The adiabatic frictional-flow assumption is appropriate to high-speed flow in short ducts. For flow in long ducts, e.g., natural-gas pipelines, the gas state more closely approximates an isothermal flow. The analysis is the same except that the isoenergetic energy equation (9.60c) is replaced by the simple relation

$$T = \text{const} \hspace{1cm} dT = 0 \hspace{3cm} (9.70)$$

Again it is possible to write all property changes in terms of the Mach number. Integration of the Mach number versus friction relation yields

$$\frac{\bar{f}L_{\text{max}}}{D} = \frac{1 - \gamma \, Ma^2}{\gamma \, Ma^2} + \ln(\gamma \, Ma^2) \hspace{2cm} (9.71)$$

which is the isothermal analog of Eq. (9.66) for adiabatic flow.

This friction relation has the interesting result that L_{max} becomes zero not at the sonic point but at $Ma_{crit} = 1/\gamma^{1/2} = 0.845$ if $\gamma = 1.4$. The inlet flow, whether subsonic or supersonic, tends downstream toward this limiting Mach number $1/\gamma^{1/2}$. If the tube length L is greater than L_{max} from Eq. (9.71), a subsonic flow will choke back to a smaller Ma_1 and mass flow and a supersonic flow will experience a normal-shock adjustment similar to Fig. 9.15.

The exit isothermal choked flow is not sonic, and so the use of the asterisk is inappropriate. Let p', ρ', and V' represent properties at the choking point $L = L_{max}$. Then the isothermal analysis leads to the following Mach-number relations for the flow properties:

$$\frac{p}{p'} = \frac{1}{Ma\,\gamma^{1/2}} \qquad \frac{V}{V'} = \frac{\rho'}{\rho} = Ma\,\gamma^{1/2} \tag{9.72}$$

The complete analysis and some examples are given in advanced texts [e.g., 8, sec. 6.4].

An interesting by-product of the isothermal analysis is an exact relation between pressure drop and duct mass flow. In contrast, it is practically impossible in adiabatic flow to eliminate the Mach number between Eqs. (9.66) and (9.68a) to get a direct relation between pressure and friction. Consequently, the common problem of predicting adiabatic duct mass flow for a given pressure drop can only be solved by a laborious Mach-number-iteration procedure.

For isothermal flow, we can substitute $V^2 = G^2/(p/RT)^2$ in Eq. (9.60b) to obtain

$$\frac{p\,dp}{G^2 RT} + \tfrac{1}{2}f\frac{dx}{D} + \frac{dV}{V} = 0 \tag{9.73}$$

But from Eqs. (9.60a) and (9.61) with $dT = 0$, we have $dV/V = -d\rho/\rho = -dp/p$, so that Eq. (9.73) becomes

$$\frac{2p\,dp}{G^2 RT} + f\frac{dx}{D} - \frac{2\,dp}{p} = 0 \tag{9.74}$$

Since $G^2 RT$ is constant for isothermal flow, these are exact differentials and can be integrated from $(x, p) = (0, p_1)$ to (L, p_2), giving

$$G^2 = \left(\frac{\dot{m}}{A}\right)^2 = \frac{p_1^2 - p_2^2}{RT[\bar{f}L/D + 2\ln(p_1/p_2)]} \tag{9.75}$$

Thus we have an explicit expression for isothermal mass flow as a function of the pressure drop in the duct. If you don't think Eq. (9.75) is beautiful, try solving the same problem for adiabatic flow with the Mach-number relations or Table B.3.

Equation (9.75) has one flaw: with the Mach number eliminated, it cannot recognize the choking phenomenon. Therefore one should check the physical realism of a solution using Eq. (9.75) by computing the exit Mach number Ma_2 to ensure that it is not greater than $1/\gamma^{1/2}$ for a subsonic inlet flow.

EXAMPLE 9.13 Air enters a pipe of 1 in diameter at subsonic velocity and $p_1 = 30\,\text{lbf/in}^2$ absolute, $T_1 = 550°\text{R}$. If the pipe is 10 ft long, $\bar{f} = 0.025$, and the exit pressure $p_2 = 20\,\text{lbf/in}^2$ absolute, compute the mass flow for (a) isothermal flow and (b) adiabatic flow.

Solution

Part (a)   For isothermal flow Eq. (9.75) applies. Compute

$$\frac{\bar{f}L}{D} + 2\ln\frac{p_1}{p_2} = \frac{0.025(10\text{ ft})}{\frac{1}{12}\text{ ft}} + 2\ln\tfrac{30}{20} = 3.0 + 0.81 = 3.81$$

$$G^2 = \frac{[30(144\text{ lbf/ft}^2)]^2 - [20(144\text{ lbf/ft}^2)]^2}{[1717\text{ ft}^2/(\text{s}^2\cdot{}^\circ\text{R})](550^\circ\text{R})(3.81)} = 2.88\ (\text{lbf}^2\cdot\text{s}^2)/\text{ft}^6$$

or $$G = 1.70\ (\text{lbf}\cdot\text{s})/\text{ft}^3 = 1.70\ \text{slugs}/(\text{s}\cdot\text{ft}^2)$$

$$A = \tfrac{1}{4}\pi D^2 = \tfrac{1}{4}\pi(\tfrac{1}{12}\text{ ft})^2 = 0.00545\text{ ft}^2$$

Hence $$\dot{m} = GA = 1.70(0.00545) = 0.00926\text{ slug/s}$$ *Ans. (a)*

Check the Mach number at inlet and exit:

$$a_1 = a_2 \approx 49(550^\circ\text{R})^{1/2} = 1149\text{ ft/s}$$

$$\rho_1 = \frac{p_1}{RT} = \frac{30(144)}{1717(550)} = 0.00457\text{ slug/ft}^3$$

$$\rho_2 = \frac{p_2}{RT} = \frac{20(144)}{1717(550)} = 0.00305\text{ slug/ft}^3$$

Then $$V_1 = \frac{G}{\rho_1} = \frac{1.70}{0.00457} = 371\text{ ft/s}$$

$$\text{Ma}_1 = \frac{V_1}{a_1} = \frac{371}{1149} = 0.323$$

$$V_2 = \frac{G}{\rho_2} = 557\text{ ft/s} \qquad \text{Ma}_2 = \frac{557}{1149} = 0.484$$

Since these are well below choking, the solution is accurate.

Part (b)   For adiabatic flow, the only solution the writer can think of is to guess Ma_1, compute $(fL^*/D)_1$, subtract $\bar{f}\,\Delta L/D = 3.0$ to get $(\bar{f}L^*/D)_2$, read Ma_2, finally compute $p_1/p_2 = (p_1/p^*)/(p_2/p^*)$, and see if it equals $30/20 = 1.5$, which is the ratio of the given pressures. If so, you have found the correct solution. If not, try another Ma_1 and iterate to find the proper solution. It is difficult to make an educated initial guess for Ma_1 without additional computation. The isothermal relation (9.75) provides an excellent guess of, say, $\text{Ma}_1 \approx 0.3$ for this problem. The adiabatic computations can then be listed as follows:

Ma_1	$(\bar{f}L^*/D)_1$	$(\bar{f}L^*/D)_2$	Ma_2	p_1/p_2
0.30	5.2993 − 3.0 = 2.2993		0.4005	1.344
0.32	4.4467 − 3.0 = 1.4467		0.4604	1.454
0.33	4.0821 − 3.0 = 1.0821		0.4985	1.531
0.326	4.2235 − 3.0 = 1.2235		0.4823	1.498

The last value is close enough: $Ma_1 = 0.326$, $p_1/p_2 \approx 1.5$.

$$V_1 = Ma_1 \, a_1 = 0.326(1149) = 375 \text{ ft/s}$$

$$\dot{m} = \rho_1 A V_1 = 0.00457(0.00545)(375)$$

$$= 0.00935 \text{ slug/s} \quad (1\% \text{ more}) \qquad \textit{Ans. (b)}$$

This was a lot of work to obtain a result only 1 percent different from the simple isothermal relation (9.75).

9.8 FRICTIONLESS DUCT FLOW WITH HEAT TRANSFER[1]

Heat addition or removal has an interesting effect on a compressible flow. Advanced texts [e.g., 8, chap. 8] consider the combined effect of heat transfer coupled with friction and area change in a duct. Here we confine the analysis to heat transfer with no friction in a constant-area duct.

Consider the elemental duct control volume in Fig. 9.16. Between sections 1 and 2 an amount of heat δQ is added (or removed) to each incremental mass δm passing through. With no friction or area change, the control-volume conservation relations are quite simple

Continuity: $\qquad\qquad \rho_1 V_1 = \rho_2 V_2 = G = \text{const} \qquad\qquad (9.76a)$

x Momentum: $\qquad\qquad p_1 - p_2 = G(V_2 - V_1) \qquad\qquad (9.76b)$

Energy: $\qquad\qquad \dot{Q} = \dot{m}(h_2 + \tfrac{1}{2}V_2^2 - h_1 - \tfrac{1}{2}V_1^2)$

or $\qquad\qquad q = \dfrac{\dot{Q}}{\dot{m}} = \dfrac{\delta Q}{\delta m} = h_{02} - h_{01} \qquad\qquad (9.76c)$

The heat transfer results in a change in stagnation enthalpy of the flow. We shall not specify exactly how the heat is transferred—combustion, nuclear reaction, evaporation, condensation, or wall heat exchange—but simply that it happened in amount q between 1 and 2. We remark, however, that wall heat exchange is not a good candidate for the theory because wall convection is inevitably coupled with wall friction, which we neglected.

To complete the analysis we use the perfect-gas and Mach-number relations

$$\frac{p_2}{\rho_2 T_2} = \frac{p_1}{\rho_1 T_1} \qquad h_{02} - h_{01} = c_p(T_{02} - T_{01})$$

$$\frac{V_2}{V_1} = \frac{Ma_2 \, a_2}{Ma_1 \, a_1} = \frac{Ma_2}{Ma_1}\left(\frac{T_2}{T_1}\right)^{1/2} \qquad\qquad (9.77)$$

For a given heat transfer $q = \delta Q/\delta m$ or, equivalently, a given change $h_{02} - h_{01}$, Eqs. (9.76) and (9.77) can be solved algebraically for the property ratios p_2/p_1,

[1] This section may be omitted without loss of continuity.

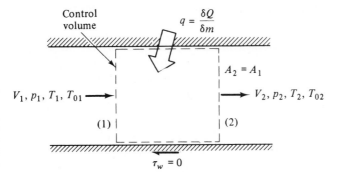

Fig. 9.16 Elemental control volume for frictionless flow in a constant-area duct with heat transfer. The length of the element is indeterminate in this simplified theory.

Ma_2/Ma_1, etc., between inlet and outlet. Note that because the heat transfer allows the entropy either to increase or decrease, the second law imposes no restrictions on these solutions.

Before writing down these property-ratio functions, we illustrate the effect of heat transfer in Fig. 9.17, which shows T_0 and T versus Mach number in the duct. Heating increases T_0, and cooling decreases it. The maximum possible T_0 occurs at $Ma = 1.0$, and we see that heating, whether the inlet is subsonic or supersonic, drives the duct Mach number toward unity. This is analogous to the effect of friction in the previous section. The temperature of a perfect gas increases from $Ma = 0$ up to $Ma = 1/\gamma^{1/2}$ and then decreases. Thus there is a peculiar—or at least unexpected—region where heating (increasing T_0) actually decreases the gas temperature, the difference being reflected in a large increase of the gas kinetic energy. For $\gamma = 1.4$ this peculiar area lies between $Ma = 0.845$ and 1.0 (interesting but not very useful information).

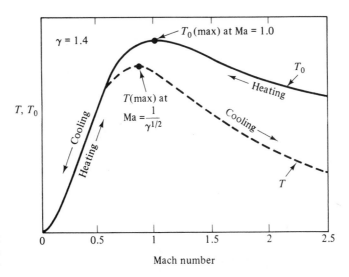

Fig. 9.17 Effect of heat transfer on Mach number.

The complete list of the effects of simple T_0 change on duct-flow properties is as follows:

	Heating		Cooling	
	Subsonic	Supersonic	Subsonic	Supersonic
T_0	Increases	Increases	Decreases	Decreases
Ma	Increases	Decreases	Decreases	Increases
p	Decreases	Increases	Increases	Decreases
ρ	Decreases	Increases	Increases	Decreases
V	Increases	Decreases	Decreases	Increases
p_0	Decreases	Decreases	Increases	Increases
s	Increases	Increases	Decreases	Decreases
T	†	Increases	‡	Decreases

† Increases up to Ma $= 1/\gamma^{1/2}$ and decreases thereafter.
‡ Decreases up to Ma $= 1/\gamma^{1/2}$ and increases thereafter.

Probably the most significant item on this list is the stagnation pressure p_0, which always decreases during heating whether the flow is subsonic or supersonic. Thus heating does increase the Mach number of a flow but entails a loss in effective pressure recovery.

Mach-Number Relations

Equations (9.76) and (9.77) can be rearranged in terms of the Mach number and the results tabulated. For convenience we specify that the outlet section is sonic, Ma $= 1$, with reference properties T_0^*, T^*, p^*, ρ^*, V^*, and p_0^*. The inlet is assumed to be at arbitrary Mach number Ma. Equations (9.76) and (9.77) then take the following form:

$$\frac{T_0}{T_0^*} = \frac{(\gamma + 1)\, \text{Ma}^2\, [2 + (\gamma - 1)\, \text{Ma}^2]}{(1 + \gamma\, \text{Ma}^2)^2} \tag{9.78a}$$

$$\frac{T}{T^*} = \frac{(\gamma + 1)^2\, \text{Ma}^2}{(1 + \gamma\, \text{Ma}^2)^2} \tag{9.78b}$$

$$\frac{p}{p^*} = \frac{\gamma + 1}{1 + \gamma\, \text{Ma}^2} \tag{9.78c}$$

$$\frac{V}{V^*} = \frac{\rho^*}{\rho} = \frac{(\gamma + 1)\, \text{Ma}^2}{1 + \gamma\, \text{Ma}^2} \tag{9.78d}$$

$$\frac{p_0}{p_0^*} = \frac{\gamma + 1}{1 + \gamma\, \text{Ma}^2} \left[\frac{2 + (\gamma - 1)\, \text{Ma}^2}{\gamma + 1} \right]^{\gamma/(\gamma - 1)} \tag{9.78e}$$

These formulas are all tabulated versus Mach number in Appendix Table B.4. The tables are very convenient if inlet properties Ma_1, V_1, etc., are given but are somewhat cumbersome if the given information centers on T_{01} and T_{02}. Let us illustrate with an example.

EXAMPLE **9.14** A fuel-air mixture, assumed thermodynamically equivalent to air with $\gamma = 1.4$, enters a duct combustion chamber at $V_1 = 250$ ft/s, $p_1 = 20$ lbf/in^2 absolute, and $T_1 = 530°$R. The heat addition by combustion is 400 Btu per pound of mixture. Compute (a) the exit properties V_2, p_2, and T_2 and (b) the total heat addition which would have caused a sonic exit flow.

Solution

Part (a)   First convert the heat addition to a change in T_0 of the gas:

$$q = 400 \text{ Btu/lb} = 10{,}020{,}000 \text{ (ft} \cdot \text{lbf)/slug}$$

$$= c_p(T_{02} - T_{01}) = [6010 \text{ (ft} \cdot \text{lbf)/(slug} \cdot °\text{R})](T_{02} - T_{01})$$

Thus $$T_{02} - T_{01} = \frac{10{,}020{,}000}{6010} = 1667°\text{R}$$

We have enough information to compute Ma_1 and T_{01}:

$$a_1 \approx 49(530°\text{R})^{1/2} = 1128 \text{ ft/s}$$

$$\text{Ma}_1 = \frac{V_1}{a_1} = \frac{250}{1128} = 0.222$$

$$T_{01} = T_1 + \frac{\frac{1}{2}V_1^2}{c_p} = 530 + \frac{\frac{1}{2}(250)^2}{6010} = 535°\text{R}$$

Now use Eq. (9.78a) or interpolate in Table B.4 to find the inlet ratio corresponding to $\text{Ma}_1 = 0.222$

$$\frac{T_{01}}{T_0^*} = 0.2084$$

Hence $$T_0^* = \frac{T_{01}}{0.2084} = \frac{535}{0.2084} = 2568°\text{R}$$

At the exit section we can now compute

$$T_{02} = T_{01} + (T_{02} - T_{01}) = 535 + 1667 = 2202°\text{R}$$

Thus we can compute the exit ratio

$$\frac{T_{02}}{T_0^*} = \frac{2202}{2568} = 0.8575$$

From Eq. (9.78a) or interpolation in Table B.4 we find that

$$\text{Ma}_2 \approx 0.638$$

With inlet and exit Mach numbers known, we can tabulate the velocity, pressure, and temperature ratios as follows:

	(1)	(2)
Ma	0.222	0.638
V/V^*	0.110	0.623
p/p^*	2.246	1.529
T/T^*	0.248	0.951

Then the exit properties are given by

$$p_2 = \frac{p_1(p_2/p^*)}{p_1/p^*} = \frac{20(1.529)}{2.246} = 13.6 \text{ lbf/in}^2 \text{ absolute} \qquad Ans.\ (a)$$

$$V_2 = \frac{V_1(V_2/V^*)}{V_1/V^*} = \frac{250(0.623)}{0.110} = 1415 \text{ ft/s} \qquad Ans.\ (a)$$

$$T_2 = \frac{T_1(T_2/T^*)}{T_1/T^*} = \frac{530(0.951)}{0.248} = 2030°\text{R} \qquad Ans.\ (a)$$

Part (b) The maximum heat addition allowed without choking would drive the exit Mach number to unity:

$$T_{02} = T_0^* = 2568°\text{R}$$

Thus

$$q_{max} = c_p(T_0^* - T_{01}) = 6010(2568 - 535)$$

$$= 12,220,000 \text{ (ft·lbf)/slug} = 488 \text{ Btu/lb} \qquad Ans.\ (b)$$

Choking Effects Due to Simple Heating

Equation (9.78a) and Table B.4 indicate that the maximum possible stagnation temperature in simple heating corresponds to T_0^*, or the sonic exit Mach number. Thus, for given inlet conditions, only a certain maximum amount of heat can be added to the flow, for example, 488 Btu/lb in Example 9.14. For a subsonic inlet there is no theoretical limit on heat addition: the flow chokes more and more as we add more heat, with the inlet velocity approaching zero. For supersonic flow, even if Ma_1 is infinite, there is a finite ratio $T_{01}/T_0^* = 0.4898$ for $\gamma = 1.4$. Thus if heat is added without limit to a supersonic flow, a normal-shock-wave adjustment is required to accommodate to the required property changes.

In subsonic flow there is no theoretical limit to the amount of cooling allowed: the exit flow just becomes slower and slower, and the temperature approaches zero. In supersonic flow only a finite amount of cooling can be allowed before the exit flow approaches infinite Mach number, with $T_{02}/T_0^* = 0.4898$ and the exit temperature equal to zero. There are very few practical applications for supersonic cooling.

EXAMPLE 9.15 What happens to the inlet flow in Example 9.14 if the heat addition is increased to 600 Btu/lb and the inlet pressure and stagnation temperature are fixed? What will the decrease in mass flow be?

Solution For $q = 600 \text{ Btu/lb} = 15,025,000 \text{ (ft·lbf)/slug}$, the flow will be choked at an exit stagnation temperature of

$$T_{02} = T_0^* = T_{01} + \frac{q}{c_p} = 535 + \frac{15,025,000}{6010} = 535 + 2500 = 3035°\text{R}$$

Hence $T_{01}/T_0^* = 535/3035 = 0.1763$, and the flow will choke down until this value corresponds to the proper subsonic Mach number. From Eq. (9.78a) or Table B.4 for $T_{01}/T_0^* = 0.1763$ read

$$Ma_{1,\text{new}} = 0.202$$

It was specified that T_{01} and p_1 remain the same. The other inlet properties will change according to Ma_1:

$$T_1 = \frac{T_{01}}{1 + 0.2(0.202)^2} = \frac{535}{1.008} = 531°\text{R}$$

$$a_1 \approx 49(531°\text{R})^{1/2} = 1129 \text{ ft/s}$$

$$V_1 = \text{Ma}_1 \, a_1 = 0.202(1129) = 228 \text{ ft/s}$$

$$\rho_1 = \frac{p_1}{RT_1} = \frac{20(144)}{1717(531)} = 0.00316 \text{ slug/ft}^3$$

Finally,

$$\frac{\dot{m}_{\text{new}}}{A} = \rho_1 V_1 = 0.00316(228) = 0.720 \text{ slug/(s} \cdot \text{ft}^2)$$

This is 9 percent less than the mass flow of 0.791 slug/(s · ft^2) in Example 9.14, due to the excess heat addition choking the flow.

Relationship to the Normal-Shock Wave

The normal-shock-wave relations of Sec. 9.5 actually lurk within the simple heating relations as a special case. From Table B.4 or Fig. 9.17 we see that for a given stagnation temperature less than T_0^* there are two flow states which satisfy the simple heating relations, one subsonic and the other supersonic. These two states have (1) the same value of T_0, (2) the same mass flow per unit area, and (3) the same value of $p + \rho V^2$. Therefore these two states are exactly equivalent to the conditions on each side of a normal-shock wave. The second law would again require that the upstream flow Ma_1 be supersonic.

To illustrate this point, take $\text{Ma}_1 = 3.0$ and from Table B.4 read $T_{01}/T_0^* = 0.6540$ and $p_1/p^* = 0.1765$. Now, for the same value $T_{02}/T_0^* = 0.6540$, use Table B.4 or Eq. (9.78a) to compute $\text{Ma}_2 = 0.4752$ and $p_2/p^* = 1.8235$. The value of Ma_2 is exactly what we read in the shock table, B.2, as the downstream Mach number when $\text{Ma}_1 = 3.0$. The pressure ratio for these two states is $p_2/p_1 = (p_2/p^*)/(p_1/p^*) = 1.8235/0.1765 = 10.33$, which again is just what we read in Table B.2 for $\text{Ma}_1 = 3.0$. This illustration is meant only to show the physical background of the simple heating relations: it would be silly to make a practice of computing normal-shock waves in this manner.

9.9 TWO-DIMENSIONAL SUPERSONIC FLOW

Up to this point we have considered only one-dimensional compressible-flow theories. This illustrated many important effects, but a one-dimensional world completely loses sight of the wave motions which are so characteristic of supersonic flow. The only "wave motion" we could muster in a one-dimensional theory was the normal-shock wave, which amounted only to a flow discontinuity in the duct.

Mach Waves

When we add a second dimension to the flow, wave motions immediately become apparent if the flow is supersonic. Figure 9.18 shows a celebrated graphical

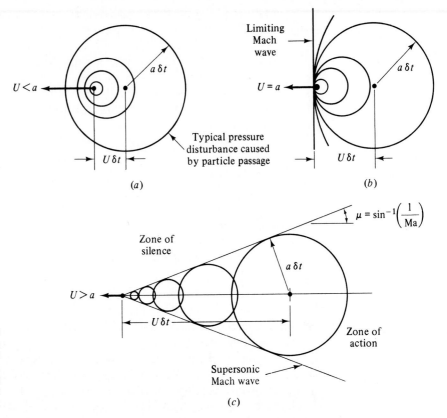

Fig. 9.18 Wave patterns set up by a particle moving at speed U into still fluid of sound velocity a: (a) subsonic, (b) sonic, and (c) supersonic motion.

construction which appears in every fluid mechanics textbook and was first presented by Ernst Mach in 1887. The figure shows the pattern of pressure disturbances (sound waves) sent out by a small particle moving at speed U through a still fluid whose sound velocity is a.

As the particle moves, it continually crashes against fluid particles and sends out spherical sound waves emanating from every point along its path. A few of these spherical disturbance fronts are shown in Fig. 9.18. The behavior of these fronts is quite different according as the particle speed is subsonic or supersonic.

In Fig. 9.18a, the particle moves subsonically, $U < a$, $\mathrm{Ma} = U/a < 1$. The spherical disturbances move out in all directions and do not catch up with each other. They move well out in front of the particle also, because they travel a distance $a\,\delta t$ during the time interval δt in which the particle has moved only $U\,\delta t$. Therefore a subsonic body motion makes its presence felt everywhere in the flow field: you can "hear" or "feel" the pressure rise of an oncoming body before it reaches you. This is apparently why that pigeon in the road, without turning around to look at you, takes to the air and avoids being hit by your car.

At sonic speed, $U = a$, Fig. 9.18b, the pressure disturbances move at exactly the speed of the particle and thus pile up on the left at the position of the particle into a sort of "front locus," which is now called a *Mach wave* after Ernst Mach. No disturbance reaches beyond the particle. if you are stationed to the left of the particle, you cannot "hear" the oncoming motion. If the particle blew its horn, you couldn't hear that either: a sonic car can sneak up on a pigeon.

In supersonic motion, $U > a$, the lack of advance warning is even more pronounced. The disturbance spheres cannot catch up with the fast-moving particle which created them. They all trail behind the particle and are tangent to a conical locus called the *Mach cone*. From the geometry of Fig. 9.18c the angle of the Mach cone is seen to be

$$\mu = \sin^{-1} \frac{a\,\delta t}{U\,\delta t} = \sin^{-1} \frac{a}{U} = \sin^{-1} \frac{1}{\text{Ma}} \qquad (9.79)$$

The higher the particle Mach number, the more slender the Mach cone; for example, $\mu = 30°$ at Ma = 2.0 and 11.5° at Ma = 5.0. For the limiting case of sonic flow, Ma = 1, $\mu = 90°$; the Mach cone becomes a plane front moving with the particle, in agreement with Fig. 9.18b.

You cannot "hear" the disturbance caused by the supersonic particle in Fig. 9.18c until you are in the *zone of action* inside the Mach cone. No warning can reach your ears if you are in the *zone of silence* outside the cone. Thus an observer on the ground beneath a supersonic airplane does not hear the *sonic boom* of the passing cone until the plane is well past.

Fig. 9.19 Supersonic wave pattern emanating from a projectile moving at Ma ≈ 2.0. The heavy lines are oblique-shock waves and the light lines Mach waves (*courtesy of U.S. Army Ballistic Research Laboratory, Aberdeen Proving Ground.*)

The Mach wave need not be a cone: similar waves are formed by a small disturbance of any shape moving supersonically with respect to the ambient fluid. For example, the "particle" in Fig. 9.18c could be the leading edge of a sharp flat plate, which would form a Mach wedge of exactly the same angle μ. Mach waves are formed by small roughnesses or boundary-layer irregularities in a supersonic wind tunnel or at the surface of a supersonic body. Look again at Fig. 9.10: Mach waves are clearly visible along the body surface downstream of the recompression shock, especially at the rear corner. Their angle is about 30°, indicating a Mach number of about 2.0 along this surface. A more complicated system of Mach waves emanates from the supersonic projectile in Fig. 9.19. The Mach angles change, indicating a variable supersonic Mach number along the body surface. There are also several stronger oblique-shock waves formed along the surface.

EXAMPLE 9.16 An observer on the ground does not hear the sonic boom caused by an airplane moving at 5 km altitude until it is 9 km past him. What is the approximate Mach number of the plane? Assume a small disturbance and neglect the variation of sound speed with altitude.

Solution A finite disturbance like an airplane will create a finite-strength oblique-shock wave whose angle will be somewhat larger than the Mach-wave angle μ and will curve downward due to the variation in atmospheric sound speed. If we neglect these effects, the altitude and distance are a measure of μ, as seen in Fig. E9.16. Thus

$$\tan \mu = \frac{5 \text{ km}}{9 \text{ km}} = 0.5556 \qquad \text{or} \qquad \mu = 29.05°$$

Hence, from Eq. (9.79),

$$\text{Ma} = \csc \mu = 2.06 \qquad\qquad Ans.$$

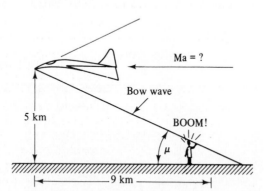

Fig. E9.16

The Oblique-Shock Wave

Figures 9.10 and 9.19 and our earlier discussion all indicate that a shock wave can form at an oblique angle to the oncoming supersonic stream. Such a wave will deflect the stream through an angle θ, unlike the normal-shock wave, for which the downstream flow is in the same direction. In essence, an oblique shock is caused by the necessity for a supersonic stream to turn through such an angle. Examples

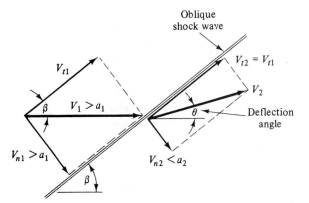

Fig. 9.20 Geometry of the flow through an oblique-shock wave.

could be a finite wedge at the leading edge of a body or a ramp in the wall of a supersonic wind tunnel.

The flow geometry of an oblique shock is shown in Fig. 9.20. As for the normal shock of Fig. 9.8, state 1 denotes the upstream conditions and state 2 is downstream. The shock angle has an arbitrary value β, and the downstream flow V_2 turns at an angle θ which is a function of β and state 1 conditions. The upstream flow is always supersonic, but the downstream Mach number $\text{Ma}_2 = V_2/a_2$ may be subsonic, sonic, or supersonic, depending upon the conditions.

It is convenient to analyze the flow by breaking it up into normal and tangential components with respect to the wave, as shown in Fig. 9.20. For a thin control volume just encompassing the wave, we then can derive the following integral relations, canceling out $A_1 = A_2$ on each side of the wave:

Continuity: $$\rho_1 V_{n1} = \rho_2 V_{n2} \tag{9.80a}$$

Normal momentum: $$p_1 - p_2 = \rho_2 V_{n2}^2 - \rho_1 V_{n1}^2 \tag{9.80b}$$

Tangential momentum: $$0 = \rho_1 V_{n1}(V_{t2} - V_{t1}) \tag{9.80c}$$

Energy: $$h_1 + \tfrac{1}{2}V_{n1}^2 + \tfrac{1}{2}V_{t1}^2 = h_2 + \tfrac{1}{2}V_{n2}^2 + \tfrac{1}{2}V_{t2}^2 = h_0 \tag{9.80d}$$

We see from Eq. (9.80c) that there is no change in tangential velocity across an oblique shock

$$V_{t2} = V_{t1} = V_t = \text{const} \tag{9.81}$$

Thus tangential velocity has as its only effect the addition of a constant kinetic energy $\tfrac{1}{2}V_t^2$ to each side of the energy equation (9.80d). We conclude that Eqs. (9.80) are identical to the normal-shock relations (9.49), with V_1 and V_2 replaced by the normal components V_{n1} and V_{n2}. All the various relations from Sec. 9.5 can be used to compute properties of an oblique-shock wave. The trick is to use the "normal" Mach numbers in place of Ma_1 and Ma_2:

$$\text{Ma}_{n1} = \frac{V_{n1}}{a_1} = \text{Ma}_1 \sin \beta$$

$$\text{Ma}_{n2} = \frac{V_{n2}}{a_2} = \text{Ma}_2 \sin(\beta - \theta) \tag{9.82}$$

Then, for a perfect gas with constant specific heats, the property ratios across the oblique shock are the analogs of Eqs. (9.55) to (9.58) with Ma_1 replaced by Ma_{n1}

$$\frac{p_2}{p_1} = \frac{1}{\gamma + 1}[2\gamma \, Ma_1^2 \sin^2 \beta - (\gamma - 1)] \tag{9.83a}$$

$$\frac{\rho_2}{\rho_1} = \frac{\tan \beta}{\tan(\beta - \theta)} = \frac{(\gamma + 1) \, Ma_1^2 \sin^2 \beta}{(\gamma - 1) \, Ma_1^2 \sin^2 \beta + 2} = \frac{V_{n1}}{V_{n2}} \tag{9.83b}$$

$$\frac{T_2}{T_1} = [2 + (\gamma - 1) \, Ma_1^2 \sin^2 \beta]\frac{2\gamma \, Ma_1^2 \sin^2 \beta - (\gamma - 1)}{(\gamma + 1)^2 \, Ma_1^2 \sin^2 \beta} \tag{9.83c}$$

$$T_{02} = T_{01} \tag{9.83d}$$

$$\frac{p_{02}}{p_{01}} = \left[\frac{(\gamma + 1) \, Ma_1^2 \sin^2 \beta}{2 + (\gamma - 1) \, Ma_1^2 \sin^2 \beta}\right]^{\gamma/(\gamma-1)}\left[\frac{\gamma + 1}{2\gamma \, Ma_1^2 \sin^2 \beta - (\gamma - 1)}\right]^{1/(\gamma-1)} \tag{9.83e}$$

$$Ma_{n2}^2 = \frac{(\gamma - 1) \, Ma_{n1}^2 + 2}{2\gamma \, Ma_{n1}^2 - (\gamma - 1)} \tag{9.83f}$$

All these are tabulated in the normal-shock Table B.2. If you wondered why that table listed the Mach numbers as Ma_{n1} and Ma_{n2}, it should be clear now that the table is also valid for the oblique-shock wave.

Thinking all of this over, we realize by hindsight that an oblique-shock wave is the flow pattern one would observe by running along a normal-shock-wave (Fig. 9.8) at a constant tangential speed V_t. Thus the normal and oblique shocks are related by a galilean, or inertial, velocity transformation and therefore satisfy the same basic equations.

If we continue with this run-along-the-shock analogy, we find that the deflection angle θ increases with speed V_t up to a maximum and then decreases. From the geometry of Fig. 9.20 the deflection angle is given by

$$\theta = \tan^{-1}\frac{V_t}{V_{n2}} - \tan^{-1}\frac{V_t}{V_{n1}} \tag{9.84}$$

If we differentiate θ with respect to V_t and set the result equal to zero, we find that the maximum deflection occurs when $V_t/V_{n1} = (V_{n2}/V_{n1})^{1/2}$. We can substitute this back into Eq. (9.84) to compute

$$\theta_{max} = \tan^{-1}r^{1/2} - \tan^{-1}r^{-1/2} \qquad r = \frac{V_{n1}}{V_{n2}} \tag{9.85}$$

For example, if $Ma_{n1} = 3.0$, from Table B.2 we find that $V_{n1}/V_{n2} = 3.8571$, the square root of which is 1.9640. Then Eq. (9.85) predicts a maximum deflection of $\tan^{-1} 1.9640 - \tan^{-1}(1/1.9640) = 36.03°$. The deflection is quite limited even for infinite Ma_{n1}: from Table B.2 for this case $V_{n1}/V_{n2} = 6.0$, and we compute from Eq. (9.85) that $\theta_{max} = 45.58°$.

This limited-deflection idea and other facts become more evident if we plot some of the solutions of Eqs. (9.83). For a given value of V_1 and a_1, assuming as usual that $\gamma = 1.4$, we can plot all possible solutions for V_2 downstream of the shock.

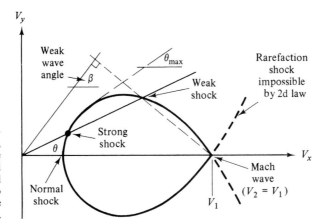

Fig. 9.21 The oblique-shock polar hodograph, showing double solutions (strong and weak) for small deflection angle and no solutions at all for large deflection.

Figure 9.21 does this in velocity-component coordinates V_x and V_y, with x parallel to V_1. Such a plot is called a *hodograph*. The heavy dark line which looks like a fat airfoil is the locus, or *shock polar*, of all physically possible solutions for the given Ma_1. The two dashed-line fishtails are solutions which increase V_2; they are physically impossible because they violate the second law.

Examining the shock polar in Fig. 9.21, we see that a given deflection line of small angle θ crosses the polar at two possible solutions: the *strong* shock, which greatly decelerates the flow, and the *weak* shock, which causes a much milder deceleration. The flow downstream of the strong shock is always subsonic, while that of the weak shock is usually supersonic but occasionally subsonic if the deflection is large. Both types of shock occur in practice: the weak shock is more prevalent, but the strong shock will occur if there is a blockage or high-pressure condition downstream.

Since the shock polar is only of finite size, there is a maximum deflection θ_{max}, shown in Fig. 9.21, which just grazes the upper edge of the polar curve. This verifies the kinematic discussion which led to Eq. (9.85). What happens if a supersonic flow is forced to deflect through an angle greater than θ_{max}? The answer is illustrated in Fig. 9.22 for flow past a wedge-shaped body.

In Fig. 9.22*a* the wedge half-angle θ is less than θ_{max}, and thus an oblique shock forms at the nose of wave angle β just sufficient to cause the oncoming supersonic stream to deflect through the wedge angle θ. Except for the usually small effect of boundary-layer growth (see, for example, Ref. 19, sec. 7-5.2), the Mach number Ma_2 is constant along the wedge surface and is given by the solution of Eqs. (9.83). The pressure, density, and temperature along the surface are also nearly constant, as predicted by Eqs. (9.83). When the flow reaches the corner of the wedge, it expands to higher Mach number and forms a wake (not shown) similar to that in Fig. 9.10.

In Fig. 9.22*b* the wedge half-angle is greater than θ_{max}, and an attached oblique shock is impossible. The flow cannot deflect at once through the entire angle θ_{max}, yet somehow the flow must get around the wedge. A detached curved shock wave forms in front of the body, discontinuously deflecting the flow through angles

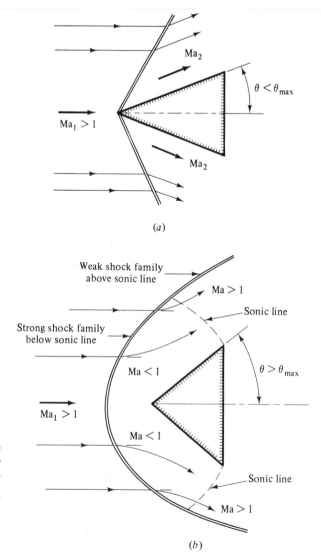

(a)

Fig. 9.22 Supersonic flow past a wedge: (a) small wedge angle, attached oblique shock forms; (b) large wedge angle, attatched shock not possible, broad curved detached shock forms.

(b)

smaller than θ_{max}. The flow then curves, expands, and deflects subsonically around the wedge, becoming sonic and then supersonic as it passes the corner region. The flow just inside each point on the curved shock exactly satisfies the oblique-shock relations (9.83) for that particular value of β and the given Ma_1. Every condition along the curved shock is a point on the shock polar of Fig. 9.21. Points near the front of the wedge are in the strong-shock family, and points aft of the sonic line are in the weak-shock family. The analysis of detached shock waves is extremely complex, and experimentation is usually needed, e.g., the shadowgraph optical technique of Fig. 9.10.

The complete family of oblique-shock solutions can be plotted or computed from Eqs. (9.83). For a given γ, the wave angle β varies only with Ma_1 and θ, from

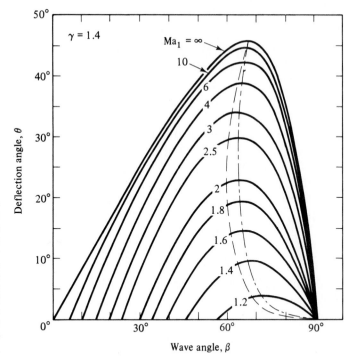

Fig. 9.23 Oblique-shock deflection versus wave angle for various upstream Mach numbers, $\gamma = 1.4$: dash-dot curve, locus of θ_{max}, divides strong (right) from weak (left) shocks; dashed curve, locus of sonic points, divides subsonic Ma_2 (right) from supersonic Ma_2 (left).

Eq. (9.83b). By using a trigonometric identity for $\tan(\beta - \theta)$ this can be rewritten in the more convenient form

$$\tan \theta = \frac{2 \cot \beta \, (Ma_1^2 \sin^2 \beta - 1)}{Ma_1^2 \, (\gamma + \cos 2\beta) + 2} \tag{9.86}$$

All possible solutions of Eq. (9.86) for $\gamma = 1.4$ are shown in Fig. 9.23. For deflections $\theta < \theta_{max}$ there are two solutions: a weak shock (small β) and a strong shock (large β), as expected. All points along the dash-dot line for θ_{max} satisfy Eq. (9.85). A dashed line has been added showing where Ma_2 is exactly sonic. We see that there is a narrow region near maximum deflection where the weak-shock downstream flow is subsonic.

For zero deflections ($\theta = 0$) the weak-shock family satisfies the wave-angle relation

$$\beta = \mu = \sin^{-1} \frac{1}{Ma_1} \tag{9.87}$$

Thus weak shocks of vanishing deflection are equivalent to Mach waves. Meanwhile the strong shocks all converge at zero deflection to the normal-shock condition $\beta = 90°$.

Two additional oblique-shock charts are given in Appendix B, where Fig. B.1 gives the downstream Mach number Ma_2 and Fig. B.2 the pressure ratio p_2/p_1, each plotted as a function of Ma_1 and θ.

Very-Weak-Shock Waves

For any finite θ the wave angle β for a weak shock is greater than the Mach angle μ. For small θ Eq. (9.86) can be expanded in a power series in $\tan \theta$ with the following linearized result for the wave angle:

$$\sin \beta = \sin \mu + \frac{\gamma + 1}{4 \cos \mu} \tan \theta + \cdots \mathcal{O}(\tan^2 \theta) \cdots \tag{9.88}$$

For Ma_1 between 1.4 and 20.0 and deflections less than $6°$ this relation predicts β to within $1°$ for a weak shock. For larger deflections it can be used as a useful initial guess for iterative solution of Eq. (9.86).

Other property changes across the oblique shock can also be expanded in a power series for small deflection angles. Of particular interest is the pressure change from Eq. (9.83a), for which the linearized result for a weak shock is

$$\frac{p_2 - p_1}{p_1} = \frac{\gamma \, Ma_1^2}{(Ma_1^2 - 1)^{1/2}} \tan \theta + \cdots \mathcal{O}(\tan^2 \theta) \cdots \tag{9.89}$$

The differential form of this relation is used in the next section to develop a theory for supersonic expansion turns. Figure 9.24 shows the exact weak-shock pressure jump computed from Eq. (9.83a). At very small deflections the curves are linear with slopes given by Eq. (9.89).

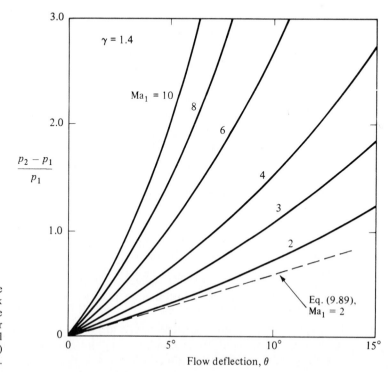

Fig. 9.24 Pressure jump across a weak oblique-shock wave from Eq. (9.83a) for $\gamma = 1.4$. For very small deflections Eq. (9.89) applies.

Finally, it is educational to examine the entropy change across a very weak shock. Using the same power-series expansion technique, we can obtain the following result for small flow deflections:

$$\frac{s_2 - s_1}{c_p} = \frac{(\gamma^2 - 1)\,\mathrm{Ma}_1^6}{12(\mathrm{Ma}_1^2 - 1)^{3/2}}\tan^3\theta + \cdots \mathcal{O}(\tan^4\theta)\cdots \qquad (9.90)$$

The entropy change is cubic in the deflection angle θ. Thus weak-shock waves are very nearly isentropic, a fact which is also used in the next section.

EXAMPLE 9.17 Air at Ma = 2.0 and $p = 10\,\mathrm{lbf/in}^2$ absolute is forced to turn through $10°$ by a ramp at the body surface. A weak oblique shock forms as in Fig. E9.17. For $\gamma = 1.4$ compute from exact oblique-shock theory (a) the wave angle β; (b) Ma_2; (c) p_2. Also use the linearized theory to estimate (d) β and (e) p_2.

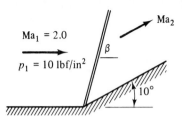

$\mathrm{Ma}_1 = 2.0$

$p_1 = 10\,\mathrm{lbf/in}^2$

Ma_2

β

$10°$

Fig. E9.17

Solution With $\mathrm{Ma}_1 = 2.0$ and $\theta = 10°$ known, we find β from Eq. (9.83b) or (9.86). From Fig. 9.23 we read $\beta \approx 39°$ within about $1°$. If we wish more accuracy, we can try 39 and $40°$ in Eq. (9.86) and see how close θ is to $10°$:

Eq. (9.86) ($\mathrm{Ma}_1 = 2.0$): $\beta = 39°$ $\theta = 9.7102°$

$\beta = 40°$ $\theta = 10.6229°$

By interpolation we have the more accurate result

$$\beta = 39.32° \qquad\qquad \textit{Ans. (a)}$$

The normal Mach number upstream is thus

$$\mathrm{Ma}_{n1} = \mathrm{Ma}_1 \sin\beta = 2.0 \sin 39.32° = 1.267$$

With Ma_{n1} we can use the normal-shock relations (Table B.2) or Fig. 9.9 or Eqs. (9.56) to (9.58) to compute

$$\mathrm{Ma}_{n2} = 0.8031 \qquad \frac{p_2}{p_1} = 1.707$$

Thus the downstream Mach number and pressure are

$$\mathrm{Ma}_2 = \frac{\mathrm{Ma}_{n2}}{\sin(\beta - \theta)} = \frac{0.8031}{\sin(39.32° - 10°)} = 1.64 \qquad\qquad \textit{Ans. (b)}$$

$$p_2 = (10\,\mathrm{lbf/in}^2\ \text{absolute})(1.707) = 17.07\,\mathrm{lbf/in}^2\ \text{absolute} \qquad\qquad \textit{Ans. (c)}$$

Notice that the computed pressure ratio agrees with Figs. 9.24 and B.2.

For the linearized theory the Mach angle is $\mu = \sin^{-1}(1/2.0) = 30°$. Equation (9.88) then estimates that

$$\sin \beta \approx \sin 30° + \frac{2.4 \tan 10°}{4 \cos 30°} = 0.622$$

or

$$\beta \approx 38.5° \qquad\qquad Ans.\ (d)$$

Equation (9.89) estimates that

$$\frac{p_2}{p_1} \approx 1 + \frac{1.4(2)^2 \tan 10°}{(2^2 - 1)^{1/2}} = 1.57$$

or

$$p_2 \approx 1.57(10 \text{ lbf/in}^2 \text{ absolute}) \approx 15.7 \text{ lbf/in}^2 \text{ absolute} \qquad Ans.\ (e)$$

These are reasonable estimates in spite of the fact that 10° is really not a "small" flow deflection.

9.10 PRANDTL-MEYER EXPANSION WAVES

The oblique-shock solution of Section 9.9 is for a finite compressive deflection θ which obstructs a supersonic flow and thus decreases its Mach number and velocity. The present section treats gradual changes in flow angle which are primarily *expansive*; i.e., they widen the flow area and increase the Mach number and velocity. The property changes accumulate in infinitesimal increments, and the linearized relations (9.88) and (9.89) are used. The local flow deflections are infinitesimal, so that the flow is nearly isentropic according to Eq. (9.90).

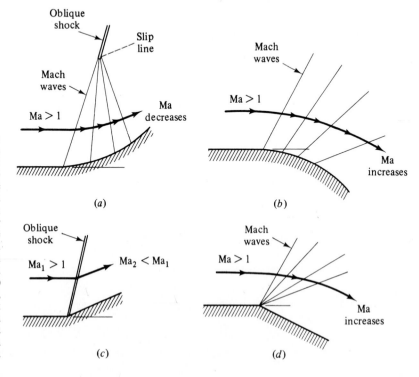

Fig. 9.25 Some examples of supersonic expansion and compression: (a) gradual isentropic compression on a concave surface, Mach waves coalesce farther out to form oblique shock; (b) gradual isentropic expansion on convex surface, Mach waves diverge; (c) sudden compression, nonisentropic shock forms; (d) sudden expansion, centered isentropic fan of Mach waves forms.

Figure 9.25 shows four examples, one of which (Fig. 9.25c) fails the test for gradual changes. The gradual compression of Fig. 9.25a is essentially isentropic, with a smooth increase in pressure along the surface, but the Mach angle decreases along the surface and the waves tend to coalesce farther out into an oblique-shock wave. The gradual expansion of Fig. 9.25b causes a smooth isentropic increase of Mach number and velocity along the surface, with diverging Mach waves formed.

The sudden compression of Fig. 9.25c cannot be accomplished by Mach waves: an oblique shock forms, and the flow is nonisentropic. This could be what you would see if you looked at Fig. 9.25a from far away. Finally, the sudden expansion of Fig. 9.25d is isentropic and forms a fan of centered Mach waves emanating from the corner. Note that the flow on any streamline passing through the fan changes smoothly to higher Mach number and velocity. In the limit as we near the corner, the flow expands almost discontinuously at the surface. The cases in Fig. 9.25a, b, and d can all be handled by the Prandtl-Meyer supersonic-wave theory of the present section, first formulated by Ludwig Prandtl and his student Theodor Meyer in 1907–1908.

Note that none of this discussion makes sense if the upstream Mach number is subsonic, since Mach wave and shock wave patterns cannot exist in subsonic flow.

The Prandtl-Meyer Perfect-Gas Function

Consider a small, nearly infinitesimal flow deflection $d\theta$ such as occurs between the first two Mach waves in Fig. 9.25a. From Eqs. (9.88) and (9.89) we have, in the limit,

$$\beta \approx \mu = \sin^{-1} \frac{1}{\text{Ma}} \tag{9.91a}$$

$$\frac{dp}{p} \approx \frac{\gamma \, \text{Ma}^2}{(\text{Ma}^2 - 1)^{1/2}} \, d\theta \tag{9.91b}$$

Since the flow is nearly isentropic, we have the frictionless differential momentum equation for a perfect gas

$$dp = -\rho V \, dV = -\gamma p \, \text{Ma}^2 \frac{dV}{V} \tag{9.92}$$

Combining Eqs. (9.91a) and (9.92) to eliminate dp, we obtain a relation between turning angle and velocity change

$$d\theta = -(\text{Ma}^2 - 1)^{1/2} \frac{dV}{V} \tag{9.93}$$

This can be integrated into a functional relation for finite turning angles if we can relate V to Ma. We do this from the definition of Mach number

$$V = \text{Ma} \, a$$

or
$$\frac{dV}{V} = \frac{d\,\text{Ma}}{\text{Ma}} + \frac{da}{a} \tag{9.94}$$

Finally, we can eliminate da/a because the flow is isentropic and hence a_0 is constant for a perfect gas

$$a = a_0[1 + \tfrac{1}{2}(\gamma - 1)\,\mathrm{Ma}^2]^{-1/2}$$

or
$$\frac{da}{a} = \frac{-\tfrac{1}{2}(\gamma - 1)\,\mathrm{Ma}\,d\,\mathrm{Ma}}{1 + \tfrac{1}{2}(\gamma - 1)\,\mathrm{Ma}^2} \tag{9.95}$$

Eliminating dV/V and da/a between Eqs. (9.93) to (9.95), we obtain a relation solely between turning angle and Mach number

$$d\theta = -\frac{(\mathrm{Ma}^2 - 1)^{1/2}}{1 + \tfrac{1}{2}(\gamma - 1)\,\mathrm{Ma}^2}\frac{d\,\mathrm{Ma}}{\mathrm{Ma}} \tag{9.96}$$

Before integrating this expression let us note that the primary application is to expansions, i.e., increasing Ma and decreasing θ. Therefore, for convenience, we define the Prandtl-Meyer angle $\omega(\mathrm{Ma})$ which increases when θ decreases and is zero at the sonic point

$$d\omega = -d\theta \qquad \omega = 0 \text{ at Ma} = 1 \tag{9.97}$$

Thus we integrate Eq. (9.96) from the sonic point to any value of Ma

$$\int_0^\omega d\omega = \int_1^{\mathrm{Ma}} \frac{(\mathrm{Ma}^2 - 1)^{1/2}}{1 + \tfrac{1}{2}(\gamma - 1)\,\mathrm{Ma}^2}\frac{d\,\mathrm{Ma}}{\mathrm{Ma}} \tag{9.98}$$

The integrals are evaluated in closed form, with the result, in radians,

$$\omega(\mathrm{Ma}) = K^{1/2}\tan^{-1}\left(\frac{\mathrm{Ma}^2 - 1}{K}\right)^{1/2} - \tan^{-1}(\mathrm{Ma}^2 - 1)^{1/2} \tag{9.99}$$

where
$$K = \frac{\gamma + 1}{\gamma - 1}$$

This is the *Prandtl-Meyer supersonic expansion function*, which is plotted in Fig. 9.26 and tabulated in Table B.5 for $\gamma = 1.4$. The angle ω changes rapidly at first and then levels off at high Mach number to a limiting value as $\mathrm{Ma} \to \infty$:

$$\omega_{\max} = \frac{\pi}{2}(K^{1/2} - 1) = 130.45° \qquad \text{if } \gamma = 1.4 \tag{9.100}$$

Thus a supersonic flow can expand only through a finite turning angle before it reaches infinite Mach number, maximum velocity, and zero temperature.

Gradual expansion or compression between finite Mach numbers Ma_1 and Ma_2, neither of which is unity, is computed by relating the turning angle $\Delta\omega$ to the difference in Prandtl-Meyer angles for the two conditions

$$\Delta\omega_{1\to2} = \omega(\mathrm{Ma}_2) - \omega(\mathrm{Ma}_1) \tag{9.101}$$

The change $\Delta\omega$ may be either positive (expansion) or negative (compression) as long as the end conditions lie in the supersonic range. Let us illustrate with an example.

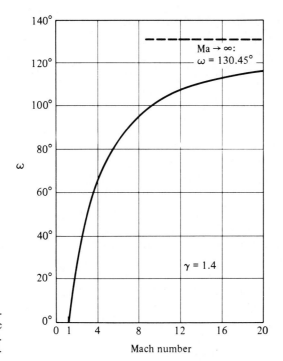

EXAMPLE 9.18   Air ($\gamma = 1.4$) flows at $Ma_1 = 3.0$ and $p_1 = 200\,kPa$. Compute the final downstream Mach number and pressure for (a) an expansion turn of $20°$ and (b) a gradual compression turn of $20°$.

Solution

Part (a)   The isentropic stagnation pressure is

$$p_0 = p_1[1 + 0.2(3.0)^2]^{3.5} = 7347\,kPa$$

and this will be the same at the downstream point. For $Ma_1 = 3.0$ we find from Table B.5 or Eq. (9.99) that $\omega_1 = 49.757°$. The flow expands to a new condition such that

$$\omega_2 = \omega_1 + \Delta\omega = 49.757° + 20° = 69.757°$$

From Eq. (9.99) or interpolation in Table B.5 this corresponds to

$$Ma_2 = 4.32 \qquad\qquad\qquad \textit{Ans. (a)}$$

The isentropic pressure at this new condition is

$$p_2 = \frac{p_0}{[1 + 0.2(4.32)^2]^{3.5}} = \frac{7347}{230.1} = 31.9\,kPa \qquad \textit{Ans. (a)}$$

Part (b)   The flow compresses to a lower Prandtl-Meyer angle

$$\omega_2 = 49.757° - 20° = 29.757°$$

Again from Eq. (9.99) or Table B.5 we compute that

$$\text{Ma}_2 = 2.125 \qquad \qquad Ans. (b)$$

$$p_2 = \frac{p_0}{[1 + 0.2(2.125)^2]^{3.5}} = \frac{7347}{9.51} = 773 \text{ kPa} \qquad Ans. (b)$$

Similarly, density and temperature changes are computed by noticing that T_0 and ρ_0 are constant for isentropic flow.

Application to Supersonic Airfoils

The oblique-shock and Prandtl-Meyer expansion theories can be used to patch together a number of interesting and practical supersonic flow fields. This marriage, called *shock-expansion theory*, is limited by two conditions: (1) except in rare instances the flow must be supersonic throughout, and (2) the wave pattern must not suffer interference from waves formed in other parts of the flow field.

A very successful application of shock-expansion theory is to supersonic airfoils. Figure 9.27 shows two examples, a flat plate and a diamond-shaped foil. In contrast to subsonic-flow designs (Fig. 8.25), these airfoils must have sharp leading edges,

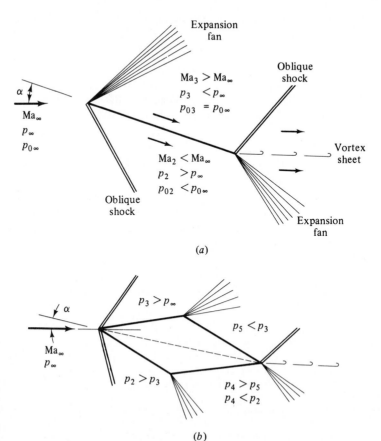

Fig. 9.27 Supersonic airfoils: (*a*) flat plate, higher pressure on lower surface, drag due to small downstream component of net pressure force; (*b*) diamond foil, higher pressures on both lower surfaces, additional drag due to body thickness.

which form attached oblique shocks or expansion fans. Rounded supersonic leading edges would cause detached bow shocks, as in Fig. 9.19 or 9.22b, greatly increasing the drag and lowering the lift.

In applying shock-expansion theory one examines each surface turning angle to see whether it is an expansion ("opening up") or compression (obstruction) to the surface flow. Figure 9.27a shows a flat-plate foil at an angle of attack. There is a leading-edge shock on the lower edge with flow deflection $\theta = \alpha$, while the upper edge has an expansion fan with increasing Prandtl-Meyer angle $\Delta\omega = \alpha$. We compute p_3 with expansion theory and p_2 with oblique-shock theory. The force on the plate is thus $F = (p_2 - p_3)Cb$, where C is the chord length and b the span width (assuming no wingtip effects). This force is normal to the plate, and thus the lift force normal to the stream is $L = F \cos \alpha$ and the drag parallel to the stream is $D = F \sin \alpha$. The dimensionless coefficients C_L and C_D have the same definitions as in low-speed flow, Eq. (7.65), except that the perfect-gas-law identity $\frac{1}{2}\rho V^2 \equiv \frac{1}{2}\gamma p \, \mathrm{Ma}^2$ is very useful here

$$C_L = \frac{L}{\frac{1}{2}\gamma p_\infty \, \mathrm{Ma}_\infty^2 \, bC} \qquad C_D = \frac{D}{\frac{1}{2}\gamma p_\infty \, \mathrm{Ma}_\infty^2 \, bC} \tag{9.102}$$

The typical supersonic lift coefficient is much smaller than the subsonic value $C_L \approx 2\pi\alpha$, but the lift can be very large because of the large value of $\frac{1}{2}\rho V^2$ at supersonic speeds.

At the trailing edge in Fig. 9.27a a shock and fan appear in reversed positions and bend the two flows back so that they are parallel in the wake and have the same pressure. They do not have quite the same velocity because of the unequal shock strengths on the upper and lower surfaces; hence a vortex sheet trails behind the wing. This is very interesting, but in the theory you ignore the trailing-edge pattern entirely, since it does not affect the surface pressures: the supersonic surface flow cannot "hear" the wake disturbances.

The diamond foil in Fig. 9.27b adds two more wave patterns to the flow. At this particular α less than the diamond half-angle, there are leading-edge shocks on both surfaces, the upper shock being much weaker. Then there are expansion fans on each shoulder of the diamond: the Prandtl-Meyer angle change $\Delta\omega$ equals the sum of the leading-edge and trailing-edge diamond half-angles. Finally, the trailing-edge pattern is similar to that of the flat plate (9.27a) and can be ignored in the calculation. Both lower-surface pressures p_2 and p_4 are greater than their upper counterparts, and the lift is nearly that of the flat plate. There is an additional drag due to thickness, because p_4 and p_5 on the trailing surfaces are lower than their counterparts p_2 and p_3. The diamond drag is greater than the flat-plate drag, but this must be endured in practice to achieve a wing structure strong enough to hold these forces.

The theory sketched in Fig. 9.27 is in good agreement with measured supersonic lift and drag as long as the Reynolds number is not too small (thick boundary layers) and the Mach number not too large (hypersonic flow). It turns out that for large Re_C and moderate supersonic Ma_∞ the boundary layers are thin and separation seldom occurs, so that the shock-expansion theory, although frictionless, is quite successful. Let us look now at an example.

EXAMPLE 9.19   A flat-plate airfoil with $C = 2$ m is immersed at $\alpha = 8°$ in a stream with $Ma_\infty = 2.5$ and $p_\infty = 100$ kPa. Compute (a) C_L and (b) C_D and compare with low-speed airfoils. Compute (c) lift and (d) drag in newtons per unit span width.

Solution   Instead of using a lot of space outlining the detailed oblique-shock and Prandtl-Meyer expansion computations, we list all pertinent results in Fig. E9.19 on the upper and lower surfaces. Using the theories of Secs. 9.9 and 9.10, you should verify every single one of the calculations in Fig. E9.19 to make sure that all details of shock-expansion theory are well understood.

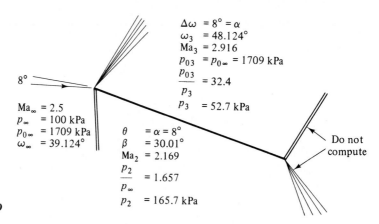

Fig. E9.19

The important final results are p_2 and p_3, from which the total force per unit width on the plate is

$$F = (p_2 - p_3)bC = (165.7 - 52.7)(\text{kPa})(1 \text{ m})(2 \text{ m}) = 226 \text{ kN}$$

The lift and drag per meter width are thus

$$L = F \cos 8° = 224 \text{ kN} \qquad\qquad \textit{Ans. (c)}$$

$$D = F \sin 8° = 31 \text{ kN} \qquad\qquad \textit{Ans. (d)}$$

These are very large forces for only 2 m² of wing area.

From Eq. (9.102) the lift coefficient is

$$C_L = \frac{224 \text{ kN}}{\frac{1}{2}(1.4)(100 \text{ kPa})(2.5)^2(2 \text{ m}^2)} = 0.256 \qquad\qquad \textit{Ans. (a)}$$

The comparable low-speed coefficient from Eq. (8.96) is $C_L = 2\pi \sin 8° = 0.874$, which is 3.4 times larger.

From Eq. (9.102) the drag coefficient is

$$C_D = \frac{31 \text{ kN}}{\frac{1}{2}(1.4)(100 \text{ kPa})(2.5)^2(2 \text{ m}^2)} = 0.036 \qquad\qquad \textit{Ans. (b)}$$

From Fig. 7.22 for the NACA 0009 airfoil C_D at $\alpha = 8°$ is about 0.009, or about 4 times smaller.

Notice that this supersonic theory predicts a finite drag in spite of assuming frictionless flow with infinite wing aspect ratio. This is called *wave drag*, and we see that the d'Alembert paradox of zero body drag does not occur in supersonic flow.

Thin-Airfoil Theory

In spite of the simplicity of the flat-plate geometry, the calculations in Example 9.19 were laborious. In 1925 Ackeret [21] developed simple yet effective expressions for the lift, drag, and center of pressure of supersonic airfoils, assuming small thickness and angle of attack.

The theory is based on the linearized expression (9.89), where $\tan \theta \approx$ the surface deflection relative to the free stream and condition 1 is the free stream, $\text{Ma}_1 = \text{Ma}_\infty$. For the flat-plate airfoil the total force F is based on

$$\frac{p_2 - p_3}{p_\infty} = \frac{p_2 - p_\infty}{p_\infty} - \frac{p_3 - p_\infty}{p_\infty}$$

$$= \frac{\gamma \, \text{Ma}_\infty^2}{(\text{Ma}_\infty^2 - 1)^{1/2}} [\alpha - (-\alpha)] \qquad (9.103)$$

Substitution into Eq. (9.102) gives the linearized lift coefficient for a supersonic flat-plate airfoil

$$C_L \approx \frac{(p_2 - p_3)bC}{\frac{1}{2}\gamma p_\infty \, \text{Ma}_\infty^2 \, bC} \approx \frac{4\alpha}{(\text{Ma}_\infty^2 - 1)^{1/2}} \qquad (9.104)$$

Computations for diamond and other finite-thickness airfoils show no first-order effect of thickness on lift. Therefore Eq. (9.104) is valid for any sharp-edged supersonic thin airfoil at a small angle of attack.

The flat-plate drag coefficient is

$$C_D = C_L \tan \alpha \approx C_L \alpha \approx \frac{4\alpha^2}{(\text{Ma}_\infty^2 - 1)^{1/2}} \qquad (9.105)$$

However, the thicker airfoils have additional thickness drag. Let the chord line of the airfoil be the x axis and let the upper surface shape be denoted by $y_u(x)$ and the lower profile by $y_l(x)$. Then the complete Ackeret drag theory (see, for example, Ref. 8, sec. 14.6, for details) shows that the additional drag depends on the mean square of the slopes of the upper and lower surfaces, defined by

$$\overline{y'^2} = \frac{1}{C} \int_0^C \left(\frac{dy}{dx}\right)^2 dx \qquad (9.106)$$

The final expression for drag [8, p. 442] is

$$C_D \approx \frac{4}{(\text{Ma}_\infty^2 - 1)^{1/2}} [\alpha^2 + \tfrac{1}{2}(\overline{y_u'^2} + \overline{y_l'^2})] \qquad (9.107)$$

These are all in reasonable agreement with more exact computations, and their extreme simplicity makes them attractive alternatives to the laborious but accurate shock-expansion theory. Consider the following example.

EXAMPLE 9.20   Repeat parts (*a*) and (*b*) of Example 9.19 using the linearized Ackeret theory.

Solution   From Eqs. (9.104) and (9.105) we have, for $Ma_\infty = 2.5$ and $\alpha = 8° = 0.1396$ rad,

$$C_L \approx \frac{4(0.1396)}{(2.5^2 - 1)^{1/2}} = 0.244 \qquad C_D = \frac{4(0.1396)^2}{(2.5^2 - 1)^{1/2}} = 0.034 \qquad \textit{Ans.}$$

These are only 5 percent lower than the more exact computations of Example 9.19.

A further result of the Ackeret linearized theory is an expression for the position x_{CP} of the center of pressure of the force distribution on the wing:

$$\frac{x_{CP}}{C} = 0.5 + \frac{S_u - S_l}{2\alpha C^2} \qquad (9.108)$$

where S_u is the cross-sectional area between the upper surface and the chord and S_l the area between the chord and the lower surface. For a symmetric airfoil ($S_l = S_u$) we obtain x_{CP} at the half-chord point, in contrast with the low-speed airfoil result of Eq. (8.98), where x_{CP} is at the quarter-chord.

The difference in difficulty between the simple Ackeret theory and shock-expansion theory is even greater for a thick airfoil, as the following example shows.

EXAMPLE 9.21   By analogy with Example 9.19 analyze a diamond, or double-wedge, airfoil of 2° half-angle and $C = 2$ m at $\alpha = 8°$ and $Ma_\infty = 2.5$. Compute C_L and C_D by (*a*) shock-expansion theory and by (*b*) Ackeret theory. Pinpoint the difference from Example 9.19.

Solution
Part (*a*)   Again we omit the details of shock-expansion theory and simply list the properties computed on each of the four airfoil surfaces in Fig. E9.21. Assume $p_\infty = 100$ kPa. There are both a force F normal to the chord line and a force P parallel to the chord. For the normal force the pressure difference on the front half is $p_2 - p_3 = 186.4 - 65.9 = 120.5$ kPa, and on the rear half it is $p_4 - p_5 = 146.9 - 48.8 = 98.1$ kPa. The average pressure difference is $\frac{1}{2}(120.5 + 98.1) = 109.3$ kPa, so that the normal force is

$$F = (109.3 \text{ kPa})(2 \text{ m}^2) = 218.6 \text{ kN}$$

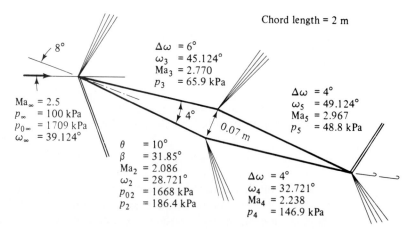

Fig. E9.21

For the chordwise force P the pressure difference on the top half is $p_3 - p_5 = 65.9 - 48.8 = 17.1$ kPa and on the bottom half is $p_2 - p_4 = 186.4 - 146.9 = 39.5$ kPa. The average difference is $\frac{1}{2}(17.1 + 39.5) = 28.3$ kPa, which when multiplied by the frontal area (maximum thickness times 1 m width) gives

$$P = (28.3 \text{ kPa})(0.07 \text{ m})(1 \text{ m}) = 2.0 \text{ kN}$$

Both F and P have components in the lift and drag directions. The lift force normal to the free stream is

$$L = F \cos 8° - P \sin 8° = 216.2 \text{ kN}$$

and

$$D = F \sin 8° + P \cos 8° = 32.4 \text{ kN}$$

For computing the coefficients, the denominator of Eq. (9.102) is the same as in Example 9.19: $\frac{1}{2}\gamma p_\infty \text{ Ma}_\infty^2 \, bC = \frac{1}{2}(1.4)(100 \text{ kPa})(2.5)^2(2 \text{ m}^2) = 875$ kN. Thus, finally, shock-expansion theory predicts

$$C_L = \frac{216.2 \text{ kN}}{875 \text{ kN}} = 0.247 \qquad C_D = \frac{32.4 \text{ kN}}{875 \text{ kN}} = 0.0370 \qquad \text{Ans. } (a)$$

Part (b) Meanwhile, by Ackeret theory, C_L is the same as in Example 9.20:

$$C_L = \frac{4(0.1396)}{(2.5^2 - 1)^{1/2}} = 0.244 \qquad \text{Ans. } (b)$$

This is 1 percent less than the shock-expansion result above. For the drag we need the mean-square slopes from Eq. (9.106)

$$\overline{y_u'^2} = \overline{y_l'^2} = \tan^2 2° = 0.00122$$

Then Eq. (9.107) predicts the linearized result

$$C_D = \frac{4}{(2.5^2 - 1)^{1/2}} [(0.1396)^2 + \tfrac{1}{2}(0.00122 + 0.00122)] = 0.0362 \qquad \text{Ans. } (b)$$

This is 2 percent lower than shock-expansion theory. We could judge Ackeret theory to be "satisfactory." Ackeret theory predicts $p_2 = 167$ kPa $(-11$ percent), $p_3 = 60$ kPa $(-9$ percent), $p_4 = 140$ kPa $(-5$ percent), and $p_5 = 33$ kPa $(-6$ percent).

Three-Dimensional Supersonic Flow

We have gone about as far as we can go in an introductory treatment of compressible flow. Of course there is much more, and you are invited to study further in the references at the end of the chapter.

Three-dimensional supersonic flows are highly complex, especially if they concern blunt bodies, which therefore contain embedded regions of subsonic and transonic flow, e.g., Fig. 9.10. Some flows, however, yield to accurate theoretical treatment, like flow past a cone at zero incidence, as shown in Fig. 9.28. The exact theory of cone flow is discussed in advanced texts [e.g., 8, chap. 17], and extensive tables of such solutions have been published [22, 23]. There are similarities between cone flow and the wedge flows illustrated in Fig. 9.22: an attached oblique shock, a thin turbulent boundary layer, and an expansion fan at the rear corner.

Fig. 9.28
Shadowgraph of flow past an 8° half-angle cone at $Ma_\infty = 2.0$. The turbulent boundary layer is clearly visible. The Mach lines curve slightly, and the Mach number varies from 1.98 just inside the shock to 1.90 at the surface. (*Courtesy of U.S. Army Ballistic Research Center, Aberdeen Proving Ground.*)

However, the conical shock deflects the flow through an angle less than the cone half-angle, unlike the wedge shock. As in the wedge flow, there is a maximum cone angle above which the shock must detach, as in Fig. 9.22b. For $\gamma = 1.4$ and $Ma_\infty = \infty$ the maximum cone half-angle for an attached shock is about 57°, compared with the maximum wedge angle of 45.6° (see Ref. 23).

For more complicated body shapes one usually resorts to experimentation in a supersonic wind tunnel. Figure 9.29 shows a wind-tunnel study of supersonic flow past a model of an interceptor aircraft. The many junctions and wingtips and shape changes make theoretical analysis very difficult. Here the surface-flow patterns, which indicate boundary-layer development and regions of flow separation, have been visualized by the smearing of oil drops placed on the model surface before the test.

As we shall see in the next chapter, there is an interesting analogy between gas-dynamic shock waves and the surface water waves which form in an open-channel flow. Chapter 11 of Ref. 14 explains how a water channel can be used in an inexpensive simulation of supersonic flow experiments.

Like most other areas of fluid mechanics, the visual aspects of compressible flow are so striking and educational that they should not be missed. The films 18 to 22 on compressible flow in Appendix C are highly recommended and useful to both novices and experts.

Fig. 9.29 Wind-tunnel test of the Cobra P-530 supersonic interceptor. The surface flow patterns are visualized by the smearing of oil droplets. (*Courtesy of Northrup Corporation.*)

SUMMARY

This chapter is a brief introduction to a very broad subject, compressible flow, sometimes called *gas dynamics*. The primary parameter is the Mach number V/a, which is large and causes the fluid density to vary significantly. This means that the continuity and momentum equations must be coupled to the energy relation and the equation of state to solve for the four unknowns (p, ρ, T, **V**).

The chapter reviews the thermodynamic properties of an ideal gas and derives a formula for the speed of sound of a fluid. The analysis is then simplified to one-dimensional steady adiabatic flow without shaft work, for which the stagnation enthalpy of the gas is constant. A further simplification to isentropic flow enables formulas to be derived for high-speed gas flow in a variable-area duct. This reveals the phenomenon of sonic-flow *choking* (maximum mass flow) in the throat of a nozzle. At supersonic velocities there is the possibility of a normal-shock wave, where the gas discontinuously reverts to subsonic conditions. The normal shock explains the effect of back pressure on the performance of converging-diverging nozzles.

To illustrate nonisentropic flow conditions, there is a brief study of constant-area duct flow with friction and with heat transfer, both of which lead to choking of the exit flow.

The chapter ends with a discussion of two-dimensional supersonic flow, where oblique-shock waves and Prandtl-Meyer (isentropic) expansion waves appear.

With a proper combination of shocks and expansions one can analyze supersonic airfoils.

REFERENCES

1. A. Y. Pope, "Aerodynamics of Supersonic Flight," 2d ed., Pitman, New York, 1958.
2. A. B. Cambel and B. H. Jennings, "Gas Dynamics," McGraw-Hill, New York, 1958.
3. F. Cheers, "Elements of Compressible Flow," Wiley, New York, 1963.
4. J. E. John, "Gas Dynamics," 2d ed., Allyn & Bacon, Boston, 1984.
5. A. J. Chapman and W. F. Walker, "Introductory Gas Dynamics," Holt, New York, 1971.
6. B. W. Imric, "Compressible Fluid Flow," Halstead, New York, 1974.
7. R. Courant and K. O. Friedrichs, "Supersonic Flow and Shock Waves," Interscience, New York, 1948; reprinted by Springer-Verlag, New York, 1977.
8. A. H. Shapiro, "The Dynamics and Thermodynamics of Compressible Fluid Flow," 2 vols., Ronald, New York, 1953.
9. H. W. Liepmann and A. Roshko, "Elements of Gasdynamics," Wiley, New York, 1957.
10. R. von Mises, "Mathematical Theory of Compressible Fluid Flow," Academic, New York, 1958.
11. J. A. Owczarek, "Gas Dynamics," International Textbook, Scranton, Pa., 1964.
12. W. G. Vincenti and C. Kruger, "Introduction to Physical Gas Dynamics," Wiley, New York, 1965.
13. W. D. Hayes and R. F. Probstein, "Hypersonic Flow Theory," vol. 1, Academic, New York, 1966.
14. P. A. Thompson, "Compressible Fluid Dynamics," McGraw-Hill, New York, 1972.
15. J. H. Keenan et al., "Steam Tables: Thermodynamic Properties of Water Including Vapor, Liquid, and Solid Phases," Wiley-Interscience, New York, 1978.
16. J. H. Keenan and J. Kaye, "Gas Tables," 2d ed., Wiley, New York, 1980.
17. W. C. Reynolds and H. C. Perkins, "Engineering Thermodynamics," McGraw-Hill, New York, 1977.
18. K. Wark, "Thermodynamics," 4th ed., McGraw-Hill, New York, 1983.
19. F. M. White, "Viscous Fluid Flow," McGraw-Hill, New York, 1974.
20. J. H. Keenan and E. P. Neumann, Measurements of Friction in a Pipe for Subsonic and Supersonic Flow of Air, *J. Appl. Mech.*, vol. 13, no. 2, p. A-91, 1946.
21. J. Ackeret, Air Forces on Airfoils Moving Faster than Sound Velocity, *NACA Tech. Mem.* 317, 1925.
22. Z. Kopal, Tables of Supersonic Flow around Cones, *M.I.T. Center Anal. Tech. Rep.* 1, 1947 (see also *Tech. Rep.* 3 and 5, 1947).
23. J. L. Sims, Tables for Supersonic Flow around Right Circular Cones at Zero Angle of Attack, NASA SP-3004, 1964 (see also NASA SP-3007).
24. Ames Research Staff, Equations, Tables, and Charts for Compressible Flow, *NACA Rep.* 1135, 1953.
25. L. Rosenhead, "A Selection of Graphs for Use in Calculations of Compressible Flow," Clarendon Press, London, 1954.
26. S. M. Yahya, "Fundamentals of Compressible Flow," Wiley Eastern, New Delhi, 1982.
27. S. Schreier, "Compressible Flow," Wiley, New York, 1982.
28. M. A. Saad, "Compressible Fluid Flow," Prentice-Hall, Englewood Cliffs, N.J., 1985.
29. A. Y. Pope and K. L. Goin, "High Speed Wind Tunnel Testing," Wiley, New York, 1965.
30. W. Bober and R. A. Kenyon, "Fluid Mechanics," Wiley, New York, 1980.
31. J. D. Anderson, "Modern Compressible Flow: With Historical Perspective," McGraw-Hill, New York, 1982.

RECOMMENDED FILMS

Films numbered 10 and 18 to 22 in Appendix C.

Problems

PROBLEM DISTRIBUTION

Section	Topic	Problems
9.1	Introduction	9.1–9.7
9.2	The speed of sound	9.8–9.18
9.3	Adiabatic and isentropic flows	9.19–9.30
9.4	Isentropic flow with area changes	9.31–9.45
9.5	The normal-shock wave	9.46–9.54
9.6	Converging and diverging nozzles	9.55–9.70
9.7	Duct flow with friction	9.71–9.83
9.8	Frictionless duct flow with heat transfer	9.84–9.89
9.9	Mach waves	9.90–9.92, 9.95
9.9	The oblique-shock wave	9.93, 9.96–9.105, 9.107
9.10	Prandtl-Meyer expansion waves	9.94, 9.97, 9.103, 9.106
9.10	Supersonic airfoils	9.108–9.114

9.1 A gas flows adiabatically through a duct. At section 1 $p_1 = 200\,lbf/in^2$ absolute, $T_1 = 500°F$, and $V_1 = 250\,ft/s$, while farther downstream $V_2 = 1100\,ft/s$, $p_2 = 40\,lbf/in^2$ absolute. Calculate T_2 in degrees Fahrenheit and $s_2 - s_1$ in Btu/(lb · °R) if the gas is (a) air ($\gamma = 1.4$), (b) argon ($\gamma = 1.67$).

9.2 Solve Prob. 9.1 if the gas is steam. Use two approaches: (a) an ideal gas from Table A.4 and (b) a real gas from the steam tables [15].

9.3 If 6 kg of oxygen in a closed tank at 250 kPa is heated from 100 to 350°C, calculate the new pressure, the heat added, and the change in entropy.

9.4 Steam at 400°F and 60 lbf/in² absolute is compressed isentropically to 100 lbf/in² absolute. What is the new temperature in degrees Fahrenheit?

9.5 Carbon dioxide ($\gamma = 1.28$) enters a constant-area duct at 400°F, 100 lbf/in² absolute, and 500 ft/s. Farther downstream the properties are $V_2 = 1000\,ft/s$ and $T_2 = 900°F$. Compute (a) p_2, (b) the heat added between sections, (c) the entropy change between sections, and (d) the mass flow per unit area. *Hint*: This problem requires the continuity equation.

9.6 Atmospheric air at 20°C enters and fills an insulated tank which is initially evacuated. Using a control-volume analysis from Eq. (3.82), compute the tank air temperature when it is full.

9.7 Steam enters a duct at $p_1 = 50\,lbf/in^2$ absolute, $T_1 = 360°F$, $V_1 = 200\,ft/s$ and leaves at $p_2 = 100\,lbf/in^2$ absolute, $T_2 = 800°F$, and $V_2 = 1200\,ft/s$. How much heat was added in Btu per pound?

9.8 A certain aircraft flies at the same Mach number regardless of its altitude. It flies 100 km/h slower at 10,000 m standard altitude than at sea level. What is its Mach number?

9.9 At 250°C and 1 atm compute the speed of sound of (a) air, (b) oxygen, (c) hydrogen, (d) steam, (e) carbon monoxide, and (f) $^{238}UF_6$ ($\gamma \approx 1.06$).

9.10 Assuming that water follows the liquid equation of state (1.38) with $n = 7$ and $B = 3000$, compute the bulk modulus in pounds force per square inch absolute and speed of sound in feet per second at (a) 1 atm and (b) 1100 atm (the deepest part of the ocean).

9.11 The measured value of the speed of sound of water at 20°C and 9000 atm is 2650 m/s (A. H. Smith and A. W. Lawson, *J. Chem. Phys.*, vol. 22, 1954, p. 351). Compare this with the value computed by the analysis of Prob. 9.10.

9.12 Mercury at 1 atm has a bulk modulus of about 2.8×10^{10} Pa. It has also $n \approx 7$ in Eq. (1.38). What value of B in Eq. (1.38) best fits the measured bulk modulus? Estimate the bulk modulus and speed of sound of mercury at 2000 atm.

9.13 The properties of a dense gas (high pressure and low temperature) are often approximated by van der Waals' equation of state [17, p. 250, or 18, p. 532]:

$$p = \frac{\rho RT}{1 - b_1 \rho} - a_1 \rho^2$$

where the constants a_1 and b_1 can be found from the critical temperature and pressure:

$$a_1 = \frac{27 R^2 T_c^2}{64 p_c} = 9.0 \times 10^5 \ (\text{lbf} \cdot \text{ft}^4)/\text{slug}^2$$

for air, and

$$b_1 = \frac{RT_c}{8 p_c} = 0.65 \ \text{ft}^3/\text{slug}$$

for air. Find an analytic expression for the speed of sound of a van der Waals gas. Assuming $\gamma = 1.4$, compute the speed of sound of air in feet per second at $-100°F$ and 20 atm for (a) a perfect gas and (b) a van der Waals gas. What percentage higher density does the van der Waals relation predict?

9.14 Why do (a) water and mercury from Table 9.1 and (b) aluminum and steel have nearly equal speeds of sound in spite of the fact that the second material of each pair is much heavier? Can this behavior be predicted from molecular theory?

9.15 An airplane flies at 350 m/s through air at $-20°C$ and 40 kPa. Is the airplane supersonic?

9.16 A weak pressure wave (sound wave) with a pressure change $\Delta p = 40$ Pa propagates through air at $20°C$ and 1 atm. Estimate (a) the density change, (b) the temperature change, and (c) the velocity change across the wave.

9.17 A weak pressure pulse Δp propagates through still air. Discuss the type of reflected pulse which occurs and the boundary conditions which must be satisfied when the wave strikes normal to, and is reflected from, (a) a solid wall and (b) a free liquid surface.

9.18 A submarine at a depth of 800 m sends a sonar signal and receives the reflected wave back from a similar submerged object in 15 s. Using Prob. 9.10 as a guide, estimate the distance to the other object.

9.19 A high-speed gas flow has $V = 800$ m/s, $p = 140$ kPa, and $T = 200°C$. Compute the pressure and temperature which the gas would achieve if brought isentropically to rest for (a) air, (b) argon, and (c) steam.

9.20 The Concorde airplane flies at Ma = 2.2 at 10,000 m standard altitude. Estimate the Celsius air

temperature at the front stagnation point. At what Mach number would it fly to have a front stagnation temperature of $400°C$?

9.21 A gas flows at $V = 300$ m/s, $p = 100$ kPa, and $T = 150°C$. Compute the maximum velocity attainable by adiabatic expansion of this gas for (a) air, (b) helium.

9.22 Air expands isentropically through a duct from $p_1 = 125$ kPa and $T_1 = 100°C$ to $p_2 = 80$ kPa and $V_2 = 325$ m/s. Compute (a) T_2, (b) Ma_2, (c) T_0, (d) p_0, (e) V_1, (f) Ma_1.

9.23 Given the pitot stagnation temperature and pressure and the static-pressure measurements in Fig. P9.23, estimate the air velocity V assuming (a) incompressible flow and (b) compressible flow.

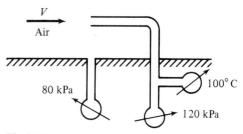

Fig. P9.23

9.24 For low-speed flow the stagnation pressure can be computed from Bernoulli's incompressible equation, $p_0 = p + \frac{1}{2}\rho V^2$. For high-speed perfect-gas flow Eq. (9.28a) should be used. Show that Eq. (9.28a) can be expanded into a series as follows:

$$p_0 = p + \tfrac{1}{2}\rho V^2 \left(1 + \tfrac{1}{4} \text{Ma}^2 + \frac{2 - \gamma}{24} \text{Ma}^4 + \cdots \right)$$

For $\gamma = 1.4$, at what Mach number will the difference $p_0 - p$ from the incompressible formula be in error by 5 percent? Suppose that a pitot tube measures $p_0 - p$ in subsonic flow and that the velocity is computed from the approximation $V^2 = 2(p_0 - p)/\rho_0$. At what Mach number will this calculation be in error by 2 percent?

9.25 Air flows isentropically in a duct. At section 1 $Ma_1 = 0.6$, $T = 250°C$, and $p_1 = 300$ kPa. At section 2 $Ma_2 = 3.1$. Compute (a) p_2, (b) T_2, and (c) p_{02}.

9.26 If it is known that the air velocity in the duct is 750 ft/s, use the mercury-manometer measurement in Fig. P9.26 to estimate the static pressure in the duct in pounds force per square inch absolute.

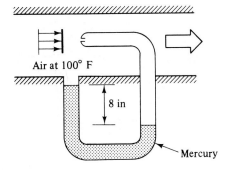

Fig. P9.26

Air at 100° F

8 in

Mercury

9.27 Show that for isentropic flow of a perfect gas, if a pitot-static probe measures p_0, p, and T_0, the gas velocity can be calculated from

$$V^2 = 2c_p T_0 \left[1 - \left(\frac{p}{p_0} \right)^{(\gamma - 1)/\gamma} \right]$$

What would be a source of error if a shock wave is formed in front of the probe?

9.28 At the exit of a nozzle flow the air temperature is 50°C and the velocity is 360 m/s. What is the Mach number and stagnation temperature there? If the flow is adiabatic, what is the Mach number upstream, where $T = 100°C$?

9.29 The large compressed-air tank in Fig. P9.29 exhausts from a nozzle at an exit velocity of 235 m/s. The mercury manometer reads $h = 30$ cm. Assuming isentropic flow, compute the pressure (a) in the tank and (b) in the atmosphere. (c) What is the exit Mach number?

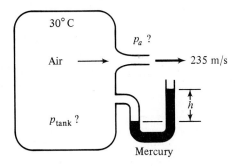

30° C

Air

p_a ?

235 m/s

p_{tank} ?

h

Mercury

Fig. P9.29

9.30 Air flows adiabatically through a duct. At one section $V_1 = 400$ ft/s, $T_1 = 200°F$, and $p_1 = 35$ lbf/in^2

absolute, while farther downstream $V_2 = 1100$ ft/s and $p_2 = 18$ lbf/in^2 absolute. Compute (a) Ma$_2$, (b) U_{max}, and (c) p_{02}/p_{01}.

9.31 Air flows isentropically from a reservoir, where $p = 300$ kPa and $T = 500$ K, to section 1 in a duct, where $A_1 = 0.2$ m^2 and $V_1 = 600$ m/s. Compute (a) Ma$_1$, (b) T_1, (c) p_1, (d) $\dot{m}$, and (e) A^*. Is the flow choked?

9.32 Steam in a tank at 450°F and 100 lbf/in^2 absolute exhausts through a converging nozzle of 0.1 in^2 throat area to a 1-atm environment. Compute the initial mass flow (a) for an ideal gas and (b) from the steam tables [15].

9.33 Air in a tank at 700 kPa and 20°C exhausts through a converging nozzle of throat area 0.65 cm^2 to 1-atm environment. Compute the initial mass flow in kilograms per second.

9.34 In Prob. 9.33, if the tank volume is 1.5 m^3, how long will it take the tank air pressure to blow down from 700 to 500 kPa absolute? Assume that the tank stagnation temperature is constant. How long will it blow down before the nozzle flow ceases being choked?

9.35 Make an exact control-volume analysis of the blowdown process in Fig. P9.35, assuming an insulated tank with negligible kinetic and potential energy within. Assume critical flow at the exit and show that both p_0 and T_0 decrease during blowdown. Set up first-order differential equations for $p_0(t)$ and $T_0(t)$ and reduce and solve as far as you can.

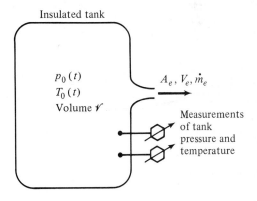

Insulated tank

$p_0(t)$
$T_0(t)$
Volume $\mathcal{V}$

A_e, V_e, $\dot{m}_e$

Measurements of tank pressure and temperature

Fig. P9.35

9.36 Prob. 9.35 makes an ideal senior project or combined laboratory and computer problem, as described in Ref. 30, sec. 8.6. In Bober and Kenyon's lab experiment the tank had a volume of 0.0352 ft^3 and was initially filled with air at 50 lb/in² gage and 72°F. Atmospheric pressure was 14.5 lb/in² absolute, and the nozzle exit diameter was 0.05 in. After 2 s of blowdown, the measured tank pressure was 20 lb/in² gage and the tank temperature was −5°F. Compare these values with the theoretical analysis of Prob. 9.35.

9.37 In Prob. 9.29, if $p_a = 1$ atm, what would the exit velocity and the manometer reading h be if the exit flow were critical?

9.38 Air is to be expanded isentropically from $p_0 = 150$ lbf/in² absolute and $T_0 = 200°F$ to a section where $A_1 = 3.5$ in² and $p_1 = 4$ lbf/in² absolute. Compute (a) Ma_1, (b) V_1, (c) the throat area, and (d) $\dot{m}$. At section 2 between the throat and section 1 the area is 1.5 in². What is Ma_2?

9.39 In Prob. 3.34 we knew nothing about compressible flow at the time so merely assumed exit conditions p_2 and T_2 and computed V_2 as an application of the continuity equation. Suppose that the throat diameter is 3 in. For the given stagnation conditions in the rocket chamber in Fig. P3.34 and assuming $\gamma = 1.4$ and a molecular weight of 26, compute the actual exit velocity, pressure, and temperature according to one-dimensional theory. If $p_a = 14.7$ lbf/in² absolute, compute the thrust from the analysis of Prob. 3.68. This thrust is entirely independent of the stagnation temperature (check this by changing T_0 to 2000°R if you like). Why?

9.40 Repeat Prob. 9.31 if the gas is argon (Table A.4).

9.41 At a point upstream of the throat of a converging-diverging nozzle the properties are $V_1 = 200$ m/s, $T_1 = 300$ K, and $p_1 = 125$ kPa. If the exit flow is supersonic, compute, from isentropic theory, (a) $\dot{m}$ and (b) A_1. The throat area is 35 cm².

9.42 In transonic wind tunnel testing the small area decrease caused by model blockage can be significant. Suppose the test-section Mach number is 1.08 and the section area is 1 m² for air with $T_0 = 20°C$. Find, according to one-dimensional theory, the percentage change in test-section velocity caused by a model of cross section 0.005 m² (a 0.5 percent blockage).

9.43 In flow of air in a converging-diverging duct with a supersonic exit, the throat area is 10 cm² and the throat pressure is 315 kPa. Find the pressure on either side of the throat where $A = 29$ cm².

9.44 A force $F = 2500$ N pushes a piston of diameter 12 cm through an insulated cylinder containing air at 20°C, as in Fig. P9.44. The exit diameter is 3 mm and $p_a = 1$ atm. Estimate (a) V_e, (b) V_p, and (c) $\dot{m}_e$.

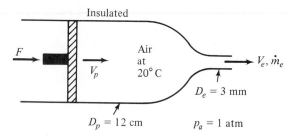

Fig. P9.44

(d) Under what conditions is the piston velocity independent of the force?

9.45 Air flows steadily from a reservoir at 20°C through a nozzle of exit area 20 cm² and strikes a vertical plate as in Fig. P9.45. The flow is subsonic throughout. A force of 135 N is required to hold the plate stationary. Compute (a) V_e, (b) Ma_e, and (c) p_0.

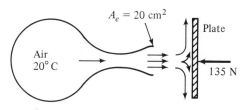

Fig. P9.45

9.46 For flow of air through a normal shock the upstream conditions are $V_1 = 600$ m/s, $T_{01} = 500$ K, $p_{01} = 700$ kPa. Compute the downstream conditions Ma_2, V_2, T_2, p_2, and p_{02}.

9.47 Air flows through a converging-diverging nozzle from a reservoir, where $p_0 = 400$ kPa. The throat area is 10 cm². A normal shock stands in the duct where $A = 15$ cm². Compute the pressure just downstream of this shock.

9.48 In Prob. 9.47 compute the pressure and Mach number at a point downstream of the normal shock, where $A = 30$ cm².

9.49 Air from a reservoir at 20°C and 500 kPa flows through a duct and forms a normal shock downstream of a throat of area 10 cm². By an odd coincidence it is found that the stagnation pressure downstream of this shock exactly equals the throat pressure. What is the area where the shock wave stands?

9.50 Air passes through a normal shock with upstream conditions $V_1 = 800$ m/s, $p_1 = 100$ kPa, and

$T_1 = 300$ K. What are the downstream conditions V_2 and p_2? What pressure p_2 would result if the same velocity change V_1 to V_2 were accomplished isentropically?

9.51 When a pitot tube such as Fig. 6.29 is placed in a supersonic flow, a normal shock will stand in front of the probe. Suppose the probe reads $p_0 = 190$ kPa and $p = 150$ kPa. If the stagnation temperature is 400 K, estimate the (supersonic) Mach number and velocity upstream of the shock.

9.52 Repeat Prob. 9.49 except this time let the odd coincidence be that the *static* pressure downstream of the shock exactly equals the throat pressure. What is the area where the shock stands?

9.53 An atomic explosion propagates into still air at 14.7 lbf/in² absolute and 520°R. The pressure just inside the shock is 5000 lbf/in² absolute. Assuming $\gamma = 1.4$, what is the speed C of the shock and the velocity V just inside the shock?

9.54 The normal-shock wave from an explosion propagates at 1500 m/s into still air at 20°C and 101 kPa. What are the pressure, velocity, and temperature just inside the shock?

9.55 Air in a large tank at 100°C and 150 kPa exhausts to the atmosphere through a converging nozzle with a 5-cm² throat area. Compute the exit mass flow if the atmospheric pressure is (*a*) 100 kPa, (*b*) 60 kPa, and (*c*) 30 kPa.

9.56 Air flows through a converging-diverging nozzle between two large reservoirs, as shown in Fig. P9.56. A mercury manometer between the throat and the downstream reservoir reads $h = 20$ cm. Estimate the downstream reservoir pressure. Is there a normal shock in the flow? If so, does it stand in the exit plane or further upstream?

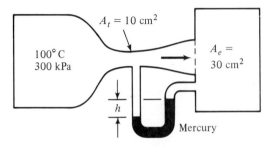

Fig. P9.56

9.57 Air in a tank at 120 kPa and 300 K exhausts to the atmosphere through a 5-cm²-throat converging nozzle at a rate of 0.12 kg/s. What is the atmospheric pressure? What is the maximum mass flow possible at low atmospheric pressure?

9.58 With reference to Prob. 3.68, show that the thrust of a rocket engine exhausting into a vacuum is given by

$$F = \frac{p_0 A_e (1 + \gamma\, \mathrm{Ma}_e^2)}{\left(1 + \dfrac{\gamma - 1}{2}\, \mathrm{Ma}_e^2\right)^{\gamma/(\gamma - 1)}}$$

where A_e = exit area

Ma_e = exit Mach number

p_0 = stagnation pressure in combustion chamber

Note that stagnation temperature does not enter into the thrust.

9.59 A double tank system in Fig. P9.59 has two identical converging nozzles of 1 in² throat area. Tank 1 is very large and tank 2 is small enough to be in steady-flow equilibrium with the jet from tank 1. Nozzle flow is isentropic, but entropy changes between 1 and 3 due to jet dissipation in tank 2. Compute the mass flow. (If you give up, Ref. 14, pp. 288–290, has a good discussion.)

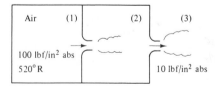

Fig. P9.59

9.60 Air flows isentropically through a nozzle of throat area 6 cm² and exit area 24 cm². If $p_0 = 600$ kPa and $T_0 = 200$°C, compute the mass flow, exit pressure, and exit Mach number for (*a*) subsonic and (*b*) supersonic flow.

9.61 For the nozzle of Prob. 9.60, allowing for non-isentropic flow, what is the range of exit tank pressures p_b for which (*a*) the diverging nozzle flow is fully supersonic, (*b*) the exit flow is subsonic, (*c*) the mass flow is independent of p_b, (*d*) exit plane pressure p_e is independent of p_b, and (*e*) $p_e < p_b$?

9.62 Suppose the nozzle flow of Prob. 9.60 is not isentropic but instead has a normal shock at the position where area is 20 cm^2. Compute the resulting mass flow, exit pressure, and exit Mach number.

9.63 A supersonic nozzle with an exit area of 8 cm^2 discharges air at $\text{Ma} = 2.5$, $p = 101 \text{ kPa}$, and $T = 300 \text{ K}$. Compute (a) exit velocity, (b) p_0, (c) T_0, (d) throat area, and (e) mass flow.

9.64 A perfect gas (not air) expands isentropically through a supersonic nozzle with an exit area 5 times its throat area. The exit Mach number is 3.8. What is the specific-heat ratio of the gas? What might this gas be? If $p_0 = 300 \text{ kPa}$, what is the exit pressure of the gas?

9.65 The orientation of a hole can make a difference. Consider holes A and B in Fig. P9.65, which are identical but reversed. For the given air properties on either side, compute the mass flow through each hole and explain why they are different.

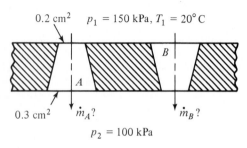

0.2 cm^2 $p_1 = 150 \text{ kPa}, T_1 = 20° \text{C}$

B

A

0.3 cm^2 $\dot{m}_A?$ $\dot{m}_B?$

$p_2 = 100 \text{ kPa}$

Fig. P9.65

9.66 Air with $p_0 = 300 \text{ kPa}$ and $T_0 = 500 \text{ K}$ flows through a converging-diverging nozzle with throat area of 1 cm^2 and exit area of 4 cm^2 into a receiver tank. The mass flow is 195.2 kg/h. For what range of receiver pressure is this mass flow possible?

9.67 An automobile tire contains 3 ft^3 of air at 45 lbf/in^2 absolute and $60°\text{F}$. It is punctured by a 0.025-in-diameter hole which resembles a convergent nozzle. Estimate the time required to discharge the tire down to 30 lbf/in^2 absolute if $p_a = 14.7 \text{ lbf/in}^2$.

9.68 A spaceship rocket engine has a thrust of 1 million pounds when operating at its design point $(p_e = p_a)$. If the chamber pressure and temperature are 500 lbf/in^2 absolute and $5000°\text{R}$ and the gas approximates air with $\gamma = 1.4$, compute the throat diameter of the engine in feet.

9.69 Air at $T_0 = 600 \text{ K}$ flows through a supersonic nozzle with $A_t = 1 \text{ cm}^2$ and $A_e = 3 \text{ cm}^2$. The mass flow is 148.5 kg/h. A pitot-static probe placed in the exit plane reads $p_0 = 200 \text{ kPa}$ and $p = 191.5 \text{ kPa}$. What is

the exit velocity? Is there a normal shock in the duct? If so, what is the Mach number just upstream of this shock?

9.70 Air flows through a duct as in Fig. P9.70, where $A_1 = 25 \text{ cm}^2$, $A_2 = 20 \text{ cm}^2$, and $A_3 = 30 \text{ cm}^2$. A normal shock stands at section 2. Compute (a) the mass flow, (b) the Mach number, and (c) the stagnation pressure at section 3.

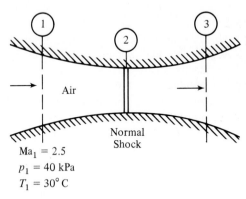

Air

Normal
Shock

$\text{Ma}_1 = 2.5$
$p_1 = 40 \text{ kPa}$
$T_1 = 30° \text{C}$

Fig. P9.70

9.71 Air enters a 3-cm-diameter pipe 15 m long at $V_1 = 73 \text{ m/s}$, $p_1 = 550 \text{ kPa}$, and $T_1 = 60°\text{C}$. The friction factor is 0.022. Compute V_2, p_2, T_2, and p_{02} at the end of the pipe. How much additional pipe length would cause the exit flow to be sonic?

9.72 Air enters a duct of $L/D = 40$ at $V_1 = 200 \text{ m/s}$ and $T_1 = 300 \text{ K}$. The flow at the exit is choked. What is the average friction factor in the duct for adiabatic flow?

9.73 Air in a tank at $p_0 = 100 \text{ lbf/in}^2$ absolute and $T_0 = 520°\text{R}$ flows through a converging nozzle into pipe of 1 in diameter. What will be the mass flow through the pipe if its length is (a) 0 ft, (b) 1 ft, and (c) 10 ft? Assume $\bar{f} = 0.025$. The pressure outside the duct exit is negligibly small.

9.74 Hydrogen (Table A.4) enters a 5-cm-diameter pipe at $p_1 = 500 \text{ kPa}$, $V_1 = 300 \text{ m/s}$, and $T_1 = 20°\text{C}$. The friction factor is 0.023. How long is the duct if the flow is choked? What is the exit pressure?

9.75 Air enters a 5- by 5-cm square duct at $V_1 = 900 \text{ m/s}$ and $T_1 = 300 \text{ K}$. The friction factor is 0.018. For what length duct will the flow exactly decelerate to $\text{Ma} = 1.0$? If the duct length is 2 m, will there be a normal shock in the duct? If so, at what Mach number will it occur?

9.76 A compressor forces air through a smooth pipe 20 m long and 4 cm in diameter, as in Fig. P9.76. The air leaves at 101 kPa and 200°C. The compressor data for pressure rise versus mass flow are shown in the figure. Using the Moody chart to estimate $\bar{f}$, compute the resulting mass flow.

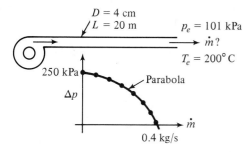

Fig. P9.76

9.77 Air enters a 4-cm-diameter duct at $p_0 = 150$ kPa, $T_0 = 400$ K, and $V_1 = 120$ m/s, with $\bar{f} = 0.025$. For adiabatic flow compute (a) the maximum duct length for these conditions, (b) the mass flow and exit stagnation pressure if the duct length is 5 m, and (c) the reduced mass flow if $L = 20$ m.

9.78 Air enters a 0.5-in-diameter pipe subsonically at $p_1 = 60$ lbf/in^2 absolute and $T_1 = 600°$R. The pipe length is 20 ft, $\bar{f} = 0.022$, and the receiver pressure outside the pipe exit is 20 lbf/in^2 absolute. Compute the mass flow in the pipe, assuming (a) isothermal flow and (b) adiabatic flow.

9.79 Air at 550 kPa and 100°C enters a smooth 1-m-long pipe and then passes through a second smooth pipe to a 30-kPa reservoir, as in Fig. P9.79. Using the Moody chart to compute $\bar{f}$, estimate the mass flow through this system. Is the flow choked?

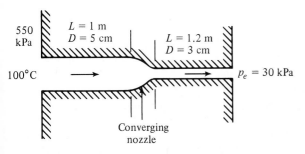

Fig. P9.79

9.80 Oxygen (Table A.4) enters a 120-m-long smooth pipe at 250 kPa and 60°C. The pipe exits into a low-pressure reservoir. The desired mass flow is 0.25 kg/s. Using the Moody chart to compute $\bar{f}$, estimate the minimum pipe diameter to transport this flow and the resulting exit pressure and temperature.

9.81 We were unable to find an adiabatic-flow formula analogous to Eq. (9.75) relating mass flow to pressure drop, but one can derive the following formula relating densities in adiabatic duct flow:

$$\rho_1^2 = \rho_2^2 + \rho^{*2}\left(\frac{2\gamma}{\gamma+1}\frac{\bar{f}L}{D} + 2\ln\frac{\rho_1}{\rho_2}\right)$$

Verify that this formula is correct. What is the disadvantage of using it to predict mass flow versus density change?

9.82 Air enters a 1-in-diameter cast-iron duct 20 ft long at $V_1 = 200$ ft/s, $p_1 = 40$ lbf/in^2 absolute, and $T_1 = 520°$R. Compute the exit pressure and mass flow using the Moody chart to predict the friction factor.

9.83 Air at 320 K is to be transported through a duct 40 m long. If $\bar{f} = 0.025$, what is the minimum diameter of a duct which can carry the flow without choking for $V_1 = (a)$ 50 m/s, (b) 100 m/s, (c) and 400 m/s? Assume adiabatic flow.

9.84 A fuel-air mixture, assumed equivalent to air, enters a duct combustion chamber at $V_1 = 100$ m/s and $T_1 = 400$ K. What amount of heat addition in kilojoules per kilogram will cause the exit flow to be choked? What will be the exit Mach number and temperature if 1000 kJ/kg is added during combustion?

9.85 What happens to the inlet flow of Prob. 9.84 if the combustion yields 2500 kJ/kg heat addition and p_{01} and T_{01} remain the same? How much is the mass flow reduced?

9.86 Air enters a 4-in-diameter duct at 15 lbf/in^2 absolute, 70°F, and 200 ft/s. What frictionless heat addition in Btu per pound and heating rate in Btu per second are necessary for the exit temperature to be 1300°F? What will be the exit pressure, velocity, and Mach number?

9.87 A jet engine at 7000 m altitude takes in 45 kg/s of air and adds 500 kJ/kg in the combustion chamber. The chamber cross section is 0.5 m^2, and the air enters the chamber at 80 kPa and 5°C. After combustion the air expands through an isentropic converging nozzle to exit at atmospheric pressure. Estimate (a) the nozzle throat diameter, (b) nozzle exit velocity, and (c) the thrust produced by the engine.

9.88 Air enters a combustion chamber at 100 m/s, 300 K, and 80 kPa. Combustion releases 500 kJ/kg to the air. What are the exit velocity and Mach number and the stagnation pressure loss? What heat addition would choke the flow?

9.89 Air at 110 kPa and 20°C enters a 12- by 18-cm rectangular duct at $V_1 = 86$ m/s. After frictionless heat addition, the flow exits at $V_2 = 200$ m/s. Compute (a) the exit pressure, (b) the exit temperature, and (c) the rate of heat addition in watts. How much heat addition would cause the exit flow to be choked?

9.90 An observer at sea level does not hear an aircraft flying at 20,000 ft overhead until it is 8 statute miles past her. Estimate the Mach number and speed of the aircraft.

9.91 A particle moving at uniform velocity in sea-level standard air creates the two disturbance spheres shown in Fig. P9.91. Compute the particle velocity and Mach number.

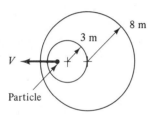

Fig. P9.91

9.92 The particle in Fig. P9.92 is moving supersonically in sea-level standard air. From the two given disturbance spheres compute the particle Mach number, velocity, and Mach angle.

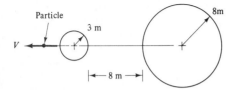

Fig. P9.92

9.93 Supersonic air takes a 5° compression turn, as in Fig. P9.93. Compute the downstream pressure and Mach number and the wave angle and compare with small-disturbance theory.

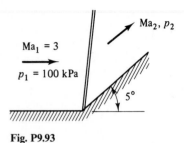

Fig. P9.93

9.94 Supersonic airflow takes a 5° expansion turn, as in Fig. P9.94. Compute the downstream Mach number and pressure and compare with small-disturbance theory.

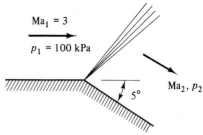

Fig. P9.94

9.95 A needlelike thermistor probe inserted in a supersonic airstream reads a static temperature of -10°C. The probe creates a conical disturbance wave of half-angle 20°. Estimate the Mach number, velocity, and stagnation temperature of the stream.

9.96 Do the Mach waves upstream of an oblique-shock wave intersect with the shock? Assuming supersonic downstream flow, do the downstream Mach waves intersect the shock? Show that for small deflections the shock-wave angle β lies halfway between μ_1 and $\mu_2 + \theta$ for any Mach number.

9.97 A supersonic flow at $Ma_1 = 3.5$ and $p_1 = 100$ kPa undergoes a compression shock and an isentropic expansion of 25° flow deflection each. Compute the final Mach number and pressure (a) if the shock is followed by the expansion and (b) if the expansion is followed by the shock. Do the final conditions match the initial flow?

9.98 Air flows at supersonic speed toward a compression ramp, as in Fig. P9.98. A scratch on the wall at point a creates a wave of 30° angle, while the oblique shock created has a 50° angle. What is (a) the ramp angle θ and (b) the wave angle ϕ caused by a scratch at b?

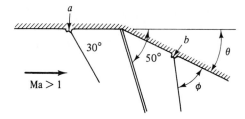

Fig. P9.98

9.99 Air flows at Ma = 3 and $p = 10\,\text{lbf/in}^2$ absolute toward a wedge of 16° angle at zero incidence in Fig. P9.99. If the pointed edge is forward, what will be the pressure at point A? If the blunt edge is forward, what will be the pressure at point B?

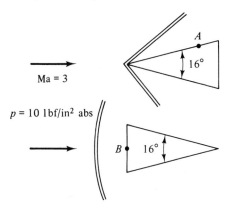

Fig. P9.99

9.100 For the wedge with its pointed edge forward in the upper part of Fig. P9.99, let the base be 6 in high and have an average base pressure of $2.5\,\text{lbf/in}^2$. Compute the drag coefficient of this wedge and the drag force per foot of depth.

9.101 When an oblique shock strikes a solid wall, it reflects as a shock of sufficient strength to cause the exit flow Ma_3 to be parallel to the wall, as in Fig. P9.101. For air flow with $\text{Ma}_1 = 2.5$ and $p_1 = 100\,\text{kPa}$, compute Ma_3, p_3, and the angle ϕ.

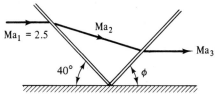

Fig. P9.101

9.102 A bend in the bottom of a supersonic duct flow induces a shock wave which reflects from the upper wall, as in Fig. P9.102. Compute the Mach number and pressure in region 3.

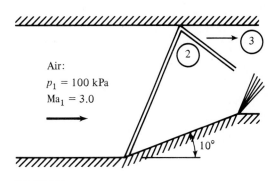

Fig. P9.102

9.103 Airflow at $\text{Ma}_1 = 3.0$ passes through a 20° oblique-shock deflection. What isentropic expansion turn is required to bring the flow back to (a) Ma_1 and (b) pressure p_1?

9.104 If the body in Fig. 9.28 were a wedge instead of a cone, what would be (a) the shock angle and (b) the surface Mach number? Use shock theory to obtain accurate numerical results.

9.105 The supersonic nozzle of Fig. P9.105 is overexpanded (case G of Fig. 9.12) with $A_e/A_t = 3.0$ and stagnation pressure of 350 kPa. If the jet edge makes a 4° angle with the nozzle centerline, what is the back pressure p_r in kilopascals?

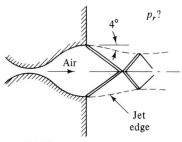

Fig. P9.105

9.106 A converging-diverging nozzle with a 4:1 exit-area ratio and $p_0 = 500\,\text{kPa}$ operates in an underexpanded condition (case I of Fig. 9.12) as in Fig. P9.106. The receiver pressure is $p_a = 10\,\text{kPa}$, which is less than exit pressure, so that expansion waves form outside the

$p_a = 10$ kPa

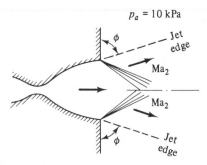

Fig. P9.106

exit. For the given conditions, what will the Mach number Ma_2 and the angle ϕ of the edge of the jet be? Assume $\gamma = 1.4$ as usual.

9.107 Airflow at Ma = 2.2 takes a compression turn of 12° and then another turn of angle θ in Fig. P9.107. What is the maximum value of θ for the second shock to be attached? Will the two shocks intersect for any θ less than θ_{max}?

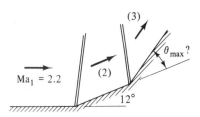

Fig. P9.107

9.108 Air flows at $Ma_\infty = 2.8$ and $p_\infty = 100$ kPa past a flat-plate airfoil at $\alpha = 7°$. Compute the lift and drag coefficients by (a) shock-expansion theory and (b) Ackeret theory.

9.109 A flat-plate airfoil with $C = 3$ ft is to have a lift of 1500 lbf per foot of width when flying in air with $Ma_\infty = 2.0$ and $p_\infty = 8$ lbf/in^2 absolute. What should the angle of attack be? What will the drag per foot for this condition be?

9.110 Repeat Example 9.21 for an angle of attack of 4°. Is the lift coefficient linear with α in this range of 0 to 8°? Is the drag coefficient parabolic in α in this range?

9.111 Repeat Example 9.21 except that the airfoil is taken to be an unsymmetrical diamond shape with a leading-edge included angle of 5° and a trailing-edge included angle of 3°.

9.112 Air flows at Ma = 2.5 past a half-wedge airfoil whose angles are 4°, as in Fig. P9.112. Compute the lift and drag coefficient at $\alpha = (a)$ 0° and (b) 6°.

Fig. P9.112

9.113 A supersonic airfoil has a parabolic symmetrical shape for upper and lower surfaces

$$y_{u,l} = \pm 2t\left(\frac{x}{C} - \frac{x^2}{C^2}\right)$$

such that the maximum thickness is t at $x = \frac{1}{2}C$. Compute the drag coefficient at zero incidence by Ackeret theory and compare with a symmetrical double wedge of the same thickness.

9.114 Prove from Ackeret theory that for a given supersonic airfoil shape with sharp leading and trailing edges and a given thickness, the minimum thickness drag occurs for a symmetrical double-wedge shape.

Open-Channel Flow

10.1 INTRODUCTION

Simply stated, open-channel flow is the flow of a liquid in a conduit with a free surface. There are many practical examples, both artificial (flumes, spillways, canals, weirs, drainage ditches, culverts) and natural (streams, rivers, estuaries, floodplains). This chapter introduces the elementary analysis of such flows, which are dominated by the effects of gravity.

The presence of the free surface, which is essentially at atmospheric pressure, both helps and hurts the analysis. It helps because the pressure can be taken constant along the free surface, which therefore is equivalent to the hydraulic grade line (HGL) of the flow. Unlike flow in closed ducts, pressure gradient is not a direct factor in open-channel flow, where the balance of forces is confined to gravity and friction.[1] But the free surface complicates the analysis because its shape is a priori unknown: the depth profile changes with conditions and must be computed as part of the problem, especially in unsteady problems involving wave motion.

Before proceeding we remark, as usual, that whole books have been written on open-channel hydraulics [1–4]. There are also specialized texts devoted to wave motion [5, 6] and to engineering aspects of coastal free-surface flows [7]. The present chapter is only an introduction to broader and more detailed treatments.

The One-Dimensional Approximation

An open channel always has two sides and a bottom, where the flow satisfies the no-slip condition. Therefore even a straight channel has a three-dimensional velocity distribution. Some measurements of straight-channel velocity contours are

---

[1] Surface tension is rarely important because open channels are normally quite large and have a very large Weber number.

shown in Fig. 10.1. The profiles are quite complex, with maximum velocity typically occurring in the midplane about 20 percent below the surface. In very broad shallow channels the maximum velocity is near the surface, and the velocity profile is nearly logarithmic from the bottom to the free surface, as in Eq. (6.93). In noncircular channels there are also secondary motions similar to Fig. 6.18 for closed-duct flows. If the channel curves or meanders, the secondary motion intensifies due to centrifugal effects, with high velocity occurring near the outer radius of the bend. Curved natural channels are subject to strong bottom erosion and deposition effects.

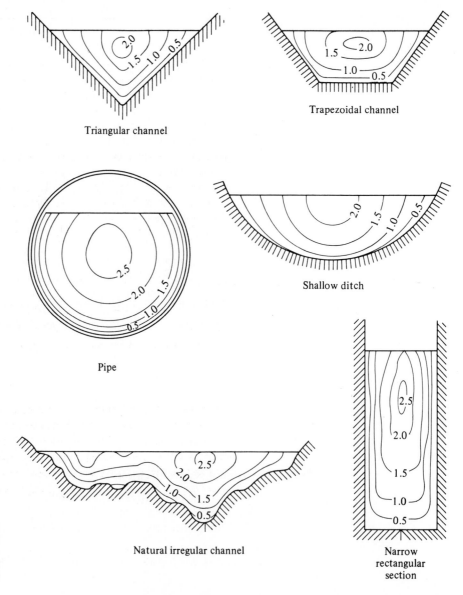

Fig. 10.1 Measured isovelocity contours in typical straight open-channel flows. (*From Ref. 3.*)

Little theoretical work has been done on velocity distributions like Fig. 10.1. Instead, the engineering approach is to assume one-dimensional flow with an average velocity $V(x)$ at each cross-sectional area $A(x)$, where x is distance along the channel. Since density variations are negligible in low-speed liquid flow, for steady flow the volume flow Q is constant along the channel from continuity

$$Q = V(x)A(x) = \text{const} \tag{10.1}$$

A second relation between velocity and depth is the Bernoulli equation including friction losses. If points 1 (upstream) and 2 (downstream) are on the free surface, $p_1 = p_2 = p_a$ and we have, for steady flow,

$$\frac{V_1^2}{2g} + z_1 = \frac{V_2^2}{2g} + z_2 + h_f \tag{10.2}$$

where h_f is the friction-head loss. The wall friction in channel flow is quite similar to that in steady duct flow and can be correlated adequately with the Moody-diagram (Fig. 6.13) formula for turbulent flow with a rough surface

$$\frac{1}{f^{1/2}} = -2.0 \log\left(\frac{\varepsilon/D_h}{3.7} + \frac{2.51}{\text{Re}_{Dh}f^{1/2}}\right) \tag{10.3}$$

where $f = 8\tau_w/\rho V^2$ and D_h, the hydraulic diameter, is equal to $4A/P$. Most open-channel analyses use the hydraulic radius R_h instead, which is one-fourth of the hydraulic diameter

$$R_h = \tfrac{1}{4}D_h = \frac{A}{P} \tag{10.4}$$

The quantity P is the *wetted perimeter*, which includes the sides and bottom of the channel but not the free surface and of course not the parts of the sides above the water level. For example, if a rectangular channel is b wide and h high and contains water to depth y, the wetted perimeter is

$$P = b + 2y \tag{10.5}$$

not $2b + 2h$.

Open channels are usually large and deep and contain low-viscosity water; hence they are almost always turbulent. The only laminar channel flows of practical importance are the thin sheet flows which occur as rainwater drains from streets and airport runways.

Flow Classification by Depth Variation

The most common method of classifying open-channel flows is by the rate of change of the free-surface depth. The simplest and most widely analyzed case is *uniform flow*, where the depth (hence the velocity in steady flow) remains constant. Uniform-flow conditions are approximated by long straight runs of constant-slope and constant-area channel. A channel in uniform flow is said to be moving at its *normal depth* y_n, which is an important design parameter.

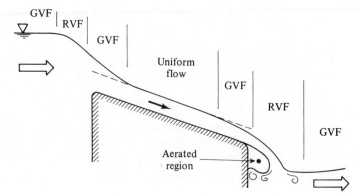

If the channel slope or cross section changes or there is an obstruction in the flow, the depth changes and the flow is said to be *varied*. The flow is *gradually varying* if the one-dimensional approximation is valid and *rapidly varying* if not. Some examples of this method of classification are shown in Fig. 10.2. The classes can be summarized as follows:

1. Uniform flow (constant depth and slope)
2. Varied flow
    a. Gradually varied (one-dimensional)
    b. Rapidly varied (multidimensional)

Typically uniform flow is separated from rapidly varying flow by a region of gradually varied flow. Gradually varied flow can be analyzed by a first-order differential equation (Sec. 10.6), but rapidly varying flow usually requires experimentation or three-dimensional potential theory.

Flow Classification by Froude Number

A second and very interesting classification is by dimensionless Froude number, which for a rectangular or very wide channel takes the form $Fr = V/(gy)^{1/2}$, where y is the water depth. The three flow regimes are

$$Fr < 1.0 \qquad \text{subcritical flow}$$
$$Fr = 1.0 \qquad \text{critical flow} \qquad\qquad (10.6)$$
$$Fr > 1.0 \qquad \text{supercritical flow}$$

The Froude number for irregular channels is defined in Sec. 10.4. As mentioned in Sec. 9.10, there is a strong analogy here with the three compressible-flow regimes of Mach number: subsonic (Ma < 1), sonic (Ma = 1), and supersonic (Ma > 1). We shall pursue the analogy in Sec. 10.4.

The Froude-number denominator $(gy)^{1/2}$ is the speed of an infinitesimal shallow-water surface wave. We can derive this with reference to Fig. 10.3a, which shows a wave of height δy propagating at speed c into still liquid. To achieve a

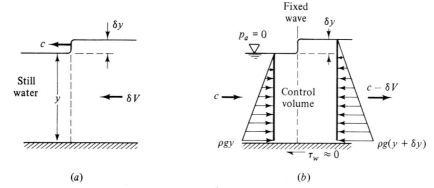

Fig. 10.3 Analysis of a small surface wave propagating into still shallow water: (*a*) moving wave, noninertial frame; (*b*) fixed wave, inertial frame of reference.

steady-flow inertial frame of reference, we fix the coordinates on the wave as in Fig. 10.3*b*, so that the still water moves to the right at velocity *c*. Figure 10.3 is exactly analogous to Fig. 9.1, which analyzed the speed of sound in a fluid.

For the control volume of Fig. 10.3*b*, the one-dimensional continuity relation is, for channel width *b*,

$$\rho c y b = \rho(c - \delta V)(y + \delta y)b$$

or

$$\delta V = c\,\frac{\delta y}{y + \delta y} \tag{10.7}$$

This is analogous to Eq. (9.10); the velocity change δV induced by a surface wave is small if the wave is "weak," $\delta y \ll y$. If we neglect bottom friction in the short distance across the wave in Fig. 10.3*b*, the momentum relation is a balance between net hydrostatic pressure force and momentum

$$-\tfrac{1}{2}\rho g b[(y + \delta y)^2 - y^2] = \rho c b y(c - \delta V - c)$$

or

$$g\left(1 + \frac{\tfrac{1}{2}\delta y}{y}\right)\delta y = c\,\delta V \tag{10.8}$$

This is analogous to Eq. (9.12). By eliminating δV between Eqs. (10.7) and (10.8) we obtain the desired expression for wave-propagation speed

$$c^2 = gy\left(1 + \frac{\delta y}{y}\right)\left(1 + \frac{\tfrac{1}{2}\delta y}{y}\right) \tag{10.9}$$

The "stronger" the wave height δy the faster the wave speed *c*, by analogy with Eq. (9.13). In the limit of an infinitesimal wave height $\delta y \to 0$, the speed becomes

$$c_0^2 = gy \tag{10.10}$$

This is the surface-wave equivalent of fluid sound speed *a*, and thus the Froude number in channel flow $\text{Fr} = V/c_0$ is the analog of Mach number.

As in gas dynamics, a channel flow can accelerate from subcritical to critical to supercritical flow and then return to subcritical flow through a sort of normal

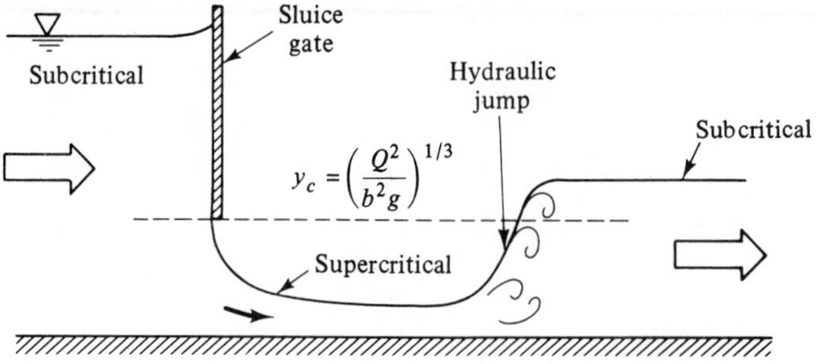

Fig. 10.4 Flow under a sluice gate accelerates from subcritical to critical to supercritical flow and then jumps back to subcritical flow.

shock called a *hydraulic jump* (Sec. 10.5). This is illustrated in Fig. 10.4. The flow upstream of the sluice gate is subcritical. It then accelerates to critical and supercritical flow as it passes under the gate, which serves as a sort of "nozzle." Further downstream the flow "shocks" back to subcritical flow because the downstream "receiver" height is too high to maintain supercritical flow. The reader should note the similarity with the nozzle gas flows of Fig. 9.12.

The critical depth $y_c = (Q^2/b^2 g)^{1/3}$ is sketched as a dashed line in Fig. 10.4 for reference. Like the normal depth y_n, y_c is an important parameter in characterizing open-channel flow (see Sec. 10.4).

An excellent discussion of the various regimes of open-channel flow is given in Ref. 8.

10.2 UNIFORM FLOW; THE CHÉZY FORMULA

Uniform flow can occur in long straight runs of constant slope and constant channel cross section. The water depth is constant at $y = y_n$, and the velocity is constant at $V = V_0$. Let the slope be $S_0 = \tan \theta$, where θ is the angle the bottom makes with the horizontal, considered positive for downhill flow. Then Eq. (10.2), with $V_1 = V_2 = V_0$, becomes

$$h_f = z_1 - z_2 = S_0 L \qquad (10.11)$$

where L is the channel length between sections 1 and 2. The head loss thus balances the loss in height of the channel. The flow is essentially fully developed, so that the Darcy-Weisbach relation, Eq. (6.30), holds

$$h_f = f \frac{L}{D_h} \frac{V_0^2}{2g} \qquad D_h = 4R_h \qquad (10.12)$$

with $D_h = 4A/P$ used to accommodate noncircular channels. The geometry and notation for open-channel flow analysis are shown in Fig. 10.5.

By combining Eqs. (10.11) and (10.12) we obtain an expression for flow velocity in uniform channel flow

$$V_0 = \left(\frac{8g}{f}\right)^{1/2} R_h^{1/2} S_0^{1/2} \qquad (10.13)$$

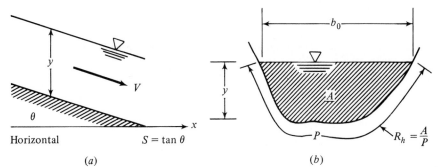

Fig. 10.5 Geometry and notation for open-channel flow: (*a*) side view; (*b*) cross section. All these parameters are constant in uniform flow.

For a given channel shape and bottom roughness, the quantity $(8g/f)^{1/2}$ is constant and can be denoted by C. Equation (10.13) becomes

$$V_0 = C(R_h S_0)^{1/2} \qquad Q = CA(R_h S_0)^{1/2} \qquad (10.14)$$

These are called the *Chézy formulas*, first developed by the French engineer Antoine Chézy in conjunction with his experiments on the Seine River and the Courpalet Canal in 1769. The quantity C, called the *Chézy coefficient*, varies from about 60 $\text{ft}^{1/2}/\text{s}$ for small rough channels to 160 $\text{ft}^{1/2}/\text{s}$ for large smooth channels (30 to 90 $\text{m}^{1/2}/\text{s}$ in SI units).

Over the past century a great deal of hydraulics research [17] has been devoted to the correlation of the Chézy coefficient with the roughness, shape, and slope of various open channels. The most popular correlations are due to Ganguillet and Kutter in 1869, Manning in 1889, Bazin in 1897, and Powell in 1950 [9]. All these formulations are discussed in delicious detail in Ref. 3, chap. 5. Here we confine our treatment to Manning's correlation, the most popular.

The Manning Roughness Correlation

Basically, the friction loss in uniform channel flow cannot be very different from fully developed turbulent pipe flow, Eq. (10.3). Channels are typically quite rough and flow at Reynolds numbers in excess of 10^6, so that Re can be neglected in Eq. (10.3) for a fully rough approximation

Fully rough flow: $$\frac{1}{f^{1/2}} \approx 2.0 \log \frac{3.7 D_h}{\varepsilon} \qquad (10.15)$$

Equation (10.15) can be approximated by the straight-line power-law fit

$$f \approx 0.18 \left(\frac{\varepsilon}{D_h}\right)^{1/3} = 0.113 \left(\frac{\varepsilon}{R_h}\right)^{1/3} \qquad (10.16)$$

which is very accurate in the typical channel-roughness range $0.001 < \varepsilon/D_h < 0.05$.

If we adopt Eq. (10.16) as a simple curve fit, the Chézy coefficient becomes

$$C \approx (8.4 g^{1/2} \varepsilon^{-1/6}) R_h^{1/6} \qquad (10.17)$$

This is exactly the type of variation which Manning found by correlating channel resistance data in 1889.

Manning expressed his correlation in the form

$$C = \frac{1.49}{n} [R_h \text{ (ft)}]^{1/6} = \frac{1.0}{n} [R_h \text{ (m)}]^{1/6} \tag{10.18}$$

where n is called *Manning's roughness coefficient* and is taken to be dimensionless; i.e., it is given the same value whether SI or BG are used, hence the change in the numerator from 1.0 (SI) to $1.49 = (3.2808 \text{ ft/m})^{1/3}$ for BG units.[1]

By comparison of Eqs. (10.17) and (10.18) we find that Manning's coefficient n is equivalent to wall-roughness height to the one-sixth power

$$n \approx 0.0313[\varepsilon \text{ (ft)}]^{1/6} = 0.0382[\varepsilon \text{ (m)}]^{1/6} \tag{10.19}$$

Thus Manning's formula for uniform-flow velocity

$$V_0 \text{ (ft/s)} \approx \frac{1.49}{n} [R_h \text{ (ft)}]^{2/3} S_0^{1/2} \qquad Q = V_0 A \tag{10.20}$$

and the friction-factor formulation, Eq. (10.13), can be used interchangeably with similar accuracy. For SI units, Manning's uniform-flow formula becomes

$$V_0 \text{ (m/s)} \approx \frac{1.0}{n} [R_h \text{ (m)}]^{2/3} S_0^{1/2} \tag{10.21}$$

Some values of Manning's n and the equivalent channel-roughness height are given for typical channel surfaces in Table 10.1. There is a factor of 15 variation in roughness from a smooth glass channel ($n = 0.01$) to a tree-lined floodplain ($n = 0.15$). There is a large scatter in channel resistance due to variations in slope, cross section, bank configuration, and vegetation, especially in natural channels. The scatter bands in Table 10.1 should be taken seriously.

Note that Manning's approximation is accurate in an intermediate range of roughness ratios: it predicts unrealistically low friction and high discharge for both deep smooth and shallow rough channels, for which the friction-factor formulation would be preferred. Thus a natural channel can have a variable n depending upon water depth. For example, the Mississippi River near Memphis, Tennessee, has $n = 0.032$ at 40-ft flood depth, $n = 0.030$ at normal 20-ft depth, and $n = 0.040$ at 5-ft low-stage depth. Seasonal vegetation growth and factors like bottom erosion can also affect the value of n.

EXAMPLE 10.1   A finished-concrete channel 8 ft wide has a slope of 0.5° and a water depth of 4 ft. Predict the uniform flow rate Q in cubic feet per second by (*a*) Manning's formula and (*b*) the friction-factor analysis.

---

[1] An interesting discussion of the history and "dimensionality" of Manning's formula is given in Ref. 3, pp. 98–99. Recall that we gave this formula as a warning about units in Example 1.4.

Table 10.1

EXPERIMENTAL VALUES OF MANNING'S n FACTOR†

	n	Average roughness height ε	
		ft	mm
Artificial lined channels:			
Glass	0.010 ± 0.002	0.0011	0.3
Brass	0.011 ± 0.002	0.0019	0.6
Steel, smooth	0.012 ± 0.002	0.0032	1.0
Painted	0.014 ± 0.003	0.0080	2.4
Riveted	0.015 ± 0.002	0.012	3.7
Cast iron	0.013 ± 0.003	0.0051	1.6
Cement, finished	0.012 ± 0.002	0.0032	1.0
Unfinished	0.014 ± 0.002	0.0080	2.4
Planed wood	0.012 ± 0.002	0.0032	1.0
Clay tile	0.014 ± 0.003	0.0080	2.4
Brickwork	0.015 ± 0.002	0.012	3.7
Asphalt	0.016 ± 0.003	0.018	5.4
Corrugated metal	0.022 ± 0.005	0.12	37
Rubble masonry	0.025 ± 0.005	0.26	80
Excavated earth channels:			
Clean	0.022 ± 0.004	0.12	37
Gravelly	0.025 ± 0.005	0.26	80
Weedy	0.030 ± 0.005	0.8	240
Stony, cobbles	0.035 ± 0.010	1.5	500
Natural channels:			
Clean and straight	0.030 ± 0.005	0.8	240
Sluggish, deep pools	0.040 ± 0.010	3	900
Major rivers	0.035 ± 0.010	1.5	500
Floodplains:			
Pasture, farmland	0.035 ± 0.010	1.5	500
Light brush	0.05 ± 0.02	6	2000
Heavy brush	0.075 ± 0.025	15	5000
Trees	0.15 ± 0.05	?	?

† A more complete list is given in Ref. 3, pp. 110–113.

Solution

Part (*a*)   From Table 10.1, for finished concrete, $n \approx 0.012$ and $\varepsilon \approx 0.0032$ ft. The slope is $S = \tan 0.5° = 0.00873$. For depth $y = 4$ ft, the geometric properties are

$$A = by = (8 \text{ ft})(4 \text{ ft}) = 32 \text{ ft}^2 \quad P = b + 2y = 8 + 2(4) = 16 \text{ ft}$$

$$R_h = \frac{A}{P} = \frac{32 \text{ ft}^2}{16 \text{ ft}} = 2.0 \text{ ft} \qquad D_h = 4R_h = 8.0 \text{ ft}$$

From Manning's formula (10.20) in BG units, the flow rate is

$$Q = \frac{1.49}{n} A R_h^{2/3} S^{1/2} = \frac{1.49}{0.012} (32\ \text{ft}^2)(2.0)^{2/3}(0.00873)^{1/2}$$

$$= 589\ \text{ft}^3/\text{s} \qquad\qquad\qquad Ans.\ (a)$$

Part (b) For the friction-factor formulation, compute $\varepsilon/D_h = (0.0032\ \text{ft})/(8\ \text{ft}) = 0.0004$. The friction factor from Eq. (10.15) is

$$f^{-1/2} = 2.0\ \log \frac{3.7}{0.0004} = 7.93 \qquad f = 0.0159$$

$$C = \left(\frac{8g}{f}\right)^{1/2} = \left[\frac{8(32.2)}{0.0159}\right]^{1/2} = 127\ \text{ft}^{1/2}/\text{s}$$

whence

$$Q = A C R_h^{1/2} S^{1/2} = (32)(127)(2.0)^{1/2}(0.00873)^{1/2}$$

$$= 538\ \text{ft}^3/\text{s}\ (9\ \text{percent less}) \qquad\qquad Ans\ (b)$$

The accuracy is comparable, the friction-factor computation being more accurate but also more cumbersome.

With the normal depth given as in Example 10.1, computation of Q is quite straightforward. However, if Q is given, the computation of normal depth y_n usually requires trial and error because R_h is a complicated function of y_n. Consider the following example.

EXAMPLE 10.2 The asphalt-lined trapezoidal channel in Fig. E10.2 carries 300 ft³/s of water under uniform flow conditions when $S = 0.0015$. What is the normal depth y_n?

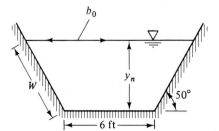

Fig. E10.2

Solution From Table 10.1, for asphalt, $n \approx 0.016$ and $\varepsilon \approx 0.018$ ft. The area and hydraulic radius are functions of y_n, which is unknown

$$b_0 = 6\ \text{ft} + 2y_n \cot 50° \qquad A = \tfrac{1}{2}(6 + b_0)y_n = 6y_n + y_n^2 \cot 50°$$

$$P = 6 + 2W = 6 + 2y_n \csc 50°$$

From Manning's formula (10.20) with a known $Q = 300$ ft³/s, we have

$$300 = \frac{1.49}{0.016} (6y_n + y_n^2 \cot 50°)\left(\frac{6y_n + y_n^2 \cot 50°}{6 + 2y_n \csc 50°}\right)^{2/3} (0.0015)^{1/2}$$

or

$$(6y_n + y_n^2 \cot 50°)^{5/3} = 83.2(6 + 2y_n\ \text{scs}\ 50°)^{2/3}$$

to be solved for y_n. Similarly, from the friction-factor analysis, we compute f from ε/D_h, then set

$$Q = 300 \text{ ft}^3/\text{s} = \left(\frac{8g}{f}\right)^{1/2} A(R_h S_0)^{1/2}$$

Probably the simplest procedure is to guess a range of values of y_n, compute Q, and see whether it is near 300 ft^3/s. the following are some tabulated guesses:

	Q, ft^3/s	
y_n, ft	Manning	Friction factor
4.0	234	230
4.5	291	285
5.0	354	347
4.6	303	297

The two formulations are nearly the same because $\varepsilon/D_h \approx 0.004$, which is the right range for Manning's formula. By interpolation we estimate the normal depth to be

$$y_n \approx \begin{cases} 4.58 \text{ ft} & \text{Manning} \\ 4.63 \text{ ft} & \text{friction factor} \end{cases} \qquad \textit{Ans.}$$

If we change n for the same flow rate by X percent, the normal depth will change by about two-thirds of X percent. For example, for this problem, if we reduce n by 25 percent to 0.012, the normal depth for 300 ft^3/s is 3.92 ft from Manning's formula, or about 15 percent less than the above computation. From now on we shall use Manning's formula exclusively in solving problems.

10.3 EFFICIENT UNIFORM-FLOW CHANNELS

The simplicity of Manning's formulation (10.20) enables us to analyze channel flows to determine the most efficient low-resistance sections for given conditions. The most common problem is that of maximizing R_h for a given flow area and discharge. Since $R_h = A/P$, maximizing R_h for given A is the same as minimizing the wetted perimeter P. There is no general solution for arbitrary cross sections, but an analysis of the trapezoid section will show the basic results.

Consider the generalized trapezoid of angle θ in Fig. 10.6. For a given side angle θ, the flow area is

$$A = by + \alpha y^2 \qquad \alpha = \cot \theta \tag{10.22}$$

The wetted perimeter is

$$P = b + 2W = b + 2y(1 + \alpha^2)^{1/2} \tag{10.23}$$

Eliminating b between (10.22) and (10.23) gives

$$P = \frac{A}{y} - \alpha y + 2y(1 + \alpha^2)^{1/2} \tag{10.24}$$

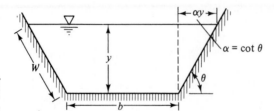

Fig. 10.6 Geometry of a trapezoidal channel section.

To minimize P, evaluate dP/dy for constant A and α and set equal to zero. The result is

$$A = y^2[2(1 + \alpha^2)^{1/2} - \alpha] \qquad P = 4y(1 + \alpha^2)^{1/2} - 2\alpha y \qquad R_h = \tfrac{1}{2}y \qquad (10.25)$$

The last result is very interesting: for any angle θ, the most efficient cross section for uniform flow occurs when the hydraulic radius is half the depth.

Since a rectangle is a trapezoid with $\alpha = 0$, the most efficient rectangular section is such that

$$A = 2y^2 \qquad P = 4y \qquad R_h = \tfrac{1}{2}y \qquad b = 2y \qquad (10.26)$$

To find the correct depth y, these relations must be solved in conjunction with Manning's flow-rate formula (10.20) for the given discharge Q.

Best Trapezoid Angle

Equations (10.25) are valid for any value of α. What is the best value of α for a given depth and area? To answer this question, evaluate $dP/d\alpha$ from Eq. (10.24) with A and y held constant. The result is

$$2\alpha = (1 + \alpha^2)^{1/2} \qquad \alpha = \cot\theta = \frac{1}{3^{1/2}}$$

or

$$\theta = 60° \qquad (10.27)$$

Thus the very best trapezoid section is half a hexagon.

Similar calculations with a circular channel section running partially full show best efficiency for a semicircle, $y = \tfrac{1}{2}D$. In fact, the semicircle is the best of all possible channel sections (minimum wetted perimeter for a given flow area). The percentage improvement over, say, half a hexagon is very slight however.

EXAMPLE 10.3

What are the best dimensions for a rectangular brick channel designed to carry 5 m³/s of water in uniform flow with $S_0 = 0.001$?

Solution

From Eq. (10.26) $A = 2y^2$ and $R_h = \tfrac{1}{2}y$. Manning's formula (10.21) in SI units gives, with $n \approx 0.015$ from Table 10.1,

$$Q = \frac{1.0}{n} AR_h^{2/3}S_0^{1/2} \qquad \text{or} \qquad 5\text{m}^3/\text{s} = \frac{1.0}{0.015}(2y^2)(\tfrac{1}{2}y)^{2/3}(0.001)^{1/2}$$

which can be solved for

$$y^{8/3} = 1.882 \text{ m}^{8/3}$$

$$y = 1.27 \text{ m} \qquad \qquad Ans.$$

The proper area and width are

$$A = 2y^2 = 3.214 \text{ m}^2 \qquad b = \frac{A}{y} = 2.535 \text{ m}$$ *Ans.*

It is constructive to see what flow rate a half-hexagon and semicircle would carry for the same area of 3.214 m².

For the half-hexagon, with $\alpha = 1/3^{1/2} = 0.577$, Eq. (10.25) predicts

$$A = y_{HH}^2[2(1 + 0.577^2)^{1/2} - 0.577] = 1.732y_{HH}^2 = 3.214$$

or $y_{HH} = 1.362$ m, whence $R_h = \frac{1}{2}y = 0.681$ m. The half-hexagon flow rate is thus

$$Q = \frac{1.0}{0.015} (3.214)(0.681)^{2/3}(0.001)^{1/2} = 5.25 \text{ m}^3/\text{s}$$

or about 5 percent more than the rectangle.

For a semicircle, $A = 3.214 \text{ m}^2 = \pi D^2/8$, or $D = 2.861$ m, whence $P = \frac{1}{2}\pi D = 4.494$ m and $R_h = A/P = 3.214/4.484 = 0.715$ m. The semicircle flow rate will thus be

$$Q = \frac{1.0}{0.015} (3.214)(0.715)^{2/3}(0.001)^{1/2} = 5.42 \text{ m}^3/\text{s}$$

or about 8 percent more than the rectangle and 3 percent more than the half-hexagon.

10.4 SPECIFIC ENERGY; CRITICAL DEPTH

As suggested by Bakhmeteff [1] in 1911, the specific energy E is a useful parameter in channel flow

$$E = y + \frac{V^2}{2g}$$ (10.28)

where y is the water depth. It is seen from Fig. 10.7a that E is the height of the energy grade line (EGL) above the channel bottom. For a given flow rate, there are usually two states possible for the same specific energy.

Fig. 10.7 Specific-energy considerations: (*a*) illustration sketch; (*b*) depth versus E from Eq. (10.29), showing minimum specific energy occurring at critical depth.

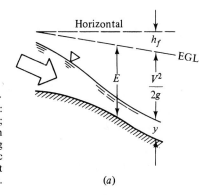

(*a*)

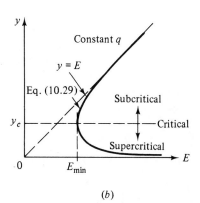

(*b*)

Rectangular Channels

Consider the possible states at a given location. Let $q = Q/b = Vy$ be the discharge per unit width of a rectangular channel. Then, with q constant, Eq. (10.28) becomes

$$E = y + \frac{q^2}{2gy^2} \qquad q = \frac{Q}{b} \tag{10.29}$$

Figure 10.7b is a plot of y versus E for constant q from Eq. (10.29). There is a minimum value of E at a certain value of y called the *critical depth*. By setting $dE/dy = 0$ at constant q we find that E_{min} occurs at

$$y = y_c = \left(\frac{q^2}{g}\right)^{1/3} = \left(\frac{Q^2}{b^2 g}\right)^{1/3} \tag{10.30}$$

The associated minimum energy is

$$E_{min} = E(y_c) = \tfrac{3}{2}y_c \tag{10.31}$$

The depth y_c corresponds to channel velocity equal to the shallow-water wave-propagation speed C_0 from Eq. (10.10). To see this, rewrite Eq. (10.30) as

$$q^2 = gy_c^3 = (gy_c)y_c^2 = V_c^2 y_c^2 \tag{10.32}$$

By comparison it follows that the critical channel velocity is

$$V_c = (gy_c)^{1/2} = C_0 \qquad \text{Fr} = 1 \tag{10.33}$$

For $E < E_{min}$ no solution exists in Fig. 10.7b, and thus such a flow is impossible physically. For $E > E_{min}$ two solutions are possible: (1) large depth with $V < V_c$, called *subcritical*, and (2) small depth with $V > V_c$, called *supercritical*. In subcritical flow, disturbances can propagate upstream because wave speed $C_0 > V$. In supercritical flow waves are swept downstream: upstream is a zone of silence, and a small obstruction in the flow will create a wedge-shaped wave exactly analogous to the Mach waves in Fig. 9.18c.[1] The angle of these waves must be

$$\mu = \sin^{-1}\frac{C_0}{V} = \sin^{-1}\frac{(gy)^{1/2}}{V} \tag{10.34}$$

The wave angle and the depth can thus be used as a simple measurement of supercritical-flow velocity.

Uniform critical flow can occur if the slope is the value S_c for which the velocity is critical. From the Chézy formula (10.14), with $R_h = y$ for a wide rectangular channel, the uniform critical-flow rate would be

$$q = Cy_c(y_c S_c)^{1/2} = (gy_c^3)^{1/2}$$

or

$$S_c = \frac{g}{C^2} = \frac{f}{8} \approx \frac{gn^2}{\xi y_c^{1/3}} \tag{10.35}$$

---

[1] This is the basis of the water-channel analogy for supersonic gas-dynamics experimentation [10, chap. 11].

where $\xi = 1.0$ for SI units and 2.208 for BG units. For fully rough channels the critical slope varies only from about 0.002 to 0.006 using (10.15) (Moody formula):

ε/R_h	0.001	0.01	0.1
S_c	0.0018	0.0031	0.0066

For narrow channels $R_h \neq y$, and Eq. (10.35) needs modification.

Note from Fig. 10.7b that small changes in E near E_{min} cause a large change in depth y, by analogy with small changes in duct area near the sonic point in Fig. 9.7. Thus critical flow is neutrally stable and is often accompanied by waves and undulations in the free surface. Channel designers should avoid long runs of near-critical flow.

EXAMPLE 10.4 A wide rectangular clean-earth channel has a flow rate $q = 50 \text{ ft}^3/(\text{s} \cdot \text{ft})$. (a) What is the critical depth? (b) What type of flow exists if $y = 3$ ft? (c) What is the critical slope?

Solution

Part (a) From Table 10.1, $n \approx 0.022$ and $\epsilon \approx 0.12$ ft. The critical depth follows from Eq. (10.30)

$$y_c = \left(\frac{q^2}{g}\right)^{1/3} = \left(\frac{50^2}{32.2}\right)^{1/3} = 4.27 \text{ ft} \qquad Ans. \ (a)$$

Part (b) If the actual depth is 3 ft, which is less than y_c, the flow must be *supercritical*. *Ans. (b)*

Part (c) The critical slope follows from Eq. (10.35), assuming $y = y_c$

$$S_c \approx \frac{gn^2}{\xi y_c^{1/3}} = \frac{32.2(0.022)^2}{2.208(4.27)^{1/3}} = 0.00435$$

$$\theta = 0.25° \qquad Ans. \ (c)$$

Nonrectangular Channels

If the channel width varies with y, the specific energy must be written in the form

$$E = y + \frac{Q^2}{2gA^2} \qquad (10.36)$$

The critical point of minimum energy occurs where $dE/dy = 0$ at constant Q. Since $A = A(y)$, Eq. (10.36) yields, for $E = E_{min}$,

$$\frac{dA}{dy} = \frac{gA^3}{Q^2} \qquad (10.37)$$

But $dA = b_0 \, dy$, where b_0 is the channel width at the free surface. Therefore Eq. (10.37) is equivalent to

$$A_c = \left(\frac{b_0 Q^2}{g}\right)^{1/3} \qquad (10.38a)$$

$$V_c = \frac{Q}{A_c} = \left(\frac{gA_c}{b_0}\right)^{1/2} \qquad (10.38b)$$

For a given channel shape $A(y)$ and $b_0(y)$ and a given Q, Eq. (10.38) has to be solved by trial and error or by plotting to find the critical area A_c, from which V_c can be computed.

By comparing the actual depth and velocity with the critical values we can determine the local flow condition

$$y > y_c, \ V < V_c: \qquad \text{Subcritical flow (Fr < 1)}$$
$$y < y_c, \ V > V_c: \qquad \text{Supercritical flow (Fr > 1)} \tag{10.39}$$

For uniform flow the critical slope is given by combining Eq. (10.38a) and the Chézy formula

$$Q^2 = \frac{gA_c^3}{b_0} = C^2 A_c^2 R_h S_c$$

or

$$S_c = \frac{f}{8}\frac{P}{b_0} = \frac{gn^2}{\xi R_h^{1/3}}\frac{P}{b_0} \tag{10.40}$$

where again $\xi = 1.0$ (2.208) for SI (BG) units. The quantity P/b_0 is unity for a wide rectangular channel.

EXAMPLE 10.5   The 50° triangular channel in Fig. E10.5 has a flow rate $Q = 16 \ \text{m}^3/\text{s}$. Compute (a) y_c, (b) V_c, and (c) S_c if $n = 0.018$.

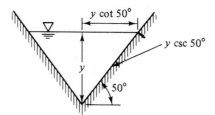

Fig. E10.5

Solution

Part (a)   This is an easy cross section because all geometric quantities can be written directly in terms of depth y

$$P = 2y \csc 50° \qquad A = y^2 \cot 50°$$
$$R_h = \tfrac{1}{2}y \cos 50° \qquad b_0 = 2y \cot 50° \tag{1}$$

The critical-flow condition satisfies Eq. (10.38a)

$$gA_c^3 = b_0 Q^2$$

or

$$g(y_c^2 \cot 50°)^3 = (2y_c \cot 50°)Q^2$$

whence

$$y_c = \left(\frac{2Q^2}{g \cot^2 50°}\right)^{1/5} = \left[\frac{2(16)^2}{9.81(0.839)^2}\right]^{1/5} = 2.37 \ \text{m} \qquad Ans. \ (a)$$

Part (b) With y_c known, from Eqs. (1) we compute $P_c = 6.18$ m, $R_{hc} = 0.760$ m, $A_c = 4.70$ m², and $b_{0c} = 3.97$ m. The critical velocity from Eq. (10.38b) is

$$V_c = \frac{Q}{A_c} = \frac{16 \text{ m}^3/\text{s}}{4.70 \text{ m}^2} = 3.41 \text{ m/s} \qquad\qquad Ans.\ (b)$$

Part (c) With $n = 0.018$, we compute from Eq. (10.40) a critical slope

$$S_c = \frac{gn^2P}{\xi R_h^{1/3}b_0} = \frac{9.81(0.018)^2(6.18)}{1.0(0.760)^{1/3}(3.97)} = 0.00542 \qquad\qquad Ans.\ (c)$$

The Moody approach gives $f = 0.0276$ and $S_c = 0.00537$.

10.5 THE HYDRAULIC JUMP

In open-channel flow a supercritical flow can change quickly back to a subcritical flow by passing through a hydraulic jump, as in Fig. 10.4. The upstream flow is fast and shallow, and the downstream flow is slow and deep, analogous to the normal shock wave of Fig. 9.8. Unlike the infinitesimally thin normal shock, the hydraulic jump is quite thick, ranging in length from 4 to 6 times the downstream depth y_2 [11].

Being extremely turbulent and agitated, the hydraulic jump is a very effective energy dissipator and is a feature of stilling-basin and spillway applications [12]. Figure 10.8 shows the jump formed at the bottom of a dam spillway in a model test.

Fig. 10.8 Spillway flow (right to left) and hydraulic-jump formation in a model of Lower Granite Dam, Snake River, Washington. Model scale 1:42.5. (*Courtesy of North Pacific Division Hydraulic Laboratory, U.S. Army Corps of Engineers.*)

It is very important that such jumps be located on specially designed firm aprons; otherwise the channel bottom will be badly scoured by the agitation. Jumps also mix fluids very effectively and have application to sewage- and water-treatment designs.

Classification

The principal parameter affecting hydraulic-jump performance is the upstream Froude number $Fr_1 = V_1/(gy_1)^{1/2}$. The Reynolds number and channel geometry have only a secondary effect. As detailed in Ref. 11, the following ranges of operation can be outlined, as illustrated in Fig. 10.9:

$Fr_1 < 1.0$: jump impossible, violates second law of thermodynamics.

$Fr_1 = 1.0$ to 1.7: standing-wave, or *undular, jump* about $4y_2$ long; low dissipation, less than 5 percent.

$Fr_1 = 1.7$ to 2.5: smooth surface rise with small rollers, known as a *weak jump*; dissipation 5 to 15 percent.

$Fr_1 = 2.5$ to 4.5: unstable, *oscillating jump*; each irregular pulsation creates a large wave which can travel downstream for miles, damaging earth banks and other structures. Not recommended for design conditions. Dissipation 15 to 45 percent.

$Fr_1 = 4.5$ to 9.0: stable, well-balanced, *steady jump*; best performance and action, insensitive to downstream conditions. Best design range. Dissipation 45 to 70 percent.

$Fr_1 > 9.0$: rough, somewhat intermittent *strong jump*, but good performance. Dissipation 70 to 85 percent.

Further details can be found in Ref. 11 and Ref. 3, chap. 15.

Theory for a Horizontal Jump

A jump which occurs on a steep channel slope can be affected by the difference in water-weight components along the flow. The effect is small, however, so that the classic theory assumes that the jump occurs on a horizontal bottom.

You will be pleased to know that we have already analyzed this problem in Sec. 10.1. A hydraulic jump is exactly equivalent to the strong fixed wave in Fig. 10.3b, where the change in depth δy is not neglected, If V_1 and y_1 upstream are known, V_2 and y_2 are computed by applying continuity and momentum across the wave, as in Eqs. (10.7) and (10.8). Equation (10.9) is therefore the correct solution for a jump if we interpret C and y in Fig. 10.3b as upstream conditions V_1 and y_1, with $C - \delta V$ and $y + \delta y$ being the downstream conditions V_2 and y_2, as in Fig. 10.9b. Equation (10.9) becomes

$$V_1^2 = \tfrac{1}{2}gy_1\eta(\eta + 1) \tag{10.41}$$

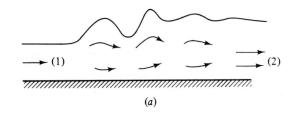

(a)

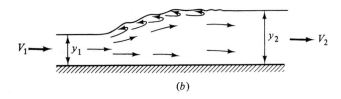

(b)

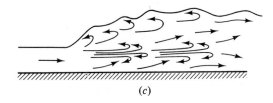

(c)

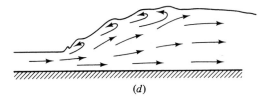

Fig. 10.9 Classification of hydraulic jumps: *(a)* Fr = 1.0 to 1.7: undular jumps; *(b)* Fr = 1.7 to 2.5: weak jump; *(c)* Fr = 2.5 to 4.5: oscillating jump; *(d)* Fr = 4.5 to 9.0: steady jump; *(e)* Fr > 9.0: strong jump. (*Adapted from Ref. 11.*)

(d)

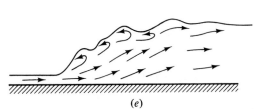

(e)

where $\eta = y_2/y_1$. Introducing the Froude number $\mathrm{Fr}_1 = V_1/(gy_1)^{1/2}$ and solving this quadratic equation for η, we obtain

$$\frac{2y_2}{y_1} = -1 + (1 + 8\,\mathrm{Fr}_1^2)^{1/2} \qquad (10.42)$$

With y_2 thus known, V_2 follows from the wide-channel continuity relation

$$V_2 = \frac{V_1 y_1}{y_2} \qquad (10.43)$$

Finally, we can evaluate the dissipation head loss across the jump from the steady-flow energy equation

$$h_f = E_1 - E_2 = \left(y_1 + \frac{V_1^2}{2g}\right) - \left(y_2 + \frac{V_2^2}{2g}\right) \tag{10.44}$$

Introducing y_2 and V_2 from Eqs. (10.42) and (10.43), we find after considerable algebraic manipulation that

$$h_f = \frac{(y_2 - y_1)^3}{4y_1 y_2} \tag{10.45}$$

Equation (10.45) shows that the dissipation loss is positive only if $y_2 > y_1$, which is a requirement of the second law of thermodynamics. Equation (10.42) then requires that $Fr_1 > 1.0$; that is, the upstream flow must be supercritical. Finally, Eq. (10.43) shows that $V_2 < V_1$ and the downstream flow is subcritical. All these results agree with our previous experience analyzing the normal-shock wave.

The present theory is for hydraulic jumps in wide horizontal channels. For the theory of prismatic or sloping channels see advanced texts [e.g., 3, chaps. 15 and 16].

EXAMPLE 10.6 Water flows in a wide channel at $q = 10 \text{ m}^3/(\text{s} \cdot \text{m})$ and $y_1 = 1.25$ m. If the flow undergoes a hydraulic jump, compute (a) y_2, (b) V_2, (c) Fr_2, (d) h_f, (e) the percentage dissipation, (f) the power dissipated per unit width, and (g) the temperature rise due to dissipation if $c_p = 4200 \text{ J}/(\text{kg} \cdot \text{K})$.

Solution

Part (a) The upstream velocity is

$$V_1 = \frac{q}{y_1} = \frac{10 \text{ m}^3/(\text{s} \cdot \text{m})}{1.25 \text{ m}} = 8.0 \text{ m/s}$$

The upstream Froude number is therefore

$$Fr_1 = \frac{V_1}{(gy_1)^{1/2}} = \frac{8.0}{[9.81(1.25)]^{1/2}} = 2.285$$

From Fig. 10.9 this is a weak jump. The depth y_2 is obtained from Eq. (10.42):

$$\frac{2y_2}{y_1} = -1 + [1 + 8(2.285)^2]^{1/2} = 5.54$$

or $\qquad y_2 = \tfrac{1}{2}y_1(5.54) = \tfrac{1}{2}(1.25)(5.54) = 3.46 \text{ m}$ *Ans. (a)*

Part (b) From Eq. (10.43) the downstream velocity is

$$V_2 = \frac{V_1 y_1}{y_2} = \frac{8.0(1.25)}{3.46} = 2.89 \text{ m/s} \qquad \textit{Ans. (b)}$$

Part (c) The downstream Froude number is

$$Fr_2 = \frac{V_2}{(gy_2)^{1/2}} = \frac{2.89}{[9.81(3.46)]^{1/2}} = 0.496 \qquad \textit{Ans. (c)}$$

Part (d)   As expected, Fr_2 is subcritical. From Eq. (10.45) the dissipation loss is

$$h_f = \frac{(3.46 - 1.25)^3}{4(3.46)(1.25)} = 0.625 \text{ m} \qquad \qquad \textit{Ans. (d)}$$

Part (e)   The percentage dissipation relates h_f to upstream energy

$$E_1 = y_1 + \frac{V_1^2}{2g} = 1.25 + \frac{(8.0)^2}{2(9.81)} = 4.51 \text{ m}$$

Hence $$\text{Percentage loss} = (100)\frac{h_f}{E_1} = \frac{100(0.625)}{4.51} = 14 \text{ percent} \qquad \textit{Ans. (e)}$$

Part (f)   The power dissipated per unit width is

$$\text{Power} = \rho g q h_f = (9800 \text{ N/m}^3)[10 \text{ m}^3/(\text{s} \cdot \text{m})](0.625 \text{ m})$$

$$= 61.3 \text{ kW/m} \qquad \qquad \textit{Ans. (f)}$$

Part (g)   Finally the mass flow rate is $\dot{m} = \rho q = (1000 \text{ kg/m}^3)[10 \text{ m}^3/(\text{s} \cdot \text{m})] = 10,000 \text{ kg/(s} \cdot \text{m})$, and the temperature rise from the steady-flow energy equation is

$$\text{Power} = \dot{m} c_p \, \Delta T$$

or $$61,300 \text{ W/m} = [10,000 \text{ kg/(s} \cdot \text{m})][4200 \text{ J/(kg} \cdot \text{K})] \, \Delta T$$

from which

$$\Delta T = 0.0015 \text{ K} \qquad \qquad \textit{Ans. (g)}$$

The dissipation is large, but the temperature rise is negligible.

10.6 GRADUALLY VARIED FLOW[1]

In practical channel flows both the bottom slope and the water depth change with position, as in Fig. 10.2. An approximate analysis is possible if the flow is gradually varied, i.e., if the slopes are small and changes not too sudden. The basic assumptions are

1. Slowly changing bottom slope
2. Slowly changing water depth (no hydraulic jumps)
3. Slowly changing cross section
4. One-dimensional velocity distribution
5. Pressure distribution approximately hydrostatic

The flow then satisfies the continuity relation (10.1) plus the energy equation with bottom friction losses included. The two unknowns for steady flow are velocity $V(x)$ and water depth $y(x)$, where x is distance along the channel.

---

[1] This section may be omitted without loss of continuity.

Basic Differential Equation

Consider the length of channel dx illustrated in Fig. 10.10. All the terms which enter the steady-flow energy equation are shown, and the balance between x and $x + dx$ is

$$\frac{V^2}{2g} + y + S_0 \, dx = S \, dx + \frac{V^2}{2g} + d \, \frac{V^2}{2g} + y + dy$$

or

$$\frac{dy}{dx} + \frac{d}{dx}\left(\frac{V^2}{2g}\right) = S_0 - S \tag{10.46}$$

where S_0 is the slope of the channel bottom (positive as shown in Fig. 10.10) and S is the slope of the EGL (which drops due to wall friction losses).

To eliminate the velocity derivative, differentiate the continuity relation

$$\frac{dQ}{dx} = 0 = A \, \frac{dV}{dx} + V \, \frac{dA}{dx} \tag{10.47}$$

But $dA = b_0 \, dy$, where b_0 is the channel width at the surface. Eliminating dV/dx between Eqs. (10.46) and (10.47), we obtain

$$\frac{dy}{dx}\left(1 - \frac{V^2 b_0}{gA}\right) = S_0 - S \tag{10.48}$$

Finally, recall from Eq. (10.38) that $V^2 b_0/gA$ is the square of the Froude number of the local channel flow. The final desired form of the gradually varied flow equation is

$$\frac{dy}{dx} = \frac{S_0 - S}{1 - \mathrm{Fr}^2} \tag{10.49}$$

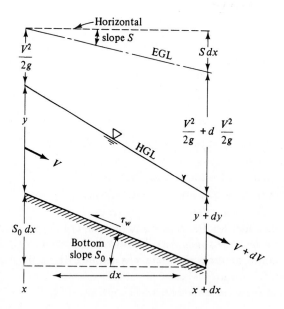

Fig. 10.10 Energy balance between two sections in a gradually varied open-channel flow.

This equation changes sign according as the Froude number is subcritical or supercritical and is analogous to the one-dimensional gas-dynamic area-change formula (9.40).

The numerator of Eq. (10.49) changes sign according as S_0 is greater or less than S, which is the slope equivalent to uniform flow at the same discharge Q

$$S = S_{0n} = \frac{dh}{dx} f = \frac{f}{D_h} \frac{V^2}{2g} = \frac{V^2}{R_h C^2} \tag{10.50}$$

where C is the Chézy coefficient. The behavior of Eq. (10.49) thus depends upon the relative magnitude of the local bottom slope $S_0(x)$, compared with (1) uniform flow, $y = y_n$ and (2) critical flow, $y = y_c$.

Classification of Solutions

It is customary to compare the actual channel slope S_0 with the critical slope S_c for the same Q from Eq. (10.4). There are five classes for S_0, giving rise to twelve distinct types of solution curves, all of which are illustrated in Fig. 10.11:

Slope class	Slope notation	Solution curves
$S_0 > S_c$	Steep	S-1, S-2, S-3
$S_0 = S_c$	Critical	C-1, C-3
$S_0 < S_c$	Mild	M-1, M-2, M-3
$S_0 = 0$	Horizontal	H-2, H-3
$S_0 < 0$	Adverse	A-2, A-3

The solution letters S, C, M, H, and A obviously denote the names of the five types of slope. The numbers 1, 2, 3 relate to the position of the initial point on the solution curve with respect to the normal depth y_n and the critical depth y_c. In type 1 solutions the initial point is above both y_n and y_c, and in all cases the water-depth solution $y(x)$ becomes even deeper and farther away from y_n and y_c. In type 2 solutions, the initial point lies between y_n and y_c, and if there is no change in S_0 or roughness, the solution tends asymptotically toward the lower of y_n or y_c. In type 3 cases, the initial point lies below both y_n and y_c, and the solution curve tends asymptotically toward the lower of these.

Figure 10.11 shows the basic character of the local solutions, but in practice, of course, S_0 varies with x and the overall solution patches together the various cases to form a continuous depth profile $y(x)$ compatible with a given initial condition and a given discharge Q. There is a fine discussion of various composite solutions in Ref. 3, chap. 9; see also Ref. 13, sec. 12.5.

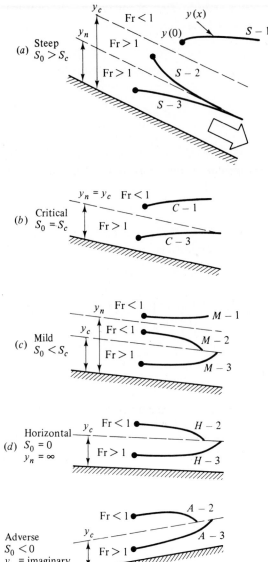

Fig. 10.11 Gradually varied flow for five classes of channel slope, showing the twelve basic solution curves.

Numerical Solution

For practical analysis of gradually varied flow profiles, numerical analysis is usually required. A simple but effective numerical scheme is to write Eq. (10.46) in finite-difference form between two depths y and $y + \Delta y$

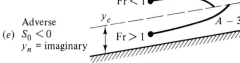

$$\Delta x \approx \frac{E(y + \Delta y) - E(y)}{(S_0 - S)_{av}} \tag{10.51}$$

where $E = y + V^2/2g$ varies only with y because Q is assumed constant. The EGL average slope S_{av} can be estimated from Eq. (10.50), using either the Moody or the Manning approach

$$S_{av} \approx \frac{f_{av} V^2_{av}}{8g R_{h,av}} \approx \frac{n^2 V^2_{av}}{\xi R_{h,av}^{4/3}} \tag{10.52}$$

where $\xi = 1.0$ for SI and 2.208 for BG units. The average velocity and hydraulic radius can be estimated from

$$V_{av} \approx \tfrac{1}{2}[V(y) + V(y + \Delta y)] \qquad R_{h,av} \approx \tfrac{1}{2}[R_h(y) + R_h(y + \Delta y)] \tag{10.53}$$

Alternately, S can be computed at y and at $y + \Delta y$ and *then* averaged. Finally, the average bottom slope would be

$$S_{0,av} \approx \tfrac{1}{2}[S_0(x) + S_0(x + \Delta x)] \tag{10.54}$$

Since Δx is not known until Eq. (10.5) has been evaluated, it may be necessary to iterate a bit to establish $S_{0,av}$ accurately. One should select the interval Δy small enough to ensure that $S_{0,av} \approx S_0(x)$.

Beginning at some initial value y_0 at $x = 0$, Eq. (10.51) can be computed either upstream or downstream, utilizing small values of Δy and computing the resulting Δx, after which $x = \sum \Delta x$ and the profile $y(x)$ can be constructed. Let us illustrate with an example. Further details of the computation are given in Ref. 3, chap. 10, and a short computer program for Eq. (10.51) is given in Ref. 14, p. 498. Physically, you should know on what class of solution profile you lie (M-2, C-3, S-1, etc.) before beginning the computation, but mathematically this is not necessary.

EXAMPLE 10.7 Let us extend the data of Example 10.4 to compute a portion of the profile shape. Given a wide channel with $n = 0.022$, $S_0 = 0.0048$, and $q = 50$ ft^3/(s · ft). If $y_0 = 3$ ft at $x = 0$, how far along the channel $x = L$ does it take the depth to rise to $y_L = 4$ ft? Use increments $\Delta y = 0.2$ ft and Manning's formulation. Is the 4-ft depth position upstream or downstream in Fig. E10.7.?

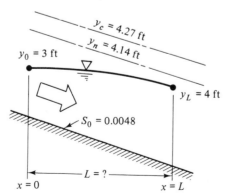

Fig. E10.7

Solution Since Example 10.4 gave the result $y_c = 4.27$ ft, we know that the flow is supercritical; and the given $S_0 = 0.0048$ is greater than $S_c = 0.00435$ from Example 10.4, so that the slope is steep and we must be on an S profile.

The normal depth is found by setting $q = 50$ in Manning's Chézy formula (10.20) with $R_h = y_n$

$$q = \frac{1.49}{0.022} [y_n(1 \text{ ft})] y_n^{2/3}(0.0048)^{1/2} = 50$$

$$y_n = 4.14 \text{ ft}$$

Thus y_0 and y_L are less than y_n, which is less than y_c and we *must* be on an S-3 curve, as in Fig. 10.11a. For the numerical solution, we tabulate $y = 3.0$ to 4.0 in intervals of 0.2, computing six values of $V = q/y$, $E = y + V^2/2g$, and $S = n^2V^2/2.208y^{4/3}$, from which S_{av} and ΔX follow from Eq. (10.51). The slope $S_0 = 0.0048$ is constant.

y	$V = 50/y$	$E = y + V^2/2g$	S	S_{av}	Δx	$x = \sum \Delta x$
3.0	16.67	7.313	0.01407			0.0
				0.01271	40.7	
3.2	15.62	6.991	0.01135			40.7
				0.01031	42.3	
3.4	14.71	6.758	0.00927			83.0
				0.00847	44.4	
3.6	13.89	6.595	0.00766			127.4
				0.00703	48.0	
3.8	13.16	6.488	0.00640			175.4
				0.00590	56.4	
4.0 ft	12.50	6.426	0.00539			$L = 231.8$ ft

For this depth increment of 0.2 ft, gradually-varied-flow theory predicts that a length of about 232 ft is required for the depth to rise from 3 to 4 ft in this supercritical flow. Using 10 increments by reducing Δy to 0.1 ft would give an estimate of $L = 235.2$ ft (calculations not shown).

It should be clear that the $y = 4$ ft position is *downstream*. *Ans.*

Note that the computed Δx's increase as the depth increases in the table. As $y \to y_n = 4.14$ ft, $\Delta x \to \infty$; that is, the profile asymptotically approaches normal depth.

An alternative to the use of Eq. (10.51) is to rewrite Eq. (10.49) in finite-difference form, substituting for S from Eq. (10.52) and for Fr from Eq. (10.38). The resulting algorithm is

$$\Delta x \approx \frac{1 - V^2 b_0/gA}{S_0 - n^2V^2/\xi R_h^{4/3}} \Delta y \qquad \xi = \begin{cases} 1.0 & \text{SI} \\ 2.208 & \text{BG} \end{cases} \qquad (10.55)$$

This expression can easily be programmed for a personal computer and used to step numerically upstream or downstream from a known position x where the depth y is known. We shall use this method in Example 10.9, where a dam in a river creates a known condition (x, y).

10.7 FLOW MEASUREMENT BY WEIRS[1]

A weir is an obstruction in the bottom of a channel which the flow must deflect over. For certain simple geometries the channel discharge Q correlates with the blockage height H to which the upstream flow is deflected by the presence of the weir. Thus a weir is an elementary but effective open-channel flowmeter.

Figure 10.12 shows two common weirs, sharp-crested, and broad-crested, assumed to be very wide. In both cases the flow upstream is subcritical, accelerates to critical near the top of the weir, and spills over in a supercritical *nappe*, which splashes into the downstream flow. In both cases, the discharge q per unit width is proportional to $H^{3/2}$, where H is the height of the upstream flow above the crest of the weir. The correlation changes if the nappe is unventilated and attaches to the weir wall downstream. (The spillway of Fig. 10.8 is a sort of unventilated weir, which can also be calibrated in the form $q = cH^{3/2}$.)

Analysis of Sharp-Crested Weirs

One can analyze weir flow by neglecting friction and using potential theory (Chap. 8) with an unknown free surface. The solutions are complicated but accurate [e.g., 15].

A very simple frictionless one-dimensional analysis is credited to the French engineer J. V. Boussinesq in 1907. As sketched in Fig. 10.12a, one estimates the velocity distribution $V_2(h)$ above the weir from the Bernoulli equation related to point 1 upstream. From Fig. 10.12a,

$$\frac{V_1^2}{2g} + H + Y \approx \frac{V_2^2}{2g} + H + Y - h$$

or
$$V_2^2 = 2gh + V_1^2$$

Then the volume rate of flow over the weir is approximately

$$q = \int_{H/3}^{H} V_2\, dh = \int_{H/3}^{H} (2gh + V_1^2)^{1/2}\, dh$$

$$= \tfrac{2}{3}(2g)^{1/2}\left[\left(H + \frac{V_1^2}{2g}\right)^{3/2} - \left(\frac{H}{3} + \frac{V_1^2}{2g}\right)^{3/2}\right] \qquad (10.56)$$

where we have taken without proof that the nappe is only about $2H/3$ high above the weir. Normally the upstream velocity head $V_1^2/2g$ is neglected, so that Eq. (10.56) becomes

$$q \approx 0.81(\tfrac{2}{3})(2g)^{1/2}H^{3/2} \qquad (10.57)$$

The formula is functionally correct, but the coefficient 0.81 is too high because of end contractions, friction, and surface-tension effects not considered in this simple

---

[1] This section may be omitted without loss of continuity.

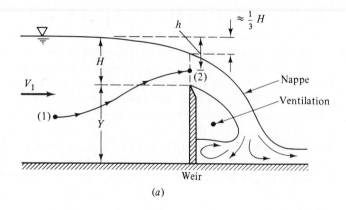

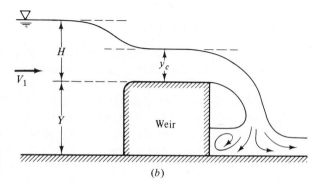

Fig. 10.12 Flow over wide, well-ventilated weirs: (*a*) sharp-crested; (*b*) broad-crested.

theory. The accepted formula for flow metering uses a weir coefficient C_w correlated from experiments by T. Rehbock in 1929

$$q = \tfrac{2}{3}C_w(2g)^{1/2}H^{3/2} \tag{10.58}$$

where

$$C_w \approx 0.611 + \frac{0.075H}{Y}$$

Analysis of Broad-Crested Weirs

The broad-crested weir of Fig. 10.12*b* is very easy to analyze because it creates a short run of nearly one-dimensional critical flow over its top. The Bernoulli equation from upstream to the top of the weir gives

$$\frac{V_1^2}{2g} + H + Y \approx \frac{V_c^2}{2g} + y_c + Y \tag{10.59}$$

Since $V_c^2 = gy_c$ from Eq. (10.33), we can solve for

$$y_c \approx \frac{2H}{3} + \frac{V_1^2}{3g} \approx \frac{2H}{3} \tag{10.60}$$

which is the estimate we used without proof in Eq. (10.56). The flow rate is thus given by

$$q = (gy_c^3)^{1/2} = g^{1/2}\left(\frac{2H}{3} + \frac{V_1^2}{3g}\right)^{3/2} \tag{10.61}$$

If we neglect the upstream velocity head, this reduces to

$$q \approx (3^{-(1/2)})(\tfrac{2}{3})(2g)^{1/2}H^{3/2} \tag{10.62}$$

By comparison with Eq. (10.58), the broad-crested weir has a theoretical weir coefficient $C_w = 3^{-(1/2)} = 0.577$. The formula recommended in Ref. 16 varies with weir height

$$C_w = \frac{0.65}{(1 + H/Y)^{1/2}} \tag{10.63}$$

See Ref. 3, pp. 52–53, for a description of test data.

EXAMPLE **10.8**   A weir in a horizontal channel is 12 ft wide and 4 ft high. The upstream water depth is 5.2 ft. Estimate the discharge if the weir is (a) sharp-crested; (b) broad-crested.

Solution

Part (a)   We are given $Y = 4$ ft and $Y + H \doteq 5.2$ ft; hence $H = 1.2$ ft. The sharp-crested-weir coefficient from Eq. (10.58) is

$$C_{w,\mathrm{SC}} = 0.611 + \frac{0.075(1.2\ \mathrm{ft})}{4\ \mathrm{ft}} = 0.634$$

so that the flow rate per unit width is

$$q = \tfrac{2}{3}(0.634)[2(32.2)]^{1/2}(1.2)^{3/2} = 4.46\ \mathrm{ft}^3/(\mathrm{s \cdot ft})$$

The total discharge is

$$Q_{\mathrm{SC}} = qb = [4.46\ \mathrm{ft}^3/(\mathrm{s \cdot ft})](12\ \mathrm{ft}) = 53.5\ \mathrm{ft}^3/\mathrm{s} \qquad Ans.\ (a)$$

Part (b)   End contractions [3, p. 362] reduce this by about 2 percent. The broad-crested-weir coefficient from Eq. (10.63) is

$$C_{w,\mathrm{BC}} = \frac{0.65}{(1 + 1.2/4.0)^{1/2}} = 0.570$$

Since everything else is the same, we simply scale down the sharp-crested result

$$Q_{\mathrm{BC}} = \frac{Q_{\mathrm{SC}} C_{w,\mathrm{BC}}}{C_{w,\mathrm{SC}}} = \frac{53.5(0.570)}{0.634} = 48.1\ \mathrm{ft}^3/\mathrm{s} \qquad Ans.\ (b)$$

End contractions are negligible. We can correct for upstream velocity head by computing $V_1 \approx q/(H + Y) = (48.1/12)/5.2 = 0.77$ ft/s. Hence $H' = H + V_1^2/2g = 1.209$ ft, and use of H' instead of H in Eq. (10.62) gives 48.7 ft³/s, or 1 percent more.

Backwater Curves

A weir is a flow barrier which not only alters the local flow over the weir but also modifies the flow depth distribution far upstream. Any strong barrier in an open-channel flow creates a *backwater curve*, which can be computed by the gradually varied flow methods of Sec. 10.6. Analysis of the flow near the barrier determines the local water depth just upstream, which then is used as input for numerical analysis of Eq. (10.51) or (10.55). For a dam or weir, dam height Y and flow rate Q are known, whence Eq. (10.58) can be solved for H, so that the total upstream water depth is $y = Y + H$. Civil engineers refer to a barrier as a *control point* where water depth can be specified as a function of flow rate. These are the starting points for numerical analysis of floodwater profiles in rivers as made, for example, by the U.S. Army Corps of Engineers [18].

We can illustrate backwater-curve computation with the following example.

EXAMPLE 10.9
A rectangular channel 8 m wide, with a flow rate of 30 m³/s, encounters a 4-m-high sharp-edged dam, as shown in Fig. E10.9. Determine the water depth 2 km upstream if the channel slope is $S_0 = 0.0004$ and $n = 0.025$.

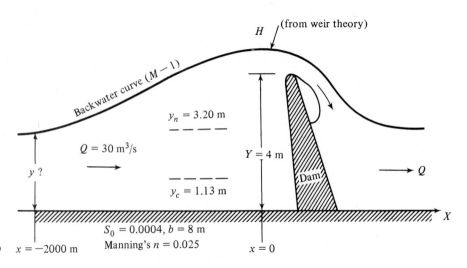

Fig. E10.9 $x = -2000$ m

Solution
The water depth just upstream of the dam (see Fig. E10.9) is determined by weir theory, Eq. (10.58)

$$q = \frac{Q}{b} = \frac{30}{8} = \frac{2}{3}\left(0.611 + 0.075\,\frac{H}{4.0}\right)[2(9.81)]^{1/2}H^{3/2}$$

By trial and error determine $H = 1.58$ m, whence $y = Y + H$ or 5.58 m at $x = 0$. This is our initial point for applying gradually varied theory.

We should verify that the upstream flow is subcritical, so that weir theory is valid. Supercritical flow approaching a barrier will create quite a different effect, such as a hydraulic jump or other rapidly varied flow. Equation (10.30) predicts that the critical depth for this flow rate will be

$$y_c = \left(\frac{Q}{b^2 g}\right)^{1/3} = \left[\frac{(30)^2}{(8)^2(9.81)}\right]^{1/3} = 1.13 \text{ m}$$

This is shown in Fig. E10.9, and since it is less than $y = 5.58$ m, the flow is subcritical.

We could now plunge right into gradually varied theory, but it is better to check the normal depth to see what type of flow profile the backwater curve is. From Eq. (10.21) for the given S_0

$$Q = 30 = \frac{1}{n} AR_h^{2/3}S_0^{1/2} = \frac{1}{0.025}(8y)\left(\frac{8y}{8+2y}\right)^{2/3}(0.0004)^{1/2}$$

By trial and error, solve for $y_n = 3.20$ m, which is also shown in Fig. E10.9. Since y is greater than both y_n and y_c, the backwater curve is of the M-1 type, as in Fig. 10.11c, and y will decrease upstream. Let us apply the alternative numerical method, Eq. (10.55), with $\xi = 1.0$ for SI units and $A/b_0 = y$ for a rectangular channel

$$\frac{\Delta y}{\Delta x} = \frac{S_0 - n^2V^2/R_h^{4/3}}{1 - V^2/gy} \tag{1}$$

Since we want to compute y at $x = -2000$ m (recall that x decreases upstream), we shall apply Eq. (1) in four equal increments $\Delta x = -500$ m. The numerical results are as follows:

x	y	V	$1 - \dfrac{V^2}{gy}$	R_h	$S_0 - \dfrac{n^2V^2}{R_h^{4/3}}$	$\dfrac{\Delta y}{\Delta x}$	$\Delta y = \dfrac{\Delta y}{\Delta x}(-500)$
0	5.580	0.672	0.992	2.33	0.000309	0.000311	
							-0.156
-500	5.424	0.691	0.991	2.30	0.000302	0.000302	
							-0.152
-1000	5.272	0.711	0.990	2.27	0.000294	0.000297	
							-0.149
-1500	5.123	0.732	0.989	2.25	0.000286	0.000289	
							-0.145
-2000	4.978 _m_						

The theory thus predicts that the water depth 2 km upstream is about 4.98 m. The change in depth is very small because the slope is very mild and the Froude number also very small (about 0.1). If S_0, n, and b remain constant upstream, the flow will approach the normal depth, $y_n = 3.20$ m, at about $x = -16$ km.

SUMMARY

This chapter is an introduction to open-channel flow analysis, limited to steady, one-dimensional flow conditions. The basic analysis combines the continuity equation with the extended Bernoulli equation including friction losses.

Open-channel flows are classified either by depth variation or by Froude number, the latter being analogous to the Mach number in compressible duct flow (Chap. 9). Flow at constant slope and depth is called uniform flow and satisfies the classical Chézy equation (10.20). Straight prismatic channels can be optimized to find the cross section which gives maximum flow rate with minimum friction losses.

As slope and flow velocity increase, the channel reaches a *critical* condition of Froude number unity, where velocity equals the speed of a small-amplitude surface wave in the channel. Every channel has a critical slope which varies with flow rate and roughness. If the flow becomes supercritical (Fr > 1), it may undergo a hydraulic jump to a greater depth and lower (subcritical) velocity, analogous to a normal-shock wave.

The analysis of gradually varied flow leads to a differential equation (10.49) which can be solved by finite-difference methods. The chapter ends with a discussion of the flow over a dam or weir, where total flow rate can be correlated with upstream water depth.

REFERENCES

1. B. A. Bakhmeteff, "Hydraulics of Open Channels," McGraw-Hill, New York, 1932.
2. S. M. Woodward and C. J. Posey, "Steady Flow in Open Channels," Wiley, New York, 1941.
3. Ven Te Chow, "Open Channel Hydraulics," McGraw-Hill, New York, 1959.
4. F. M. Henderson, "Open Channel Flow," Macmillan, New York, 1966.
5. J. J. Stoker, "Water Waves," Interscience, New York, 1957.
6. B. Kinsman, "Wind Waves: Their Generation and Propagation on the Ocean Surface," Prentice-Hall, Englewood Cliffs, N.J., 1965; Dover, New York, 1984.
7. A. T. Ippen, "Estuary and Coastline Hydrodynamics," McGraw-Hill, New York, 1966.
8. J. M. Robertson and H. Rouse, The Four Regimes of Open Channel Flow, *Civ. Eng.*, vol. 11, no. 3, pp. 169–171, March 1941.
9. R. W. Powell, Resistance to Flow in Rough Channels, *Trans. Am. Geophys. Union*, vol. 31, no. 4, pp. 575–582, August 1950.
10. P. A. Thompson, "Compressible-Fluid Dynamics," McGraw-Hill, New York, 1972.
11. U.S. Bureau of Reclamation, Research Studies on Stilling Basins, Energy Dissipators, and Associated Appurtenances, *Hydraul. Lab. Rep.* Hyd-399, June 1, 1955.
12. E. A. Elevatorski, "Hydraulic Energy Dissipators," McGraw-Hill, New York, 1959.
13. R. M. Olson, "Essentials of Engineering Fluid Mechanics," 4th ed., Harper & Row, New York, 1980.
14. J. K. Vennard and R. L. Street, "Elementary Fluid Mechanics," 6th ed., Wiley, New York, 1982.
15. J. S. McNown et al., Applications of the Relaxation Technique in Fluid Mechanics, *Proc. ASCE*, vol. 120, pp. 650–686, July 1953.
16. H. A. Doeringsfeld and C. L. Barker, Pressure-Momentum Theory Applied to the Broad-Crested Weir, *ASCE Trans.*, vol. 106, pp. 934–946, 1941.
17. Friction Factors in Open Channels, Report of the Committee on Hydromechanics, *ASCE J. Hydraul. Div.*, March 1963, pp. 97–143.
18. U.S. Army Corps of Engineers, HEC-2 Water Surface Profiles, Users Manual, *Hydrolog. Eng. Cen., Davis, Calif. Rep.* 723-02A, November 1976.
19. A. T. Lenz, Viscosity and Surface Tension Effects on V-Notch Weir Coefficients, *ASCE Trans.*, vol. 108, pp. 759–802, 1943.

RECOMMENDED FILMS

Films numbered 7, 12, 24, 25, 27 to 33, 36, and 41 in Appendix C.

Problems

PROBLEM DISTRIBUTION

10.1 The formula for shallow-water wave-propagation speed, Eq. (10.9) or (10.10), is independent of the physical properties of the liquid, i.e., density, viscosity, or surface tension. Does this mean that waves propagate at the same speed in water, mercury, gasoline, and glycerin? Explain.

10.2 A shallow-water wave 1 in high propagates into still water of depth 2 ft. Compute the wave speed c and the velocity δV induced by the wave.

10.3 The water-channel flow in Fig. P10.3 has a free surface in three places. Does it qualify as an open-channel flow? Explain. What does the dashed line represent?

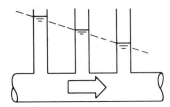

Fig. P10.3

10.4 Water flows rapidly in a flat wide channel 30 cm deep. Pebbles dropped successively in the water at the same spot create two circular ripples which are swept downstream as shown from above in Fig. P10.4. What is the current speed V?

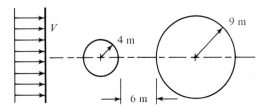

Fig. P10.4

10.5 Water flows in a wide flat channel 50 cm deep. Pebbles dropped successively at the same spot create two circular ripples as shown from above in Fig. P10.5. What is the current speed V?

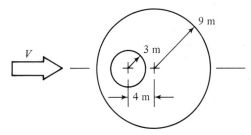

Fig. P10.5

10.6 Consider flow in a wide channel over a bump, as in Fig. P10.6. One can estimate the water depth change or *transition* with frictionless flow. Use continuity and the Bernoulli equation to show that

$$\frac{dy}{dx} = -\frac{dh/dx}{1 - V^2/gy}$$

Is the drawdown of the water surface realistic in Fig. P10.6? Explain under what conditions the surface might rise above its upstream position y_0.

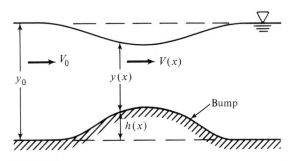

Fig. P10.6

10.7 In Fig. P10.6 suppose that $y_0 = 80$ cm and the flow rate is 1.2 m^3/s per meter of channel width. If the bump has $h_{max} = 10$ cm, compute the velocity and water depth above the peak of the bump and comment about the Froude-number effects and other details.

10.8 In Fig. P10.6 suppose that $y_0 = 80$ cm and $Q = 3.6$ m^3/s per meter of channel width. If the bump has $h_{max} = 10$ cm, compute V and y at the peak of the bump and comment about the results.

10.9 Given the flow of a channel of large width b under a sluice gate, as in Fig. P10.9. Assuming frictionless steady flow with negligible upstream kinetic energy, derive a formula for the dimensionless flow ratio $Q^2/y_1^3 b^2 g$ as a function of the ratio y_2/y_1. Show by differentiation that the maximum flow rate occurs at $y_2 = 2y_1/3$.

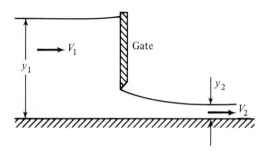

Fig. P10.9

10.10 For the sluice gate in Fig. P10.9 take the special case $y_1 = 90$ cm and $y_2 = 60$ cm, with $b = 10$ m. Compute the flow rate Q if the upstream kinetic energy is (a) neglected and (b) considered. Compare the downstream Froude numbers and comment.

10.11 For laminar draining of a wide thin sheet of water of depth y_0 on pavement of small slope S_0, show that the flow rate per unit width is $q = gy_0^3 S_0/3v$. (Recall Prob. 4.35.)

10.12 For Prob. 10.11 by comparison with the Chézy formula, show that the friction factor is $f = 24/Re$, where the Reynolds number is defined by $Re = V_{av} y_0/v = q/v$.

10.13 For laminar sheet draining as in Prob. 10.11 the flow may become turbulent if $Re = 500$. If $S_0 = 0.002$, what is the maximum sheet thickness y_0 to ensure laminar flow of water at 20°C?

10.14 A rectangular channel is 3 m wide and contains water 2 m deep. If the slope is 0.6° and the lining is unfinished cement, compute the discharge for uniform flow by the formulas of (a) Manning and (b) Moody.

10.15 A riveted-steel channel slopes at 1:400 and has a V shape with an included angle of 70°. Find the water depth for uniform flow at 800 m^3/h.

10.16 A trapezoidal channel similar to Fig. 10.6 has $b = 10$ ft, $\theta = 40°$, and $y = 3$ ft. Compute the uniform-flow discharge for a clean earth channel with $S_0 = 0.0003$.

10.17 For the channel of Prob. 10.16 compute the normal depth y_n if the flow rate is 250 ft^3/s.

10.18 In smooth channels the effective value of Manning's n can increase significantly with channel size. For roughness $\epsilon = 0.001$ ft, use the Moody formula to compute Manning's n for R_h equal to (a) 1 ft and (b) 8.75 ft.

10.19 A circular corrugated-metal storm drain is flowing half-full over a slope 5 ft per stature mile. Estimate the discharge if the drain diameter is 7 ft.

10.20 The Chézy formula (10.14) is independent of the density and viscosity of the liquid. Does this mean that water, mercury, gasoline, and glycerin will all flow at the same rate down a given open channel? Explain.

10.21 A trapezoid aqueduct (Fig. 10.6) has $b = 6$ m and $\theta = 35°$ and carries 50 m^3/s water when $y = 3$ m. If $n = 0.014$, what is the required elevation drop per kilometer?

10.22 For the aqueduct of Prob. 10.21, what will the water depth be if the slope is 0.0002 and the discharge is 25 m^3/s?

10.23 Uniform water flow in a wide brick channel of slope 0.02° moves over a 10-cm bump as in Fig. P10.23. A slight depression in water surface results. If the minimum water depth over the bump is 50 cm, compute (a) the velocity over the bump and (b) the flow rate per meter of width.

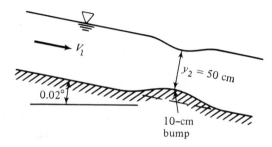

Fig. P10.23

10.24 A triangular channel (Fig. E10.5) has 45° sides and a slope of 0.0006 and is lined with asphalt. What will the normal depth be when the discharge is (*a*) 150 ft³/s and (*b*) 15 ft³/s?

10.25 A brick rectangular channel with $S_0 = 0.002$ is designed to carry 230 ft³/s of water in uniform flow. There is an argument over whether the channel width should be 4 or 8 ft. Which design needs fewer bricks? By what percentage?

10.26 In flood stage a natural channel often consists of a deep main channel plus two floodplains, as in Fig. P10.26. The floodplains are often shallow and rough. If the channel has the same slope everywhere, how would you analyze this situation for the discharge? Suppose that $y_1 = 20$ ft, $y_2 = 5$ ft, $b_1 = 40$ ft, $b_2 = 100$ ft, $n_1 = 0.020$, $n_2 = 0.040$, with a slope of 0.0002. Estimate the discharge in cubic feet per second.

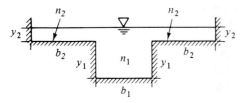

Fig. P10.26

10.27 For the flood stage in Fig. P10.26 compute the total flow rate if the main channel is 8-by-8 m clean earth and each flood plain is 1-m-deep farmland 100 m wide. All channels have a slope of 0.1°.

10.28 A 1-m-diameter clay tile sewer pipe runs half full on a slope of 0.2°. Compute the flow rate using (*a*) Manning's formula and (*b*) the Moody chart.

10.29 Four of the sewer pipes from Prob. P10.28 empty into a single finished-cement pipe, also sloping at 0.2°. If the large pipe is also to run half full, what should its diameter be? Use the Manning correlation.

10.30 For the circular channel of Fig. P10.30, if $n = 0.016$, $D = 3$ m, and the slope is 0.2°, find the normal depth y_n for a discharge of 20 m³/s.

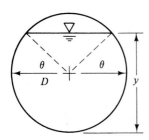

Fig. P10.30

10.31 A rectangular channel has $b = 3$ m and $y = 1$ m. If n and S_0 are the same, what is the diameter of a semicircular channel that will have the same discharge? Compare the two wetted perimeters.

10.32 A trapezoidal channel has $n = 0.022$ and $S_0 = 0.0003$ and is made in the shape of a half-hexagon for maximum efficiency. What should the length of the side of the hexagon be if the channel is to carry 225 ft³/s of water? What is the discharge of a semicircular channel of the same cross-sectional area and the same S_0 and n?

10.33 For the circular channel flowing partly full, as in Fig. P10.30, show that for a given n and S_0 the maximum average velocity in uniform flow occurs when $y = 0.813D$. Show also that the maximum discharge occurs when $y = 0.938D$.

10.34 Prove from the geometry of Fig. P10.30 that the most efficient circular channel (maximum R_h for a given A) is a semicircle.

10.35 What are the most efficient dimensions for a planed-wood rectangular channel to carry 3.5 m³/s at $S_0 = 0.0006$?

10.36 What is the most efficient depth for an asphalt trapezoidal channel, with sides sloping at 45°, to carry 4 m³/s at $S_0 = 0.0004$? How much additional discharge is carried by a semicircle with the same area?

10.37 Determine the most efficient value of θ for the V-shaped channel of Fig. P10.37.

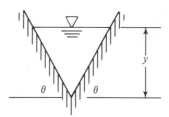

Fig. P10.37

10.38 Suppose that the side angles of the trapezoidal channel in Prob. 10.32 are reduced to $\theta = 20°$ to avoid earth slides. If the bottom flat width is 8 ft, what should the normal depth of this new channel be? Compare the wetted perimeter with the perimeter of 24.07 ft in Prob. 10.32. (Do not reveal this answer to friends still struggling with Prob. 10.32.)

10.39 Replot Fig. 10.7*b* in the form of q versus y for constant E. Does the maximum q occur at the critical depth?

10.40 A wide, clean-earth river has a flow rate $q = 135 \ \text{ft}^3/(\text{s} \cdot \text{ft})$. What is the critical depth? If the actual depth is 15 ft, what is the Froude number of the river? Compute the critical slope by (a) Manning's formula and (b) the Moody chart.

10.41 Find the critical depth of the brick channel in Prob. 10.25 for both the 4- and 8-ft widths. Are the normal flows sub- or supercritical?

10.42 Interpret the flow under a sluice gate of Fig. P10.9 in terms of points on Fig. 10.7b. Is the flow at constant E?

10.43 Interpret the flow over a bump in Fig. P10.23 in terms of points on Fig. 10.7b. Is the flow at constant E? Explain.

10.44 A thumbnail piercing the surface of a channel flow creates a wedgelike wave of half-angle 25°, as seen from above in Fig. P10.44. If the water depth is 20 cm, what is the flow velocity?

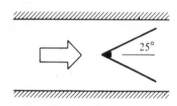

Fig. P10.44

10.45 Suppose that the wave of Fig. P10.44 is seen instead on the surface of water flowing half full in a circular channel of diameter 75 cm. What is the flow rate if the surface is finished cement?

10.46 Suppose that the wave of Fig. P10.44 is seen instead on the surface of water flowing full in a half-hexagon of side length 30 cm. What is the flow rate if the sides are of planed wood?

10.47 For the given flow rate of 50 m³/s in the trapezoidal channel of Prob. 10.21, find (a) the critical depth and (b) the critical slope.

10.48 For the river flow of Prob. 10.40, find the depth y_2 which has the same specific energy as the given depth $y_1 = 15$ ft. These are called *conjugate depths*. What is Fr_2?

10.49 A clay-tile triangular channel has sides sloping at 60°. If the channel carries 10 m³/s, compute (a) the critical depth, (b) the critical velocity, and (c) the critical slope.

10.50 A riveted-steel triangular duct flows partly full as in Fig. P10.50. If the critical depth is 60 cm, compute (a) the critical flow rate and (b) the critical slope.

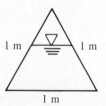

Fig. P10.50

10.51 For the triangular duct of Prob. 10.50, if the critical flow rate is 1.0 m³/s, compute (a) the critical depth and (b) the critical slope.

10.52 A rectangular channel 2 m wide and 1 m deep is found to have a critical slope of 0.008. Estimate (a) Manning's n, (b) the surface roughness height, and (c) the critical flow rate.

10.53 Water flows in a rectangular unfinished-cement channel 10 m wide laid on a slope of 0.2°. Find the depth and discharge if the channel flow is critical.

10.54 A circular corrugated-metal channel 6 ft in diameter is found to have a Froude number of 0.5 in uniform half-full flow. Compute (a) the channel slope and (b) the flow rate.

10.55 Show that the Froude number downstream of a hydraulic jump will be given by $\text{Fr}_2 = 8^{1/2} \ \text{Fr}_1/[(1 + 8 \ \text{Fr}_1^2)^{1/2} - 1]^{3/2}$. Does the formula remain correct if we reverse subscripts 1 and 2? Why?

10.56 For what Froude number Fr_1 is the energy dissipation in a hydraulic jump exactly 50 percent according to theory?

10.57 Water flows in a wide channel at $q = 20 \ \text{ft}^3/(\text{s} \cdot \text{ft})$, $y_1 = 1$ ft and then undergoes a hydraulic jump. Compute y_2, V_2, Fr_2, h_f, the percentage dissipation, and the horsepower dissipated per unit width. What is the critical depth?

10.58 A wide-channel flow at depth 50 cm passes through a hydraulic jump and emerges at a depth of 2.8 m. Compute the velocities upstream and downstream of the jump. What is the critical depth for this flow?

10.59 The flow downstream of a hydraulic jump in a rectangular channel is 9 m deep and has a velocity of 3.5 m/s. What are the velocity and depth upstream of the jump? What is the critical depth for the flow?

10.60 Water in a horizontal channel accelerates smoothly over a bump and then undergoes a hydraulic jump, as in Fig. P10.60. If $y_1 = 1$ m and $y_3 = 40$ cm, estimate (a) V_1, (b) V_3, (c) y_4, and (d) the bump height h.

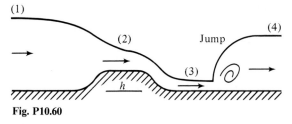

Fig. P10.60

10.61 Repeat Prob. 10.60 if $y_1 = 2$ ft and $y_3 = 6$ in.

10.62 For the conditions of Prob. 10.60 explain the reasoning by which the bump height is estimated. Also estimate the depth and velocity at section 2 on top of the bump.

10.63 At the bottom of a spillway is a hydraulic jump with water depths 2 ft upstream and 11 ft downstream. If the spillway is 100 ft wide, how much horsepower is being dissipated in the jump? What percentage dissipation occurs?

10.64 At one point in a rectangular channel 8 ft wide the depth is 3 ft and the flow rate is 450 ft³/s. If a hydraulic jump occurs, will it be upstream or downstream of this point?

10.65 A *bore* is a hydraulic jump which propagates upstream into a slower-moving or still fluid, as in Fig. 10.3a. Suppose the still fluid is 3 m deep and the water behind the bore is 5 m deep. What will the propagation speed of the bore be?

10.66 Repeat Prob. 10.65 assuming that the water upstream is flowing toward the bore at 5 m/s ground speed. What will the ground speed of propagation of the bore be?

10.67 Consider the flow under the sluice gate of Fig. P10.67. If $y_1 = 10$ ft and all losses are neglected except

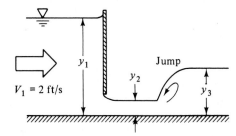

Fig. P10.67

the dissipation in the jump, calculate y_2 and y_3 and the percentage dissipation, and sketch the flow to scale with the EGL included. The channel is horizontal and wide.

10.68 In the sluice-gate flow of Fig. P10.67 measurement of y_1 and y_3 can serve as a crude estimate of flow rate. Suppose that V_1 is unknown but $y_1 = 10$ ft and $y_3 = 6$ ft. Estimate the flow rate q in ft³/(s·ft) and criticize this measurement scheme.

10.69 A 10-cm-high bump in a wide horizontal water channel creates a hydraulic jump just upstream and the flow pattern in Fig. P10.69. Neglecting losses except in the jump, for the case $y_3 = 30$ cm, estimate (a) V_4, (b) y_4, (c) V_1, and (d) y_1.

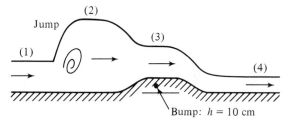

Fig. P10.69

10.70 Show what simplifications occur in Eq. (10.46) if the channel flow is uniform.

10.71 At a section of a rectangular channel 10 ft wide with $n = 0.015$ and $S_0 = 0.001$, $Q = 300$ ft³/s. Compute the normal depth and critical depth. Is the slope steep or mild? What type of gradually varied profile are we on from Fig. 10.11 if $y = $ (a) 2 ft, (b) 4 ft, (c) 6 ft?

10.72 Repeat Prob. 10.71 if the slope is $S_0 = 0.007$.

10.73 If bottom friction is included in the sluice-gate flow of Fig. P10.67, the depths (y_1, y_2, y_3) will vary with x. What type of solution curves will we have in regions (1, 2, 3) from Fig. 10.11?

10.74 Water flows in a wide clean-earth channel at $S_0 = 0.007$ and $q = 10$ m³/(m·s). Is this a steep slope? For what range of depths will the flow be on a type 1, 2, or 3 curve?

10.75 Repeat Prob. 10.74 if the channel slope is $S_0 = 0.0015$.

10.76 Water flows at 125 ft³/s in an asphalt triangular channel with 45° sloping sides and $S_0 = 0.009$. Is the slope steep or mild? For what range of depths will the flow be on a type 1, 2, or 3 curve?

10.77 A circular riveted-steel duct is 2 m in diameter and laid on a slope of 4 ft per statute mile. The duct is half full of water flowing at 2.5 m³/s. Is this a mild or steep slope? Is the flow on a type 1, 2, or 3 curve?

10.78 Consider the gradual change from the profile beginning at point a in Fig. P10.78 on a mild slope S_{01} to a mild but steeper slope S_{02} downstream. Sketch and label the curve $y(x)$ expected.

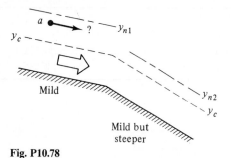

Fig. P10.78

10.79 Consider the wide-channel flow in Fig. P10.79, which changes from a mild to a steep slope. Beginning at point a, sketch the water surface profile $y(x)$ which is expected for gradually varied flow.

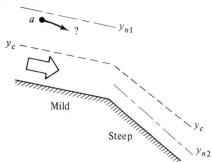

Fig. P10.79

10.80 In Fig. P10.80 the channel slope changes from steep to less steep. Beginning at point a, sketch and

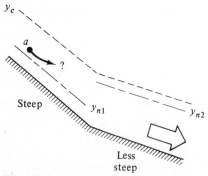

Fig. P10.80

label the expected surface curve $y(x)$ from gradually varied flow theory.

10.81 Repeat Example 10.7 to find the length L using only two depth increments of 0.5 ft each. Is the accuracy sufficient?

10.82 Let the value y_2 computed from frictionless theory in Prob. 10.67 occur just downstream of the gate at $x = 0$. If the channel is horizontal with $n = 0.015$ and there is no hydraulic jump, compute from the theory for gradually varied flow the downstream distance $x = L$, where $y = 2.0$ ft.

10.83 A wide-channel flow with $n = 0.018$ is proceeding up an adverse slope with $S_0 = -0.001$. Estimate from the theory for gradually varied flow the distance required for the depth to drop from 2.5 to 2.0 m if $q = 4$ m³/(s · m).

10.84 Figure P10.84 illustrates a free overfall or *dropdown* flow pattern, where a channel flow accelerates down a slope and falls freely over an abrupt edge. As shown, the flow reaches critical just before the overfall. Between y_c and the edge the flow is rapidly varied and does not satisfy gradually varied theory. Suppose that the flow rate is $q = 1.1$ m³/(s · m) and the surface is unfinished cement. Use Eq. (10.51) to estimate the water depth 300 m upstream as shown. Do not use Eq. (10.55) because of the singularity at the critical point $y = y_c$.

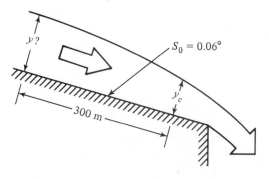

Fig. P10.84

10.85 The clean-earth channel in Fig. P10.85 is 6 m wide and slopes at 0.3°. Water flows at 30 m³/s in the channel and enters a reservoir so that the channel depth is 3 m just before the entry. Assuming gradually varied flow, how far is the distance L to a point in the channel where $y = 2$ m? What type of curve is the water surface?

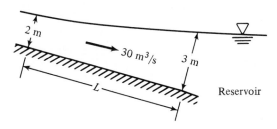

Fig. P10.85

10.86 Figure P10.86 shows a channel width change accompanied by a water-depth change, sometimes called a *venturi flume* because measurements of y_1 and y_2 can be used to meter the flow rate. Suppose that $b_1 = 5$ m, $b_2 = 3$ m, $y_1 = 2$ m, and $y_2 = 1.5$ m. Assuming no losses, compute the flow rate Q.

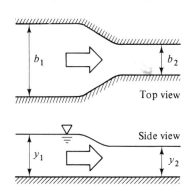

Fig. P10.86

10.87 In Prob. 10.86 for what value of y_2 will the exit flow V_2 be critical? What will the critical flow rate be?

10.88 In Prob. 10.86 suppose the flow rate is 10 m³/s and y_1 is unknown. Find the proper value of upstream depth y_1. Are there two solutions?

10.89 For the general flow pattern sketched in Fig. P10.86, show that if losses are neglected, the flow rate is given by

$$Q = \left[\frac{2g(y_1 - y_2)}{1/b_2^2 y_2^2 - 1/b_1^2 y_1^2} \right]^{1/2}$$

Verify for the numerical values of Prob. 10.86 that the flow rate should be 15.8 m³/s.

10.90 In the flume transition of Fig. P10.86, suppose that $b_1 = 4$ ft, $b_2 = 3$ ft, $y_1 = 2$ ft, and $Q = 30$ ft³/s. Compute y_2 and V_2 assuming no losses. Why is there no solution for $Q = 35$ ft³/s?

10.91 A weir in a horizontal channel is 7 m wide and 1.2 m high. If the upstream depth is 2 m, estimate the channel discharge for (*a*) a sharp-edged weir and (*b*) a broad-crested weir.

10.92 Using an analysis similar to Fig. 10.12*a*, show that the discharge of the V-notch weir in Fig. P10.92 is given by

$$Q = C_w \tfrac{8}{15} (2g)^{1/2} \tan \alpha \ H^{5/2}$$

Experiments show that C_w varies from 0.65 at $\alpha = 5°$ to 0.58 at $\alpha = 50°$.

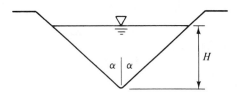

Fig. P10.92

10.93 Data by Lenz [19] show an effect of very low heads on the discharge coefficient of the V-notch weir in Fig. P10.92. For example, for $\alpha = 20°$, the data are

H, ft	C_w
0.2	0.663
0.4	0.624
0.6	0.612
0.8	0.606
1.0	0.601

Suggest a curve-fit expression for these results. What fundamental parameter (Froude number, etc.) might be involved? What head H results when the flow rate is 2 ft³/s?

10.94 Rehbock's correlation of weir coefficients, Eq. (10.58), actually suggested an additional dimensional term,

$$C_w \approx 0.611 + 0.075 H/Y + 1/305 H(\text{ft})$$

What dimensionless parameter involving water properties could account for this term?

10.95 Water flows at 500 ft³/s in a rectangular channel 20 ft wide with $n = 0.025$. A dam increases the depth to 15 ft as in Fig. P10.95. Using gradually varied theory, estimate the distance L upstream at which the water depth will be 10 ft if the channel slope is $S_0 = 0.0015$. What is the solution curve type from Fig. 10.11?

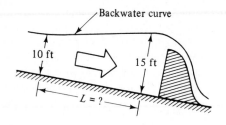

Fig. P10.95

10.96 The Tupperware dam on the Blackstone River is 12 ft high, 100 ft wide, and sharp-edged. It creates a backwater similar to Fig. P10.95. Assume that the river is a weedy-earth rectangular channel 100 ft wide with a flow rate of 800 ft^3/s. Using increments of $\frac{1}{2}$ mile, estimate the water depth 2 statute miles upstream of the dam if $S_0 = 0.001$.

10.97 A rectangular channel 4 m wide is blocked by a broad-crested weir 2 m high, as in Fig. P10.97. The channel is horizontal for 200 m upstream and then slopes at 0.7° as shown. The flow rate is 12 m^3/s and $n = 0.02$. Using 50-m increments, compute the water depth y at 300 m upstream from gradually varied theory.

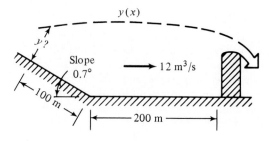

Fig. P10.97

Turbomachinery

11.1 INTRODUCTION AND CLASSIFICATION

A turbomachine is a device which adds energy to a fluid or extracts energy from it. If the machine adds energy to the fluid, it is commonly called a *pump*; if it extracts energy, it is called a *turbine*. The prefix *turbo-* is a Latin word meaning "spin" or "whirl," implying that turbomachines rotate in some way. This is generally true for turbines, but there are several types of pumps which do not rotate.

The pump is probably the oldest fluid-energy-transfer device known. At least two types date before Christ: (1) the undershot-bucket waterwheels, or *norias*, used in Asia and Africa 3000 years ago and still common today, and (2) Archimedes' screw pump (around 250 B.C.), still being manufactured today to handle solid-liquid mixtures.

A device which pumps liquids is simply called a pump, but if it pumps gases, three different terms may be used depending upon the pressure rise achieved. Up to about 1 lbf/in² pressure rise a gas pump is called a *fan*; between 1 and 40 lbf/in² absolute it is called a *blower*; and above 40 lbf/in² absolute it is called a *compressor*.

Classification of Pumps

There are two basic types of pumps, positive-displacement and dynamic, or momentum-change, pumps. There are several billion of each type in use in the world today.

Positive-displacement pumps (PDPs) have a moving boundary which forces the fluid along by volume changes. A cavity opens, and the fluid is admitted through an inlet. The cavity is then closed and the fluid is squeezed through an outlet. The classic example is the mammalian heart, but mechanical versions are in wide use.

They may be classified as follows:

A. Reciprocating
  1. Piston or plunger
  2. Diaphragm
B. Rotary
  1. Single rotor
    a. Sliding vane
    b. Flexible tube or lining
    c. Screw
    d. Peristaltic (wave contraction)
  2. Multiple rotor
    a. Gear
    b. Lobe
    c. Screw
    d. Circumferential piston

All PDPs deliver a pulsating or periodic flow as the cavity volume opens, traps, and squeezes the fluid. Their great advantage is the delivery of any fluid regardless of viscosity.

Figure 11.1 shows schematics of the operating principles of seven of these PDPs. It is rare for such devices to be run backward, so to speak, as turbines or energy extractors, the steam engine (reciprocating piston) being a classic exception.

Dynamic pumps simply add momentum to the fluid by means of fast-moving blades or vanes or certain special designs. There is no closed volume: the fluid increases momentum while moving through open passages and then converts its high velocity into a pressure increase by exiting into a diffuser section. Dynamic pumps can be classified as follows:

A. Rotary
  1. Centrifugal or radial exit flow
  2. Axial flow
  3. Mixed flow (between radial and axial)
B. Special designs
  1. Jet pump or ejector
  2. Electromagnetic pumps for liquid metals
  3. Fluid-actuated: gas-lift or hydraulic-ram

We shall concentrate in this chapter on the rotary designs, sometimes called *rotodynamic pumps*. Other designs of both PDP and dynamic pumps are discussed in specialized texts [e.g., 25].

Dynamic pumps generally provide a higher flow rate than PDPs and a much steadier discharge but are ineffective in handling high-viscosity liquids. Dynamic pumps also generally need *priming*; i.e., if they are filled with gas, they cannot suck up a liquid from below into their inlet. The PDP, on the other hand, is self-priming

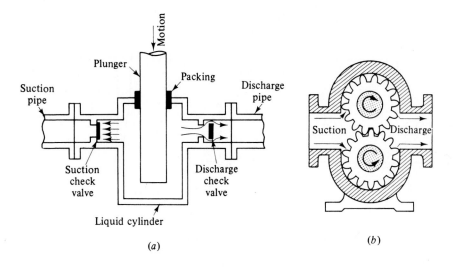

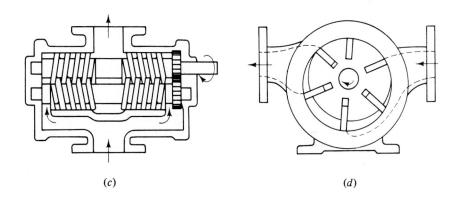

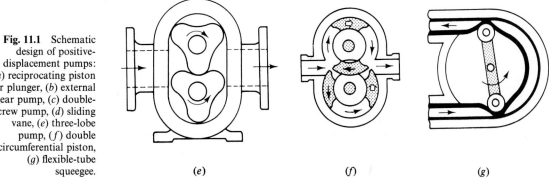

Fig. 11.1 Schematic design of positive-displacement pumps: (*a*) reciprocating piston or plunger, (*b*) external gear pump, (*c*) double-screw pump, (*d*) sliding vane, (*e*) three-lobe pump, (*f*) double circumferential piston, (*g*) flexible-tube squeegee.

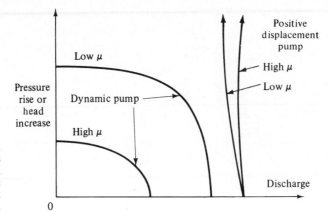

for almost any application. A PDP is appropriate for high pressure rise and low flow rate, for example, 500 lbf/in^2 and 1000 gal/min (2.2 ft^2/s), while a dynamic pump provides high flow rate (up to 300,000 gal/min) with low pressure rise, for example, 100 lbf/in^2 or less.

Figure 11.2 shows an interesting operational difference between the two types of pump. The PDP provides nearly constant flow rate over a wide range of pressure rises, while the dynamic pump gives uniform pressure rise over a range of flow rates. Note the effect of viscosity on performance.

As usual—and for the last time—we must remind the reader that this is merely an introductory chapter. There are whole books written on turbomachines: generalized treatments [1–6], texts specializing in pumps [7–14] and turbines [15–20], books giving engineering design applications [21–23], and several excellent handbooks [24–28]. There is also at least one elementary textbook [29] with a very comprehensive treatment of turbomachines. Refer to these sources for further details.

11.2 THE CENTRIFUGAL PUMP

Let us begin our brief look at rotodynamic machines by examining the characteristics of the centrifugal pump. As sketched in Fig. 11.3, this pump consists of an impeller rotating within a casing. Fluid enters axially through the *eye* of the casing,

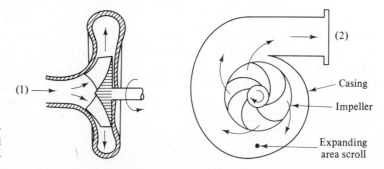

Fig. 11.3 Cutaway schematic of a typical centrifugal pump.

is caught up in the impeller blades, and is whirled tangentially and radially outward until it leaves through all circumferential parts of the impeller into the diffuser part of the casing. The fluid gains both velocity and pressure while passing through the impeller. The doughnut-shaped diffuser, or *scroll*, section of the casing decelerates the flow and further increases the pressure.

The impeller blades are usually *backward-curved*, as in Fig. 11.3, but there are also radial and forward-curved blade designs, which slightly change the output pressure. The blades may be *open*, i.e., separated from the front casing only by a narrow clearance, or *closed*, i.e., shrouded from the casing on both sides by an impeller wall. The diffuser may be *vaneless*, as in Fig. 11.3, or fitted with fixed vanes to help guide the flow toward the exit.

Basic Output Parameters

Assuming steady flow, the pump basically increases the Bernoulli head of the flow between point 1, the eye, and point 2, the exit. From Eq. (3.88), neglecting viscous work and heat transfer, this change is denoted by H:

$$H = \left(\frac{p}{\rho g} + \frac{V^2}{2g} + z \right)_2 - \left(\frac{p}{\rho g} + \frac{V^2}{2g} + z \right)_1 = h_s - h_f \qquad (11.1)$$

where h_s is the pump head supplied and h_f the losses. The net head H is a primary output parameter for any turbomachine. Since Eq. (11.1) is for incompressible flow, it must be modified for gas compressors with large density changes.

Usually V_2 and V_1 are about the same, $z_2 - z_1$ is no more than a meter or so, and the net pump head is essentially equal to the change in pressure head

$$H \approx \frac{p_2 - p_1}{\rho g} = \frac{\Delta p}{\rho g} \qquad (11.2)$$

The power delivered to the fluid simply equals the specific weight times the discharge times the net head change

$$P_w = \rho g Q H \qquad (11.3)$$

This is traditionally called the *water horsepower*. The power required to drive the pump is the *brake horsepower*

$$\text{bhp} = \omega T \qquad (11.4)$$

where ω is the shaft angular velocity and T the shaft torque. If there were no losses, P_w and brake horsepower would be equal, but of course P_w is actually less and the efficiency η of the pump is defined as

$$\eta = \frac{P_w}{\text{bhp}} = \frac{\rho g Q H}{\omega T} \qquad (11.5)$$

The chief aim of the pump designer is to make η as high as possible over as broad a range of discharge Q as possible.

The efficiency is basically composed of three parts, volumetric, hydraulic, and mechanical. The volumetric efficiency is

$$\eta_v = \frac{Q}{Q + Q_L} \tag{11.6}$$

where Q_L is the loss of fluid due to leakage in the impeller-casing clearances. The hydraulic efficiency is

$$\eta_h = 1 - \frac{h_f}{h_s} \tag{11.7}$$

where h_f has three parts: (1) *shock* loss at the eye due to imperfect match between inlet flow and the blade entrances, (2) *friction* losses in the blade passages, and (3) *circulation* loss due to imperfect match at the exit side of the blades.

Finally, the mechanical efficiency is

$$\eta_m = 1 - \frac{P_f}{\text{bhp}} \tag{11.8}$$

where P_f is the power loss due to mechanical friction in the bearings, packing glands, and other contact points in the machine.

By definition, the total efficiency is simply the product of its three parts

$$\eta \equiv \eta_v \eta_h \eta_m \tag{11.9}$$

The designer has to work in all three areas to improve the pump.

Elementary Pump Theory

You may have thought that Eqs. (11.1) to (11.9) were pump-theory relations. Not so; they are merely definitions of pump parameters and cannot be used in any predictive sense. Actually, in spite of the long history of pump development, pump theory is still rather tentative and sketchy, and most design improvements still come from testing and experience. After all, turbomachines have complex three-dimensional geometries with strong viscous effects which preclude any simple theory. The recent development of the laser anemometer, Fig. 6.28h, now allows detailed measurements to be made of the internal flow patters [30]. Within the next 10 years such data will probably be successfully coupled to three-dimensional computer solutions for predictive capability. Japikse [31] gives a review of recent progress of numerical analysis of flow in turbomachines.

To construct an elementary theory of pump performance, we assume one-dimensional flow and combine idealized fluid-velocity vectors through the impeller with the angular-momentum theorem for a control volume, Eq. (3.71).

The idealized velocity diagrams are shown in Fig. 11.4. The fluid is assumed to enter the impeller at $r = r_1$ with velocity component w_1 tangent to the blade angle β_1 plus circumferential speed $u_1 = \omega r_1$ matching the tip speed of the impeller. Its absolute entrance velocity is thus the vector sum of w_1 and u_1, shown as V_1. Similarly, the flow exits at $r = r_2$ with component w_2 parallel to the blade angle β_2 plus tip speed $u_2 = \omega r_2$, with resultant velocity V_2.

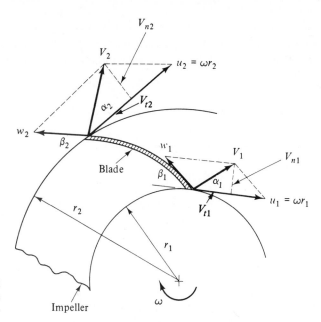

Fig. 11.4 Inlet and exit velocity diagrams for an idealized pump impeller.

We applied the angular-momentum theorem to a turbomachine in Example 3.18 (Fig. 3.16) and arrived at a result for the applied torque T

$$T = \rho Q(r_2 V_{t2} - r_1 V_{t1}) \tag{11.10}$$

where V_{t1} and V_{t2} are the absolute circumferential velocity components of the flow. The power delivered to the fluid is thus

$$P_w = \omega T = \rho Q(u_2 V_{t2} - u_1 V_{t1})$$

or

$$H = \frac{P_w}{\rho g Q} = \frac{1}{g}(u_2 V_{t2} - u_1 V_{t1}) \tag{11.11}$$

These are the *Euler turbomachine equations*, showing that the torque, power, and ideal head are functions only of the rotor tip velocities $u_{1,2}$ and the absolute fluid tangential velocities $V_{t1,2}$, independent of the axial velocities (if any) through the machine.

Additional insight is gained by rewriting these relations in another form. From the geometry of Fig. 11.4

$$V^2 = u^2 + w^2 - 2uw \cos \beta \qquad w \cos \beta = u - V_t$$

or

$$uV_t = \tfrac{1}{2}(V^2 + u^2 - w^2) \tag{11.12}$$

Substituting this into Eq. (11.11) gives

$$H = \frac{1}{2g}[(V_2^2 - V_1^2) + (u_2^2 - u_1^2) - (w_2^2 - w_1^2)] \tag{11.13}$$

Thus the ideal head relates to the absolute plus the relative kinetic-energy change of the fluid minus the rotor-tip kinetic-energy change. Finally, substituting for H from its definition in Eq. (11.1) and rearranging, we obtain the classic relation

$$\frac{p}{\rho g} + z + \frac{w^2}{2g} - \frac{r^2 \omega^2}{2g} = \text{const} \tag{11.14}$$

This is the *Bernoulli equation in rotating coordinates* and applies to either two- or three-dimensional ideal incompressible flow.

For a centrifugal pump, the power can be related to the radial velocity $V_n = V_t \tan \alpha$ and the continuity relation

$$P_w = \rho Q(u_2 V_{n2} \cot \alpha_2 - u_1 V_{n1} \cot \alpha_1) \tag{11.15}$$

where

$$V_{n2} = \frac{Q}{2\pi r_2 b_2} \quad \text{and} \quad V_{n1} = \frac{Q}{2\pi r_1 b_1}$$

and where b_1 and b_2 are the blade widths at inlet and exit. With the pump parameters $r_1, r_2, \beta_1, \beta_2$, and ω known, Eqs. (11.11) or (11.15) are used to compute idealized power and head versus discharge. The "design" flow rate Q^* is commonly estimated by assuming that the flow enters exactly normal to the impeller

$$Q = Q^*: \qquad \qquad \alpha_1 = 90° \qquad V_{n1} = V_1 \tag{11.16}$$

We can expect this simple analysis to yield estimates within ± 25 percent for the head, water horsepower, and discharge of a pump. Let us illustrate with an example.

EXAMPLE 11.1   Given the following data for a commercial centrifugal water pump: $r_1 = 4$ in, $r_2 = 7$ in, $\beta_1 = 30°$, $\beta_2 = 20°$, speed $= 1440$ r/min. Estimate (a) design-point discharge, (b) water horsepower, and (c) head if $b_1 = b_2 = 1.75$ in.

Solution

Part (a)   The angular velocity is $\omega = 2\pi$ r/s $= 2\pi(1440/60) = 150.8$ rad/s. Thus the tip speeds are $u_1 = \omega r_1 = 150.8(4/12) = 50.3$ ft/s, and $u_2 = \omega r_2 = 150.8(7/12) = 88.0$ ft/s. From the inlet-velocity diagram, Fig. E11.1a, with $\alpha_1 = 90°$ for design point, we compute

$$V_{n1} = u_1 \tan 30° = 29.0 \text{ ft/s}$$

whence the discharge is

$$Q = 2\pi r_1 b_1 V_{n1} = 2\pi \frac{4}{12} \frac{1.75}{12} (29.0) = (8.87 \text{ ft}^3/\text{s})(60 \text{ s/min}) \left(\frac{1728}{231} \text{ gal/ft}^3 \right)$$

$$= 3980 \text{ gal/min} \qquad \qquad \qquad \textit{Ans. (a)}$$

(The actual pump produces about 3500 gal/min.)

Part (b)   The outlet radial velocity follows from Q

$$V_{n2} = \frac{Q}{2\pi r_2 b_2} = \frac{8.87}{2\pi(7/12)(1.75/12)} = 16.6 \text{ ft/s}$$

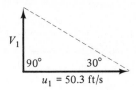

Fig. E11.1a

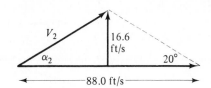

Fig. E11.1b

This enables us to construct the outlet-velocity diagram as in Fig. E11.1b, given $\beta_2 = 20°$. The tangential component is

$$V_{t2} = u_2 - V_{n2} \cot \beta_2 = 88.0 - 16.6 \cot 20° = 42.4 \text{ ft/s}$$

$$\alpha_2 = \tan^{-1} \frac{16.6}{42.4} = 21.4°$$

The power is then computed from Eq. (11.11) with $V_{t1} = 0$ at the design point

$$P_w = \rho Q u_2 V_{t2} = (1.94 \text{ slugs/ft}^3)(8.87 \text{ ft}^3/\text{s})(88.0 \text{ ft/s})(42.4 \text{ ft/s})$$

$$= \frac{64{,}100 \text{ (ft} \cdot \text{lbf)/s}}{550} = 117 \text{ hp} \qquad \qquad Ans. (b)$$

(The actual pump delivers about 125 water horsepower, requiring 147 bhp at 85 percent efficiency.)

Part (c)   Finally, the head is estimated from Eq. (11.11)

$$H \approx \frac{P_w}{\rho g Q} = \frac{64{,}100 \text{ (ft} \cdot \text{lbf)/s}}{(62.4 \text{ lbf/ft}^3)(8.87 \text{ ft}^3/\text{s})} = 116 \text{ ft} \qquad \qquad Ans. (c)$$

(The actual pump develops about 140 ft head.) Improved methods for obtaining closer estimates are given in advanced references [e.g., 6, 14, and 25].

Effect of Blade Angle on Pump Head

The simple theory above can be used to predict an important blade-angle effect. If we neglect inlet angular momentum, the theoretical water horsepower is

$$P_w = \rho Q u_2 V_{t2} \qquad \qquad (11.17)$$

where $$V_{t2} = u_2 - V_{n2} \cot \beta_2 \qquad V_{n2} = \frac{Q}{2\pi r_2 b_2}$$

Then the theoretical head from Eq. (11.11) becomes

$$H \approx \frac{u_2^2}{g} - \frac{u_2 \cot \beta_2}{2\pi r_2 b_2 g} Q \qquad \qquad (11.18)$$

The head varies linearly with discharge Q, having a shutoff value u_2^2/g, where u_2 is the exit blade-tip speed. The slope is negative if $\beta_2 < 90°$ (backward-curved blades) and positive for $\beta_2 > 90°$ (forward-curved blades). This effect is shown in Fig. 11.5 and is accurate only at low flow rates.

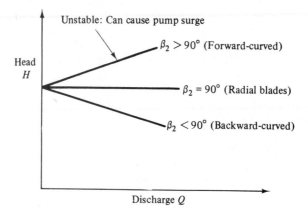

Fig. 11.5   Theoretical
effect of blade exit angle
on pump head versus
discharge.

The positive-slope condition can be unstable and cause pump *surge*, an oscillatory condition where the pump "hunts" for the proper operating point. Surge may cause only rough operation in a liquid pump, but it can be a major problem in gas-compressor operation. For this reason a backward-curved or radial blade design is generally preferred. A survey of the problem of pump stability is given by Greitzer [37].

11.3   PUMP PERFORMANCE CURVES AND SIMILARITY RULES

Since the theory of the previous section is rather qualitative, the only solid indicator of a pump's performance lies in extensive testing. For the moment let us discuss the centrifugal pump in particular. The general principles and the presentation of data are exactly the same for mixed-flow and axial-flow pumps and compressors.

Performance charts are almost always plotted for constant shaft-rotation speed n (in revolutions per minute, usually). The basic independent variable is taken to be discharge Q (in gallons per minute usually for liquids and cubic feet per minute for gases). The dependent variables, or "output," are taken to be head H (pressure rise Δp for gases), brake horsepower (bhp), and efficiency η.

Figure 11.6 shows typical performance curves for a centrifugal pump. The head is approximately constant at low discharge and then drops to zero at $Q = Q_{max}$. At this speed and impeller size, the pump cannot deliver any more fluid than Q_{max}. The positive-slope part of the head is shown dashed; as mentioned earlier, this region can be unstable and cause compressor surge.

The brake horsepower provided by the pump motor in Fig. 11.6 rises monotonically with discharge and then typically drops off slightly near Q_{max}. The dropoff region is shown dashed because it is also potentially unstable and can cause motor overload during a transient. Actually, the overload danger is much more serious in an axial-flow pump, where the brake-horsepower dropoff is much steeper and occurs at low flow rate.

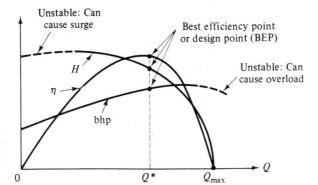

Fig. 11.6 Typical centrifugal-pump performance curves at constant impeller-rotation speed.

Finally, the efficiency in Fig. 11.6 rises to a maximum at about 60 percent of Q_{max}. This is the design flow rate Q^* or best-efficiency point (BEP), $\eta = \eta_{max}$. The head and horsepower at the BEP will be termed H^* and bhp*, respectively. Note that η is zero at the origin (no flow) and at Q_{max} (no head). Note also that the η curve is not independent but is simply calculated from the $H(Q)$ and bhp(Q) data according to the relation in Eq. (11.5), $\eta = \rho gQH/$bhp.

Measured Performance Curves

Figure 11.7 shows actual performance data for a commercial centrifugal pump. Figure 11.7a is for a basic casing size with three different impeller diameters. The head curves $H(Q)$ are shown, but the horsepower and efficiency curves have to be inferred from the contour plots. Maximum discharges are not shown, being far outside the normal operating range near the BEP. Everything is plotted raw, of course [feet, horsepower, gallons per minute (1 U.S. gal = 231 in^3)] since it is to be used directly by designers. Figure 11.7b is the same pump design with a 20 percent larger casing, a lower speed, and three larger impeller diameters. Comparing the two pumps may be a little confusing: the larger pump produces exactly the same discharge but only half the horsepower and half the head. This will be readily understood from the scaling or similarity laws we are about to formulate.

A point often overlooked is that raw curves like Fig. 11.7 are strictly applicable to a fluid of a certain density and viscosity, in this case water. If the pump were used to deliver, say, mercury, the brake horsepower would be about 13 times higher while Q, H, and η would be about the same. But in that case H should be interpreted as feet of *mercury*, not feet of water. If the pump were used for SAE 30 oil, *all* data would change (brake horsepower, Q, H, and η) due to the large change in viscosity (Reynolds number). Again this should become clear with the similarity rules.

Net Positive-Suction Head

In the top of Fig. 11.7 is plotted the net positive-suction head (NPSH), which is the head required at the pump inlet to keep the liquid from cavitating or boiling. The

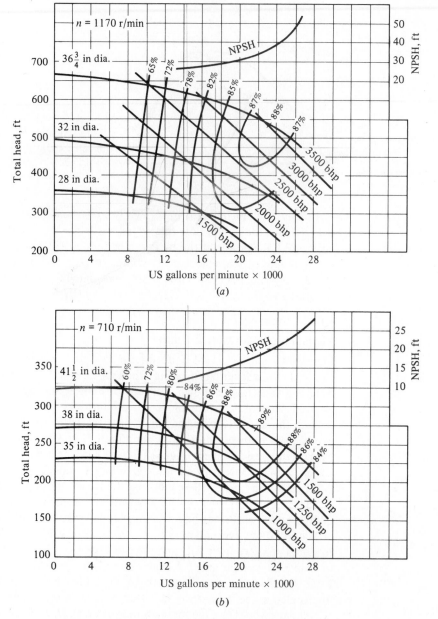

Fig. 11.7 Measured-performance curves for two models of a centrifugal water pump; (*a*) basic casing with three impeller sizes; (*b*) 20 percent larger casing with three larger impellers at slower speed. (*Ingersoll-Rand Corporation, Cameron Pump Division.*)

pump inlet or suction side is the low-pressure point where cavitation will first occur. The NPSH is defined as

$$\text{NPSH} = \frac{p_i}{\rho g} + \frac{V_i^2}{2g} - \frac{p_v}{\rho g} \tag{11.19}$$

where p_i and V_i are the pressure and velocity at the pump inlet and p_v is the vapor pressure of the liquid. Given the left-hand side, NPSH, from the pump performance curve, we must ensure that the right-hand side is equal or greater in the actual system to avoid cavitation.

If the pump inlet is placed at a height Z_i above a reservoir whose free surface is at pressure p_a, we can use Bernoulli's equations to rewrite NPSH as

$$\text{NPSH} = \frac{p_a}{\rho g} - Z_i - h_{fi} - \frac{p_v}{\rho g} \tag{11.20}$$

where h_{fi} is the friction-head loss between the reservoir and the pump inlet. Knowing p_a and h_{fi}, we can set the pump at a height Z_i which will keep the right-hand side greater than the "required" NPSH plotted in Fig. 11.7.

If cavitation does occur, there will be pump noise and vibration, pitting damage to the impeller, and a sharp dropoff in pump head and discharge. In some liquids this deterioration starts before actual boiling, as dissolved gases and light hydrocarbons are liberated.

EXAMPLE 11.2 The 32-in pump of Fig. 11.7a is to pump 24,000 gal/min of water at 1170 r/min from a reservoir whose surface is at 14.7 lbf/in² absolute. If head loss from reservoir to pump inlet is 6 ft, where should the pump inlet be placed to avoid cavitation for water at (a) 60°F, $p_v = 0.26$ lbf/in² absolute, SG = 1.0 and (b) 200°F, $p_v = 11.52$ lbf/in² absolute, SG = 0.9635?

Solution For either case read from Fig. 11.7a at 24,000 gal/min that the required NPSH is 40 ft.
Part (a) For this case $\rho g = 62.4$ lbf/ft³. From Eq. (11.20) it is necessary that

$$\text{NPSH} \leq \frac{p_a - p_v}{\rho g} - Z_i - h_{fi}$$

or
$$40 \text{ ft} \leq \frac{(14.7 - 0.26)(144)}{62.4} - Z_i - 6.0$$

or
$$Z_i \leq 27.3 - 40 = -12.7 \text{ ft} \qquad \qquad Ans. (a)$$

The pump must be placed at least 12.7 ft *below* the reservoir surface to avoid cavitation.
Part (b) For this case $\rho g = 62.4(0.9635) = 60.1$ lbf/ft³. Equation (11.20) applies again with the higher p_v

$$40 \text{ ft} \leq \frac{(14.7 - 11.52)(144)}{60.1} - Z_i - 6.0$$

or
$$Z_i \leq 1.6 - 40 = -38.4 \text{ ft} \qquad \qquad Ans. (b)$$

The pump must now be placed at least 38.4 ft below the reservoir surface. These are unusually stringent conditions because a large, high-discharge pump requires a large NPSH.

Dimensionless Pump Performance

For a given pump design, the output variables H and brake horsepower should be dependent upon discharge Q, impeller diameter D, and shaft speed n, at least. Other possible parameters are fluid density ρ, viscosity μ, and surface roughness ϵ. Thus the performance curves in Fig. 11.7 are equivalent to the following assumed functional relations:[1]

$$gH = f_1(Q, D, n, \rho, \mu, \epsilon) \qquad \text{bhp} = f_2(Q, D, n, \rho, \mu, \epsilon) \qquad (11.21)$$

This is a straightforward application of dimensional-analysis principles from Chap. 5. As a matter of fact, it was given as an exercise (Probs. 5.19 and 5.20). For each function in Eq. (11.21) there are seven variables and three primary dimensions (M, L, and T); hence we expect $7 - 3 = 4$ dimensionless pi's, and that is what we get. You can verify as an exercise that appropriate dimensionless forms for Eqs. (11.21) are

$$\frac{gH}{n^2 D^2} = g_1\left(\frac{Q}{nD^3}, \frac{\rho n D^2}{\mu}, \frac{\epsilon}{D}\right)$$

$$\frac{\text{bhp}}{\rho n^3 D^5} = g_2\left(\frac{Q}{nD^3}, \frac{\rho n D^2}{\mu}, \frac{\epsilon}{D}\right) \qquad (11.22)$$

The quantities $\rho n D^2/\mu$ and ϵ/D are recognized as the Reynolds number and roughness ratio, respectively. Three new pump parameters have arisen:

$$\text{Capacity coefficient } C_Q = \frac{Q}{nD^3}$$

$$\text{Head coefficient } C_H = \frac{gH}{n^2 D^2} \qquad (11.23)$$

$$\text{Power coefficient } C_P = \frac{\text{bhp}}{\rho n^3 D^5}$$

Note that only the power coefficient contains fluid density, the parameters C_Q and C_H being kinematic types.

Figure 11.7 gives no warning of viscous or roughness effects. The Reynolds numbers are from 0.8 to 1.5×10^7, or fully turbulent flow in all passages probably. The roughness is not given and varies greatly among commercial pumps. But at such high Reynolds numbers we expect more or less the same percentage effect on all these pumps. Therefore it is common to assume that the Reynolds number and the roughness ratio have a constant effect, so that Eqs. (11.23) reduce to, approximately,

$$C_H \approx C_H(C_Q) \qquad C_P \approx C_P(C_Q) \qquad (11.24)$$

For geometrically similar pumps, we expect head and power coefficients to be (nearly) unique functions of the capacity coefficient. We have to watch out that the pumps are geometrically similar or nearly so because (1) manufacturers put different sized impellers in the same casing, thus violating geometric similarity, and (2) large pumps have smaller ratios of roughness and clearances to impeller diameter than small pumps. In addition, the more viscous liquids will have

---

[1] We adopt gH as a variable instead of H for dimensional reasons.

significant Reynolds-number effects; e.g., a factor of 3 or more viscosity increase causes a clearly visible effect on C_H and C_P.

The efficiency η is already dimensionless and is uniquely related to the other three. It varies with C_Q also

$$\eta \equiv \frac{C_H C_Q}{C_P} = \eta(C_Q) \tag{11.25}$$

We can test Eqs. (11.24) and (11.25) from the data of Fig. 11.7. The impeller diameters of 32 and 38 in are approximately 20 percent different in size, and so their ratio of impeller to casing size is the same. The parameters C_Q, C_H, and C_P are computed with n in revolutions per second, Q in cubic feet per second (gallons per minute $\times$ 2.23 $\times$ 10^{-3}), H and D in feet, $g = 32.2$ ft/s^2, and brake horsepower in horsepower times 550 (ft $\cdot$ lbf)/(s $\cdot$ hp). The nondimensional data are then plotted in Fig. 11.8. A dimensionless suction-head coefficient is also defined

$$C_{HS} = \frac{g(\text{NPSH})}{n^2 D^2} = C_{HS}(C_Q) \tag{11.26}$$

The coefficients C_P and C_{HS} are seen to correlate almost perfectly into a single function of C_Q, while η and C_H data deviate by a few percent. The last two parameters are more sensitive to slight discrepancies in model similarity; since the

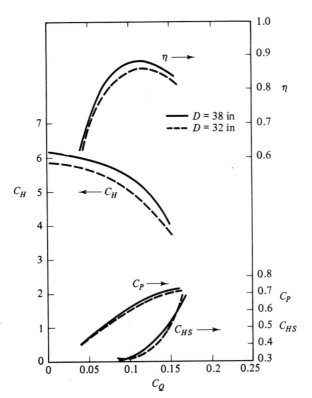

Fig. 11.8
Nondimensional plot of the pump performance data from Fig. 11.7. These numbers are not representative of other pump designs.

larger pump has smaller roughness and clearance ratios and a 40 percent larger Reynolds number, it develops slightly more head and is more efficient. The overall effect is a resounding victory for dimensional analysis.

The best efficiency point in Fig. 11.8 is approximately

$$\eta_{max} \approx 0.88: \qquad\qquad C_{Q*} \approx 0.115 \qquad C_{P*} \approx 0.65$$
$$C_{H*} \approx 5.0 \qquad C_{HS*} \approx 0.37 \qquad\qquad (11.27)$$

These values can be used to estimate the BEP performance of any size pump in this geometrically similar family. In like manner, the shutoff head is $C_H(0) \approx 6.0$, and by extrapolation the shutoff power is $C_P(0) \approx 0.25$ and the maximum discharge is $C_{Q,max} \approx 0.23$. Note, however, that Fig. 11.8 gives no reliable information about, say, the 28- or 35-in impellers in Fig. 11.7, which have a different impeller-to-casing-size ratio and thus must be correlated separately.

By comparing values of n^2D^2, nD^3, and n^3D^5 for two pumps in Fig. 11.7 we can see readily why the large pump had the same discharge but less power and head:

	D, ft	n, r/s	Discharge nD^3, ft^3/s	Head n^2D^2/g, ft	Power $\rho n^3 D^5/550$, hp
Fig. 11.7a	$\dfrac{32}{12}$	$\dfrac{1170}{60}$	370	84	3527
Fig. 11.7b	$\dfrac{38}{12}$	$\dfrac{710}{60}$	376	44	1861
Ratio			1.02	0.52	0.53

Discharge goes as nD^3, which is about the same for both pumps. Head goes as n^2D^2 and power as n^3D^5 for the same ρ (water), and these are about half as much for the larger pump. The NPSH goes as n^2D^2 and is also half as much for the 38-in pump.

EXAMPLE 11.3

A pump from the family of Fig. 11.8 has $D = 21$ in and $n = 1500$ r/min. Estimate (a) discharge, (b) head, (c) pressure rise, and (d) brake horsepower of this pump for water at 60°F and best efficiency.

Solution

Part (a)

In BG units take $D = 21/12 = 1.75$ ft and $n = 1500/60 = 25$ r/s. At 60°F, ρ of water is 1.94 slugs/ft^3. The BEP parameters are known from Fig. 11.8 or Eq. (11.27). The BEP discharge is thus

$$Q^* = C_{Q*}nD^3 = 0.115(25)(1.75)^3 = (15.4 \text{ ft}^3/\text{s})(448.8) = 6900 \text{ gal/min} \qquad Ans. (a)$$

Part (b)

Similarly, the BEP head is

$$H^* = \frac{C_{H*}n^2D^2}{g} = \frac{5.0(25)^2(1.75)^2}{32.2} = 300 \text{ ft water} \qquad Ans. (b)$$

Part (c)

Since we are not given elevation or velocity-head changes across the pump, we neglect them and estimate

$$\Delta p \approx \rho g H = 1.94(32.2)(300) = 18,600 \text{ lbf/ft}^2 = 129 \text{ lbf/in}^2 \qquad Ans. (c)$$

Part (d) Finally, the BEP power is

$$P* = C_{P*}\rho n^3 D^5 = 0.65(1.94)(25)^3(1.75)^5$$

$$= \frac{323{,}000 \text{ (ft} \cdot \text{lbf)/s}}{550} = 590 \text{ hp} \qquad\qquad Ans. (d)$$

EXAMPLE 11.4 We want to build a pump from the family of Fig. 11.8, which delivers 3000 gal/min water at 1200 r/min at best efficiency. Estimate (a) the impeller diameter, (b) the maximum discharge, (c) the shut-off head, and (d) the NPSH at best efficiency.

Solution

Part (a) 3000 gal/min = 6.68 ft³/s   and   1200 r/min = 20 r/s.
At BEP we have

$$Q* = C_{Q*}nD^3 = 6.68 \text{ ft}^3/\text{s} = (0.115)(20)D^3$$

or

$$D = \left[\frac{6.68}{0.115(20)}\right]^{1/3} = 1.43 \text{ ft} = 17.1 \text{ in} \qquad\qquad Ans. (a)$$

Part (b) The max Q is related to Q* by a ratio of capacity coefficients

$$Q_{\max} = \frac{Q*C_{Q,\max}}{C_{Q*}} \approx \frac{3000(0.23)}{0.115} = 6000 \text{ gal/min} \qquad\qquad Ans. (b)$$

Part (c) From Fig. 11.8 we estimated the shutoff-head coefficient to be 6.0. Thus

$$H(0) \approx \frac{C_H(0)n^2 D^2}{g} = \frac{6.0(20)^2(1.43)^2}{32.2} = 152 \text{ ft} \qquad\qquad Ans. (c)$$

Part (d) Finally, from Eq. (11.27), the NPSH at BEP is approximately

$$\text{NPSH}* = \frac{C_{HS*}n^2 D^2}{g} = \frac{0.37(20)^2(1.43)^2}{32.2} = 9.4 \text{ ft} \qquad\qquad Ans. (d)$$

Since this a small pump, it will be less efficient than the pumps in Fig. 11.8, probably about 82 percent maximum.

Similarity Rules

The success of Fig. 11.8 in correlating pump data leads to simple rules for comparing pump performance. If pump 1 and pump 2 are from the same geometric family and are operating at homologous points (the same dimensionless position on a chart such as Fig. 11.8), their flow rates, heads, and powers will be related as follows:

$$\frac{Q_2}{Q_1} = \frac{n_2}{n_1}\left(\frac{D_2}{D_1}\right)^3 \qquad \frac{H_2}{H_1} = \left(\frac{n_2}{n_1}\right)^2\left(\frac{D_2}{D_1}\right)^2$$

$$\frac{P_2}{P_1} = \frac{\rho_2}{\rho_1}\left(\frac{n_2}{n_1}\right)^3\left(\frac{D_2}{D_1}\right)^5 \qquad\qquad (11.28)$$

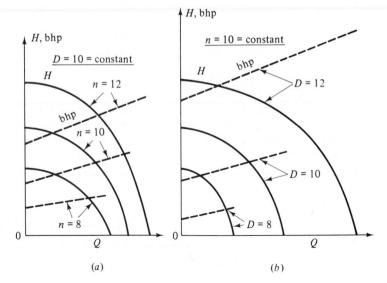

Fig. 11.9 Effect of changes in size and speed on homologous-pump performance: (a) 20 percent change in speed at constant size; (b) 20 percent change in size at constant speed.

These are the similarity rules, which can be used to estimate the effect of changing fluid, speed, or size on any dynamic turbomachine, pump, or turbine within a geometrically similar family. Strictly speaking, we would also expect $\eta_2 = \eta_1$ for perfect similarity, but we have seen that the larger pumps are more efficient, having a higher Reynolds number and lower roughness and clearance ratios. An empirical formula for estimating the change in efficiency due to size was given by Moody [32]

$$\frac{1 - \eta_2}{1 - \eta_1} \approx \left(\frac{D_1}{D_2}\right)^{1/4} \tag{11.29}$$

This formula, first developed for turbines, is widely used for both pumps and turbines when efficiency data are lacking.

A graphic display of the similarity rules implied by Eqs. (11.28) is given in Fig. 11.9, showing the effect of speed and diameter changes on pump performance. In Fig. 11.9a the size is held constant and the speed varied 20 percent, while Fig. 11.9b shows a 20 percent size change with constant speed. The curves are plotted to scale but with arbitrary units. The head and brake horsepower change magnitude, but the shape stays the same. Although the speed effect (Fig. 11.9a) is substantial, the size effect is even more dramatic (Fig. 11.9b), especially on brake horsepower, which varies as D^5. Generally we see that a given pump family can be adjusted in size and speed to fit a large variety of system characteristics.

Effect of Viscosity

Centrifugal pumps are often used to pump oils and other viscous liquids up to 1000 times the viscosity of water. But the Reynolds numbers become low-turbulent or even laminar, with a strong effect on performance. Figure 11.10 shows typical test

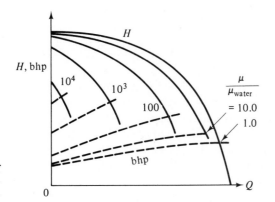

Fig. 11.10 Effect of viscosity on centrifugal-pump performance.

curves of head and brake horsepower versus discharge. High viscosity causes a dramatic drop in head and discharge and increases in power requirements. The efficiency also drops substantially according to the following typical results:

μ/μ_{water}	1.0	10.0	100	1000
η_{max}, %	85	76	52	11

Beyond about $300\mu_{water}$ the deterioration in performance is so great that a positive-displacement pump is recommended.

11.4 MIXED- AND AXIAL-FLOW PUMPS: THE SPECIFIC SPEED

We have seen from the previous section that the modern centrifugal pump is a formidable device, able to deliver very high heads and reasonable flow rates with excellent efficiency. It can match many system requirements. But basically the centrifugal pump is a high-head, low-flow machine, whereas there are many applications requiring low head and high discharge. To see that the centrifugal design is not convenient for such systems, consider the following example.

EXAMPLE 11.5   We want to use a centrifugal pump from the family of Fig. 11.8 to deliver 100,000 gal/min of water at 60°F with a head of 25 ft. What should be (*a*) the pump size and speed and (*b*) brake horsepower assuming operation at best efficiency?

Solution

Part (a)   Enter the known head and discharge into the BEP parameters from Eq. (11.27):

$$H^* = 25 \text{ ft} = \frac{C_{H*}n^2D^2}{g} = \frac{5.0n^2D^2}{32.2}$$

$$Q^* = 100,000 \text{ gal/min} = 222.8 \text{ ft}^3/\text{s} = C_{Q*}nD^3 = 0.115nD^3$$

The two unknowns are n and D. Solve simultaneously for

$$D = 12.4 \text{ ft} \qquad n = 1.03 \text{ r/s} = 62 \text{ r/min} \qquad \qquad Ans. \text{ (}a\text{)}$$

Part (b)   The most efficient horsepower is then, from Eq. (11.27),

$$\text{bhp*} \approx C_{P*}\rho n^3 D^5 = \frac{0.65(1.94)(1.03)^3(12.4)^5}{550} = 720 \text{ hp} \qquad Ans. (b)$$

The solution to Example 11.5 is mathematically correct but results in a grotesque pump: an impeller more than 12 ft in diameter, rotating so slowly one can visualize oxen walking in a circle turning the shaft.

There are other dynamic-pump designs which do provide low head and high discharge. For example, there is a type of 38-in, 710 r/min pump, e.g., with the same input parameters as Fig. 11.7b, which will deliver the 25-ft head and 100,000 gal/min flow rate called for in Example 11.5. This is done by allowing the flow to pass through the impeller with an axial-flow component and less centrifugal component. The passages can be opened up to the increased flow rate with very little size increase, but the drop in radial outlet velocity decreases the head produced. These are the mixed-flow (part radial, part axial) and axial flow (propeller-type) families of dynamic pump. Some vane designs are sketched in Fig. 11.11, which introduces an interesting new "design" parameter, the specific speed, N_S or N_s'.

The Specific Speed

Most pump applications involve a known head and discharge for the particular system, plus a speed range dictated by electric-motor speeds or cavitation requirements. The designer then selects the best size and shape (centrifugal, mixed, axial) for the pump. To help this selection, we need a dimensionless parameter involving speed, discharge, and head but not size. This is accomplished by eliminating diameter between C_Q and C_H, applying the result only to the BEP. This ratio is called the *specified speed* and has both a dimensionless form and a somewhat lazy, practical form:

Rigorous form:
$$N_s' = \frac{C_{Q*}^{1/2}}{C_{H*}^{3/4}} = \frac{n(Q^*)^{1/2}}{(gH^*)^{3/4}} \qquad (11.30a)$$

Lazy but common:
$$N_s = \frac{(\text{r/min})(\text{gal/min})^{1/2}}{[H \text{ (ft)}]^{3/4}} \qquad (11.30b)$$

In other words, practicing engineers do not bother to change n to revolutions per second or Q^* to cubic feet per second or include gravity with head, although the latter would be necessary for, say, a pump on the moon. The conversion factor is

$$N_s = 17182 N_s'$$

Note that N_s is applied only to BEP; thus a single number characterizes an entire family of pumps. For example, the family of Fig. 11.8 has $N_s' \approx (0.115)^{1/2}/(5.0)^{3/4} = 0.1014$, $N_s = 1740$, regardless of size or speed.

It turns out that specific speed is directly related to the most efficient pump design, as shown in Fig. 11.11. Low N_s means low Q and high H, hence a centrifugal

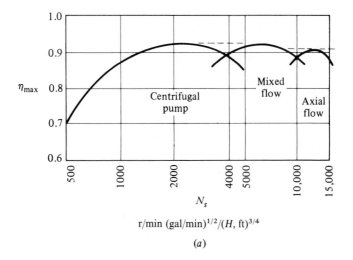

r/min (gal/min)$^{1/2}$/(H, ft)$^{3/4}$

(a)

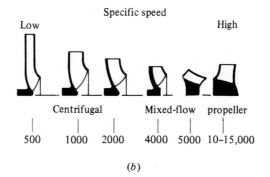

Fig. 11.11 Optimum efficiency (a) and vane design (b) of dynamic-pump families as a function of specific speed.

(b)

pump, and large N_s implies an axial pump. The centrifugal pump is best for N_s between 500 and 4000, the mixed-flow pump between 4000 and 10,000, and the axial-flow pump above 10,000. Note the changes in impeller shape as N_s increases.

Suction Specific Speed

If we use NPSH rather than H in Eq. (11.30), the result is called *suction specific speed*

Rigorous

$$N'_{ss} = \frac{nQ^{1/2}}{(g \text{ NPSH})^{3/4}}$$

(11.31a)

Lazy:

$$N_{ss} = \frac{(\text{r/min})(\text{gal/min})^{1/2}}{[\text{NPSH (ft)}]^{3/4}}$$

(11.31b)

where NPSH denotes the available suction head of the system. Data from Wislicenus [3] show that a given pump is in danger of inlet cavitation if

$$N'_{ss} \geq 0.47 \qquad N_{ss} \geq 8100$$

In the absence of test data this relation can be used, given n and Q, to estimate the minimum required NPSH.

Axial-Flow Pump Theory

A multistage axial-flow geometry is shown in Fig. 11.12a. The fluid essentially passes almost axially through alternate rows of fixed *stator* blades and moving *rotor* blades. The incompressible flow assumption is frequently used even for gases, because the pressure rise per stage is usually small.

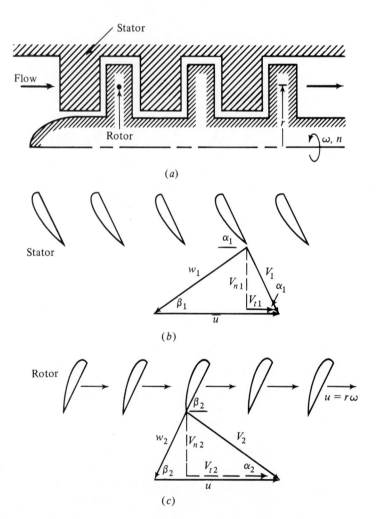

Fig. 11.12 Analysis of an axial-flow pump: (*a*) basic geometry; (*b*) stator blades and exit-velocity diagram; (*c*) rotor blades and exit-velocity diagram.

The simplified vector-diagram analysis assumes that the flow is one-dimensional and leaves each blade row at a relative velocity exactly parallel to the exit blade angle. Figure 11.12b shows the stator blades and their exit velocity diagram. Since the stator is fixed, ideally the absolute velocity V_1 is parallel to the trailing edge of the blade. After vectorially subtracting the rotor tangential velocity u from V_1, we obtain the velocity w_1 relative to the rotor, which ideally should be parallel to the rotor leading edge.

Figure 11.12c shows the rotor blades and their exit-velocity diagram. Here the relative velocity w_2 is parallel to the blade trailing edge, while the absolute velocity V_2 should be designed to enter smoothly the next row of stator blades.

The theoretical power and head are given by Euler's turbine relation (11.11). Since there is no radial flow, the inlet and exit rotor speeds are equal, $u_1 = u_2$, and one-dimensional continuity requires that the axial velocity component remain constant

$$V_{n1} = V_{n2} = V_n = Q/A = \text{const} \tag{11.32}$$

From the geometry of the velocity diagrams, $V_{t1} = V_{n1} \cot \alpha_1$ and $V_{t2} = u - V_{n2} \cot \beta_2$. Euler's relation (11.11) becomes

$$gH = uV_n(\cot \alpha_2 - \cot \alpha_1)$$
$$= u^2 - uV_n(\cot \alpha_1 + \cot \beta_2) \tag{11.33}$$

the preferred form because it relates to the blade angles α_1 and β_2. The shutoff or no-flow head is seen to be $H_0 = u^2/g$, just as in Eq. (11.18) for a centrifugal pump. The blade-angle parameter $\cot \alpha_1 + \cot \beta_2$ can be designed to be negative, zero, or positive, corresponding to a rising, flat, or falling head curve, as in Fig. 11.5.

Strictly speaking, Eq. (11.33) applies only to a single stream tube of radius r, but it is a good approximation for very short blades if r denotes the average radius. For long blades it is customary to sum Eq. (11.33) in radial strips over the blade area. Such complexity may not be warranted since theory, being idealized, neglects losses and usually predicts head and power larger than those in actual pump performance.

Performance of an Axial-Flow Pump

At high specific speeds, the most efficient choice is an axial-flow, or propeller, pump, which develops high flow rate and low head. A typical dimensionless chart for a propeller pump is shown in Fig. 11.13. Note, as expected, the higher C_Q and lower C_H compared with Fig. 11.8. The head curve drops sharply with discharge, so that a large system-head change will cause a mild flow change. The power curve drops with head also, which means a possible overloading condition if the system discharge should suddenly decrease. Finally, the efficiency curve is rather narrow and triangular, as opposed to the broad, parabolic-shaped centrifugal-pump efficiency (Fig. 11.8).

By inspection of Fig. 11.13, $C_{Q*} \approx 0.55$, $C_{H*} \approx 1.07$, $C_{P*} \approx 0.70$, and $\eta_{\max} \approx 0.84$. From this we compute $N'_s \approx (0.55)^{1/2}/(1.07)^{3/4} = 0.705$, $N_s = 12{,}000$. The relatively low efficiency is due to small pump size: $d = 14$ in, $n = 690$ r/min, $Q^* = 4400$ gal/min.

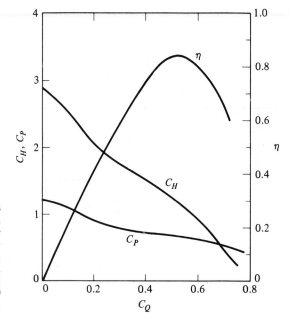

Fig. 11.13
Dimensionless
performance curves for
a typical axial-flow
pump, N_s = 12,000.
Constructed from data
given by Stepanoff [14]
for a 14-in pump at 690
r/min.

A repetition of Example 11.5 using Fig. 11.13 would show that this propeller pump family can provide a 25-ft head and 100,000 gal/min discharge if D = 46 in and n = 430 r/min, with bhp = 750; this is a much more reasonable design solution, with improvements still possible at larger N_s conditions.

Pump Performance versus Specific Speed

Specific speed is such an effective parameter that it is used as an indicator of both performance and efficiency. Figure 11.14 shows a correlation of the optimum efficiency of a pump as a function of specific speed and capacity. Because the dimensional parameter Q is a rough measure of both size and Reynolds number, η increases with Q. When this type of correlation was first published by Wislicenus [3] in 1947, it became known as *the* pump curve: a challenge to all manufacturers. We can check that the pumps of Figs. 11.7 and 11.13 fit the correlation very well.

Figure 11.15 shows the effect of specific speed on the shape of the pump performance curves, normalized with respect to the BEP point. The numerical values shown are representative but somewhat qualitative. The high-specific-speed pumps ($N_s \approx$ 10,000) have head and power curves which drop sharply with discharge, implying overload or start-up problems at low flow. Their efficiency curve is very narrow.

A low-specific-speed pump (N_s = 600) has a broad efficiency curve, a rising power curve, and a head curve which "droops" at shutoff, implying possible surge or hunting problems.

Although turbomachinery designs have traditionally been accomplished through extensive experimental programs, the theoretical techniques are catching

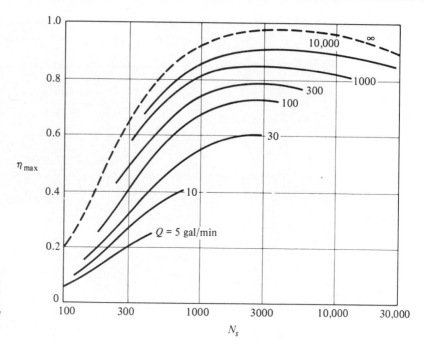

Fig. 11.14 Optimum efficiency of pumps versus capacity and specific speed. (*Adapted from Refs. 3 and 25.*)

up. Modern turbomachine design is a combination of sophisticated measurements [35] and theories which make use of boundary-layer theory (Sec. 7.5), numerical analysis (Sec. 8.9), compressible-flow analysis (Chap. 9), and dimensionless parameters (Chap. 5). Computer graphics help visualize the complex three-dimensional geometries. These modern concepts are illustrated in Figure 11.16. A recent international symposium [38] was devoted entirely to the performance of turbomachinery.

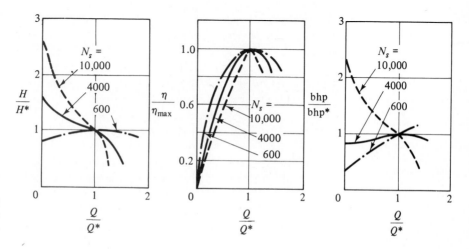

Fig. 11.15 Effect of specific speed on pump performance curves.

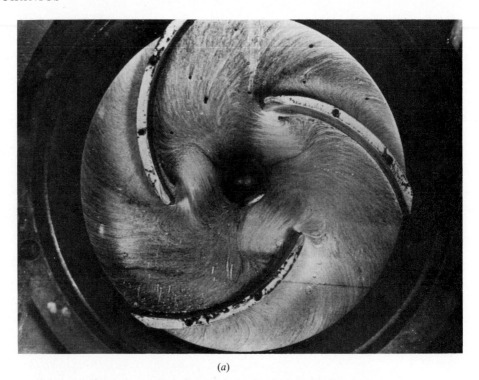

(a)

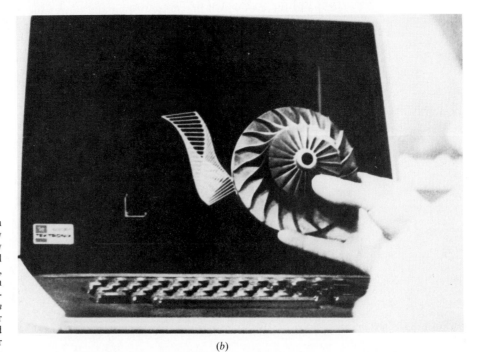

Fig. 11.16 Modern turbomachinery analysis uses a variety of techniques: (a) oil surface-flow pattern, showing inlet section blockage, on a three-bladed impeller (*from Ref. 36*); (b) impeller design curves visualized on a personal computer (*Concepts ETI Inc.*).

(b)

11.5 MATCHING PUMPS TO SYSTEM CHARACTERISTICS

The ultimate test of a pump is its match with the operating-system characteristics. Physically, the system head must match the head produced by the pump, and this intersection should occur in the region of best efficiency.

The system head will probably contain a static-elevation change $z_2 - z_1$ plus friction losses in pipes and fittings

$$H_{\text{syst}} = (z_2 - z_1) + \frac{V^2}{2g}\left(\sum \frac{fL}{D} + \sum K\right) \qquad (11.34)$$

where $\sum K$ denotes minor losses and V is the flow velocity in the principal pipe. Since V is proportional to pump discharge Q, Eq. (11.34) represents a system-head curve $H_s(Q)$. Three examples are shown in Fig. 11.17: a static head $H_s = a$, static head plus laminar friction $H_s = a + bQ$, and static head plus turbulent friction $H_s = a + cQ^2$. The intersection of the system curve with the pump-performance curve $H(Q)$ defines the operating point. In Fig. 11.17 the laminar-friction operating point is at maximum efficiency while the turbulent and static curves are off design. This may be unavoidable if system variables change, but the pump should be changed in size or speed if its operating point is consistently off design. Of course a perfect match may not be possible because commercial pumps have only certain discrete sizes and speeds. Let us illustrate these concepts with an example.

EXAMPLE 11.6   We want to use the 32-in pump of Fig. 11.7a at 1170 r/min to pump water at 60°F from one reservoir to another 120 ft higher through 1500 ft of 16-in-ID pipe with friction factor $f = 0.030$. (a) What will the operating point and efficiency be? (b) To what speed should the pump be changed to operate at the BEP?

Solution

Part (a)   For reservoirs the initial and final velocities are zero; thus the system head is

$$H_s = z_2 - z_1 + \frac{V^2}{2g}\frac{fL}{D} = 120 \text{ ft} + \frac{V^2}{2g}\frac{0.030(1500 \text{ ft})}{\frac{16}{12}\text{ ft}}$$

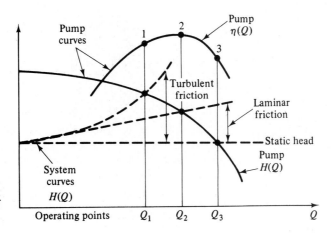

Fig. 11.17 Illustration of pump operating points for three types of system-head curves.

From continuity in the pipe, $V = Q/A = Q/[\frac{1}{4}\pi(\frac{16}{12}\text{ ft})^2]$, and so we substitute for V above to get

$$H_s = 120 + 0.269Q^2 \qquad Q \text{ in ft}^3/\text{s} \tag{1}$$

Since Fig. 11.7a uses thousands of gallons per minute for the abscissa, we convert Q in Eq. (1) to this unit:

$$H_s = 120 + 1.335Q^2 \qquad Q \text{ in } 10^3 \text{ gal/min} \tag{2}$$

We can plot Eq. (2) on Fig. 11.7a and see where it intersects the 32-in pump head curve, as in Fig. E11.6. A graphical solution gives approximately

$$H \approx 430 \text{ ft} \qquad Q \approx 15{,}000 \text{ gal/min}$$

The efficiency is about 82 percent, slightly off design.

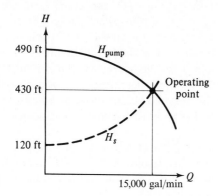

Fig. E11.6

An analytic solution is possible if we fit the pump-head curve to a parabola, which is very accurate

$$H_{\text{pump}} \approx 490 - 0.26Q^2 \qquad Q \text{ in } 10^3 \text{ gal/min} \tag{3}$$

Equations (2) and (3) must match at the operating point:

$$490 - 0.26Q^2 = 120 + 1.335Q^2$$

or
$$Q^2 = \frac{490 - 120}{0.26 + 1.335} = 232$$

$$Q = 15.2 \times 10^3 \text{ gal/min} = 15{,}200 \text{ gal/min} \qquad \qquad \textit{Ans. (a)}$$

$$H = 490 - 0.26(15.2)^2 = 430 \text{ ft} \qquad \qquad \textit{Ans. (a)}$$

Part (b) To move the operating point to BEP, we change n, which changes both $Q \propto n$ and $H \propto n^2$. From Fig. 11.7a, at BEP, $H^* \approx 386$ ft; thus for any n, $H^* = 386(n/1170)^2$. Also read $Q^* \approx 20 \times 10^3$ gal/min; thus for any n, $Q^* = 20(n/1170)$. Match H^* to the system characteristics, Eq. (2),

$$H^* = 386\left(\frac{n}{1170}\right)^2 \approx 120 + 1.335\left(20\,\frac{n}{1170}\right)^2 \qquad \qquad \textit{Ans. (b)}$$

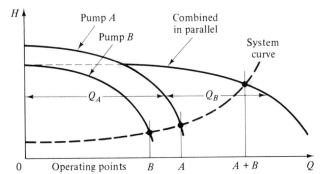

Fig. 11.18
Performance and
operating points of two
pumps singly and
combined in parallel.

which gives $n^2 < 0$?. Thus it is impossible to operate at maximum efficiency with this particular system and pump.

Pumps Combined in Parallel

If a pump provides the right head but too little discharge, a possible remedy is to combine two similar pumps in parallel, i.e., sharing the same suction and inlet conditions. A parallel arrangement is also used if delivery demand varies, so that one pump is used at low flow and the second pump started up for higher discharges. Both pumps should have check valves to avoid backflow when one is shut down.

The two pumps in parallel need not be identical. Physically, their flow rates will sum for the same head, as illustrated in Fig. 11.18. If pump A has more head than pump B, pump B cannot be added in until the operating head is below the shutoff head of pump B. Since the system curve rises with Q, the combined delivery Q_{A+B} will be less than the separate operating discharges $Q_A + Q_B$ but certainly greater than either one. For a very flat (static) curve two similar pumps in parallel will deliver nearly twice the flow. The combined brake horsepower is found by adding brake horsepower for each of pumps A and B at the same head as the operating point. The combined efficiency equals $\rho g(Q_{A+B})(H_{A+B})/(550 \text{ bhp}_{A+B})$.

If pumps A and B are not identical, as in Fig. 11.18, pump B should not be run and cannot even be started up if the operating point is above its shutoff head.

Pumps Combined in Series

If a pump provides the right discharge but too little head, consider adding a similar pump in series, with the output of pump B fed directly into the suction side of pump A. As sketched in Fig. 11.19, the physical principle for summing in series is that the two heads add at the same flow rate to give the combined-performance curve. The two need not be identical at all, since they merely handle the same discharge; they may even have different speeds although normally both are driven by the same shaft.

The need for a series arrangement implies that the system curve is steep, i.e., requires higher head than either pump A or B can provide. The combined operating-point head will be more than either A or B separately but not as great as

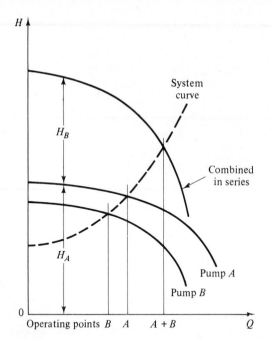

Fig. 11.19
Performance of two
pumps combined in
series.

their sum. The combined power is the sum of brake horsepower for A and B at the operating point flow rate. The combined efficiency is

$$\rho g (Q_{A+B})(H_{A+B})/550 \, \text{bhp}_{A+B}$$

similar to parallel pumps.

Whether pumps are used in series or in parallel, the arrangement will be uneconomical unless both pumps are operating near their best efficiency.

Multistage Pumps

For very high heads in continuous operation, the solution is a multistage pump, with the exit of one impeller feeding directly into the eye of the next. Centrifugal, mixed-flow, and axial-flow pumps have all been grouped in as many as 50 stages, with heads up to 8000 ft of water and pressure rises up to 5000 lbf/in² absolute. Figure 11.20 shows a section of a seven-stage centrifugal propane compressor which develops 300 lbf/in² rise at 40,000 ft³/min and 35,000 bhp.

EXAMPLE 11.7   Investigate extending Example 11.6 by using two 32-in pumps in parallel to deliver more flow. Is this efficient?

Solution   Since the pumps are identical, each delivers $\frac{1}{2}Q$ at the same 1170 r/min speed. The system curve is the same, and the balance-of-head relation becomes

$$H = 490 - 0.26(\tfrac{1}{2}Q)^2 = 120 + 1.335Q^2$$

or $$Q^2 = \frac{490 - 120}{1.335 + 0.065} \qquad Q = 16,300 \, \text{gal/min} \qquad \textit{Ans.}$$

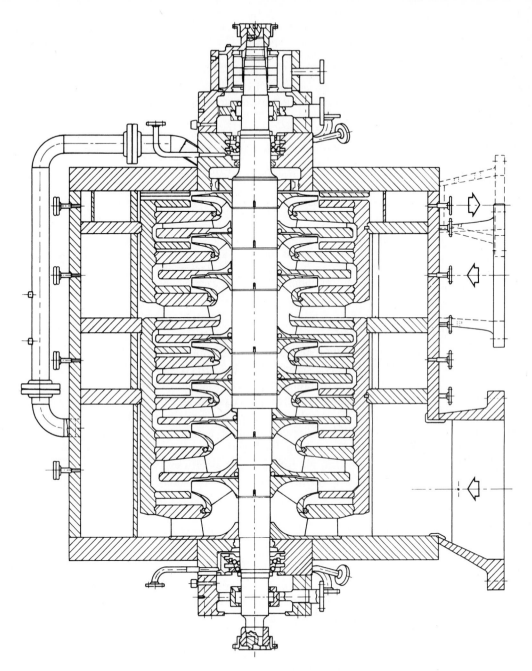

Fig. 11.20 Cross section of a seven-stage centrifugal propane compressor which delivers 40,000 ft³/min at 35,000 bhp and a pressure rise of 300 lbf/in². Note the second inlet at stage 5 and the varying impeller designs. (*DeLaval-Stork V.O.F.*, *Centrifugal Compressor Division.*)

This is only 7 percent more than a single pump. Each pump delivers $\frac{1}{2}Q = 8130$ gal/min, for which the efficiency is only 60 percent. The total brake horsepower required is 3200, whereas a single pump used only 2000 bhp.

EXAMPLE 11.8   Suppose the elevation change in Example 11.6 is raised from 120 to 500 ft, greater than a single 32-in pump can supply. Investigate using 32-in pumps in series at 1170 r/min.

Solution   Since the pumps are identical, the total head is twice as much and the constant 120 in the system-head curve is replaced by 500. The balance of heads becomes

$$H = 2(490 - 0.26Q^2) = 500 + 1.335Q^2$$

or
$$Q^2 = \frac{980 - 500}{1.335 + 0.52} \qquad Q = 16.1 \times 10^3 \text{ gal/min} \qquad\qquad\qquad Ans.$$

The operating head is $500 + 1.335(16.1)^2 = 845$ ft, or 97 percent more than a single pump in Example 11.5. Each pump is operating at 16.1×10^3 gal/min, which from Fig. 11.7a is 83 percent efficient, a pretty good match to the system. To pump at this operating point requires 4100 bhp, or about 2050 bhp for each pump.

11.6   TURBINES

A turbine extracts energy from a fluid which possesses high head, but it is fatuous to say a turbine is a pump run backward. Basically there are two types, reaction and impulse, the difference being the manner of head conversion. In the reaction turbine, the fluid fills the blade passages, and the head change or pressure drop occurs within the impeller. Reaction designs are of the radial-flow, mixed-flow, and axial-flow types and are essentially dynamic devices designed to admit the high-energy fluid and extract its momentum. An impulse turbine first converts the high head through a nozzle into a high-velocity jet, which then strikes the blades at one position as they pass by. The impeller passages are not fluid-filled, and the jet flow past the blades is essentially at constant pressure.

Reaction Turbines

Reaction turbines are low-head, high-flow devices. The flow is opposite that in a pump, entering at the larger-diameter section and discharging through the eye after giving up most of its energy to the impeller. Early designs were very inefficient because they lacked stationary guide vanes at the entrance to direct the flow smoothly into the impeller passages. The first efficient inward-flow turbine was built in 1849 by James B. Francis, an American engineer [34], and all radial- or mixed-flow designs are now called *Francis turbines*. At still lower heads, a turbine can be designed more compactly with purely axial flow and is termed a *propeller turbine*. The propeller may be either fixed-blade or adjustable (Kaplan type), the latter being complicated mechanically but much more efficient at low power settings. Figure 11.21 shows sketches of runner designs for Francis radial, Francis mixed-flow, and propeller-type turbines.

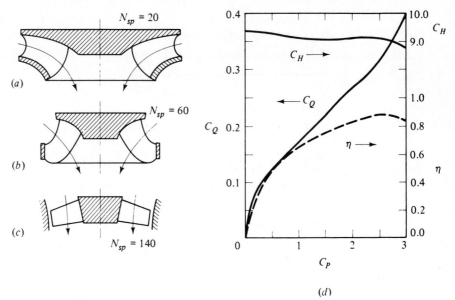

Fig. 11.21 Reaction turbines: (*a*) Francis (radial type); (*b*) Francis (mixed-flow); (*c*) propeller (axial-flow); (*d*) performance curves for a Francis turbine, $n = 600$ r/min, $D = 2.25$ ft, $N_{sp} = 29$.

Power Specific Speed

Turbine parameters are similar to those of a pump, but the dependent variable is the output brake horsepower, which depends upon the inlet flow rate Q, the available head H, the impeller speed n, and diameter D. The efficiency is the output brake horsepower divided by the available water horsepower $\rho g Q H$. The dimensionless forms are C_Q, C_H, and C_P, defined just as for a pump, Eqs. (11.23). If we neglect Reynolds-number and roughness effects, the functional relationships are written with C_P as the independent variable

$$C_H = C_H(C_P) \qquad C_Q = C_Q(C_P) \qquad \eta = \frac{\text{bhp}}{\rho g Q H} = \eta(C_P) \qquad (11.35)$$

Figure 11.21*d* shows typical performance curves for a small Francis radial turbine. The maximum efficiency point is called the *normal power*, and the values for this particular turbine are

$$\eta_{\max} = 0.89 \qquad C_{P*} = 2.70 \qquad C_{Q*} = 0.34 \qquad C_{H*} = 9.03 \qquad (11.36)$$

A parameter which compares the output power with the available head, independent of size, is found by eliminating diameter between C_H and C_P. It is called the *power specific speed*:

Rigorous form: $$N'_{sp} = \frac{C_P^{*1/2}}{C_H^{*5/4}} = \frac{n(\text{bhp})^{1/2}}{\rho^{1/2}(gH)^{5/4}} \qquad (11.37a)$$

Lazy but common: $$N_{sp} = \frac{(\text{r/min})(\text{bhp})^{1/2}}{[H\ (\text{ft})]^{5/4}} \qquad (11.37b)$$

For water, $\rho = 1.94$ slugs/ft^3 and $N_{sp} = 273.3 N'_{sp}$. The various turbine designs divide up nicely according to the range of power specific speed, as follows:

Turbine type	N_{sp} range	C_H range
Impulse	1–10	15–50
Francis	10–110	5–25
Propeller:		
Water	100–250	1–4
Gas, steam	25–300	10–80

Note that N_{sp}, like N_s for pumps, is defined only with respect to the BEP and has a single value for a given turbine family. In Fig. 11.21d, $N_{sp} = 273.3(2.70)^{1/2}/(9.03)^{5/4} = 29$, regardless of size.

Like pumps, turbines of large size are generally more efficient, and the Moody step-up formula (11.29) can be used as an estimate when data are lacking.

The design of a complete large-scale power-generating turbine system is a major engineering project, involving inlet and outlet ducts, trash racks, guide vanes, wicket gates, spiral cases, generator with cooling coils, bearings and transmission gears, runner blades, draft tubes, and automatic controls. Some typical large-scale reaction turbine designs are shown in Fig. 11.22. The reversible pump-and-turbine design of Fig. 11.22d requires special care for adjustable guide vanes to be efficient for flow in either direction.

The largest (1000-MW) hydropower designs are awesome when viewed on a human scale, as shown in Fig. 11.23. The economic advantages of small-scale model testing are evident from this photograph of the Francis turbine units at Grand Coulee Dam.

Impulse Turbines

For high head and relatively low power, i.e., low N_{sp}, a reaction turbine would not only require too high a speed but also the high pressure in the runner would require a massive casing thickness. The impulse turbine of Fig. 11.24 is ideal for this situation. Since N_{sp} is low, n will be low and the high pressure is confined to the small nozzle, which converts the head into an atmospheric pressure jet of high velocity V_j. The jet strikes the buckets and imparts a momentum change similar to that in our control-volume analysis for a moving vane in Example 3.10 or Prob. 3.57. The buckets have an elliptical split-cup shape, as in Fig. 11.24b. They are named *Pelton wheels*, after Lester A. Pelton (1829–1908), who produced the first efficient design.

From Example 3.10 the force and power delivered to a Pelton wheel are theoretically

$$F = \rho Q(V_j - u)(1 - \cos \beta)$$

$$P = Fu = \rho Q u(V_j - u)(1 - \cos \beta)$$

$$(11.38)$$

Fig. 11.22 Large-scale reaction turbine designs depend upon available head and flow rate and operating conditions: (*a*) Francis (radial); (*b*) Kaplan (propeller); (*c*) bulb mounting with propeller runner; (*d*) reversible pump-turbine with radial runner. (*Allis-Chalmers Fluid Products Company.*)

Fig. 11.23 Interior view of the 1.1-million-horsepower (820-MW) turbine units on the Grand Coulee Dam of the Columbia River, showing the spiral case, the outer fixed vanes ("stay ring"), and the inner adjustable vanes ("wicket gates"). (*Allis-Chalmers Fluid Products Company.*)

where $u = 2\pi n r$ is the bucket linear velocity and r is the *pitch radius*, or distance to the jet centerline. A bucket angle $\beta = 180°$ gives maximum power but is physically impractical. In practice, $\beta \approx 165°$, or $1 - \cos \beta \approx 1.966$ or only 2 percent less than maximum power.

From Eq. (11.38) the theoretical power of an impulse turbine is parabolic in bucket speed u and is maximum when $dP/du = 0$, or

$$u^* = 2\pi n^* r = \tfrac{1}{2} V_j \tag{11.39}$$

For a perfect nozzle, the entire available head would be converted into jet velocity, $V_j = (2gH)^{1/2}$. Actually, since there are 2 to 8 percent nozzle losses, a velocity coefficient C_v is used

$$V_j = C_v(2gH)^{1/2} \qquad 0.92 \le C_v \le 0.98 \tag{11.40}$$

By combining Eqs. (11.35) and (11.40), the theoretical impulse-turbine efficiency becomes

$$\eta = 2(1 - \cos \beta)\phi(C_v - \phi) \tag{11.41}$$

where
$$\phi = \frac{u}{(2gH)^{1/2}} = \text{peripheral-velocity factor}$$

Maximum efficiency occurs at $\phi = \tfrac{1}{2}C_v \approx 0.47$.

Figure 11.25 shows Eq. (11.41) plotted for an ideal turbine ($\beta = 180°$, $C_v = 1.0$) and for typical working conditions ($\beta = 160°$, $C_v = 0.94$). The latter case predicts

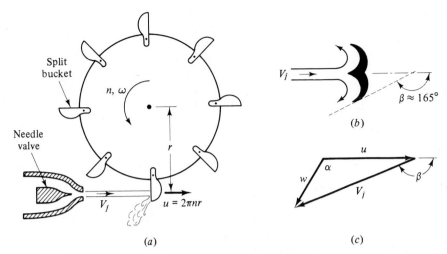

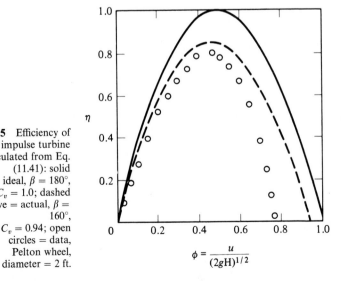

Fig. 11.24 Impulse turbine: (*a*) side view of wheel and jet; (*b*) top view of bucket; (*c*) typical velocity diagram.

$\eta_{\max} = 85$ percent at $\phi = 0.47$, but the actual data for a 24-in Pelton-wheel test are somewhat less efficient due to windage, mechanical friction, backsplashing, and nonuniform bucket flow. For this test $\eta_{\max} = 80$ percent, and, generally speaking, an impulse turbine is not quite as efficient as the Francis or propeller turbines at their BEP.

Figure 11.26 shows the optimum efficiency of the three turbine types, showing the importance of the power specific speed N_{sp} as a selection tool for the designer. These efficiencies are optimum and are obtained in careful design of large machines.

The water power available to a turbine may vary due to either net-head or flow-rate changes, both of which are common in field installations such as hydroelectric plants. The demand for turbine power also varies from light to heavy, and the

Fig. 11.25 Efficiency of an impulse turbine calculated from Eq. (11.41): solid curve = ideal, $\beta = 180°$, $C_v = 1.0$; dashed curve = actual, $\beta = 160°$, $C_v = 0.94$; open circles = data, Pelton wheel, diameter = 2 ft.

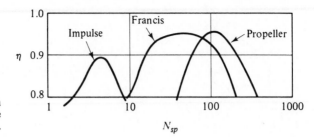

Fig. 11.26 Optimum efficiency of turbine designs.

operating response is a change in the flow rate by adjustment of a gate valve or needle valve (Fig. 11.24a). As shown in Fig. 11.27, all three turbine types achieve fairly uniform efficiency as a function of the level of power being extracted. Especially effective is the adjustable-blade (Kaplan-type) propeller turbine, while the poorest is a fixed-blade propeller. The term *rated power* in Fig. 11.27 is the largest power delivery guaranteed by the manufacturer, as opposed to *normal power*, which is delivered at maximum efficiency.

For further details of design and operation of turbomachinery the readable and interesting treatment in Ref. 29 is especially recommended. The feasibility of microhydropower is discussed in Noyes [43].

EXAMPLE 11.25 Investigate the possibility of using (a) a Pelton wheel similar to Fig. 11.25 or (b) the Francis-turbine family of Fig. 11.21d to deliver 30,000 bhp from a net head of 1200 ft.

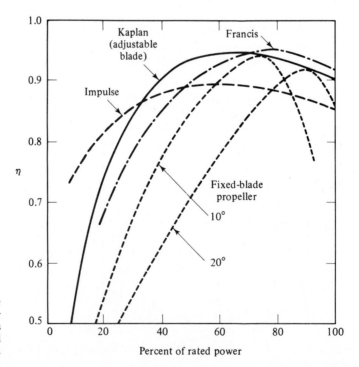

Fig. 11.27 Efficiency versus power level for various turbine designs at constant speed and head.

Solution

Part (a) From Fig. 11.26, the most efficient Pelton wheel occurs at about

$$N_{sp} \approx 4.5 = \frac{(\text{r/min})(30{,}000 \text{ bhp})^{1/2}}{(1200 \text{ ft})^{1.25}}$$

or

$$n = 183 \text{ r/min} = 3.06 \text{ r/s}$$

From Fig. 11.25 the best operating point is

$$\phi \approx 0.47 = \frac{\pi D(3.06 \text{ r/s})}{[2(32.2)(1200)]^{1/2}}$$

or

$$D = 13.6 \text{ ft} \hspace{3cm} \textit{Ans. (a)}$$

This Pelton wheel is perhaps a little slow and a trifle large. You could reduce D and increase n by increasing N_{sp} to, say, 6 or 7 and accepting the slight reduction in efficiency. Or you could use a double-hung, two-wheel configuration, each delivering 15,000 bhp, which changes D and n by the factor $2^{1/2}$:

Double wheel: $\qquad n = (183)2^{1/2} = 260 \text{ r/min} \qquad D = \dfrac{(13.6)}{2^{1/2}} = 9.6 \text{ ft} \hspace{1.5cm} \textit{Ans. (a)}$

Part (b) The Francis wheel of Fig. 11.21d must have

$$N_{sp} = 29 = \frac{(\text{r/min})(30{,}000 \text{ bhp})^{1/2}}{(1200 \text{ ft})^{1.25}}$$

or

$$n = 1183 \text{ r/min} = 19.7 \text{ r/s}$$

Then the optimum power coefficient from Eq. (11.36) is

$$C_{P*} = 2.70 = \frac{P}{\rho n^3 D^5} = \frac{30{,}000(550)}{(1.94)(19.7)^3 D^5}$$

or

$$D^5 = 412 \qquad D = 3.33 \text{ ft} = 40 \text{ in} \hspace{3cm} \textit{Ans. (b)}$$

This is a faster speed than normal practice, and the casing would have to withstand 1200 ft of water or about 520 lbf/in² internal pressure, but the 40-in size is extremely attractive. Francis turbines are now being operated at heads up to 1500 ft.

Wind Turbines

Wind energy has long been used as a source of mechanical power. The familiar four-bladed windmills of Holland, England, and the Greek islands have been used for centuries to pump water, grind grain, and saw wood. Modern research concentrates on the ability of wind turbines to generate electric power. Koeppl [39] stresses the potential for propeller-type machines, while Inglis [40] gives a more general analysis of wind turbines compared with alternative energy sources. Jarass [41] gives a detailed discussion of the technical and economic feasibility of large-scale electric/power generation by wind.

Some examples of wind turbine designs are shown in Fig. 11.28. The familiar American multiblade farm windmill (Fig. 11.28a) is of low efficiency, but thousands are in use as a rugged, reliable, and inexpensive way to pump water. A more

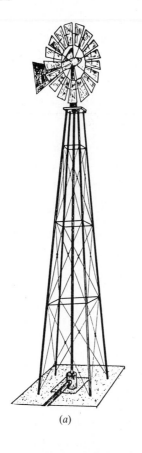

(a)

(c)

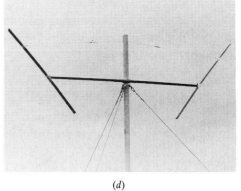

(d)

Fig. 11.28 Wind turbine designs: (a) the American multiblade farm HAWT; (b) propeller HAWT (*Grumman Areospace Corp.*); (c) the Darrieus VAWT (*National Research Council, Canada*); (d) modified straight-blade Darrieus VAWT (*Reading University*).

(b)

efficient design is the propeller mill in Fig. 11.28*b*, similar to the pioneering Smith-Putnam 1250-kW two-bladed system which operated on Grampa's Knob, 12 mi west of Rutland, Vermont, during 1941 to 1945. The Smith-Putnam design broke because of inadequate blade strength, but it withstood winds up to 115 mi/h and its efficiency was amply demonstrated [39].

The Dutch, American multiblade, and propeller mills are examples of *horizontal-axis wind turbines* (HAWTs), which are efficient but somewhat awkward in that they require extensive bracing and gear systems when combined with an electric generator. Thus a competing family of vertical-axis wind turbines (VAWTs) has been proposed which simplifies gearing and strength requirements. Figure 11.28*c* shows the "eggbeater" VAWT invented by G. J. M. Darrieus in 1925, now used in several government-sponsored demonstration systems [38]. To minimize centrifugal stresses, the twisted blades of the Darrieus turbine follow a *troposkien* curve formed by a chain anchored at two points on a spinning vertical rod.

An alternative VAWT, simpler to construct than the troposkien, is the straight-bladed Darrieus-type turbine in Fig. 11.28*d*. This design, proposed by Reading University in England, has blades which pivot due to centrifugal action as wind speeds increase, thus limiting bending stresses.

Idealized Wind-Turbine Theory

The ideal, frictionless efficiency of a propeller windmill was predicted by A. Betz in 1920, using the simulation shown in Fig. 11.29. The propeller is represented by an *actuator disk* which creates across the propeller plane a pressure discontinuity of area A and local velocity V. The wind is represented by a streamtube of approach velocity V_1 and a slower downstream wake velocity V_2. The pressure rises to p_b just

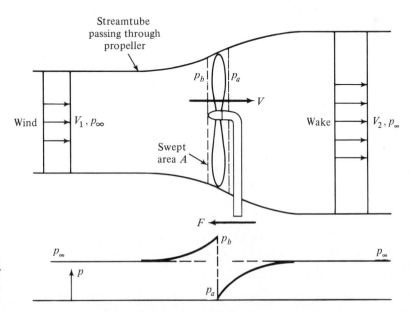

Fig. 11.29 Idealized actuator-disk and streamtube analysis of flow through a windmill.

before the disk and drops to p_a just after, returning to free-stream pressure in the far wake. To hold the propeller rigid when it is extracting energy from the wind, there must be a leftward force F on its support, as shown.

A control-volume–horizontal-momentum relation applied between sections 1 and 2 gives

$$\sum F_x = -F = \dot{m}(V_2 - V_1)$$

A similar relation for a control volume just before and after the disk gives

$$\sum F_x = -F + (p_b - p_a)A = \dot{m}(V_a - V_b) = 0$$

Equating these two yields the propeller force

$$F = (p_b - p_a)A = \dot{m}(V_1 - V_2) \tag{11.42}$$

Assuming ideal flow, the pressures can be found by applying the incompressible Bernoulli relation up to the disk

From 1 to b: $\qquad\qquad p_\infty + \tfrac{1}{2}\rho V_1^2 = p_b + \tfrac{1}{2}\rho V^2$

From a to 2: $\qquad\qquad p_a + \tfrac{1}{2}\rho V^2 = p_\infty + \tfrac{1}{2}\rho V_2^2$

Subtracting these and noting that $\dot{m} = \rho A V$ through the propeller, we can substitute for $p_b - p_a$ in Eq. (11.42) to obtain

$$p_b - p_a = \tfrac{1}{2}\rho(V_1^2 - V_2^2) = \rho V(V_1 - V_2)$$

or $\qquad\qquad\qquad\qquad V = \tfrac{1}{2}(V_1 + V_2) \tag{11.43}$

Continuity and momentum thus require that the velocity V through the disk equal the average of the wind and far-wake speeds.

Finally, the power extracted by the disk can be written in terms of V_1 and V_2 by combining Eqs. (11.42) and (11.43)

$$P = FV = \rho A V^2(V_1 - V_2) = \tfrac{1}{4}\rho A(V_1^2 - V_2^2)(V_1 + V_2) \tag{11.44}$$

For a given wind speed V_1, we can find the maximum possible power by differentiating P with respect to V_2 and setting equal to zero. The result is

$$P = P_{max} = \tfrac{8}{27}\rho A V_1^3 \qquad \text{at} \qquad V_2 = \tfrac{1}{3}V_1 \tag{11.45}$$

which corresponds to $V = 2V_1/3$ through the disk.

The maximum available power to the propeller is the mass flow through the propeller times the total kinetic energy of the wind

$$P_{available} = \tfrac{1}{2}\dot{m}V_1^2 = \tfrac{1}{2}\rho A V_1^3$$

Thus the maximum possible efficiency of an ideal frictionless wind turbine is usually stated in terms of the *power coefficient*

$$C_P = \frac{P}{\tfrac{1}{2}\rho A V_1^3} \tag{11.46}$$

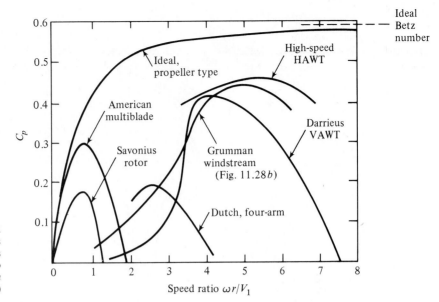

Fig. 11.30 Estimated performance of various wind turbine designs as a function of blade-tip speed ratio. (*From Ref. 42.*)

Equation (11.45) states that the total power coefficient is

$$C_{p,\max} = \frac{16}{27} = 0.593 \qquad (11.47)$$

This is called the *Betz number* and serves as an ideal with which to compare the actual performance of real windmills.

Figure 11.30 shows the measured power coefficients of various wind-turbine designs. The independent variable is not V_2/V_1 (which is artificial and convenient only in the ideal theory) but the ratio of blade-tip speed ωr to wind speed. Note that the tip can move much faster than the wind, a fact disturbing to the laity but familiar to engineers in the behavior of iceboats and sailing vessels. The Darrieus VAWT and the high-speed propeller HAWTs are quite efficient. The Darrieus has the many advantages of a vertical axis but has little torque at low speeds and also rotates slower at maximum power than a propeller, thus requiring a higher gear ratio for the generator. The Savonius rotor (Fig. 6.28b) has been suggested as a VAWT design [40] because it produces power at very low wind speeds, but it is inefficient and susceptible to storm damage because it cannot be feathered in high winds.

As shown in Fig. 11.31, there are many areas of the world where wind energy is an attractive alternative. Greenland, Newfoundland, Argentina, Chile, New Zealand, Iceland, Ireland, and the United Kingdom have the highest prevailing winds, but Australia, for example, with only moderate winds, has the potential to generate half its electricity with wind turbines [42]. In addition, since the ocean is generally even windier than the land, there are many island areas of high potential for wind power. There have also been proposals [39] for ocean-based floating windmill farms. The inexhaustible availability of the winds, coupled with improved low-cost turbine designs, indicates a bright future for this alternative.

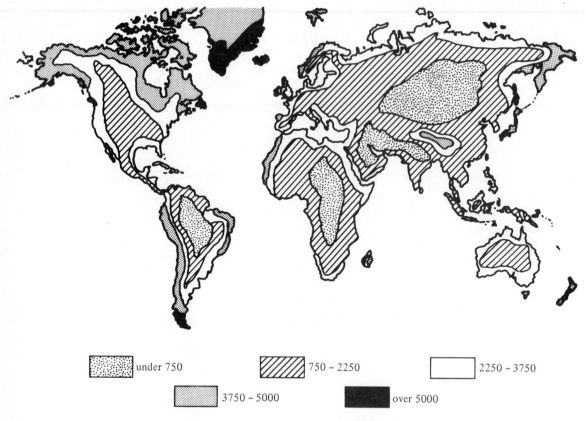

�	under 750	⧄	750 – 2250	□	2250 – 3750
⧈	3750 – 5000	■	over 5000		

Fig. 11.31 World availability of landbased wind energy: estimated annual electric output in kilowatthours per kilowatt of a wind turbine rated at 11.2 m/s (25 mi/h). (*From Ref. 44.*)

SUMMARY

Turbomachinery design is perhaps the most practical and most active application of the principles of fluid mechanics. There are billions of pumps and turbines in use in the world, and thousands of companies seeking improvements. This chapter discusses both positive-displacement devices and, more extensively, rotodynamic machines. With the centrifugal pump as an example, the basic concepts of torque, power, head, flow rate, and efficiency are developed for a turbomachine. Nondimensionalization leads to the pump similarity rules and some typical dimensionless performance curves for axial and centrifugal machines. The single most useful pump parameter is found to be the specific speed, which delineates the type of design needed. An interesting design application is the theory of pumps combined in series and in parallel.

Turbines extract energy from flowing fluids and are of two types: impulse turbines, which convert the momentum of a high-speed stream, and reaction turbines, where the pressure drop occurs within the blade passages in an internal flow. By analogy with pumps, the power specific speed is important for turbines and is used to classify them into impulse, Francis, and propeller-type designs. A special case of reaction turbine with unconfined flow is the wind turbine. Several types of windmills are discussed and their relative performance compared.

REFERENCES

1. S. L. Dixon, "Fluid Mechanics of Turbomachinery," 3d ed., Pergamon, Oxford, 1978.
2. A. Betz, "Introduction to the Theory of Flow Machines," Pergamon, New York, 1966.
3. G. F. Wislicenus, "Fluid Mechanics of Turbomachinery," 2d ed., McGraw-Hill, New York, 1965.
4. G. T. Csanady, "Theory of Turbomachines," McGraw-Hill, New York, 1964.
5. O. E. Balje, "Turbomachines: A Guide to Design, Selection, and Theory," Wiley, New York, 1981.
6. D. G. Shepherd, "Principles of Turbomachinery," Macmillan, New York, 1956.
7. A. J. Stepanoff, "Pumps and Blowers: Two-Phase Flow," Wiley, New York, 1965.
8. T. B. Ferguson, "The Centrifugal Compressor Stage," Butterworth, London, 1963.
9. I. J. Karassick and R. Carter, "Centrifugal Pumps," Dodge, New York, 1960.
10. J. H. Horlock, "Axial Flow Compressors," Butterworth, London, 1958.
11. R. A. Wallis, "Axial Flow Fans and Ducts," Wiley, New York, 1983.
12. F. A. Kristal and F. A. Annett, "Pumps," 2d ed., McGraw-Hill, New York, 1953.
13. H. Addison, "Centrifugal and Rotodynamic Pumps," Chapman & Hall, London, 1948.
14. A. J. Stepanoff, "Centrifugal and Axial Flow Pumps," 2d ed., Wiley, New York, 1957.
15. J. H. Horlock, "Axial Flow Turbines," Butterworth, London, 1966.
16. A. W. Judge, "Small Gas Turbines," Chapman & Hall, London, 1960.
17. J. F. Lee, "Theory and Design of Steam and Gas Turbines," McGraw-Hill, New York, 1954.
18. H. Constant, "Gas Turbines and Their Problems," Todd, London, 1953.
19. W. J. Kearton, "Steam Turbine Theory and Practice," 6th ed., Pitman, 1951.
20. W. W. Bathe, "Fundamentals of Gas Turbines," Wiley, New York, 1984.
21. T. G. Hicks and T. W. Edwards, "Pump Application Engineering," McGraw-Hill, New York, 1971.
22. R. A. Wallis, "Axial Flow Fans: Design and Practice," Newnes, London, 1961.
23. T. G. Hicks, "Pump Operation and Maintenance," McGraw-Hill, New York, 1958.
24. "Pump Selector for Industry," Worthington Pump, Mountainside, N.J., 1977.
25. I. J. Karassick et al., "Pump Handbook," 2d ed., McGraw-Hill, New York, 1985.
26. Engineering Equipment Users Associated, "Guide to the Selection of a Rotodynamic Pump," Constable, London, 1972.
27. R. Walker, "Pump Selection," 2d ed., Butterworth, London, 1979.
28. Hydraulic Institute, "Standards of the Hydraulic Institute," 11th ed., New York, 1965.
29. R. L. Daugherty and J. B. Franzini, "Fluid Mechanics," McGraw-Hill, 7th ed., New York, 1977.
30. D. Eckardt, Detailed Flow Investigations within a High Speed Centrifugal Compressor Impeller, *J. Fluids Eng.*, September 1976, pp. 390–402.
31. D. Japikse, Review: Progress in Numerical Turbomachinery Analysis, *J. Fluid Eng.*, December 1976, pp. 592–606.
32. L. F. Moody, The Propeller Type Turbine, *ASCE Trans.*, vol. 89, p. 628, 1926.
33. H. Gartmann (ed.), "DeLaval Engineering Handbook," 3d ed., McGraw-Hill, New York, 1970.
34. J. B. Francis, "Lowell Hydraulic Experiments," 5th ed., Van Nostrand, Princeton, N.J., 1909.

35. B. Lakshminarayana and P. Runstadler Jr. (eds.), Measurement Methods in Rotating Components of Turbomachinery, *Proc. ASME Symp. New Orleans, 1980*, vol. no. I00130.

36. M. Murakami, K. Kikuyama, and E. Asakura, Velocity and Pressure Distributions in the Impeller Passages of Centrifugal Pumps, *J. Fluids Eng.*, vol. 102, pp. 420–426, December 1980.

37. E. M. Greitzer, The Stability of Pumping Systems: The 1980 Freeman Scholar Lecture, *J. Fluids Eng.*, vol. 103, pp. 193–242, June 1981.

38. W. L. Swift et al. (eds), Performance Characteristics of Hydraulic Turbines and Pumps, *Proc. ASME Symp., Boston, 1983*, Book no. H00280.

39. G. W. Koeppl, "Putnam's Power from the Wind," 2d ed., Van Nostrand Reinhold, New York, 1982.

40. D. R. Inglis, "Wind Power," University of Michigan Press, Ann Arbor, 1978.

41. L. Jarass, "Wind Energy: An Assessment of the Technical and Economic Potential," Springer-Verlag, Berlin, 1981.

42. M. L. Robinson, The Darrieus Wind Turbine for Electrical Power Generation, *Aeronaut. J.*, June 1981, pp. 244–255.

43. R. Noyes (ed.), Small and Micro Hydroelectric Power Plants, Noyes Data Corp., Park Ridge, N.J., 1980.

44. D. F. Warne and P. G. Calnan, Generation of Electricity from the Wind, *IEE Rev.*, vol. 124, no. 11R, pp. 963–985, November 1977.

45. L. A. Haimerl, The Crossflow Turbine, *Waterpower*, January 1960, pp. 5–13; see also *ASME Symp. Small Hydropower Fluid Mach.*, vol. 1, 1980, and vol. 2, 1982.

Problems

PROBLEM DISTRIBUTION

Section	Topic	Problems
11.1	Introduction and classification	11.1–11.11
11.2	Centrifugal pump theory	11.12–11.17
11.3	Pump performance and similarity rules	11.18–11.35
11.4	Specific speed: mixed and axial pumps	11.36–11.46
11.5	Matching pumps to systems	11.47–11.55
11.6	Reaction and impulse turbines	11.56–11.67
11.6	Windmills	11.68–11.70

11.1 Describe the geometry and operation of a peristaltic positive-displacement pump which occurs in nature.

11.2 What would be the technical classification of the following turbomachines: (*a*) a household fan, (*b*) a windmill, (*c*) an aircraft propeller, (*d*) a fuel pump in a car, (*e*) an eductor, (*f*) a fluid-coupling transmission, and (*g*) a power-plant steam turbine?

11.3 A PDP can deliver almost any fluid, but there is always a limiting very high viscosity for which performance will deteriorate. Can you explain the probable reason?

11.4 Figure 11.2*c* shows a double-screw pump. Sketch a single-screw-pump and explain its operation. How did Archimedes' screw pump operate?

11.5 What type of pump is shown in Fig. P11.5? How does it operate?

Fig. P11.5

11.6 A piston PDP has a 7-in diameter and a 3-in stroke. Its crankshaft rotates at 350 r/min. How many gallons per minute does it deliver at 100 percent volumetric efficiency?

11.7 If the PDP of Prob. 11.6 delivers water against a total head of 30 ft, what horsepower is required at 75 percent efficiency?

11.8 A pump delivers 1500 L/min of water at 20°C against a pressure rise of 300 kPa. Kinetic- and potential-energy changes are negligible. If the driving motor supplies 9 kW, what is the overall efficiency?

11.9 Which is better, a PDP or a centrifugal pump, for (a) high flow rate, (b) high pressure rise, (c) self-priming, (d) low wear on moving parts, (e) steady outlet stream, (f) delivering high-viscosity fluids?

11.10 A 20-hp pump delivers 400 gal/min of gasoline at 20°C with 80 percent efficiency. What head and pressure rise result across the pump?

11.11 A pump delivers gasoline at 20°C and 12 m³/h. At the inlet $p_1 = 100$ kPa, $z_1 = 1$ m, and $V_1 = 2$ m/s. At the exit $p_2 = 500$ kPa, $z_2 = 4$ m, and $V_2 = 3$ m/s. How much power is required if the motor efficiency is 80 percent?

11.12 A lawn sprinkler can be used as a simple turbine. As shown in Fig. P11.12, flow enters normal to the paper in the center and splits evenly into $Q/2$ and V_{rel} leaving each nozzle. The arms rotate at angular velocity ω and do work on a shaft. Draw the velocity diagram for this turbine. Neglecting friction, find an expression for the power delivered to the shaft. Find the rotation rate for which the power is a maximum.

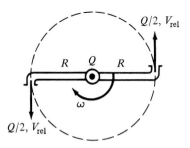

Fig. P11.12

11.13 Apply the analysis of Prob. 11.12 to the specific case of $R = 15$ cm, $Q = 16$ m³/h, and water flow with a nozzle exit diameter of 1 cm. Compute the maximum power in watts which can be delivered.

11.14 A centrifugal pump has $r_1 = 6$ in, $r_2 = 12$ in, $b_1 = 3$ in, $b_2 = 2$ in, $\beta_1 = 20°$, and $\beta_2 = 10°$ and rotates at 1200 r/min. If the fluid is water at 60°F, estimate the theoretical Q, H, and P_w.

11.15 A centrifugal pump has $r_2 = 8$ in, $b_2 = 1.5$ in, and $\beta_2 = 30°$. It delivers water at 2700 gal/min when rotating at 1500 r/min. Estimate the theoretical head rise and water horsepower.

11.16 A jet of velocity V strikes a vane which moves to the right at speed V_c, as in Fig. P11.16. The vane has a turning angle θ. Derive an expression for the power delivered to the vane by the jet. For what vane speed is the power maximum?

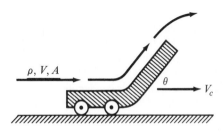

Fig. P11.16

11.17 A centrifugal water pump impeller has uniform blade width $b = 6$ cm, $D_1 = 12$ cm, and $D_2 = 40$ cm. Blade angles are $\beta_1 = 40°$ and $\beta_2 = 10°$. If the impeller rotates at 900 r/min and the efficiency is 85 percent, estimate (a) the flow rate, (b) the pressure rise, and (c) the power required.

11.18 A centrifugal pump delivers 900 gal/min of water at 1500 r/min with a head rise of 120 ft. What will the homologous discharge and head be when the pump is run at 1300 r/min?

11.19 If the 28-in-diameter water pump of Fig. 11.7a is run at 900 r/min, what head and brake horsepower will result when the discharge is 12,000 gal/min?

11.20 If the 35-in-diameter pump of Fig. 11.7b is used to deliver 20°C gasoline at 900 r/min and 25,000 gal/min, what head and brake horsepower will result?

11.21 At what speed in revolutions per minute should the 35-in-diameter pump of Fig. 11.7b be run to produce a head of 400 ft at a discharge of 20,000 gal/min? What brake horsepower will be required? *Hint:* Fit $H(Q)$ to a formula.

11.22 Tests by the Byron Jackson Co. of a 14.62-in-diameter centrifugal water pump at 2134 r/min yield the following data:

Q, ft^3/s	0	2	4	6	8	10
H, ft	340	340	340	330	300	220
bhp	135	160	205	255	330	330

What is the BEP? What is the specific speed? Estimate the maximum discharge possible.

11.23 For the pump of Prob. 11.22 for what speed will the shutoff head be 220 ft? For what speed will Q^* be 6.5 ft^3/s? For what speed will shutoff conditions require 200 bhp?

11.24 The pump of Prob. 11.22 is run at 2134 r/min to deliver water at 20°C through 70 m of 12-cm-diameter cast-iron pipe. All other system losses are neglected. What flow rate and brake horsepower result?

11.25 A pump from the same family as that of Prob. 11.22 is built with $D = 18$ in and $n = 1500$ r/min. Find the discharge, water pressure rise, and brake horse-power at the BEP.

11.26 You are asked to build a pump geometrically similar to that of Prob. 11.22 to deliver 8000 gal/min at 900 r/min and best efficiency. What should the impeller diameter be? What head and brake horsepower will result for water flow? Estimate the maximum efficiency.

11.27 A 5-in-diameter centrifugal pump, running at 1800 r/min, yields the following performance data:

Q, ft^3/min	1	3	5	7	9
H, ft	35	39	40	36	25
η, %	17	64	82	81	30

When it runs at 2400 r/min and 6 ft^3/min, the measured head is 66 ft. Compare this with the similarity rule estimate and compute the horsepower required at this new point.

11.28 Plot the dimensionless performance curves for the pump of Prob. 11.27 and compare with Fig. 11.8. Find the appropriate diameter in inches and speed in revolutions per minute for a geometrically similar pump to deliver 400 gal/min against a head of 200 ft. What brake horsepower would be required?

11.29 The 36.75-in pump in Fig. 11.7a at 1170 r/min is used to pump water at 60°F from a reservoir through 1000 ft of 12-in-ID galvanized-iron pipe to a point 200 ft above the reservoir surface. What flow rate and brake horsepower will result? If there is 40 ft of pipe upstream of the pump, how far below the surface should the pump inlet be placed to avoid cavitation?

11.30 In Prob. 11.29 the operating point is off design at an efficiency of only 77 percent. Is it possible, with the similarity rules, to change the pump rotation speed to deliver the water near BEP? Explain your results.

11.31 The efficiency curve for a centrifugal pump can be approximated by the curve fit $\eta \approx aQ - bQ^3$, where a and b are constants. For this approximation what is the ratio of Q^* at best efficiency to Q_{max}? If the maximum efficiency is 90 percent, what is η at (a) $Q = \frac{1}{2}Q_{max}$; (b) $Q = \frac{1}{2}Q^*$?

11.32 A typical household basement sump pump provides a discharge of 10 gal/min against a head of 20 ft. What is the minimum horsepower required to drive such a pump?

11.33 Suppose that the 32-in pump of Fig. 11.7a is used to deliver air with inlet conditions of 14.7 lbf/in^2 absolute and 60°F. What pressure rise, discharge in cubic feet per minute, and brake horsepower would result if the pump is run at 1800 r/min?

11.34 The Allis-Chalmers D30LR centrifugal compressor delivers 32,850 ft^3/min of SO_2 with a pressure change from 14.0 to 18.0 lbf/in^2 absolute using an 800-hp motor at 3550 r/min. What is the overall efficiency? What will the flow rate and Δp be at 3000 r/min? Estimate the diameter of the impeller.

11.35 An 8-in model pump delivering 180°F water at 800 gal/min and 2400 r/min begins to cavitate when the inlet pressure and velocity are 12 lbf/in^2 absolute and 20 ft/s, respectively. Find the required NPSH of a prototype which is 4 times larger and runs at 1000 r/min.

11.36 An axial-flow pump delivers 40 ft^3/s of air which enters at 20°C and 1 atm. The flow passage has a 10-in outer radius and an 8-in inner radius. Blade angles are $\alpha_1 = 60°$ and $\beta_2 = 70°$, and the rotor runs at 1800 r/min. For the first stage compute (a) the head rise and (b) the power required.

11.37 Compute the specific speed of the pump in Prob. 11.27. Is it appropriate for the type of pump?

11.38 If the axial-flow pump of Fig. 11.13 is used to deliver 80,000 gal/min of water at 1800 r/min, what is the proper impeller diameter? What are the head and brake horsepower at shutoff? What is Δp^*?

11.39 The Colorado River Aqueduct uses Worthington Corp. pumps which deliver 200 ft³/s at 450 r/min against a head of 440 ft. What type of pumps are these? Estimate the impeller diameter.

11.40 We want to pump 70°C water at 20,000 gal/min and 1800 r/min. Estimate the type of pump, the horsepower required, and the impeller diameter if the required pressure rise for one stage is (a) 170 kPa and (b) 1350 kPa.

11.41 Suppose the pump of Fig. 11.13 is used to satisfy the conditions of Prob. 11.40. How many stages would be needed to provide a pressure rise of (a) 170 kPa and (b) 1350 kPa?

11.42 We want to pump 50,000 gal/min of gasoline at 60°F against a head of 120 ft. Find the appropriate impeller size, speed, and brake horsepower needed for the pump families of (a) Fig. 11.8 and (b) Fig. 11.13. Which is the better design?

11.43 Performance data for a 21-in-diameter air blower running at 3550 r/min are:

Δp, inH$_2$O	29	30	28	21	10
Q, ft³/min	500	1000	2000	3000	4000
bhp	6	8	12	18	25

Note the fictitious expression of pressure rise in terms of water rather than air. What is the specific speed? How does the performance compare with Fig. 11.8? What are C_Q^*, C_H^*, and C_P^*?

11.44 The Worthington Corp. model A-12251 water pump, operating at maximum efficiency, produces 53 ft of head at 3500 r/min, 1.1 bhp at 3200 r/min, and 60 gal/min at 2940 r/min. What type of pump is this? What is its efficiency, and how does this compare with Fig. 11.14? Estimate the impeller diameter.

11.45 Estimate the specific speed and approximate impeller diameter of a single stage of the compressor in Fig. 11.20, assuming an inlet pressure of 15 lbf/in² absolute and a rotor speed of 1750 r/min. The molecular weight of propane is 44.06.

11.46 Suppose that the blower family of Prob. 11.43 is used to move 7000 ft³/min of air at 1 atm and 20°C with a single-stage pressure rise of 10 inH$_2$O. What are the proper impeller diameter and speed for this task? Is the

axial-flow family of Fig. 11.13 better suited to this flow? If not, what type of pump is best for a single stage?

11.47 A 24-in pump is homologous to the 32-in pump in Fig. 11.7a. At 1400 r/min this pump delivers 12,000 gal/min of water from one reservoir through a long pipe to another 50 ft higher. What will the flow rate be if the pump speed is increased to 1750 r/min? Assume no change in pipe friction factor or efficiency.

11.48 The 32-in pump in Fig. 11.7a is used at 1170 r/min in a system whose head curve is H_s (ft) = $100 + 1.5Q^2$, with Q in thousands of gallons per minute. Find the discharge and brake horsepower required for (a) one pump; (b) two pumps in parallel; (c) two pumps in series. Which configuration is best?

11.49 The 38-in pump of Fig. 11.7b is to be used in series at 710 r/min to lift water 4000 ft through 5000 ft of 18-in-ID cast-iron pipe. How many pumps in series would give most efficient operation? Neglect minor losses and cavitation.

11.50 It is proposed to run the pump of Prob. 11.27 at 1800 r/min to pump water at 20°C through the system in Fig. P11.50. The pipe is 3-cm-diameter commercial steel. What flow rate in cubic feet per minute will result? Is this an efficient application?

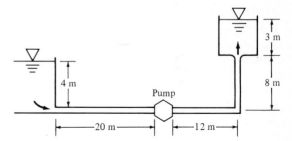

Fig. P11.50

11.51 Repeat Prob. 11.49 if the pumps are run at 1200 r/min.

11.52 Two 32-in pumps from Fig. 11.7a are combined in parallel to deliver water at 60°F through 1500 ft of horizontal pipe. If $f = 0.025$, what pipe diameter will ensure a flow rate of 35,000 gal/min for $n = 1170$ r/min?

11.53 Two pumps of the type tested in Prob. 11.27 are to be used at 1800 r/min to pump water at 20°C vertically through 60 ft of commercial steel pipe. Should they be in series or in parallel? What is the proper pipe diameter for most efficient operation?

11.54   Two 32-in pumps from Fig. 11.7a are to be used in series at 1170 r/min to lift water through 500 ft of vertical cast-iron pipe. What should the pipe diameter be for most efficient operation? Neglect minor losses.

11.55   It is proposed to use one 32- and one 28-in pump from Fig. 11.7a in parallel to deliver water at 60°F. The system-head curve is $H_s = 50 + 0.3Q^2$, with Q in thousands of gallons per minute. What will the head and delivery be if both pumps run at 1170 r/min? If the 28-in pump is reduced below 1170 r/min, at what speed will it cease to deliver?

11.56   Show that if the net head H is varied for a turbine operating at a given valve opening and efficiency, the speed will vary as $H^{1/2}$ and the output power as $H^{3/2}$.

11.57   Turbines are to be installed where the net head is 400 ft and the flow rate 250,000 gal/min. Discuss the type, number, and size of turbine which might be selected if the generator selected is (a) 48-pole, 60 cycle ($n = 150$ r/min); and (b) 8-pole ($n = 900$ r/min). Why are at least two turbines desirable from a planning point of view?

11.58   Turbines at the Conowingo Plant on the Susquehanna River each develop 54,000 bhp at 82 r/min under a head of 89 ft. What type of turbines are these? Estimate the flow rate and impeller diameter.

11.59   The Tupperware hydroelectric plant on the Blackstone River has four 36-in-diameter turbines, each providing 447 kW at 200 r/min and 205 ft³/s for a head of 30 ft. What type of turbines are these? How does their performance compare with Fig. 11.21?

11.60   Figure P11.60 shows a cutaway of a *crossflow* turbine [45], which resembles a squirrel cage with slotted curved blades. The flow enters at about 2 o'clock, passes through the center, and leaves at about 8 o'clock. Report to the class on the operation and advantages of this design, including the velocity diagrams.

11.61   A certain turbine in Switzerland delivers 25,000 bhp at 500 r/min under a net head of 5330 ft. What type of turbine is this? Estimate the approximate discharge and size.

11.62   A Pelton wheel of 12-ft pitch diameter operates under a net head of 2000 ft. Estimate the speed, power output, and flow rate for best efficiency if the nozzle exit diameter is 4 in.

11.63   How can the rotor and stator flows of Fig. 11.12 be applied to an axial-flow *turbine*? What are the

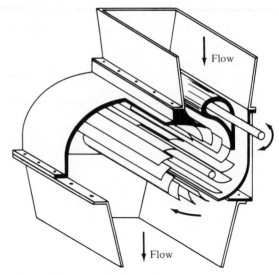

Fig. P11.60

proper blade angles? Sketch a suitable blade arrangement. For further details see chap. 8 of Ref. 20.

11.64   At a certain proposed turbine installation the net head is 1000 ft and the flow rate 100 ft³/s. Discuss the size, speed, and number of turbines which might be suitable for (a) a Pelton wheel, Fig. 11.25 and (b) a Francis wheel, Fig. 11.21d.

11.65   It is planned to use the Francis-turbine family of Fig. 11.21d at an installation with a head of 600 ft and a flow rate of 200 ft³/s. What are the proper impeller diameter and the optimum speed and power produced?

11.66   Francis and Kaplan turbines are often provided with *draft tubes*, which lead the exit flow into the tailwater region, as in Fig. P11.66. Explain at least two advantages in using a draft tube.

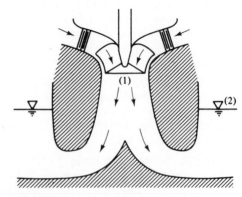

Fig. P11.66

11.67 Turbines can also cavitate when the pressure at point 1 in Fig. P11.66 drops too low. With NPSH defined by Eq. (11.20), the empirical criterion given by Wislicensus [3] for cavitation is

$$N_{ss} = \frac{(r/min)(gal/min)^{1/2}}{[NPSH (ft)]^{3/4}} \geq 11,000$$

Use this criterion to compute how high $z_1 - z_2$, the impeller eye in Fig. P11.66, can be placed for a Francis turbine with a head of 300 ft, $N_{sp} = 40$, and $p_a = 14$ lbf/in^2 absolute before cavitation occurs in 60°F water.

11.68 One of the largest wind generators in operation today is the ERDA/NASA two-blade propeller HAWT in Sandusky, Ohio. The blades are 125 ft in diameter and reach maximum power in 19 mi/h winds. For this condition estimate (*a*) the power generated in kilowatts (*b*) the rotor speed in revolutions per minute, and (*c*) the velocity V_2 behind the rotor.

11.69 A Darrieus VAWT in operation in Lumsden, Saskatchewan, that is 32 ft high and 20 ft in diameter sweeps out an area of 432 ft^2. Estimate (*a*) the maximum power and (*b*) the rotor speed if it is operating in 16 mi/h winds.

11.70 An American 6-ft-diameter multiblade HAWT is used to pump water to a height of 10 ft through 3-in-diameter cast-iron pipe. If the winds are 12 mi/h, estimate the rate of water flow in gallons per minute.

APPENDIX A
Physical Properties of Fluids

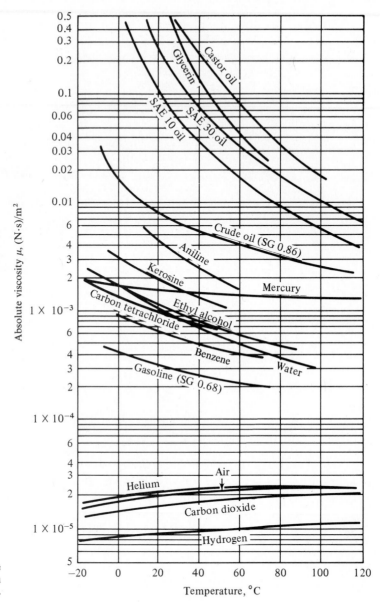

Fig. A.1 Absolute viscosity of common fluids at 1 atm.

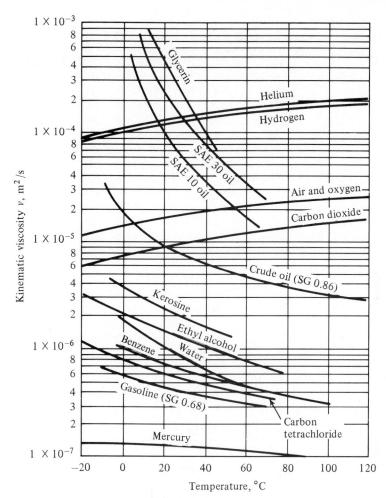

Fig. A.2 Kinematic viscosity of common fluids at 1 atm.

Table A.1
VISCOSITY AND DENSITY OF WATER AT 1 atm

T, °C	ρ, kg/m³	μ, (N·s)/m²	v, m²/s	T, °F	ρ, slug/ft³	μ, (lb·s)/ft²	v, ft²/s
0	1000	1.788 E−3	1.788 E−6	32	1.940	3.73 E−5	1.925 E−5
10	1000	1.307 E−3	1.307 E−6	50	1.940	2.73 E−5	1.407 E−5
20	998	1.003 E−3	1.005 E−6	68	1.937	2.09 E−5	1.082 E−5
30	996	0.799 E−3	0.802 E−6	86	1.932	1.67 E−5	0.864 E−5
40	992	0.657 E−3	0.662 E−6	104	1.925	1.37 E−5	0.713 E−5
50	988	0.548 E−3	0.555 E−6	122	1.917	1.14 E−5	0.597 E−5
60	983	0.467 E−3	0.475 E−6	140	1.908	0.975 E−5	0.511 E−5
70	978	0.405 E−3	0.414 E−6	158	1.897	0.846 E−5	0.446 E−5
80	972	0.355 E−3	0.365 E−6	176	1.886	0.741 E−5	0.393 E−5
90	965	0.316 E−3	0.327 E−6	194	1.873	0.660 E−5	0.352 E−5
100	958	0.283 E−3	0.295 E−6	212	1.859	0.591 E−5	0.318 E−5

Table A.2
VISCOSITY AND DENSITY OF AIR AT 1 atm

T, °C	ρ, kg/m³	μ, (N·s)/m²	v, m²/s	T, °F	ρ, slug/ft³	μ, (lb·s)/ft²	v, ft²/s
−40	1.52	1.51 E−5	0.99 E−5	−40	2.94 E−3	3.16 E−7	1.07 E−4
0	1.29	1.71 E−5	1.33 E−5	32	2.51 E−3	3.58 E−7	1.43 E−4
50	1.09	1.95 E−5	1.79 E−5	122	2.12 E−3	4.08 E−7	1.93 E−4
100	0.946	2.17 E−5	2.30 E−5	212	1.84 E−3	4.54 E−7	2.47 E−4
150	0.835	2.38 E−5	2.85 E−5	302	1.62 E−3	4.97 E−7	3.07 E−4
200	0.746	2.57 E−5	3.45 E−5	392	1.45 E−3	5.37 E−7	3.71 E−4
250	0.675	2.75 E−5	4.08 E−5	482	1.31 E−3	5.75 E−7	4.39 E−4
300	0.616	2.93 E−5	4.75 E−5	572	1.20 E−3	6.11 E−7	5.12 E−4
400	0.525	3.25 E−5	6.20 E−5	752	1.02 E−3	6.79 E−7	6.67 E−4
500	0.457	3.55 E−5	7.77 E−5	932	0.89 E−3	7.41 E−7	8.37 E−4

Table A.3
PROPERTIES OF COMMON LIQUIDS AT 1 atm AND 20°C (68°F)

Liquid	ρ, kg/m³	μ, (N·s)/m²	Υ, N/m†	p_v, N/m²	Bulk modulus, N/m²
Ammonia	608	2.20 E−4	2.13 E−2	9.10 E+5	
Benzene	881	6.51 E−4	2.88 E−2	1.01 E+4	1.05 E+9
Carbon tetrachloride	1,590	9.67 E−4	2.70 E−2	1.20 E+4	9.65 E+8
Ethanol	789	1.20 E−3	2.28 E−2	5.7 E+3	8.96 E+8
Gasoline	680	2.92 E−4	2.16 E−2	5.51 E+4	9.58 E+8
Glycerin	1,260	1.49	6.33 E−2	1.4 E−2	4.34 E+9
Kerosine	804	1.92 E−3	2.8 E−2	3.11 E+3	1.43 E+9
Mercury	13,550	1.56 E−3	4.84 E−1	1.1 E−3	2.55 E+10
Methanol	791	5.98 E−4	2.25 E−2	1.34 E+4	8.27 E+8
SAE 10 oil	917	1.04 E−1	3.6 E−2		1.31 E+9
SAE 30 oil	917	2.90 E−1	3.5 E−2		1.38 E+9
Water	998	1.00 E−3	7.28 E−2	2.34 E+3	2.19 E+9
Seawater	1,025	1.07 E−3	7.28 E−2	2.34 E+3	2.28 E+9

† In contact with air.

Table A.4

PROPERTIES OF COMMON GASES AT 1 atm AND 20°C (68°F)

Gas	Molecular weight	R, m²/(s²·K)	ρg, N/m³	μ, (N·s)/m²	Specific heat ratio
H_2	2.016	4124	0.822	9.05 E−6	1.41
He	4.003	2077	1.63	1.97 E−5	1.66
H_2O	18.02	461	7.35	1.02 E−5	1.33
Ar	39.944	208	16.3	2.24 E−5	1.67
Dry air	28.96	287	11.8	1.80 E−5	1.40
CO_2	44.01	189	17.9	1.48 E−5	1.30
CO	28.01	297	11.4	1.82 E−5	1.40
N_2	28.02	297	11.4	1.76 E−5	1.40
O_2	32.00	260	13.1	2.00 E−5	1.40
NO	30.01	277	12.1	1.90 E−5	1.40
N_2O	44.02	189	17.9	1.45 E−5	1.31
Cl_2	70.91	117	28.9	1.03 E−5	1.34
CH_4	16.04	518	6.54	1.34 E−5	1.32

Table A.5

SURFACE TENSION, VAPOR PRESSURE, AND SOUND SPEED OF WATER

T, °C	Υ, N/m	p_v, kPa	a, m/s
0	0.0756	0.611	1402
10	0.0742	1.227	1447
20	0.0728	2.337	1482
30	0.0712	4.242	1509
40	0.0696	7.375	1529
50	0.0679	12.34	1542
60	0.0662	19.92	1551
70	0.0644	31.16	1553
80	0.0626	47.35	1554
90	0.0608	70.11	1550
100	0.0589	101.3	1543
120	0.0550	198.5	1518
140	0.0509	361.3	1483
160	0.0466	617.8	1440
180	0.0422	1,002	1389
200	0.0377	1,554	1334
220	0.0331	2,318	1268
240	0.0284	3,344	1192
260	0.0237	4,688	1110
280	0.0190	6,412	1022
300	0.0144	8,581	920
320	0.0099	11,274	800
340	0.0056	14,586	630
360	0.0019	18,651	370
374 †	0.0 †	22,090 †	0 †

† Critical point.

Table A.6

PROPERTIES OF THE STANDARD ATMOSPHERE

z, m	T, K	p, Pa	ρ, kg/m^3
−500	291.41	107,508	1.2854
0	288.16	101,350	1.2255
500	284.91	95,480	1.1677
1,000	281.66	89,889	1.1120
1,500	278.41	84,565	1.0583
2,000	275.16	79,500	1.0067
2,500	271.91	74,684	0.9570
3,000	268.66	70,107	0.9092
3,500	265.41	65,759	0.8633
4,000	262.16	61,633	0.8191
4,500	258.91	57,718	0.7768
5,000	255.66	54,008	0.7361
5,500	252.41	50,493	0.6970
6,000	249.16	47,166	0.6596
6,500	245.91	44,018	0.6237
7,000	242.66	41,043	0.5893
7,500	239.41	38,233	0.5564
8,000	236.16	35,581	0.5250
8,500	232.91	33,080	0.4949
9,000	229.66	30,723	0.4661
9,500	226.41	28,504	0.4387
10,000	223.16	26,416	0.4125
10,500	219.91	24,455	0.3875
11,000	216.66	22,612	0.3637
11,500	216.66	20,897	0.3361
12,000	216.66	19,312	0.3106
12,500	216.66	17,847	0.2870
13,000	216.66	16,494	0.2652
13,500	216.66	15,243	0.2451
14,000	216.66	14,087	0.2265
14,500	216.66	13,018	0.2094
15,000	216.66	12,031	0.1935
15,500	216.66	11,118	0.1788
16,000	216.66	10,275	0.1652
16,500	216.66	9,496	0.1527
17,000	216.66	8,775	0.1411
17,500	216.66	8,110	0.1304
18,000	216.66	7,495	0.1205
18,500	216.66	6,926	0.1114
19,000	216.66	6,401	0.1029
19,500	216.66	5,915	0.0951
20,000	216.66	5,467	0.0879

APPENDIX B
Compressible-Flow Tables

Table B.1
ISENTROPIC FLOW OF A PERFECT GAS, $\gamma = 1.4$

Ma	p/p_0	ρ/ρ_0	T/T_0	A/A^*	Ma	p/p_0	ρ/ρ_0	T/T_0	A/A^*
0.0	1.0	1.0	1.0	∞	0.9	0.5913	0.6870	0.8606	1.0089
0.02	0.9997	0.9998	0.9999	28.9421	0.92	0.5785	0.6764	0.8552	1.0056
0.04	0.9989	0.9992	0.9997	14.4815	0.94	0.5658	0.6658	0.8498	1.0031
0.06	0.9975	0.9982	0.9993	9.6659	0.96	0.5532	0.6551	0.8444	1.0014
0.08	0.9955	0.9968	0.9987	7.2616	0.98	0.5407	0.6445	0.8389	1.0003
0.1	0.9930	0.9950	0.9980	5.8218	1.0	0.5283	0.6339	0.8333	1.0000
0.12	0.9900	0.9928	0.9971	4.8643	1.02	0.5160	0.6234	0.8278	1.0003
0.14	0.9864	0.9903	0.9961	4.1824	1.04	0.5039	0.6129	0.8222	1.0013
0.16	0.9823	0.9873	0.9949	3.6727	1.06	0.4919	0.6024	0.8165	1.0029
0.18	0.9776	0.9840	0.9936	3.2779	1.08	0.4800	0.5920	0.8108	1.0051
0.2	0.9725	0.9803	0.9921	2.9635	1.1	0.4684	0.5817	0.8052	1.0079
0.22	0.9668	0.9762	0.9904	2.7076	1.12	0.4568	0.5714	0.7994	1.0113
0.24	0.9607	0.9718	0.9886	2.4956	1.14	0.4455	0.5612	0.7937	1.0153
0.26	0.9541	0.9670	0.9867	2.3173	1.16	0.4343	0.5511	0.7879	1.0198
0.28	0.9470	0.9619	0.9846	2.1656	1.18	0.4232	0.5411	0.7822	1.0248
0.3	0.9395	0.9564	0.9823	2.0351	1.2	0.4124	0.5311	0.7764	1.0304
0.32	0.9315	0.9506	0.9799	1.9219	1.22	0.4017	0.5213	0.7706	1.0366
0.34	0.9231	0.9445	0.9774	1.8229	1.24	0.3912	0.5115	0.7648	1.0432
0.36	0.9143	0.9380	0.9747	1.7358	1.26	0.3809	0.5019	0.7590	1.0504
0.38	0.9052	0.9313	0.9719	1.6587	1.28	0.3708	0.4923	0.7532	1.0581
0.4	0.8956	0.9243	0.9690	1.5901	1.3	0.3609	0.4829	0.7474	1.0663
0.42	0.8857	0.9170	0.9659	1.5289	1.32	0.3512	0.4736	0.7416	1.0750
0.44	0.8755	0.9094	0.9627	1.4740	1.34	0.3417	0.4644	0.7358	1.0842
0.46	0.8650	0.9016	0.9594	1.4246	1.36	0.3323	0.4553	0.7300	1.0940
0.48	0.8541	0.8935	0.9559	1.3801	1.38	0.3232	0.4463	0.7242	1.1042
0.5	0.8430	0.8852	0.9524	1.3398	1.4	0.3142	0.4374	0.7184	1.1149
0.52	0.8317	0.8766	0.9487	1.3034	1.42	0.3055	0.4287	0.7126	1.1262
0.54	0.8201	0.8679	0.9449	1.2703	1.44	0.2969	0.4201	0.7069	1.1379
0.56	0.8082	0.8589	0.9410	1.2403	1.46	0.2886	0.4116	0.7011	1.1501
0.58	0.7962	0.8498	0.9370	1.2130	1.48	0.2804	0.4032	0.6954	1.1629
0.6	0.7840	0.8405	0.9328	1.1882	1.5	0.2724	0.3950	0.6897	1.1762
0.62	0.7716	0.8310	0.9286	1.1656	1.52	0.2646	0.3869	0.6840	1.1899
0.64	0.7591	0.8213	0.9243	1.1451	1.54	0.2570	0.3789	0.6783	1.2042
0.66	0.7465	0.8115	0.9199	1.1265	1.56	0.2496	0.3710	0.6726	1.2190
0.68	0.7338	0.8016	0.9153	1.1097	1.58	0.2423	0.3633	0.6670	1.2344
0.7	0.7209	0.7916	0.9107	1.0944	1.6	0.2353	0.3557	0.6614	1.2502
0.72	0.7080	0.7814	0.9061	1.0806	1.62	0.2284	0.3483	0.6558	1.2666
0.74	0.6951	0.7712	0.9013	1.0681	1.64	0.2217	0.3409	0.6502	1.2836
0.76	0.6821	0.7609	0.8964	1.0570	1.66	0.2151	0.3337	0.6447	1.3010
0.78	0.6690	0.7505	0.8915	1.0471	1.68	0.2088	0.3266	0.6392	1.3190
0.8	0.6560	0.7400	0.8865	1.0382	1.7	0.2026	0.3197	0.6337	1.3376
0.82	0.6430	0.7295	0.8815	1.0305	1.72	0.1966	0.3129	0.6283	1.3567
0.84	0.6300	0.7189	0.8763	1.0237	1.74	0.1907	0.3062	0.6229	1.3764
0.86	0.6170	0.7083	0.8711	1.0179	1.76	0.1850	0.2996	0.6175	1.3967
0.88	0.6041	0.6977	0.8659	1.0129	1.78	0.1794	0.2931	0.6121	1.4175

Table B.1 (*Cont.*)

Ma	p/p_0	ρ/ρ_0	T/T_0	A/A^*	Ma	p/p_0	ρ/ρ_0	T/T_0	A/A^*
1.8	0.1740	0.2868	0.6068	1.4390	2.7	0.0430	0.1056	0.4068	3.1830
1.82	0.1688	0.2806	0.6015	1.4610	2.72	0.0417	0.1033	0.4033	3.2440
1.84	0.1637	0.2745	0.5963	1.4836	2.74	0.0404	0.1010	0.3998	3.3061
1.86	0.1587	0.2686	0.5910	1.5069	2.76	0.0392	0.0989	0.3963	3.3695
1.88	0.1539	0.2627	0.5859	1.5308	2.78	0.0380	0.0967	0.3928	3.4342
1.9	0.1492	0.2570	0.5807	1.5553	2.8	0.0368	0.0946	0.3894	3.5001
1.92	0.1447	0.2514	0.5756	1.5804	2.82	0.0357	0.0926	0.3860	3.5674
1.94	0.1403	0.2459	0.5705	1.6062	2.84	0.0347	0.0906	0.3827	3.6359
1.96	0.1360	0.2405	0.5655	1.6326	2.86	0.0336	0.0886	0.3794	3.7058
1.98	0.1318	0.2352	0.5605	1.6597	2.88	0.0326	0.0867	0.3761	3.7771
2.0	0.1278	0.2300	0.5556	1.6875	2.9	0.0317	0.0849	0.3729	3.8498
2.02	0.1239	0.2250	0.5506	1.7160	2.92	0.0307	0.0831	0.3696	3.9238
2.04	0.1201	0.2200	0.5458	1.7451	2.94	0.0298	0.0813	0.3665	3.9993
2.06	0.1164	0.2152	0.5409	1.7750	2.96	0.0289	0.0796	0.3633	4.0763
2.08	0.1128	0.2104	0.5361	1.8056	2.98	0.0281	0.0779	0.3602	4.1547
2.1	0.1094	0.2058	0.5313	1.8369	3.0	0.0272	0.0762	0.3571	4.2346
2.12	0.1060	0.2013	0.5266	1.8690	3.02	0.0264	0.0746	0.3541	4.3160
2.14	0.1027	0.1968	0.5219	1.9018	3.04	0.0256	0.0730	0.3511	4.3990
2.16	0.0996	0.1925	0.5173	1.9354	3.06	0.0249	0.0715	0.3481	4.4835
2.18	0.0965	0.1882	0.5127	1.9698	3.08	0.0242	0.0700	0.3452	4.5696
2.2	0.0935	0.1841	0.5081	2.0050	3.1	0.0234	0.0685	0.3422	4.6573
2.22	0.0906	0.1800	0.5036	2.0409	3.12	0.0228	0.0671	0.3393	4.7467
2.24	0.0878	0.1760	0.4991	2.0777	3.14	0.0221	0.0657	0.3365	4.8377
2.26	0.0851	0.1721	0.4947	2.1153	3.16	0.0215	0.0643	0.3337	4.9304
2.28	0.0825	0.1683	0.4903	2.1538	3.18	0.0208	0.0630	0.3309	5.0248
2.3	0.0800	0.1646	0.4859	2.1931	3.2	0.0202	0.0617	0.3281	5.1210
2.32	0.0775	0.1609	0.4816	2.2333	3.22	0.0196	0.0604	0.3253	5.2189
2.34	0.0751	0.1574	0.4773	2.2744	3.24	0.0191	0.0591	0.3226	5.3186
2.36	0.0728	0.1539	0.4731	2.3164	3.26	0.0185	0.0579	0.3199	5.4201
2.38	0.0706	0.1505	0.4688	2.3593	3.28	0.0180	0.0567	0.3173	5.5234
2.4	0.0684	0.1472	0.4647	2.4031	3.3	0.0175	0.0555	0.3147	5.6286
2.42	0.0663	0.1439	0.4606	2.4479	3.32	0.0170	0.0544	0.3121	5.7358
2.44	0.0643	0.1408	0.4565	2.4936	3.34	0.0165	0.0533	0.3095	5.8448
2.46	0.0623	0.1377	0.4524	2.5403	3.36	0.0160	0.0522	0.3069	5.9558
2.48	0.0604	0.1346	0.4484	2.5880	3.38	0.0156	0.0511	0.3044	6.0687
2.5	0.0585	0.1317	0.4444	2.6367	3.4	0.0151	0.0501	0.3019	6.1837
2.52	0.0567	0.1288	0.4405	2.6865	3.42	0.0147	0.0491	0.2995	6.3007
2.54	0.0550	0.1260	0.4366	2.7372	3.44	0.0143	0.0481	0.2970	6.4198
2.56	0.0533	0.1232	0.4328	2.7891	3.46	0.0139	0.0471	0.2946	6.5409
2.58	0.0517	0.1205	0.4289	2.8420	3.48	0.0135	0.0462	0.2922	6.6642
2.6	0.0501	0.1179	0.4252	2.8960	3.5	0.0131	0.0452	0.2899	6.7896
2.62	0.0486	0.1153	0.4214	2.9511	3.52	0.0127	0.0443	0.2875	6.9172
2.64	0.0471	0.1128	0.4177	3.0073	3.54	0.0124	0.0434	0.2852	7.0471
2.66	0.0457	0.1103	0.4141	3.0647	3.56	0.0120	0.0426	0.2829	7.1791
2.68	0.0443	0.1079	0.4104	3.1233	3.58	0.0117	0.0417	0.2806	7.3135

Table B.1 (*Cont.*)

ISENTROPIC FLOW OF A PERFECT GAS, $\gamma = 1.4$

Ma	p/p_0	ρ/ρ_0	T/T_0	A/A^*	Ma	p/p_0	ρ/ρ_0	T/T_0	A/A^*
3.6	0.0114	0.0409	0.2784	7.4501	4.32	0.0043	0.0205	0.2113	14.1984
3.62	0.0111	0.0401	0.2762	7.5891	4.34	0.0042	0.0202	0.2098	14.4456
3.64	0.0108	0.0393	0.2740	7.7305	4.36	0.0041	0.0198	0.2083	14.6965
3.66	0.0105	0.0385	0.2718	7.8742	4.38	0.0040	0.0194	0.2067	14.9513
3.68	0.0102	0.0378	0.2697	8.0204	4.4	0.0039	0.0191	0.2053	15.2099
3.7	0.0099	0.0370	0.2675	8.1691	4.42	0.0038	0.0187	0.2038	15.4724
3.72	0.0096	0.0363	0.2654	8.3202	4.44	0.0037	0.0184	0.2023	15.7388
3.74	0.0094	0.0356	0.2633	8.4739	4.46	0.0036	0.0181	0.2009	16.0092
3.76	0.0091	0.0349	0.2613	8.6302	4.48	0.0035	0.0178	0.1994	16.2837
3.78	0.0089	0.0342	0.2592	8.7891	4.5	0.0035	0.0174	0.1980	16.5622
3.8	0.0086	0.0335	0.2572	8.9506	4.52	0.0034	0.0171	0.1966	16.8449
3.82	0.0084	0.0329	0.2552	9.1148	4.54	0.0033	0.0168	0.1952	17.1317
3.84	0.0082	0.0323	0.2532	9.2817	4.56	0.0032	0.0165	0.1938	17.4228
3.86	0.0080	0.0316	0.2513	9.4513	4.58	0.0031	0.0163	0.1925	17.7181
3.88	0.0077	0.0310	0.2493	9.6237	4.6	0.0031	0.0160	0.1911	18.0178
3.9	0.0075	0.0304	0.2474	9.7990	4.62	0.0030	0.0157	0.1898	18.3218
3.92	0.0073	0.0299	0.2455	9.9771	4.64	0.0029	0.0154	0.1885	18.6303
3.94	0.0071	0.0293	0.2436	10.1581	4.66	0.0028	0.0152	0.1872	18.9433
3.96	0.0069	0.0287	0.2418	10.3420	4.68	0.0028	0.0149	0.1859	19.2608
3.98	0.0068	0.0282	0.2399	10.5289	4.7	0.0027	0.0146	0.1846	19.5828
4.0	0.0066	0.0277	0.2381	10.7188	4.72	0.0026	0.0144	0.1833	19.9095
4.02	0.0064	0.0271	0.2363	10.9117	4.74	0.0026	0.0141	0.1820	20.2409
4.04	0.0062	0.0266	0.2345	11.1077	4.76	0.0025	0.0139	0.1808	20.5770
4.06	0.0061	0.0261	0.2327	11.3068	4.78	0.0025	0.0137	0.1795	20.9179
4.08	0.0059	0.0256	0.2310	11.5091	4.8	0.0024	0.0134	0.1783	21.2637
4.1	0.0058	0.0252	0.2293	11.7147	4.82	0.0023	0.0132	0.1771	21.6144
4.12	0.0056	0.0247	0.2275	11.9234	4.84	0.0023	0.0130	0.1759	21.9700
4.14	0.0055	0.0242	0.2258	12.1354	4.86	0.0022	0.0128	0.1747	22.3306
4.16	0.0053	0.0238	0.2242	12.3508	4.88	0.0022	0.0125	0.1735	22.6963
4.18	0.0052	0.0234	0.2225	12.5695	4.9	0.0021	0.0123	0.1724	23.0671
4.2	0.0051	0.0229	0.2208	12.7916	4.92	0.0021	0.0121	0.1712	23.4431
4.22	0.0049	0.0225	0.2192	13.0172	4.94	0.0020	0.0119	0.1700	23.8243
4.24	0.0048	0.0221	0.2176	13.2463	4.96	0.0020	0.0117	0.1689	24.2109
4.26	0.0047	0.0217	0.2160	13.4789	4.98	0.0019	0.0115	0.1678	24.6027
4.28	0.0046	0.0213	0.2144	13.7151	5.0	0.0019	0.0113	0.1667	25.0000
4.3	0.0044	0.0209	0.2129	13.9549					

Table B.2

NORMAL-SHOCK RELATIONS FOR A PERFECT GAS, $\gamma = 1.4$

Ma_{n1}	Ma_{n2}	p_2/p_1	$V_1/V_2 = \rho_2/\rho_1$	T_2/T_1	p_{02}/p_{01}	A_2^*/A_1^*
1.0	1.0000	1.0000	1.0000	1.0000	1.0000	1.0000
1.02	0.9805	1.0471	1.0334	1.0132	1.0000	1.0000
1.04	0.9620	1.0952	1.0671	1.0263	0.0999	1.0001
1.06	0.9444	1.1442	1.1009	1.0393	0.9998	1.0002
1.08	0.9277	1.1941	1.1349	1.0522	0.9994	1.0006
1.1	0.9118	1.2450	1.1691	1.0649	0.9989	1.0011
1.12	0.8966	1.2968	1.2034	1.0776	0.9982	1.0018
1.14	0.8820	1.3495	1.2378	1.0903	0.9973	1.0027
1.16	0.8682	1.4032	1.2723	1.1029	0.9961	1.0040
1.18	0.8549	1.4578	1.3069	1.1154	0.9946	1.0055
1.2	0.8422	1.5133	1.3416	1.1280	0.9928	1.0073
1.22	0.8300	1.5698	1.3764	1.1405	0.9907	1.0094
1.24	0.8183	1.6272	1.4112	1.1531	0.9884	1.0118
1.26	0.8071	1.6855	1.4460	1.1657	0.9857	1.0145
1.28	0.7963	1.7448	1.4808	1.1783	0.9827	1.0176
1.3	0.7860	1.8050	1.5157	1.1909	0.9794	1.0211
1.32	0.7760	1.8661	1.5505	1.2035	0.9758	1.0249
1.34	0.7664	1.9282	1.5854	1.2162	0.9718	1.0290
1.36	0.7572	1.9912	1.6202	1.2290	0.9676	1.0335
1.38	0.7483	2.0551	1.6549	1.2418	0.9630	1.0384
1.4	0.7397	2.1200	1.6897	1.2547	0.9582	1.0436
1.42	0.7314	2.1858	1.7243	1.2676	0.9531	1.0492
1.44	0.7235	2.2525	1.7589	1.2807	0.9476	1.0552
1.46	0.7157	2.3202	1.7934	1.2938	0.9420	1.0616
1.48	0.7083	2.3888	1.8278	1.3069	0.9360	1.0684
1.5	0.7011	2.4583	1.8621	1.3202	0.9298	1.0755
1.52	0.6941	2.5288	1.8963	1.3336	0.9233	1.0830
1.54	0.6874	2.6002	1.9303	1.3470	0.9166	1.0910
1.56	0.6809	2.6725	1.9643	1.3606	0.9097	1.0993
1.58	0.6746	2.7458	1.9981	1.3742	0.9026	1.1080
1.6	0.6684	2.8200	2.0317	1.3880	0.8952	1.1171
1.62	0.6625	2.8951	2.0653	1.4018	0.8877	1.1266
1.64	0.6568	2.9712	2.0986	1.4158	0.8799	1.1365
1.66	0.6512	3.0482	2.1318	1.4299	0.8720	1.1468
1.68	0.6458	3.1261	2.1649	1.4440	0.8639	1.1575
1.7	0.6405	3.2050	2.1977	1.4583	0.8557	1.1686
1.72	0.6355	3.2848	2.2304	1.4727	0.8474	1.1801
1.74	0.6305	3.3655	2.2629	1.4873	0.8389	1.1921
1.76	0.6257	3.4472	2.2952	1.5019	0.8302	1.2045
1.78	0.6210	3.5298	2.3273	1.5167	0.8215	1.2173
1.8	0.6165	3.6133	2.3592	1.5316	0.8127	1.2305
1.82	0.6121	3.6978	2.3909	1.5466	0.8038	1.2441
1.84	0.6078	3.7832	2.4224	1.5617	0.7948	1.2582
1.86	0.6036	3.8695	2.4537	1.5770	0.7857	1.2728
1.88	0.5996	3.9568	2.4848	1.5924	0.7765	1.2877
1.9	0.5956	4.0450	2.5157	1.6079	0.7674	1.3032

Table B.2 (*Cont.*)
NORMAL-SHOCK RELATIONS FOR A PERFECT GAS, $\gamma = 1.4$

Ma_{n1}	Ma_{n2}	p_2/p_1	$V_1/V_2 = \rho_2/\rho_1$	T_2/T_1	p_{02}/p_{01}	A_2^*/A_1^*
1.92	0.5918	4.1341	2.5463	1.6236	0.7581	1.3191
1.94	0.5880	4.2242	2.5767	1.6394	0.7488	1.3354
1.96	0.5844	4.3152	2.6069	1.6553	0.7395	1.3522
1.98	0.5808	4.4071	2.6369	1.6713	0.7302	1.3695
2.0	0.5774	4.5000	2.6667	1.6875	0.7209	1.3872
2.02	0.5740	4.5938	2.6962	1.7038	0.7115	1.4054
2.04	0.5707	4.6885	2.7255	1.7203	0.7022	1.4241
2.06	0.5675	4.7842	2.7545	1.7369	0.6928	1.4433
2.08	0.5643	4.8808	2.7833	1.7536	0.6835	1.4630
2.1	0.5613	4.9783	2.8119	1.7705	0.6742	1.4832
2.12	0.5583	5.0768	2.8402	1.7875	0.6649	1.5039
2.14	0.5554	5.1762	2.8683	1.8046	0.6557	1.5252
2.16	0.5525	5.2765	2.8962	1.8219	0.6464	1.5469
2.18	0.5498	5.3778	2.9238	1.8393	0.6373	1.5692
2.2	0.5471	5.4800	2.9512	1.8569	0.6281	1.5920
2.22	0.5444	5.5831	2.9784	1.8746	0.6191	1.6154
2.24	0.5418	5.6872	3.0053	1.8924	0.6100	1.6393
2.26	0.5393	5.7922	3.0319	1.9104	0.6011	1.6638
2.28	0.5368	5.8981	3.0584	1.9285	0.5921	1.6888
2.3	0.5344	6.0050	3.0845	1.9468	0.5833	1.7144
2.32	0.5321	6.1128	3.1105	1.9652	0.5745	1.7406
2.34	0.5297	6.2215	3.1362	1.9838	0.5658	1.7674
2.36	0.5275	6.3312	3.1617	2.0025	0.5572	1.7948
2.38	0.5253	6.4418	3.1869	2.0213	0.5486	1.8228
2.4	0.5231	6.5533	3.2119	2.0403	0.5401	1.8514
2.42	0.5210	6.6658	3.2367	2.0595	0.5317	1.8806
2.44	0.5189	6.7792	3.2612	2.0788	0.5234	1.9105
2.46	0.5169	6.8935	3.2855	2.0982	0.5152	1.9410
2.48	0.5149	7.0088	3.3095	2.1178	0.5071	1.9721
2.5	0.5130	7.1250	3.3333	2.1375	0.4990	2.0039
2.52	0.5111	7.2421	3.3569	2.1574	0.4911	2.0364
2.54	0.5092	7.3602	3.3803	2.1774	0.4832	2.0696
2.56	0.5074	7.4792	3.4034	2.1976	0.4754	2.1035
2.58	0.5056	7.5991	3.4263	2.2179	0.4677	2.1381
2.6	0.5039	7.7200	3.4490	2.2383	0.4601	2.1733
2.62	0.5022	7.8418	3.4714	2.2590	0.4526	2.2093
2.64	0.5005	7.9645	3.4937	2.2797	0.4452	2.2461
2.66	0.4988	8.0882	3.5157	2.3006	0.4379	2.2835
2.68	0.4972	8.2128	3.5374	2.3217	0.4307	2.3218
2.7	0.4956	8.3383	3.5590	2.3429	0.4236	2.3608
2.72	0.4941	8.4648	3.5803	2.3642	0.4166	2.4005
2.74	0.4926	8.5922	3.6015	2.3858	0.4097	2.4411
2.76	0.4911	8.7205	3.6224	2.4074	0.4028	2.4825
2.78	0.4896	8.8498	3.6431	2.4292	0.3961	2.5246
2.8	0.4882	8.9800	3.6636	2.4512	0.3895	2.5676
2.82	0.4868	9.1111	3.6838	2.4733	0.3829	2.6115

Table B.2 (*Cont.*)

Ma_{n1}	Ma_{n2}	p_2/p_1	$V_1/V_2 = \rho_2/\rho_1$	T_2/T_1	p_{02}/p_{01}	A_2^*/A_1^*
2.84	0.4854	9.2432	3.7039	2.4955	0.3765	2.6561
2.86	0.4840	9.3762	3.7238	2.5179	0.3701	2.7017
2.88	0.4827	9.5101	3.7434	2.5405	0.3639	2.7481
2.9	0.4814	9.6450	3.7629	2.5632	0.3577	2.7954
2.92	0.4801	9.7808	3.7821	2.5861	0.3517	2.8436
2.94	0.4788	9.9175	3.8012	2.6091	0.3457	2.8927
2.96	0.4776	10.0552	3.8200	2.6322	0.3398	2.9427
2.98	0.4764	10.1938	3.8387	2.6555	0.3340	2.9937
3.0	0.4752	10.3333	3.8571	2.6790	0.3283	3.0456
3.02	0.4740	10.4738	3.8754	2.7026	0.3227	3.0985
3.04	0.4729	10.6152	3.8935	2.7264	0.3172	3.1523
3.06	0.4717	10.7575	3.9114	2.7503	0.3118	3.2072
3.08	0.4706	10.9008	3.9291	2.7744	0.3065	3.2630
3.1	0.4695	11.0450	3.9466	2.7986	0.3012	3.3199
3.12	0.4685	11.1901	3.9639	2.8230	0.2960	3.3778
3.14	0.4674	11.3362	3.9811	2.8475	0.2910	3.4368
3.16	0.4664	11.4832	3.9981	2.8722	0.2860	3.4969
3.18	0.4654	11.6311	4.0149	2.8970	0.2811	3.5580
3.2	0.4643	11.7800	4.0315	2.9220	0.2762	3.6202
3.22	0.4634	11.9298	4.0479	2.9471	0.2715	3.6835
3.24	0.4624	12.0805	4.0642	2.9724	0.2668	3.7480
3.26	0.4614	12.2322	4.0803	2.9979	0.2622	3.8136
3.28	0.4605	12.3848	4.0963	3.0234	0.2577	3.8803
3.3	0.4596	12.5383	4.1120	3.0492	0.2533	3.9483
3.32	0.4587	12.6928	4.1276	3.0751	0.2489	4.0174
3.34	0.4578	12.8482	4.1431	3.1011	0.2446	4.0877
3.36	0.4569	13.0045	4.1583	3.1273	0.2404	4.1593
3.38	0.4560	13.1618	4.1734	3.1537	0.2363	4.2321
3.4	0.4552	13.3200	4.1884	3.1802	0.2322	4.3062
3.42	0.4544	13.4791	4.2032	3.2069	0.2282	4.3815
3.44	0.4535	13.6392	4.2178	3.2337	0.2243	4.4581
3.46	0.4527	13.8002	4.2323	3.2607	0.2205	4.5361
3.48	0.4519	13.9621	4.2467	3.2878	0.2167	4.6154
3.5	0.4512	14.1250	4.2609	3.3151	0.2129	4.6960
3.52	0.4504	14.2888	4.2749	3.3425	0.2093	4.7780
3.54	0.4496	14.4535	4.2888	3.3701	0.2057	4.8614
3.56	0.4489	14.6192	4.3026	3.3978	0.2022	4.9461
3.58	0.4481	14.7858	4.3162	3.4257	0.1987	5.0324
3.6	0.4474	14.9533	4.3296	3.4537	0.1953	5.1200
3.62	0.4467	15.1218	4.3429	3.4819	0.1920	5.2091
3.64	0.4460	15.2912	4.3561	3.5103	0.1887	5.2997
3.66	0.4453	15.4615	4.3692	3.5388	0.1855	5.3918
3.68	0.4446	15.6328	4.3821	3.5674	0.1823	5.4854
3.7	0.4439	15.8050	4.3949	3.5962	0.1792	5.5806
3.72	0.4433	15.9781	4.4075	3.6252	0.1761	5.6773
3.74	0.4426	16.1522	4.4200	3.6543	0.1731	5.7756

Table B.2 (*Cont.*)

NORMAL-SHOCK RELATIONS FOR A PERFECT GAS, $\gamma = 1.4$

Ma_{n1}	Ma_{n2}	p_2/p_1	$V_1/V_2 = \rho_2/\rho_1$	T_2/T_1	p_{02}/p_{01}	A_2^*/A_1^*
3.76	0.4420	16.3272	4.4324	3.6836	0.1702	5.8755
3.78	0.4414	16.5031	4.4447	3.7130	0.1673	5.9770
3.8	0.4407	16.6800	4.4568	3.7426	0.1645	6.0801
3.82	0.4401	16.8578	4.4688	3.7723	0.1617	6.1849
3.84	0.4395	17.0365	4.4807	3.8022	0.1589	6.2915
3.86	0.4389	17.2162	4.4924	3.8323	0.1563	6.3997
3.88	0.4383	17.3968	4.5041	3.8625	0.1536	6.5096
3.9	0.4377	17.5783	4.5156	3.8928	0.1510	6.6213
3.92	0.4372	17.7608	4.5270	3.9233	0.1485	6.7348
3.94	0.4366	17.9442	4.5383	3.9540	0.1460	6.8501
3.96	0.4360	18.1285	4.5494	3.9848	0.1435	6.9672
3.98	0.4355	18.3138	4.5605	4.0158	0.1411	7.0861
4.0	0.4350	18.5000	4.5714	4.0469	0.1388	7.2069
4.02	0.4344	18.6871	4.5823	4.0781	0.1364	7.3296
4.04	0.4339	18.8752	4.5930	4.1096	0.1342	7.4542
4.06	0.4334	19.0642	4.6036	4.1412	0.1319	7.5807
4.08	0.4329	19.2541	4.6141	4.1729	0.1297	7.7092
4.1	0.4324	19.4450	4.6245	4.2048	0.1276	7.8397
4.12	0.4319	19.6368	4.6348	4.2368	0.1254	7.9722
4.14	0.4314	19.8295	4.6450	4.2690	0.1234	8.1067
4.16	0.4309	20.0232	4.6550	4.3014	0.1213	8.2433
4.18	0.4304	20.2178	4.6650	4.3339	0.1193	8.3819
4.2	0.4299	20.4133	4.6749	4.3666	0.1173	8.5227
4.22	0.4295	20.6098	4.6847	4.3994	0.1154	8.6656
4.24	0.4290	20.8072	4.6944	4.4324	0.1135	8.8107
4.26	0.4286	21.0055	4.7040	4.4655	0.1116	8.9579
4.28	0.4281	21.2048	4.7135	4.4988	0.1098	9.1074
4.3	0.4277	21.4050	4.7229	4.5322	0.1080	9.2591
4.32	0.4272	21.6061	4.7322	4.5658	0.1062	9.4131
4.34	0.4268	21.8082	4.7414	4.5995	0.1045	9.5694
4.36	0.4264	22.0112	4.7505	4.6334	0.1028	9.7280
4.38	0.4260	22.2151	4.7595	4.6675	0.1011	9.8889

Table B.2 (*Cont.*)

Ma_{n1}	Ma_{n2}	p_2/p_1	$V_1/V_2 = \rho_2/\rho_1$	T_2/T_1	p_{02}/p_{01}	A_2^*/A_1^*
4.4	0.4255	22.4200	4.7685	4.7017	0.0995	10.0522
4.42	0.4251	22.6258	4.7773	4.7361	0.0979	10.2179
4.44	0.4247	22.8325	4.7861	4.7706	0.0963	10.3861
4.46	0.4243	23.0402	4.7948	4.8053	0.0947	10.5567
4.48	0.4239	23.2488	4.8034	4.8401	0.0932	10.7298
4.5	0.4236	23.4583	4.8119	4.8751	0.0917	10.9054
4.52	0.4232	23.6688	4.8203	4.9102	0.0902	11.0835
4.54	0.4228	23.8802	4.8287	4.9455	0.0888	11.2643
4.56	0.4224	24.0925	4.8369	4.9810	0.0874	11.4476
4.58	0.4220	24.3058	4.8451	5.0166	0.0860	11.6336
4.6	0.4217	24.5200	4.8532	5.0523	0.0846	11.8222
4.62	0.4213	24.7351	4.8612	5.0882	0.0832	12.0136
4.64	0.4210	24.9512	4.8692	5.1243	0.0819	12.2076
4.66	0.4206	25.1682	4.8771	5.1605	0.0806	12.4044
4.68	0.4203	25.3861	4.8849	5.1969	0.0793	12.6040
4.7	0.4199	25.6050	4.8926	5.2334	0.0781	12.8065
4.72	0.4196	25.8248	4.9002	5.2701	0.0769	13.0117
4.74	0.4192	26.0455	4.9078	5.3070	0.0756	13.2199
4.76	0.4189	26.2672	4.9153	5.3440	0.0745	13.4310
4.78	0.4186	26.4898	4.9227	5.3811	0.0733	13.6450
4.8	0.4183	26.7133	4.9301	5.4184	0.0721	13.8620
4.82	0.4179	26.9378	4.9374	5.4559	0.0710	14.0820
4.84	0.4176	27.1632	4.9446	5.4935	0.0699	14.3050
4.86	0.4173	27.3895	4.9518	5.5313	0.0688	14.5312
4.88	0.4170	27.6168	4.9589	5.5692	0.0677	14.7604
4.9	0.4167	27.8450	4.9659	5.6073	0.0667	14.9928
4.92	0.4164	28.0741	4.9728	5.6455	0.0657	15.2284
4.94	0.4161	28.3042	4.9797	5.6839	0.0647	15.4672
4.96	0.4158	28.5352	4.9865	5.7224	0.0637	15.7092
4.98	0.4155	28.7671	4.9933	5.7611	0.0627	15.9545
5.0	0.4152	29.0000	5.0000	5.8000	0.0617	16.2032

Table B.3

ADIABATIC FRICTIONAL FLOW IN A CONSTANT-AREA
DUCT FOR $\gamma = 1.4$

Ma	$\bar{f}L^*/D$	p/p^*	T/T^*	$\rho^*/\rho = V/V^*$	p_0/p_0^*
0.0	∞	∞	1.2000	0.0	∞
0.02	1778.4500	54.7701	1.1999	0.0219	28.9421
0.04	440.3520	27.3817	1.1996	0.0438	14.4815
0.06	193.0310	18.2508	1.1991	0.0657	9.6659
0.08	106.7180	13.6843	1.1985	0.0876	7.2616
0.1	66.9216	10.9435	1.1976	0.1094	5.8218
0.12	45.4080	9.1156	1.1966	0.1313	4.8643
0.14	32.5113	7.8093	1.1953	0.1531	4.1824
0.16	24.1978	6.8291	1.1939	0.1748	3.6727
0.18	18.5427	6.0662	1.1923	0.1965	3.2779
0.2	14.5333	5.4554	1.1905	0.2182	2.9635
0.22	11.5961	4.9554	1.1885	0.2398	2.7076
0.24	9.3865	4.5383	1.1863	0.2614	2.4956
0.26	7.6876	4.1851	1.1840	0.2829	2.3173
0.28	6.3572	3.8820	1.1815	0.3043	2.1656
0.3	5.2993	3.6191	1.1788	0.3257	2.0351
0.32	4.4467	3.3887	1.1759	0.3470	1.9219
0.34	3.7520	3.1853	1.1729	0.3682	1.8229
0.36	3.1801	3.0042	1.1697	0.3893	1.7358
0.38	2.7054	2.8420	1.1663	0.4104	1.6587
0.4	2.3085	2.6958	1.1628	0.4313	1.5901
0.42	1.9744	2.5634	1.1591	0.4522	1.5289
0.44	1.6915	2.4428	1.1553	0.4729	1.4740
0.46	1.4509	2.3326	1.1513	0.4936	1.4246
0.48	1.2453	2.2313	1.1471	0.5141	1.3801
0.5	1.0691	2.1381	1.1429	0.5345	1.3398
0.52	0.9174	2.0519	1.1384	0.5548	1.3034
0.54	0.7866	1.9719	1.1339	0.5750	1.2703
0.56	0.6736	1.8975	1.1292	0.5951	1.2403
0.58	0.5757	1.8282	1.1244	0.6150	1.2130
0.6	0.4908	1.7634	1.1194	0.6348	1.1882
0.62	0.4172	1.7026	1.1143	0.6545	1.1656
0.64	0.3533	1.6456	1.1091	0.6740	1.1451
0.66	0.2979	1.5919	1.1038	0.6934	1.1265
0.68	0.2498	1.5413	1.0984	0.7127	1.1097
0.7	0.2081	1.4935	1.0929	0.7318	1.0944
0.72	0.1721	1.4482	1.0873	0.7508	1.0806
0.74	0.1411	1.4054	1.0815	0.7696	1.0681
0.76	0.1145	1.3647	1.0757	0.7883	1.0570
0.78	0.0917	1.3261	1.0698	0.8068	1.0471
0.8	0.0723	1.2893	1.0638	0.8251	1.0382
0.82	0.0559	1.2542	1.0578	0.8433	1.0305
0.84	0.0423	1.2208	1.0516	0.8614	1.0237
0.86	0.0310	1.1889	1.0454	0.8793	1.0179
0.88	0.0218	1.1583	1.0391	0.8970	1.0129

Table B.3 (*Cont.*)

Ma	$\bar{f}L^*/D$	p/p^*	T/T^*	$\rho^*/\rho = V/V^*$	p_0/p_0^*
0.9	0.0145	1.1291	1.0327	0.9146	1.0089
0.92	0.0089	1.1011	1.0263	0.9320	1.0056
0.94	0.0048	1.0743	1.0198	0.9493	1.0031
0.96	0.0021	1.0485	1.0132	0.9663	1.0014
0.98	0.0005	1.0238	1.0066	0.9833	1.0003
1.0	0.0000	1.0000	1.0000	1.0000	1.0000
1.02	0.0005	0.9771	0.9933	1.0166	1.0003
1.04	0.0018	0.9551	0.9866	1.0330	1.0013
1.06	0.0038	0.9338	0.9798	1.0492	1.0029
1.08	0.0066	0.9133	0.9730	1.0653	1.0051
1.1	0.0099	0.8936	0.9662	1.0812	1.0079
1.12	0.0138	0.8745	0.9593	1.0970	1.0113
1.14	0.0182	0.8561	0.9524	1.1126	1.0153
1.16	0.0230	0.8383	0.9455	1.1280	1.0198
1.18	0.0281	0.8210	0.9386	1.1432	1.0248
1.2	0.0336	0.8044	0.9317	1.1583	1.0304
1.22	0.0394	0.7882	0.9247	1.1732	1.0366
1.24	0.0455	0.7726	0.9178	1.1879	1.0432
1.26	0.0517	0.7574	0.9108	1.2025	1.0504
1.28	0.0582	0.7427	0.9038	1.2169	1.0581
1.3	0.0648	0.7285	0.8969	1.2311	1.0663
1.32	0.0716	0.7147	0.8899	1.2452	1.0750
1.34	0.0785	0.7012	0.8829	1.2591	1.0842
1.36	0.0855	0.6882	0.8760	1.2729	1.0940
1.38	0.0926	0.6755	0.8690	1.2864	1.1042
1.4	0.0997	0.6632	0.8621	1.2999	1.1149
1.42	0.1069	0.6512	0.8551	1.3131	1.1262
1.44	0.1142	0.6396	0.8482	1.3262	1.1379
1.46	0.1215	0.6282	0.8413	1.3392	1.1501
1.48	0.1288	0.6172	0.8344	1.3520	1.1629
1.5	0.1361	0.6065	0.8276	1.3646	1.1762
1.52	0.1433	0.5960	0.8207	1.3770	1.1899
1.54	0.1506	0.5858	0.8139	1.3894	1.2042
1.56	0.1579	0.5759	0.8071	1.4015	1.2190
1.58	0.1651	0.5662	0.8004	1.4135	1.2344
1.6	0.1724	0.5568	0.7937	1.4254	1.2502
1.62	0.1795	0.5476	0.7869	1.4371	1.2666
1.64	0.1867	0.5386	0.7803	1.4487	1.2836
1.66	0.1938	0.5299	0.7736	1.4601	1.3010
1.68	0.2008	0.5213	0.7670	1.4713	1.3190
1.7	0.2078	0.5130	0.7605	1.4825	1.3376
1.72	0.2147	0.5048	0.7539	1.4935	1.3567
1.74	0.2216	0.4969	0.7474	1.5043	1.3764
1.76	0.2284	0.4891	0.7410	1.5150	1.3967
1.78	0.2352	0.4815	0.7345	1.5256	1.4175

Table B.3 (*Cont.*)

ADIABATIC FRICTIONAL FLOW IN A CONSTANT-AREA DUCT FOR $\gamma = 1.4$

Ma	$\bar{f}L^*/D$	p/p^*	T/T^*	$\rho^*/\rho = V/V^*$	p_0/p_0^*
1.8	0.2419	0.4741	0.7282	1.5360	1.4390
1.82	0.2485	0.4668	0.7218	1.5463	1.4610
1.84	0.2551	0.4597	0.7155	1.5564	1.4836
1.86	0.2616	0.4528	0.7093	1.5664	1.5069
1.88	0.2680	0.4460	0.7030	1.5763	1.5308
1.9	0.2743	0.4394	0.6969	1.5861	1.5553
1.92	0.2806	0.4329	0.6907	1.5957	1.5804
1.94	0.2868	0.4265	0.6847	1.6052	1.6062
1.96	0.2929	0.4203	0.6786	1.6146	1.6326
1.98	0.2990	0.4142	0.6726	1.6239	1.6597
2.0	0.3050	0.4082	0.6667	1.6330	1.6875
2.02	0.3109	0.4024	0.6608	1.6420	1.7160
2.04	0.3168	0.3967	0.6549	1.6509	1.7451
2.06	0.3225	0.3911	0.6491	1.6597	1.7750
2.08	0.3282	0.3856	0.6433	1.6683	1.8056
2.1	0.3339	0.3802	0.6376	1.6769	1.8369
2.12	0.3394	0.3750	0.6320	1.6853	1.8690
2.14	0.3449	0.3698	0.6263	1.6936	1.9018
2.16	0.3503	0.3648	0.6208	1.7018	1.9354
2.18	0.3556	0.3598	0.6152	1.7099	1.9698
2.2	0.3609	0.3549	0.6098	1.7179	2.0050
2.22	0.3661	0.3502	0.6043	1.7258	2.0409
2.24	0.3712	0.3455	0.5989	1.7336	2.0777
2.26	0.3763	0.3409	0.5936	1.7412	2.1153
2.28	0.3813	0.3364	0.5883	1.7488	2.1538
2.3	0.3862	0.3320	0.5831	1.7563	2.1931
2.32	0.3911	0.3277	0.5779	1.7637	2.2333
2.34	0.3959	0.3234	0.5728	1.7709	2.2744
2.36	0.4006	0.3193	0.5677	1.7781	2.3164
2.38	0.4053	0.3152	0.5626	1.7852	2.3593
2.4	0.4099	0.3111	0.5576	1.7922	2.4031
2.42	0.4144	0.3072	0.5527	1.7991	2.4479
2.44	0.4189	0.3033	0.5478	1.8059	2.4936
2.46	0.4233	0.2995	0.5429	1.8126	2.5403
2.48	0.4277	0.2958	0.5381	1.8192	2.5880
2.5	0.4320	0.2921	0.5333	1.8257	2.6367
2.52	0.4362	0.2885	0.5286	1.8322	2.6865
2.54	0.4404	0.2850	0.5239	1.8386	2.7372
2.56	0.4445	0.2815	0.5193	1.8448	2.7891
2.58	0.4486	0.2781	0.5147	1.8510	2.8420
2.6	0.4526	0.2747	0.5102	1.8571	2.8960
2.62	0.4565	0.2714	0.5057	1.8632	2.9511
2.64	0.4604	0.2682	0.5013	1.8691	3.0073
2.66	0.4643	0.2650	0.4969	1.8750	3.0647
2.68	0.4681	0.2619	0.4925	1.8808	3.1233

Table B.3 (*Cont.*)

Ma	$\bar{f}L^*/D$	p/p^*	T/T^*	$\rho^*/\rho = V/V^*$	p_0/p_0^*
2.7	0.4718	0.2588	0.4882	1.8865	3.1830
2.72	0.4755	0.2558	0.4839	1.8922	3.2440
2.74	0.4791	0.2528	0.4797	1.8978	3.3061
2.76	0.4827	0.2498	0.4755	1.9033	3.3695
2.78	0.4863	0.2470	0.4714	1.9087	3.4342
2.8	0.4898	0.2441	0.4673	1.9140	3.5001
2.82	0.4932	0.2414	0.4632	1.9193	3.5674
2.84	0.4966	0.2386	0.4592	1.9246	3.6359
2.86	0.5000	0.2359	0.4552	1.9297	3.7058
2.88	0.5033	0.2333	0.4513	1.9348	3.7771
2.9	0.5065	0.2307	0.4474	1.9398	3.8498
2.92	0.5097	0.2281	0.4436	1.9448	3.9238
2.94	0.5129	0.2256	0.4398	1.9497	3.9993
2.96	0.5160	0.2231	0.4360	1.9545	4.0763
2.98	0.5191	0.2206	0.4323	1.9593	4.1547
3.0	0.5222	0.2182	0.4286	1.9640	4.2346
3.02	0.5252	0.2158	0.4249	1.9686	4.3160
3.04	0.5281	0.2135	0.4213	1.9732	4.3989
3.06	0.5310	0.2112	0.4177	1.9777	4.4835
3.08	0.5339	0.2090	0.4142	1.9822	4.5696
3.1	0.5368	0.2067	0.4107	1.9866	4.6573
3.12	0.5396	0.2045	0.4072	1.9910	4.7467
3.14	0.5424	0.2024	0.4038	1.9953	4.8377
3.16	0.5451	0.2002	0.4004	1.9995	4.9304
3.18	0.5478	0.1981	0.3970	2.0037	5.0248
3.2	0.5504	0.1961	0.3937	2.0079	5.1210
3.22	0.5531	0.1940	0.3904	2.0120	5.2189
3.24	0.5557	0.1920	0.3872	2.0160	5.3186
3.26	0.5582	0.1901	0.3839	2.0200	5.4201
3.28	0.5607	0.1881	0.3807	2.0239	5.5234
3.3	0.5632	0.1862	0.3776	2.0278	5.6286
3.32	0.5657	0.1843	0.3745	2.0317	5.7358
3.34	0.5681	0.1825	0.3714	2.0355	5.8448
3.36	0.5705	0.1806	0.3683	2.0392	5.9558
3.38	0.5729	0.1788	0.3653	2.0429	6.0687
3.4	0.5752	0.1770	0.3623	2.0466	6.1837
3.42	0.5775	0.1753	0.3594	2.0502	6.3007
3.44	0.5798	0.1736	0.3564	2.0537	6.4198
3.46	0.5820	0.1718	0.3535	2.0573	6.5409
3.48	0.5842	0.1702	0.3507	2.0607	6.6642
3.5	0.5864	0.1685	0.3478	2.0642	6.7896
3.52	0.5886	0.1669	0.3450	2.0676	6.9172
3.54	0.5907	0.1653	0.3422	2.0709	7.0471
3.56	0.5928	0.1637	0.3395	2.0743	7.1791
3.58	0.5949	0.1621	0.3368	2.0775	7.3135

Table B.3 (*Cont.*)

ADIABATIC FRICTIONAL FLOW IN A CONSTANT-AREA DUCT FOR $\gamma = 1.4$

Ma	$\bar{f}L^*/D$	p/p^*	T/T^*	$\rho^*/\rho = V/V^*$	p_0/p_0^*
3.6	0.5970	0.1606	0.3341	2.0808	7.4501
3.62	0.5990	0.1590	0.3314	2.0840	7.5891
3.64	0.6010	0.1575	0.3288	2.0871	7.7305
3.66	0.6030	0.1560	0.3262	2.0903	7.8742
3.68	0.6049	0.1546	0.3236	2.0933	8.0204
3.7	0.6068	0.1531	0.3210	2.0964	8.1691
3.72	0.6087	0.1517	0.3185	2.0994	8.3202
3.74	0.6106	0.1503	0.3160	2.1024	8.4739
3.76	0.6125	0.1489	0.3135	2.1053	8.6302
3.78	0.6143	0.1475	0.3111	2.1082	8.7891
3.8	0.6161	0.1462	0.3086	2.1111	8.9506
3.82	0.6179	0.1449	0.3062	2.1140	9.1148
3.84	0.6197	0.1436	0.3039	2.1168	9.2817
3.86	0.6214	0.1423	0.3015	2.1195	9.4513
3.88	0.6231	0.1410	0.2992	2.1223	9.6237
3.9	0.6248	0.1397	0.2969	2.1250	9.7990
3.92	0.6265	0.1385	0.2946	2.1277	9.9771
3.94	0.6282	0.1372	0.2923	2.1303	10.1581
3.96	0.6298	0.1360	0.2901	2.1329	10.3420
3.98	0.6315	0.1348	0.2879	2.1355	10.5289
4.0	0.6331	0.1336	0.2857	2.1381	10.7188

Table B.4

FRICTIONLESS DUCT FLOW WITH HEAT TRANSFER FOR $\gamma = 1.4$

Ma	T_0/T_0^*	p/p^*	T/T^*	$\rho^*/\rho = V/V^*$	p_0/p_0^*
0.0	0.0	2.4000	0.0	0.0	1.2679
0.02	0.0019	2.3987	0.0023	0.0010	1.2675
0.04	0.0076	2.3946	0.0092	0.0038	1.2665
0.06	0.0171	2.3880	0.0205	0.0086	1.2647
0.08	0.0302	2.3787	0.0362	0.0152	1.2623
0.1	0.0468	2.3669	0.0560	0.0237	1.2591
0.12	0.0666	2.3526	0.0797	0.0339	1.2554
0.14	0.0895	2.3359	0.1069	0.0458	1.2510
0.16	0.1151	2.3170	0.1374	0.0593	1.2461
0.18	0.1432	2.2959	0.1708	0.0744	1.2406
0.2	0.1736	2.2727	0.2066	0.0909	1.2346
0.22	0.2057	2.2477	0.2445	0.1088	1.2281
0.24	0.2395	2.2209	0.2841	0.1279	1.2213
0.26	0.2745	2.1925	0.3250	0.1482	1.2140
0.28	0.3104	2.1626	0.3667	0.1696	1.2064
0.3	0.3469	2.1314	0.4089	0.1918	1.1985
0.32	0.3837	2.0991	0.4512	0.2149	1.1904
0.34	0.4206	2.0657	0.4933	0.2388	1.1822
0.36	0.4572	2.0314	0.5348	0.2633	1.1737
0.38	0.4935	1.9964	0.5755	0.2883	1.1652
0.4	0.5290	1.9608	0.6151	0.3137	1.1566
0.42	0.5638	1.9247	0.6535	0.3395	1.1480
0.44	0.5975	1.8882	0.6903	0.3656	1.1394
0.46	0.6301	1.8515	0.7254	0.3918	1.1308
0.48	0.6614	1.8147	0.7587	0.4181	1.1224
0.5	0.6914	1.7778	0.7901	0.4444	1.1141
0.52	0.7199	1.7409	0.8196	0.4708	1.1059
0.54	0.7470	1.7043	0.8469	0.4970	1.0979
0.56	0.7725	1.6678	0.8723	0.5230	1.0901
0.58	0.7965	1.6316	0.8955	0.5489	1.0826
0.6	0.8189	1.5957	0.9167	0.5745	1.0753
0.62	0.8398	1.5603	0.9358	0.5998	1.0682
0.64	0.8592	1.5253	0.9530	0.6248	1.0615
0.66	0.8771	1.4908	0.9682	0.6494	1.0550
0.68	0.8935	1.4569	0.9814	0.6737	1.0489
0.7	0.9085	1.4235	0.9929	0.6975	1.0431
0.72	0.9221	1.3907	1.0026	0.7209	1.0376
0.74	0.9344	1.3585	1.0106	0.7439	1.0325
0.76	0.9455	1.3270	1.0171	0.7665	1.0278
0.78	0.9553	1.2961	1.0220	0.7885	1.0234
0.8	0.9639	1.2658	1.0255	0.8101	1.0193
0.82	0.9715	1.2362	1.0276	0.8313	1.0157
0.84	0.9781	1.2073	1.0285	0.8519	1.0124
0.86	0.9836	1.1791	1.0283	0.8721	1.0095
0.88	0.9883	1.1515	1.0269	0.8918	1.0070

Table B.4 (*Cont.*)

FRICTIONLESS DUCT FLOW WITH HEAT TRANSFER FOR $\gamma = 1.4$

Ma	T_0/T_0^*	p/p^*	T/T^*	$\rho^*/\rho = V/V^*$	p_0/p_0^*
0.9	0.9921	1.1246	1.0245	0.9110	1.0049
0.92	0.9951	1.0984	1.0212	0.9297	1.0031
0.94	0.9973	1.0728	1.0170	0.9480	1.0017
0.96	0.9988	1.0479	1.0121	0.9658	1.0008
0.98	0.9997	1.0236	1.0064	0.9831	1.0002
1.0	1.0000	1.0000	1.0000	1.0000	1.0000
1.02	0.9997	0.9770	0.9930	1.0164	1.0002
1.04	0.9989	0.9546	0.9855	1.0325	1.0008
1.06	0.9977	0.9327	0.9776	1.0480	1.0017
1.08	0.9960	0.9115	0.9691	1.0632	1.0031
1.1	0.9939	0.8909	0.9603	1.0780	1.0049
1.12	0.9915	0.8708	0.9512	1.0923	1.0070
1.14	0.9887	0.8512	0.9417	1.1063	1.0095
1.16	0.9856	0.8322	0.9320	1.1198	1.0124
1.18	0.9823	0.8137	0.9220	1.1330	1.0157
1.2	0.9787	0.7958	0.9118	1.1459	1.0194
1.22	0.9749	0.7783	0.9015	1.1584	1.0235
1.24	0.9709	0.7613	0.8911	1.1705	1.0279
1.26	0.9668	0.7447	0.8805	1.1823	1.0328
1.28	0.9624	0.7287	0.8699	1.1938	1.0380
1.3	0.9580	0.7130	0.8592	1.2050	1.0437
1.32	0.9534	0.6978	0.8484	1.2159	1.0497
1.34	0.9487	0.6830	0.8377	1.2264	1.0561
1.36	0.9440	0.6686	0.8269	1.2367	1.0629
1.38	0.9391	0.6546	0.8161	1.2467	1.0701
1.4	0.9343	0.6410	0.8054	1.2564	1.0777
1.42	0.9293	0.6278	0.7947	1.2659	1.0856
1.44	0.9243	0.6149	0.7840	1.2751	1.0940
1.46	0.9193	0.6024	0.7735	1.2840	1.1028
1.48	0.9143	0.5902	0.7629	1.2927	1.1120
1.5	0.9093	0.5783	0.7525	1.3012	1.1215
1.52	0.9042	0.5668	0.7422	1.3095	1.1315
1.54	0.8992	0.5555	0.7319	1.3175	1.1419
1.56	0.8942	0.5446	0.7217	1.3253	1.1527
1.58	0.8892	0.5339	0.7117	1.3329	1.1640
1.6	0.8842	0.5236	0.7017	1.3403	1.1756
1.62	0.8792	0.5135	0.6919	1.3475	1.1877
1.64	0.8743	0.5036	0.6822	1.3546	1.2002
1.66	0.8694	0.4940	0.6726	1.3614	1.2131
1.68	0.8645	0.4847	0.6631	1.3681	1.2264
1.7	0.8597	0.4756	0.6538	1.3746	1.2402
1.72	0.8549	0.4668	0.6445	1.3809	1.2545
1.74	0.8502	0.4581	0.6355	1.3870	1.2692
1.76	0.8455	0.4497	0.6265	1.3931	1.2843
1.78	0.8409	0.4415	0.6176	1.3989	1.2999

Table B.4 (*Cont.*)

Ma	T_0/T_0^*	p/p^*	T/T^*	$\rho^*/\rho = V/V^*$	p_0/p_0^*
1.8	0.8363	0.4335	0.6089	1.4046	1.3159
1.82	0.8317	0.4257	0.6004	1.4102	1.3324
1.84	0.8273	0.4181	0.5919	1.4156	1.3494
1.86	0.8228	0.4107	0.5836	1.4209	1.3669
1.88	0.8185	0.4035	0.5754	1.4261	1.3849
1.9	0.8141	0.3964	0.5673	1.4311	1.4033
1.92	0.8099	0.3895	0.5594	1.4360	1.4222
1.94	0.8057	0.3828	0.5516	1.4408	1.4417
1.96	0.8015	0.3763	0.5439	1.4455	1.4616
1.98	0.7974	0.3699	0.5364	1.4501	1.4821
2.0	0.7934	0.3636	0.5289	1.4545	1.5031
2.02	0.7894	0.3575	0.5216	1.4589	1.5246
2.04	0.7855	0.3516	0.5144	1.4632	1.5467
2.06	0.7816	0.3458	0.5074	1.4673	1.5693
2.08	0.7778	0.3401	0.5004	1.4714	1.5924
2.1	0.7741	0.3345	0.4936	1.4753	1.6162
2.12	0.7704	0.3291	0.4868	1.4792	1.6404
2.14	0.7667	0.3238	0.4802	1.4830	1.6653
2.16	0.7631	0.3186	0.4737	1.4867	1.6908
2.18	0.7596	0.3136	0.4673	1.4903	1.7168
2.2	0.7561	0.3086	0.4611	1.4938	1.7434
2.22	0.7527	0.3038	0.4549	1.4973	1.7707
2.24	0.7493	0.2991	0.4488	1.5007	1.7986
2.26	0.7460	0.2945	0.4428	1.5040	1.8271
2.28	0.7428	0.2899	0.4370	1.5072	1.8562
2.3	0.7395	0.2855	0.4312	1.5104	1.8860
2.32	0.7364	0.2812	0.4256	1.5134	1.9165
2.34	0.7333	0.2769	0.4200	1.5165	1.9476
2.36	0.7302	0.2728	0.4145	1.5194	1.9794
2.38	0.7272	0.2688	0.4091	1.5223	2.0119
2.4	0.7242	0.2648	0.4038	1.5252	2.0451
2.42	0.7213	0.2609	0.3986	1.5279	2.0789
2.44	0.7184	0.2571	0.3935	1.5306	2.1136
2.46	0.7156	0.2534	0.3885	1.5333	2.1489
2.48	0.7128	0.2497	0.3836	1.5359	2.1850
2.5	0.7101	0.2462	0.3787	1.5385	2.2218
2.52	0.7074	0.2427	0.3739	1.5410	2.2594
2.54	0.7047	0.2392	0.3692	1.5434	2.2978
2.56	0.7021	0.2359	0.3646	1.5458	2.3370
2.58	0.6995	0.2326	0.3601	1.5482	2.3770
2.6	0.6970	0.2294	0.3556	1.5505	2.4177
2.62	0.6945	0.2262	0.3512	1.5527	2.4593
2.64	0.6921	0.2231	0.3469	1.5549	2.5018
2.66	0.6896	0.2201	0.3427	1.5571	2.5451
2.68	0.6873	0.2171	0.3385	1.5592	2.5892

Table B.4 (*Cont.*)

FRICTIONLESS DUCT FLOW WITH HEAT TRANSFER FOR $\gamma = 1.4$

Ma	T_0/T_0^*	p/p^*	T/T^*	$\rho^*/\rho = V/V^*$	p_0/p_0^*
2.7	0.6849	0.2142	0.3344	1.5613	2.6343
2.72	0.6826	0.2113	0.3304	1.5634	2.6802
2.74	0.6804	0.2085	0.3264	1.5654	2.7270
2.76	0.6781	0.2058	0.3225	1.5673	2.7748
2.78	0.6760	0.2030	0.3186	1.5693	2.8235
2.8	0.6738	0.2004	0.3149	1.5711	2.8731
2.82	0.6717	0.1978	0.3111	1.5730	2.9237
2.84	0.6696	0.1953	0.3075	1.5748	2.9752
2.86	0.6675	0.1927	0.3039	1.5766	3.0278
2.88	0.6655	0.1903	0.3004	1.5784	3.0813
2.9	0.6635	0.1879	0.2969	1.5801	3.1359
2.92	0.6615	0.1855	0.2934	1.5818	3.1914
2.94	0.6596	0.1832	0.2901	1.5834	3.2481
2.96	0.6577	0.1809	0.2868	1.5851	3.3058
2.98	0.6558	0.1787	0.2835	1.5867	3.3646
3.0	0.6540	0.1765	0.2803	1.5882	3.4245
3.02	0.6522	0.1743	0.2771	1.5898	3.4854
3.04	0.6504	0.1722	0.2740	1.5913	3.5476
3.06	0.6486	0.1701	0.2709	1.5928	3.6108
3.08	0.6469	0.1681	0.2679	1.5942	3.6752
3.1	0.6452	0.1660	0.2650	1.5957	3.7408
3.12	0.6435	0.1641	0.2620	1.5971	3.8076
3.14	0.6418	0.1621	0.2592	1.5985	3.8756
3.16	0.6402	0.1602	0.2563	1.5998	3.9449
3.18	0.6386	0.1583	0.2535	1.6012	4.0154
3.2	0.6370	0.1565	0.2508	1.6025	4.0871
3.22	0.6354	0.1547	0.2481	1.6038	4.1602
3.24	0.6339	0.1529	0.2454	1.6051	4.2345
3.26	0.6324	0.1511	0.2428	1.6063	4.3101
3.28	0.6309	0.1494	0.2402	1.6076	4.3871
3.3	0.6294	0.1477	0.2377	1.6088	4.4655
3.32	0.6280	0.1461	0.2352	1.6100	4.5452
3.34	0.6265	0.1444	0.2327	1.6111	4.6263

Table B.4 (*Cont.*)

Ma	T_0/T_0^*	p/p^*	T/T^*	$\rho^*/\rho = V/V^*$	p_0/p_0^*
3.36	0.6251	0.1428	0.2303	1.6123	4.7089
3.38	0.6237	0.1412	0.2279	1.6134	4.7929
3.4	0.6224	0.1397	0.2255	1.6145	4.8783
3.42	0.6210	0.1381	0.2232	1.6156	4.9652
3.44	0.6197	0.1366	0.2209	1.6167	5.0536
3.46	0.6184	0.1351	0.2186	1.6178	5.1435
3.48	0.6171	0.1337	0.2164	1.6188	5.2350
3.5	0.6158	0.1322	0.2142	1.6198	5.3280
3.52	0.6145	0.1308	0.2120	1.6208	5.4226
3.54	0.6133	0.1294	0.2099	1.6218	5.5188
3.56	0.6121	0.1280	0.2078	1.6228	5.6167
3.58	0.6109	0.1267	0.2057	1.6238	5.7162
3.6	0.6097	0.1254	0.2037	1.6247	5.8173
3.62	0.6085	0.1241	0.2017	1.6257	5.9201
3.64	0.6074	0.1228	0.1997	1.6266	6.0247
3.66	0.6062	0.1215	0.1977	1.6275	6.1310
3.68	0.6051	0.1202	0.1958	1.6284	6.2390
3.7	0.6040	0.1190	0.1939	1.6293	6.3488
3.72	0.6029	0.1178	0.1920	1.6301	6.4605
3.74	0.6018	0.1166	0.1902	1.6310	6.5739
3.76	0.6008	0.1154	0.1884	1.6318	6.6893
3.78	0.5997	0.1143	0.1866	1.6327	6.8065
3.8	0.5987	0.1131	0.1848	1.6335	6.9256
3.82	0.5977	0.1120	0.1830	1.6343	7.0466
3.84	0.5967	0.1109	0.1813	1.6351	7.1696
3.86	0.5957	0.1098	0.1796	1.6359	7.2945
3.88	0.5947	0.1087	0.1779	1.6366	7.4215
3.9	0.5937	0.1077	0.1763	1.6374	7.5505
3.92	0.5928	0.1066	0.1746	1.6381	7.6816
3.94	0.5918	0.1056	0.1730	1.6389	7.8147
3.96	0.5909	0.1046	0.1714	1.6396	7.9499
3.98	0.5900	0.1036	0.1699	1.6403	8.0873
4.0	0.5891	0.1026	0.1683	1.6410	8.2269

Table B.5

PRANDTL-MEYER SUPERSONIC EXPANSION FUNCTION FOR $\gamma = 1.4$

Ma	ω, deg	Ma	ω, deg	Ma	ω, deg	Ma	ω, deg
1.00	0.0						
1.05	0.49	3.05	50.71	5.05	77.38	7.05	91.23
1.10	1.34	3.10	51.65	5.10	77.84	7.10	91.49
1.15	2.38	3.15	52.57	5.15	78.29	7.15	91.75
1.20	3.56	3.20	53.47	5.20	78.73	7.20	92.00
1.25	4.83	3.25	54.35	5.25	79.17	7.25	92.24
1.30	6.17	3.30	55.22	5.30	79.60	7.30	92.49
1.35	7.56	3.35	56.07	5.35	80.02	7.35	92.73
1.40	8.99	3.40	56.91	5.40	80.43	7.40	92.97
1.45	10.44	3.45	57.73	5.45	80.84	7.45	93.21
1.50	11.91	3.50	58.53	5.50	81.24	7.50	93.44
1.55	13.38	3.55	59.32	5.55	81.64	7.55	93.67
1.60	14.86	3.60	60.09	5.60	82.03	7.60	93.90
1.65	16.34	3.65	60.85	5.65	82.42	7.65	94.12
1.70	17.81	3.70	61.60	5.70	82.80	7.70	94.34
1.75	19.27	3.75	62.33	5.75	83.17	7.75	94.56
1.80	20.73	3.80	63.04	5.80	83.54	7.80	94.78
1.85	22.16	3.85	63.75	5.85	83.90	7.85	95.00
1.90	23.59	3.90	64.44	5.90	84.26	7.90	95.21
1.95	24.99	3.95	65.12	5.95	84.61	7.95	95.42
2.00	26.38	4.00	65.78	6.00	84.96	8.00	95.62
2.05	27.75	4.05	66.44	6.05	85.30	8.05	95.83
2.10	29.10	4.10	67.08	6.10	85.63	8.10	96.03
2.15	30.43	4.15	67.71	6.15	85.97	8.15	96.23
2.20	31.73	4.20	68.33	6.20	86.29	8.20	96.43
2.25	33.02	4.25	68.94	6.25	86.62	8.25	96.63
2.30	34.28	4.30	69.54	6.30	86.94	8.30	96.82
2.35	35.53	4.35	70.13	6.35	87.25	8.35	97.01
2.40	36.75	4.40	70.71	6.40	87.56	8.40	97.20
2.45	37.95	4.45	71.27	6.45	87.87	8.45	97.39
2.50	39.12	4.50	71.83	6.50	88.17	8.50	97.57
2.55	40.28	4.55	72.38	6.55	88.47	8.55	97.76
2.60	41.41	4.60	72.92	6.60	88.76	8.60	97.94
2.65	42.53	4.65	73.45	6.65	89.05	8.65	98.12
2.70	43.62	4.70	73.97	6.70	89.33	8.70	98.29
2.75	44.69	4.75	74.48	6.75	89.62	8.75	98.47
2.80	45.75	4.80	74.99	6.80	89.90	8.80	98.64
2.85	46.78	4.85	75.48	6.85	90.17	8.85	98.81
2.90	47.79	4.90	75.97	6.90	90.44	8.90	98.98
2.95	48.78	4.95	76.45	6.95	90.71	8.95	99.15
3.00	49.76	5.00	76.92	7.00	90.97	9.00	99.32

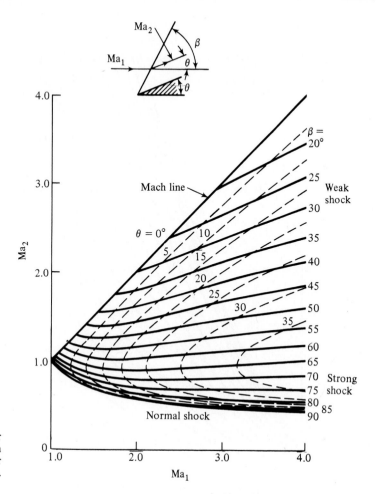

Fig. B.1 Mach number downstream of an oblique shock for $\gamma = 1.4$.

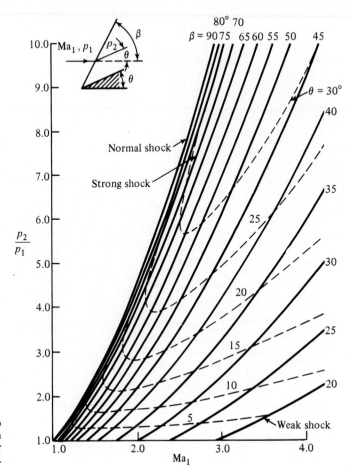

Fig. B.2 Pressure ratio downstream of an oblique shock for $\gamma = 1.4$.

Films on Fluid Mechanics

A comprehensive list of fluid mechanics films is given in the ASME Film Catalog, pages 151 to 155 of the June 1976 issue of the *Journal of Fluids Engineering*. All films are stocked for sale and loan by the Engineering Societies Library, United Engineering Center, 345 East 47th Street, New York, NY 10017.

Films cited in this book can also be purchased or borrowed from the agencies which produced them. The following is a brief list.

Encyclopedia Brittanica Educational Corp., 425 North Michigan Avenue, Chicago IL 60611:

1. The Fluid Dynamics of Drag (119 min in four parts)
2. Vorticity (44 min in two parts)
3. Flow Visualization (31 min)
4. Deformation of Continuous Media (38 min)
5. Pressure Fields and Fluid Acceleration (30 min)
6. Surface Tension in Fluid Mechanics (29 min)
7. Waves in Fluids (33 min)
8. Secondary Flow (30 min)
9. Boundary Layer Control (25 min)
10. Channel Flow of a Compressible Fluid (29 min)
11. Low Reynolds Number Flows (33 min)
12. Flow Instabilities (27 min)
13. Cavitation (27 min)
14. Fundamentals of Boundary Layers (24 min)
15. Turbulence (29 min)
16. Eulerian and Lagrangian Descriptions in Fluid Mechanics (27 min)
17. Rheological Behavior of Fluids (22 min)

The Shell Film Library is now closed. Many colleges own the following five color movies about compressible flow, which are highly recommended for students:

18. Approaching the Speed of Sound (27 min)
19. Transonic Flight (20 min)
20. Beyond the Speed of Sound (19 min)
21. Schlieren (20 min)
22. High Speed Flight (20 min)

University of Iowa, Media Library, Audiovisual Center, Iowa City, IA 52242:

23. Introduction to the Study of Fluid Motion (25 min)
24. Fundamental Principles of Flow (23 min)
25. Fluid Motion in a Gravitational Field (24 min)
26. Characteristics of Laminar and Turbulent Flow (26 min)

St. Anthony Falls Hydraulic Laboratory, Mississippi River at 3d Avenue S.E., Minneapolis, MN 55414:

27. Hydraulic Model Studies (20 min)
28. Some Phenomena of Open Channel Flow (33 min)
29. Flow in Culverts (20 min)
30. Spillway Design, Guayabo Dam, El Salvador (15 min)
31. Surface Waves (24 min)

U.S. Army Engineers Waterways Experiment Station, P.O. Box 631, Vicksburg, MS 39180:

32. Speaking of Models (20 min)

Association Sterling Films, 600 Grand Avenue, Ridgefield, NJ 07657:

33. Four Experiments in Hydraulics (17 min)

Engineering Societies Library, 345 East 47th Street, New York, NY 10017:

34. Turbulent Flow (An Introductory Lecture) (35 min)
35. Flow Separation and Formation of Vortices (16 min)
36. Hydraulic Analogy for Flow Studies (24 min)
37. Visual Cavitation Studies of Mixed Flow Pump Impellers (22 min)

NASA Lewis Research Center, Photographic Branch, Mail Stop 5-2, 21000 Brookpark Road, Cleveland OH 44135:

38. Vorticity Amplification in Stagnation Flow about a Circular Cylinder (12 min)
39. Flow Visualization in Experimental Combustion Chambers (8 min)

40. Laser Doppler Velocimetry (13 min)

41. Spinning Water Waves in a Circular Pool (5 min)

42. Stability of Cylindrical Bubbles in a Vertical Pipe (9 min)

43. Visual Observations of Flow through a Radial-Bladed Centrifugal Impeller (22 min)

44. Some Visual Observations of Cavitation in Rotating Machinery (17 min)

45. Hydrodynamic Seals (14 min)

46. Compressor Flow Patterns Observed with a Hot-Wire Anemometer (11 min)

47. Smoke Study of Nozzle Secondary Flows in a Low-Speed Turbine (20 min)

48. Computer-Generated Flow-Visualization Motion Pictures (12 min)

APPENDIX D

Conversion Factors

During this period of transition there is a constant need for conversions between BG and SI units (see Table 1.2). Some additional conversions are given here. Conversion factors are given inside the front and back covers.

Length	Volume
1 ft = 12 in = 0.3048 m 1 mi = 5280 ft = 1609.344 m 1 nautical mile (nmi) = 6076 ft = 1852 m 1 yard = 3 ft	1 ft^3 = 0.028317 m^3 1 U.S. gal = 231 in^3 = 0.0037854 m^3 1 L = 0.001 m^3 = 0.035315 ft^3

Mass	Density
1 slug = 32.174 lbm = 14.594 kg 1 lbm = 0.4536 kg	1 slug/ft^3 = 515.38 kg/m^3 1 lbm/ft^3 = 16.0185 kg/m^3

Velocity	Acceleration
1 ft/s = 0.3048 m/s 1 mi/h = 1.466666 ft/s = 0.44704 m/s 1 knot = 1 nmi/h = 1.6878 ft/s = 0.5144 m/s	1 ft/s^2 = 0.3048 m/s^2

Mass flow	Volume flow
1 slug/s = 14.594 kg/s 1 lbm/s = 0.4536 kg/s	1 gal/min = 0.002228 ft^3/s = 0.06309 L/s 1 × 10^6 gal/day = 1.5472 ft^3/s = 0.04381 m^3/s

Pressure	Force
1 lbf/ft^2 = 47.88 Pa 1 lbf/in^2 = 144 lbf/ft^2 = 6895 Pa 1 atm = 2116.2 lbf/ft^2 = 14.696 lbf/in^2 = 101,325 Pa	1 lbf = 4.448222 N = 16 oz 1 kgf = 2.2046 lbf = 9.80665 N 1 U.S. (short) ton = 2000 lbf

Energy	Power
1 ft·lbf = 1.35582 J 1 Btu = 252 cal = 1055.056 J = 778.17 ft·lbf	1 hp = 550 (ft·lbf)/s = 745.7 W 1 (ft·lbf)/s = 1.3558 W

Specific weight	Density
1 lbf/ft^3 = 157.09 N/m^3	1 slug/ft^3 = 515.38 kg/m^3 1 lbm/ft^3 = 16.0185 kg/m^3

Viscosity	Kinematic viscosity
1 slug/(ft·s) = 47.88 kg/(m·s) 1 poise (p) = 1 g/(cm·s) = 0.1 kg/(m·s)	1 ft^2/h = 0.000025806 m^2/s 1 stoke (st) = 1 cm^2/s = 0.0001 m^2/s

Temperature scale readings

$$T_F = \tfrac{9}{5}T_C + 32 \qquad T_C = \tfrac{5}{9}(T_F - 32) \qquad T_R = T_F + 459.69 \qquad T_K = T_C + 273.16$$

where subscripts F, C, R, and K refer to readings on the Fahrenheit, Celsius, Kelvin, and Rankine scales, respectively

Specific heat or gas constant†	Thermal conductivity†
1 (ft·lbf)/(slug·°R) = 0.16723 (N·m)/(kg·K) 1 Btu/(lb·°R) = 4186.8 J/(kg·K)	1 Btu/(h·ft·°R) = 1.7307 W/(m·K)

† Although the absolute (Kelvin) and Celsius temperature scales have different starting points, the intervals are the same size: 1 kelvin = 1 Celsius degree. The same holds true for the nonmetric absolute (Rankine) and Fahrenheit scales: 1 Rankine degree = 1 Fahrenheit degree. It is customary to express temperature differences in absolute-temperature units.

Equations of Motion in Cylindrical Coordinates

The equations of motion of an incompressible newtonian fluid with constant μ, k, and c_v are given here in cylindrical coordinates (r, θ, z), which are related to cartesian coordinates (x, y, z) as in Fig. 4.2:

$$x = r \cos \theta \qquad y = r \sin \theta \qquad z = z \tag{E.1}$$

The velocity components are v_r, v_θ, and v_z. The equations are:

Continuity:

$$\frac{1}{r} \frac{\partial}{\partial r}(rv_r) + \frac{1}{r} \frac{\partial}{\partial \theta}(v_\theta) + \frac{\partial}{\partial z}(v_z) = 0 \tag{E.2}$$

Convective time derivative:

$$\mathbf{V} \cdot \mathbf{V} = v_r \frac{\partial}{\partial r} + \frac{1}{r} v_\theta \frac{\partial}{\partial \theta} + v_z \frac{\partial}{\partial z} \tag{E.3}$$

Laplacian operator:

$$\nabla^2 = \frac{1}{r} \frac{\partial}{\partial r}\left(r \frac{\partial}{\partial r}\right) + \frac{1}{r^2} \frac{\partial^2}{\partial \theta^2} + \frac{\partial^2}{\partial z^2} \tag{E.4}$$

The r-momentum equation:

$$\frac{\partial v_r}{\partial t} + (\mathbf{V} \cdot \mathbf{V})v_r - \frac{1}{r} v_\theta^2 = -\frac{1}{\rho} \frac{\partial p}{\partial r} + g_r + \nu\left(\nabla^2 v_r - \frac{v_r}{r^2} - \frac{2}{r^2} \frac{\partial v_\theta}{\partial \theta}\right) \tag{E.5}$$

The θ-momentum equation:

$$\frac{\partial v_\theta}{\partial t} + (\mathbf{V} \cdot \mathbf{V})v_\theta + \frac{1}{r} v_r v_\theta = -\frac{1}{\rho r} \frac{\partial p}{\partial \theta} + g_\theta + \nu\left(\nabla^2 v_\theta - \frac{v_\theta}{r^2} + \frac{2}{r^2} \frac{\partial v_r}{\partial \theta}\right) \tag{E.6}$$

The z-momentum equation:

$$\frac{\partial v_z}{\partial t} + (\mathbf{V} \cdot \nabla) v_z = -\frac{1}{\rho} \frac{\partial p}{\partial z} + g_z + \nu \nabla^2 v_z \tag{E.7}$$

The energy equation:

$$\rho c_v \left[\frac{\partial T}{\partial t} + (\mathbf{V} \cdot \nabla) T \right] = k \nabla^2 T + \mu [2(\epsilon_{rr}^2 + \epsilon_{\theta\theta}^2 + \epsilon_{zz}^2) + \epsilon_{\theta z}^2 + \epsilon_{rz}^2 + \epsilon_{r\theta}^2] \tag{E.8}$$

where

$$\epsilon_{rr} = \frac{\partial v_r}{\partial r} \qquad \epsilon_{\theta\theta} = \frac{1}{r} \left(\frac{\partial v_\theta}{\partial \theta} + v_r \right)$$

$$\epsilon_{zz} = \frac{\partial v_z}{\partial z} \qquad \epsilon_{\theta z} = \frac{1}{r} \frac{\partial v_z}{\partial \theta} + \frac{\partial v_\theta}{\partial z}$$

$$\epsilon_{rz} = \frac{\partial v_r}{\partial z} + \frac{\partial v_z}{\partial r} \qquad \epsilon_{r\theta} = \frac{1}{r} \left(\frac{\partial v_r}{\partial \theta} - v_\theta \right) + \frac{\partial v_\theta}{\partial r}$$

Angular-velocity components:

$$\omega_r = \frac{1}{r} \frac{\partial v_z}{\partial \theta} - \frac{\partial v_\theta}{\partial z}$$

$$\omega_\theta = \frac{\partial v_r}{\partial z} - \frac{\partial v_z}{\partial r}$$

$$\omega_z = \frac{1}{r} \frac{\partial}{\partial r} (r v_\theta) - \frac{1}{r} \frac{\partial v_r}{\partial \theta}$$

Index

725

CONVERSION FACTORS FROM SI TO BG UNITS

	To convert from	To	Multiply by
Acceleration	m/s^2	ft/s^2	3.2808
Area	m^2	ft^2	1.0764 E + 1
		mi^2	3.8610 E − 7
		acres	2.4711 E − 4
Density	kg/m^3	$slug/ft^3$	1.9403 E − 3
		lbm/ft^3	6.2428 E − 2
Energy	J	$ft \cdot lbf$	7.3756 E − 1
		Btu	9.4782 E − 4
		cal	2.3884 E − 1
Force	N	lbf	2.2481 E − 1
		kgf	1.0197 E − 1
Length	m	ft	3.2808
		in	3.9370 E + 1
		mi (statute)	6.2137 E − 4
		nmi (nautical)	5.3996 E − 4
Mass	kg	slug	6.8522 E − 2
		lbm	2.2046
Mass flow	kg/s	slug/s	6.8522 E − 2
		lbm/s	2.2046
Power	W	$(ft \cdot lbf)/s$	7.3756 E − 1
		hp	1.3410 E − 3